ESPRIT '90

ESPRIT '90

Proceedings of the Annual ESPRIT Conference,
Brussels, November 12–15, 1990

Edited by

COMMISSION OF THE EUROPEAN COMMUNITIES
Directorate-General TELECOMMUNICATIONS,
INFORMATION INDUSTRIES and INNOVATION

KLUWER ACADEMIC PUBLISHERS
DORDRECHT / BOSTON / LONDON

ISBN 978-94-010-6803-1   e-ISBN-13: 978-94-009-0705-8
DOI: 10.1007/978-94-009-0705-8

Publication arrangements by
Commission of the European Communities
Directorate-General Telecommunications, Information Industries and Innovation,
Scientific and Technical Communications Service, Luxembourg

EUR 13148
© 1990 ECSC, EEC, EAEC, Brussels and Luxembourg
Softcover reprint of the hardcover 1st edition 1990

Published by Kluwer Academic Publishers,
P.O. Box 17, 3300 AA Dordrecht, The Netherlands.

Kluwer Academic Publishers incorporates the publishing programmes of
D. Reidel, Martinus Nijhoff, Dr W. Junk and MTP Press.

Sold and distributed in the U.S.A. and Canada
by Kluwer Academic Publishers,
101 Philip Drive, Norwell, MA 02061, U.S.A.

In all other countries, sold and distributed
by Kluwer Academic Publishers Group,
P.O. Box 322, 3300 AH Dordrecht, The Netherlands.

Printed on acid-free paper

Foreword

The 1990 ESPRIT Conferene is being held in Brussels from the 12th November to the 15th November. Well over 1700 participants from all over Europe and overseas are expected to attend the various events. The Conference will offer the opportunity to be updated on the results of the ESPRIT projects and Basic Research actions and to develop international contacts with colleagues, both within a specific branch of Information Technology and across different branches.

The first three days of the Conference are devoted to presentations of Esprit projects and Basic Research actions structured into plenary and parallel sessions; the scope of the Conference has been broadened this year by the inclusion of several well-known international speakers. All areas of Esprit work are covered: Microelectronics, Information Processing Systems, Office and Business Systems, Computer Integrated Manufacturing, Basic Research and aspects of the Information Exchange System.

During the IT Forum on Thursday November 15th, major European industrial and political decision-makers will address the audience in the morning. In the afternoon, a Round Table will discuss the impact of Information Technology on society.

More than 100 projects and actions will display their major innovations and achievements at the Esprit Exhibition which will be, for the first time, open to the general public. The exhibition offers a unique opportunity to acquire in a short time first hand, complete and comprehensive knowledge about the results obtained in all areas of the Esprit programme. These exhibits also show the pervasive role of IT and the contribution it makes to the economic and social life of everyone in the Community.

I would like to congratulate and thank everyone who has contributed to the Conference: the authors and reviewers of papers and reports; the chairmen, the speakers, the panelists and workshop participants at the Conference; the project teams which set up the demonstrations. The success of this Conference is due to all their efforts.

J.M. Cadiou
Director
Information Technologies - ESPRIT

CONTENTS

Information Processing Systems

Computer Integrated Manufacturing

Office and Business Systems

Information Exchange System

Basic Research

Indexes

PLENARY SESSION

Project No. 2589

Assessment, Methodology and Standardisation in Multilingual Speech Technology

SAM[1]

INTRODUCTION

The necessary use within the European Community of the languages of the member states has important implications for the development and use of speech information systems. All widely used technology has to be produced with reference to accepted standards concerning its manufacture and its assessment. Speech technology similarly requires standards to be developed and specified and these must be applicable across languages within the context of the European Community.

The SAM ['Speech Assessment Methods'] Project (2589) is dedicated to the definition and application of these multi-lingual EC standards. At present the project is based on the collaboration of twenty-eight laboratories in eight countries, six within the EC and two from EFTA. The SAM project is now at the half-way stage of its three-year ESPRIT II Main Phase. This follows a preliminary 'Definition Phase' (ESPRIT 1541) in which the state of the art, and the requirements in Europe and the rest of the world were investigated, and a 'bridging' 'Extension Phase', in which preparatory work for the Main Phase was undertaken. Current work is in progress in three inter-connected working areas:

I	Speech Recognition Assessment (Input)
II	Speech Synthesis Assessment (Output)
III	Enabling Technology and Research (ETR)

At the very beginning of the SAM Project, the need to ensure a practical basis for ready collaboration between so many different laboratories in different countries was met by the definition of a reference, standard, workstation - SESAM. The minimum hardware requirements for SESAM are an IBM pc-at or compatible computer, an analogue interface board, OROS-AU21 or AU22, 1 Mbyte of extended memory, and means for accessing speech data eg CD-ROM reader. C is used as the common programming language. Each one of the three workgroups, above, has made use of this simple reference standard so that software, data and assessment results can be interchanged. This has proved to be very successful both between project members and in the provision of data and support for other laboratories across Europe - all the work of the SAM Project is designed to be readily

---

1 prepared for the SAM project by Adrian Fourcin & colleagues at UCL and across Europe

available within the European Community.

I. Input Assessment

In recognition assessment, the simple reference standard workstation has been implemented and tested in multi-lingual, multi-laboratory trials. It comprises a DBMS interface, for automatically feeding the recognition items to a recogniser; a recogniser assessment control module with standard recogniser-to-workstation interface which is structured to minimise the amount of work needed to develop recogniser-specific drivers; a scoring module for scoring and statistically processing the recogniser response,(including a software module which has been developed to include NIST scoring criteria, providing a degree of compatibility with DARPA assessment work).

In order to provide a flexible tool for recogniser assessment, the component software packages are designed as separate modules which can be independently developed by different laboratory groupings within the project. The first package, PAOSAM, is designed to be capable of managing the information associated with the standard SAM format speech databases. The second package, EURPAC, primarily controls the interaction between the assessment system and the recogniser itself and the third package. This last package, SAM_SCOR, provides a series of performance measures. All three software modules are interconnected via ASCII files, and all programs are in C using the microsoft 5.1 compiler, and executable on the SESAM workstation running MS-DOS as the operating system (Version 3 or later).

Database Management

The PAOSAM program has been developed to cater for the major needs of data retrieval and data archiving for all languages and all speakers in the SAM project. A commercially available DataBase Management System (ORACLE) is used as the basic building block. The management structure has been designed to allow the integration of all of the characteristics of both present and future SAM speech databases. Effectively, PAOSAM enables the user to specify the characteristic assessment aspects to be targetted in terms, for example, of language, speaker and speech types, and for an automatic procedure to be utilised for the composition of training and testfiles in the assessment of a defined recogniser.

Control Module

The EURPAC program is designed to operate from this basis in controlling the assessment of isolated or connected word recognisers. The assessment session can be controlled by information given in a separate control file, defined by the user, and giving details of the unique serial number of the test run, the identification of the recogniser, and the names of the configuration-, training-, test- and response-files. An important aspect of the design of this particular software module is that it uses resident drivers to control individual recognisers. In this way, the greater part of the software is quite independent of the analogue interface board which is utilised, and it is easier to develop new recogniser drivers which can have separate communication protocols.

Scoring

The SAM_SCOR program provides a range of recognition performance measures - hit; miss; substitution; correct rejection; false alarm. In addition, at the isolated word level, confusion matrices, confidence analyses, and the application of the McNemar test are standard facilities. For connected word and continuous speech recognisers string matching at the orthographic level is available employing NIST scoring routines which have been made executable on the SESAM workstation. The output of this scoring software is designed to provide uniform presentations of the assessment results that are easy to understand and cross compare. SAM_SCOR generates a file which can subsequently be fed back into the DBMS to make it possible to relate speech material characteristics to recogniser performance measures.

Applications

More than 10 EC laboratories in the Project have been involved in the application of recogniser assessments so far for six commercially-available or in-house recognisers. Considerable use has been made of the first SAM CD-ROM speech database - EUROM 0 - which gives 5 hours from 20 speakers in five languages. This cross laboratory single and multi-language testing of equivalent recognisers has provided the foundation for the setting up of a basic calibration procedure for the SESAM input assessment workstation. Work is currently in progress to define a common method for standard reference calibration and hardware setting up protocols.

In collaboration with the ETR Group, a new multi-lingual speech database has been designed and is in the process of being recorded. The contents of the database have been defined to meet the present and near future need for the development of diagnostic and predictive assessment methodologies. The database is divided into two sets: a 'Many Speaker set' and a 'Few Speaker set'. The vocabulary of the 'Many Speaker set' contains a list of selected numbers between zero to nine thousand nine hundred and ninety nine covering all the phonotactic possibilities of the languages' number systems, and blocks of five sentences giving continuous speech with paragraph prosody rather than individual sentences. The vocabulary of the 'Few Speaker set' is expanded with a CVC list and more repetitions per item.

Future Input Assessment Activities

The present availability of this suite of software packages operating together with standard data bases and a set of common hardware facilities has provided an essential set of practical tools which we have shown to be usable in many different laboratory settings in Europe. We are faced, however, with the need to develop a more fundamental and comprehensive approach towards the investigation and the evaluation of input assessment methodologies when the needs have to be met not only of operating with a range of speakers, accents and dialects, but also languages. A main prospective area of our future work is directed towards the development of reliable language-independent predictive assessment methods. For this purpose several analytic approaches are under consideration:

RAMOS (Recogniser Assessment by means of Manipulation Of Speech) an approach developed at TNO, Netherlands. This is a diagnostic method based on a test vocabulary of CVC-words and resynthesis. The CVC-list of the new database is designed to be used with this method.

RSA (Recogniser Sensitivity Analysis) investigated in the UK for English in the Alvey STA-project. This is potentially complementary to RAMOS and can also be based on the "Few Speaker set" of the new database, since the speakers of the "Few Speaker set" are carefully selected from the "Many Speaker set" providing a controlled variability of performance related factors.

EVC (Effective Vocabulary Capacity method) could also be of interest, since the method has shown evidence that it produces estimates, which are relatively independent of the size and composition of the test vocabulary, with only one to two hundred utterances.

A complementary aspect of this analytic approach towards recogniser assessment involves the determination of the speech production factors applicable in our present eight European languages. This then can be correlated with their influence on the performance of individual recognisers. For this purpose a speech parameter extractor has been developed from work in the UK STA project using a software package, SAM-SPEX produced in Denmark. Currently six speaker-dependent parameters are measured: speaking rate; energy; larynx frequency; 'voice quality'; vocal tract area estimate; and finally, (temporal) pattern congruence. The aim is to make it possible for speech databases eventually to be calibrated in terms of factors which show a sensitivity in respect of recogniser performance.

Finally, we have arranged our data collection so that post-production factors can be incorporated by simulation into the (originally anechoically recorded) speech data, and this also has been the object of concerted activity within the project.

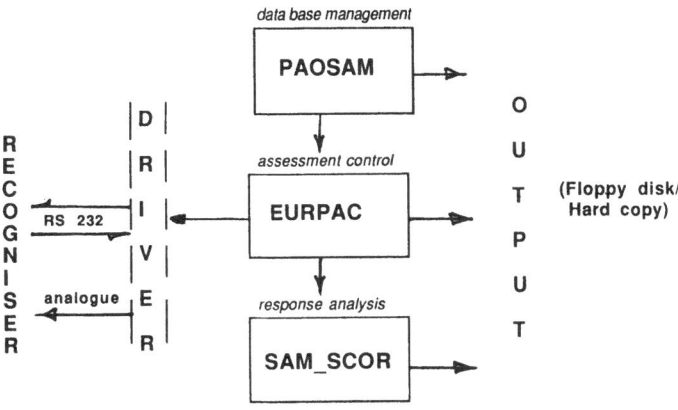

SESAM WORKSTATION FOR RECOGNISER ASSESSMENT

II. Output Assessment

Standard word-level and sentence-level segmental multi-lingual intelligibility tests have already been defined. They can be automatically generated on the SESAM workstation in the languages of the project using phonotactic and word frequency constraints. Compatible software provides for response collection, collation and scoring.

Segmental Structures

The SAM segmental test contains guidelines for the automatic generation of nonsense-word lists for all eight partner languages, using a set of fixed word structures and phoneme lists. The test material is language specific in that phoneme combinations respect phonotactic constraints for the languages in which they are prepared. The SAM group has chosen to use nonsense words in its definition of this standard with an open response set in order to get an intelligibility score which is not influenced by contextual information or semantically restricted answer choices. This type of material is the most relevant when an analysis of phoneme confusions is required, and in application, for instance, to synthesis material where error patterns may be quite device-specific. The SAM Segmental test consists of two parts: a first "core test" containing structures common to all languages of the consortium, and which cover consonants in initial, medial and final positions: VCV, VC (+ fixed final V for Italian) and CV. In all cases, the full inventory of consonants is used with only a sub-set of vowels. This sub-test cannot be considered as a *full* diagnostic test, but it is substantially diagnostic for consonants. The "full test" will include more extensive language-specific and even synthesiser-specific sub-tests with complex structures such as CVC, CVVC, VCCV and possibly CnVC, CVCn. Phonemes are presented in equal numbers per list so that an equal probability score

8

will be obtained. This score can then be weighted according to phoneme-frequency-of-occurrence counts to obtain scores which reflect phonemic balance.

Segmental Assessment

A system to support the automatic segmental assessment of synthesisers has been implemented on the SESAM workstation. The present form of this system, called SOAS, consists of three modules: the test module, the pre-processing module, and the scoring module. The test software controls the playing of sampled synthetic speech tokens in a sequence determined by a definable test file. The subject responds using the keyboard (a mouse-driven subject interface will soon be also available) and these responses are stored together with details describing the test. The pre-processing module applies a SAM-defined protocol to interpret these responses and produces a file containing the subject's responses converted to an unambiguous SAMPA form which is listener independent. These results are then automatically scored, using the third module, to produce percentage scores, confusion matrices, analysis in terms of certain types of phonetic feature (eg place of articulation and voicing), and the effect of vowel environment together with other statistics all in a form suitable for hard copy output. Further developments to this system will include a test manager to control all modules in a user-friendly environment and a test file generator for automatically producing randomised test sequences.

Assessment of words in context

A test of word intelligibility in sentence context has also been developed for SESAM, using semantically unpredictable sentences (SUS). Grammatical structures and word lists are defined for all the languages of the consortium to permit the generation of an unlimited number of test sentences. The SUS test material has already found wide acceptance outside ESPRIT, and has been recommended to the CCITT. The standard form of the SUS test is to be finalised by the end of 1990. The importance of word-intelligibility at this sentence level is in regard to the fuller information it provides on the quality of distributional and contextual variants. These are key factors in overall intelligibility, and deficiencies in this area of synthesis probably contribute to the reduced comprehension of synthesised messages in noise.

Prosody and Quality

Work on prosodic assessment is now under way. Pilot tests have been devised and are in progress to examine the acceptability of:

a) intonational structure
b) intonational function

All tests are designed for use with linguistically naïve subjects, an important factor if they are to be generally applicable.

Multi-lingual application has required the definition of structural tests which are formulated in general terms, independent of language-specific intonation. The functional tests are specified in terms of pragmatic speech-act categories, which also guarantees comparability across the languages.

In addition to intelligibility tests at word- and sentence-level, there is a need for tests which provide a global comparison between systems. A number of psycho-physical scaling procedures have been used for this purpose in the estimation of speech quality; these include categorical scaling, and magnitude estimation. Within the SAM Output group, various methodological parameters have been systematically estimated to establish how best to reduce the context effects (range of systems) and subject effects which might affect the rating of a system. It is being investigated whether the introduction of a reference system (natural undistorted speech) common to all experiments could help reduce these context/group effects. The two techniques - magnitude estimation (ME) and categorical estimations (CE) are being compared in terms of their 'resistance' to these effects and a major study has been completed [***].

III. Enabling Technology and Research

The core SAM workstation, SESAM, has been specified and implemented for data collection, following standard protocols, database management, and speech signal labelling. A phonemic notational system for all European languages, SAMPA, has been developed and is in use both for manual labelling and, currently, for semi-automatic label alignment. Phonemic level structural constraints across the languages of the project have been compiled and are used in corpus definition. Broader descriptors are being investigated for multi-lingual application. Other, physical, levels of description are being quantified as a contribution to analytic methods of assessment. Information on cross-language lexica is being compiled.

SESAM

Hardware (see the INTRODUCTION above) and software specifications are now well established and widely applied in regard, for example, to: the structure and code normalisation of software; the formatting of data and organisation of data-bases; and the provision of interfaces.

Two, key, software packages are central to the use of the workstation within the project. The first is EUROPEC, which is designed to provide for the realisation of large speech databases. Two-channel acquisition (eg for microphone and laryngo-graph signals) and monitoring is now possible with visual prompting for the speaker which may be manually controlled or automatically triggered as a function of signal level. Automatic end-point detection facilitates the handling and recording of large organised corpora. This is also substantially assisted by the automatic inclusion in

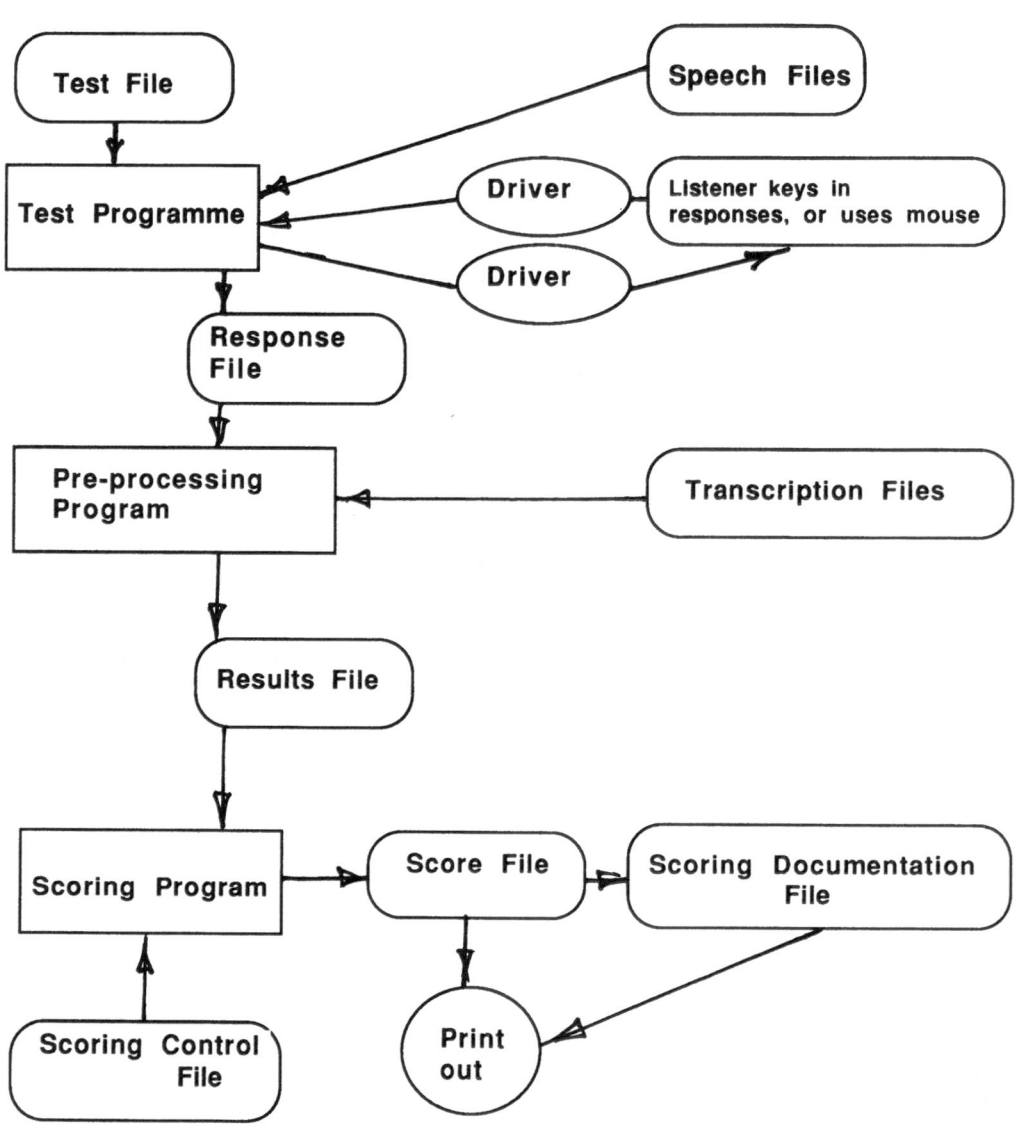

Speech Output Assessment System

the database of description text files in standard form with header and body, so that the orthographic prompt can be routinely incorporated together with complete sessional and recording item and condition information.

A complementary package, VERIPEC, is designed to give ready access to these standard data and text files, making it possible to display the orthographic prompts, access and monitor recorded items and show the label files.

The second important package, PTS, is designed fo operate from the data acquired via EUROPEC. Its primary function is to enable the labelling of speech data files with either SAMPA (see below) or IPA notations, using window based displays of waveform and spectrograms of the signal. A range of manipulation functions has been incorporated for viewing and editing (eg cut, paste, copy, save), measurement, monitoring and saving signal and label files.

SESAM was originally conceived as a common tool giving test-bed facilities and acting as interface in networking with more practical, more powerful, mainframes. It has emerged, however, both as a means of fulfilling these reference facilities and as a workstation in its own right. It is being adopted as a common tool not only within the SAM laboratories and in other ESPRIT projects but also as a bridge between US and EC speech standards. SAM software ('CONVERT') is incorporated into DARPA CD-ROMs so that American speech material can be used on SESAM in the same way as SAM-standard files.

Data

The first SAM database, EUROM 0, was distributed on a single CD-ROM and contained five hours of speech material recorded using a condenser microphone in anechoic rooms from four single accent speakers in each of five languages (with 16 kHz sampling). NATO single and triple digit sequences were obtained with only the speech signal, and a continuous speech passage, with a common numeric theme across languages, was recorded using two channels - with both speech and laryngographic inputs.

A new database is now in preparation using the same standard format with sixty speakers in each of eight languages. Nonsense words, number sequences up to 9999, (both phonotactically balanced) and situationally linked sentence blocks are being recorded. Anechoic condenser microphone recordings will permit the subsequent imposition of post-production effects. A small subset of data will have two channel representation, as above. Two CD-ROMs are planned for each language - using 20 kHz sampling.

SAMPA

The SAM Phonetic Alphabet (SAMPA), which defines a standard keyboard based notation (ASCII) corresponding to the relevant International Phonetic Association symbols for each of the languages represented in the project, was agreed very early in the project, and has now been extended to cover all the major European languages. It has also been adopted by a number of ESPRIT projects and both the British and German national speech databases. This consensus for the representation of phonemic contrasts in all the languages of the group provides a common labelling basis for cross-comparison and for a structured multi-lingual approach to database specification in the development of standard methods of assessment. The basic SAM transcription system was originally intended to evolve as a multi-tier labelling tool and work is currently directed towards the introduction of prosodic and acoustic element levels of description.

Labelling

Multi-lingual labelling, in which phoneme categories are assigned to successive regions of the speech signal, has always been an important part of the SAM group's activity. This is because overall assessment, detailed evaluation and the processes of training themselves ultimately depend on an accurate definition of speech which can be given in phonetic and orthographic terms. So, although the precise assignment of discrete categories, for different sound classes, to the continuous speech signal is an impossible task - since the subjective level of labelling is not compatible with any physical set of exact temporal stretches of the signal - the consistent correlation is of real value. The SESAM workstation is designed to support this work, and manual labelling in all the languages of the project has provided essential reference material.

A further development of this work currently involves a semi-automatic approach, using label alignment. In this way the larger quantities of speech material generated by current database gathering, and which are in need of labelling, can be accommodated without imposing an impossibly large manual labelling task - simply by using an ordinary transcription made without reference to signal details. This is currently being done for texts in Norwegian, French, Danish and English as an exercise across the whole of the SAM project. Three different methods have been selected for implementation. The first, from France is based on speech knowledge rules, the second, from Norway, on Hidden Markov modelling, and the third, from Denmark, uses neural network techniques in conjunction with phonetic feature transformations. Automatic scoring software (ELSA) has been written and, although it is still necessary for an expert labeller to check the setting of the label time boundaries, this provides a common basis for evaluation.

Speech and Language Material Complexity

The increasing development of databases and of detail in speech technology processing itself, leads to the need for the specification of speech and language material along additional dimensions to those discussed above. Work in hand and planned in the SAM project addresses some of these issues in regard to: the further collation of phonotactic information; the measurement of basic physical speech parameters - using SAM_SPEX for example - to tie in with the analytic procedures

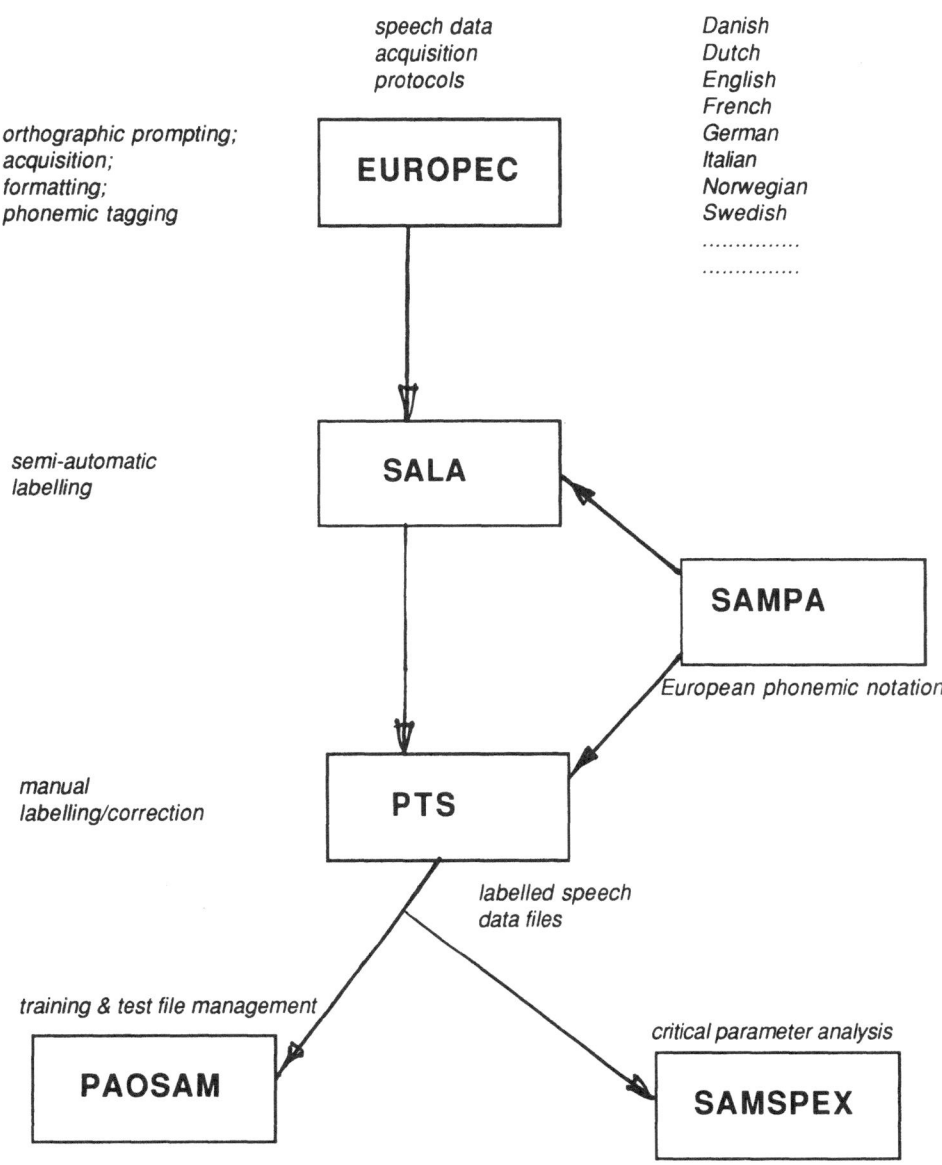

speech data
acquisition
protocols

Danish
Dutch
English
French
German
Italian
Norwegian
Swedish
....
....

orthographic prompting;
acquisition;
formatting;
phonemic tagging

EUROPEC

semi-automatic
labelling

SALA

SAMPA

European phonemic notation

manual
labelling/correction

PTS

labelled speech
data files

training & test file management

critical parameter analysis

PAOSAM

SAMSPEX

SESAM WORKSTATION FOR SPEECH DATA ACQUISITION

14

discussed in I above; the use of more representative, mildly pathological, speech; and the gathering of information on computer compatible lexica.

In Conclusion

The SAM Project (ESPRIT 2589: Multi-lingual Speech Input/Output Assessment, Methodology and Standardisation) is well placed to promote the aims that have been identified as central to the ESPRIT initiative:

- to provide the European Information Technology industry with the technologies it needs to meet the competitive requirements of the 1990s;
- to promote European cooperation in Information Technology;
- to contribute towards the development and implementation of international standards.

In common with all ESPRIT projects, it is concerned to fulfil the first two aims:

1. Common assessment methods and common tools can give a basis for the valid cross comparison of recogniser and synthesiser performance.
2. In addressing the problem of speech-technology assessment across 8 languages, with 8 partner-countries and 28 laboratories the SAM project is able to put the principle of cooperation into practice more extensively than most projects. The need for a multi-lingual approach to every aspect of assessment calls for parallel rather than distributed solutions. SAM cannot be a project based on division of labour; each partner country has to be actively involved in integrating the requirements for their language into a common approach.
   At another level, too, SAM may go further in collaboration and cooperation, in that the inclusion of two EFTA countries as fully active, self-financing partners, points the way to even wider European collaboration.
3. But in regard to the third aim the SAM project has as a **main goal** to provide, and to help set, standards within the EC. Its direct objectives are to supply standardised methods for the assessment of speech recognition and speech synthesis systems. In doing this within Europe it is already giving a foundation for broader collaboration internationally.

REFERENCES

The most relevant sources of information on the SAM project come from its progress reports. The first of these is available in book form from the publishers:

Ellis Horwood Ltd, Chichester, 1989 [ISBN 0-7458-0651-1] Speech Input and Output Assessment; Multi-lingual Methods and Standards
Two further progress reports may be obtained directly from:
Kate Jones, Phonetics and Linguistics Department, University College London, UK.

Project No. 2202

MANUFACTURING SYSTEMS PLANNING AND PROGRAMMING IN A CIM ENVIRONMENT

R. Bernhardt
Fraunhofer-Institute for Production Systems and Design Technology
Head.: o. Prof. Dr.-Ing. Drs. h.c. G. Spur
Main Department of Robot Systems Technology
Director: o. Prof. Dr.-Ing. G. Duelen
Pascalstraße 8-9, D - 1000 Berlin 10

Abstract

This paper reports on two ESPRIT projects (623 and 2202). The first one (started in 1985) was finished in 1990. Results and benefits of this project therefore can be presented in detail. Additionally the application spectrum of the realised systems and tools is outlined. For the second project which was started im May 1989 the objectives and the state of affairs after the first project year are described.

1. Introduction

Information techniques have initiated a structural change of the manufacturing industry. Productivity, flexibility, quality and reliability will a level which cannot be realized on the basis of conventional production structures. The very differentiated technological requirements for products lead to an increased product variation and a quick product substitution. These alterations of the market situation demand automation of the highest flexibility and productivity. This can be reached by computer integrated, automated and flexible manufacturing, whereby information techniques take over a key function.

In production technology it was comprehended at a very early stage that computers can be important components. Stations of these developments were the NC technique, CNC controls up to the development of FMS with robots. In this context it can be stated that robots play an important role as the most flexible automation components. But their effective use and their integration into a CIM environment requires the availability of powerful tools for planning and programming of robotised manufacturing cells. In this area R&D work has been done within the ESPRIT project 623:

Operational Control for Robot System Integration Into CIM.

Encouraged by the results reached in this project a further ESPRIT project (2202) has been launched:

CIM System Planning Toolbox (CIM-PLATO).

While the first project was dedicated specifically to robots the R&D contents of the latter one addresses manufacturing systems in general. Robots and the tools for their integration in a CIM environment are regarded as mere components but nevertheless important ones among many others.

For these tasks computer-aided tools like CAD, data base, expert, simulation, off-line programming and communication systems are either already available but still have to be adopted to todays needs of production industries or they are in development. This shows also that there are many relations between informatics and production technology. On the one hand, methods and results from informatics are used and on the other hand the needs from production technology have initiated research and development in informatics. These interrelations have meanwhile opened a big market for the information industry.

2. Robot System Integration into CIM

2.1 Objectives and Approaches

Robots are important components for flexible automation. The enlargement of their application area as well as their integration into CIM requires the availability of computer-aided planning and off-line programming means. Therefore a project (ESPRIT 623) was started in 1985 and finished in 1990. The general objective was to specify and build means to demonstrate the integration of robots into CIM systems /1/. This included two closely interrelated fields of research, i.e. computer-aided planning and an off-line programming system for robots to be integrated into CIM systems. The project had three branches of work: realisation of an explicit (motion oriented) programming and simulation system, realisation of an implicit (task oriented) programming system and realisation of a planning system for robotised work cells.

Work in these three areas proceeded simultaneously. In all cases, definition and specification of software modules was followed by their realisation and application/demonstration in real or simulated industrial environments.

In April 1988 project groups for realising demonstrator systems were organised to integrate components and subsystems for planning and programming of robotised work cells. The main objective for realising demonstrator systems by integrating modules of different partners was to show the increased functionality and efficiency of an integrated planning and programming procedure.

For the realisation different principles like automatic or interactive planning functions and explicit or implicit programming procedures were considered. The integration was achieved primarily through information exchange and management, performed via a relational data base. Additionally realistic industrial applications or well-known benchmark tests were selected which covered a broad spectrum of problems and showed their solution. The realised systems or parts of them were also used by the project partners for a variety of industral projects. In the meantime products based on the developments of this ESPRIT project have been introduced on the market by the project partners KUKA and Renault.

2.2 Industrial Applications

The realised planning and off-line programming tools have been used by the involved partners for a broad spectrum of applications. This is enabled because of the modular system structure allowing the exchange of user and task specific modules. Subsequently some examples are presented /2/.

2.2.1 Handling of Car Seats

An important area of use for planning systems are robot applications in which complex operations within restricted spatial conditions have to be performed. An example concerning the handling of car seats is given by fig. 2.2-1 (simulation) and fig. 2.2-2 which shows the realized workcell.

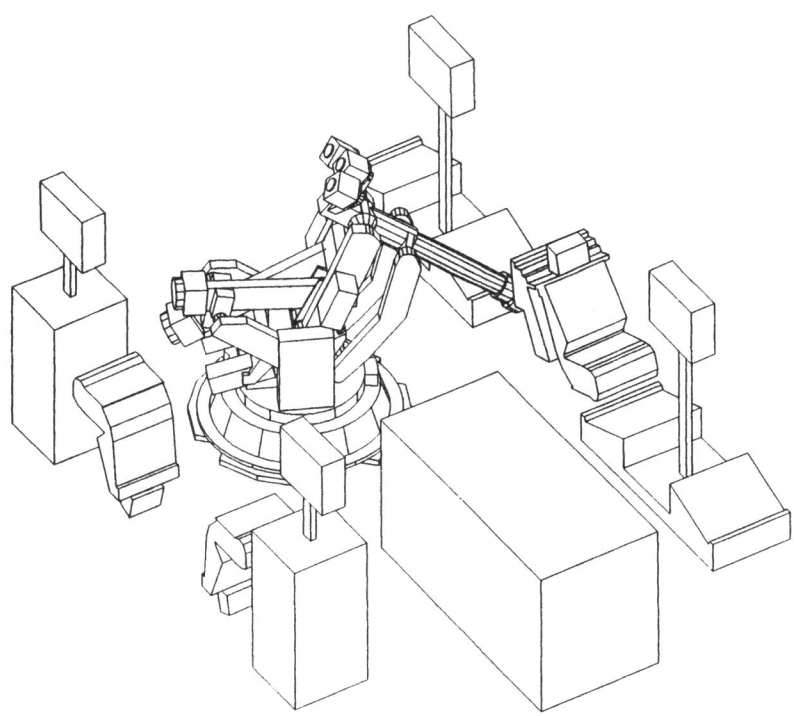

Fig. 2.2-1 Complex Robot Application (KUKA)

18

The robot transports car seats from different positions with a specific gripper to a measuring station. The actual seat position is used for the calculation of tool correction data. After measuring the seats are transported to the next workstation for further processing. In this project the simulation system has been used to analyze the reachability.

Fig. 2.2-2 Realised Workcell (KUKA)

2.2.2 Laser Cutting

A laser cutting method has been developed which consists of

- a curve definition on the part,
- a curve discretion according to different criteria,
- the laser trajectory definition including the laser torch adjustment and speed control, particularily in the windings of the laser path,
- the choice and positioning of the robot,
- the simulation and trajectory improvement.

This method is used particularily for sheet metal laser cutting application for the cutting of holes in pre-production cars where it is difficult to build specific dies. An example of such an application is given in the figures 2.2.-3 and 2.2.-4 where first a solid model representation of the simulation is shown and then the realised laser cutting cell is presented.

Laser cutting is also important outside the automotive industry. In short batch production there are a lot of products which need various holes differing from one variant of a part to another. For this reason such holes must be cut in the final step of the manufacturing process which is generally done manually. A laser is sometimes used but the part must be flat (2D curve) and the shape of the hole simple.

The developed system enables the use of this technology for complex parts and for complex contours of the hole once the part has been represented in the CAD system. From the application point of view, two approaches are followed:

- The first consists of a detemination on what kind of product the technology may be used (e.g. bath tubs, yacht hulls...).
- The second is an extension of this approach to another cutting system: the high pressure water jet cutting.

2.2.3 Inspection of Underwater Structures

The overall objective of the so-called OSIRIS project is to make available an underwater working robot in connection with a suitable carrier. The main tasks are cleaning, measuring and inspection of welding seams at underwaters structures without diver assistance. This requires the planning and programming of the submersible vehicles's motion as well as the task execution of the robot itself. Therefore the existing system has been enlarged by the additional module "VehicleProgramming". As the first step of the programming procedure the application program for the submersible vehicle is generated. The motion execution is tested via the simulation system. In the next step the same procedure is applied to generate and test the robot program. A testbed was built in a laboratory consisting of an industrial robot, a tube intersection of an off-shore platform and a tool exchange system including cleaning, measuring and inspection tools. Off-line generated and

20

Fig. 2.2-3   Solid Model Representation (Simulation Renault)

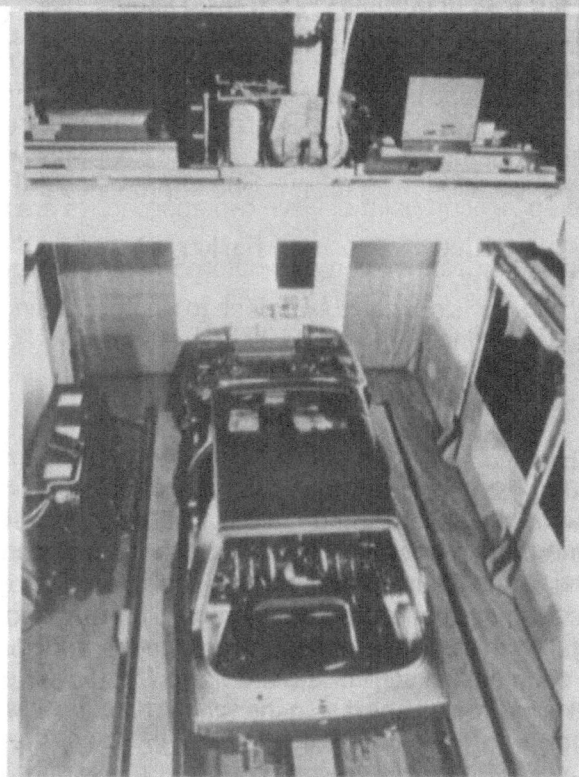

Fig. 2.2-4   Realised Laser Cutting Cell (Renault

tested robot programs were transfered to the robot control and executed by the robot. The project is conducted by the partners Interatom, GKSS and IPK Berlin /3/.

2.2.4 Automation and Robotics in Space

For the German space-lab mission envisaged for 1991 a robot technology experiment is in development. Thereby a robot in an experiment box has to fulfill different tasks after a calibration procedure. The required robot programs have to be planned, generated and tested by simulation in the gound station and transferred for execution to orbit. The project is financed by the German Ministry of Research and Techology with Dornier as prime contractor and IPK Berlin as a subcontractor of Dornier /4/. Futhermore at the IPK Berlin a lab for automation and robotics (A&R) in space has been built which provides the development environment for automation procedures and components for space applications. This lab allows the development and test of experiments under realistic conditions. The project is partly supported by the Senate of Berlin. Two research institutions and five Berlin SMEs participate in it. As a first experiment, a robotised work cell (5 axes robot mounted on two external linear axes) for the exchange of samples from a melting stove has been realised. Thereby the task execution is planned and tested by the off-line programming and simulation system.

2.3 Results and Benefits

As the partners who were concerned in the project are not only from the industry but also a number of academic partners were involved, it must be differentiated between results and benefits seen mainly from industrial points of view and those which more concern research institutions /5, 6/.

First the results and benefits of the industrial partners within the project are presented. Computer aided tools were developed to support the various planning and programming activities in the companies. These prototype products resulting from the project work were customised and extended by each partner. They then were introduced into the planning and programming departments and applied to many industrial projects. The yield was higher productivity on the suppliers side and better quality of the delivered manufacturing systems.

Two kinds of products were developed in order to serve the planning and programming needs. First, there are integrated planning and off-line programming systems that are used in the office and run on work stations or minicomputers. The work is done by using simulation models. The second category are application support systems that run on small, portable computers. They are used for adapting the off-line created programs to the specific requirements of the real manufacturing system on the shop floor.

To support planning tasks, robot simulation is the most important method. The most frequent kinds of application are

– restricted spatial conditions,

– tasks with critical time specifications,
– complex robot operations with multi-purpose endeffectors and
– complex kinematic structures with more than six joints involved.

Also for off-line programming industrial applications have been carried out. Hitherto, three categories have been identified where the use of these techniques was most advantageous for the industry:

– Programming of regular (e.g. symmetric) workpieces is considerably facilitated by applying program generation or manipulation functions (e.g. mirror functions).
– High accuracy applications (e.g. laser beam cutting) requiring joint level program optimisation.
– Transfer and adaptation of existing programs to similar applications (e.g. adaptation of a transfer line to a new car model).

To evaluate the results and benefits from an industrial point of view, the most important qualitative advantages are mentioned:

– Cost reduction:
  Personnel costs for design, optimisation and programming are saved and costs for workshop tests and subsequent alterations of the systems are significantly reduced.
– Faster results:
  At a very early stage in a project, highly accurate information is available, e.g. with regard to the configuration of the layout.
– Time saving:
  User solutions are developed faster, also within the tender preparation phase.
– Optimal engineering products:
  Several alternative solutions can be analysed and compared, resulting in a decisive improvement of the quality of a solution.

To quantify the results and benefits, the industrial partners have carried out inquiries. To give an example, some results of an inquiry by the project partner KUKA are presented in the figures 2.3-1 to 2.3-3. It may also be mentioned that project results have been presented on 25 international fairs.

In close cooperation with the industrial partners two approaches were followed by the academic partners during the course of the project. In the first place, existing techniques were used for test and integration into prototype systems.

These systems were installed and evaluated together with the industrial partners. The second approach was aimed at developing new technologies and showing their usefulness in experiments. In this way a proper input to the development of prototypes was ensured.

The objective of the research and the development activities was to demonstrate the applicability of advanced technologies in the area of planning and programming for operational robot control. Thereby two major areas were addressed. The first

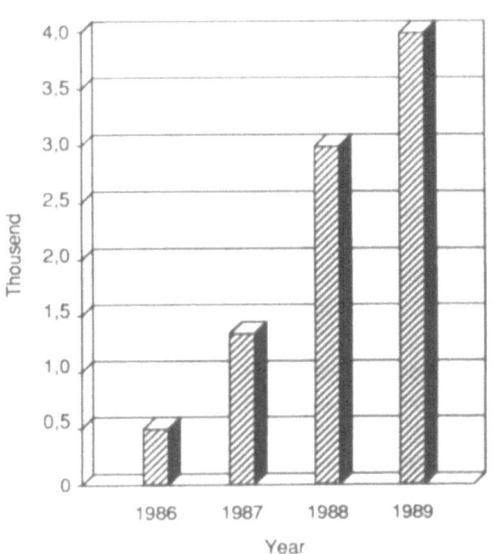

Fig. 2.3-1: Robot Simulation Working Hours (KUKA)

Fig. 2.3-2: Time Saving by Appling Robot Simulation (KUKA)

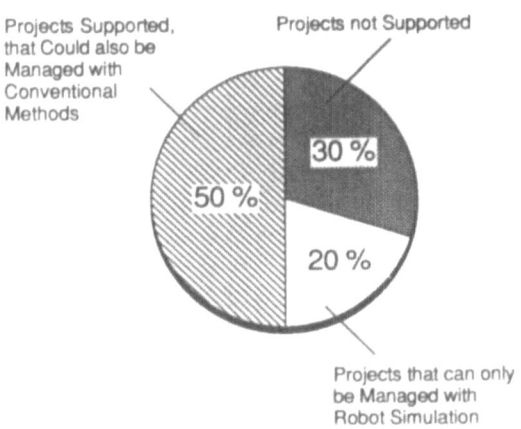

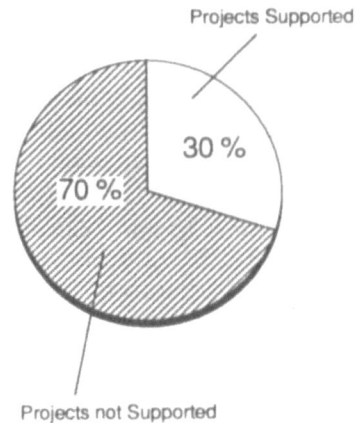

Fig. 2.3-3: Application of Robot Simulation (KUKA)

concerned the development of an automatic planning and programming system and the gradual replacement of interactive by automatic systems. Secondly information integration aspects were studied in detail. In particular in the later phases of the project the research efforts were directed to this second area. Throughout the project close contact with other research groups outside the consortium was kept.

In total 30 projects and services resulted from the academic partners. From these 20 products and services were deliverd to external institutions and industries. In particular the research experiences were used for other application areas such as tele-robotics and underwater applications. The academic partners were able to start 14 new industrial cooperations and 26 public projects in national and European programs.

An important aspect for the academic partners is the dissemination of knowledge through courses, publications and conferences. New courses were set up in the areas of CIM systems, robot programming and also in new subjects of sensors and artificial intelligence. Over a 100 publications were edited by the partners and more than 50 conferences were attended with a contribution. In particular during the last period, the established academic and industrial cooperations within the consortium presented themselves together in special sessions centred around on of the topics of the project. These research cooperations will also be used to address the new challenges in robot application.

The basic research efforts of the academic partners are now focussing on the new technologies emerging. These include the application of expert systems techniques and advanced programming approaches like object oriented formalisms.

3. CIM System Planning Tools

3.1 Project Objectives

Encouraged by the results reached in the ESPRIT 623 project a further ESPRIT project (2202) has been launched: CIM System Planning Toolbox (CIM-PLATO). The overall objective of the CIM-PLATO project is the development of an industrial toolbox prototype of computer-based procedures and tools which support the design, planning and installation of FMS and FAS systems in a CIM environment /7/. To reach this goal, three fields of R&D can be stated. These are the manufacturing system planning, the process execution planning and the provision of all necessary information to fulfill these tasks as well as the integration into a factory information system. In fig. 3-1 a functional reference model is presented which shows how the manufacturing system design tasks are embedded in a CIM environment.

The manufacturing system design is part of the a priori planning area. Its output is the design of the manufacturing system, the control design, the generation of all required executable programs and strategies for the manufacturing control valid for the class of products envisaged to be produced in the plant. This represents the base for the manufacturing control process in the executive area which has to control actual order-related production of the required products in an optimal manner.

In the case where events occur in the executive area which have not been considered during the a priori planning process and which cannot be handled by the means of the manufacturing control, a feedback to the planning area is required, and a planning process taking these situations into account has to be realized.

The a priori planning area can be regarded as a sequence of planning subtasks which are fairly non-interacting in the forward direction. This does not exclude iterations between sequential planning steps. The main inputs for the manufacturing system design are coming from the product program planning and the product design and development area: In a first step the manufacturing system design can be further decomposed into the subtasks

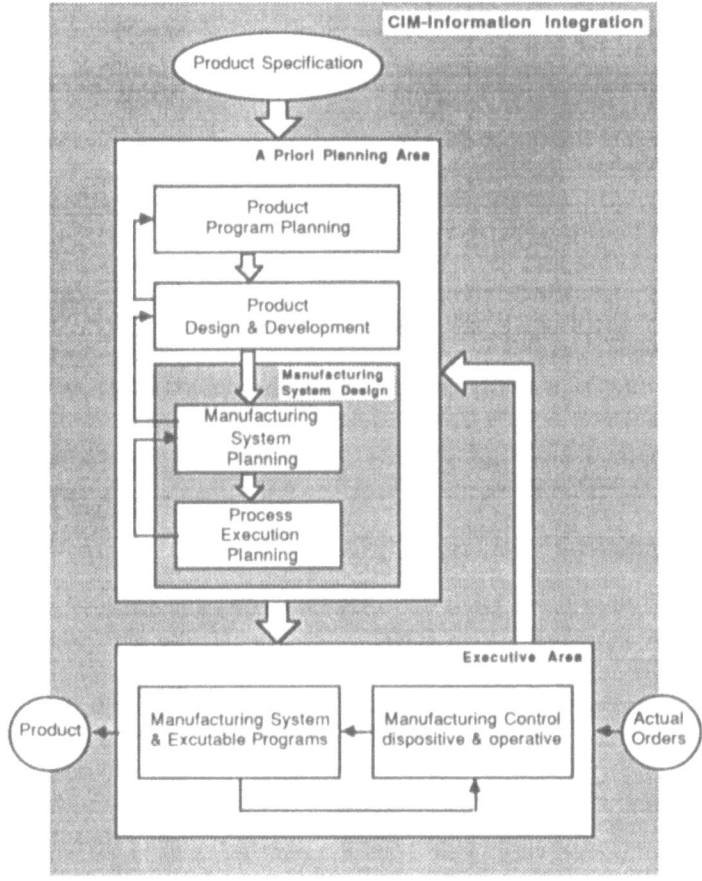

Fig. 3-1 Functional Reference Model

- manufacturing system planning and
- process execution planning.

Planning of a manufacturing system requires initially a system configuration which can be subdivided into a selection of components and the planning of layouts. As a second step of the planning process, a scheduling has to be performed which is an assignment of resources and times to manufacturing operations. For the completion of the planning process a verification has to be carried out. The resultant information is a formal representation of the layout as well as a description of the production task and this constitutes the input for the following more detailed planning phase in the overall goal of getting a manufacturing system into operation.

Process execution planning is based on the delivered information produced by manufacturing system planning. Using additional technological and geometrical information the process execution planning as well as the exceptional case planning has to be carried out. Process execution planning contains event and motion oriented planning, technological process planning as well as application oriented verification, simulation and test procedures. Furthermore optimization procedures related to task sequencing and trajectories have to be considered. Another important item is related to the conversion and transfer of application programs to the real manufacturing system.

The integration of manufacturing system planning and process execution planning for manufacturing system design purposes is mainly a problem of information integration. Therefore the information flow within both systems, information exchange between both and their integration into a CIM structure have to be considered. This requires the elaboration of an information system architecture, the development of suitable (and possibly general valid) knowledge/information models as well as the use of standards for modelling and information exchange. This includes to guarantee a safe data handling and to reach a most flexible system structure related to the integration and management of the tools as well as to a quick adaption to user needs.

3.2 Approaches and Intermediate Results

Within the frame of the project tools will be realized or already existing tools will be adapted and integrated to a toolbox "library". Tools are designed as far as possible and required to be applicable as "stand-alone units" (single tools) for specific industrial applications (Fig. 3-2). Additionally specific toolboxes can be configured from the toolbox library for applications requiring the cooperation of different tools, e.g. planning and programming of a robotized assembly cell. This principle approach adopted in the project is outlined in Fig. 3-3.

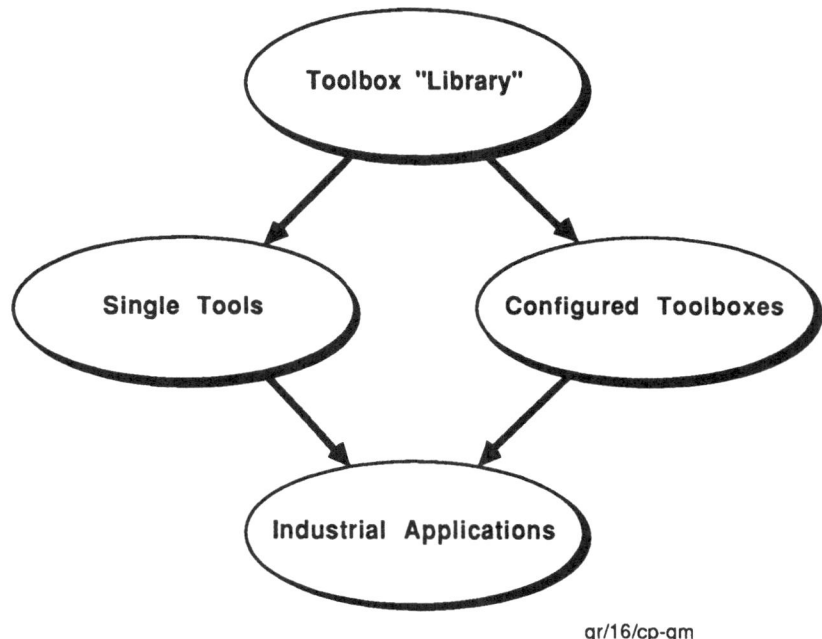

gr/16/cp-gm

Fig. 3-2: Principle Approach

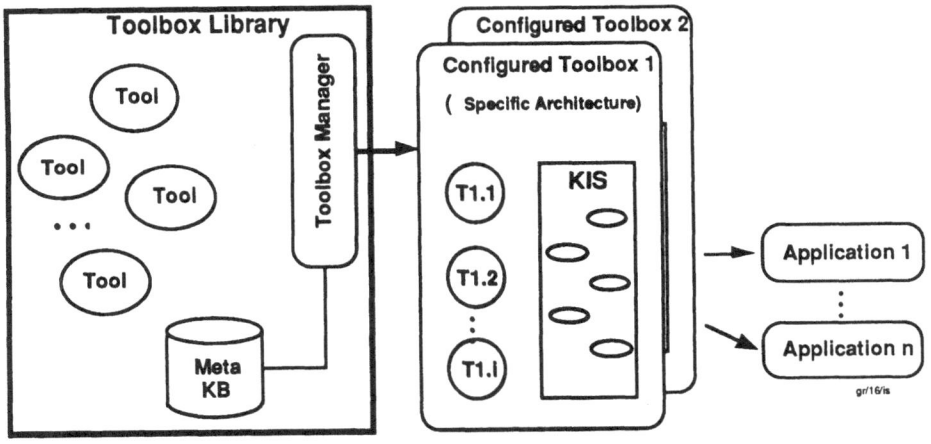

KIS : Knowledge-based Information System

Fig. 3-3: Basic Project Structure and Principle Tasks

To ensure an effective cooperation between partners, a subgroup structure has been installed. There are two subgroups for the configuration of examplary tool-boxes to evaluate and to prove the functionality and effectivity of the tools to be developed as well as to demonstrate the benefits and advantages reached through integration of tools to a toolbox. Additionally a subgroup has been installed working in the areas of toolbox management system and information integration.

The first year of this four year project was mainly a definition phase. The tasks to be done in this phase were the elaboration of a functional description (first report) and a functional specification of each tool (second report) to be realised in the project. Additionally the software design for some tools was started already. On the one hand this concerns already existing tools which have to be adapted . On the other hand the software design for new tools was started and first prototypes are now available, e.g. a technology planning tool for seam welding. Further activities which were addressed in the definition phase concern the specification of demonstrator systems. It is envisaged to realise four different demonstrator systems via the integration of tools from various partners. Furthermore it was proposed during the definition phase to build a toolbox manager which supports a user in selecting tools to fulfill a specific task concerning planning and programming of manufacturing systems. As a first step towards realisation a task force was established to define the user requirements. This task force consists of the industrial project partners to ensure the consideration of different backgrounds and experiences from various industries (e.g. computer, car, robot, textile industry, FMS/FAS system builders).

Beside the editing of two interim reports and the elaboration of a variety of working papers for internal information exchange within the project, also twelve articles were published by the consortium.

The ongoing work in the project is splitted in two directions: The first one is dedicated to the realisation of individual tools, the second one to the specification of demonstrator systems which integrate tools of different project partners.

4. Summary

The article gives a brief survey of the two ESPRIT projects. The project 623 "Operational Control for Robot System Integration Into CIM" was started in1985 and finished by the end of May 1990. As a final event of this project a workshop organised by the consortium was held in Berlin to present the project results to persons involved in the technical and research management from companies, universities and administrations. The different demonstrations and presentations as well as the interest of the audience clarly showed the industrial relevance of the work done and the scientific/technical value.

The project 2202 "CIM System Planning Toolbox" (CIM-PLATO) was started in 1989 and is now in the software design phase. Furthermore first fast prototypes of some tools have already been realised and demonstrated. Due to the importance of standardisation to ensure the cooperation of the tools to be realised, a workshop dealing with these aspects was held in May 1990. As speakers, experts from industry and research insitutions working in international standardisation boards were invited. Due to the broad spectrum of presentations during the workshop reaching from product modelling (STEP) to robot codes (ICR) the project partners got very compact information about the activities and state of the art in this area.

Due to the fact that speakers invited to the workshop are also involved in other ESPRIT projects, areas of common interest could be identified. As a result a number of cooperations have been agreed. Together with the CIM-OSA project a seminar is planned to discuss topics like integrating infrastructure, information modelling, engineering tools set and implementation reference models. With the NIRO project information and experiences will be exchanged concerning the formal description of manufacturing tasks, the industrial robot language (IRL), and the intermediate code for robots (ICR). Another cooperation has been launched with the IMPPACT project in the area of STEP. Of specific interest to the CIM PLATO project is the standardisation of kinematic and mechanic parameters.

Furthermore the CIM-PLATO consortium is involved in the organisation of a workshop which mainly aims at intensifying the information exchange among ESPRIT projects. This CIM Europe supported workshop is being organised by the consortia of the projects 2165, 2202, 2434 and 2527 and will be held at 3-4 December 1990 in Saarbrücken, Germany.

Finally it may be mentioned that all activities for cooperation and information exchange had not been planned at the beginning of the project. In so far they cause additional work and costs. On the other hand it is a way to effectively get informed about methods, approaches and results of other projects working in similar fields or the state of the art in related standardisation areas. In this sense projects can benefit from each other even economically provided that the cooperative activities are thoroughly planned and well prepared.

5. References

1. Spur et. al., Planning and Programming of Robot Integrated Production Cells Proceedings ESPRIT Technical Conference, Sept. 1987, Brussels

2. R. Bernhardt, Integrated Planning and Off-Line Programming System for Robotised Work Cells, Proceedings ESPRIT Technical Conference, Nov. 1989, Brussels

3. G. Duelen, V. Katschinski, Programmierung und Simulation eines mobilen Inspektionssystems fur Off-Shore-Anlagen, Vortragsband des 5. Fachgespraches über Autonome Mobile Systeme, November 1989, München, Bayrisches Forschungszentrum für wissensbasierte Systeme

4. G. Duelen, Th. Seidl, D-2 Technologieexperiment "ROTEX Ground Programming", Statusseminar des BMFT zu "Automatisierungstechnologien für die Raumfahrt" am 13. und 14.2.1990 in Bad Honnef

5. Operational Control for Robot Systems Into CIM, Results and Benefits, 10th Interim Report of the ESPRIT 623 Project, Nov. 1989

6. Robot System Integration Into CIM, Workshop Proceedings, March 1990, Berlin

7. R. Bernhardt, V. Katschinski, G. Schreck, CIM System Plannibg Toolbox: CIM-PLATO, Proceedings CIM Europe Conference, May 1990, Lisbon

6. Acknowledgements

The following partners were involved in the ESPRIT project 623:

Fraunhofer-Institute for Production Systems and Design Technolgy (IPK), Berlin, Germany
KUKA Schweißanlagen und Roboter GmbH, Augsburg, Germany
Renault Automation D.T.A.A., Paris, France
University College Galway, Galway, Republic of Ireland
FIAR Spa, Milano, Italy
University of Karlsruhe, Karlsruhe, Germany
Universidade Nova de Lisboa, Lisbon, Portugal
Universidad Politecnica de Madrid, Madrid, Spain
Universiteit Amsterdam, Amsterdam, The Netherlands
Gesellschaft fur Prozeßsteuerungs- und Informationssysteme GmbH
(PSI), Berlin, Germany
Consiglio Nationale delle Richerche (LADSEB-CNR), Padova, Italy
Politecnico di Milano, Milano, Italy

In the ESPRIT project 2202 the following additional partners are involved:

BULL France, Angers, France
Investronica, Madrid, Spain

The author in his capacity as the project manager of both projects reports about the work done by the researchers of the companies and institutions mentioned above.

Project No 2705

The ITHACA Technology
A Landscape for Object-Oriented Application Development

Martin Ader
Bull S.A.

Oscar Nierstrasz
Université de Genève

Stephen McMahon
Gerhard Müller
Anna-Kristin Pröfrock
Siemens Nixdorf Informationssysteme
AG

Abstract

The ITHACA environment[*] offers an application support system which incorporates advanced technologies in the fields of object-oriented programming in general and programming languages, database technologies, user interface systems and software development tools in particular. ITHACA provides an integrated and open-ended toolkit which exploits the benefits of object-oriented technologies for promoting reusability, tailorability and integratability, factors which are crucial for ensuring software quality and productivity. Industrial applications from the fields of office automation, public administration, finance/insurance and chemical engineering are developed in parallel and used to evaluate the suitability of the system.

1. Introduction

The ITHACA project aims to build a development environment for the construction of large-scale applications. Compared with the current situation, the applications are expected to be more reliable and more convenient to handle and to run. The manner in which they interact will, on the whole, be more integrated. At the same time, it is necessary to considerably reduce development costs, minimise the overall development risk and ensure the maintainability of the resulting applications. Moreover, applications built within the environment must be extensible with only a moderate amount of effort, thus allowing them to be adapted to meet changing requirements. This latter point is probably the major drawback of current hardwired application systems. The application domains we envisage in the ITHACA context

---

[*] The ITHACA project is funded as a Technology Integration Project by the Commission of the European Communities as part of the ESPRIT II programme. The partners involved are Siemens Nixdorf Informationssysteme AG (Germany), the prime contractor, with the associated partners Trinity College Dublin, University of Zurich, Altair (France) and SQL Datenbanksysteme GmbH (Germany); Bull S.A. with the associated partners Delphi S.p.A. (Italy), INRIA (France) and CMSU (Greece); Datamont S.p.A. (Italy) with Politecnico di Milano and Università di Milano as subcontractors; TAO (Spain); F.O.R.T.H. (Greece) and the University of Geneva (Switzerland).

belong to the domain of information systems, office systems and C* systems (CAD, CAM, CIM, CASE etc.).

It is widely accepted that the object-oriented paradigm promises best to be a basis for meeting the above requirements [Tsichritzis and Nierstrasz '88]. In spite of this, however, object-oriented programming is not yet used for the industrial development of systems in the foreseen domains. To reason this neglect we may draw from our experience in putting object-oriented programming environments to practical use and may identify some major lacks of present-day environments [Müller and Pröfrock '89].

The reasons for this are manifold. First, object-oriented programming is not supported by any adequate life-cycle model. Existing life-cycle models are insufficient for they do not support any notion of reusability - the core advantage of the object-oriented paradigm. Second (and most probably due to the missing life cycle model), tools for supporting the early phases of object-oriented programming are still in their infancy. Finally, object-oriented programming does not support any notion of persistence and thus limits the use of object-oriented programming to the development of system software, tools and user interface systems.

The ITHACA project aims to contribute towards providing solutions for these problem areas. The first section of this paper gives an integrated overview of the entire system and the approach adopted. This is followed by a more detailed description of the persistent programming and storage environment (HooDS). Subsequently, an appropriate object-oriented development life-cycle is introduced together with the tools required to support this. A brief description of selected pilot applications follows, including an explanation of end-user assistance and the support provided by groupware for cooperative processes. We conclude by outlining the minimum hardware and software platform on which the system runs and describe the status of development work so far.

2. The ITHACA Approach

As mentioned above, the objective of ITHACA is to provide an environment for supporting the development of large-scale application software systems [Pröfrock et al. '90]. The general approach to large-scale system design and development is to move the emphasis towards engineering rather than programming. To do so, we focus on the reuse of components on a macro-scale, a technique which is understood well and which is common practice in traditional engineering sectors, but not in software engineering, despite its name.

Complexity in engineering design is managed by hierarchical decomposition. This entails the repeated application of the following design cycle: Specifications are laid down and then, using knowledge of existing lower-level components, their properties and possible methods of interconnection, a configuration is proposed. This configuration usually undergoes a number of improvements in order to meet the specifications and criteria which may have been omitted or unquantified intially. The initial configuration together with subsequent improvements includes two major steps. First, the choice of component *types* and their layout and interconnection (topology of the design). Second, the choice of *specific* components (in terms of characteristic parameter values) to meet these specifications. We have assumed that the design is composed of available components. However, there may be a need for some *new* component(s). In this case, a design cycle at a lower level will be initiated to provide the missing component (or components), configuration of which again being attempted using existing components at a lower level still. For any given class of designs, systems and subsystems are constructed using a defined set of building blocks, the primitive components. The primitive components are not universal: gates are the basic building blocks for some designs, although they are themselves configured from more elementary components. At lower design levels, the primitive components are usually standardised. As a field of application matures, standardisation propagates upwards until it reaches final products.

It should be noted that the described approach to engineering is quite different from the stepwise refinement offered by software engineering. Stepwise refinement leads the software engineer to split a complex problem into less complex subproblems by means of iteration. Reuse of existing components is not considered to be a goal of the process and, if at all, is achieved by accident. In general, stepwise refinement results in recoding of similar programs.

Another, more intuitive example can be drawn from an architect's approach to building a house for a specific client. First of all, the architect draws up details of the client's requirements and the budget restrictions. Based on those conditions, the architect decides if he can offer his client a prefab. If not, he has to make a customised design in the manner described above. The more he cannot use standard components, the client's budget is burdened exponentially. He will thus try to find standard kitchen designs, common bathroom layouts etc. in order to remain within his budget limitations. Such a "generic kitchen frame" can be specialised by concrete objects which are differentiated with regard to material, colour, form and quality.

Object-oriented programming allows this approach to be followed because it adheres to the philosophy of using prefabricated components. Here, the term "component" covers more than just basic code; it also includes the context in which a piece of software is used. Thus, reuse of components also means the reuse of designs, models, requirements and even experience.

The following scenario can be used to apply the principles of engineering design to software engineering. Well-engineered components are described in catalogs and stored in software information bases in the form of application frames. An application frame is a prefabricated application and constitutes a network which links the application's model, requirements, design, experience, realisation and documentation. The application developer gathers the client's requirements. He navigates through his prefabricated applications with the aim of finding either a solution which satisfies the demands or pieces which fit into a specific layout. This cycle is applied recursively until no specific layout is left. In addition to the support of object-oriented programming and application configuration, the developer is assisted by means of dedicated construction tools (e.g. for user interfaces) and high-level system components (e.g. adequate database systems).

In order to support a manufacturing approach to software development, ITHACA provides an application support system which places an additional, object-oriented layer between the application and the underlying operating system.

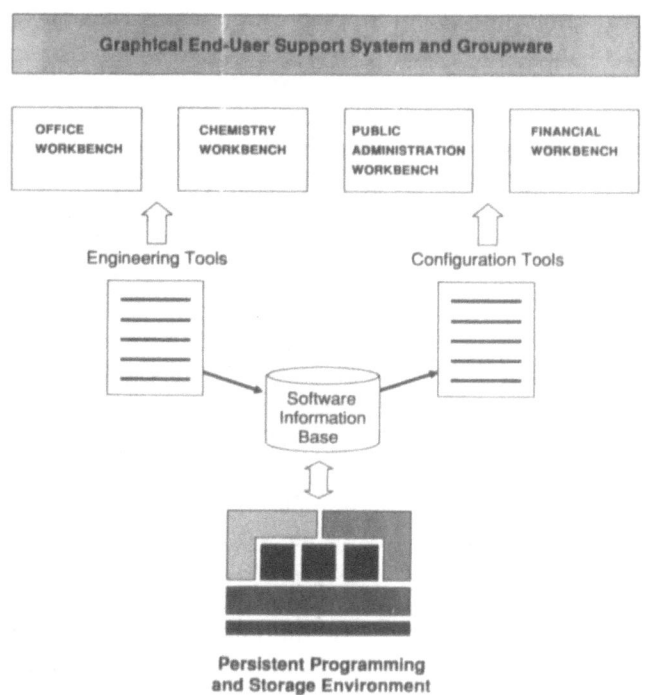

Fig. 2.1 : The ITHACA approach to application engineering

For the purpose of application programming, ITHACA provides a persistent programming and storage environment which serves as the kernel of the system. This kernel, called HooDS, includes an object-oriented database system, adequate programming languages and the appropriate basic programming environment. In addition, ITHACA presupposes an object-oriented design style. This design style is supported by dedicated tools for engineering and configuring applications. These tools are based on a software information base with appropriate navigation facilities. For end-user support, ITHACA offers a Motif-based user interface and groupware for assisting cooperative work between end-users. In order to have an early evaluation of the technology provided and to allow user participation from a very early stage onwards, four complex "real-world" applications are being designed and developed within the ITHACA framework. In addition to spanning the scope of the application areas we aim to serve, all four applications have a high generic nature and are expected to be relevant to the market in the near future.

3. The Persistent Object Environment HooDS

HooDS supports object-oriented programming in general and persistent objects in particular. As a programming environment, HooDS provides:

- a multi-programming language approach,
- the basic programming tools for programming in the small and
- an optional desktop environment for convenient access. The desktop can be used by applications as an application interface, as well as by programmers as a system interface to the system.

As an environment which supports persistent objects, HooDS provides the functionality of a fully-fledged object-oriented database system. Thus, HooDS supports persistent objects, multi-user access to shared objects, concurrency control, (closed) nested transactions, object retrieval by both navigation and association and, last but not least, recovery in the case of software failures. Note that in the context of ITHACA, persistence of objects is considered as a feature for those applications which need database support and not as an end in itself in order to store development data - although this would be equally possible and is in fact planned for future versions.

The following diagram shows the component structure of HooDS.

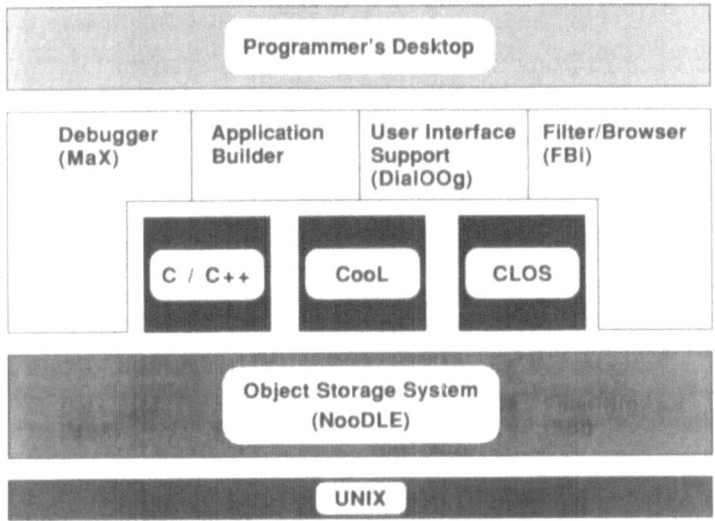

Fig. 3.1 : The general system architecture of HooDS

On the highest level of abstraction, the kernel consists of a set of object-oriented programming languages. In order to realise persistent objects, these languages are connected to an object storage subsystem, called NooDLE.

To support the practical needs of programmers, a runtime monitoring and debugging facility, called MaX is supported [Brodde '89], [Brodde '90]. In a world where the reuse of software components written by other programmers is the standard, a facility such as this constitutes more than just one of life's luxuries.

· A specialised application builder is under development [Krickstadt '90] in order to ease the process of building a specific application system by connecting class declarations, interconnect this application to the appropriate user interface objects and finally generate the necessary meta-information for the management of persistent objects in NooDLE (i.e. the schema). Due to the internal complexity of object-oriented application systems (e.g. the resolution of inheritance) the standard UNIX® *make* facility is only of limited use in this context. To nevertheless ensure conformance with the standard, the application builder is built on top of the *make* facility and thus generates standard *make* files.

The design and implementation of appropriate graphical user interfaces is an expensive task. Based on meta-information kept internally, the dialog management system DialOOg is able to generate simple form-oriented, graphical user interfaces on the basis of OSF/Motif for a majority of standard representations.

Finally, in order to retrieve objects interactively the kernel provides a filter/browser system called FBi which acts on both the storage subsystem to retrieve the data of an object, as well as on the languages able to execute the methods of an object.

Unlike current approaches to object-oriented database system development, HooDS focuses on achieving an open, extensible and configurable architecture.

A number of research projects are currently under way which focus on the development of prototypes for extensible database systems [Batory et al. '86], [Carey et al. '86], [Thompson et al. '89]. The general idea of extensible database systems is to modularise the system's architecture by factoring out functionalities into modules. This idea is quite simple, but nevertheless contradicts the common approach to database system design which centres on monolithic systems with a layered architecture.

We may intuitively draw some clear advantages of extensible architectures. First, an extensible architecture allows systems to be configured at installation time in accordance with the needs of different application domains - if a specific functionality is not needed then it is not configured at all. Thus, HooDS permits any number of configurations to be installed in a system related to the different needs of application domains. Nevertheless, HooDS guarantees that applications based on different configurations may also freely co-operate and exchange information.

Second, extensible architectures are better suited for staying abreast of the progress of technology and for absorbing improvements more frequently, faster and mostly without the need to rewrite major portions of the system.

---

* UNIX is a registered trademark of AT&T

Last but not least, extensible architectures provide a better basis for the construction of open systems. The vital case for openness with respect to object-oriented database systems is the ease with which additional programming languages can be bound into them. Current object-oriented database systems are implemented as huge, monolithic systems, which closely combine a specific programming language with a storage management system for handling persistent objects. Due to the fact that (object-oriented) languages do not support any notion of persistence, the favoured language is usually extended by some model of persistence. The implications of this approach are that the overall conceptual model of persistence is both language-dependent and model-specific. In practice, this approach thus implies that it is neither possible to integrate any additional language into the system nor to introduce any different model of persistence. In fact, current implementations are noticeably closed solutions and thus resemble database system developments of the late 1960s.

In order to compensate these drawbacks of current implementations, HooDS supports a formal data model and a related algebra called NO2. This model serves as the explicit binding interface between languages and the storage subsystem NooDLE. The NO2 data model is comparatively general and allows the developer of a specific language binding to decide more or less freely how to extend his favoured (object-oriented or imperative) language by concepts of persistence. It should be noted that the use of a formal data model not only fomalises language binding, but also ensures that different languages (and language extensions) may interoperate within the HooDS system.

In order to facilitate implementation of different language bindings and to increase the flexibility of binding, NooDLE supports different levels of interfaces. At the highest interface level, NooDLE supports an object-oriented extension of SQL which allows embedded oSQL approaches to be built as used in the conventional relational approach. On a medium level, NooDLE provides a converter interface which permits language binding via a call interface. At the lowest level, NooDLE maintains a heap interface which corresponds to an abstract low-level heap as used in most programming languages. Unlike other approaches, this heap is also used internally by NooDLE as a database buffer, hence constituting some sort of stable heap [Kolodner et al. '89]. This interface minimises any overhead and is thus extremely useful when language binders have direct access to the internals of the language's compiler.

NO2 Data Model

The NO2 data model [Elsholtz '89] [Dittrich et al. '90] [Geppert et al. '90] is based conceptually on the O2 data model [Lecluse and Richard '89], but extends this model by ideas from the ENF2 model [Pistor and Anderson '86]. Thus, the NO2 model is equally well suited to serve both object-oriented and imperative programming languages as a storage medium.

NO2 clearly distinguishes between objects and values. Objects have an identity and exist by themselves independent of their values, while values exist only by belonging to an object. An object is thus a pair comprising a surrogate and a value. Objects are always instances of an object type. Complex objects are constructed by orthogonally including objects in values of objects. As with objects, values are instances of value sets. Constructors for tuples, arrays, lists and sets are offered in order to construct structured value sets. To increase the flexibility of data modelling, NO2 also supports the composition of objects, in addition to aggregation of objects. Composition models the IS-PART-OF relationship between objects. Subobjects of a composite object may be owned exclusively on the existence of their parent object. Composition is first of all a value of its own for application domains which have to handle hierarchies of objects as a whole. Moreover, composition provides a basis for the design of specialised (and more efficient) storage and retrieval strategies.

Language Bindings

The initial version of HooDS supports C + +, CLOS and CooL as programming languages. Two successive approaches are used for binding C + +. In a first attempt, a persistent class library is provided which allows application programmers to achieve persistence for their actual class by subclassing from persistent classes. In a second step, which is currently under design, an extension to C + + in the direction of O + + [Agrawal and Gehani '89] is planned. These extensions, which are few in number, are processed by a preprocessor and translated to calls to the converter interface. For CLOS (the de-facto standard extension of LISP by object-oriented concepts [Attardi et al. '89]), persistence is transparent to the programmer and is achieved through direct binding on the level of the heap interface.

CooL is a newly designed object-oriented programming language which closely combines the object-oriented approach to programming with current database technology.

CooL

CooL [CooL '90] is especially designed with the objective of giving programmers an adequate language for building large and complex application systems. Large-scale system building does not usually take place in "splendid isolation", but has to consider existing software. Thus, CooL is fully compatible with C and allows any existing C software to be integrated without additional effort. Despite this compatibility, CooL does not inherit the insecurity of C, but rather offers a high level of reliablity by incorporating strong typing, data encapsulation, controlled inheritance and exception handling. Of course, CooL is an object-oriented language which supports the usual features of object-orientation. Nevertheless, the design is not religious and thus extends the language for the benefit of professional programmers by a number of concepts which are not necessarily a 'must' for object-oriented language systems. An example of such an extension is that CooL supports a rich set of type constructors, including sets with a computational complete set algebra.

Support for persistent objects is an integral concept of CooL. For programmers, persistence is transparent and uniform. This implies that they may use any construct together with any object, regardless of whether its property is persistent or volatile. CooL provides transactions, read/write locks and elaborated iterators in order to manage persistent objects. A runtime overhead for persistence only has to be paid if persistent objects are actually used in an application. Otherwise CooL is at least as efficient as C + +.

Under the present-day practical conditions encountered in industry, object-oriented database systems only constitute small islands in the world of relational database systems. To succeed commercially integration to SQL systems is thus essential. Hence CooL supports a type constructor relation which is mapped directly onto a SQL system and which allows persistent (complex) objects with tuples from a SQL system to be merged within a CooL program.

NooDLE

The object storage system NooDLE constitutes the core of HooDS and provides a so-called structurally object-oriented storage system. NooDLE supports the NO2 data model and implements the IS-A hierachy based on multiple inheritance, (closed) nested transactions, concurrency control based on read/write locking, query processing (also on type hierarchies), recovery based on before images and a n-client/1-server architecture with dedicated cache management. Of course, NooDLE is in itself again extensible and hence configurable.

The following figure illustrates the architecture.

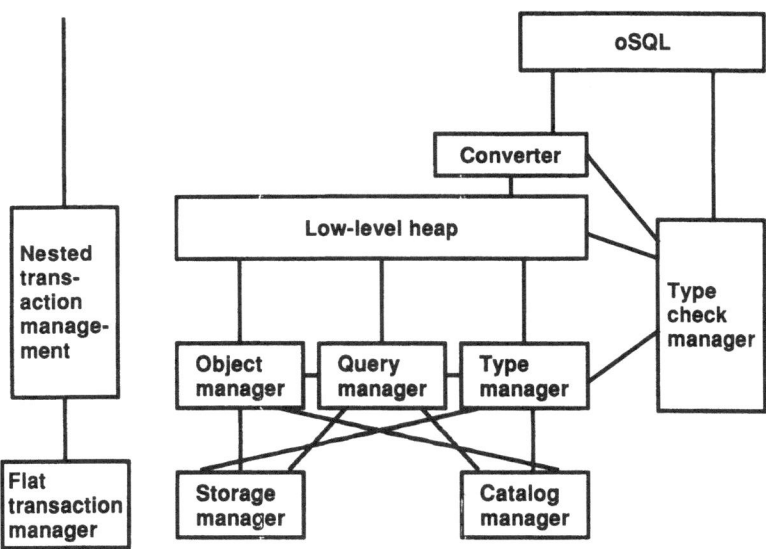

Fig. 3.2 : General architecture of the storage manager

In the current implementation, the flat transaction manager and the catalog manager are based on a relational database system. However, dedicated severs are under evaluation.

MaX

MaX provides comfortable monitoring and debugging facilities to the programmer of HooDS. In this paper, we restrict ourselves to emphasising the debugging facility of MaX. The MaX debugger provides multi-language support for language implementations based on the COFF format. MaX debugs at source code level, providing all code printouts in the original programming language. As a debugger suited for object-oriented languages, it also solves inheritance by listing the inherited code when it is executed. Similarly, MaX is able to cope with exception handling and displays raised exceptions and the name of the exception to the programmer. Besides common debugging facilities, the MaX runtime environment supports conditional and unconditional breakpoints and trace points, full variable access (persistent and volatile) and all data types. On the basis of standard UNIX(R) facilities, MaX is able to debug programs written in different languages, thus providing CooL source code whenever CooL is used, for example, and switching to C source code when C code is called.

Filter/Browser

The filter/browser system constitutes an end-user tool which allows the objects of an application to be retrieved interactively. The retrieval process is carried out in two steps. The filter provides a query-by-example interface [Zloof '82] to the user which allows objects to be queried in a natural way by describing patterns for the results of the query. The evaluation of the query yields a so-called result set. The user can navigate through this result set by means of the browser. It should be noted that the filter/browser is a standard tool for which the information is generated internally by the system. It need not be implemented by the programmer for a specific application.

DialOOg

This tool is currently undergoing design and will provide a dialog manager which generates user interfaces based on meta-information kept internally within the system. These interfaces will be form-oriented and will incorporate as a default the entire dialog behaviour to permit object creation, modification, display and method activation. References to objects will be marked and may be dereferenced by navigation. This provides an implicit browsing mechanism within the object base conforming to a hyper-object approach. DialOOg is based on OSF/Motif and can be tailored at several levels for supporting specific interface layouts.

4. Application Development Tools

The goal of the ITHACA Application Development Environment is to reduce the long-term costs of application development and maintenance for standard applications in selected application domains. By "standard" applications we mean classes of similar applications that share concepts, domain knowledge, functionality and software classes.

The key requirement is that applications developed using the environment be flexible and open-ended: it should be possible not only to develop applications quickly and flexibly, but it should be relatively easy to reconfigure applications to adapt to evolving requirements.

The key assumption of the approach is that one must be able to adequately characterise the selected application domains in order that individual application can be constructed largely from standard object-oriented software components. Achieving reusability of not just software but of previous experience is therefore an essential activity within the approach. This, in turn, implies the need for a different kind of software life-cycle in which the long-term development and evolution of reusable software proceeds in tandem with the short-term development of specific applications.

The ITHACA approach requires the development of "application workbenches" consisting of reusable application domain specific software components and software information. This information is stored in a *Software Information Base* (SIB) and is accessed by an application developer either interactively by a Selection Tool, or indirectly through the use of the other development tools. We distinguish between *application development*, which refers to the development of specific applications by means of the application workbenches and the ITHACA development tools, and *application engineering*, which refers to the activity of preparing the contents of the SIB, that is, developing the application workbenches themselves.

Application engineering is expected to be an incremental, long-term activity, since reusable software can only be developed on the basis of experience gained from the development of specific applications (including existing "precursor" applications). The benefits are to be realised during application development, since one can reuse existing requirements models, existing (generic) designs and existing software. Most of the activity in application development should ideally take place in the task of matching user requirements to generic designs, and in configuring running applications from available components. Since mostly tried and tested software will be used, less time should be spent on detailed debugging, and since construction of applications from existing parts should rapidly lead to evolutionary prototypes, more time can be spent ensuring that the client's needs are properly met. Except for unusual applications, little effort should be spent capturing exceptional requirements, re-engineering existing designs or programming new custom software. Clearly this scenario depends on the extent to which one can capture application domains, but even non-standard applications can benefit from the approach if the special requirements can be localised to particular subcomponents or to incremental modifications of existing software objects.

The raw material with which the application developer works is the software information contained in the SIB. This information is organised into *Generic Application Frames* (GAFs), which describe how specific applications can be constructed from the available components. In order to ensure that it will be possible to map the application requirements to a GAF, it is essential that requirements collection and specification start by using the SIB as a basis for specifying the application. Matching requirements to existing application domain knowledge starts *immediately* as requirements are collected. The *Requirements Collection And Specification Tool* (RECAST) effectively provides a "guided tour" of the SIB, attempting to construct a specific application frame (SAF) on the basis of generic information and user requirements. Software components selected and identified at this stage are then tailored and composed to construct the running application by a combination of programming and *scripting*. The *Visual Scripting Tool* (VISTA) supports the interactive construction of applications by graphically editing and connecting visual representations of application and user interface objects. The tools are tightly integrated to permit, for example, simultaneously development of parts of the application during requirements collection, re-examination and refinement of requirements specifications during development, and access to programming tools during scripting. We shall briefly present the functionality and current status of each of these tools. The following diagram provides a simplified view of the tools interaction.

The SIB provides the underlying mechanisms for storing and representing descriptions. Descriptions encapsulate properties of software components and knowledge concerning application domains for use by the other tools. Since descriptions may refer to one another, the contents of the SIB can be seen as a semantic network, sharing properties of both object-oriented (design) databases and of hypertext systems. A prototype of the SIB has been built using the Telos knowledge representation language [Koubarikis et al. '89]. Descriptions are thus represented internally as Telos propositions, but externally may appear as, for example, software templates, requirements collection forms, or application designs. The uniform internal representation permits advanced queries to be posed to the SIB and evaluated by the Telos inferencing mechanisms.

The Selection Tool provides the means to retrieve software descriptions from the SIB and to navigate through GAFs. The Selection Tool may be used either interactively or indirectly via the other tools. The two modes supported are (1) querying, in which a thesaurus and a set of filters are used to reduce the focus of interest, and (2) browsing, in which the client can navigate through the software information network. A prototype has been built using the **Labyrinth** system (Laby), a constraint-oriented graphics engine which runs on top of X. Laby is used to display various hierarchical views of the SIB during browsing.

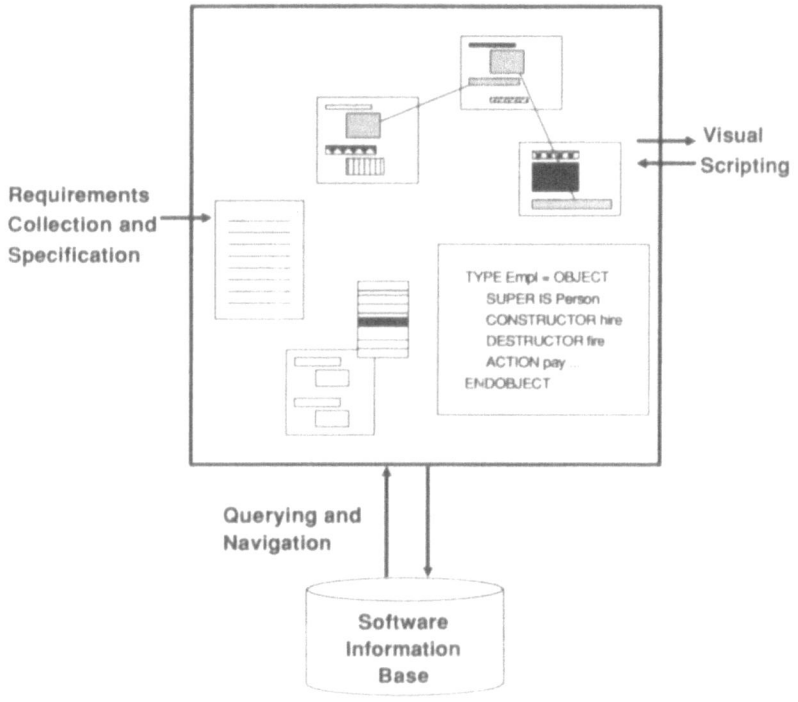

Fig. 4.1 : Simplified view of user/tools interaction

RECAST directs the application developer in the task of collecting requirements and matching them to existing designs and specifications by providing a "guided tour" of the SIB. RECAST is based on the Objects with Roles Model (ORM), an object-oriented specification model which considers that objects undergo a life-cycle in which they may play a variety of roles, each of which may associate different properties, states, methods, usage rules and constraints with objects [Pernici '90]. A first prototype of RECAST guides the application developer along a series of choices associated with different parameters of a generic application. Current work on RECAST is concerned with refining this specification technique and with the flexible interplay between RECAST and the other tools.

VISTA is an interactive tool for scripting running applications from previously programmed components [Nierstrasz et al. '90]. The term "scripting" refers to the idea that components should completely encapsulate some well-defined behaviour, like character actors in a play, whereas the interaction between components and the precise roles they should play should be specified separately, that is, within a script. A script, then, may bind parameters of generic classes, specify which instances of object classes are needed in various parts of an application, and introduce objects to one another. Both user interface components and internal application components are visually presented to the user, and graphical editing facilities are used to link together the components. Scripts encapsulate the bindings between objects, and so may themselves be reused as components within other scripts. Furthermore, a script, with the help of the other tools, can viewed as a generic design: once an application has been built as a script, it is a simple matter to unpack it and use it as a prototype for a new application, obtaining suggestions for design alternatives via RECAST and the Selection Tool.

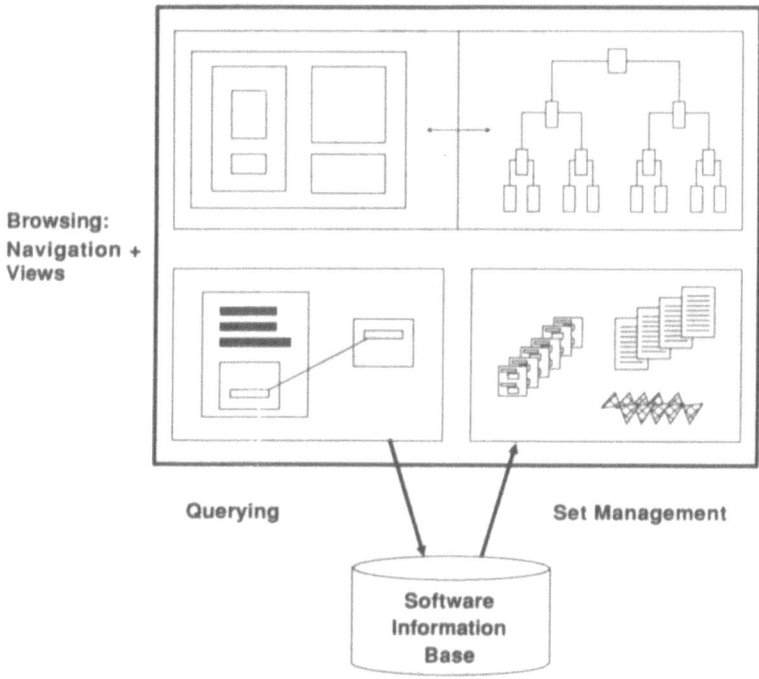

Fig. 4.2 : SIB management

Scripting depends heavily on the definition of standard interfaces between components and on the existence of a rich library of components that conform to these interfaces. Object-oriented techniques are essential here as they provide the mechanisms by which subclass hierarchies can share a common abstract interface. A "scripting model" is a description of these standard interfaces and the allowable interconnections supported by components for a given application domain. For example, a scripting model for a user interface kit might specify which methods are to be used to propagate events between application and user interface objects, and which events are valid to pass to which classes of objects.

A proof-of-concept prototype based on a UNIX® scripting model has been implemented [Stadelmann et al. '90]. An object-oriented visual scripting tool (VISTA) based on Laby and the MOTIF toolkit is now under development, and an initial version is expected to be released within the project by the end of 1990. Work on scripting models for various application domains is ongoing.

5. The Applications

Demonstrator applications which reflect real-world situations and common problem areas facing industrial software production are used to test and ensure the suitability of the environment developed in ITHACA. In addition to specifying their requirements, these 'workbenches' are also developing new domain architectures and innovative techniques oriented to their respective, non-IT-specific spheres of activity.

Office applications constitute a real challenge for testing the suitability of ITHACA technology in the commercial/industrial scenario. Offices in general form an application scenario which requires a wide range of different tools and facilities in order to meet the requirements of the different "styles" of office encountered (size, type of work etc.). The general nature of these tools allows them to be used in a variety of other applications to a greater or lesser extent.

Due to the fact that the ITHACA Office Workbench provides a general concept in terms of a reference office model and several basic technologies, it is described in some detail here. The components it includes can be applied to the other workbenches as necessary.

5.1. The Office Workbench

Any office application must (1) deal with a complex environment (distribution, client/server architecture, large objects), (2) integrate existing pieces of software (editors, spreadsheet, printing and filing services, electronic mail services), and (3) provide a very simple interface to non-expert end-users. The major application target selected is computer-supported cooperative work in all its possible variations: from rigid to flexible procedures, from a few instances to hundreds of thousands of instances, from a few steps to one hundred steps, with a total duration ranging from one day to several months, and involving two to dozens of actors.

In order to meet this goal, several interrelated components are provided that can be used together or separately. A CooL Office Model provides an object representation of organisation, information and facilities. A set of CooL Office Operators enable activation of classical office operations (such as editing, filing, mailing, printing) on these objects and the definition of more elaborate actions by assembling them in scripts using the scripting tool. A coordination procedure application provides the means to express procedures in the form of a network of activities which encapsulate actions. A desktop application provides the user with a workstation interface to the overall application. In addition, a Budget Management System provides a decision-making tool using constraint-based artificial intelligence techniques for analysing and optimising budgets in large organisations. This tool will be able to access information provided by commercial information services with access to an information services programmatic interface.

The **Office Model** provides a set of CooL objects representing office concepts in the fields of organisation, facilities and information [Ang '90]. The organisation model includes three main object types: grouping, actor and role. It constitutes the basis for authorisation and access control. The facilities model represents objects to organise the user space in the form of folders, repositories, trash-can, in-tray, out-tray, printer and file servers. The information model represents objects, such as forms, documents, messages, drawings, spreadsheets etc. The various parts of the model are interrelated. A folder can contain a letter whose "to" and "copy" instance variables can have sets of actors or roles as their value. Together, objects of the office model represent the application domain on which office operators can be applied and combined into scripts to form actions linked together into procedures by using the COP tool.

Office Operators represent an abstraction of most commonly used actions in an office environment, such as editing, mailing, filing and printing. They comprise a combination of objects and methods which encapsulate existing pieces of software and which take into account the objects of the office model. For example, a print operation will involve the information object to be printed, a printer server object representing the available printer server, a trading object able to negotiate the applicability of the service to print the information object, and a conversion object which, if required, is able to convert the format of the content portion of the object to a format acceptable to the printer.

COP (COordination Procedure) is a tool for supporting the application engineer and the end-user in modelling and executing cooperative task processing [Tueni et al. '90]. It is designed to support a large scope of applications, from highly repetitive, highly structured administrative procedures (workflow processing) to highly unstructured and flexible cooperative sessions between office actors (cooperative computer-supported work). COP relies on a language providing constructs to express dynamic behaviour of the coordination procedure, an engine which interprets structures generated by the compiler for supporting the distributed execution of the procedure, and support tools for graphical programming, scripting actions, debugging and for using the COP applications. The principal COP elements are activity, input and output, pre- and post-condition, procedure and object. The language supports the expression of sequences, alternatives, parallelism, loop and rendezvous, composition and refinement, aggregation and specialisation, and an unlimited level of abstractions. The COP engine offers resume, suspension, correction, cancellation, jumping, exception, help, history, event triggering and explanation.

The general **Desktop** is configured to present the user with a metaphor of her/his current working environment that must be easy to use and as flexible as possible to incorporate all personal applications available to her/him [Brady '90]. In order to facilitate speedy access, the Desktop presents all documents which have just been created by the user, or which the user wants to edit immediately and which are therefore available locally on the workstation. The Desktop is used to start all local applications, such as editors for texts, graphics and images. In addition, the Desktop application allows the user to access all office services, such as electronic mail, central filer, printer service etc. Documents no longer required can be stored in a trash-can which is also a component of the desktop. The Desktop is to be implemented in C and CooL, with OSF/Motif for the user interface.

The **Budget Management System** supports the functions of a company top management budgeting process, including budget preparation and budget evaluation using financial analysis [Bicard-Mandel et al. '89]. It incorporates a data model and appropriate knowledge-based tools. The data model provides the means for viewing the budget elements at several levels of abstraction and/or factors of analysis (e.g. company departments, chart of accounts, products, time). The knowledge-based tools provide constraint formalisms tailored to the budgeting domain. These formalisms are used for expressing consistency goals for the data model. Tools are provided for building the semantic network representing a company's budget, where nodes are budget entities (accounts, products, departments, production factors) and where links have budget-oriented semantics with associated methods for regulating information flow through the system.

Access to information services provides a CooL programmatic interface for implementing access to information in a client/server mode. This programmatic interface acts as a client for the service and as a server for the client application or user interface. The services integrated in this way could act as sources of up-to-date information, which could then be transferred to and exploited in local databases, spreadsheets and word processors. A specific access can be integrated as an action part in an activity network controlled by the COP engine.

5.2. Financial Workbench

The activities of the Financial Workbench are organised around the development of both a specific and a generic application for the sale of financial products, in particular insurance [Bruni et al. '89]. Key marketing issues are the development of new insurance services and products, as well as the extension and enhancement of the distribution network. In this context, a sales support tool is being developed which collects and uses all relevant information related to consultancy on the subject of insurance products geared specifically to customer needs.

Activities centered on developing the specific insurance application represent a key step for acquiring the necessary knowledge and expertise concerning the domain to be abstracted in the application frame.

5.3. Public Administration Workbench

The Public Administration Workbench demonstrates the applicability of the ITHACA environment in relation to a real setting within a public administration setup

[Garcia et al. '89]. The application is concerned with workflow automation, where a citizen's request must be processed within the administrative setup by different services according to precise rules and at different locations.

5.4. Chemistry Workbench

The Chemistry Workbench belongs to the CIM domain and involves the development of a graphical control desk for controlling plant and machinery in the chemicals production process. This control system is linked to a planning system for solving the logistic problems specific to this particular sector.

6. An Industrial Approach

From the very outset, the ITHACA project chose an industrial approach by selecting standards to be applied for development and by defining a common hardware platform to ensure full integration [Konstantas '90]. The goals of the project already correspond to the established main line of activity of the development units of the main industrial partners involved.

The standards and de facto standards which have been adopted within the project include the following:
- UNIX®, conforming to the X/Open portability guide,
- C compiler and GNU C++ compiler,
- X/Windows and OSF/Motif toolkits for user interfaces,
- TCP/IP and NFS,
- SQL for database interface,
- ODA, ODIFF, DFR, X.400 and X.500 for office automation.

Besides ensuring compatibility between the various organisations involved in the project, the consistent use of these standards will also ensure conformance with the market requirements when finished products are ready to be launched.

In addition, both Siemens Nixdorf Informationssysteme AG and Bull S.A. participate actively in the same worldwide organisations for the main elements of the architecture: X/Open for UNIX(R) portability, the Open Software Foundation for UNIX® extensions, and the Object Management Group for future object-oriented standards.

The ITHACA common hardware platform is an Intel 80386-based workstation delivered either by Bull S.A. or Siemens Nixdorf Informationssysteme AG. The workstations run the UNIX(R) operating system and conform to the 386 application binary interface. The workstation provides full support of a selection of components, including C, GNU C++, X/Windows System, OSF/Motif, TCP/IP, NFS, INET and ORACLE. All software components produced in ITHACA are qualified on this platform and have to run there before they are classified as accepted and able to be officially delivered to each partner.

7. Conclusions

In order to gain as much exposure to the market as possible, the ITHACA environment will be provided to external companies and universities for evaluation purposes on the basis of evaluation licence contracts as soon as individual components are available. For example, the first version of the CooL compiler has been available since June 1990 and the first version of the NooDLE/CooL integrated environment will be available in October 1990. An office application prototype written in CooL will be on show at the exhibition accompanying the ESPRIT Conference 1990.

Prototypes of each of the application development tools were produced during the first year of the project. Redesign and reimplementation of the tools is now under way. A C + + /X version of Labyrinth has been made available to partners and is now being used as a graphical front-end to VISTA. A first version of VISTA is being made available to the partners as of October 1990.

References

[1] R. Agrawal and N.H. Gehani, "Rationale for the Design of Persistence and Query Processing Facilities in the Database Programming Language O++", Proceedings of the Second International Workshop on Database Programming Languages, Morgan Kaufmann Publishers, Inc., San Mateo, California, 1989.

[2] J. Ang, "A Comprehensive Office Modeling Framework for Ithaca", Technical Report, Bull S.A., Paris, 1989, ITHACA.BULL.89.D8.1a.

[3] G. Attardi, C. Bonini, M.R. Boscotrecase, T. Flagella and M. Gaspari, "Metalevel Programming in CLOS", in Proceedings ECOOP '89, ed. S. Cook, British Computer Society Workshop Series, 1989.

[4] D.S. Batory, J. Barnett, J. Garza, K. Smith, K. Tsukuda, B. Twichell and T. Wise, "Genesis: A configur- able database management system", Technical Report TR- 86-07, Department of Computer Science, University of Texas at Austin, 1986.

[5] J. Bicard-Mandel (Ed.), C. Chen, G. Vlodakis and M. Androulakis, "User Requirements and Architecture Specifications for the Budget Management System", Technical Report, Bull & NTUA-CMSU, 1989, ITHACA.BULL.89.D10.1.

[6] M. Brady, S. Kienapfel and G. Zschoche, "An Office Desktop Approach", Technical Report, Nixdorf Micropro- cessor Engineering GmbH, Berlin, 1989, ITHACA.NIX-DORF.89.D3#1.

[7] E.Brodde, "MaX - Monitoring and X-Ray Tool for ITHACA", Technical Report, Nixdorf Microprocessor Engineering GmbH, Berlin, July 1989, ITHACA.NIXDORF.89.E4.#1.

[8] E.Brodde, "MaX User Handbook", User Handbook, Nixdorf Microprocessor Engineering GmbH, Berlin, August 1990, ITHACA.NIXDORF.90.E4.4.1#1.

[9] G. Bruni, C. Cardigno, M. Damiani and G. Seminati, "Final Report on Insurance Domain Requirement Analysis", Technical Report, DATAMONT R&D, Milano, 1989, ITHACA.DATA-MONT.89.D.7.#3.

[10] M. Carey, D. DeWitt, D. Frank, G. Graefe, M. Muralikr- ishna, J. Richardson and E.

Shekita, "The Architecture of the EXODUS extensible DBMS", Proceedings of Object-Oriented Database Workshop, 1986.

[11] CooL Development Team, "CooL/0 Language Description", Reference Manual, Nixdorf Microprocessor Engineering GmbH, Berlin, May 1990, ITHACA.NIXDORF.90.L2.#2.

[12] K.R. Dittrich, A. Geppert and V. Goebel, "The Data Definition Language of NO2", Technical Report, Univer- sity of Zürich, March 1990, ITHACA.ZUERICH.90.X.4#2.

[13] A.Elsholtz, "NooDLE, a New Object-Oriented Database System for Advanced Pro-gramming Language Environments", Technical Report, Nixdorf Microprocessor Engineer-ing GmbH, Berlin, November 1989, ITHACA.NIXDORF.89.X.4#1.

[14] J. Garcia, J. Lopez, J. Mongiou and R. Sole, "Design Description of the Administration Workbench. First Draft.", Technical Report, TAO, Barcelona, 1989, ITHACA.TAO.89.D.6.#8.

[15] A. Geppert, K.R. Dittrich and V. Goebel, "An Algebra for the NO2 Data Model", Technical Report, University of Zürich, July 1990, ITHACA.ZUERICH.90.X.4#4.

[16] Brian W. Kernighan and Dennis Ritchie, The C Program- ming Language, Prentice-Hall, Inc., Englewood Cliffs, New Jersey 07632, 1978.

[17] E. Kolodner, B. Liskov and W. Wheil, "Atomic Garbage Collection: Managing a Stable Heap", in Proceedings of the ACM SIGMOD International Conference on the Manage- ment of Data, 1989.

[18] D. Konstantas, "The Ithaca UNIX Development Platform", Technical Report, D-Tech, 1990, ITHACA.D-TECH.90.X.#4.

[19] M. Koubarikis, J. Mylopoulos, M. Stanley and A. Bor- gida, "Telos: Features and Formalization", Technical Report CSI 1989/018, FORTH, Herklion, Crete, 1989.

[20] T. Krickstadt, CAKE: Towards an Object-Oriented Appli- cation Builder, Nixdorf Microprocessor Engineering GmbH, Berlin, September 1990, ITHACA.NIXDORF.90.E1.#1.

[21] C. Lecluse and P. Richard, "The O2 Data Model", Techni- cal Report, pp. 39-89, Altair, Le Chesnaey Cedex, 1989.

[22] G. Müller and A.-K. Pröfrock, "Four Steps and a Rest in Putting an Object-Oriented Programming Environment to Practical Use", in Proceedings ECOOP '89, ed. S. Cook, British Computer Society Workshop Series, 1989.

[23] O.M. Nierstrasz, L. Dami, V. de Mey, M. Stadelmann, D.C. Tsichritzis and J. Vitek, "Visual Scripting -- Towards Interactive Construction of Object-Oriented Applications", in Object Management, ed. D.C. Tsi- chritzis, pp. 315-331, Centre Universitaire d'Infor-matique, University of Geneva, July 1990.

[24] B. Pernici, "Objects with Roles", Proceedings ACM-IEEE Conference of Office Infor-mation Systems (COIS), Bos- ton, April 1990.

[25] P. Pistor and F. Andersen, "Designing a Generalized NF2 Model with a SQL-Type Language Interface", Proceedings of the Twelfth International Conference on Very Large Data Bases, IBM Wissenschaftliches Zentrum, Kyoto, August 1986.

[26] A.-K. Pröfrock, M. Ader, G. Müller and D. Tsichritzis, "ITHACA: An Overview", Proceed-ings of the Spring 1990 EUUG Conference, pp. 99-105, 1990.

[27] M. Stadelmann, G. Kappel and J. Vitek, "VST: A Script- ing Tool Based on the UNIX Shell", in Object Manage- ment, ed. D.C. Tsichritzis, pp. 333-344, Centre Universitaire d'Informatique, University of Geneva, July 1990.

[28] B. Stroustrup, The C++ Programming Language, Addison- Wesley, Reading, Mass., 1986.

[29] C. Thompson et al., "Open Architecture for Object- Oriented Database Systems", Technical Report 89-12-01, Texas Instruments, 1986.

[30] D.C. Tsichritzis and O.M. Nierstrasz, "Application Development Using Objects", Proceedings of the First European Conference on Information Technology for Organisational Systems, pp. 15-232, North Holland, Athens, May 1988.

[31] M. Tueni (Ed.), J. Alsina, A. Graffigna, J. Li, G. de Michelis, J. Monnguio and H. Wiegmann, "Towards a Com- mon Activity Coordination System", Technical Report, Bull S.A., Nixdorf Computer AG, TAO, University of Milano, 1989, ITHACA.BULL.89.U2.#1.

[32] M.M. Zloof, "Office-By-Example: a business language that unifies data and word processing and electronic mail", IBM Syst. Journal, vol. 21, no. 3, pp. 272 - 304, 1982.

Action, No. BR 3175
DEFAULT REASONING
AND DYNAMIC INTERPRETATION OF NATURAL LANGUAGE

FRANK VELTMAN
Department of Philosophy
University of Amsterdam
15 Nieuwe Doelenstraat
Amsterdam, The Netherlands

EWAN KLEIN & MARC MOENS
Centre for Cognitive Science
2 Buccleuch Place
University of Edinburgh
Edinburgh EH8 9LW, Scotland, UK

Abstract: We present a proposal for treating default reasoning from the perspective of a dynamic approach to semantics, where meaning is a mapping between information states. Information states are identified with sets of possible worlds—the epistemic possibilities which those states admit. Generic rules, like *On weekdays, Giles normally gets up at 8.00* are then taken to induce a pre-order on possible worlds, where worlds complying with the rules are less exceptional than those which go against the rules. Thus, a particular weekday on which Giles gets up at 8.00 is less exceptional than one on which he stays in bed till noon. Unlike many other approaches to nonmonotonicity, we draw a distinction at the level of the object language between defeasible and indefeasible conclusions.

1 Introduction

The ESPRIT Basic Research Action DYANA —*Dynamic Interpretation of Natural Language*—is concerned with developing a formal theory of language interpretation and processing which models human cognitive abilities but is at the same time mathematically precise and admits computational interpretation. An important goal of DYANA is to go beyond particular, isolated problems occurring at individual levels of interpretation and to study the way these levels of interpretation interact in an integrated theory. The work programme is divided into three interdependent components:

- grammar development, speech and prosody
- meaning, discourse and reasoning
- logic and computation

For a more detailed discussion of the work in each of these areas, and for a general overview of the project, the reader is referred to Klein and Moens [9]. In this paper we will describe recent DYANA work on default reasoning. In § 2, we will sketch why this work is relevant to natural language understanding and show how it fits into the DYANA project. In § 3, we discuss some basic notions of update semantics, while § 4 presents a key distinction between stable and non-stable sentences. § 5 sketches the mechanisms by which default rules induce a preference order on epistemic states. Finally, in § 6, we briefly discuss relations between our approach and that of semantic networks, and point to some directions for future research.

2 Partiality, Dynamics and Nonmonotonicity

DYANA focuses on two important themes, namely the dynamics of natural language interpretation, and theories of partial information. The two themes are connected: interpretation is dynamic since it involves the constant manipulation of information which is extracted, transduced and modified at all levels of representation—phonological, syntactic, semantic, and pragmatic. Meaning thus becomes a **dynamic** notion: at all levels of representation, it can be defined as a function from information states to information states.[1] Both the domain and the range of this function will be states of partial information, since complete information states hardly need an update. But partiality also arises as a result of the dynamics of the interpretation process itself: ambiguities and other indeterminacies will be encountered at each stage of the interpretation process. The result is that it is not always possible to pass complete and reliable information between levels of representation in a predefined way.

Defeasibility plays a pervasive role in natural language understanding. At the most global level, understanding a discourse involves the integration of new incoming information into an existing body of beliefs, assumptions and commitments. It is hardly surprising that information states evolve in a nonmonotonic fashion—assumptions which are plausible at one stage become rendered untenable later on, and even deeply-held commitments may have to be abandoned in the face of new, conflicting facts. However, defeasibility infuses the very texture of the human processing mechanisms which map linguistic input (whether speech or written text) into some kind of discourse representation. We briefly list below just a few examples where defeasible inferences are drawn on the basis of partial information:

Semantics:

- Generic sentences: *Tigers have four legs. Shere Khan is a tiger... Shere Khan has lost a leg.*

- Quantifier scope preferences: *Every student here speaks a foreign language... It is French.*

- Tense and aspect: *Lee was crossing the street... Unfortunately, she was hit by a truck before she reached the other side.*

- Lexical semantics: *This is a flower... In fact, it's a plastic flower.*

Morphology:

1. English verbs take a past tense by suffixing -*d* (e.g. *bake* ~ *baked*).

2. Verbs with roots of the form *Xing* (e.g. *sing*) take a past tense by changing to *Xang* (e.g. *sang*)

3. *bring* is a verb of the form *Xing* and takes a past tense *brought*

Morphophonology:

1. The masculine singular form of the French lexeme BEAU is *bo*\.

2. The masculine singular form of the lexeme BEAU is *bεl*\ if the following word starts with a vowel.

Notice that our morphology example is analogous to the following set of statements:

1. Birds (normally) fly.

2. Penguins (which are birds) waddle (but don't fly).

3. Max is penguin which hops (but doesn't waddle).

---

[1] For a synoptic discussion of the dynamic perspective to logic, see van Benthem [1].

Formulated in terms of rule application within the framework of generative grammar, we would say that more specific rules (like rule (3) for *bring*) are deemed to take precedence over less specific ones (such as rule (2) for verbs of the form *Xing*, or rule (1) for verbs in general). The most general rule is said to be the 'elsewhere' case.

The problem of formalizing nonmonotonic inference is an active research topic in the area of Artificial Intelligence. Reasoning devices are supposed to derive conclusions that follow logically from the facts and rules stored in their databases. However, it has often been noted that reasoning devices are sometimes expected to draw conclusions that are not necessarily true but nevertheless seem reasonable given the circumstances. The Artificial Intelligence literature contains many examples of this type of default reasoning in domains other than natural language, and offers a plethora of techniques for the formalisation of the nonmonotonic behaviour of reasoning systems (see, e.g., Reiter [12]; Shoham [13]).

One branch of DYANA work on nonmonotonicity, carried out by Morreau [11], studies the dynamics of information states which support or contain, in the form of conditional sentences, meta-information about their own response to revision. In a second branch of research, Veltman [14] develops a modal semantics for default reasoning and, to the extent that these express default rules, for generic sentences. As such it moves on territory which is familiar from the Artificial Intelligence literature on the subject, while importing into it techniques which originated in philosophical logic. Given limitations of space, we will not try to discuss Morreau's work further here, but instead give a brief overview of Veltman's results. Before entering into more details, it is worth drawing attention right away to a central feature of our approach: the notion of default reasoning is captured by drawing a distinction between defeasible and non-defeasible conclusions at the level of the **object language**. As a result, our task is provide an adequate semantics for special kinds of sentences, namely those which express default rules and defeasible conclusions.

3 Dynamic Interpretation

According to the standard explication of logical validity, an argument is valid if its premises cannot all be true without its conclusion being true as well. Crucial to this approach is the specification of truth conditions. The heart of the theory presented below consists instead in a specification of **update conditions**. That is, you know the meaning $[\phi]$ of a sentence ϕ if you know the change that it brings about in the information state of anyone who accepts the news conveyed by ϕ. Thus, as we suggested above, $[\phi]$ is an operation for updating the information states of an idealised agent.

Let σ be an information state and ϕ a sentence with meaning $[\phi]$. Then we write

$$\sigma[\phi]$$

for the information state that results when σ is updated with $[\phi]$. In most cases $\sigma[\phi]$ will be different from σ, but it is possible that the information conveyed by ϕ is already subsumed by σ; thus, updating σ by $[\phi]$ will simply result in σ. In such a case, i.e. when

$$\sigma[\phi] = \sigma,$$

we say that ϕ is **accepted in** σ, and write this as

$$\sigma \Vdash \phi.$$

It may be helpful to the reader if we give a preliminary example to show what updating rules look like. To begin with, let us be more specific about how to characterise an information state. For the sake of simplicity, an information state σ can be identified with a subset of the set W of possible worlds. Intuitively, σ represents everything that an agent takes to be true at a given time, and thus contains those worlds which may yet turn out to be the real one. If the agent happens to know nothing at all, then any world may be the actual one, and σ is just W. We shall use '1' to represent this minimal information state. As the agent's information increases, σ shrinks, until—in the limit—it just consists of a single world. Thus, the growth of information is understood as the elimination of possibilities. Moreover, we also admit an absurd information state '0', identified with the empty set. Thus, we will have $\sigma[\phi] = 0$ when ϕ is inconsistent with σ. We can now introduce some additional terminology. If $\sigma[\phi] \neq 0$, then we say that ϕ is **acceptable in** σ, whereas if $\sigma[\phi] = 0$, then ϕ is **not acceptable in** σ.

Notice finally that we do not assume information states to be 'veridical', in the sense that they must contain the actual world. We admit $\sigma[\phi] = 0$ when in fact ϕ is true, and equally we allow $\sigma[\phi] = \sigma$ when in fact ϕ is false. Suppose however that $\sigma[\phi] = 0$, for a true sentence ϕ. In this case, an agent cannot refuse to accept ϕ when confronted with the facts; rather, she should **revise** her information state in such a manner that ϕ becomes acceptable. However, we shall not attempt to say anything here about how such a revision is carried out.

Let us take as given a finite set $\mathcal{A}$ of atomic sentences, and let $L(\mathcal{A})$ be a propositional language based on $\mathcal{A}$, whose sentences are built in the usual way. We can think of such sentences as expressing the kind of descriptive content which constitutes an information state. In addition, we add to the language a one-place sentential operator *might*. This can be prefixed to any sentence ϕ which does not already contain occurrences of *might*. *Might* ϕ sentences should not be thought of as expressing descriptive propositions. Rather, they have a meta-semantic character which tells us something about our current information state, namely whether ϕ is acceptable given what we already take to be true.

As pointed out by van Benthem [1], a dynamic approach to semantics makes it natural to postulate various **modes** of operating on an information state σ. For example:

Update: make a transition from σ to a new state σ' which extends the information in σ.

Downdate: revise an unsatisfactory information σ to produce a new state σ' which eliminates certain (mis)information in σ.

Test: check whether a given proposition is accepted in σ, and leave σ unchanged.

In terms of this taxonomy, descriptive sentences will perform updates, whereas sentences such as *might* ϕ carry out tests. In a fuller treatment, we would also need to allow an information state to be revised by a downdate statement—however, we will not consider problems of revision here.

Before turning to some explicit clauses to illustrate the dynamic approach, it is a useful technical detail to identify a possible world w with the set of atomic sentences from $\mathcal{A}$ which are true in w; hence W is the powerset of $\mathcal{A}$. With this clarification, the evaluation clauses for our language can now be stated as follows:

(1)   a $\sigma[p] = \sigma \cap \{w \in W \mid p \in w\}$, for any atom $p \in \mathcal{A}$

     b $\sigma[\neg\phi] = \sigma \setminus \sigma[\phi]$

     c $\sigma[\phi \wedge \psi] = \sigma[\phi] \cap \sigma[\psi]$

     d $\sigma[\phi \vee \psi] = \sigma[\phi] \cup \sigma[\psi]$

     e $\sigma[might\ \phi] = \sigma$ if $\sigma[\phi] \neq 0$

       $\sigma[might\ \phi] = 0$ if $\sigma[\phi] = 0$

The analysis of *might* is motivated by the following considerations: an agent will accept *might* ϕ just in case ϕ is consistent with what she takes to be true. As pointed out above, clause (1e) tests σ rather than updated it; if ϕ is acceptable in σ, then you have to accept *might* ϕ, while remaining in information state σ.

Although other notions of logical validity are possible in this context (cf. [14] for discussion), the one we shall employ here goes as follows: Let 1 be the minimal information state, where all epistemic possibilities are open. Then an argument is **valid** iff updating 1 with the premises

$\psi_1 \dots \psi_n$, in that order, yields an information state σ in which ϕ is accepted. Formally:

$$\psi_1 \dots \psi_n \models \phi \text{ iff } \mathbf{1}[\psi_1] \dots [\psi_n] \Vdash \phi$$

4 Stability

An important motivation of classical logic has been to abstract away from the context-dependence of ordinary discourse, with the goal of formalising arguments whose validity does not shift according to their position in a discourse. In particular, conclusions should be **stable** in the sense that if they are true at one stage, then they remain true regardless of what ensues subsequently. Equally, much attention has been paid within the framework of logic programming to find declarative formulations of data operations which abstract away from details of implementation, such as the order in which operations are carried out. Yet given that one of our goals is to formally model the context-dependence of natural language discourse, we wish to find ways of explicitly capturing the procedural aspects of informal argumentation. In pursuit of this, we will indeed allow into our formal system **non-stable** information—information which may become obsolete when more facts are acquired.

The distinction between stability and non-stability is one that we shall draw at the level of the object language. Thus, we say that some sentences ϕ are not stable, in the following sense:

Definition 1 (Stability) *A sentence ϕ is* stable *just in case for any σ and $\psi_1 \dots \psi_n$ if $\sigma \Vdash \phi$ then $\sigma[\psi_1] \dots [\psi_n] \Vdash \phi$.* [2]

Sentences involving *might* provide a simple example of non-stability. In the minimal information state $\mathbf{1}$, *it might be raining* is accepted, since *it is raining* is certainly acceptable in $\mathbf{1}$. Suppose we now learn that it isn't raining, and update our information state accordingly. Then *It might be raining* becomes unacceptable. Reading p as *it is raining*, we have $\mathbf{1} \Vdash might\ p$ by (1e), since $\mathbf{1}[p] \neq \mathbf{0}$, but we do not have $\mathbf{1}[\neg p] \Vdash might\ p$, since $\mathbf{1}[\neg p][p] = \mathbf{0}$.

There are other important epistemic operators which create non-stability. The one we wish to look at is *presumably*. Thus consider the following argument:

(2) Adults are normally employed
Wim is an adult
——————————————
Presumably Wim is employed

According to the semantics developed in [14], this argument is valid, in the sense defined above. Notice that we do not conclude that Wim **is** employed—only that he presumably is. This qualification makes explicit that an unstable, and therefore defeasible, conclusion has been drawn.

The argument (2) can remain valid as one learns more about Wim, so long as there is no evidence that the new information is relevant to the conclusion:

(3) 1: Adults are normally employed
2: Wim is an adult and a student
——————————————
Presumably Wim is employed

However, if we now adopt the default rule *Students normally aren't employed*, the argument is no longer valid. Thus, we can draw no conclusions about whether or not Wim is employed from the following premises:

(4) 1: Students normally aren't employed
2: Adults are normally employed
3: Wim is an adult and a student

---

[2]Notice that we are assuming that $\psi_1 \dots \psi_n$ onlyh express updates or tests on σ; if we were to admit downdates as well, then a different definition would be required.

Adding a fourth premise may make the balance tip. Thus, in (5), we draw a conclusion that is the opposite of what we previously inferred:

(5) 1: Students are normally adults
     2: Students normally aren't employed
     3: Adults are normally employed
     4: Wim is an adult and a student
     Presumably Wim isn't employed

In the presence of premise 1, the apparent incommensurability between Wim's being a student and his being an adult is lost, and premise 2 takes precedence over 3.

It should now be evident what we meant by our claim that defeasibility is made explicit at the level of the object language. In other theories, one may well infer from the premises in (2) that Wim is employed, only this time it is a different kind of inference (i.e. an nonmonotonic one). Default reasoning, we claim, is not a special kind of reasoning with ordinary sentences, but rather **ordinary reasoning with a special kind of sentence.**

5 Rules with Exceptions

Although there is no space to give an elaborated presentation of the intended semantics for sentences like those discussed in the preceding section, we shall attempt to sketch the basic mechanism by which default sentences are interpreted. The theory arose out of an attempt to give a dynamic twist to the theory developed by Delgrande [3], who in turn took Lewis's [10] study of counterfactuals as his starting point.

When an agent adopts a sentence of the form *normally* ϕ, she adopts certain expectations: worlds where ϕ holds are less surprising than those where it doesn't. To capture this idea, we need to give more structure to an information state. It must not only contain encode the set of epistemic alternatives, as before, but also an **expectation pattern** which makes explicit what an agent would expect to happen in the absence of complete information. Of course, the dynamics of interpretation now includes two kinds of change on an information state σ:

- modifying σ's descriptive content
- modifying σ's expectation pattern

Operations available in the language can avail themselves of just one of these options, or both.

We formalise the notion of expectation pattern in terms of a pre-order $\leq$ (i.e. a reflexive, transitive relation of 'preference'), where $w \leq w'$ just in case w is at least as normal as w'. When an agent updates her information state σ with a default statement such as

(6)   On weekdays, Giles normally gets up at 8.00,

her expectation pattern, encoded as the pre-order, will be modified in such a way that worlds in which the sentence holds are considered more normal than those in which it fails. Thus, given (6), a Monday on which Giles gets up at 8.00 is less exceptional than one on which he stays in bed till noon.

We increment our language with two new unary operators, *normally* and *presumably*. Again, we forbid iteration: sentences of the form *normally* ϕ and *presumably* ϕ are not allowed to contain further occurences of any of the epistemic operators. *Presumably* ϕ performs a simple test on an information state σ to determine whether ϕ can be (defeasibly) concluded. *Normally* ϕ, which expresses a default rule, effects a subtle change on σ's expectation pattern.

An information state now involves a pair $\langle \leq, s \rangle$, where s is again a subset of W and $\leq$ is a pre-order on W. We will call $\leq$ an **expectation pattern** on W. If $w \leq w'$ and $w' \leq w$, we write

$$w \equiv w'$$

58

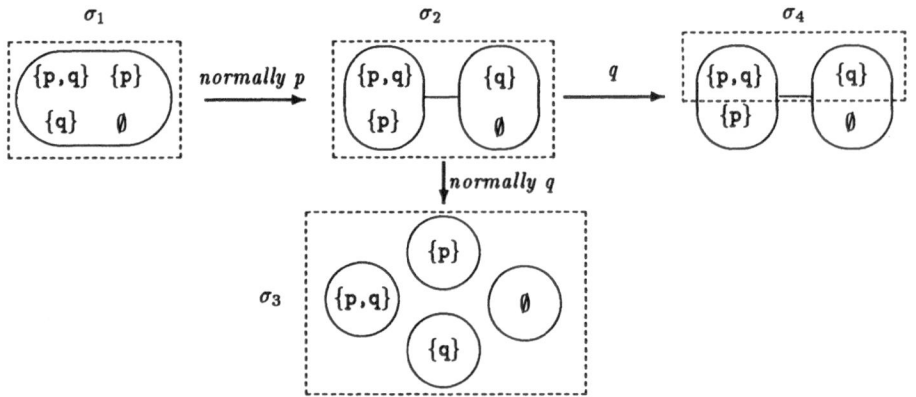

Figure 1: Expectation Patterns

Clearly, $\equiv$ is an equivalence relation.

Definition 2 *Let $\leq$ be an expectation pattern on W. Then*

1. *w is a normal world if $w \leq w'$ for every w' in W.*
2. *$\leq$ is coherent iff there exist some normal worlds.*
3. *$NORM_\leq$ is the set of all normal worlds relative to $\leq$.*

That is, a pattern is considered to be coherent if there is at least one possible world in which every proposition that expresses how things should normally be does in fact hold. This does not mean, however, that the real world must satisfy all the default rules accepted by an agent—by definition, default rules allow for exceptions.

Definition 3 *Let W be as before. Then σ is an* information state *iff $\sigma = \langle \leq, s \rangle$ and one of the following conditions is satisfied:*

1. *$\leq$ is a coherent pattern on W and s is a nonempty subset of W, or*
2. *$\leq = \{\langle w, w \rangle \mid w \in W\}$ and $s = \emptyset$.*

We now have:
$1 = \langle W \times W, W \rangle$ *is the* minimal *information state,*
$0 = \langle \{\langle w, w \rangle \mid w \in W\}, \emptyset \rangle$ *is the* absurd *information state.*

In order to grasp the semantic treatment we are proposing, it is helpful to consider illustrations like those in Figure 1. We assume in Figure 1 that $W = \{w_0, w_1, w_2, w_3\}$ where $w_0 = \emptyset, w_1 = \{p\}, w_2 = \{q\}$, and $w_3 = \{p, q\}$. If two worlds belong to the same $\equiv$ equivalence class, they are placed within the same oval, and if $w < w'$, then w is pictured to the left of w'. Finally, the worlds belonging to s are drawn within a dashed rectangle.

The information state $\sigma_1 = \langle \leq_1, s_1 \rangle$ on the left is the minimal state, 1: the expectation pattern treats all worlds as equally normal. When σ_1 is updated to a new state σ_2 with the sentence *normally p*, the expectation pattern is **refined**, with the result that worlds where p does not hold are judged to be more exceptional than those where it does. In particular, the worlds w_2 and w_0 no longer stand in the 'as normal as' relation to w_3 and w_1, and hence the corresponding pairs are removed from the new expectation pattern $\leq_2$ which arises after the update. A similar refinement occurs when σ_2 is updated by *normally q*. Now the most normal world in σ_3 is w_3, where both p and q hold.

As mentioned above, the dashed rectangles in Figure 1 demarcate the sets of worlds which are held to be true by the agent. It will be observed that default sentences only affect the expectation pattern in the information state; thus $s_3 = s_2 = s_1 = W$. However, when the information state σ_2 is updated with a descriptive sentence q, the s_2 is reduced accordingly, so that worlds w_1 and w_3 are excluded from the resulting set s_4.

We now define the notion of refinement.

Definition 4 (Refinement) *Let $\leq_0, \leq_1$ be expectation patterns on W and let $X \subset W$.*

1. *$\leq_1$ is a refinement of $\leq_0$ iff $\leq_1 \subset \leq_0$.*
2. *$\leq \circ X = \{\langle w, w' \rangle \in \leq \mid w \in X$ or $w' \notin X\}$*

As we already observed, refinement is brought into play when a new default rule is acquired. Suppose that we are currently in an information state σ where w is at least as normal as w'. That means that w satisfies at least as many default rules as w'. What happens when σ is updated with a new default rule *normally ϕ*? Let us define

$$\|\phi\|$$

to be the set of ϕ-worlds; i.e. worlds in which ϕ holds. Then only worlds within $\|\phi\|$ can satisfy all the default rules. Assume that ϕ is compatible with all the preceding rules, and thus that $NORM_{\leq}[\phi] \neq \emptyset$. Then we have to refine $\leq$ to the new pattern $\leq \circ \|\phi\|$ by excluding from $\leq$ any pairs which render a non ϕ-world at least as normal as a ϕ-world; i.e. any pairs of the form $\langle w, w' \rangle$ such that $w \notin \|\phi\|$ but $w' \in \|\phi\|$. Consider, for example, $NORM_{\leq_2}$ of σ_2 in Figure 1. This will contain the pair $\langle \{p\}, \{p, q\} \rangle$. But $\{p\} \notin \|q\| = \{\{p, q\}, \{q\}\}$, though $\{p, q\}$ is. Consequently $\langle \{p\}, \{p, q\} \rangle$ is removed from $\leq_2 \circ \|q\|$, as required by the update rule for *normally q*.

Although the expectation pattern $\leq_4$ of σ_4 is the same as that in σ_3, there is an obvious sense in which one of the normal worlds, namely w_1, is no longer relevant. The expectation patterns of an agent in information state σ_4 will be determined by the set $\{w_3\}$ of **optimal** worlds which result when $\leq_4$ is restricted to the set s_4. We define this notion as follows:

Definition 5 *Let $\leq$ be a pattern on W and let $s \subseteq W$.*

1. *w is optimal in $\langle \leq, s \rangle$ iff $w \in s$ and for every $w' \in s$, if $w' \leq w$ then $w' \equiv w$.*
2. *OPT is an optimal set in $\langle \leq, s \rangle$ iff there is some optimal w in $\langle \leq, s \rangle$ such that $OPT = \{w' \in s \mid w \equiv w'\}$.*

Optimality, we see, is relative to two considerations: the default rules which are accepted in an information state, and the set s of worlds which constitute the current epistemic alternatives. Worlds which are less than optimal at one point become important when expectations have to be readjusted. As one's knowledge increases, and more and more alternatives are eliminated, the optimal worlds may disappear, and the best among the less than optimal worlds take over their role.

We bring this section to a close by stating update clauses for *normally* and *presumably*. Thus, if $\sigma = \langle \leq, s \rangle$ and ϕ is in $L_2(\mathcal{A})$, $\sigma[\phi]$ is defined as follows:

(7)   a   If ϕ is a sentence of $L_0(\mathcal{A})$, then

      1. $\sigma[\phi] = 0$ if $s[\phi] = \emptyset$.

      2.   Otherwise, $\sigma[\phi] = \langle \leq, s[\phi] \rangle$.

   b   If ϕ is *normally* ψ, then

      1. $\sigma[\phi] = 0$ if $NORM_\leq[\psi] = \emptyset$.

      2.   Otherwise, $\sigma[\phi] = \langle \leq \circ \|\psi\|, s \rangle$.

   c   If ϕ is *presumably* ψ, then

      1. $\sigma[\phi] = \sigma$ if $OPT[\psi] = OPT$ for every optimal set OPT in $\langle \leq, s \rangle$.

      2.   Otherwise, $\sigma[\phi] = 0$.

As we noted before, *presumably* ϕ resembles *might* ϕ in being an invitation to perform a test on σ rather than updating it. If the proposition expressed by ϕ holds in all optimal worlds in $\sigma = \langle \leq, s \rangle$, then *presumably* ϕ must be accepted, and σ is left unchanged. Unlike *normally* sentences, sentences of the form *presumably* ϕ are not in general stable. Even if it is a rule that *normally* ϕ, it may be wrong to expect to expect that ϕ. Such non-stability can be illustrated schematically by the following example:

(8) $1[normally\ p]$ $\|\!\!-$ *presumably* p
       $1[normally\ p]\ [\neg p]$ $\|\!\!\!/\!-$ *presumably* p

6   Conclusion

Important requirements for any approach to nonmonotonicity are the following:

- the defeasibility of conclusions drawn by default;
- scepticism in the face of conflicting defaults;
- the priority of more specific information before more general information.

The only other theory of nonmonotonic reasoning which seems to do justice to these requirements is the sceptical theory of nonmonotonic semantic networks due to Thomason and his colleagues [8,2]. In a sense it is to that theory that the present one is most closely related: semantic nets may be regarded as sets of statements belonging to a proper sublanguage of the languages whose semantics is explicated in Veltman's [14]. Broadly speaking, a link where P inherits from Q is formalised as a sentence of the form

$$P \leadsto Q$$

which we read as 'P normally implies Q. Unfortunately there is not space here to present the semantics of this new binary operator; however, it essentially selects an expectation pattern that is appropriate for the property P. This semantics does not correspond exactly to the inference principles proposed for semantic networks, but the fit is quite close. In fact, it appears that the logic defined by [14] is preferable to that characterized by those inference principles. Cycles of defaults, for example, are excluded in the Thomason *et al.* theory for technical reasons, but pose no special problem for the current modal theory of default reasoning by update, like Veltman's. It is an interesting but still open question whether the low computational complexity that has been claimed for inferencing on semantic nets is lost when we move to the richer languages introduced in [14].

The DYANA work described above is intended initially as a contribution to the theories of nonmonotonic reasoning taking shape within philosophy and Artificial Intelligence. However, it does not address directly the more specific problems of nonmonotonic processing that arise in the context of computational linguistics. In itself this is not objectionable at the end of this first phase of the research into nonmonotonic reasoning, for much is still needed by way of general clarification of the various aspects of nonmonotonic reasoning and revision in the face of inconsistency. But ultimately it is one of the central purposes of DYANA to relate such general insights to the processing of language. It is linguistic applications such as the inheritance of morphological and phonological properties and the structure of the lexicon that we hope to address in future work.

References

[1] van Benthem, J. [1990] 'General Dynamics,' paper presented to Workshop on Semantics and Computation, CIS, Universität München, July 1990. To appear in *Theoretical Linguistics*.

[2] Carpenter, B. and Thomason, R.H. [1989] 'Inheritance Theory and Path-Based Reasoning: an Introduction', Ms., Intelligent Systems Program, Pittsburgh, PA: University of Pittsburgh.

[3] Delgrande, J. [1988] 'An Approach to Default Reasoning Based on a First-Order Conditional Logic: Revised Report', *Artificial Intelligence*, 36, 63–90.

[4] Horty, J.F., Thomason, R.H. and Touretzky, D.S. [1987] 'A Skeptical Theory of Inheritance in Nonmonotonic Semantic Networks', *Proceedings of the 6th National Conference on Artificial Intelligence*, Seattle, Washington, pp358-363.

[5] Klein, E. and Moens, M. [1989] 'The Dynamic Interpretation of Natural Language', in ESPRIT '89: Proceedings of the 6th Annual ESPRIT Conference, Brussels, 27 November-1 December, 1989, pp1100- 1107.

[6] Lewis, D. [1973] *Counterfactuals*, Basil Blackwell, Oxford.

[7] Morreau, M. [1990] 'Epistemic Semantics for Counterfactuals', in Kamp, H. (ed.) *Conditionals, Defaults and Belief Revision*. Edinburgh: DYANA Deliverable R2.5.A, pp1-27. January 1990.

[8] Reiter, R. [1987] 'Nonmonotonic Reasoning', *Annual Review of Computer Science*, 2, 147-187.

[9] Shoham, Y. [1987] 'Nonmonotonic Logics: Meaning and Utility', IJCAI87, 388-393.

[10] Veltman, F. [1990] 'Defaults in Update Semantics', in Kamp, H. (ed.) *Conditionals, Defaults and Belief Revision*. Edinburgh: DYANA Deliverable R2.5.A, pp28-64. January 1990.

MICROELECTRONICS AND PERIPHERAL TECHNOLOGIES

COMPACT MODELLING FOR ANALOGE CIRCUIT DESIGN

H.C. DE GRAAFF, W.J. KLOOSTERMAN AND M. VERSLEYEN
Philips Research Laboratories
P.O. Box 80.000
5600 JA Eindhoven
The Netherlands

ABSTRACT

Compact modelling is part of a wide simulation activity, containing also techno-logical process and device simulation. It forms the essential link to the circuit simulation. In the Esprit projekt 2016 (TIPBASE) there is a special interest in analogue circuit design, which is reflected in the demands on compact modelling: high accuracy, special effects (Early effect, non-linear distortion etc.). Various models are briefly discussed, compared to each other and confronted with measurements up to 18 GHz.

1. Introduction

The design of integrated electronic circuits is unthinkable today without the use of computer simulations of the electrical circuits [1]. Circuit analysis programs like SPICE and PANACEA are used for this purpose; these programs contain mathe-matical descriptions of the electrical behaviour of the network elements (transistors, resistors, capacitors). Such descriptions are formulated from electrical measure-ment results, or, if these are not available, form the results of device and process simulations. We will name the set of mathematical descriptions for the electrical behaviour of the network elements a compact model. This compact model forms the crucial link between the circuit performance on the one hand and the process quantities on the other hand. In fig. 1 is sketched how the compact model is positioned in the total chain of simulation tools. In the very beginning we have the flowchart data as input for the process simulator. Its output is a device structure (doping profile, geometry), that serves as input for the device simulator. The output of the device simulator consists of electrical characteristics; they form a simulation alternative of what can also be obtained from electrical measurements of real hardware devices.

A set of defining equations is not sufficient for making a compact model, we also need the numerical values of the parameters of these equations. The parameter values are obtained from the electrical characteristics (measured or simulated) by means of a parameter extraction program, using in most cases a "least squares" algorithm [2].

Flowchart data

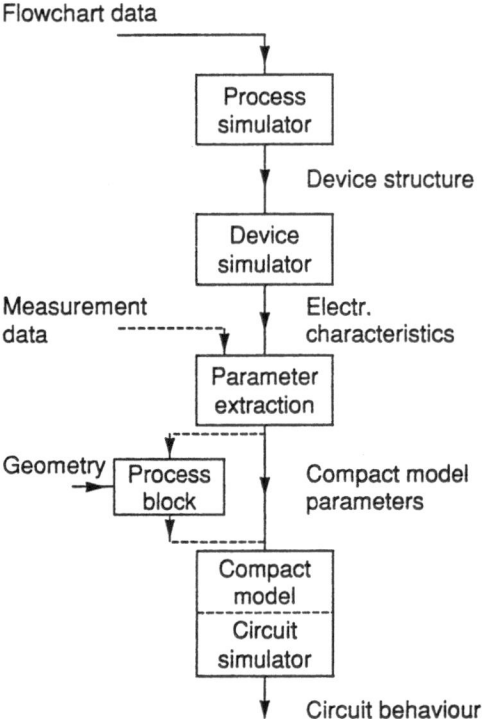

Fig. 1. Chain of coupled simulation tools.

In section 3 we will point out the possibility of generating compact model parameters with the help of a so-called "process block", which is a powerful design tool.

The different types of compact models are discussed in section 2, the important aspects for analogue circuit design in section 3 (one of these aspects, the use of the process block, was already mentioned).

In section 4 several models, known from publications, will be treated. We can distinguish between two classes of models, the charge-controlled models and the carrier-controlled models. The last section (5) will show some recent results of modern compact models, with examples taken from bipolar microwave devices, as they are fabricated and studied in the Esprit Project 2016 ("TIPBASE"). Only active devices are considered, resistors and capacitors are omitted here, although their modelling at high frequencies is not so easy.

2. Various types of compact models

We can distinguish three different types of compact models.

A. Physical models: the set of model equations are directly based on the device

physics. The model equations are preferably explicit, analytical functions.
The advantages of physical transistors models are the following:
- the model parameters have physical significance and can be used for checking the results of the parameter extraction
- they have forecasting properties
- geometrical scaling rules can be easily derived from device physics
- the statistical correlation between the model parameters can be found from device physics

Physical transistor models also have drawbacks; these are:
- the analytical functions used for the model definition are often simplifying approximations and therefore inaccurate
- the development time takes several manyears
- each new device structure may require major model revisions

B. Empirical models: here too the model equations are analytical functions, however, not based on device physics, but of a more curve-fitting nature. This is usually detrimental to the predictive power, the geometrical scalability, the physical significance and the correlation of the model parameters. The advantage is found in the shorter development time.
Purely empirical models are not found in practice, but each physical device model uses to some extent empirical formulas for the more complex device phenomena (e.g. in bipolar transistors the bias- dependent transit times).

C. Table models: the measured or simulated points of all the important electrical characteristics are stored in the computer's memory in table form. The required data that are not measured, are obtained by means of interpolation.
The development time for table models is relatively short, but the drawbacks are many: poor possibilities for geometrical scaling, statistical correlation, predictive power outside the measured range. Because analogue applications require in general a high accuracy in a wide range of operation modes (d.c., a.c. small-signal, transient, non-linear) the amount of data storage becomes prohibitive.

The compact models, developed and studied in the Esprit Project 2016 (TIP-BASE) all belong to the physical type, because the great advantages of this type outweigh the disadvantages.

3. Special model requirements for analogue circuit design

Analogue circuit design has special demands that are not found with digital circuit design: a high accuracy, an explicit description of certain device phenomena and the possibility to generate device parameters with a process block.
A high model accuracy is needed because in analogue design it is not only a matter of switching at the highest possible speed in the transient mode, but even more so a matter of signal handling in the frequency domain, together with a good

description of non-linear distortion and noise behaviour.
Device phenomena of special importance for analogue applications are the

- Early effect [3]:
  this effect increases the output conductance of the (I_c, V_{ce})-characteristics, see
  fig. 2. It can be characterized by an Early voltage V_A. For analogue applications

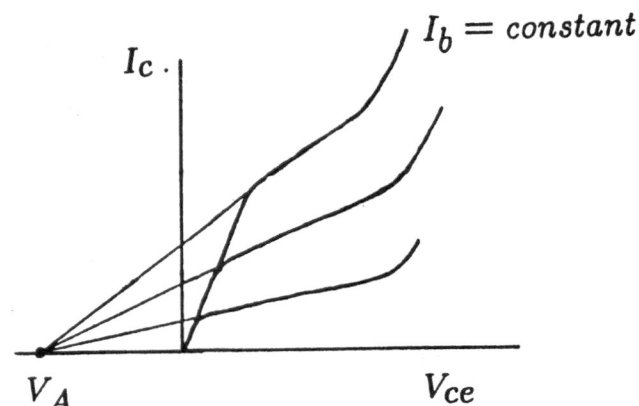

Fig. 2. Sketch of (I_c, V_{ce}) characteristics and indication of Early voltage V_A.l

this V_A value should be sufficiently high. In simple compact models V_A is taken
as a constant and it is a model parameter [4]. In reality V_A is bias-dependent; it
is sometimes modelled by taking the ratio of the collector depletion charge (Q_{TC})
to the fixed base charge (Q_{bo}) [4]. See fig. 3.

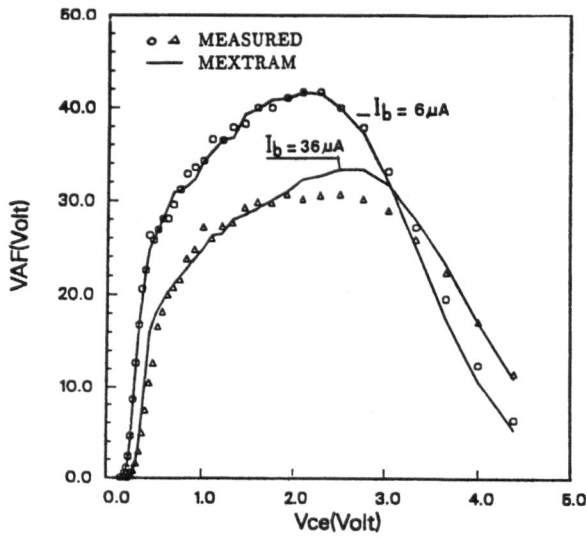

Fig. 3.Bias dependence of
the forward Early voltage

$$V_{AF} = I_c \Big/ \frac{dI_c}{dV_{ce}} - V_{ce}$$

- Quasi-saturation:
  To realize a sufficiëntly high Early voltage the collector epilayer should be lightly doped and relatively thick which results in a high collector series resistance.
  Due to voltage drop in the collector region the base-collector junction might become forward biased at high currents (al)though the external base-collector voltage remains reversely biased. This internal voltage drop arises in the active part of the transistor, under the emitter. We will call this effect quasi-saturation and it limits strongly the maximum achievable current and speed.
- Avalanche multiplication:
  this occurs at the collector-base junction. It limits the useful collector voltage range when the device is current driven (I_b is constant). At the sustaining voltage BVCEO the collector current I_c goes to infinity! This effect can be modelled with a "weak" avalanche model: the generated avalanche current is supposed to remain small enough so that it does not influence the collector field strength distribution [5]. See also fig. 4.
- Emitter-base current crowding:
  the base current gives a lateral voltage drop along the e-b junction. When this voltage drop becomes larger than kT/q volts we get an inhomogeneous carrier injection from the emitter. The main effect is that this makes the base resistance under the emitter bias-dependent. This bias dependence is modified by capacitive effects at high frequencies [6].

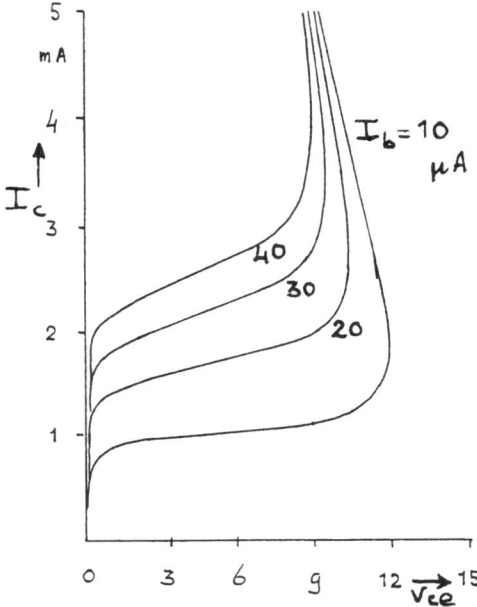

Fig. 4.Influence of avalanche multiplication on the (I_c, V_{ce}) characteristics: a. model calculations.

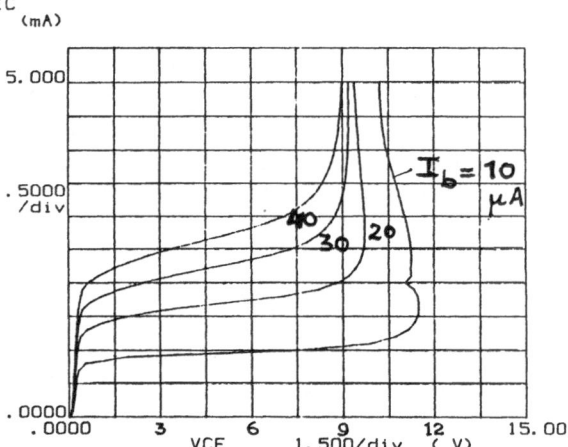

b. measurements.

- Non-linear disortion:

  non-linear network elements like transistors will produce higher harmonics and all sorts of intermodulation products, giving rise to various spurious signals. A description of these distortions in the frequency domain is usually limited to the 2nd an 3rd order terms for the sake of simplicity. Even this puts strong demands on the model, for it means that not only the analytic model functions must have sufficient accuracy, but also their 2nd and 3rd derivatives. Fig. 5 shows the 3rd harmonic output signal at the collector as a function of the bias. The input signal frequency is 10 KHz. The first maxima of the measured curves coincide with the onset of quasi-saturation of the device. Today there is no compact model known to us that can calculate these curves adequately. To find an improved model that can do this job, is one of goals in the Esprit Project 2016.

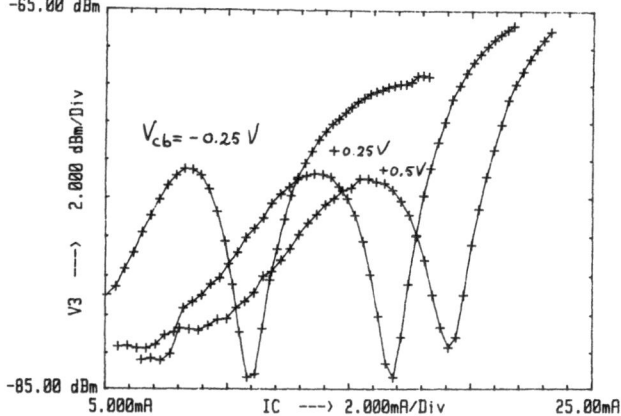

Fig. 5. Collector signal of the 3rd harmonic distortion as a function of I$_c$ at various V$_{cb}$ values.

Other phenomena, worth to be mentioned, are the 1/f and the white noise, the transit times (important for the cut-off frequency of the device) and the excess phase shifts, when the signal frequency becomes higher than the inverse of the total transit time.

The process block has proven to be a very useful design tool [7]. It is based on the idea that in principle each compact model parameter P can be split into several components:

$$P = P_b\,HL + 2\,P_{si}\,L + 2\,P_{ox}\,H + 4\,P_c. \qquad (1)$$

We have assumed here a rectangular pn junction with H and L as dimensions.

$P_b HL$ is the contribution of the bottom part, $4\,P_c$ that of the corners. The sidewall components are $2\,P_{si}\,L$ for the silicon edges and $2\,P_{ox}\,H$ for the oxide edges. If the parameter P represents e.g., the collector saturation current I_s and HxL is the emitter geometry then the quantities P_b, P_{si} and P_{ox} are current densities with dimensions A/cm^2 and A/cm respectively. In a first approximation (H and L > 0.4 m) we may neglect the corner contributions ($4P_c$) and we can put that P_b, P_{si} and P_{ox} only depend on the vertical doping profile, that is on the process only. In this way we have separated the influence of the process from that of the geometry. Eq.(1) in fact gives the geometrical scaling rule for the parameter P as well as the dependence on the process. The latter can be accomplished by establishing the correlation between the density quantities P_b, P_{si} and P_{ox} and certain process quantities like sheet resistances and breakdown voltages. Such process quantities are easy to measure, also for statistical purposes.

The process block consists of equations like eq.(1); for a given process we can then for each geometry generate the required compact model parameters. If the statistical data spread of the process is also known, we may use the process block then for statistical modelling and design centering.

4. Short review of existing bipolar models

We will distinguish two classes of models in this section:
- Charge - controlled models
- Carrier - controlled models

- The charge-controlled models have the currents directly coupled to the related charges. The Ebers-Moll (EM) and Gummel-Poon (GP) [8,9] belong to this class of models.
- In the EM model the forward active collector current, If, depends exponentially on the base-emitter voltage V_{be}. The stored diffusion charge is then given by the charge control relation;

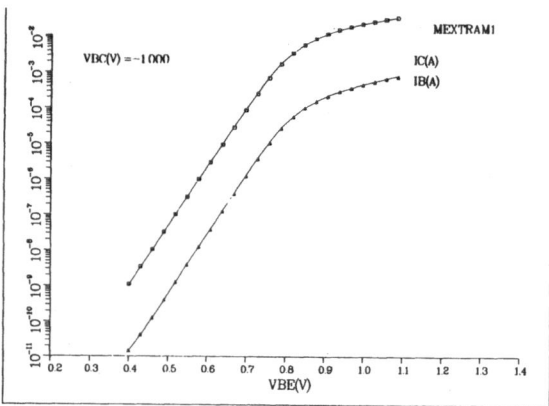

Fig. 6.

Measured and modelled
characteristics of a V₁
process device of Tip
Base.

a. Gummel plot,

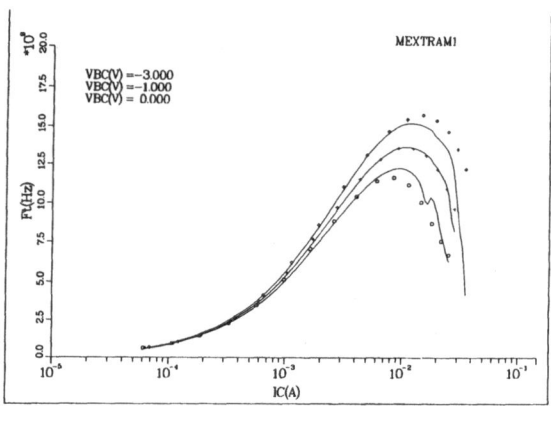

b. cut-off frequency.

$$Q_f + T_f \cdot I_f$$

In the above equation T_f is a transistor parameter. The Early effect is modelled by means of a fixed Early voltage. The EM model only models ideal transistor behaviour, i.e. constant current gain an no cut-off frequency fall-off, so the model is only of limited use in circuit simulation. The GP model is a more complex charge-controlled model which, at the cost of extra parameters, also models high injection effects by using a more complete I_c (V_{be}) relation resulting in current gain fall-off at high current densities. Furthermore, the forward transit time can be made current and voltage dependent [10] to account empirically for cut-off frequency fall-off. The GP model is fairly accurate in simulating digital circuitry. However, for analogue circuits the accuracy often is insufficient.

– More recently introduced types of models are the carrier-controlled models [11, 12, 13]. In these models both current and charge are separately described as a function of the injected carrier concentration. The injected carrier concentration is related to the junction voltage via the well known pn product. In this way the bias dependence of the base transit time is obtained naturally, from the equations for the current and charge.

In the MEXTRAM [11] model the Early effect is modelled by means of the emitter and collector depletion charge, which are voltage-dependent, thus resulting in bias-dependent Early voltages. In the models of Jeong-Fossum [12] and Michaels-Strojwas [13] the boundaries of the quasi-neutral base region are introduced in the collector current expression. The Early effect is here represented by the movement of these boundaries. Furthermore the MEXTRAM model includes an also carrier based epilayer model which models quasi-saturation effects. Emitter base current crowding, weak avalanche currents and several parasitics, sidewalls and extrinsic transistor regions are also modeled. With the above features the MEXTRAM model is very suitable for both digital and analogue circuit simulation.

5. Recent developments and results

First we will show a few results of the Mextram model, applied to a V1 process version device, as developed in the Esprit Project 2016. The device under test was an npn transistor with emitter dimensions 1.5 x 58.5 m (on silicon). Fig.6a shows the Gummel plot, measured and calculated with Mextram. Fig. 6b gives the measured and calculated cut-off frequency f_T. The parameter fit is done for the curve at V_{cb} = 1V, the other curves are model predictions.

In Fig.7 a comparison is made between the Mextram model and the widely used Gummel-Poon model (see also section 4). Confrontation with measured I_c (V_{ce}) characteristics show the far superior performance of the Mextram model in the quasi-saturation region. The tested device here was an npn device with a 2.5 x 12 μm emitter.

The a.c. small-signal behaviour of the transistors is of paramount importance for analogue circuit applications. For the characterization of this behaviour the so-called Y parameters [14] are conveniently used. In Fig. 8a the transconductance (Y 21) parameters of an npn transistor is measured up to 18 GHz at a given bias point (V_{cb} = 1V, I_c = 0.93 mA), where f_T = 4.5 GHz. The calculations are done with Ebers-Moll, Gummel-Poon and Mextram models. In Fig. 8b the same transconductance is measured at a fixed frequency (300 MHz) as a function of the d.c. bias current. In all cases the worst results are obtained with the EM-model, the best with the Mextram model. The GP model occupies an intermediate position. These results are in accordance with the increasing model complexity: EM is the simplest model, Mextram the most complex.

6. Conclusions

Compact transitor models for circuit simulation are a necessary design tool. Especially in analogue circuit design a high degree of accuracy is required. This

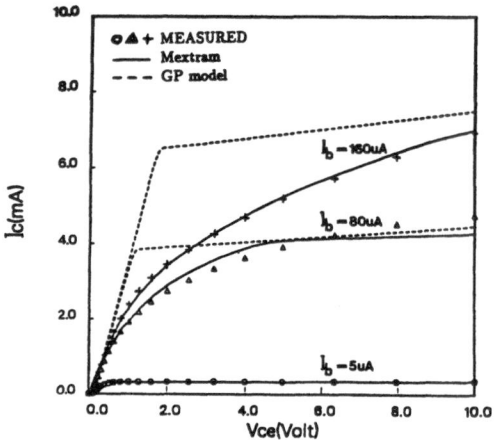

Fig. 7.Comparison of d.c. characteristics between Gummel-Poon and Mextram model and measurements.

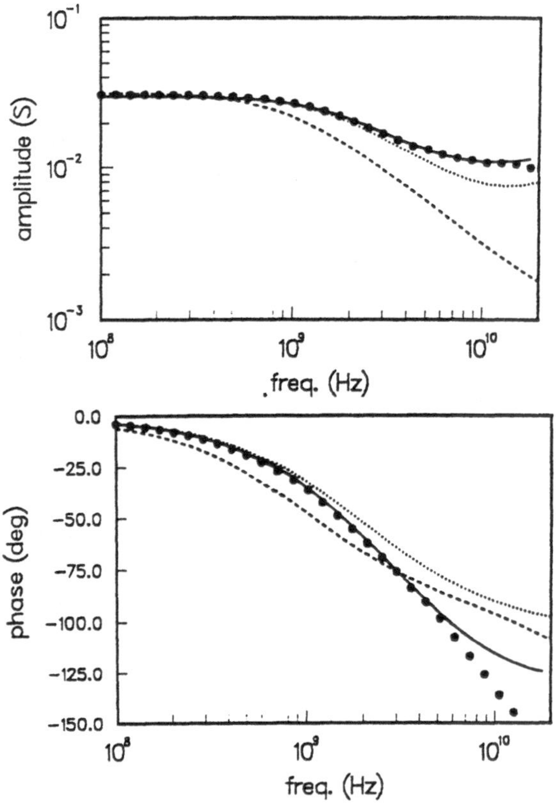

Fig. 8. Transconductance (Y21) as a function of frequency (a)

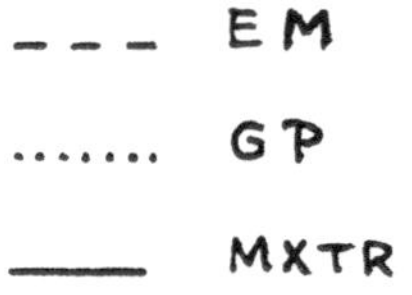

- - - E M

.... G P

——— MXTR

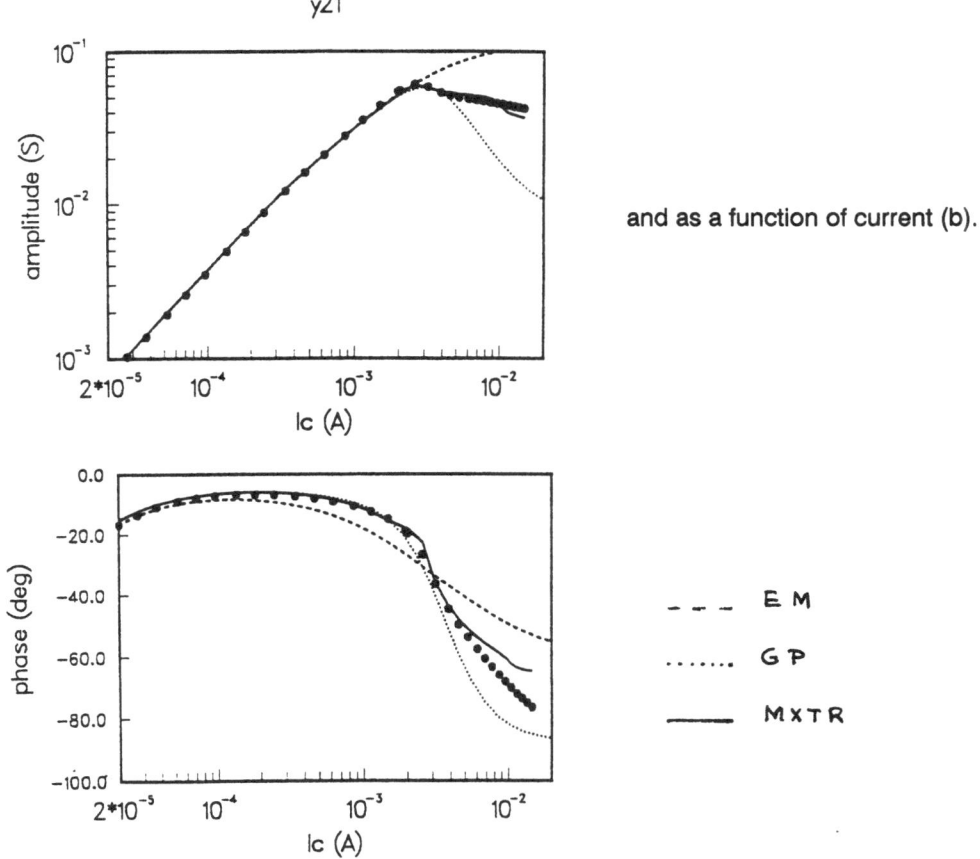

y21

and as a function of current (b).

also demands the inclusion of special effects (Early effect, avalanche multiplication in the models. The possibility of using process blocks is a highly valued feature.

The existing Ebers-Moll and Gummel-Poon models are usually not accurate enough, especially in the high-frequency range (GHz). In Esprit Project 2016 attention is paid to the improvement of the bipolar models in this respect and the first results are shown in this paper.

REFERENCES

1. L.W. Nagel: SPICE2-A Computer Program to simulate Semiconductor Circuits. Electr. Res. Lab Memo ERL-M520, University of California, Berkeley (1975)

2. K. Doganis, D.L. Scharfetter: General Optimization and Extraction of IC Device Model Parameters. IEEE Trans. Electr. Dev. ED-30, 1219 (1983)

3. J.M. Early: Effects of Space-charge layer widening in Junction Transistors. Proc. IRE 40, 1401 (1952).

4. H.C. de Graaff, F.M. Klaassen: Compact Transistor Modelling for Circuit Design, ch 3,

Springer-Verlag Wien, New York 1990.

5. W.J. Kloosterman, H.C. de Graaff: Avalanche Multiplication in a Compact Bipolar Transistor Model for Circuit Simulation. Trans. Electr. Dev. EC-36, 1376 (1989)

6. G. Rey: Effects de la Défocalisation sur le Comportement des Transistors à Jonction. Solid-St. Electr. 12, 645 (1969).

7. P.J. Rankin: Statistical Modelling for Integrated Circuits. Proc. IEEE 129, 186 (1982)

8. J.J. Ebers, J.L. Moll: Large Signal Behaviour of Junction Transistors Proc. I.R.E. 42, 1761 (1954)

9. H.K. Gummel, H.C. Poon: An integral Charge-Control Model for bipolar Transistors. Bell Syst. Tech. J. 49, 827 (1970)

10. M. Schröter, H.M. Rein: Transit time of high speed bipolar Transistors in dependence on operating point, technogical parameters and temperature. Proc. 1989 Bip. Circ. Techn. Meeting, 250 (1989)

11. H.C. de Graaff, W.J. Kloosterman: New formulation of the Current and Charge relations in bipolar Transitor Modeling for CACD Purposes. IEEE Trans. Electr. Dev. ED-32, 2415 (1985)

12. H. Jeong, J.G. Fossum: A charge-based large-signal bipolar transistor model for device and circuit simulation. IEEE Trans. Electr. Dev. ED-36, 124 (1989)

13. K.W. Michaels, A.J. Strojwas: A compact physically based Bipolar Transistor Model. Proc. 1989 Bip. Circ. Techn. Meeting 242 (1989)

14. J. Lindmayer, Ch,Y Wrigley: Fundamentals of Semiconductor Devices. D.V. Nostrand Comp., Princeton 1965

Project No. 2016
Bipolar Advanced Silicon for Europe TIP BASE

INTERNAL GETTERING AS A FUNCTION OF THE WAFER POSITION IN THE ORIGINAL CRYSTAL ROD

K. Graff
TELEFUNKEN electronic
D-7100 Heilbronn, FRG

Abstract.

This paper reports measured efficiencies of internal gettering in CZ-silicon wafers before and after the application of pre-anneal and buried layer processes as a function of the wafer position in the original crystal rod. The purpose of the investigations was to clarify whether inhomogeneities in internal gettering which were observed previously can be explained by the respective wafer position in the crystal.

The results exhibit drastic variations of the gettering efficiencies as a function of the wafer position, which change after each process. They can be explained by the overlapping of two different gettering mechanisms. The first, effective in as-received wafers and after application of lower temperature-processes, is caused by still unidentified gettering nuclei. The second which is effective after prolonged high temperature-processes is caused by bulk stacking faults. The knowledge about gettering mechanisms and behaviour is important for future processes where diffused phosphorus gettering will be strongly reduced due to phosphorus implantation and rapid processing.

1. Introduction

Since the early sixties gettering has become a common method to reduce unintentional metal impurities in semiconductor devices. The first method investigated was the extrinsic gettering of impurities by means of phosphorus diffusion [2]. Since that time a variety of different gettering techniques have been reported which recently have been reviewed [3]. One of the modern techniques is the so-called internal or intrinsic gettering. Gettering centres are created intentionally during preannealing processes generating oxygen precipitates in Czochralski-grown silicon samples. This technique is investigated in the present paper since it will be applied in future bipolar integrated circuit production and is not yet well understood.

For all investigations on gettering processes the definition of gettering should be kept in mind [3]:

Gettering is the dissolution of unwanted metal impurities followed by their diffusion and precipitation in an area of the semiconductor sample where they do not deteriorate the device performance.

From this definition the following general conclusions can be deduced [3]:

1. For gettering a high-temperature process is needed to dissolve precipitated impurities and to diffuse them to the respective gettering sites. The solubility of the impurity at this temperature must exceed the impurity concentration in the sample.

2. In most gettering processes the precipitation of the impurities takes place during the cooling period due to the supersaturation of the impurities at lower temperatures. As a consequence the cooling rate is an important parameter to characterize gettering together with the diffusivity of the respective impurity during the cooling period.

3. Gettering is based on a heterogeneous nucleation process of the impurities at nucleation centres which must be formed earlier and which must be located in regions where the precipitates do not deteriorate the electrical function of the device. In general these nucleation centres are situated either in a highly doped zone (phosphorus gettering), or in the bulk of the sample beneath a denuded zone (internal gettering), or at the reverse side of the sample (backside gettering). In the case of backside gettering the impurities have to diffuse from the electrically active zone on the front side to the reverse side of the wafer where they precipitate. Since this diffusion in general is performed during the cooling period of the sample the cooling rate must be adapted to the diffusivity of the impurity at the respective sample temperature.

The problems concerning gettering of slowly diffusing transition metals were discussed recently [3,4]. Furthermore problems can arise for the determination of the gettering effeciencies which can differ for different impurity metals [9] for instance due to their different diffusivities. In many publications gettering was characterized by an increase in carrier lifetime due to the reduction of unintentional and unknown impurities which were accidentally present in the semiconductor sample. For further characterizations the determination of the yield, measurements of the breakdown voltage or of other device parameters were used. An intentional contamination with a defined impurity of known concentration was rarely performed. In general the reduction of this impurity concentration due to gettering was then measured by means of expensive techniques such as neutron activation analysis or Rutherford backscattering

To overcome this problem two techniques were developed which can be applied rather easily for routine control since they are fast and cheap. The first one, the so-called palladium-test [5] is based on an intentional contamination of the wafer with palladium. The gettering effectiveness is examined by the inspection of the palladium haze formed during a defined temperature process and followed by a preferential etching of the sample. This method is a more qualitative test. It has been extended recently to other haze-forming metals [9]. The second technique, the so-called iron-test [1], is based on the intentional contamination of the sample with iron. In contrast to palladium iron belongs to the main impurities in device production. The concentration of the iron is adjusted by its solubility at the diffusion temperature and it can be measured before and after gettering by means of DLTS.

By this way the gettering efficiency can be determined. It is defined in the following way [3]:

The gettering efficiency is the difference of the impurity concentration before and after the application of the gettering process. It is coupled to a defined impurity and to a defined cooling rate.

Because of the possibility to obtain quantitative results for the effectiveness of gettering this method is suitable to compare different techniques applied on the same material on one hand and different materials applied to the same gettering process on the other hand. This technique was already applied to study internal gettering during bipolar processes [1], gettering by polysilicon on the backside of a wafer, and to study the homogeneity of internal gettering across a diameter of the wafer [4]. The investigations of internal gettering as function of the wafer position in the crystal rod were started previously [3,4], and are continued in the present paper. These investigations should help to understand variations in the efficiencies of internal gettering which were observed previously by the application of the palladium- and iron-test. Variations in gettering were observed in two directions, in radial direction on one wafer and in axial direction of the original boule between different wafers. In this paper only the position axially along the boule is analysed.

This paper in the first section, presents experimental details of the wafer preparation for internal gettering and for the iron-tests. The following section presents the results of the iron-tests which were applied to the as-received wafers before any temperature process was performed, after the high-low-high preannealing which was applied to activate internal gettering, and finally after a simulated buried layer diffusion process which has been performed in addition. In a final discussion a tentative interpretation of the results is given together with a comparison with other results published in the recent literature.

2. Experiments

The wafers with known positions in the original crystal rod were placed at our disposal for the respective measurements by Wacker-Chemitronic, FRG. The polished wafers originated from a 4" boron-doped p-type silicon crystal with resistivities between 10 and 12 Ωcm, Czochralski-grown and (111)-oriented. The oxygen concentration amounted to $(6-7)10^{17}$ cm^{-3}.

The sample preparation for the iron-test was performed as follows: The wafer was etched in a mixture of nitric acid and hydrofluoric acid to remove the native oxide and surface contaminations. Iron was sputtered on the reverse side of the wafer and diffused at 980C for 30 minutes in an argon atmosphere. During diffusion any cross contamination of the sample was avoided as far as possible. After the diffusion the sample was withdrawn from the furnace tube by means of an automatic loading system applying a pull velocity of 9oo mm/min. For this cooling rate the efficiency of internal gettering was found to be independent of the cooling rate [4]. The sample was etched again and magnesium Schottky contacts 2 mm in diameter were evaporated for DLTS measurements. For a most accurate eva-luation of the recorded spectra the correction factors published by Pons [6] were applied.

For internal gettering the following heat treatments were applied: Formation of the denuded zone at 1100°C for 12 hours in a N_2/O_2-atmosphere, nucleation annealing at 650°C for 4 hours in N_2-atmosphere followed by a ramping up of the temperature to 1100°C for the second high-temperature process with a duration of 3.25 hours in a H_2/O_2-atmosphere. The ramp was adjusted to 2°C/min. The simulated buried layer diffusion was performed on one half of the wafers separated after the application of the preannealing. It was carried out at a temperature of 1110C for 4 hours in a N_2/O_2-atmosphere.

Before starting the sample preparations the oxygen concentrations of all wafers were determined with high accuracy by means of IR absorption spectroscopy following the new guidelines of DIN 50438 Part I (1990) and applying a calibration factor of 2.4510^{17} cm^{-2} (DIN or new ASTM). The oxygen concentrations were remeasured after internal gettering preannealing to determine the amount of precipitated oxygen due to this heat treatment.

The iron-tests on the as-received wafers were performed on one wafer of each position. A second wafer of each position was used to obtain the results after preannealing (first half) and after the subsequent buried layer diffusion (second half). A third wafer of positions 2, 3, and 7 was selected to control the repeatability of the results after preannealing and subsequent diffusion. Finally a fourth wafer of position 3 was measured after preannealing and after diffusion because of the large variations observed in the gettering efficiency results.

3. Experimental results

The results for the oxygen concentrations are shown in Fig.1 (right hand scale). In agreement with previous experience on silicon crystals with specified oxygen content the oxygen concentrations decreased with increasing distance from the seed end of the crystal and slightly increase once more in the second half of the crystal. The variation of the oxygen concentration within the crystal is low and ranges from 6 to 710^{17} cm^{-3}. The crystal exhibits a rather low oxygen content for use in internal gettering. The variations from wafer to wafer of one position were small compared to those between the different positions and therefore they were ne-glected in the figure.

The gettering efficiencies measured on the as-received wafers are shown in the same Fig.1 for comparison (full line, left hand scale). The gettering efficiencies are the differences between the iron concentration without gettering (fixed value = 20.510^{13} cm^{-3} measured repeatedly on float-zone silicon crystals) and the measured iron concentrations in the as-received wafers. The indicated results are mean values and originate from averaging 12 results obtained on the whole diameter of the wafer. These different results scattered more or less with maximum deviations of about 30%. The wafers at the seed end and in the middle region (positions 1, 4, and 5) exhibit a vanishing gettering efficiency. The other 4 wafers (positions 2,3,6,7) exhibit gettering efficiencies between 4 and 810^{13} Fe-atoms/cm^3.

Gettering effects of this order of magnitude in as-received wafers were already observed previously [1] but they never have been measured as a function of the wafer position.

The mean gettering efficiencies as a function of the respective wafer position obtained after applying internal gettering preannealing are indicated in the same Fig.1 (dashed line). For the positions 2, 3, and 7 the control measurements were also taken into account. Only the remeasurement for position 3 disagreed substantially as shown in Fig.2. Whereas the first wafer resulted in a very low gettering efficiency the second and the third sample yielded very high values. The reason for this discrepancy is not known but it is not caused by measurement errors. This was verified by counting the bulk stacking fault densities revealed by angle polishing and preferential etching. Whereas the first sample of position 3 exhibited low mean bulk stacking fault densities of about 1000 cm^{-2} the sample of position 1 exhibited much higher densities of about 7000-8000 cm^{-2}. As pointed out previously the gettering effect after applying our preannealing process is caused essentially by bulk stacking faults and not by oxygen precipitates [1]. After averaging all three results for position 3 as indicated in Fig.1 the gettering efficiencies after preannealing as a function of the wafer position show a decrease with increasing distance from the seed end of the crystal. Only at the tang end the gettering efficiency increases again considerably.

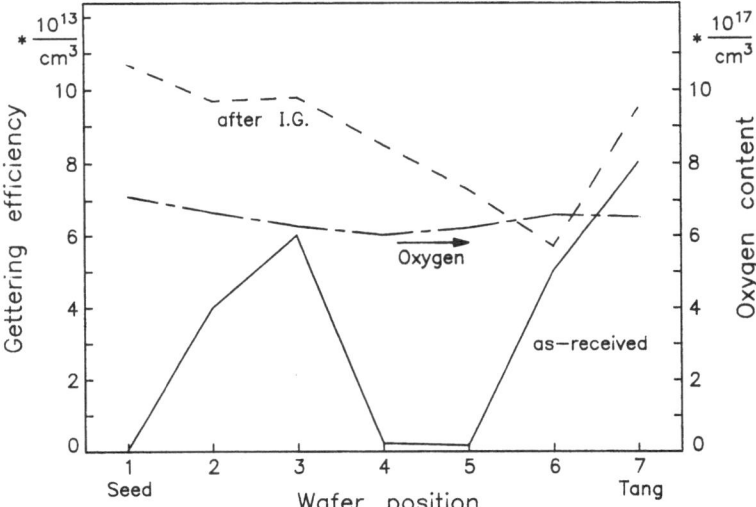

Fig.1: Iron gettering efficiencies as a function of the wafer position (left hand scale), pull velocity 900 mm/min:
1. On as-received wafers (lower full line)
2. On wafers after internal gettering preanneal (upper broken line)
3. Oxygen concentration as a function of the wafer position (dashed line and right hand scale).

The mean gettering efficiencies obtained after applying an additional simulated buried layer diffusion (without doping material) are indicated in Fig.3 together with the results obtained on the as-received wafers and the results obtained after preannealing the wafers. In the curve obtained after preannealing position 3 now exhibits the mean of the two remeasurements neglecting the result of the first measurement since it better fits the trend observed in all three curves showing maximum values at position 3 and 7 and minimum values near position 5. In general the values after diffusion exceed the gettering efficiencies obtained after preannealing.

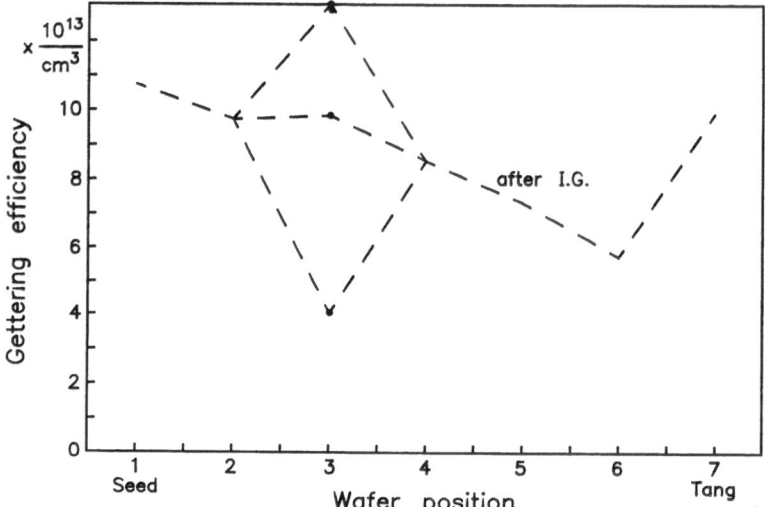

Fig.2: Iron gettering efficiencies as a function of the wafer position after internal gettering preanneal including different measurement points for position 3.

Fig.4 presents all measurement points of the gettering efficiencies after preannealing as a function of the decrease in oxygen content during the preannealing process which is the amount of the precipitated oxygen content. Although the calculated linear regression exhibits a slightly increasing tendency with increasing amount of oxygen precipitates there is no direct correlation of both parameters if the large deviations from the linear regression are regarded. These deviations do not originate from high measurement errors for the determination of the oxygen content. This is verified in Fig.5 which presents the decrease in oxygen content as a function of the original oxygen content in the as-received wafers. This curve agrees with the lower part of the well-known S-shaped functional dependence reported repeatedly in the literature [7]. Because of the rather low oxygen contents in our samples only this lower part is revealed. Within the expected measurement error all measurement points touch the indicated mean curve.

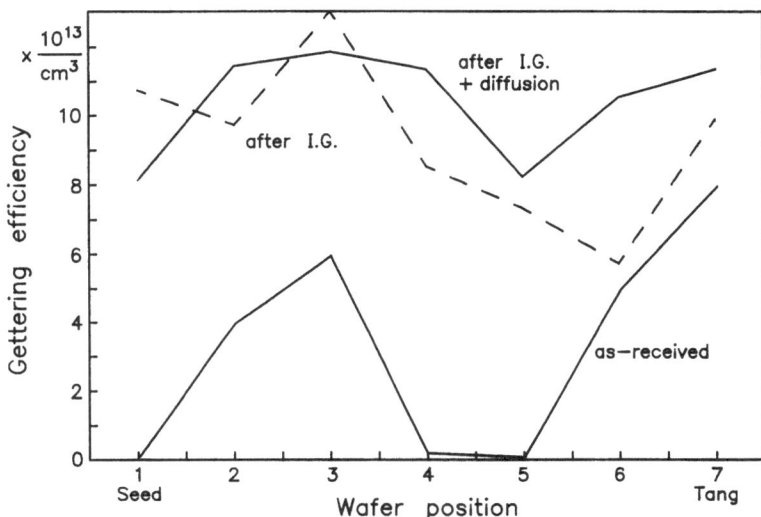

Fig.3: Simultaneous presentation of iron gettering efficiencies as a function of the wafer position before and after preannealing and after a subsequent diffusion. (Exclusion of the first measurement point from averaging position 3 after preannealing).

4. Discussion

The existence of gettering effects in as-received wafers is surprising although it is in agreement with previous experience. The dependence of the gettering efficiency upon the wafer position shown in Fig.1 has not yet been proved by remeasurements on second samples or by measurements of other authors. The increasing-decreasing-increasing gettering efficiencies do not correlate either with the oxygen content or with the amount of oxygen precipitates or with the thermal history of the crystal during growth. The respective gettering centres are not identified and therefore their variation with the wafer position in the crystal rod can not be explained. So far the only explanation for their variation in axial direction is the variation of the crystal growing parameters. Further investigations are required to understand the behaviour of these centres.

The gettering efficiencies after applying the preannealing processes result in an almost continuously decreasing curve with increasing distance from the seed end of the crystal (Fig.1). The values at the tang end of the crystal, however, increase again. This curve was obtained by using the average of all three measurement points for position 3. With the exception of the values at the tang end, this curve can be explained by the influence of the thermal history of the crystal during its growth. The slope of this curve correlates qualitatively with the concentration of thermal donors formed during the cooling period of the crystal. As mentioned before the gettering effect after preannealing of the wafers is essentially due to the gettering of bulk stacking faults. This has been found previously [1] and was verified once more by

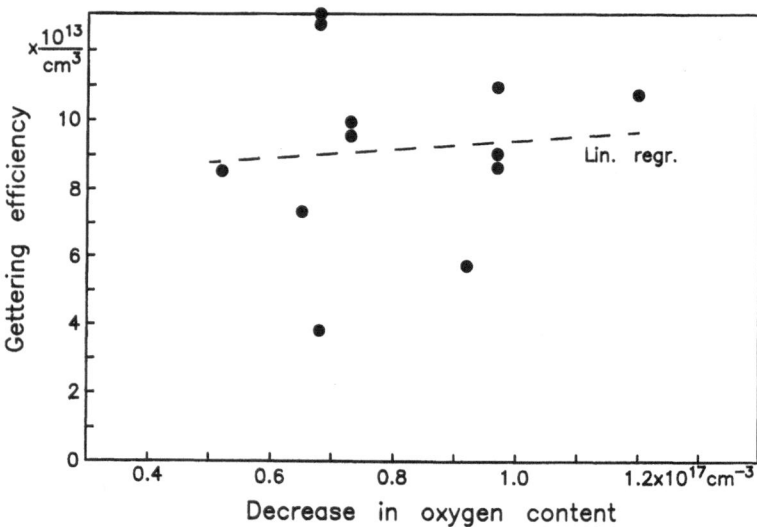

Fig.4: Iron gettering efficiencies after internal gettering preanneal as a function of the precipitated oxygen content and linear regression (dashed line).

counting the bulk stacking faults densities in the sample of position 1 and the first sample of position 3. From the mentioned correlation to the density of thermal donors it could be deduced that the nuclei for the formation of bulk stacking faults are found in the thermal donors which are rearranged by the annealing process performed by the wafer supplier to kill their electrical activity. The formation mechanism of bulk stacking faults is still unknown in detail. During the formation of the denuded zone which was the first high-temperature process, however, the thermal donor-related nuclei were disturbed or they became ineffective due to the lack of higher oxygen concentrations since bulk stacking faults are not found within the denuded zone. In conclusion the correlation of the gettering efficiency after preannealing to the thermal history of the crystal seems evident but is still not well understood.

The increasing gettering efficiency at the tang end of the crystal after applying the preannealing could be caused by an enhanced impurity content at the tang end of the crystal as it is commonly observed in Czochralski-grown crystals. It has been observed that for instance copper which precipitates by a homogeneous nucleation mechanism can serve as a nucleation centre for the precipitation of iron. The precipitation of iron is favoured since it absorbs silicon selfinterstitials to form iron silicide whereas copper emits silicon selfinterstitals during the formation of silicide [8]. The interference by copper and nickel on the results of the iron-test must be regarded during sample preparation. Probably it can be excluded for the present results obtained on the as-received wafers since they have been surface-etched before preparation and the only heat treatment was the diffusion of iron which was

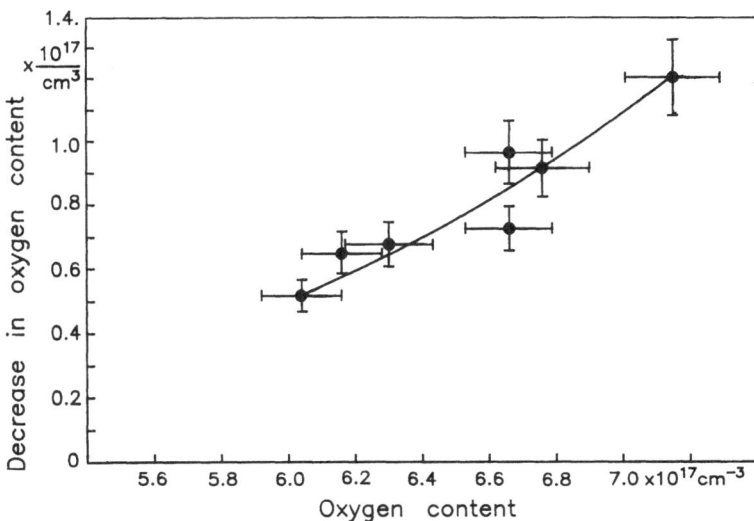

Fig.5: Precipitated oxygen contents after preannealing as a function of the respective original oxygen content in the as-received wafers including measurement errors.

performed maintaining utmost clean conditions. These impurities may, however, effect the results after preannealing and after the subsequent diffusion. To reduce the influence of impurities on our results all wafers have been heat treated simulta- neously with the exception of the second and third wafers to control the repeatability of the results. It is assumed that impurities if they were present would contaminate all wafers of the same run to a similar degree. Almost all the results of the control measurements agreed, however, with the first results within the expected measure- ment error ($\pm$10 %) although these wafers were heat treated separately several months later. Therefore it is assumed that the interference of other unintentionally added impurities may be unimportant in the present investigations.

In general the gettering efficiencies after the additional application of a simulated buried layer diffusion are more or less enhanced. At the same time their variations with the wafer positions are reduced from 46% to 30%. Within a standard deviation of $\pm$14 % all values agree. For the technical application of internal gettering the efficiencies should be equal for all wafers at the end of the last high temperature process and they should be sufficiently high to getter all impurities. Both conditions seem to be accomplished applying the present technique. The gettering efficiencies exceed 810^{13} Fe atoms/cm^3 which may exceed common impurity concentrations at the end of all processes.

In order to determine the gettering mechanism which is effective in our samples the gettering efficiencies measured after preannealing were plotted as a function of the amount of the precipitated oxygen in Fig.4. Although the gettering efficiencies vary by about a factor of 3 and the amount of oxygen precipitates by a factor of two a correlation of both parameters can not be observed. On the other hand these gettering efficiencies correlate with the density of bulk stacking faults which was observed previously [1] and confirmed by comparing the density of stacking faults in the samples of position 1 and the first sample of position 3. A difference in the density of bulk stacking faults of about a factor of 7 causes a difference in the gettering efficiency of about a factor of 3.

From the results reported here it is evident that internal gettering is not a simple process which can be applied without care to any wafer source followed by any sequence of processes for device production. This was also shown by Falster and Bergholz [9] who investigated internal gettering at different stages of a simulated CMOS process by applying the palladium-test which was extended to other haze-forming transition metals (Cu,Ni,Co,Fe). In contrast to our investigations Falster and Bergholz did not start the sequence of processes with the high-low-high-temperature preannealing which is commonly used to enhance internal gettering. As a consequence their results are quite different. Gettering was observed by the authors in CZ-silicon wafers with oxygen contents exceeding 6.010^{17} cm^{-3} even if the amount of precipitated oxygen was as low as 0.2510^{17} cm^{-3} and in general stacking faults could not be detected. Furthermore they observed a decreasing tendency of the gettering activity after prolonged process times and increased oxygen precipitation. Therefore it may be assumed that this gettering activity is caused by the same centres which were observed in our as-received wafers.

This assumption can lead to a modified interpretation of our gettering efficiency-curve after preannealing (Fig.1,2). Now it would make more sense to eliminate the exceptional result obtained on the first sample of position 3 from averaging and replace the mean by the higher mean value remeasured twice. A strongly marked maximum of the gettering efficiency is then observed located at the wafer position 3 which coincides with the strongly marked maximum obtained from measurements on the as-received wafers and with that obtained after diffusion as shown in Fig.3. A second coincidence of maximum gettering efficiencies is observed at the tang end of the crystal. Now a correlation of the gettering efficiencies before and after preannealing and after diffusion seems evident.

After preannealing this correlation is superposed by a continuous decrease of the gettering efficiencies from the seed end to the tang end of the crystal. This can be explained by the superposition of two different gettering effects due to the unidentified centres in the as-received wafers on one hand and the bulk stacking faults on the other hand. The density of bulk stacking faults after preannealing exhibits a decreasing tendency due to the influence of the thermal history of the crystal. This influence is reduced during prolonged high-temperature processes. So the decreasing tendency disappears after an additional diffusion process. Simultaneously the share of the unidentified gettering centres is further reduced and both maxima due to superposition of two gettering activities flatten but still exist (Fig.3).

Additional experiments are required to verify this new explanation of internal gettering. The influence of low temperature processes upon the gettering efficiencies of the unidentified centres in as-received wafers must be investigated. The repeatability of the gettering efficiencies as function of the wafer position in as-received wafers must be controlled and explained. Furthermore the unidentified gettering centres must be studied and their density variations must be correlated to the variations in the pulling parameters during crystal growth in order to improve the homogeneity of the gettering effect.

For the application of internal gettering in device production it would be very important to develop this alternative gettering mechanism since it is effective without the application of high-temperature processes. In future so far common elongated high-temperature processes will be replaced more and more by shorter processes at lower temperatures.

So many investigations are required where the iron-test can be successfully applied to study the repeatability of the present results and to optimize modified gettering processes for future device production. For this purpose additional wafers with known positions in the original crystal rod were ordered from different suppliers and will be available in near future.

REFERENCES:

1 Graff,K., Hefner,H.-A., and Hennerici,W. (1988) 'Monitoring of Internal Gettering during Bipolar Processes', J.Electrochem.Soc.135, 952-957.

2 Goetzberger, A., and Shockley, W. (1960) 'Metal Precipitates in Silicon p-n Junctions', J.Appl.Phys. 31 1821

3 Graff, K. (1990) 'Gettering of Transition Metals in Silicon', Technical Proceedings SEMICON/EUROPA 90, Semi. Equipment and Materials Internat. Zürich , 2-9.

4 Graff, K. (1989) 'Transition Metals in Silicon and their Gettering Behaviour', Mat. Science and Engin. 4 63-69.

5 Graff, K., Hefner, H.-A., and Pieper, H. (1985) 'Palladium-Test: A Tool to Evaluate Gettering Efficiency', M.R.S. Proc. 36 19-24.

6 Pons, D. (1980) 'Determination of the Free Energy Level of Deep Centers, with Application to GaAs', Appl.Phys.Lett. 37 413-415.

7 Chiou,H.-D. (1987) 'Oxygen Precipitation Behavior and Control in Silicon Crystals', Sol.St.Techn.(3) 77-81.

8 Seibt, M., and Schröter, W. (1989) 'Lokalisierung und Identifizierung von Mikrodefekten mittels Transmissionselektronenmikroskopie', Final report for the Ministry of Research and Technology of the FRG, 1-45.

9 Falster, R., and Bergholz, W. (1990) 'The Gettering of Transition Metals by Oxygen Related Defects in Silicon Part I', J.Electrochem.Soc. 137 1548-1559.

Project No. 2016
Bipolar Advanced Silicon for Europe TIP BASE

PREAMORPHIZATION TECHNIQUES FOR SHALLOW JUNCTIONS IN SI

J. IMSCHWEILER, H. A. HEFNER
TELEFUNKEN electronic GmbH
D 7100 HEILBRONN
GERMANY

Abstract.

Due to dopant channeling during ion implantation into crystalline silicon, shallow junctions are difficult to achieve. The scope of this study was to investigate the preamorphization technique as a tool to prevent channeling during transistor fabrication. Implantation of heavy ions such as Si^+ and Ge^+ offers the possibility of amorphizing a desired region in a highly controllable way. The preamorphization experiments were carried out using these species of atoms. In both cases boron channeling was found to be completely suppressed. In the case of Si^+ preamorphization transistors show significant leakage currents due to a high density of residual stacking faults and dislocation loops inside the active region. In the case of Ge^+ preamorphization complete recrystallization of the amorphized layer was obtained. No stacking faults or dislocation loops could be observed by TEM analysis after a two-step low temperature annealing process. A slight improvement of the peak value of the transit frequency was obtained for the preamorphized transistors.

1. Introduction

Double polysilicon selfaligning structures are used in state-of-the-art bipolar integrated circuits that offer a substantial increase of speed as well as reduced power consumption compared with conventional processing directly forming junctions in the monocrystalline silicon.

One of the key features to produce high speed bipolar transistors is the formation of shallow doping junctions. Crowder et al. [1] and Seidel et al. [2] have first shown that preamorphizing of single crystal silicon with ion implantation prior to junction formation is a promising technique.

The amorphization of the substrate prior to boron ion implantation by implanting heavier ions such as Si^+ [1, 3] and Ge^+ [2, 4] suppresses channeling and offers the potential of reducing the junction depth.

An alternative technique is the boron ion implantation through oxide or oxide/nitride layers. This requires relatively thick oxides and corresponding higher implant energies which can result in "knock on" effects and higher sheet resistance values [5].

A further simple approach to get shallow junction is the decrease of ion energy. This method, however, leads to a growing influence of channeling [6].

A method to circumvent the channeling effect by diffusion of the active base out of the polysilicon suffers from an insufficient base link. For this case an optimization of the spacer thickness is absolutely necessary [7].

In this study we address the issue of preamorphization via Si^+ and Ge^+ [8] implantation for junctions in transistors shallower than 150 nm which are formed by BF_2^+ and As^+ implantation. The goal was to investigate the structural and electrical properties of these junctions with the intention to produce a high performance bipolar transistor.

2. Experimental

The preamorphization experiments were carried out using two species of ions: Si and Ge.

Silicon ions were implanted using a conventional BALZERS MPB 200 with a cold cathode source (SiF_4) and wafers at ambient temperature. In order to avoid the formation of a buried amorphous layer the amorphization was accomplished by triple Si^+ ion implantation at three different energies.

Ge ions were implanted in a VARIAN model EXTRION type 20020 AF ion implanter using a GeH_4/H_2 mixture as a source. Wafers were hold at ambient temperature during implantation. Either ^{70}Ge or ^{74}Ge was used as amorphizing species.

The impurity concentration profiles were obtained by secondary mass spectroscopy (SIMS); amorphization and recrystallization behaviour were analyzed by transmission electron microscopy (TEM). Temperature treatment of the wafers was carried out either in a conventional furnace for low temperature steps or in a rapid thermal annealer (RTA; EATON ROA 400).

3. Results and Discussion

3.1. Preamorphization using silicon implantation

In order to obtain an amorphized layer extending from the surface the amorphization was accomplished by triple Si^+ ion implantation at three different energies ranging from 40 keV to 200 keV. Annealing of the samples was carried out in a furnace at temperatures from 400 °C to 600 °C and using RTA with cycles ranging from 5 to 35 seconds at temperatures from 900 °C to 1200 °C in Ar ambient.

The basic investigations were carried out using bare wafers for implantation. BF_2 ions were implanted at an energy of 30 keV to a dose of 8E13 cm^{-2} to simulate the base formation. Bipolar transistors were fabricated to test the electrical characteristics.

Fig. 1 shows the as implanted boron and fluorine concentration profiles after implantation of BF_2 at an energy of 30 keV into crystalline silicon. The SIMS data show significant channeling resulting in a boron penetration depth of 0.2 microns (measured at a concentration level of 2E16 cm^{-2}).

Fig. 2 shows the as implanted profiles of F and B after implantation into preamorphized silicon. Amorphization was carried out by triple Si$^+$ implantation at energies of 60, 120 and 180 keV to doses of 4E14 cm^{-2}, 5E14 cm^{-2} and 6E14 cm^{-2}, respectively. The ion channeling tails of both species are completely eliminated, consequently the boron penetration depth is reduced to about 0.1 micron at a BF2 implantation energy of 30 keV.

The influence of the preamorphization on the static and dynamic characteristics was determined using specific bipolar transistors. The process sequence of the transistors is based on a well established LOCOS process. Special test stripes on the wafer were used for SIMS and TEM analyses.

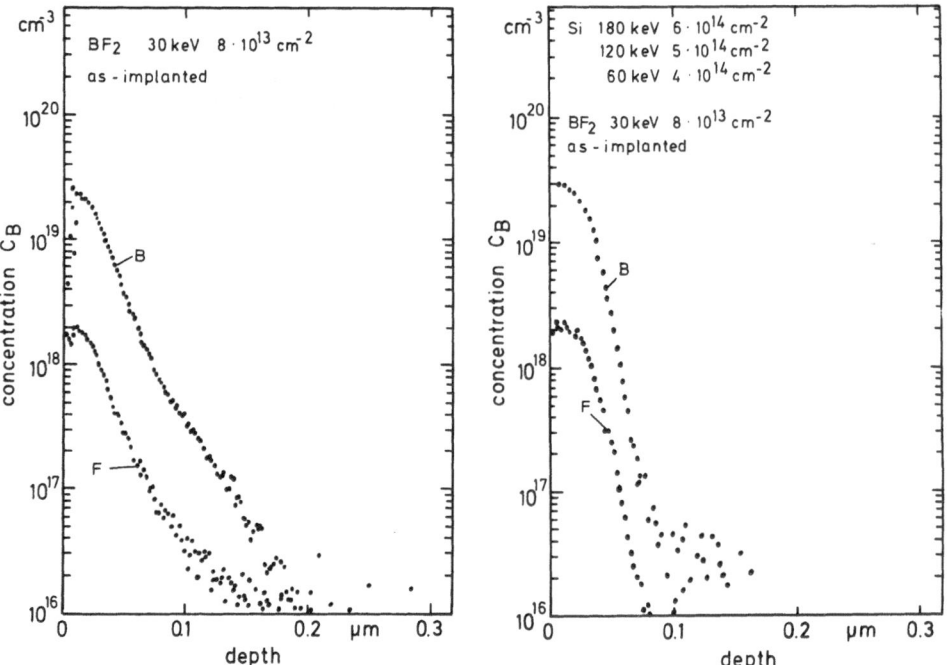

Figure 1. Concentration profile of BF2 as implanted at 30 keV into crystalline silicon.

Figure 2. Concentration profile of BF2 as implanted at 30 keV into Si$^+$-preamorphized silicon.

Extensive TEM analyses of the preamorphized regions revealed a major problem of the silicon self-preamorphization with wafers at room temperature: the residual defect bands within the active area of the transistors. In fig. 3 three defect bands are sketched along with the corresponding B and As concentration profiles. The sample was preamorphized at the same conditions described above. The BF2 implantation was carried out at 50 keV in this example.

The first layer around 0.02 nm consists of a damage type, which could not be specified by TEM. Possibly it is an effect of the surface treatment. The second layer of dislocation loops centered about 65 nm below the surface is likely to be associated with the direct arsenic implantation (this layer is absent if the process involves a polysilicon deposition and diffusion of arsenic out of polysilicon). The third layer consisting of dislocation loops and stacking faults extends from 200 nm to 300 nm with its high density center at 260 nm, the location of the end of the amorphized region. This residual damage layer lies in the critical region of the base-collector junction.

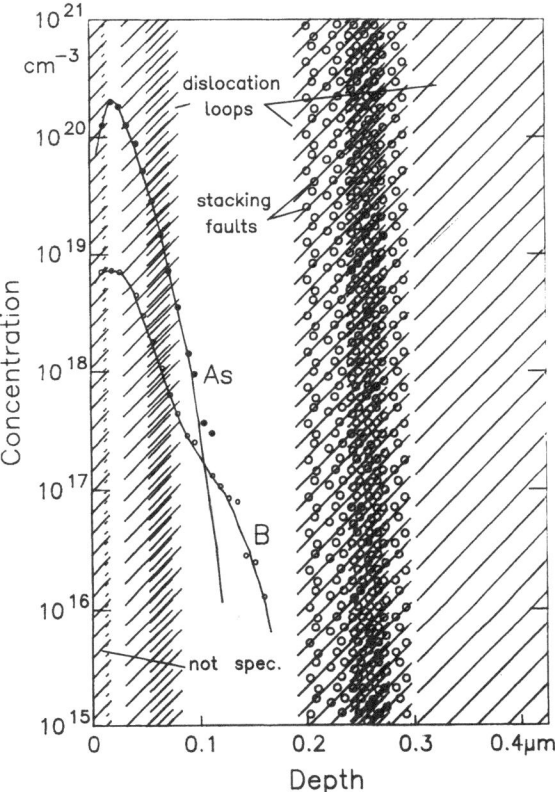

Figure 3. Residual defect bands after recrystallization obtained after TEM analysis of a Si$^+$-preamorphized transistor cross section.

Consequently all transistors with self-preamorphized base regions showed very poor electrical behaviour. Gummel plots revealed excessive leakage currents for IB and IC.

As already mentioned, the implantation was carried out with wafers at ambient temperature, which results in an incomplete recrystallization behaviour due to partly in situ annealing during implantation. To avoid this problem wafers should be at cryogenic temperature during Si$^+$ implantation, a condition not practicable with common commercial implanters.

The investigations have shown that Si$^+$ preamorphization avoids boron channeling. However the remaining defects result in excess leakage currents.

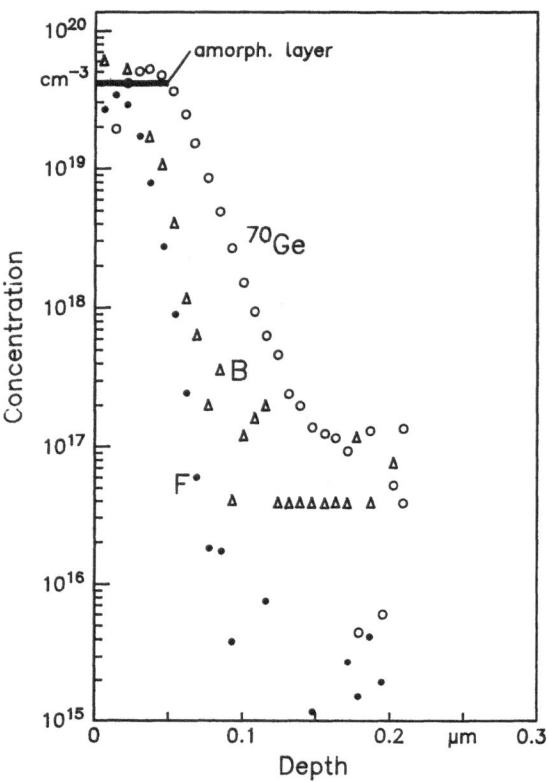

Figure 4. As implanted SIMS profile of a sample preamorphized with ^{70}Ge (60 keV) and
implanted with BF$_2$ (30 keV). The depth of the amorphous layer is indicated as obtained by TEM analysis.

3.2. Preamorphization using Ge$^+$ ion implantation

The implantation energies were varied between 60 keV and 150 keV and the
implantation doses were varied between 2E14 cm^{-2} and 9E14 cm^{-2} to determine
the optimum amorphization parameters.

Figs. 4 and 5 show the data of the SIMS analyses of two ^{70}Ge and BF^2 (30 keV)
implanted samples. Dots, triangles and open circles represent the fluorine, boron
and ^{70}Ge concentrations, respectively, where the fluorine concentration is not
calibrated. In the figures the depths of the corresponding amorphous layers are
indicated by the bars as obtained from the TEM analyses. As can be seen from the
150 keV preamorphized sample the boron profile lies well inside the amorphized
region which suppresses channeling effectively. In the 60 keV Ge sample however,

the boron depth exceeds the amorphous depth below a concentration of 8E18 cm^{-2}, consequently a certain amount of channeling must be expected. Due to the poor boron resolution of the SIMS measurement in the 1E17 cm^{-2} range, boron channeling can not be proved from the SIMS plot.

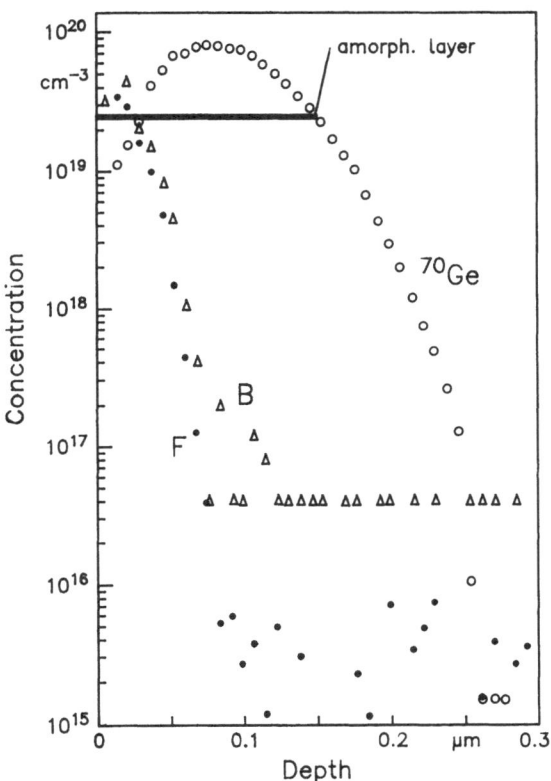

Figure 5. As implanted SIMS profile of a sample preamorphized with ^{70}Ge (150 keV) and implanted with BF_2 (30 keV). The depth of the amorphous layer is indicated as obtained by TEM analysis.

The investigations have shown, that an implantation energy of 70 keV is suitable to suppress channeling during BF_2 implantation at energies of 25 keV to 30 keV. A minimum implantation dose of 3E14 cm^{-2} is necessary for a complete amorphization.

Fig. 6 shows a TEM micrograph of an as amorphized sample. The amorphization parameters were: $^{70}Ge^+$ ion implantation at 70 keV to a dose of 6E14 cm^{-2}. The amorphized layer extends to about 85 nm below the surface. A relative broad roughness of the amorphous/crystalline (a/c) interface of 15 nm is obtained consisting of a mixture of crystalline and amorphous islands. No indication for the existence of stacking faults are found after this process step. A high residual defect density at the location of the original a/c interface was observed if the heat treatment process for solid phase epitaxial (SPE) realignment of the preamorphized layer was not optimized. Consequently the role of the a/c interface roughness and the effects of heat treatment steps on the recrystallization were studied in great detail by TEM.

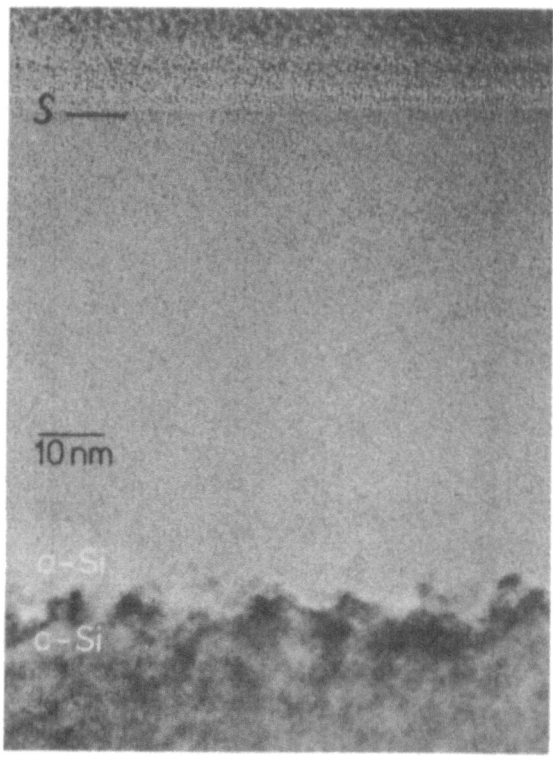

Figure 6. Cross sectional TEM micrograph of an as preamorphized sample; implantation parameters: ^{70}Ge$^+$ at 70 keV to a dose of 6E14 cm^{-2} (S: surface of the sample, a-Si: amorphous layer, c-Si: crystalline region).

Fig. 7 (see next page) shows a cross sectional TEM micrograph of the sample described above (fig. 6) after furnace anneal at 550 °C, 40 min in N$_2$ atmosphere (note the different scales between figs. 6 and 7). The amorphized layer has recrystallized whereas a layer of stacking faults originates at the location of the former a/c interface (indicated as "D"), with a stacking fault area density of 9E9 cm^{-2}.

TEM micrographs of preamorphized samples after furnace treatment at 450 °C, 40 min in N$_2$ atmosphere showed that recrystallization has not still started at this temperature. However a significant smoothing of the a/c interface and a decreasing of the lattice distortions have taken place. TEM analyses of preamorphized samples which were treated with a combination of a 450 °C, 40 min step followed by a 550 °C, 40 min step showed complete recrystallization of the amorphized layer. Furthermore no stacking faults could be determined above the lower detection limit of 5E6 cm^{-2} of the TEM analysis.

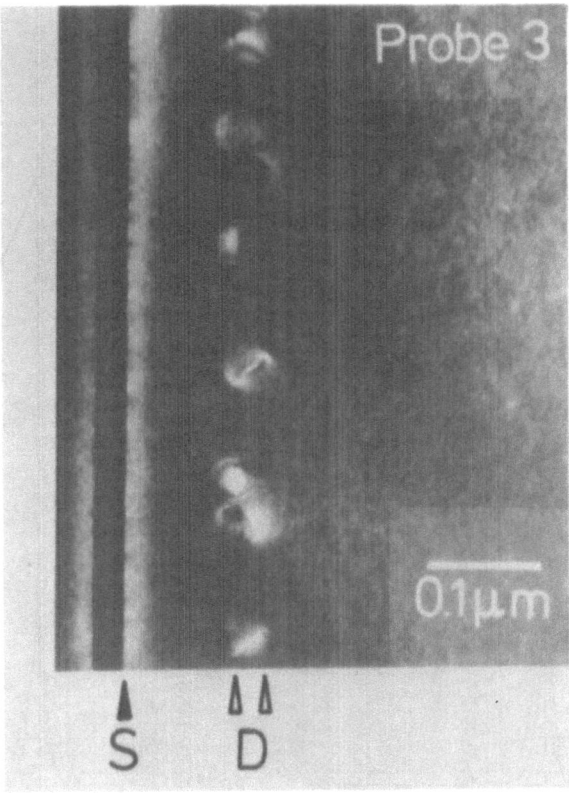

Figure 7. Cross sectional TEM micrograph of the sample of fig. 6 (different scale) after 550 C, 40 min furnace anneal (S: surface of the sample, D: residual layer of stacking faults).

In addition to the systematic studies on preamorphization and recrystallization bipolar HF transistors were fabricated. The base region was preamorphized after a LOCOS process. Amorphization parameters were: ^{70}Ge implantation 70 keV, $6E14$ cm^{-2}. After BF_2 implantation for intrinsic base formation the recrystallization of the base region was carried out using the two-step low temperature process described above. A 0.27 μm thick polysilicon layer was deposited, patterned and implanted by As$^+$ to form polysilicon emitters. Emitter drivein and dopant activation were accomplished by rapid thermal annealing.

Fig. 8 shows a SIMS analysis of a preamorphized transistor cross section. It exhibits an improvement of the obtainable base depth (X_{jB}) and base width (W_B) of 200 nm and 100 nm, respectively, in comparison to the case of a not preamorphized sample (fig. 9, X_{jB} = 250 nm, W_B = 140 nm). A further reduction of base width and base depth can be expected after reduction of the BF_2 implantation energy to 20 keV or below.

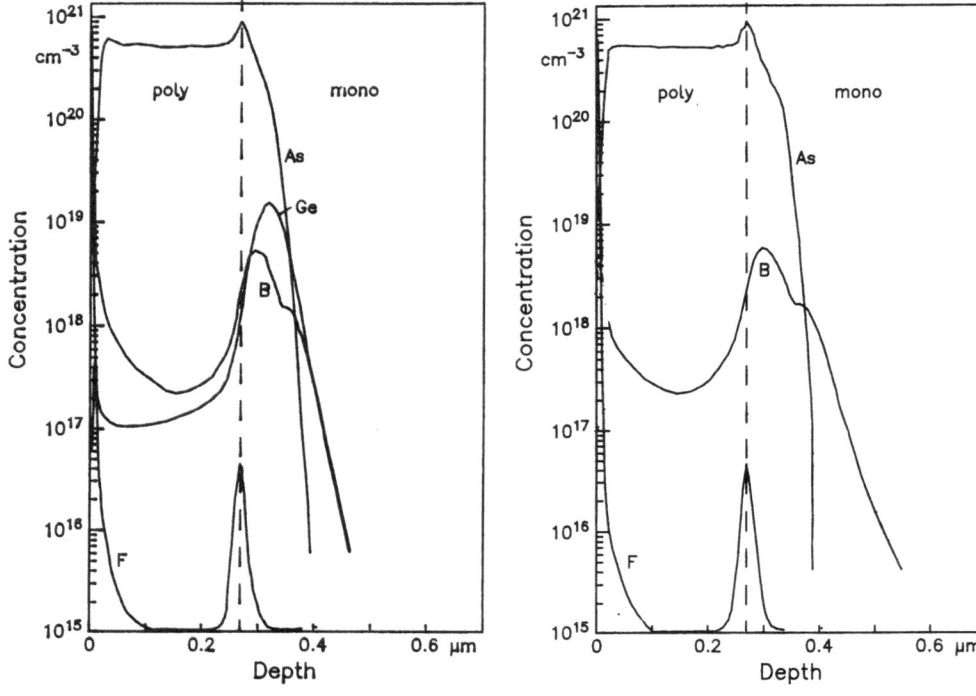

Figure 8. SIMS profile of a preamorphized transistor cross section; process parameters see text.

Figure 9. SIMS profile of a not preamorphized transistor cross section.

The transistors were electrically characterized with no significant leakage currents in dc mode.

Fig. 10 shows a sketch of the transit frequency vs. collector current at $U_{CB} = 0$ V, 1 V and 4 V for a preamorphized transistor with an emitter size of 24 μm x 1 μm.

The plot shows a peak value of the cut-off frequency f_T of 14 GHz at $U_{CB} = 4$ V, which is an improvement of > 1 GHz compared to the not preamorphized transistor.

The layout of the bipolar transistors tested in this state of the investigations was not specially optimized with regard to a high absolute f_T performance. Consequently the present result can only be a relative comparison about the improvement of the preamorphized transistors to the not preamorphized ones.

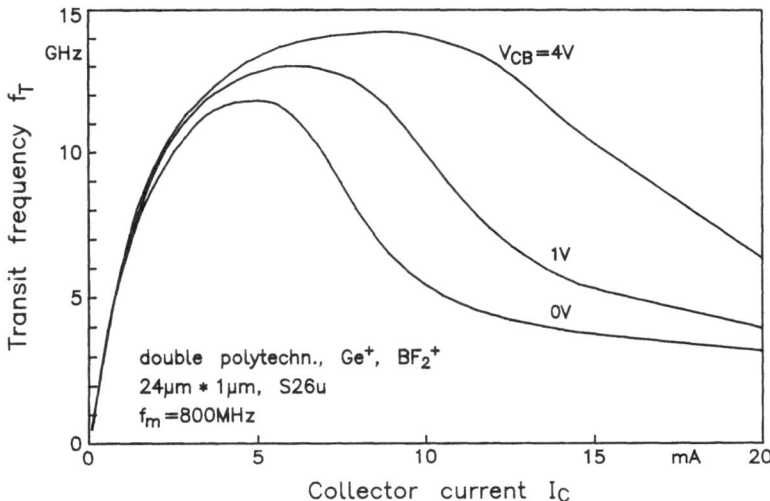

Figure 10. Transit frequency vs. collector current characteristics of a preamorphized transistor (details see text).

4. CONCLUSIONS

The investigations on Si^+ implantation have shown that boron channeling can be effectively avoided. However the remaining defects in the crystal lattice result in excess leakage current. Thus the process is not suitable for practical applications especially if wafers are implanted at ambient temperature.

The experiments on Ge^+ ion preamorphization have shown that single implantation at 70 keV, 3E14 cm^{-2} with wafers at ambient temperature produces a fully amorphized base region resulting in a complete prevention of channeling during BF_2 implantation at 25 - 30 keV. The amorphized layer effectively recrystallizes during a two-step low temperature furnace process without formation of stacking faults or dislocation loops. A base depth of 0.2 microns has been obtained for the preamorphized samples at BF_2 implantation energies of 25 - 30 keV. First ac measurements revealed a slight improvement of the peak f_T for the preamorphized transistor.

ACKNOWLEDGEMENT

The authors wish to thank Dr. M. Seibt, Universität Göttingen, for providing the TEM analyses and stimulating discussions, and B. Müller, Institut für Mikroelektronik, TU Berlin, for providing the Ge implantations.

98

REFERENCES

1 Crowder, B.L., Ziegler, J.F., and Cole, G.W. (1973) 'The Influence of the amorphous phase on boron atom distributions in ion implanted silicon', in Ion Implantation in Semiconductors and Other Materials, Plenum, New York, pp. 257 - 266

2 Seidel, T.E., Knoell, R., Poli, G., Schwartz, B., Stevie, F.A., Chu, P. (1985) 'Rapid thermal annealing of dopants implanted into preamorphized silicon', J. Appl. Phys. 58, 683 - 687

3 Tsai, M.Y. and Streetman, B.G. (1979) 'Recrystallization of implanted amorphous silicon layers. I. Electrical properties of silicon implanted with BF_2^+ or $Si^+ + B^+$', J. Appl. Phys. 50, 183 - 187

4 Ozturk, M.C., Wortman, J.J., Osburn, C.M., Ajmera, A. Rozgonyi, A., Frey, E., Chu, W.-K., and Lee, C. (1988) 'Optimization of the Germanium Preamorphization Conditions for Shallow-Junction Formation', IEEE Trans. on Electr. Dev. 35, 659 - 668

5 Ozturk, M.C., Wortman, J.J., Chu, W.-K., Rozgonyi, G.A., and Griffis, D. (1987) 'BF2 ionimplantation through surface oxides and the behaviour of the fluorine with rapid thermal annealing', Mat. Lett. 5, 311 - 314

6 Cho, K., Allen, W.R., Finstad, T.G., Chu, W.-K., Liu, J., Wortman, J.J. (1985) 'Channeling effect for low energy ion implantation in Si', Nucl. Instrum. Meth. (B) 7, 265 - 272

7 Ehinger, K., Kabza, H., Weng, J., Miura-Mattausch, M. Maier, I., Schaber, H., and Bieger, J. (1988) 'Shallow doping profiles for high-speed bipolar transistors', J. Physique 9, C4, 109 - 112

8 Imschweiler, J., Hefner, H. A., and Seibt, M., to be published

VERY HIGH SPEED BIPOLAR TECHNOLOGY AND CIRCUITS TIPBASE
BIPOLAR ADVANCED SILICON FOR EUROPE

A. J. Linssen
N.V. Philips Gloeilampen fabrieken
Eindhoven (The Netherlands)

L. Treitinger
Siemens Aktiengesellschaft
München (Germany)

R. Kaiser
TEG-Telefunken Electronic GMBH
Heilbronn (Germany)

P. Ward
SGS-Thomson Microelectronics SRL
Catania (Italy)

J. Smith
Plessey U.K. LTD
Caswell (United Kingdom)

Abstract

The contract no. 2016 has been signed between the commission of European Communities and a consortium of five partners Philips (prime contractor), Siemens, SGS-Thomson, Plessey and Telefunken.

It covers a period of three years. Effective work started on Feb. 1, 1989, the first year being a definition phase.

The project is headed by Europe's most competent partners in this field. It is firmly based on common ground partially developed within previous ESPRIT projects and positively interacts with ongoing ESPRIT and RACE projects of the CEC Research Framework Programme.

The aggressive technical goals as well as the tight timeschedule requires an additional effort of about 10% for University and Research institutes support in five European countries. The project is scheduled for three years. The manpower needed, which amounts to about 210 people has already been built up.

Strategic Goals

The project is market oriented in order to strengthen the European position in bipolar IC's being driven by the application areas consumer and communication. Thus the strategic goal of the project is to strengthen the European position in the field of high speed bipolar technology and circuits for main applications in consumer and communication. The process is also applicable to bipolar high performance circuits for data processing, instrumentation and other high performance industrial electronics.

Market perspective

TIPBASE is addressing an IC world market share estimated to be about 22,000 M$ per year in 1994 to be compared with 14,000 M$ in 1988 (see fig. 1) being an increase of about 10% per annum. Bipolar analogue (+ 12%) is growing faster than bipolar digital (- 2%). The market for high performance bipolar IC's will be 3,000 M$ in 1994.

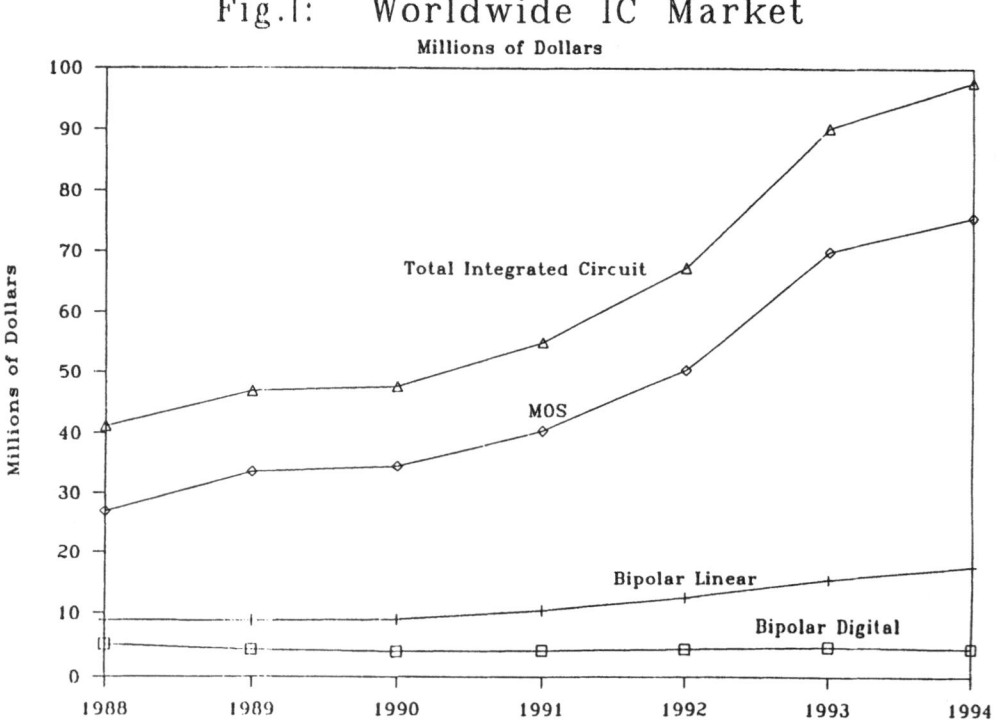

Fig.I: Worldwide IC Market
Millions of Dollars

The partners in TIPBASE today hold a market share of 13% which is to be defended and increased. Even more important than its direct impact on the semiconductor market is its impact on the high performance systems concerned with information technology with applications in the fields of consumer, communications, data processing, industrial electronics and instrumentation all demanding bipolar technology to realize key electronic functions. New systems development in Europe include satellite receivers for domestic television services, CT 2 cellular telephony and radio data system (RDS).

Technical impact

Very high speed bipolar technology fills a crucially important area between the lower performance but higher complexity per chip provided by MOS technology and the extreme device speed, but lower complexity, higher cost per function and limited availability, provided by Ga As (see fig.2).

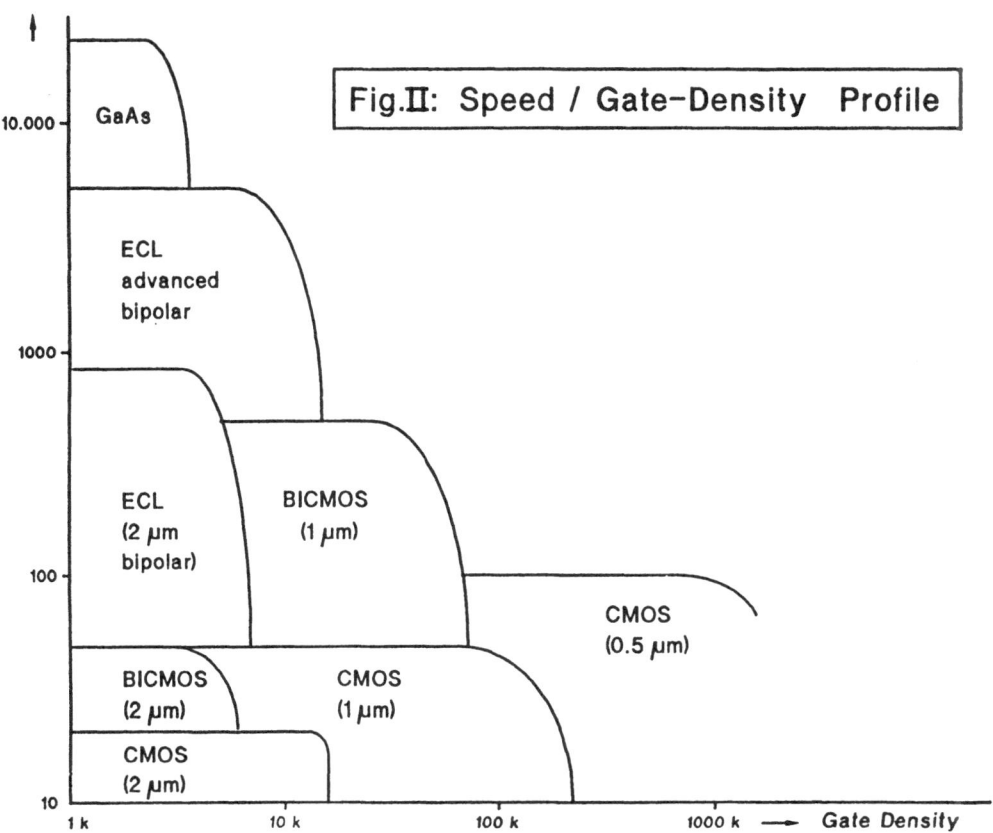

Fig.II: Speed / Gate-Density Profile

Bipolar is best suited for high precision and high speed systems because its analogue features are best, intrinsic speed is excellent, driving capability is best and availability as well as future perspectives are excellent. The bipolar technology is and will be the most important technology for high speed and high precision systems which are increasingly needed for the future applications in consumer, communication, computing and instrumentation. Higher data processing rates, broader bandwidth and lower noise applications and increased use of signal processing techniques all demand the use of new bipolar technologies.

Objectives

Fig. 3 shows the project structure of TIPBASE. The project will develop and fully integrate technology expertise in the submicron range as well as design and CAD expertise in the field of analogue, mixed analogue/digital and digital applications. The main goal is to develop a manufacturable bipolar process for high speed and high precision systems and to show its performance by circuit demonstrators.

Process/device performance

The process is characterized by the following target values:

characteristics
propagation delay	30 ps
power delay product	25 fJ
noise figure (2 GHz)	dB
packing density (ECL gates)	103/mm2

other characteristics
dimension WE (wafer)	$0.5\,\mu$m
pitch metallisation	$3.5\,\mu$m
cut-off frequency	25 GHz

For comparison, present pilot production processes are characterized by 1 to 1.5 μm emitter widths, about 100ps gate delays and three interconnect levels at about 5 μm minimum via pitch.

Technological objectives

Currently used isolation techniques are of the junction or of the recessed oxide type characterized by minimum dimensions of 2 μm. The aim within the project is to develop completely new techniques concerning the filling and etchback planarisation of the trenches with a minimum dimension of 0.8 μm.

The emitter base structures will be capable of 30ps ECL gate delay operation and will be characterized by 0.5 μm minimum emitter width, extremely shallow polysilicon contacted emitters of about 50 nm junction depth, base width of 100 nm or below and probably the use of silicides for base and emitter contacting.

The multilevel metallisation scheme will allow for up to three metal layers at a pitch of 3.5 μm with minimum restrictions on the placement of vias and tracks. Key issues will be planarisation of dielectric, via hole-filling and metal step coverage, lithography and etching and the improvement of electromigration performance, yield and reliability.

Substrate preparation and epitaxy have to continue to strive for reduction of defect densities. An important issue is the reduction of epi-layer thickness to about 0.5 μm with very narrow epi to buried layer transition regions.

Very encouraging results have recently been obtained with new device concepts and materials like hetero-devices, sidewall contacted base structures etc. Corresponding key effort will be included in this project.

103

Fig: III   PROJECTSTRUCTURE, PROJECT TIP BASE

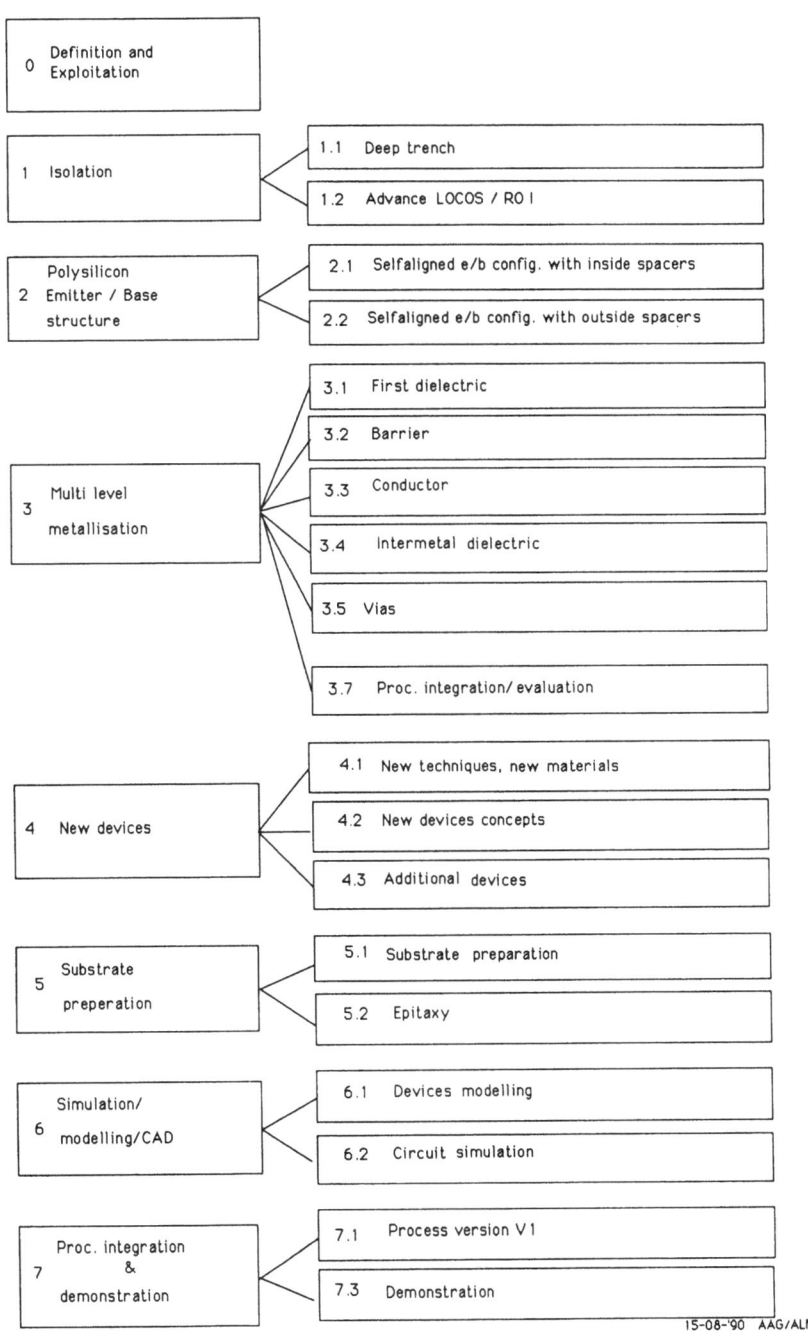

0	Definition and Exploitation

1	Isolation	1.1	Deep trench
		1.2	Advance LOCOS / RO I

2	Polysilicon Emitter / Base structure	2.1	Selfaligned e/b config. with inside spacers
		2.2	Selfaligned e/b config. with outside spacers

3	Multi level metallisation	3.1	First dielectric
		3.2	Barrier
		3.3	Conductor
		3.4	Intermetal dielectric
		3.5	Vias
		3.7	Proc. integration/ evaluation

4	New devices	4.1	New techniques, new materials
		4.2	New devices concepts
		4.3	Additional devices

5	Substrate preperation	5.1	Substrate preparation
		5.2	Epitaxy

6	Simulation/ modelling/CAD	6.1	Devices modelling
		6.2	Circuit simulation

7	Proc. integration & demonstration	7.1	Process version V 1
		7.3	Demonstration

15-08-'90 AAG/ALN

The aim is to include some of these innovative devices and techniques into the mainstream process flow at an appropriate time although others might prove not useful, or applicable only in a later process generation. A large fraction of this work will be done in collaboration with research institutions.

CAD objectives

The CAD effort will concentrate on the following areas:

- adaption of device simulation tools to reduced geometries, new effects related with shallow junctions and two dimensional simulation.
- increased frequency range of compact transistor models and also of on wafer measurement techniques.
- coupling of process-, device- and circuit simulation.

Demonstrators

The final proof of consistency and validity of all the efforts described so far will be the design and realization of demonstrator chips in analogue, mixed analogue/digital and digital applications with up to 5 GHz operating frequency and of high speed circuits for communications up to and above the 10 GHz band.
Presently available technologies allow the realization of circuits at about 100ps gate delay and some communication circuits for the 2.4 GHz band.
The demonstrators envisaged are:

- static frequency divider at 10 GHz (16:1)
- 10k mixed digital/analogue array for 10 Gbit/s systems with three interconnect layers.
- 8 bits AD converter 300 MHz, 150 MHz bandwidth (10k complexity)
- Multiplexer (14-20 Gbit/s)
- laser driver (8-10 Gbit/s)
- MUX/DEMUX 10 Gbit/s (64:1)
- amplifier + prescaler input stage (0.9 - 1.2 GHz; 1-10 mW)

These demonstrators will be delivered at the end of the project.

All partners in Tipbase already offer the possibility to design in their IC technologies in production or pilot-line production to external users, i.e. SME's, Institutes and Universities. External users could start to design in the developed technology in this project as soon as reliable design data are available.

LOW TEMPERATURE ION-ASSISTED EPITAXY OF DEPOSITED SILICON LAYERS

F. PRIOLO, C. SPINELLA and E. RIMINI
Dipartimento di Fisica, Universita di Catania,
Corso Italia 57
I95129 Catania (Italy)

Abstract.

The low temperature epitaxial crystallization of chemical vapor deposited silicon layers obtained by means of high energy ion irradiation is studied. Both the kinetics of the process and the morphology of the regrown layers are characterized. This novel procedure, in view of the small thermal budgets involved, can result interesting for possible application to the bipolar technology.

1. Introduction

The structure of deposited Si layers onto single crystal substrates plays a dominant role in bipolar transistor technology where the emitter is often made by a deposited layer. A process capable of transforming deposited amorphous layers into good quality single crystals is therefore of fundamental importance to optimize the performance of the whole device.

Solid phase epitaxial growth of amorphous Si (a-Si) layers onto single crystal substrates presents several problems, mainly related to the presence of contaminants at the deposited layer/substrate interface. Substrates cleaned by the conventional chemical procedure before deposition have at least a monolayer of native oxide and/or contaminants at the interface between the substrate and the deposited layer. Thermal annealing at low temperatures ($< 1000°C$) is usually not able to dissolve this interfacial oxide and the competitive phenomenon of random nucleation and grain growth takes place. The result is a polycrystalline layer. Single crystal layers can be obtained provided that a high temperature thermal process is used (e.g. $1100°C$ for a few seconds). In this case columns of crystal silicon grow in those regions, randomly distributed, where windows are opened in the interfacial oxide. The columnar growth is then followed by a lateral growth and, at the end, a uniform single crystal is obtained[1-2].

High temperature processing, though able of producing single crystal emitters, is in contrast with the demand for low thermal budgets, needed to avoid long range dopant diffusion and produce shallow junctions. Alternative techniques able to epitaxially recrystallize deposited silicon layers at low temperatures are therefore required.

In this paper we will present a novel process able to recrystallize deposited Si layers at temperatures as low as 400°C. The paper presents the results of a fundamental research but the possible applications to bipolar technology seem particularly promising.

2. Experimental Procedure

Si layers, 500-1000 Å thick, were deposited by silane dissociation at a temperature of 540°C, at a pressure of 250 mTorr and at a rate of 3 nm/min. A set of samples (hereafter referred to as set A) was loaded into the reactor as received from the manufacturer, a second set (hereafter referred to as set B) was instead cleaned by a simple *HF* dip prior to deposition.

The oxygen content at the interface between the deposited layer and the substrate was accurately measured by nuclear reaction analysis (NRA) using a 830 keV deuteron beam extracted from a 2.5 MeV Van de Graaff generator. The nuclear reaction $^{16}O(d,p)^{17}O^*$ was used in the analysis and the outcoming proton beam was detected at an angle of 150° through a 12μm thick mylar sheet (to avoid backscattered particles) by a solid state detector. The NRA showed that samples of set A have ~ $6.5 \times 10^{15} O/cm^2$ at the interface (which corresponds to about 15Å of native oxide) whilst those of set B have ~ $1.5 \times 10^{15} O/cm^2$ (i.e. about a monolayer). Both of these oxygen concentrations completely inhibit the occurrence of epitaxy by conventional thermal treatments at temperatures below 1000°C.

Recently it has been shown that recrystallization of implanted *a*-Si layers can also be induced by means of high energy ion beam irradiation in the temperature range 250-450°C[3-5] and at crystal-amorphous interface velocities of ~ 1Å/sec. This process is referred to as ion-beam-induced epitaxial crystallization (IBIEC). We have studied IBIEC using a 600 keV *Kr* beam scanned at an ion flux of $1 \times 10^{12}/cm^2 sec$. An interferometric technique to follow *in situ*, during the irradiation, the IBIEC process has been also mounted. The experimental apparatus is schematically shown in Fig.1. The sample is held on a resistively heated copper block whose temperature is monitored by a chromel-alumel thermocouple. A 5 mW σ-polarized *He - Ne* laser is focussed, through a window present in the scattering chamber, onto the irradiated sample and the reflected light is detected by a photodiode connected to an X-Y recorder (whose X axis is driven by a signal proportional to the *Kr* dose). The reflectivity will oscillate due to successive constructive and destructive interferences between the light reflected from the surface and that reflected from the buried, advancing crystal-amorphous interface. A simulation of the reflectivity signal as a function of the thickness of the surface amorphous layer is shown in Fig.2. For instance, as soon as the signal passes from a maximum to a successive minimum the surface amorphous layer has regrown by 33 nm. Fig.3 shows how transient reflectivity measurements can give direct informations on the ion-induced growth rate. Experimentally we measure the reflectivity vs. the irradiation dose (a), the theory gives the reflectivity vs. the amorphous thickness (b), a comparison of (a) and (b) directly yields the regrown layer vs. the irradiation dose

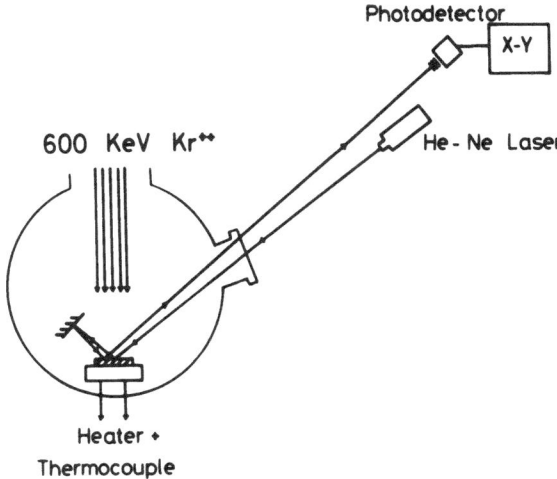

Fig.1 Experimental apparatus

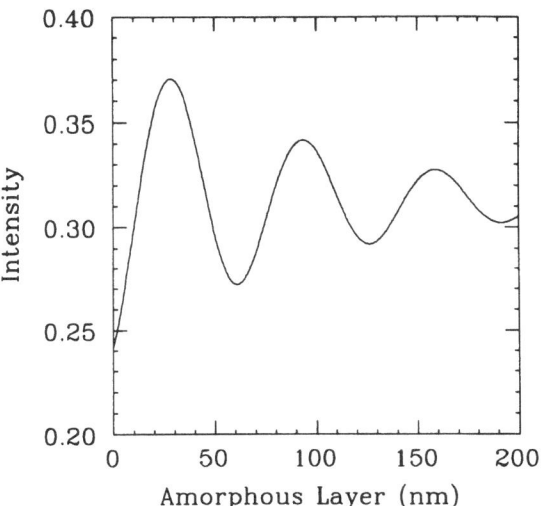

Fig.2 Calculated reflectivity vs thickness of the amorphous layer.

108

(c) and a derivative of (c) finally gives the growth rate as a function of depth (d). A computer program takes care of all of these different steps and we can obtain the ion-induced growth rate in real time.

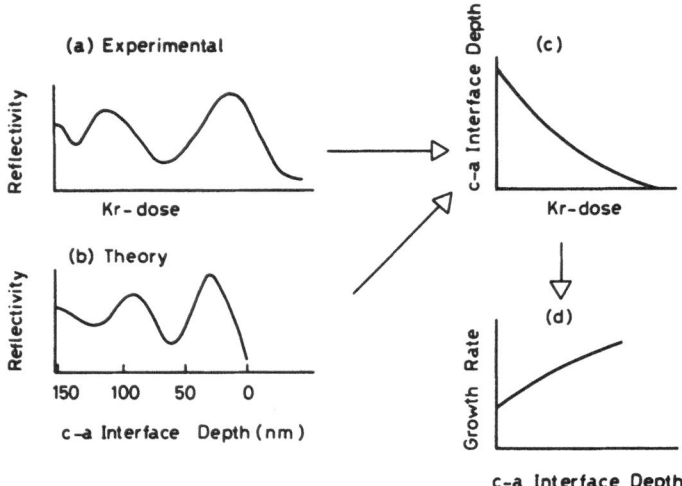

Fig.3 Schematic illustration of the use of transient reflectivity measurements to obtain the growth rate. The experiment measures the reflectivity vs the irradiation dose (a), the theory gives the reflectivity vs the amorphous thickness (b), a comparison of (a) and (b) yields the regrown layer as a function of the irradiation dose (c) and, finally, a derivative of (c) gives the growth rate vs the depth (d).

After irradiation samples were analyzed *ex situ* by 2.0 MeV He^+ Rutherford backscattering (RBS) in connection with the channeling effect and by Transmission Electron Microscopy (TEM).

3. Kinetics

The transient reflectivity technique is particularly useful when applied to study the influence of impurities on the kinetics of IBIEC. Oxygen is well known to inhibit epitaxy by conventional techniques. We have shown that its effect on IBIEC, though present, is however less dramatic. Fig.4 shows the ion-induced normalized growth rate for a 1500Å thick amorphous layer onto < 100 > Si implanted with 20 keV O to a dose of $6.2 \times 10^{15}/cm^2$ (the normalized rate is the growth rate of the O-implanted layer divided to that of an impurity-free layer). In the same figure the O profile is also reported. The rate decreases with increasing oxygen concentration and it reaches a minimum of 0.3 at the peak value of $1 \times 10^{21} O/cm^3$. The extracted dependence of the growth rate on O concentration is shown in Fig.5 where an almost linear dependence is observed. For comparison the literature data [6] showing the influence of oxygen on conventional thermal annealing at 550°C are also reported. The thermal growth rate decreases exponentially with oxygen concentration and a concentration of $4 \times 10^{20}/cm^3$ is able to completely halt the process.

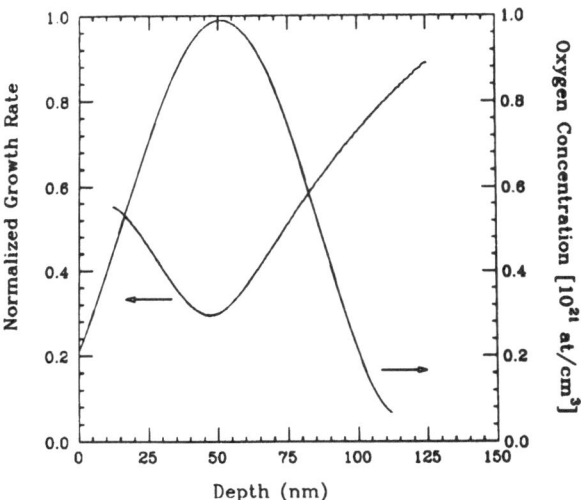

Fig.4 Ion-induced normalized growth rate as a function of depth for a 1500Å thick amorphous layer with a gaussian oxygen profile. The oxygen profile is also reported in the figure.

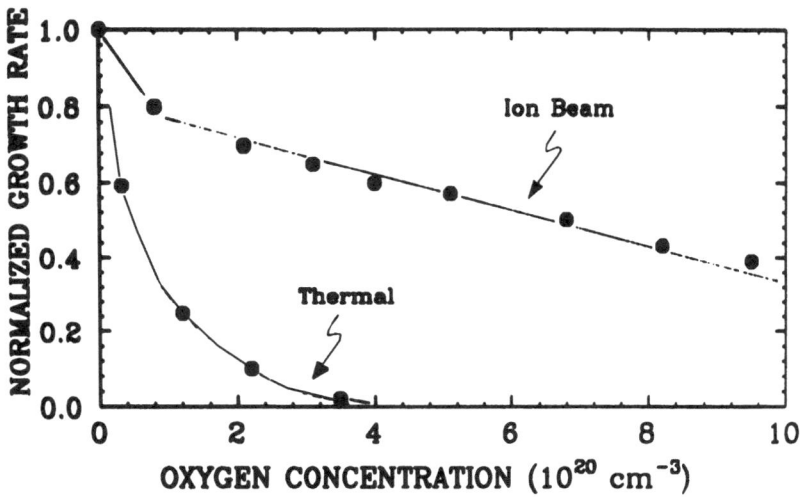

Fig.5 Normalized growth rate as a function of the oxygen concentration. They are reported for both thermal and ion-assisted treatments.

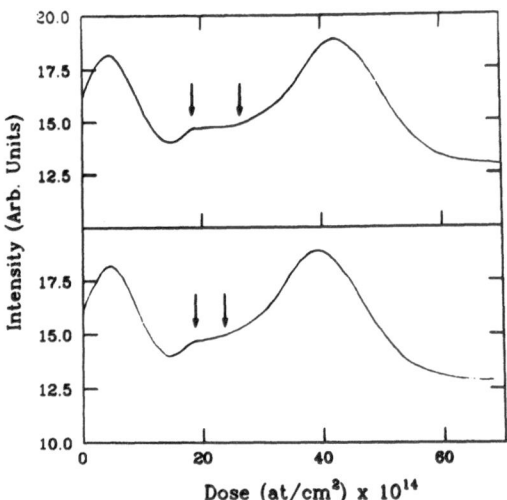

Fig.6 Transient reflectivity measurements for deposited Si layers, uncleaned (upper) and HF-cleaned (lower) prior to deposition, and recrystallized by Kr irradiation at $450°C$.

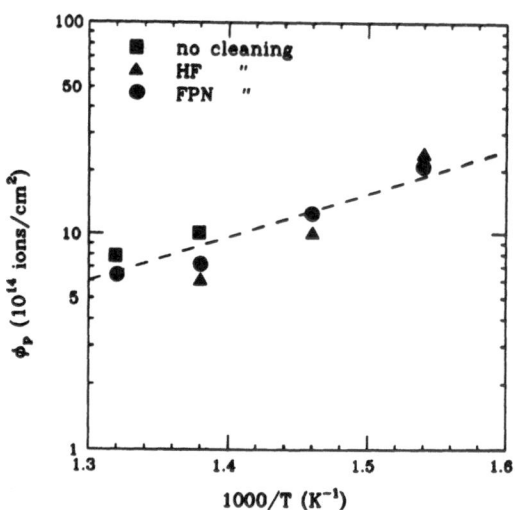

Fig.7 Incubation dose vs reciprocal temperature

The strong influence of oxygen on thermal annealing does not allow the epitaxial regrowth of deposited layers due to the presence of interfacial oxygen contamination. We have shown that IBIEC can overtake this problem and epitaxial regrowth is observed. Samples of both set A and B were implanted at room temperature by 150 keV Ge^+ ions to a dose of $5 \times 10^{14}/cm^2$. These implants were necessary to completely amorphize the deposited layers (we have indeed noticed that chemical vapor deposited layers are not really "amorphous" but contain a high density of small microcrystallites which must be destroyed to avoid the growth of randomly oriented crystal grains). These samples were subsequently irradiated with 600 keV Kr ions in the temperature range 350-480°C and recrystallization was monitored *in situ* by transient reflectivity. Fig.6 shows the reflectivity signals for sample A (upper part) and sample B (lower part) vs. Kr dose. The oscillations indicate the occurrence of epitaxial regrowth and after a dose of $\sim 6 \times 10^{15} Kr/cm^2$ the whole deposited layer is converted into single crystal material (as also confirmed by RBS). The plateau between arrows is associated with a temporary stop of the crystallization process at the deposited layer/substrate interface. After an incubation dose (which depends on the cleaning procedure performed prior to deposition) the native oxide breaks up and the epitaxial front propagates into the deposited layer. The dependence of this incubation dose on temperature and cleaning procedure is clearly illustrated in Fig.7: it increases with decreasing temperature and with increasing O content at the interface.

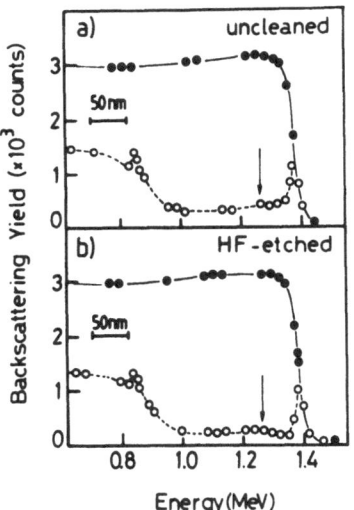

Fig.8 2.0 MeV He Rutherford Backscattering spectra of 60 nm thick deposited layers after recrystallization with Kr ions. Spectra are shown for both uncleaned (a) and HF-cleaned (b) samples. The crystal quality is better in this last case.

4. Morphology

Fig.8 shows the RBS spectra of 60 nm deposited a-Si layers after recrystallization by 600 keV Kr^{++} ions at a fluence of 6 x $10^{15}/cm^2$ and at a temperature of 450C. The spectra in Fig.8a refer to samples of set B (i.e. cleaned in HF prior to deposition) whilst those of Fig.8b refer to samples of set A (i.e. uncleaned prior to deposition). Both random (closed circles) and channeling (open circles) spectra are reported. The sharp increase in the channeling yield at energies below 0.8 MeV is associated with the end of range damage produced by the 600 keV Kr irradiation. This damage is found to start at a depth of ~ 200 nm. It should be noted, however, that inducing the recrystallization by higher energy beams (> 1.5MeV) will push this damage much deeper inside (at depths > 1.5μm), thus avoiding interferences with the active regions of the device. These high energy irradiations are currently under investigation. The main feature of Fig.8 is the indication of a good crystal quality obtained in the surface deposited layer. Samples of set B (Fig.8b) have a x_{min} (i.e. the ratio between the channeling and the random yields) of ~ 5%, comparable to that of a virgin crystal. On the other hand, samples of set A (Fig.8a) have a X_{min} of ~ 10%, demonstrating that some residual damage is present on the recrystallized layers.

In order to study in more details the damage morphology, samples have been analyzed by Transmission Electron Microscopy (TEM) both in the plan view and in the cross sectional configurations. Samples were thinned by backside etching for plan view imaging and by mechanical polishing and Ar^+ ion milling for cross sectional imaging.

Fig.9 is a cross sectional TEM (in dark field) of a sample of set A. Deep inside the end of range damage is clearly visible. Note the sharp interface present between the damage and the defect-free crystalline material. The deposited layer at the surface, though recrystallized, presents a high density of defects (bright regions). The extra spots present in the diffraction pattern of this sample (Fig.10) demonstrates that these defects are twins around the < 111 > direction.

In Fig.11 high resolution TEM cross sectional images (in bright field) are reported for both a sample of set A (Fig.11a) and a sample of set B (Fig.llb). Twins are clearly visible together with the original interface for samples of set A. A good quality single crystal is instead observed for samples of set B, though the original interface is still visible as a dark line.

The density of twinned regions has been observed by plan view TEM in dark field. Fig.12a refers to samples of set A whilst Fig.12b refers to those of set B. Bright regions are twins. These twins have lateral dimensions of ~ 10-50nm and their amount is ~ 12% of the total area for set A. Some twins are seen to be present also in samples of set B, their amount is however < 3%. These results are in agreement with the previously shown RBS spectra and demonstrate that the cleaning procedure performed prior to deposition is critical to obtain a good quality single crystal.

It should be noted that nuclear reaction analyses have shown that the amount of interfacial oxygen is 6.5 x $10^{15}/cm^2$ for samples of set A, where twins are largely present, and 1.5 x $10^{15}/cm^2$ for those of set B, where twins are almost absent (see

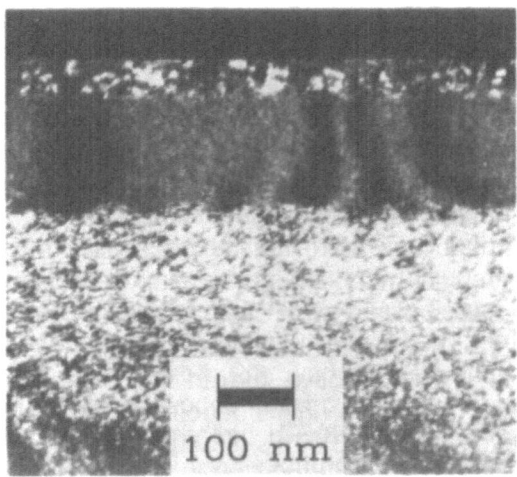

Fig.9 Cross sectional TEM of a sample of set A after recrystallization. The end of range damage is clearly visible. Twins are present at the surface.

Fig.10 Diffraction pattern of the same sample. The extra spots indicate the presence of twinned material.

also previous report). The formation of twins can therefore be associated with the presence of a high oxygen content and, in turn, with a severe retardation of the *c-a* boundary. As a matter of fact, we have noticed that twin formation occurs every time the < 100 > *c-a* interface is retarded by more than a factor of 4 with respect to its normal rate, independently of the species which produces this retardation. This behavior suggests that twin formation is simply due to kinetic reasons: as soon as the < 100 > *c-a* interface slows down, crystal growth along other orientations (such as twins) might become favorable and takes place.

It is important to stress, once more, that the problems in recrystallizing deposited *Si* layers are associated just with high interfacial oxygen content. Therefore, once the thin interfacial oxide is passed by ion beam irradiation, the epitaxial growth front should proceed unperturbed to the surface also by means of conventional thermal processing. To test this idea we have *Kr*-irradiated a sample of set B at 450°C, stopping the irradiation when the original deposited layer/substrate interface was passed (see RBS spectra in Fig.13). This treatment produced a partial recrystallization with a ~ 40*nm* surface amorphous layer left. The sample was subsequently heated in a vacuum furnace (p ~ 10^{-7}*Torr*) at 700°C for 30 minutes. The thermal treatment resulted in a complete recrystallization of the amorphous layer (circles). The possibility of using a combination of ion beam and thermal processing is of great interest in view of reducing the dose of *Kr* which induces the recrystallization and, in turn, its end of range damage.

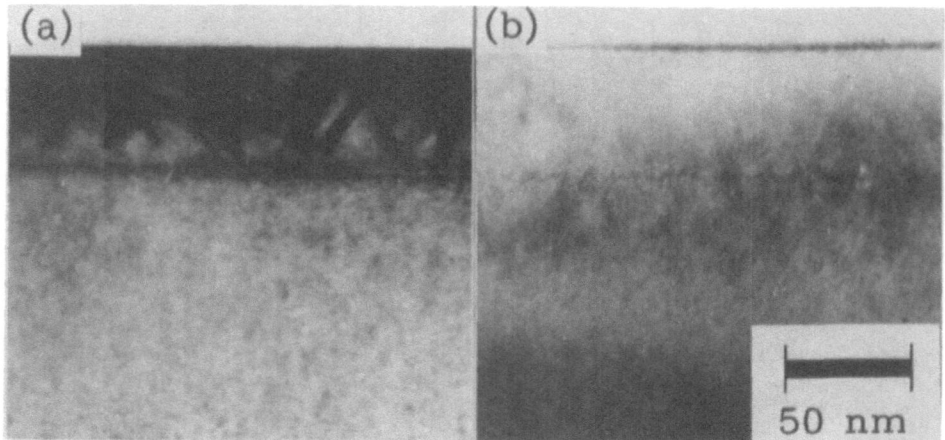

Fig.11 Cross sectional TEM images of ion-beam recrystallized deposited Si layers. The micrographs refer to samples of set A (a) and of set B (b).

5. Applications and Summary

The application of this technique to a production process, assuming that the secondary defects left by the recrystallizing beam have no effects on the electrical

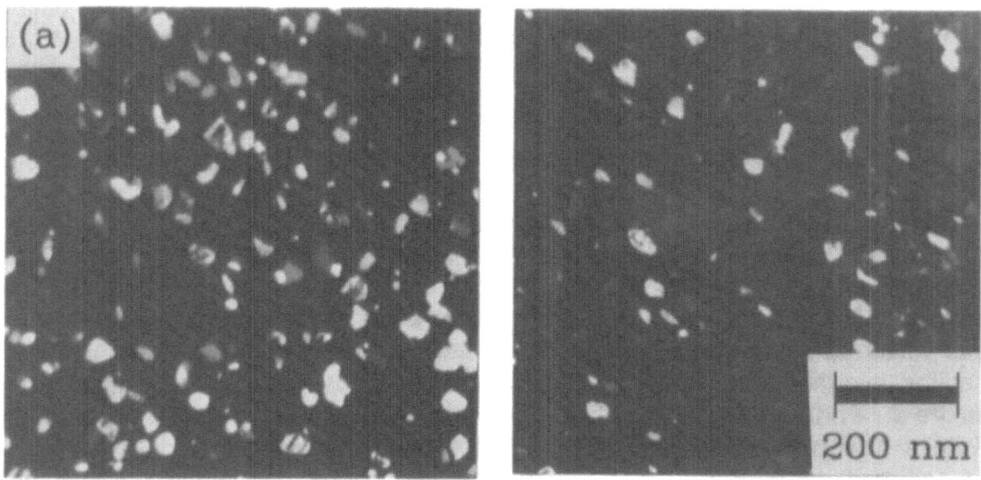

Fig.12 Plan-view TEM images, in dark field conditions, of ion-beam recrystallized deposited Si layers. The micrographs refer to samples of set A (a) and of set B (b). Bright regions are twins.

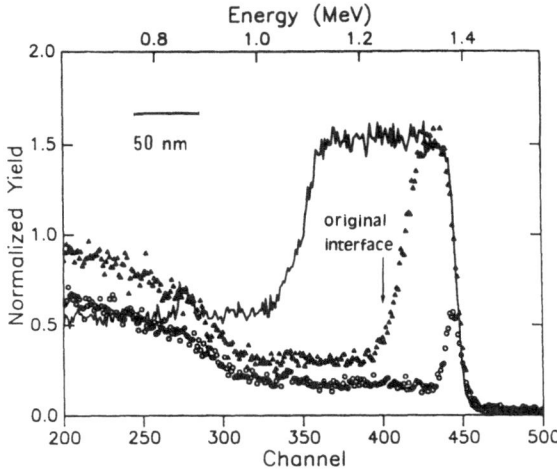

Fig.13 2.0 MeV He Rutherford Backscattering spectra in channeling conditions for a 60 nm thick deposited layer after amorphization (continuous line), after successive partial ion beam regrowth (triangles) and after a further pure thermal treatment at 700°C (circles). The original deposited layer/substrate interface is clearly indicated.

behaviour, requires an ion implanter with an heating stage in the end station. Temperatures in the 400-500°C range are required for the IBIEC as shown in the present work. This might be a serious limitation for high current implanters based on a mechanical scanning system. The huge wheel cannot be easily heated so a medium current implanter with casset to casset handling seems suitable for the process. The wafers should be processed once at a time. The required dose being in the 1×10^{15} - $1 \times 10^{16}/cm^2$ range should be compatible with all the other processes. The usual wafer dimensions can be processed.

In summary, we have shown that the epitaxy of deposited Si layers can be obtained with very low thermal budgets by using an ion beam to stimulate the process. The results show that both the kinetics and the morphology of the recrystallized layers strongly depends on the oxygen present at the original deposited layer/substrate interface. Furthermore we have demonstrated that as soon as the c - a boundary has trepassed, by ion beam irradiation, the original interface (with a high oxygen content), also a conventional thermal treatment is able to recrystallize the deposited layer. The possibility of using a combination of both ion beam and thermal processing seems particularly attractive for applications in bipolar transistor technology.

6. References

[1] B.Y. Tsaur and L.S. Hung, Appl. Phys. Lett. 37, 922 (1980)

[2] M. Delfino, J.G. Groot, K.N. Ritz and P. Maillot, J. Electrochem. Soc. 136, 215 (1989)

[3] J.S. Williams, R.G. Elliman, W.L. Brown and T.E. Seidel, Phys. Rev. Lett. 55, 1482 (1985)

[4] F. Priolo, A. La Ferla and E. Rimini, J. Mater. Res. 3, 1212 (1988)

[5] F. Priolo, C. Spinella, A. La Ferla, E. Rimini and G. Ferla, Appl. Surf. Sci.43, 178 (1989)

[6] E.F. Kennedy, L. Csepregi, J.W. Mayer and T.W. Sigmon, J. Appl. Phys. 48, 4241 (1977)

TIPBASE 2016 BIPOLAR ADVANCED SILICON FOR EUROPE ADVANCED BIPOLAR PROCESSES FOR HIGH-PERFORMANCE ANALOG APPLICATIONS

J.W.A. VAN DER VELDEN
Bipolar Technology Research & Development
Philips Research Laboratories
P.O. Box 80.000
5600 JA EINDHOVEN
The Netherlands

Abstract

Within TIPBASE a new process concept (BASIC) has been developed showing excellent analog performance. BASIC (Best Alignment with Sidewall Contact) is a super selfaligned technology which gives not only high speed but also low power because of the strong reduction in parasitic capacitances and resistances.

High f_T (17 GHz) and f_{max} (15 GHz) both at $V_{cb}=2$ V have been obtained for the npn device made in this technology.

As an analog demonstrator a 8-bit high-speed A/D converter with an analog bandwidth of 150 MHz at a sampling rate of 300 MHz will be made.

Introduction

Process capability for up to 10k gates has been demonstrated by several companies within the ESPRIT I projects with processes based on shallow ($0.2\,\mu$m) emitter base structures, polysilicon emitters and three level metallisation. Not only circuit complexity, but also high-speed and low-power performance (90 ps gate delay and 100 fJ speed power product) have been achieved which matched the most advanced values reported for high speed ECL circuits.

For the present ESPRIT II project TIPBASE a much higher performance is envisaged. Among the demonstrators ECL ring oscillators with gate delays of less than 30 ps and speed power products of less than 25 fJ will be delivered at the end of year 3 of the running project. The programme includes leading edge topics such as silicon hetero structures, molecular beam epitaxy, and new transistor structures, comparable with those in the most advanced research laboratories in the world.

Analog Demonstrator: 8 bit high-speed A/D converter

One of the company specific demonstrators that will be made by Philips within the TIPBASE consortium is a high-speed 8-bit A/D converter.

This demonstrator does not only have a fairly high complexity (10k components) but also places strong demands on the specifications with respect to analog performance.

The analog bandwidth is specified at 150 MHz with a 300 MHz sampling rate. The restriction is that at this frequency the Effective Resolution may be reduced with 0.5 least significant bit to 7.5 bits effective.

Internal bandwidth of amplifier/comparators used in this system must be at least 2 GHz (-3 dB). Matching of the components is critical because of the linearity requirements and the full scale resolution.

Application areas for this A/D converter can be found for Digital Oscilloscopes, HDTV and high speed signal processing.

BASIC/BASIC II: Advanced Analog Bipolar Processes

To obtain the required analog high-frequency performance, strong demands are placed on the bipolar process concept that can be used. One of the possible candidates is the class of the so-called sidewall contacted devices. In the context of the TIPBASE project, Philips is involved in work on such super selfaligned concepts, which are characterised by very low parasitic resistances and capacitances.

Ideally, the bipolar npn transistor contains a three base structure: a $p(++)$ extrinsic base, a $p(+)$ link-up base and a $p(-)$ intrinsic base. With the advent of selfaligned process schemes which were driven by the necessity for lower resistances (R_{bb}) and lower parasitic capacitances (e.g. C_{cb}), this $p(+)$ link-up region has become either non-existent (SICOS, ref. 1) or submicron and not separately adjustable.

Therefore, most of the recently published advanced self-aligned bipolar technologies exhibit a poor control of the emitter-base transition region (ref. 2).

The absence of an adjustable transition region between highly doped extrinsic base and lightly doped intrinsic base, results in problems such as E-B forward tunneling and low E-B reverse breakdowns (ref. 3). To study these problems, simplified process versions (e.g. SDD-2 ref. 4) without lateral base alignment have been evaluated.

Several proposals have been made to mediate this base link-up problem in advanced selfaligned schemes, but in none of them the intermediate base link-up can be tuned independently from the intrinsic transistor. For the so-called double poly process, one of the proposed solutions is the addition of a base link-up implant before spacer formation (ref. 5). Another proposal is the use of BSG spacers for outdiffusion of boron to provide a link between intrinsic base and polysilicon base boosts (ref. 6).

This base link-up problem, which is of particular importance for analog applications, is eliminated in the BASIC process by the introduction of an adjustable submicron link-up region, reproducibly defined by the lateral oxidation of polysilicon (Fig. 1a).

The final device structure, is shown in Figure 2.

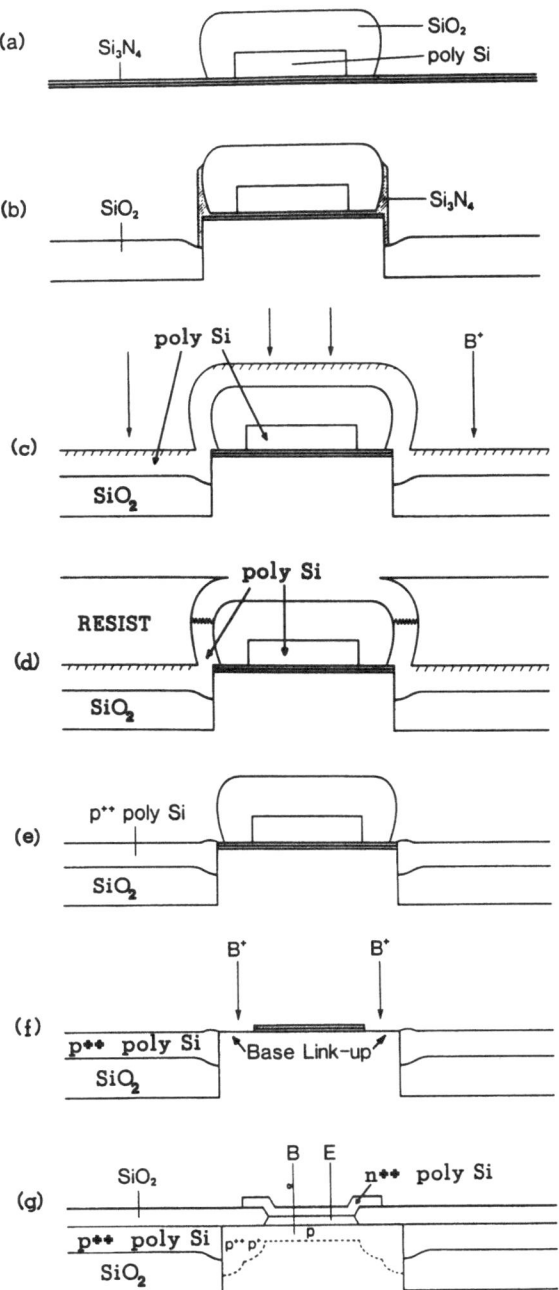

Figure 1: Fabrication steps of the BASIC transistor

120

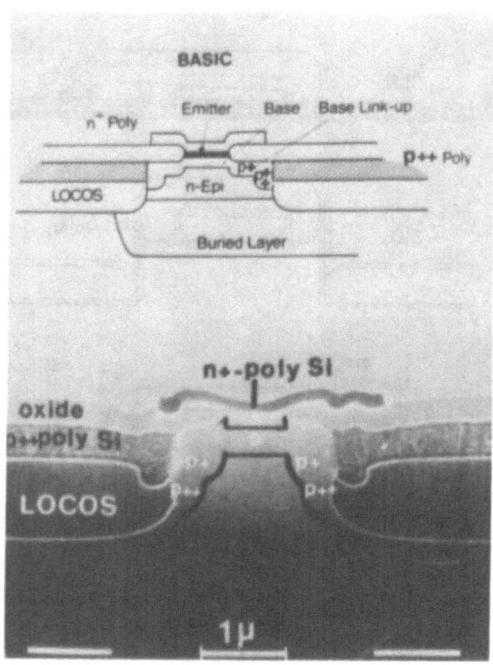

Figure 2: SEM cross section of the BASIC npn transistor together with a schematic cross-sectional view

The most thoroughly investigated process sofar giving such a sidewall base contacted structure is the SICOS process from Hitachi (ref. 1). Polysilicon base electrodes on thick silicon oxide are used to minimize parasitic effects. Intrinsic regions are formed in a mesa etched substrate and connected to polysilicon electrodes on the sidewalls. Multiple selfalignment defines - both for BASIC and SICOS - the mesa-etched isolation, base contact, intrinsic region, emitter region and emitter contact, all in one lithographic step. An extra degree of selfalignment is obtained in the BASIC process by the simultaneous definition of the submicron base link-up region. For both processes the emitter width is smaller than lithography, giving widths 0.6μm and 1.5μm smaller than minimum lithography, for SICOS and BASIC respectively. This means, that the BASIC process provides sub 0.5 m emitter widths, without having to resort to submicron lithography.

Figure 1 shows schematically the fabrication steps of the BASIC transistor.

A n-type epitaxial silicon layer of 1.2μm thickness and 0.5 Ohm.cm resistivity is grown on top of a n($+$) buried layer in a high-ohmic p-type (100) substrate. Collector plugs are defined by As($+$) implantation and outdiffusion in subsequent temperature steps. Then the process sequence is started which will result in the super selfaligned BASIC structure: after growing a stack of padox and nitride a thick polysilicon layer is deposited. This polysilicon layer is patterned by anisotropic etching and then oxidised to give oxidised polysilicon blocks, located on top of the padox/nitride stack (Fig. 1a). The lateral dimension of this oxide defines the base link-up region in the completed BASIC npn device. Since this dimension (ca. 0.8 μm) has been defined by thermal oxidation and not by underetch or spacer formation techniques, its width is not only larger than the usual emitter base separation regions, but also better controlled.

Anisotropic etching of the padox/nitride stack and subsequent shallow silicon trench etching (ca. 0.8 μm deep) provides silicon mesas. The sidewalls of these mesas are protected against thermal oxidation by deposition and subsequent anisotropic etching of a thin LPCVD nitride layer. A 1.1 μm thick field oxide is formed by wet thermal oxidation, resulting in a vertical base contact dimension on the mesa sidewall of ca. 0.5 μm. (Fig. 1b)

In the next step the nitride is removed from the sidewall by selective wet etching in hot phosphoric acid and polysilicon is conformally deposited. A simple, novel poly planarisation method to form sidewall contacted base boosts has been used. The method is based on the controlled diffusion of boron in polysilicon, followed by the selective wet etching of undoped polysilicon. The process steps used for this unique planarisation procedure are shown in Figure 1c-e. Boron is implanted into the horizontal surfaces (Fig. 1c) and using an auxiliary resistmask the exposed polysilicon on top of the silicon mesas is etched (Fig. 1d).

Subsequently, boron is outdiffused accurately to the required level in the poly-silicon base boost. "Undoped" polysilicon (with less than 6E19 B atoms/cm^3) is etched in a isopropanol/KOH mixture (ref. 7) with a high selectivity with respect to highly doped p($+ +$) polysilicon (Fig. 1e).

The next step in the fabrication sequence (Fig. 1f) is the removal of the exposed SiO_2 and the underlying Si_3N_4 by selective wet etching; also the remaining undoped polysilicon is removed. At this point a base link-up implant is done by using B($+$) ions. The dose and energy of this implant has to be fine-tuned, in combination with the subsequent thermal oxidation of the exposed polysilicon and mono silicon, to give an optimum combination of DC and AC characteristics. The remaining padox/nitride layer is removed by selective wet etching to open up emitter and collector areas simultaneously.

The intrinsic transistor is formed by low energy boron implantation and outdiffu-sion of the arsenic from n($+$) doped polysilicon, using an RTA anneal (Fig. 1g).

Figure 2 shows a SEM picture of a completed BASIC npn ($W_e = 0.2 \mu$m) together with a schematic drawing. Figure 3 shows the final perfectly planarised structure.

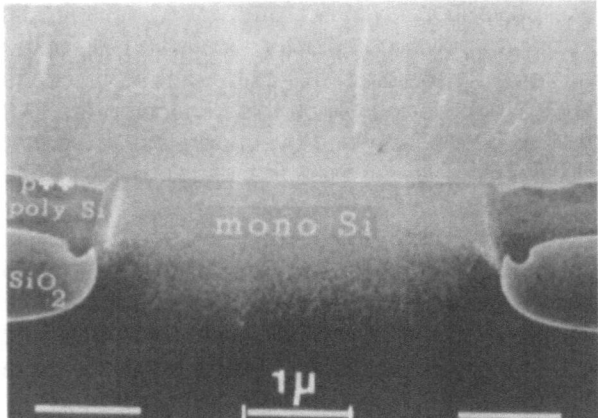

PARAMETER	
emitter area (μm^2)	1.5x28
R_E (Ohm)	2.6
R_B (Ohm)	110
$R_{bb}C_{cb}$ (ps)	7
epi thickness (μm)	1.2
epi resistivity (Ohm.cm)	0.5
x_{jE} (nm)	80
W_B (nm)	150
Rpinch (kOhm/sq)	13
h_{FE}(max)	100
V_{eaf} (V)	80
f_T(max) (V_{cb}=3 V;GHz)	17
f_{max} (V_{cb}=3 V;GHz)	9.5
BV_{ceo} (V)	7.5
BV_{ebo} (V)	7
BV_{cbo} (V)	8
C_{eb} (fF/μm^2)	3
C_{cb} (fF/μm^2;intr.)	0.3
C_{cb} (fF/μm;extr.)	1.1

Figure 3: SEM cross section of the BASIC structure after polysilicon planarisation.

Table 1: Parameters of transistor with 1.5x28 μm^2 emitter.

In Table I the most important DC and AC characteristics for the npn transistor are listed, exhibiting a combination of properties, which makes this process very suitable for analog applications.

Maximum f_T for the npn transistor is 17 GHz (fig.4); the $R_{bb}C_{cb}$ product is 7 ps for the npn transistor with double base contacts and emitter area of 1.5x28 μm^2. Other electrical properties such as high uniformity of gain (over 6-8 decades) (Fig.5), gain independency of emitter width (down to <0.5 μm;Fig. 6) and high E-B breakdown voltage (>6 V;Fig. 7) were obtained by tuning of the base link-up. The optimised doping profiles obtained from SIMS measurements and process simulations are shown in 1- and 2-dimensional plots for the three separate regions of the intrinsic transistor, base link-up and highly doped extrinsic base. (Fig.8)

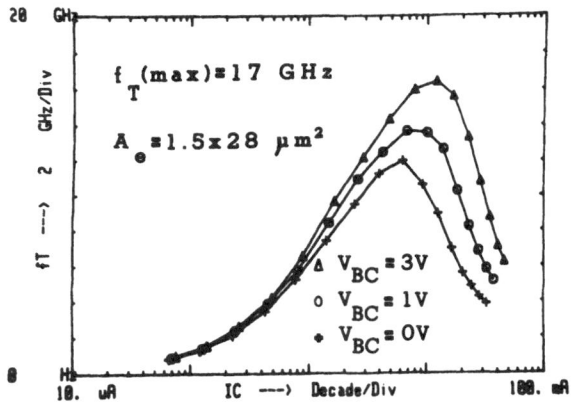

Figure 4: Cut-off frequency (f_T) versus collector current (I_C) characteristics for transistor with 1.5x28 μm^2 emitter.

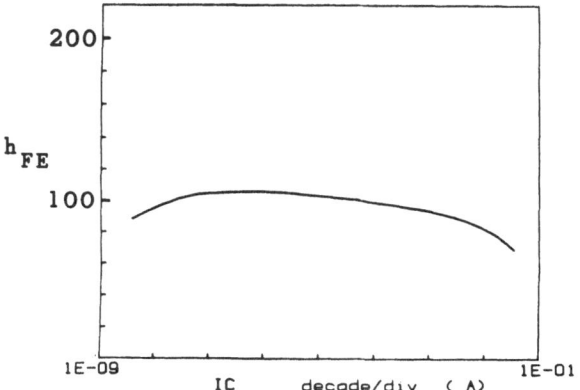

Figure 5: Current gain (h_{FE}) versus collector current (I_C) characteristics for transistor with 10x60 μm^2 emitter.

124

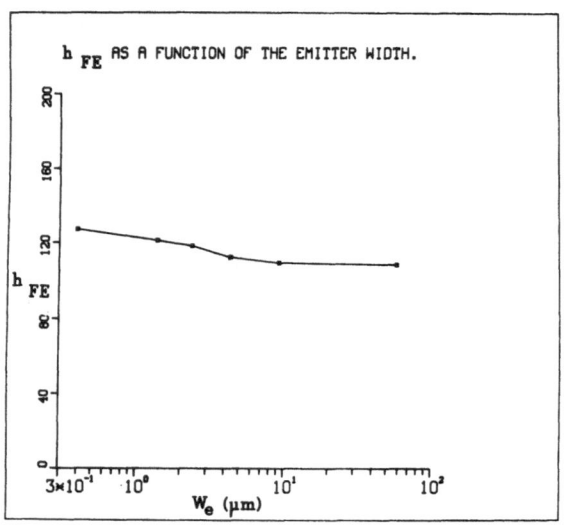

Figure 6: Current gain (h_{FE}) versus emitter width (W_E) for transistors with emitter length of 60 μm.

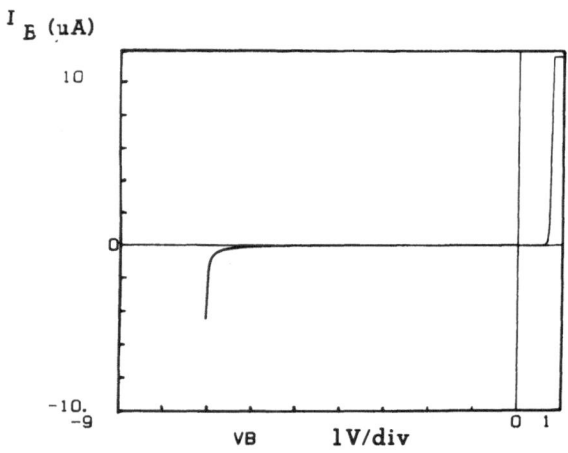

Figure 7: Base current (I_B) versus emitter-base bias (V_B) characteristics for transistor with 1.5x28 μm² emitter.

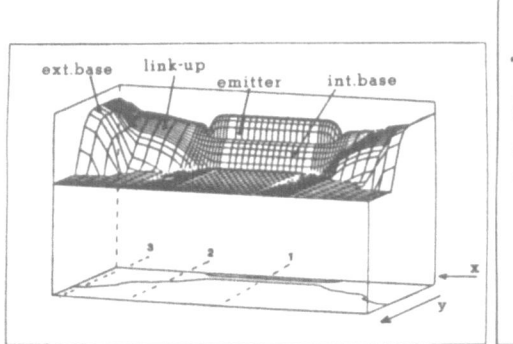 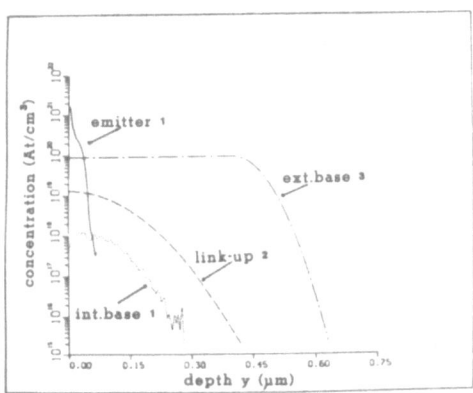

Figure 8: Dopant profiles of the BASIC transistor; two-dimensional (simulated) and one-dimensional (SIMS, curves 1 and simulated curves 2 and 3).

An extra advantage for analog applications as compared to non sidewall contacted structures is an increase in the Early voltage for narrow emitters ($W_e < 2$ μm). Measurements show an increase in the Early voltage by more than a factor of two for narrow emitters (W_e 0.5 μm, $V_{eaf} = 80$ V) as compared to wide emitters ($W_e = 10\,\mu$m), $V_{eaf} = 30$ V). Computer simulations indicate, that this phenomenon can be explained from two dimensional effects that gain importance when lateral device dimensions become comparable to depletion layer widths. Figure 9 shows a 2-dimensional computer simulation of the approximate location of the collector depletion layer for several collector-base bias conditions. Part of the intrinsic collector depletion charge at the edges of the emitter is shared by the depletion charge of the base link-up area and the $p(++)$ outdiffusion from the polysilicon electrodes on the sidewalls.

This locally increases the collector depletion width. Figure 10.a shows the electric field component perpendicular to the surface at the centerline of the transistor for several emitter widths. For wide emitters the field distribution approaches the one-dimensional case. For narrow emitters the collector depletion layer broadens; as the collector-base voltage remains constant, this results in a lower peak electric field at the metallurgical junction. Fig. 10.b and 10.c give detailed pictures of the electric field and the hole concentration at the base side of the collector-base junction. It can be observed, that since for narrow emitters the peak electric field is reduced, the neutral base widens and thus the neutral base charge increases, resulting in a higher Early voltage than might be expected from the one-dimensional case.

FIGURE 9: Location of the collector depletion area for several V_{CB} bias conditions.

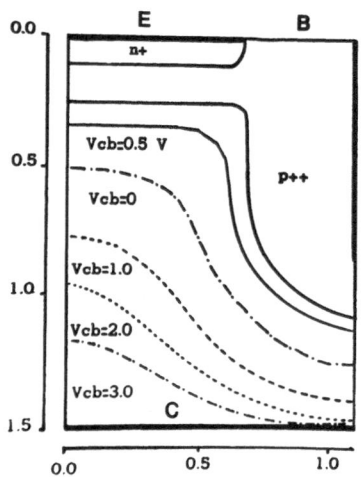

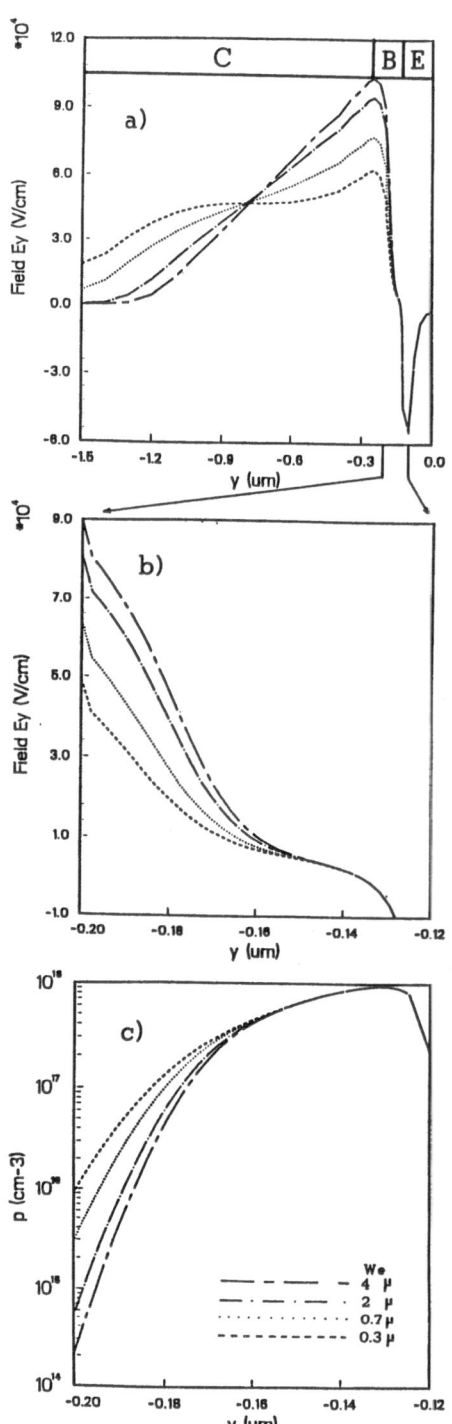

FIGURE 10: For emitter widths of 0.3, 0.7, 2.0 and 4.0 μm
A. electric field component perpendicular to the transistor surface at the centerline of the device.
B. electric field component at the centerline of the device in the base area.
C. hole concentration at the centerline of the device in the base area.

Further performance improvement has been shown to be possible by reducing the base contact area from ca. 0.5 μm to 0.1 μm, by applying BASIC-II technology. This leads to a reduced base-collector capacitance and an increased base-collector breakdown voltage.

A novel planarisation technique was merged with standard BASIC processing as described above (ref. 8) to achieve a strong reduction of the extrinsic vertical base contact dimension from 0.5 to 0.1 micron. The key process steps, which deviate from standard BASIC processing, are shown in Figure 11. Figure 11.a shows a SEM cross-section after formation of the recessed mesa-etched isolation. Locally, layers of p(+ +) poly Si are present which have been obtained by deposition of undoped polysilicon, vertical implantation of boron, annealing and subsequent removal of undoped polysilicon on the vertical sidewalls in KOH/isopropanol.

The remaining p(+ +) poly layer serves as an etch mask to remove only part of the nitride layer on the sidewall of the mesa structure yielding a very small extrinsic base contact. Secondly, it serves as a dopant source for the following deposited polysilicon layer which will function as base boost. This second polysilicon layer is planarised in a similar way as described above for the first polysilicon layer.

Now however, boron is introduced into the horizontal surfaces by both implantation and diffusion from the underlying polysilicon layer.

Again, the same dope selective etchant is used to remove undoped polysilicon from the vertical parts (Fig. 11.b).

Further processing, using similar steps as for the standard BASIC process, including base link-up implant and selfaligned emitter-base formation (ref.8) leads to a BASIC-II npn device.

The final npn BASIC-II device structure is given in Figure 12, showing a 0.6 μm wide emitter formed by RTA diffusion out of n(+) polysilicon, a 0.8 μm wide base link-up region and a 0.1 μm vertical base contact dimension.

In Table II, a comparison has been made between BASIC and BASIC-II device characteristics. Reduction of the base contact area leads to an increase in collector-base breakdown voltage from 8 to 14 V, due to an increase in distance between the p(+ +) outdiffusion of the base boost and the highly doped buried layer.

For the same reason, also the extrinsic collector-base edge capacitance is drastically reduced from 1.1 to 0.7 fF/μm.

By using the first p(+ +) polysilicon layer as an extra diffusion source for the final polysilicon base boost a reduction of p(+ +) polysilicon sheet resistance has been obtained from 300 to 150 Ohms/square. Both lead to a strong increase in f$_{MAX}$ from 9 to 15 GHz at Vcb = 2 V. Other DC characteristics such as high BVeb0 (6 V) could be maintained for this BASIC-II technology in comparison to conventional BASIC technology.

Table II: Comparison of performance of BASIC vs.
BASIC-II performance

	BASIC	BASIC-II
Emitter Area (μm^2)	1.5x58	1.5x58
BV_{ce0}	7.5	6
BV_{eb0}	7	6
BV_{cb0}	8	14
C_{eb} (fF/μm^2)	3	4
C_{cb} (fF/μm^2;intr.)	0.3	0.4
C_{cb} (fF/μm;extr.)	1.1	0.7
f_T (max) (V_{cb}=2 V;GHz)	14	17
f_{max} (V_{cb}=2 V;GHz)	9	15
h_{FE} (max)	100	65

Conclusion

In conclusion, we have presented in this abstract a process which solves the base link-up problem from which most selfaligned advanced bipolar processes suffer.

We achieved a sidewall contacted structure, with strongly reduced parasitic capacitances, resistances and transistor area, which approaches very close by the ideal one-dimensional symmetric transistor, by using the highest degree of selfalignment demonstrated sofar.

A novel, efficient method for polysilicon planarisation based on selective wet etching of undoped polysilicon, has been shown to be effective in forming sidewall contacted polysilicon base boost electrodes.

An excellent combination of DC and AC characteristics makes this process very suitable for analog applications, featuring a maximum fT of 17 GHz and an Early voltage of 80 V. An increased Early voltage is obtained by lateraldepletion of the collector region underneath the intrinsic emitter-base region.

In an improved version (BASIC-II) further enhancement of the AC and DC performance has been demonstrated with even lower parasitic capacitances and resistances. This has been achieved by strongly reducing the vertical base contact area by applying a simple polysilicon planarisation technique twice.

While still maintaining excellent analog properties, remarkable increases in both fT and fmax could be obtained (20 and 50% respectively).

A 8 bit high speed A/D converter of high complexity (> 10K components) which will be delivered at the end of year 3 of the running project, has been the driving force for the development of the high performance BASIC process.

129

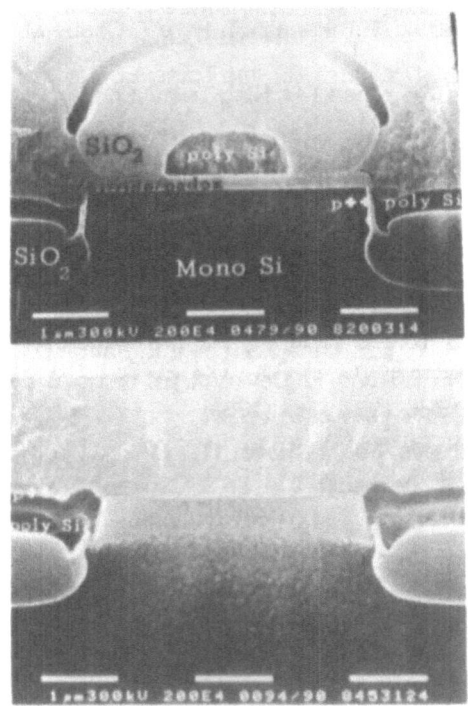

FIGURE 11: key process steps in BASIC-II technology

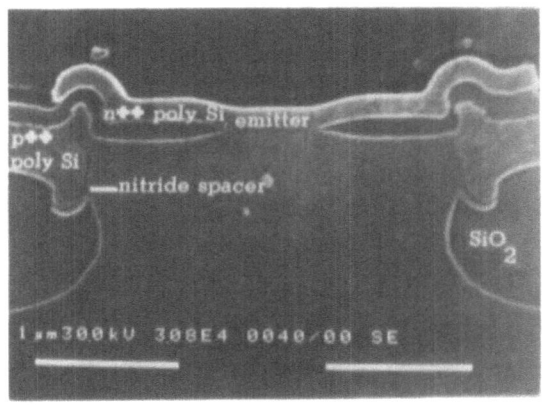

FIGURE 12: final BASIC-II npn device

References:

1: T. Nakamura,T. Miyazaki,S. Takahashi,T. Kure,T. Okabe,M. Nagata,ISSCC Dig. Techn. Pap.,p214 (1981)

2: G.P. Li,C.T. Chuang,T.C. Chen,T.H. Ning,Trans. El. Dev.,vol.35 (11), p1942 (1988);C.T. Chuang,IEEE IEDM 88,p554 (1988)

3: J.M.C. Stork,R.D. Isaac, Trans. El. Dev., vol.30 (11),p1527 (1983);K. Washio,T. Nakamura,T. Hayashida,Trans. El. Dev.,vol. 35 (10), p1596 (1988); G.P. Li,E. Hackbarth,T.C. Chen,Trans. El. Dev.,35 (1),89 (1988);Y. Tamaki, T. Shiba,K. Ikeda,T. Nakamura,N. Natsuaki,S. Ohyu,T. Hayashida,IEEE BCTM 87,p22 (1987)

4: S. Kameyama,K. Kobushi,H. Nishimura,K. Kikuchi,M. Kajiyama,H. Sakai,T. Komeda,IEEE BCTM 87,p27 (1987)

5: T. Yamaguchi,Y.C.S. Yu,E.E. Lane,J.S. Lee,E.E. Patton,D.R. Ahrendt,V.F. Drobny,T.H. Yuzuriha,V.E. Garuts,IEEE Trans. El. Dev., Vol. ED-35,no. 8 ,p1247 (1988)

6: M. Nakamae, ESSDERC 1987,p361 (1987)

7: A. Bohg, J. Electrochem. Soc.,118,p401 (1971)

8. J. van der Velden,R. Dekker,R. van Es,S. Jansen,M. Koolen,P. Kranen, H. Maas,A. Pruijmboom, IEEE IEDM 89, p. 233 (1989)

Project No. 2268

COMBINED ANALOG DIGITAL INTEGRATION: CANDI

C. MALLARDEAU, Y. DUFLOS,
J.C. MARIN, M. ROCHE
SGS-THOMSON MICROELECTRONICS
BP 217
38019 GRENOBLE Cedex, FRANCE

G. TROSTER, J. ARNDT
TELEFUNKEN ELECTRONIC, GmbH,
Theresienstrasse 2
D 7100 Heilbronn, FRG

Abstract

This paper describes the BICMOS project no 2268: CANDI, and gives its status at the end of the first phase, after 18 months of collaborative work.

The goal of the CANDI project is to develop a mixed analog digital BICMOS technology with the related CAD and design techniques. During the first phase of the project, a $1.2\mu m$ BICMOS process has been developed and optimized.; the objective of this technology is to achieve simultaneously high density CMOS logic (SV) and high performance bipolar transistors for analog applications (12V). Together with the technology, a BICMOS cell library has been implemented; and several demonstrators have been designed: a RGB digital decoder for IDTV applications, a mixed analog digital array for the paneuropean cellular radio, and a 16 bit shuffler for wide band telecommunications. The performance of the technology has just been validated on these complex demonstrators, and the next step of the project is to lead this process to a level of maturity guarantying manufacturability, yield and costs for industrial utilisation.

1. Introduction

The objective of the CANDI project is to develop an advanced merged bipolar CMOS technology with the related CAD and design techniques, and to raise them to an industrial level in order to prepare the next generation of ICs for consumer applications. Moreover, the final goal of the project is to create an analog-digital integration environment in Europe.

The use of both CMOS and bipolar structures on the same chip is the best solution to achieve simultaneously: high speed performance, high driving capability, high precision analog functions, high integration density, and low power comsumption. However, the implementation of such a complex technology can only be successful if the improved performance can counterbalance the higher costs of fabrication.

A large number of companies and research laboratories worldwide are currently working to develop advanced Bipolar/CMOS merged technologies (BICMOS) (1). In BICMOS applications, two major trends can be distinguished: high speed logic for high performance memories and gate arrays (2, 3, 4), and one chip integration of complete analog/digital system (5, 6, 7, 8).

The CANDI BICMOS project addresses the second type of application: mixed analog digital circuits with high density CMOS logic and high performance analog functions.

2. Description of the Project

The BICMOS 2268 ESPRIT project is carried out by SGS-THOMSON Microelectronics and TELEFUNKEN Electronics as the main partners. SGS-THOMSON is the prime contractor.

The project includes technology and design work, with equal repartition of manpower in each. The following associated partners contribute to the CAD and design part of the project: Alcatel SESA, THOMSON Consumer Electronics, PLESSEY Research, AEG Olympia, THOMSON CSF/LER, University of Paris-sud, DOSIS, University of Dortmund, IMS. Technology development is implemented by the two main partners.

The project is based on two phases:
- the goal of the first phase is to provide an intermediate $1.2\,\mu m$ process, adapted from the main partners technology. This process is a common process and will allow the two main partners to reach an industrial level by the end of the second year.
- the second phase will start on the 2nd year of the project and its objective is the development of a $0.8\,\mu m$ BICMOS process as well as the associated design and CAD techniques.

The project started in December 1988 and has now completed its first phase, after 18 months.

To test the full integration of technology, design and CAD, and the capacity to generate complex circuits, the main achievement of the first phase has been the realization of three demonstrators: a RGB decoder for IDTV, a mixed analog/digital array for paneuropean cellular radio, and a 16 bit shuffler for wideband telecommunications.

In the following phase, starting in June 1990, a new generation BICMOS process of $0.8\mu m$ minimum feature size is to be developed by SGS-THOMSON and TELEFUNKEN.

3. Project Organisation

The total manpower assigned to the project for 3 years is 170 man-years, with 66 manyears for SGS-THOMSON, and 62 manyears for TEG.

The project is organised in workpackages, three of them related to technology work and the last four related to design work. The manpower repartition to workpackages is given for each phase in table 1.

WORPACKAGES	First phase MXY	Second phase MXY
1. Process integration	17.9	14.7
2. 0.8 μm Modules Concept	7.9	8.2
3. Fabrication of Demonstrators	7.5	13.5
4. Design Tool Adaptation	10.8	4.1
5. Cell Library	22.1	24.5
6. Design of Demonstrators	11.7	12.8
7. Testability and testing	3.9	4.9
Management	1.5	1.5
TOTAL	83.3	84.2

Table 1- Manpower repartition to workpackages.

The technology work of the first phase consists of 1.2μm BICMOS process development and stabilisation, with a first validation on demonstrators.

The design work is divided into two main topics: first, the design of a 1.2μm BICMOS cell library with the contribution of all partners and associated partners of CANDI; second, the design of complex demonstrators to test the full capability of the technology.

4. Technology

4.1 Choice of a common 1.2μm BICMOS process

In technology, the two main partners had different backgrounds and therefore they started the project by exchanging their own experience on BICMOS process architecture.

On month 3 of the project, a common process was defined. The choice was the SGS-THOMSON process, HF3CMOS, as it was better suited for mixed-analog digital requirements: in particular, with 12 V supply voltage and high enough Early voltage for the bipolar part.

Table 2 shows the compared electrical specifications of the two processes.

		HF3 CMOS (ST)	BICMOS2 (TEG)
B I P O L A R	SUPPLY VOLT	12V/5V	10V/5V
	BVCEO	> 13.2 V	12V
	EARLY VOLT	> 50 V	
	FT	6 GHz	6 GHz
C M O S	BVDSS	> 7V	> 7V
	Gate Delay TD	0.3 ns	0.3 ns

This common 1.2 μm process is the right tool to establish the common working environment and promote the partner collaboration efficiency. It constitutes the basis for a common set of design rules, common test structures and common test methods.

The objective of this 1.2 μm BICMOS technology (refered to as HF3CMOS) is to achieve simultaneously high density CMOS Logic (5V) and high performance bipolar transistors for analog applications (12 V).

4.2 Main features of the process

The process is set up using 1.2μm design rules. Fig. 1 shows a cross sectional view of the 1.2μm BICMOS structure. The available components of the technology are NMOS, PMOS, high speed NPN bipolar and high performance vertical PNP.

The process starts from a 1.2μm CMOS technology, with five additional levels forbipolar process. The process architecture is shown in fig. 2.

The main features of the 1.2μm CMOS starting technology are the following: Psubstrate, twin well structure, mixed isolation with conventional LOCOS and junction, 1.2μm polygate with silicide (TaSi2), DDD structure for NMOS, with spacer to prevent hot electron degradation, conventional structure for PMOS, BPSG deposition and reflow to smooth down the topography. The double level metal structure consits of Ti/TiN/AlCu for metal 1, a sandwich of PECVD oxide and SOG for planarized mineral interdielectric, and AlCu for metal 2.

To add the bipolar components, the following steps have been introduced: N+ buried layer combined with N+ deep collector sink is used for low collector access resistance, P+ buried layer is added for isolation between components and PNP collector, a 3mm thick N-type epilayer is required for the NPN collector to achieve high performance analog functions. The emitter of the NPN is formed by direct As implantation into monosilicon, and the base uses an additional mask. The emitter-base drive-in is performed at 950°C.

For the vertical PNP, a N-buried layer is used to isolate the transistor from substrate. The base of the PNP is graded; it is implanted and diffused at the same time as the drain/source of the NMOS. Therefore no specific masking level is required for the PNP, besides the N- isolation layer. The intrinsic transistor profiles for both NPN and PNP are plotted in figures 3 and 4. The base width is 140nm for the NPN and 250nm for the PNP.

4.3 Device characteristics

The device parameters are given in table 3, for NMOS, PMOS, NPN and PNP.

The analog requirements like high breakdown voltage (> 13.2 V) and high Early voltage have been achieved for both bipolar transistors. Moreover the fT of the NPN was optimized by reducing the basewidth, and a maximal value of 8GHz was obtained at VCE = 5V, as plotted in figure 5. Finally excellent high frequency performances were demonstrated by the vertical PNP, as shown by the plot of

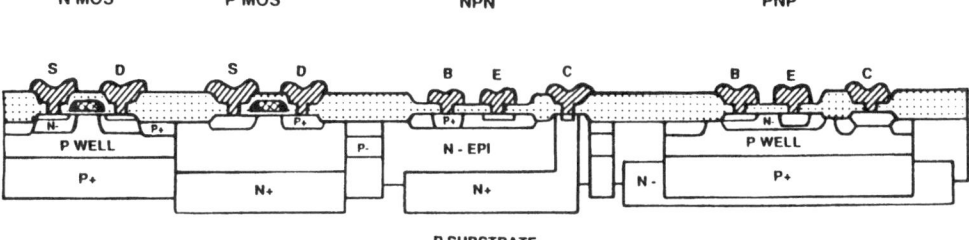

Fig.1- Cross section of the 1.2μm BICMOS technology

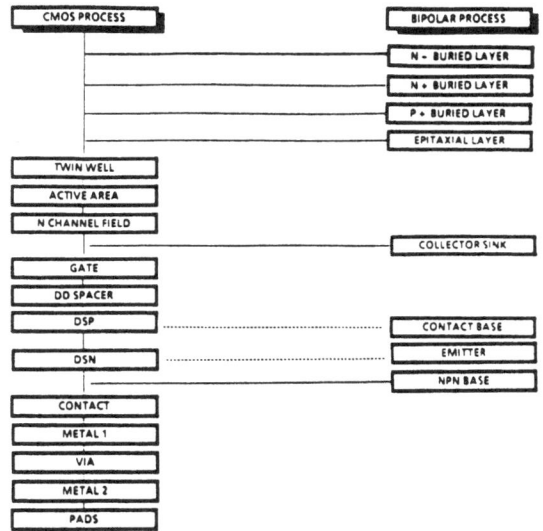

Fig.2- Process architecture

C MOS

NMOS (50 / 50) Vth   : 0.7V
PMOS (50 / 50) Vth   : 1.0V
BVDSS   : > 7V
td (fin=fout=1)   : 220 ps

NPN		PNP	
Se	: 3.6 x 3.6μm^2	Se	: 3.6 x 3.6μm^2
hfe	: 120	hfe	: 90
BVCEO	: 14V	BVCEO	: -20V
BVCBO	: 25V	BVCBO	: -30V
BVEBO	: 5V	BVEBO	: -5.4V
VEA	: 50V	VEA	: -30V
Ceb	: 37.5fF	ft	: 2.5 GHz
Ccb	: 27fF		
Ccs	: 106fF		
rbb'	: 370Ω		
ft max	: 8 GHz		

Table 3- Device parameters

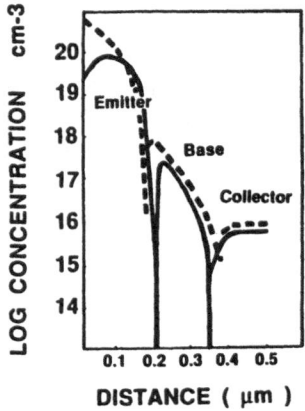

Fig.3- NPN transistor profile
  — spreading resistance
  -- simulation results

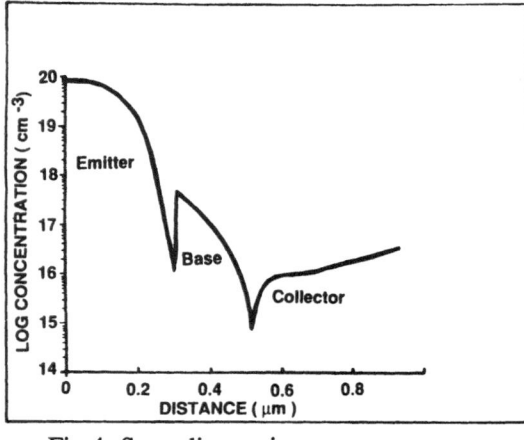

Fig.4- Spreading resistance
  Profile of PNP bipolar transistor

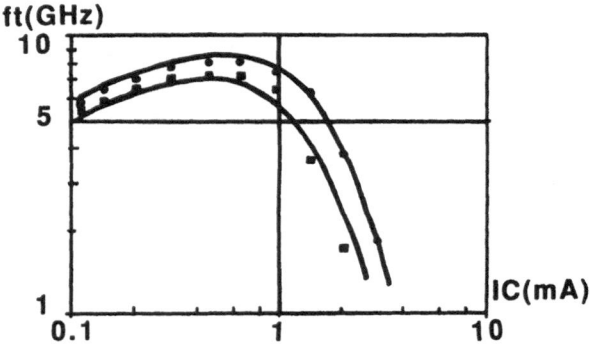

Fig.5- fT versus Ic for the NPN transistor
($Em = 3.6 \times 3.6\mu m2$)

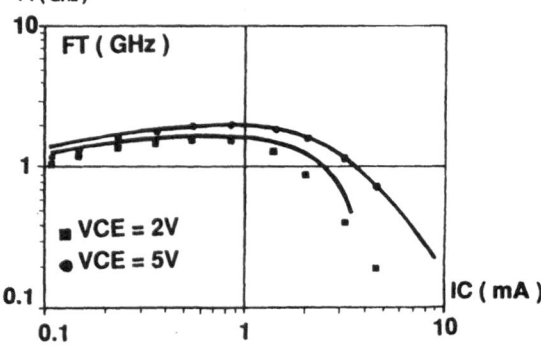

Fig.6- fT versus Ic for the PNP transistor
($Em = 3.6 \times 3.6\mu m2$)

cut-off frequency versus collector current in figure 6. At VCE = 5V, the maximum cut-off frequency is 2GHz.

High frequency and speed of NPN bipolar transistors were tested on packaged devices demonstrating performances of fmax = 890MHz on D flip-flop, and 170ps propagation delay per stage for 21-stage CML ring oscillators.

These device characteristics have been obtained at both companies: SGS-THOMSON and TELEFUNKEN. The use of the same test structure at both plants made the collaboration very efficient, as the electrical data could be directly compared.

4.4 Application to the demonstrators

The manufacturability of the process has to be proven throughout the fabrication of demonstrators. Complex demonstrators have been designed within the project, as described in paragraph 6.

5. Cell Library

The cell library is a strategic workpackage within the project. Table 1 shows that the highest amount of manpower is concentrated there. The goal of the workpackage is to create a broad basis for the design in the common 1.2μm BICMOS technology in order to best exploit the BICMOS potentials.

All CANDI partners and subcontractors contribute cells and modules to the common databook. So the different application backgrounds of the designers and companies have been included and concentrated into cell library.

The final 1.2μm databook consists of totally 101 cells divided into four main parts:

1. Analog standard cells
   basis and general purpose cells (e.g. OTAs, OPAMPs, video amplifier, sample/hold, bandgaps, ...)

2. Digital standard cells
   (e.g. (Bi)CMOS & ECL gates, counter, level translater, (CMOS-ECL), bit slice processor 2901, adder, multiplier ...)

3. Converter
   A-to-D and D-to-A converter

4. Cells for the mixed array (see workpackage 6)
   - Analog macrocells,(e.g. preamplifier, mixer, gain controlled IF amplifier, ...)
   - Digital macrocells (e.g. gates, register, counter, accumulator, ROM, ...)

The application range of the library can be characterized by following corner-stones:

analog frequency range	DC up to 1 GHz
signal levels	CMOS, ECL
supply voltage	3V to 9V
cell complexity	up to 10K devices
cell size	up to 32 mm2

In order to ensure the exchange and applicability of the cells all partners agreed upon layout standards for the standards cells. The main features of these standards are:
- power routing by abutment
- fixed cell heights in units of the characteristic grid pitch (6μm)
- three different standards for analog, (Bi)CMOS and ECL cells

For each cell four levels of description and documentation are provided in the final databook: functionnal block diagram and description, electrical characteristics, circuit schematics, layout.

For the necessary communication and exchange of cells at all design and simulation level, a common European CAD standard being supported by the companies' CAD tools is painfully missing. So a minimal set of provisional standards have been defined within CANDI:

circuit schematics	EDIF2 (if supported)
netlist	EDIF2, SPICE
layout	GDS2
device models	
bipolar	Gummel Poon
MOS	SPICE level 3

A specific cell library database has been developed containing all relevant data, which can be distributed on VAX computers. For each cell a flexible system is provided containing all electrical and physical design data.

6. Demonstrators

Three demonstrator systems have been designed and fabricated with the 1.2μm technology. Defined in agreement with the relevant product divisions the chosen demonstrators are representative of different application fields and have to confirm the project ability to meet its technology as well as design objectives.The key features ot the three demonstrators are summarized in the following table.

	Contributor	Complexity	Size
RGB decoder (digital luminance & chrominance to analog, for IDTV)	SGS-THOMSON	31k devices	35mm²
combined Analog/Digital Array, custom metallized for cellular radio receiver	TELEFUNKEN AEG Olympia	33k used devices	44mm²
16 channel bit shuffler for wide-band telecommunication multiplexer	Alcatel SESA	5.9k devices	30mm²

6.1 SGS-THOMSON demonstrator: RGB decoder

On the way leading from the pure analog TV to the future HDTV, intermediate steps are taking place, such as Improved Definition TV (IDTV), which require some degree of digital processing within the TV set. The circuit chosen by SGS-THOMSON to demontrate the possibilities of the HF3CMOS common process addresses this need,by implementing digital luminance and chrominance to analog RGB components decoding.

The function realized on the chip is shown by the block diagram in figure 7. Taking as inputs the 8-bit digital luminance signal Y and an 8-bit multiplex of the chrominance signals Cr and Cb, it provides at its outputs the three basic colour components R,G and B in analog form. In addition, an auxiliary analog input is provided for text or picture incrustation.

The BICMOS implementation allows a one chip integration of two functions which, in the former generation, were integrated in different technologies:

1 - a large digital operator (35 K MOS devices) achieving filtering and oversampling of the digital input signals, in order to ease filtering out of the aliased part of the spectrum resulting from sampling, and then dematricing the Y, Cr, Cb signals. This function stems from a pure CMOS chip originally designed in 2μm by the french Centre National d'Etudes des Télécommunications.

2 - a triple D/A converter and associated output buffer amplifier, translating the digital RGB components into analog form with 8-bit accuracy and with a high degree of matching and a low crosstalk between the channels. This part had to be designed with bipolar devices in order to allow full 8-bit linearity of the voltage output buffer amplifier under a single voltage supply of 5 volts.

The benefits resulting from BICMOS technology are:
– monolithic integration of the complete function (bipolar analog and CMOS digital parts), thus reducing system costs.
– ability to boost the operating frequency up to 54 MHz by using BICMOS buffers in the speed critical internal nodes.

Figures 8 shows the microphotograph of the chip, whose area is 35 mm2. The main characteristics of the circuit are given hereunder:
– single supply voltage: 5 volts
– digital inputs coded on 8 bits
– input data rate: 27 Mhz
– main clock at oversampled data rate: 54 Mhz
– D/A data conversion with 8 bit resolution
– differential linearity < 0.2 %
– integral linearity < 1.2 %
– voltage output buffer amplifier: 1.6 volts on 150 ohms

140

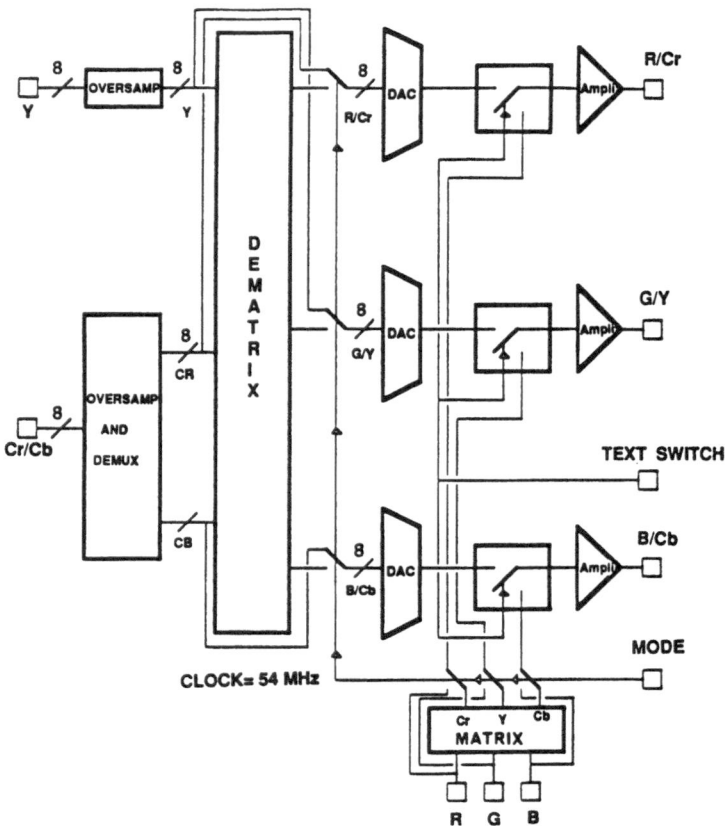

Fig.7- Block diagram of the RGB decoder

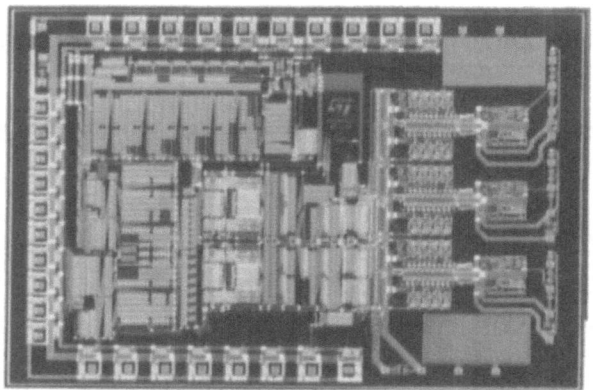

Fig.8- Microphotograph of the RGB decoder chip.

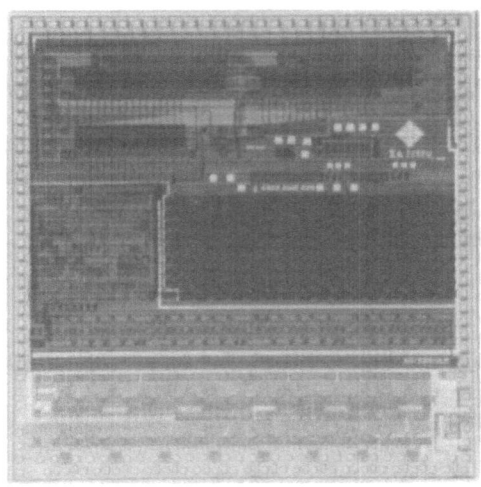

Fig.9- Microphotograph of the mixed array, custom metallized
for the cellular radio receiver.

	Analog HF	Digital SOG	Total
Die Size	10.25mm²	34.57mm²	**44.82mm²**
No.Cells	60	3625	
No.of Devices	638NPN, 80vPNP 1928 Resistors 62 Capacitors	126.375 MOS 3625 NPN	**132.208**
Supply Vol.	≤ 12V	≤ 5V	
Speed	$f_{toggle,\ ECL}$=1GHz	$t_{D,MOS}$=380ps $t_{D,BiCMOS}$=500ps	
Pads	≤ 63	≤ 82	≤ 145

Table 4 - Key features of the mixed array, custom metallized for cellular radio receiver.

The chip design was realized by SGS-THOMSON Central R & D, with the contribution of THOMSON Consumer Electronics.

6.2 TELEFUNKEN demonstrator: mixed array, customized for cellular radio receiver

The Telefunken/AEG Olympia demonstrator consists of three parts (9) :
- analog/digital array optimized for telecommunication application
- array specific CAD environment
- custom metallization for the cellular radio receiver

Analog/Digital Array

The microphotograph of Fig.9 shows that two arrays, namely a bipolar high-frequency array (on the left side) and a BICMOS Sea-of-Gate (SOG) array are combined into one masterslice. The key features of this demonstrator are summarized in table 4.

This mixed array is aimed at a 'silicon breadboard' for prototyping high-performance, combined high-frequency and complex systems, especially at the interface between analog signal ranges and digital components. The individual custom metallization is performed by four masking levels, two contact and two metallization levels.

CAD Environment

The main feature of arrays, namely the fast turnaround time can only be exhausted by appropriate design and verification tools.

Beneath the GDT ('Generator Development Tools' from 'Silicon Compiler Systems') shell a workstation was installed being specific for the SOG master and powerful enough for the realization and verification of complex systems on this masterslice. A functional model of the NPN transistor was additionally developed for the GDT's mixed-mode simulator Lsim.

The custom metallization on the analog HF array can be performed with the PARIS design system (10) which covers the areas of electrical designs of cells, performance assessment and layout generation with on-line connectivity extraction within analog circuit design.

Custom Metallization for the Cellular Radio Receiver

In the implementation phase of the paneuropean digital cellular radio net this mixed array allows an economical and flexible test as well as optimization of sensitive components. A critical module concerning the performance of the complete mobile receiver is the A-to-D converter and demodulator. This key component, whose blockdiagram is shown in fig.10 has been realized on the mixed array. The $\Delta\Sigma$ modulator converts the analog 10MHz bandpass signal X(T) in a pulse density modulated 1-bit stream. The GMSK-I/Q demodulator and decimation filter cover the complete SOG array. An essential feature of this approach is that the main characteristics of the circuit can be modified by changing only the 2nd metallization level. By varying the oversampling factor and filter coefficients this basic configuration can be adapted not only to GSM conversion, but to hetero- and homodyne receivers.

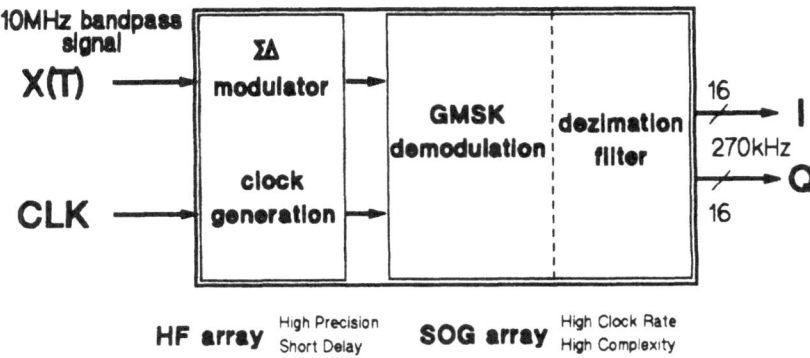

Fig.10 - Block diagram of $\Sigma\Delta$ A-to-D converter

6.3 Alcatel SESA demonstrator: 16 bit shuffler

The shuffler-scrambler is a Full Custom VLSI circuit that combines two independent functions, the reordering of data (shuffler) and descrambling (scrambler) allowing multiplexing of 16 STM-1 data channels at 155Mbit/s using the byteplexing technique according to CCITT recommendation 707, and parallel scrambling and descrambling based on CCITT recommendation 709, implemented in 1.2μm BICMOS. Both functions shuffling and scrambling, can be selected in any order by external pins controlling three integrated multiplexers giving this specific device an extra commercial value.

7. Conclusions

During the first phase of the CANDI project, a 1.2μm BICMOS process has been developed and optimized for mixed analog digital applications. Together with the technology, a BICMOS cell library has been implemented and complex demonstrators have been designed.

The performance of the technology has just been validated on a RGB decoder circuit and on a mixed analog digital array, ensuring the success of the first phase of the project.

With the second phase starting July 1990, the future goals of the CANDI project are twofold: first to lead the 1.2μm process to a level of maturity guarantying manufacturability, yield and costs for industrial utilisation; second to develop a 0.8μm BICMOS technology for the next generation of products.

Consortium:

SGS-THOMSON Microelectronics, TELEFUNKEN electronic, Alcatel SESA, THOMSON Consumer Electronics, PLESSEY, AEG Olympia, THOMSON CSF/LER, University of Paris-Sud, DOSIS, University of Dortmund, IMS .

References

[I] H.C.LIN et al Digest of Technical Papers of IEDM 1987

[2] H.IWAI et al. "1.2μm BICMOS Technology with High Performance ECL" ESSDERC 87 PP 29-32 .

[3] Electronics "Is BICMOS The Next Technology Driver" Febr. 4th, 1988, PP 55-67

[4] I.MASUDA et al . "High Performances BICMOS Technology and its Applications" Proc. of ESSCIRC, 1987, pp 53-59 .

[5] P.HOLLOWAY, "A Trimless 16b Digital Potentiometer", ISSCC Dig.Techn.Papers, 1984, pp 66-67 .

[6] FUKUSHIMA et al., "A BIMOS FET Processor for VCR Audio", ISSCC Dig. Techn.Papers, 1983, pp 243-247 .

[7] Y.KOWASE et al., "A BIMOS Analog/Digital LSI with programmable 280 bit SRAM "Proc. of IEEE Custom Integrated Circuits Conference, 1985, pp 170-173.

[8]P.A.H. Hart et al, "BICMOS, ESPRIT 86: Results and Achievements" pp 221-230

[9] G.TROSTER et al., "A BICMOS Analog/Digital array for Cellular Radio Applications.", IEEE Proc.of CICC ' 90, pp 12.6.1-12.6.4 .

[10] M.GERBERSHAGEN et al., "A Hierarchical Cell Based Engineering System for the Design of Semi-Custom Analog Integrated Circuits", IEEE Proc. of CICC'88, pp 2.2.1-2.2.4.

BICMOS: FROM DREAM TO REALITY

W.J.M.J.Josquin
Philips Research Laboratories
P.O. Box 80000
5600 JA Eindhoven
The Netherlands

H.Klose
Siemens Research Laboratory
Otto-Hahn Ring 6
8000 MGnchen 83
Germany

1. Introduction

Since end of 1988 the project 2430 "A high performance CMOS-bipolar process (BICMOS)" has focussed on the realization of circuits and circuit parts in second generation BICMOS technologies and on the development of third generation BICMOS technologies.

The realization of circuits and circuit parts is a joint effort of all the partners in the consortium, i.e. Philips and Siemens as the main partners, the University of Dublin, INESC in Lissabon and the Entwicklungszentrum fGr Mikroelektronik in Villach as associate partners and the University of Stuttgart as a subcontractor.

The development of third generation BICMOS technologies is pursued by Philips and Siemens.

2. The BICMOS history

The prime goal of any BICMOS activity is to combine the benefits of MOS and bipolar on a single chip. These benefits are well known and will not be listed here. Some of the advantages of the BICMOS combi- nation on circuit level are:

- Improved speed performance when compared to CMOS-only circuits
- Reduced power dissipation when compared to bipolar-only circuits
- Improved performance for mixed analog/digital applications

Of course a BICMOS technology is more complex than a pure CMOS or a pure bipolar technology and therefore the benefits of BICMOS should well outweigh this drawback of BICMOS.

Within the project 412/2430 (1-5) the initial emphasis has been on the development of a first generation process (BICMOS1) and according demonstrators (6,7,16).

In the second phase of the project the performance of both the bipolar devices and the MOS devices has been boosted significantly. The key issue in this phase of the development has been the extensive use of oxide sidewall spacers which were used in LDD (lightly doped drain) MOS devices and in self-aligned emitter-base structures in bipolar devices.

In this phase the demonstrator chips aimed at the edge of performance and complexity that can be reached using BICMOS technology. The demonstrators themselves will be discussed below. The targets for these demonstrators were chosen such that all the difficulties that might occur during the design of an advanced BICMOS circuit would arise, and would be solved either through modifications of the technology or through new approaches in electronic circuit design.

In the present (third) phase of the project the emphasis will be on the development of an advanced BICMOS technology as an optimum and economic answer to the electronic requirements, and based on the extensive experience in the second phase. For this purpose the third generation process should satisfy the following demands:

- a further improvement in performance of MOS and bipolar devices
- a reduction in process complexity
- a CMOS packing density that is competitive with submicron CMOS technologies.

The main feature of the demonstrators in the final phase of the project is that they are prototypes of industrial products. This implies that unlike in the second phase of the project, the demonstrators will not have extreme complexity in terms of the number of components and also that high risk electronic concepts have to be minimized. Obviously the prime goal of these prototype demonstrators is to provide a given function in silicon at a minimum cost, i.e. with a minimum number of square millimeters in a process that provides a highly efficient combination of MOS and bipolar.

3. The application areas.

In principle the application areas of BICMOS reach all the way from high performance audio functions through video circuits and industrial applications up to the edge of telecom applications, where the application area ends and fades into the regime where TIP-BASE like technologies take over.

The demonstrators throughout the project aim at covering this complete range of applications. In the first phase of the project Philips demonstrated a hifi audio processor in a BICMOS1 technology (16).

In the second and third phase of the project Philips and the University of Dublin will focus on Video applications with special attention for HDTV (HD-MAC) functions. Obviously this relates to the interest of Philips in integrated circuits for consumer electronics. This implies that the Philips technology will be optimized for this kind of applications.

In the frequency/speed range between audio and video, and beyond video INESC and the Electronic Development Center at Villach are now working on A/D conversion for industrial applications.

At the high speed side of the range Siemens has its main activities in high speed digital systems. Therefore the Siemens technology will be optimized towards this type of application like SRAMs, gate arrays and data processors.

The different optimization of the Siemens and Philips technologies allows to clarify the consequences and drawbacks of the specific optimization through design exchange: e.g. a BICMOS product for Video applications contains some digital circuit parts that cannot reach ultimate performance since the full circuit is produced in an analog-oriented environment. By exchanging that very circuit- part towards the partner with the digital-oriented technology, the loss in performance can be quantified. Obviously exactly the same can and will be done in the opposite direction: an analog circuitpart in an industrial A/D converter will not be optimum in a system and in a technology that is mainly oriented towards digital products. Again by exchanging this kind of parts towards the partner with the analog technology, the performance loss can be quantified.

The demonstrator circuits will be discussed in some more detail below.

4. The present demonstrator activities.

The high resolution video A/D converter (Philips)

In the video application range a >10 bit video A/D converter will be realized with a 13.5 MHz sampling rate and over 100k complexity. The converter uses the sigma-delta modulator principle and has an internal data rate of 432 MHz. Figure 1 shows the relation between resolution and signal bandwidth for a number of A/D converters, both commercially available products and research achievements. It is obvious that the target of the BICMOS2 demonstrator lies beyond the performance of all bipolar-only or CMOS-only solutions.

The sigma delta modulator needs an analog loop filter that should handle signal frequencies up to half the sampling frequency (0-216 MHz) with a high degree of accuracy.

The digital filter, that follows the one bit coder, performs the down sampling to 13.5 MHz together with noise reduction to below -60 dB. Apart from the accuracy, a major design goal is to limit the power dissipation to below 1 Watt so that the analog circuit parts will not be hampered too much by temperature gradients on the chip. Early 1990 a two-chip prototype of this A/D converter has been realized.

The digital filter which was realized as a separate CMOS-only chip is shown in figure 2. It has more than 100k components and has a chip area of about 70 mm2 in the BICMOS2 process. The analog part is the most critical part of the system, since it contains many new electronic design concepts. Figure 3 shows the prototype of the analog part, which was realized in a well-established 2.0 m BICMOS technology at 1/4 of the ultimate operating frequency. The prototype study has resulted in fully functional silicon, and has provided sufficient system know-how to start the design of the final BICMOS2 demonstrator chip.

The total chip size will be about 80 mm2.

148

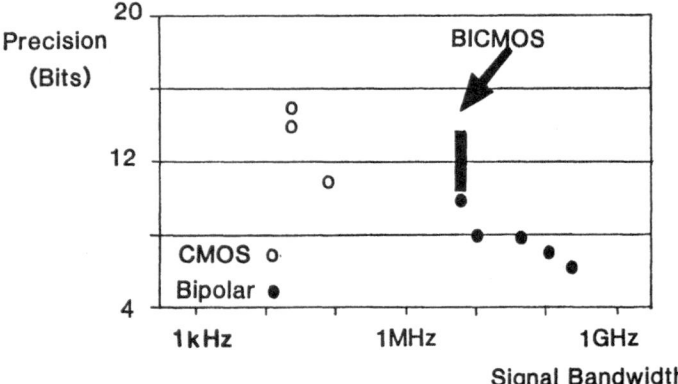

Fig. 1. **Resolution** and bandwidth of A/D converters

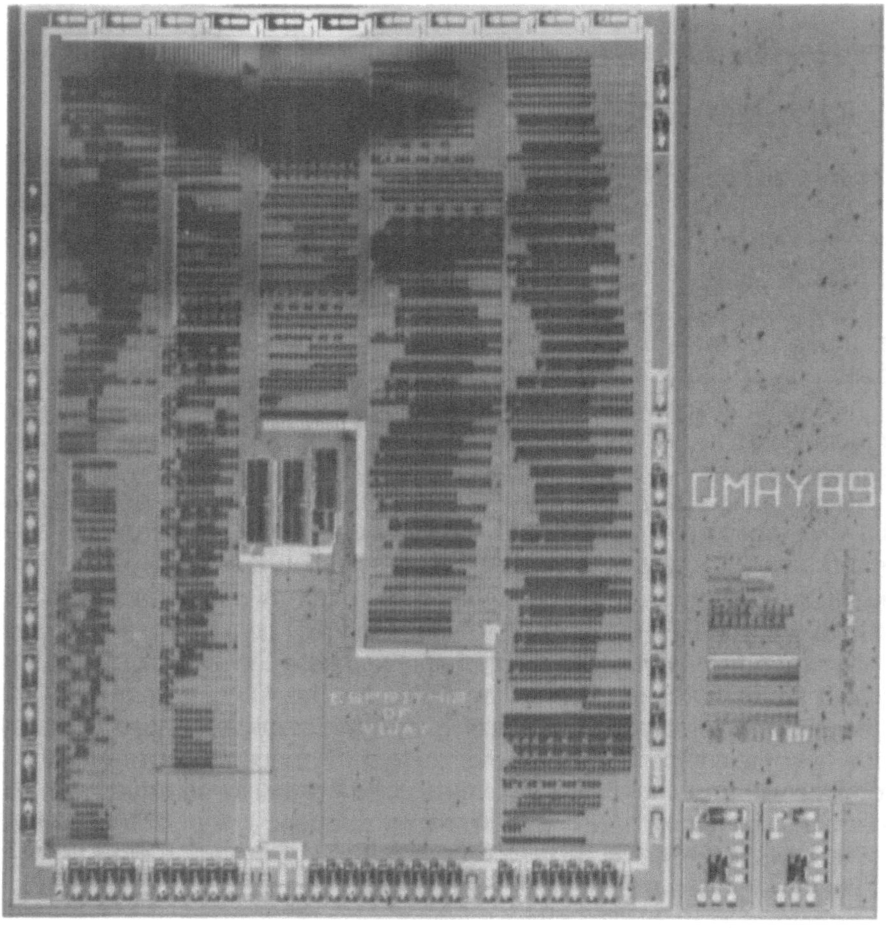

Fig. 2. Digital filter part of high resolution
video A/D converter

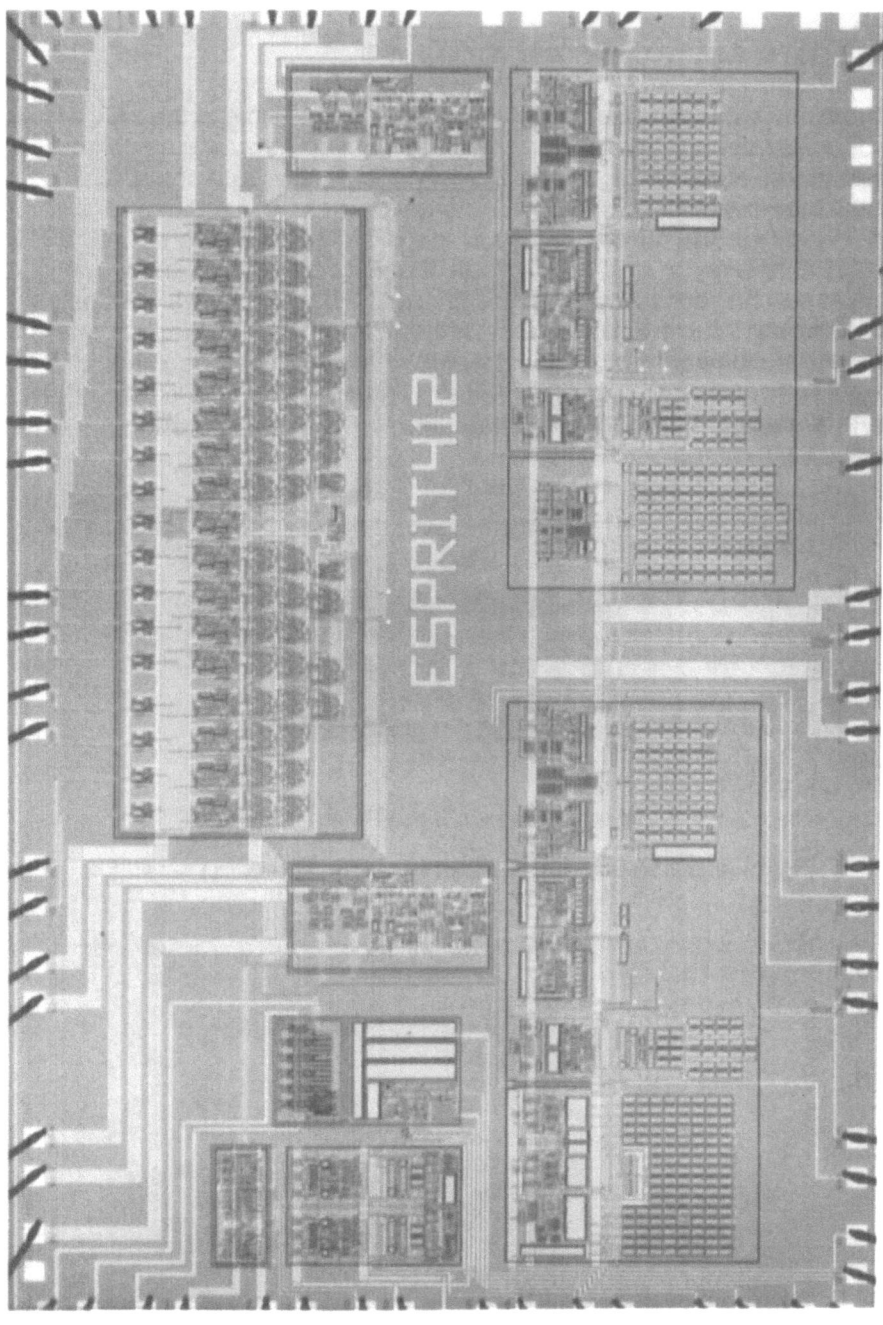

Fig. 3. Prototype analog part of high resolution
video A/D converter

The gate array with on chip SRAM (Siemens)

One of the digital applications of BICMOS Siemens aims at are ECL gate arrays with optional memory functions on chip. As CMOS devices have a 4x higher packing density than bipolar for regular arrays like SRAMs, CMOS is the natural choice to form the memory block. A two-chip demonstrator has been designed and fabricated which comprises the building blocks of this BICMOS application: a 16k SRAM and a 2k ECL gate array. A chip-micro-graph of both circuit-parts is shown in Fig. 4. The SRAM shows an access-time of 3.8ns (shown in Fig. 5), a power dissipation of 2W and an area-consumption of 23 mm^2 (14,19). The high speed macros of the ECL gate array have stage-delay times down to around 60ps for double ended stages.

This SRAM is not only being processed by Siemens, but also in a slightly modified version at Philips, to prove the exchangebility of full circuit design between both partners.

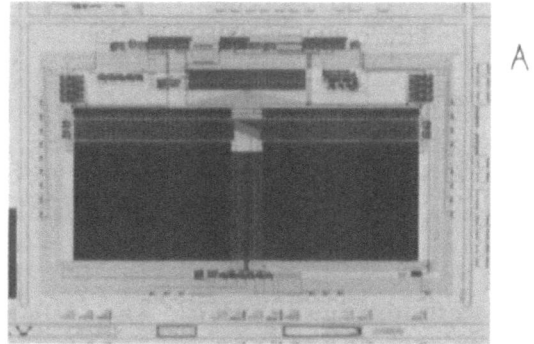

Fig. 4. Chip-micrographs of the 16k SRAM (a) and the 2k ECL gate array (b)

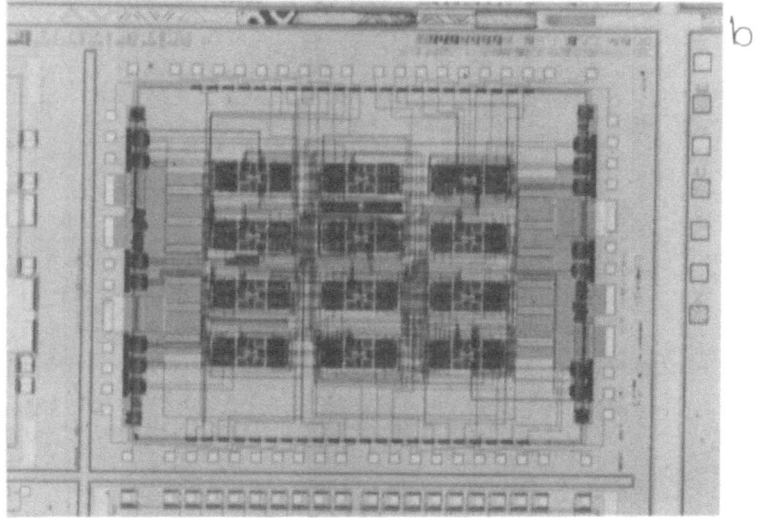

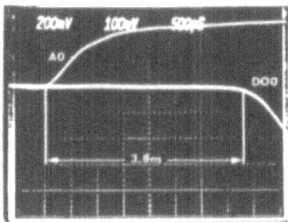

organisation	:	2kx8bit
address acces time	:	3.8 ns
power dissipation	:	1.95 W
chip size	:	23 mm^2
cell size	:	432 μm^2
I/O interface	:	ECL 100k

Fig. 5. Input/output waveforms of the 16K SRAM

An A/D converter with error correction and autocalibration (Villach)

The demonstrator designed at the "Entwicklungszentrum für Mikroelektronik/Villach" is a 200kHz bandwidth ADC. The main circuit innovation which is implemented is the selfcalibration scheme which allows for a 16bit resolution. So far the main building-block - the ADC-core without the control-unit - has been designed and is under way of processing to check the circuit modules on a hardware base, what is mandatory for analog circuit parts. A copy of the layout is sketched in Fig.6.(See next page) The full demonstrator is in the course of design and will be available in silicon towards the end of this project.

The FIR filter (University of Dublin)

The 1-D FIR filter chip can be used to perform one-, two- and three-dimensional filtering tasks in HDTV systems.The individual pixels in a continuing sequence of pictures can be located in relation to a three-dimensional set of coordinates: horizontal, vertical and time. A 1-D filter only alters the horizontal resolution. A 3-D filter shapes the bandwidth horizontally, vertically and temporally. A 2-D filter operates in any two of these three dimensions (see figure 7a).

A 1-D filter is simple enough to implement. The required number of taps are obtained by cascading an appropriate number of FIR filter chips. The 2-D filters most commonly are spatial filters and vertical-temporal filters. For general video signal processing a 2-D filter of 9*9 taps is thought to be sufficient. A spatial filter implemented using the FIR chip as the basic building block is shown in figure 7b (assuming 3 taps/chip).

The final FIR filter will contain 3 multiplier-accumulators and the structure will be cascadable. Presently a number of testcircuits are being made in the BICMOS2 process as a first step in the development of a full BICMOS2 FIR filter.

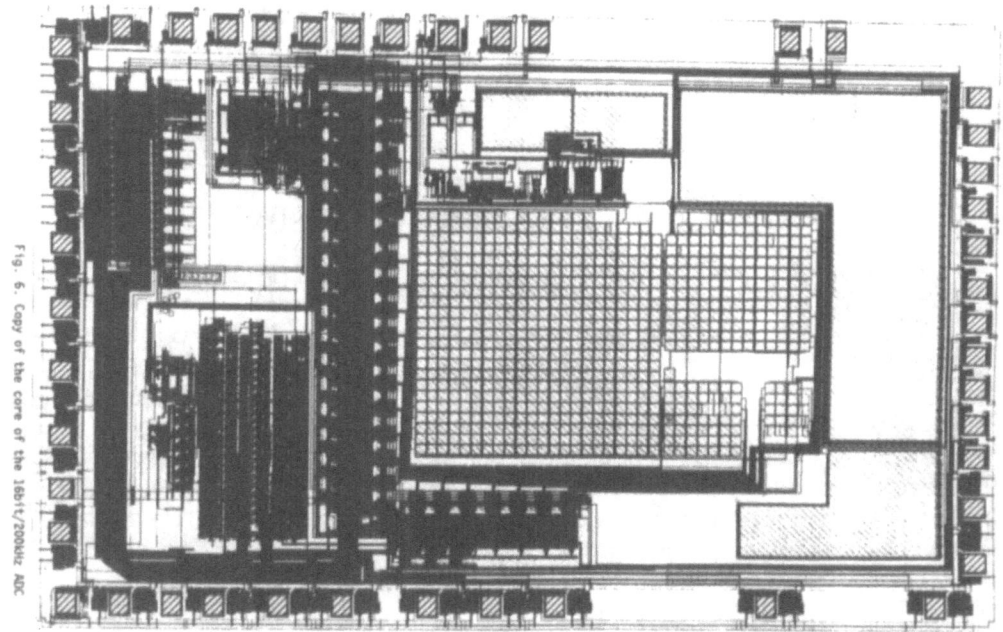

```
Circuit     : Analog Macro of 16-bit A/D-Converter
Design      : EZM VI
Technology  : BICMOS5 (ZFE SPT 22/23)
Date        : 02.10.89
```

Fig. 6. Copy of the core of the 16bit/200kHz ADC

An A/D converter with added functionality (INESC)

This demonstrator will be an 8-bit A/D converter with a high data rate of 75 MHz and will include an anti-aliasing filter. The converter aims at data rates above the 13.5 MHz used in conventional video processing. As a first step towards realization of this A/D converter in BICMOS2 a number of testcircuits are being made. Figure 8 shows a layout of the priority encoder. An innovative circuit design approach was used to achieve a minimum chip area and maximum speed.

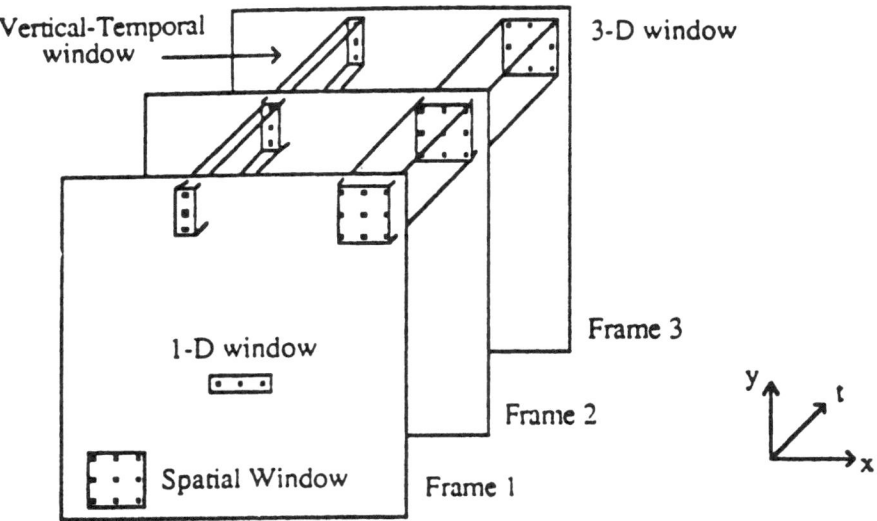

Figure 7a. Filter Windows in Video Processing.

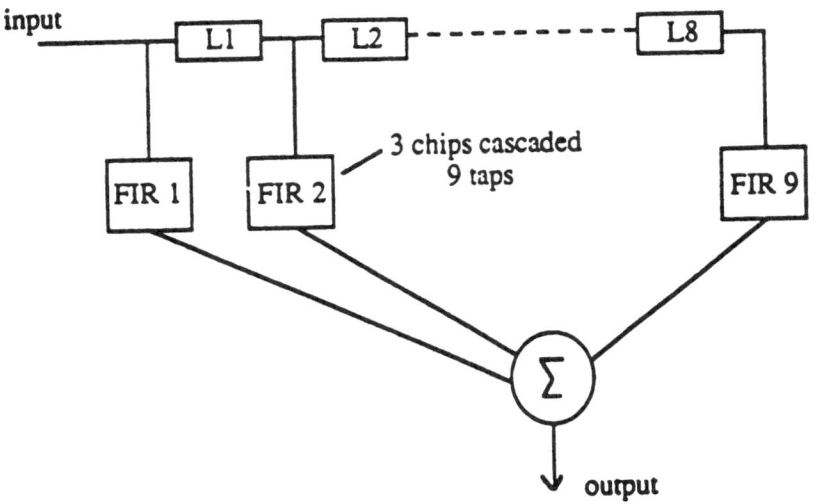

Figure 7b. Spatial Filter (9*9).

154

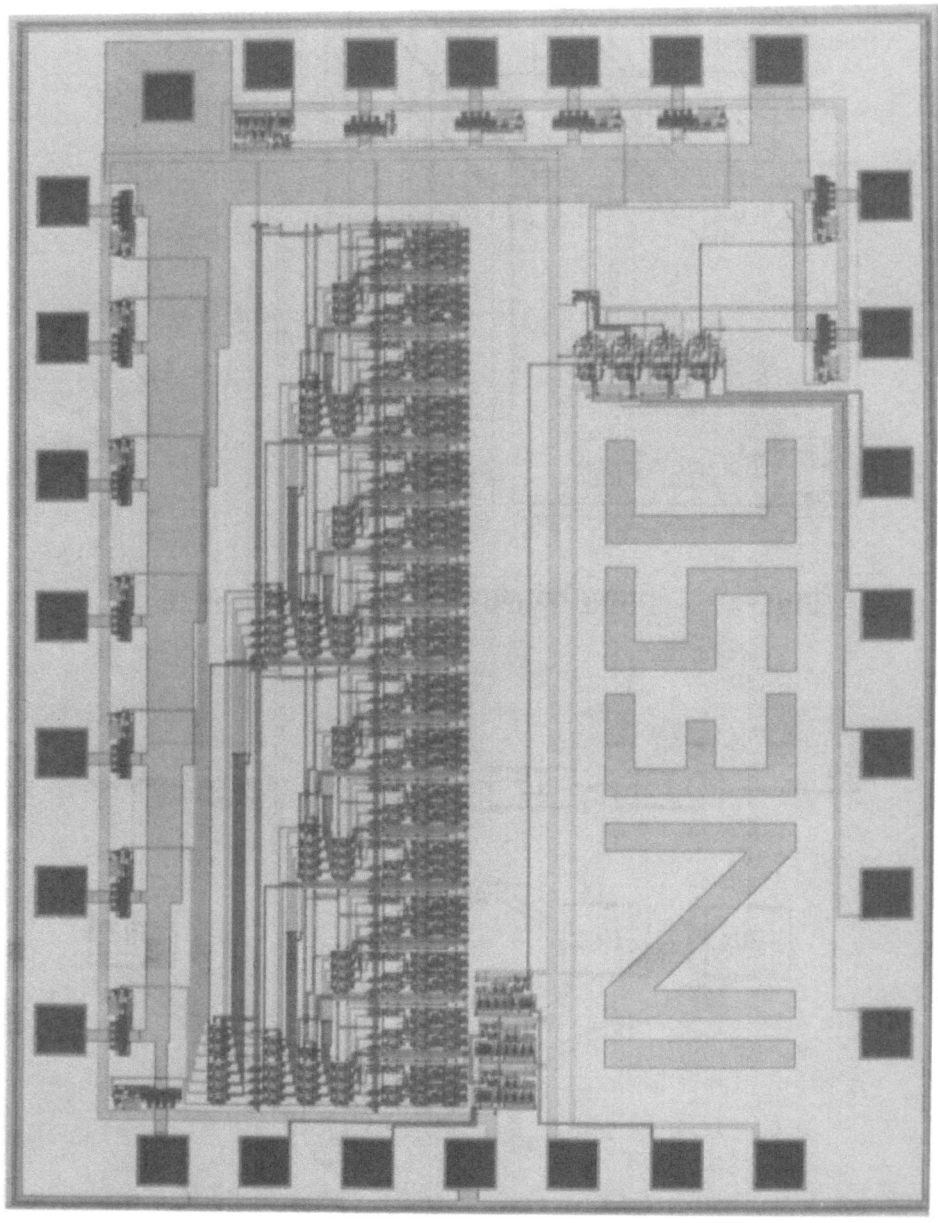

Fig. 8. Layout of priority encoder of A/D converter

5. The technology

In the second phase of the project the BICMOS2 (12,17,18) technology has been developed. The integration concept involves standard LDD MOS devices and self-aligned double-polysilicon bipolar devices. In this way an optimum use is made of a minimum set of process modules, in which the polysilicon and oxide sidewall spacers are the critical ones.

This communality of MOS and bipolar is partly due to the convergence of CMOS and bipolar technologies throughout the world, which makes the concept of the BICMOS2 process not only feasible, but also allows to arrive at highly similar technologies at Philips and Siemens: both partners use identical device structures for MOS and for bipolar, based on a common set of process modules. The differences that still remain are due to the different application backgrounds, namely the analog focus of Philips and the digital focus of Siemens.

Figure 9 gives a cross-section of the various device structures in BICMOS2.

The isolated vertical pnp transistor is only being applied by Philips in its analog systems (20). The structures of npn, NMOS and PMOS are common to all the partners in the consortium.

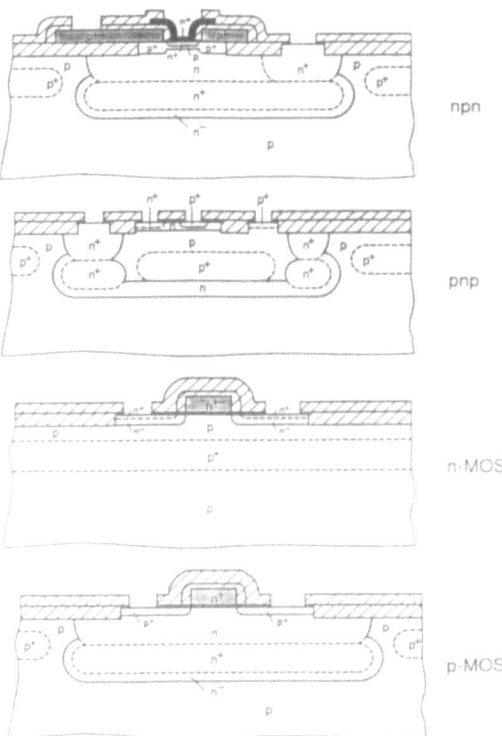

Fig. 9. Schematic device cross-sections

156

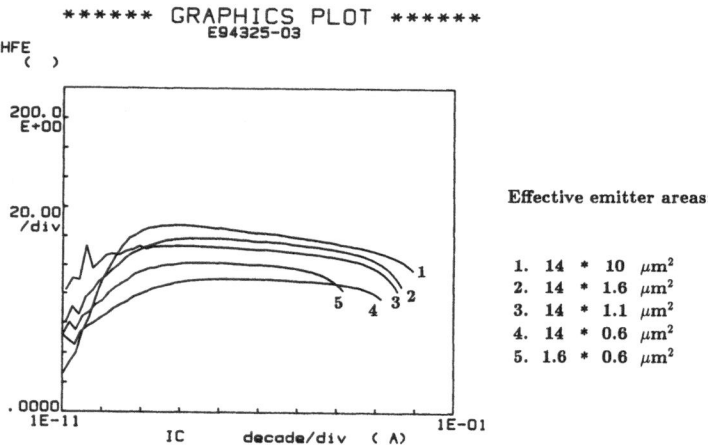

****** GRAPHICS PLOT ******
E94325-03

HFE
( )

200. 0
E+00

20. 00
/div

.0000
1E-11   IC   decade/div   (A)   1E-01

Effective emitter areas:

1. 14 * 10 μm^2
2. 14 * 1.6 μm^2
3. 14 * 1.1 μm^2
4. 14 * 0.6 μm^2
5. 1.6 * 0.6 μm^2

Fig. 10. Current gain scaling of npn transistors

Figure 10 gives an impression of the scaling of the current gain of the npn transistor: at emitter widths down to 0.6 um the current gain is constant. This indicates that this device structure is not only adequate for the BICMOS2 generation, but even for the BICMOS3 generation. The current status of the BICMOS- technology performance is documented in Fig 11 which shows the gate-delay time of ECL-stages as a function of the power-dissipation per gate. The characteristic data is a minimal stage delay time of 65ps for a stage current of 0.7mA and 250ps for 50 A respectively. For even higher speeds longer "bipolar transistors" with longer emitter-stripes can be used. E.g. a 1.0x20 m^2 bipolar transistor results in 50ps ECL-gate delay time. As expected, the Philips and Siemens technologies show similar and promising shrinking behavior.

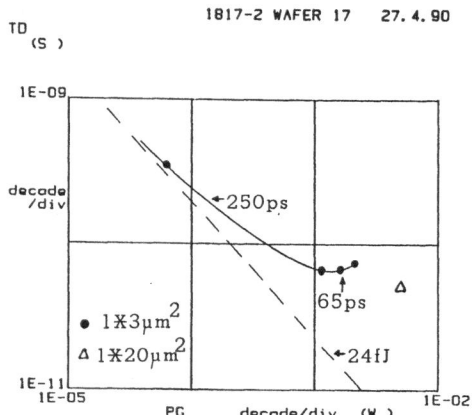

1817-2 WAFER 17   27. 4. 90

TD
(S)

1E-09

decade
/div   ←250ps

                        65ps   Δ

• 1*3μm^2

Δ 1*20μm^2 ←24fJ

1E-11
1E-05   PC   decade/div   (W)   1E-02

Fig. 11. Gate-delay time vs. power dissipation of ECL gates

In the final phase of the project the performance of MOS and bipolar will be boosted by shrinking the critical transistor dimensions and in part by introducing silicides. In fact the device structures in figure 9 will only be slightly modified, only the pnp-structure will be replaced by the complementary self-aligned structure of the npn.

Apart from the shrinking, a cost reduction will be achieved in two ways: first by increasing the packing densities, especially of the CMOS part, and second by reducing the process complexity. This reduction in process complexity involves the extensive use of self-alignment techniques on several levels in the process, and a further interweaving of MOS and bipolar process steps.

As before, the device structures for MOS and bipolar will be identical for Philips and Siemens (LDD CMOS and self-aligned double- polysilicon for bipolar), but the integration concept will show less convergence than in the BICMOS2 case. This is related to the fact that the BICMOS3 technologies are intended to serve as the base for an industrial activity, and the infrastructural background of both companies puts certain restrictions on a common approach in this phase of the development.

6. Stability issues in BICMOS applications

The application of BICMOS technology for the realization of complex analog/digital systems, has shown that special attention has to be paid to crosstalk from the digital circuitparts into the sensitive analog ciruitparts. From the experience with BICMOS2 design, a number of crosstalk-mechanisms have been identified. Presently a checklist is being defined that can serve as a handguide in future design efforts to anticipate on crosstalk problems.

The crosstalk mechanisms are listed below:

a. Groundbounce via substrate contacts
   The internal digital ground line is connected to the external ground via its own resistance. Since this line is connected to the substrate (to avoid latch-up), the analog ground will pick up digital noise via the substrate. The earth line with substrate contacts is an almost perfect noise transport ring.

b. Capacitive groundbounce
   The source-, gate- and drain-to-substrate capacitors in the NMOS devices and also the n-well to substrate capacitor of the PMOS will cause ground bounce. This bounce will be picked up by the analog part through the n-well-to-substrate capacitors.

c. Mutual inductance
   Digital power and output pins can show large current spikes. Since chip, package and printed circuit board can form loops, large magnetic fields may occur. Through magnetic coupling of input pins and chip wiring these voltage and current spikes can magnetically be coupled back into the system.

d. Minority carrier injection

In some CMOS circuits and certainly in input-protection circuits n-p junctions can become forward biased. The injected minority carriers can be collected by another n-type area elsewhere in the substrate.

e. Capacitive crosstalk via interconnect

The coupling capacitors that frequently occur in the interconnect system of the chip play an important role in crosstalk. But since many layout extractors and circuit simulators can deal with this type of phenomena, this mechanism is less likely to cause unexpected problems.

In high-performance analog-digital systems all of these crosstalk mechanisms have to be considered in order to arrive at a first time right design.

7. BICMOS: from dream to reality

Ten years ago the idea of MOS and bipolar devices on a single chip had crossed the mind of several experts in the field, but MOS and bipolar technology had so little communality at that time, that the concept of BICMOS was not too much different from GaAs and silicon devices on a single chip. And obviously this did not prevent some pioneers to start the first BICMOS developments.

The major step between dream and reality is feasibility, both in a technical sense and in an economic sense. The years from 1985 to 1990 are the years of worldwide feasibility studies of BICMOS. The convergence of MOS and bipolar technology has resulted into BICMOS technologies that have provided a definite answer on this part of the feasibility question. On the other hand electronic design in MOS and bipolar have not shown such a convergence, and this has led to an emphasis on electronic design efforts in BICMOS.

Arrived in 1990 it seems that the mastering of complex mixed systems also justifies a positive answer on the technical feasibility of electronic BICMOS design.

This is clearly reflected in the number and variety of BICMOS circuits as presented at the 1990 ISSC conference: over 10 papers discuss. BICMOS SRAMS, A/D converters, sea of gates, filters etc. The technologies typically range from 0.8 m (gate length) to 2.0 μm.

The final phase of the BICMOS project has an emphasis that relates to the economic feasibility and to the reality phase of the nineties: reduction of process complexities, increasing packing densities, and realizing electronic functions in silicon with a minimum number of devices and a minimum number of square millimeters.

At the very start of the BICMOS era, BICMOS was considered as a multi-function technology with a clear users guide: use CMOS for digital and bipolar for high-speed and analog. The reality of the nineties will show that BICMOS has an additional value as a multi-component technology: high performance MOS and bipolar devices can be used to apply efficient electronic design concepts, no matter whether the application is analog or digital.

BIBLIOGRAPHY

1.P.A.H. Hart and H. Klose, "Bipolar CMOS Esprit project", pp. 100-109, Proceedings ESPRIT Technical Week 1989.

2. P.A.H. Hart and A.W. Wieder, Bipolar CMOS Esprit project, pp 95-108. Proceedings ESPRIT Technical Week 1988.

3. P.A.H. Hart en H. Klose, BICMOS versus CMOS and Bipolar p. 327-329. Proceedings ESPRIT Technical Week 1988.

4. P.A.H. Hart, A.W. Wieder, Bipolar CMOS project BICMOS 412. Proceedings ESPRIT Technical Week Brussels 1987.

5. P.A.H. Hart, K. Bürker and A.W. Wieder, BICMOS 412, Poceedings ESPRIT Technical Week Brussels 1986.

6. F. Rausch, H. Lindeman, W.J.M.J. Josquin, D. de Lang and P.J.W. Jochems, An analog BIMOS technology. Extended Abstracts of the 18th Conference on Solid State Devices and Materials. Tokyo 1986, pp. 65-68.

7. D. de Lang and W.J.M.J. Josquin, Optimization of a 1.5 μm BICMOS process. BICMOS Symposium, Abstract no. 275. Electrochemical Soc. Philadelphia May 1987.

8. W. Josquin, D. de Lang, M. van Iersel, J. van Dijk, A. v/d Goor and E. Bladt, The integration of double-polysilicon NPN transistors in an analog BICMOS process. Extended Abstracts of the 1988 SSDM, Tokyo. Paper no.7, session A9.

9. H. Klose, T. Miester, B. Hoffmann, H. Kabza, J. Weng, Well-optimization for high speed BICMOS Technologies,ESSDERC, 1988.

10. B. Hoffmann, H. Klose, T. Miester, A 10 GHz High Performance BICMOS Technology for Mixed CMOS/ECL ICs, ESSDERC 1989.

11. H. Klose, T. Meister, B. Hoffmann, B. Pfäffel, P. Weger, Low cost and high performance BICMOS processes: A comparison, BCT Meeting 1989, pp. 178-181.

12. J. Hauenschild, H.M. Rein, P. Weger, H. Klose, A 10 Gbit/s monolithic integrated bipolar demultiplexer for optical fiber transmission systems fabricated in BICMOS Technology, Electronics Letters, Vol 25, No 12, pp. 782-783.

13. W. Heimsch, B. Hoffmann, R. Krebs, E. Müllner, B. Pfäffel, K. Ziemann, Merged CMOS/Bipolar Current Switch Logic, Proceedings of the ISSCC 1989, pp. 112-112.

14. W. Heimsch, R. Krebs, K. Zieman, A 4ns 16k BICMOS SRAM, Symposium VLSI circuits, Kyoto, 1989.

15. W. Heimsch, R. Krebs, K. Ziemann, D. Moebus, Comparing CMOS and BICMOS NOR decoder structures using Monte Carlo optimization, Presented at ESSDERC 1988.

16. P. Nuijten and K. Hart, A digitally controlled twenty bit dynamic range BICMOS stereo-audio processor. Digest ISSCC Feb. 1989.

17. H. Klose, B. Zehner, A. Wieder, "BICMOS, a Technology for High Speed/High Density ICs", Proc. 1989 IEEE Int. Conf. on Comp. Design, ICCD'89, pp. 304-309.

18. P. Weger, H. Klose, "A 7 GHz 10:1 Frequency Divider Realized in a High Performance BICMOS Technology", submitted to Electronics Letters.

19. W. Heimsch, R. Krebs, B. Pfäffel, K. Ziemann, "A 3.8 ns 16k BICMOS SRAM", IEEE-Solic State Circuits, Vol.25, No.1, pp.48-54.

20. D. de Lang, E. Bladt, A. v/d Goor, W. Josquin, "Integration of Vertical pnp Transistors in a Double-Polysilicon Bi-CMOS process", Proceedings of the 1989 BCTM, Minneapolis, 1989, p. 190-193.

INFORMATION PROCESSING SYSTEM

Project No. 311

Terminological Information Management in ADKMS

Maria Damiani, Sandro Bottarelli, Manlio Migliorati
Datamont SpA Gruppo Ferruzzi
R&D Centre
Via Restelli 1/A
I-20124 Milano

Christof Peltason
Technische Universität Berlin
Project KIT-BACK, FR 5-12
Franklinstr. 28/29
D-1000 Berlin 10

Abstract

One main goal of the ADKMS project (Advanced Data and Knowledge Management System) is the investigation of information management based on a terminological approach. We describe an application scenario in which organisational information about a large Company Group is acquired for a knowledge base. The application (developed by the project group of Datamont SpA, Milano) reveals a number of requirements which are contrasted with the service offered by the terminological knowledge representation system (developed by the project group at the Technical University Berlin). The interaction between these aspects is highlighted and the resulting techniques for appropriate knowledge base access and retrieval are sketched.

1. Introduction

The utility of an information system is critically dependant on how easily the knowledge of the application domain can be described. Appropriate description techniques allow for appropriate acquisition and access of a knowledge base. Only if classes, objects, and relations of a domain can be modelled and retrieved in a natural way is the knowledge base of any use in particular in more complex application areas. The key idea of the approach presented here is that a proper treatment of the domain terminology is a good starting point for such a task.

Work within the paradigm of terminological reasoning has been performed in the ESPRIT I Project 311, **ADKMS (Advanced Data and Knowledge Management System)** since 1985[1].

1 This work was partially supported by the Commission of the European Communities and is part of the ESPRIT Project 311 which involves the following participants: Datamont (Milan), Nixdorf (Munich), Olivetti (Pisa), Quinaty (Milan), Technische Universität Berlin, Universität Hildesheim.

In this paper we describe a joint effort carried out during the second phase of this Project (November 1988 - October 1990) which shows the interaction between two groups of the project:

- the group involved in building an application model (Datamont SpA, R&D Center, Milano),

- the group involved in the development of the terminological knowledge representation system BACK (Technical University Berlin).

The initial setting involves establishing a data/knowledge base describing the social structure of a very large Italian **Company Group.** Administrative, financial, legal and business aspects of the Group's overall organisation are to be taken into account.

It seems that for such a complex area any attempt to employ conventional design techniques is extremely problematic: Structuring of the complex interdependencies goes beyond the limitations of the relational data model, and formulating queries in an SQL-like manner requires a deep knowledge of the underlying structures.

In contrast, the representation system chosen for this task within ADKMS offers a more flexible and easy-to-use approach: The experiences made in modelling the **Company Group domain** show the usefulness of description and query techniques on a logic-oriented level. However, a number of additional requirements with respect to the expressiveness of the representational service are posed which influence the evolving design of the system.

The following study shows the interaction between application requirements and the representational service:[2]

- The application scenario is sketched. It shows the need for a knowledge base management system in the business management area.
- The requirements for the terminological model are collected. We show how conceptual dependencies require specific modelling constructs.
- The terminological model is outlined in terms of the language. of the experimental object-based representation system $BACK^3$. The main characteristics of the system and the terminological paradigm are described and the role within the overall information management system is illustrated by examples.
- The system's usage is sketched. The complexity of the representation system leads to the development of customized interface tools.

Summarizing, the cooperation between the application group and representation system group results in a common understanding of the termoinological approach, which leads to an improvement of the usability of the system but also is taken as a guiding line for the future development.

---

2 See also [DB90]

3 The Berlin Advanced Computational Knowledge Representation System

2. The Business Application Scenario

A relevant problem, quite common in very large **Company Groups,** is the maintenance of an updated view of the societary structure, evolving in time due to company acquisitions, disposals and share transfers. Such a view is fundamental for the top management in charge of defining and supervising the economic profile of the Group as well as for company shareholders.

Commonly the overall knowledge is maintained by people very close to the top management, split among several offices with different tasks, often duplicated and stored on paper. In general access to this bulk of information is not simple; complex requests often need to be decomposed into subqueries to be turned over to the competent offices.

The application we have investigated and prototyped in cooperation with the user focuses on the development of a knowledge base as the central repository of the societary structure of one of the most important private industrial groups in Italy[4].

The societary structure is quite complex, amounting to over 1500 companies operating throughout Europe, USA, South America, and Australia and involved in a wide range of businesses:

- Agribusiness
  (i.e. industrial processing of agricultural raw material; farming, commodities, trading, shipping),
- Chemicals
  (chemical products, energy, pharmaceutical),
- Engineering and Construction,
- Insurance and Financial Services.

Companies in the Group are indeed extremely heterogeneous. They operate in different countries, and are thus subject to different regulations depending on the legislation of the host nation; they are engaged in diversified businesses in several economic areas and have different internal administrative organisations in part deriving from their country-dependent juridical status.

As a consequence, legal and financial terminology as well as jargon specific to the Group is extensively used both by experts in specifying and users in accessing the domain description. In some cases terminology is precise and stable; in others it is ambiguous and not fixed.

In general, terminology acquisition is quite a complex and time consuming activity, although essential for the soundness of domain model. It leads to the identification of relevant conceptual entities in the domain and to the specification of their features including relationships with other entities. The result is the so-called terminological model.

---

4 See also [DB90] for a discussion of the application domain

3. Requirements for the Representation Service

Terminology in highly structured domains such as the one under consideration, presents specific features that it would be desirable to capture in the terminological model and thus in the representation formalism:

Description Equivalence: Often the same conceptual entity can be referred to either by using the specific terminology or through properties which unequivocally identify the object. For example the queries, taken from the domain:

1. *Which are the Italian companies listed on New York Stock Exchange?*
2. *Which are the "SpA"* [5]*listed on New York Stock Exchange?*

have an identical answer, i.e. the set of companies that are formally declared located in Italy, with capital subdivided in shares, with shares of the specific type that make them quotable on a Stock Exchange and that are listed in New York; Given *Query 1,* we can derive, reasoning on the terminology, that the answer consists only of companies with juridical status *SpA.*

In fact companies that can be listed on the Stock Exchange are only those whose capital is subdivided in shares and the Italian companies with such a property are only *SpA* companies. Vice versa, given *Query 2* and reasoning on the definition of *SpA* we can infer that the answer consists of the companies that are located in Italy with their capital subdivided in shares of a specific type. Some terms can be identified by both necessary and sufficient properties.

We can assert or retrieve a specific entity either through the proper domain term, i.e. a *SpA* company, or definitely through the distinguishing properties, i.e. an Italian company with ..., in turn relying on further terms. A knowledge base exhibiting the behaviour described above must thus incorporate both an accurate definition of domain concepts and an inferential engine for reasoning on the properties specified at the structural level.

Complex Term Interdependencies: Domain terms can be strictly interrelated. For example, to remain in the legal area, the juridical status of a company determines the form of its administrative apparatus (i.e. Board of Auditors, Secretary Board, etc.) and constrains the set of administrative/financial events the company is subject to (i.e. capital addition, share participation transfer, etc.), the financial assets, the type of business activity of the company [Gal86]. Consequently an assertion such as the following:

the company X had a capital addition of 10ML$ in ordinary shares

---

5 SpA is a specific Italian juridical form, acronym of "Società per Azioni".

is meaningful only if the company X has a juridical form for which the formally declared capital can be increased and such capital consists of shares that in turn can be ordinary. If, for instance, X is a non profit company, the assertion is wrongly formulated because by definition it doesn't make any sense.

The inconsistency can be detected only if juridical terms are properly correlated to the concepts of capital addition and, consequently, capital and share. In that hypothesis, the resulting description is a complex net of term definitions, one strictly depending on the other. For the description to be sound, overall consistency has to be guaranteed.

Taxonomic Organisation: In the domain, terminology is often inherently structured in a taxonomy. It is quite common, both in legal clauses and in the jargon of the Group, to find terms ordered in a hierarchy of abstractions. For example, with respect to the economic activities of companies, these are presented by management as grouped in increasingly specialized areas. For instance the agroindustry business one of the more relevant sectors for the Group consists of activities of transformation of agricultural products, that can in turn be further specialized on the basis of products, trading that can be further specialized depending upon the forms of trading, and so on.

Inference on Terms Definitions: For a sound representation, the positioning of a given object into the taxonomic description must respect the terminological definitions so far introduced. For example, given the economic activity of growing, defined as a type of production limited to the agricultural sector, if a company is engaged in soy bean production, it necessarily will be involved in the growing business as well. In this case the positioning of soy bean production is the result of an inference on term definitions .

Retrieval Facility and Model Scalability: The terminological model has to be accessed not only for updating but above all for querying purposes. In general, articulated queries require a complex navigation of the model and consequently powerful mechanisms of query interpretation. In this domain, the dimension of the Group (1500 companies) and its great heterogeneity entails a remarkable quantity of terminology to be modelled.

4. Modelling a Terminology in BACK

Domain modelling aims at providing a high level, conceptual representation of application specific knowledge. The different approaches to modelling mainly pursued in the Database and Artificial Intelligence fields (through, respectively, semantic data models and knowledge representation schemas) share the fundamental objective of providing highly expressive representation constructs close to the way in which reality is conceptualized [HK87,BGN89].

In general, semantic data models focus solely on the issue of representational adequacy while rarely being fully operational. Recent object-oriented DBMS [ABD$^+$89] overcome these limitations while lacking inferential features. Knowledge representation formalisms, on the other hand, stress in addition to descriptive naturalness and richness also the inferential capability of the underlying computational model. Nevertheless, representation formalisms often lack a strong formal basis as well as powerful mechanisms for uniformly accessing and manipulating domain description.

The features given in Section 3 are indeed extremely relevant for modelling the business domain under consideration. The description is conceived to be accessed by the management in charge of both business monitoring and decision activity. Application requirements suggest a model and thus a representation formalism with high structuring power, able to capture the rich terminology of the domain, formally sound, provided with a uniform and powerful high level description interface for information manipulation and querying. The chosen knowledge representation is the terminological representation system BACK [LNPS87,PSKQ89].

It is object-based in the tradition of frames and semantic nets, yet differs from these in being provided with logic-based formal semantics, making the language simpler to use and behaviourally sound and reliable. It stands in the research tradition of KL-ONE and analogously introduces a sharp distinction between terminological and assertional knowledge under the cognitivistic assumption that describing contigent facts of world knowledge is based on an "a priori" domain terminology [BS85]:

- Terminological knowledge captures the intensional aspects of the description and consists of a set of concepts and roles (concept properties) organised in subsumption taxonomies.

- Assertional knowledge consists of terminological object instances.

- A distinguishing feature of the language (and of other KL-ONE like systems) is the possibility of specifying definitional properties for concepts (i.e. properties expressing necessary and sufficient conditions for an individual to be a concept instance). Through classification inference, concepts are placed automatically in the net of terminological descriptions depending on the specified properties, while instances are automatically completed with the best, i.e. most specific, description.

Taxonomic organisation is the basic representational feature of the terminological knowledge (TBox) structured as a lattice of concepts and roles partially ordered with respect to set inclusion (subsumption) relation. At the inferential level classification identifies the correct positioning of new TBox/ABox objects; in the overall knowledge base net providing consistency checking mechanisms and deductive inferencing on term definitions.

Finally, a powerful Tell/Ask interface language is defined on both the TBox and the ABox, respectively, for knowledge assertion and retrieval [PSKQ89]. A relevant part of the interface language is the ABox Query Language (AQL), which makes the representation system particularly suitable to be used for iniormation retrieval purposes on complex descriptions[6].

We describe the approach to domain modelling with BACK through examples taken from the developed knowledge base. The implemented model encompasses the following types of information:

- juridical knowledge about the juridical status of companies;
- legal knowledge concerning the financial status as well as relationships with other companies;
- geographic knowledge;
- organisational apparatus internal to companies (bodies, charges, powers);
- knowledge about relevant events in specific companies.

A sample company description is shown in Figure 1[7].

In brief the example describes the following: the central concept is *company*, introduced as a primitive concept (i.e. only necessary conditions can be defined on it) which is a specialization of *juridical person* but disjoint from *physical person*. Relevant properties for the *company* are expressed by the following roles:

- *has_capital* with range *capital;* the role of *capital, subdivided_in* (and its related chain), expresses the property of *capital* to be subdivided into *participations* belonging to *juridical person;*
- present_in with range a *geographic locality;* this role can be further specialized in the roles: *located_in, office, registered office,* and describes the location - in its different forms - of a company.

The *Company* concept subsumes the defined concepts: (necessary and sufficient conditions can be defined on it): *limited company, Share Company, Australian company,* introduced as value restrictions of roles.

Given the above description, the insertion of a new company with, for example, its registered office in *Sydney,* listed on *New York,* and with *Montedison* as its major owner, is realized by the expression[8]:

---

6 See also [Kin90] for details of the assertional component

7 In the graphical notation an ellipse represents a concept (primitive concepts are marked with asterisks). A wide arrow denotes the subsumption link. A role is represented by a little box inside a circle with arrows from the domain concept and to the range concept. The simple boxes denote value restrictions.

8 *Sydney, New York* and *Montedison* are aliases of pre-existing ABox instances.

```
X = company
        with registered_office: 'Sydney'
        andwith listed_on: 'New York'
        andwith has_capital:
                (Y = capital with subdivided_in:
                        (Z = participation with belonging_to: 'Montedison')))
```

The above assertion entails the following set of inferences:

1. the company has a registered office in Sydney, and is therefore located in Sydney;

2. Sydney is an Australian locality and the company is thus an Australian company;

3. The Australian company is listed on a Stock Market and is therefore a listed company and, as consequence, a share company,

4. The company has a participation belonging to another company, hence it is a limited company;

5. Australian limited companies with capital subdivided in shares can only be Australian Corporation companies.

The inserted company is thus inferred to be an Australian Corporation. In this way, instances can be only partially specified and the insertion of an entity as new element of a concept is left to the knowledge base management system. Inferencing generates additional instances in the ABox as depicted in Figure 1.

Querying the ABox is done through a powerful query language [PSKQ89] which is sketched in the following simple examples:

Which are the limited companies with office in Sydney and listed on New York?

```
getall limited_company
        with listed_on: 'New York'
        andwith office: 'Sydney'
```

Which are the names of the chairmen of Italian companies engaged in both the chemical and agricultural sectors, and that have some installation in a nation in which a company participated by Montedison is also present?

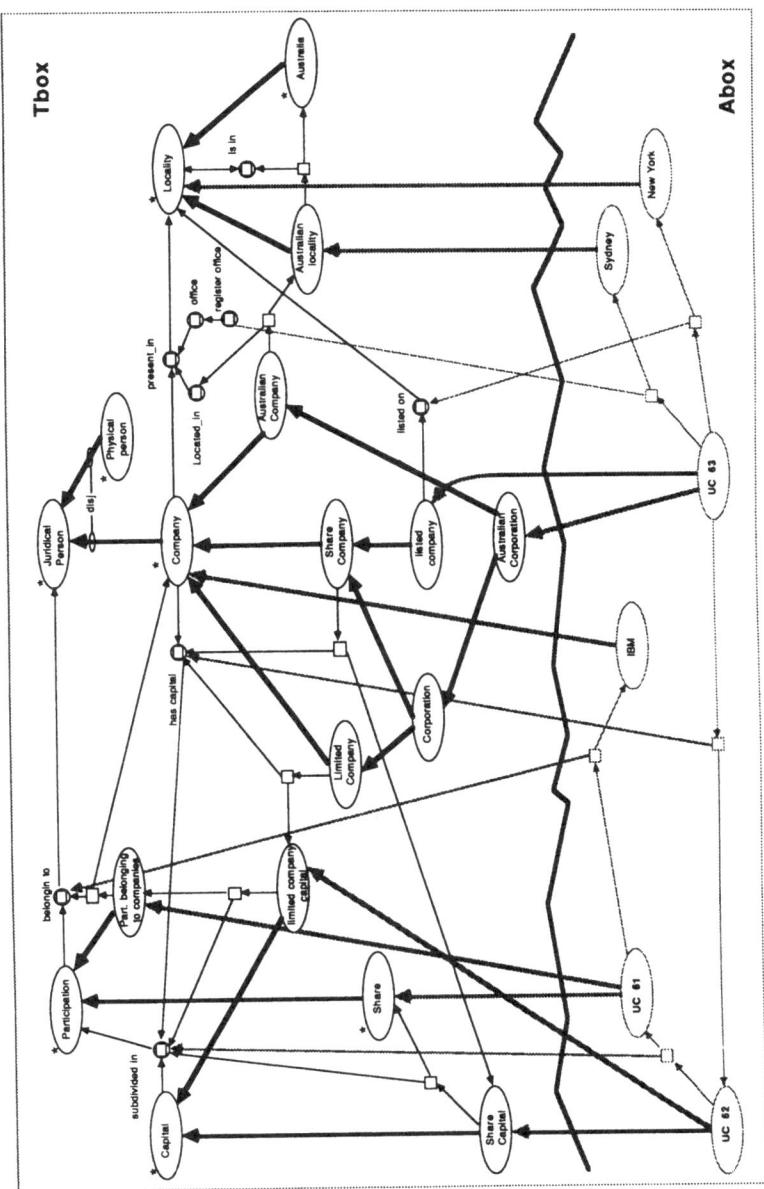

Figure 1: A detail of the **Company Group** terminology with assertions.

[rf (firstname),rf (surname)] for
getall physical_person with inv has_in_office:
    (some chairman with inv has_body:
        (some italian_company
        with engaged_in:
           (some chemical_area) and (some agricultural_area)
        andwith installation:
           (some nation with inv present_in:
              (some company with has_capital:
                (some capital with subdivided_in:
                  (some participation with belonging_to:
                    (the company with named: 'Montedison')))))))

The model we have built consists of about 500 TBox concepts/roles. The effectiveness of the underlying approach in capturing the representational features of the domain and the powerful mechanisms it provides for accessing the description have greatly simplified and sped up the realization of the model.

5. System Usability

The usability of the system is improved by three components:
- The shell environment **BIT** (BACK Interface Tool) which was specifically developed by the application group (Datamont), and which is based on the requirements emerging from the development of the knowledge base. The shell provides a set of innovative graphic based tools for knowledge base construction, access and monitoring, tailored to the specific needs of the terminological representation paradigm [BM90].
- End-user access to the knowledge base supported by an interface, which maps enduser requests on a user view of the knowledge base into sophisticated query language expressions; a first prototype has been developed in a hypermedia environment.
- Techniques for coupling the representation system with a relational database for storage of the ABox contents are currently being investigated.

5.1 The Knowledge Engineer Environment

The BIT environment is a user-friendly, graphic based shell to BACK, tailored to manage large knowledge bases. Providing both textual and graphical interaction modalities, the shell adopts the multiple windows approach, enabling the knowledge engineer to interact with the environment by several windows.

The shell, based on requirements emerging from the construction of the knowledge base, consists of three integrated environments:

Investigation Environment: In the BACK system knowledge can be structured in a rich and complex way. As a consequence it can be hard to recognize represented knowledge. To make this easier BIT provides both classic browsing tools enabling investigations considering whole knowledge base, and "ad hoc" tools, enabling the knowledge engineer to select and investigate just knowledge considered as relevant in a well defined moment, without considering unnecessary and possibly misleading details.

Input Environment: BIT provides both textual and graphical environments to define new BACK objects: in this way the knowledge engineer can choose the more comfortable one depending on the situation.

Monitoring Environment: BACK performs inferences on objects in an automatic way, making the reasoning process difficult to control from the point of view of the knowledge engineer. Monitoring tools enable the knowledge engineer to understand inferences showing the reasoning process performed by the system in a graphical and/or textual way.

The shell was developed on a SUN^{TM} hardware platform in Quintus Prolog^{TM} plus Prowindows^{TM},[9]

5.2 End User Interface

Usually, both in relational databases and in advanced representation contexts (i.e. semantic data modelling, object-oriented databases, or Knowledge Representation) little attention is paid to end-user interaction; complicated query languages are at best made available at the interface level. This constitutes a barrier for the casual end-user wishing to consult the information base, and consequently a large amount of informative potential is often wasted.

The query language sketched in this paper is still a quite sophisticated formalism incomprehensible to an end-user not already acquainted with the representation paradigm. It is therefore inconceivable for the user of our system, a manager or business professional, to interact directly with the knowledge base or, conversely, to be constrained by fixed interaction patterns, as usually happens in most information systems. To meet these needs, we are exploring a hypertextual approach to query definition and interaction. Relevant issues we are currently focussing on are:

User Views: The user does not have to be concerned with low level details of the domain description, introduced as pure implementative choices or as heuristic shortcuts. In fact, even if BACK provides high level representation constructs for domain modelling, often the actual representation of the model does not fit with the model itself. User views describe a virtual model mappable onto the actual knowledge base.

---

9 Sun is a trademark of Sun Microsystems,Inc; Quintus Prolog and ProWindows are trademarks of Quintus Computer Systems, Inc.

174

User Queries: The user must be allowed to freely query a knowledge base view. A query expression can be naturally viewed as a link of instance patterns. In the proposed approach, instance patterns are defined to constrain TBox concept extensions; links are meant to relate instance patterns through the exisiting roles. A hypertextual approach to query definition is thus indicated: the mechanism of composition is provided by navigation accross the hypertext application links. As the resulting query interrogates the user view, a mapping onto the actual knowledge base is also-specified.

Answer Navigation: The query provides a set of instances that can be further inspected according to user needs. In this case, a navigational approach is also appropriate, concerning matched instances rather the instances patterns.

A first prototype of such an end-user interface has been developed in Hypercard[TM] on the Macintosh[TM]. The Sun[TM] platform acts as the knowledge base server to this client application level.[10]

5.3 Relational Coupling

Considerable attention has been recently paid by the research community to the problem of integrating terminological formalisms with databases [BGN89]. The research is still at the initial stages, and few results can be currently reported [PS+90]. Our experience from the practical use of BACK strengthens the thesis that scalability is actually of fundamental importance for the use of representation systems in concrete applications.

The approach presented here proposes a model very suitable for replacing both relational and semantic data models when the information to be represented is highly structured, organised via complex relationships, and requiring deductive capabilities. The efficient management of information is thus the first and fundamental step towards realization of a knowledge server system as a global informative resource for client applications and users needing access to complex descriptions rather than elementary data.

At present, we are investigating the specific problem of coupling BACK with a Relational DBMS having SQL as interface for the efficient storage and retrieval of ABox instances. A first solution has been proposed in [Kin90] although not fully satisfactorily in terms of efficiency; moreover it works only on a subset of BACK, with great limitations to the formalism's expressivity. The proposal focuses on the following main aspects:

– Mapping of every TBox into a relational schema and consequently of every ABox instance into a corresponding relational instance. BACK concepts and roles have a very straighforward semantics denoting nothing but sets of individ-

---

10 HyperCard and Macintosh are trademarks of Apple Computer, Inc.; Sun is a trademark of Sun Microsystems, Inc.

uals/couples_of individuals. TBox entities can thus be *naturally* associated with relational data structures. Mapping the large amount of TBox objects necessarily entails the definition of a limited number of tables in the relational schema. How to aggregate information in order to efficiently handle retrieval and updating requests is, at the present, a major issue. To that extent, the possibility of adapting mapping techniques proposed for translating semantic data base schemas into relational schemas is currently being evaluated [NC87,LV87].

- Translation of ABox query expressions into composite SQL-based expressions with the goal of limiting the number of accesses to the database and exploiting RDBMS optimization facilities. In contrast to semantic data models, query execution entails deductive inferencing on object structures. A possible interpretation function for query expressions would access the intensional description (TBox) to infer properties on query objects, and would apply SQL expressions to manipulate extensions according to the inferred properties and query patterns.

- Translation of update operations into transactions. Update operations are extremely costly, as consistency checks need to be performed on the knowledge base to validate any new assertion. To that extent, set manipulation operations can be used for testing constraints on extensional descriptions.

6. Conclusions

We have presented a terminological knowledge base developed for a business application, and have argued the suitability of the terminological approach in highly structured domains requiring inferential capabilities. Finally, the necessity of both adequate support for enduser interaction and efficient knowledge storage for the development of real-world complex applications has been stressed.

References

[ABD[+]89] M. Atkinson, F. Bançilhon, D. DeWitt, K. Dittrich, D. Maier, and S. Zdonik. The object-oriented database system manifesto. In *Proceedings of the First International Conference on Deductive and Object-Oriented Databases, Kyoto, Japan, December 4-6, 1989,* 1989.

[BGN89] Howard W. Beck, Sunit K. Gala, and Shamkant B. Navathe. Classification as a query processing technique in the CANDIDE semantic data model. In *Proceedings of the International Data Engineering Conference, IEEE,* pages 572-581, Los Angeles, Cal., February 1989.

[BM90] C. Bagnasco and M. Migliorati. The Bit Environment: a first release. Internal Report, Datamont SpA, Milan, January 1990.

[BS85] Ronald J. Brachman and James G. Schmolze. An overview of the KL-ONE knowledge representation system. *Cognitive Science,* 9(2) 171-216, April 1985.

[DB90] M. Damiani and S. Botarelli. A terminological approach to business domain modelling. In *Proc. of the International Conference on Database and Expert Systems*

Applications, Vienna, Austria, August 1990.

[Gal86] F. Galgano. *Diritto Commerciale, Le Società.* Zanichelli, Bologna, 1986.

[HK87] R. Hull and R. King. Semantic database modelling: Survey, applications and research *issues. ACM Computing Surveys,* 19(3),1987.

[Kin90] Carsten Kindermann. Class instances in a terminological framework - An experience report -. In H. Marburger, editor, *GWAI-90. 14th German Workshop on Artificial Intelligence,* Berlin, 1990. Springer-Verlag. To appear.

[LNPS87] Kai von Luck, Bernhard Nebel, Christof Peltason, and Albrecht Schmiedel. The anatomy of the BACK system. KIT Report 41, Department of Computer Science, Technische Universität Berlin, January 1987.

[LV87] P. Lynback and V. Vianu. Mapping a semantic data model to the relational model. In *Proceedings of ACM Special Interest Group on Management of Data,* San Francisco, May 1987.

[NC87] B. Nixon and L. Chung. Implementation of a compiler for semantic data model: Experience with taxis. In *Proceedings of ACM, San Francisco,* Special Interest Group on Management of Data, May 1987.

[PS$^+$90] Peter F. Patel-Schneider et. al. Term subsumption languages in knowledge representation. *The AI Magazine,* 11(2):16-22,1990.

[PSKQ89] Christof Peltason, Albrecht Schmiedel, Carsten Kindermann, and Joachim Quantz. The BACK system revisited. KIT Report 75, Department of Computer Science, Technische Universität Berlin, September 1989.

Project No.530

COOPERATIVE DISTRIBUTED PROBLEM SOLVING IN EPSILON[*]

Günter KNIESEL, Armin B. CREMERS
University of Dortmund
Department of Computer Science 6
P.O. Box 500 500
4600 Dortmund 50
West Germany
Phone: + 49 231 755 2508
e-mail:gk@mecky.informatik.uni-dortmund.de

ABSTRACT.

EPSILON is a knowledge base management system incorporating objects, inheritance, logic (meta-)programming and databases in an uniform framework. We present an architecture for distributed problem solving in a network of EPSILON workstations, which allows different parts of one integrated application, distributed on several nodes in the network, to communicate and cooperate. Distributed problem solving, regarded as the solution of problems by cooperation of autonomous problem solving agents, is supported by a communication subsystem with control distributed among the processing nodes. The architecture allows to split one global knowledge base, allocating parts of it to the nodes where they are used most frequently, as well as to integrate existing knowledge base with different representations formalisms and their inference engines (corresponding to integration of existing expert systems from different application domains). Especially problems related to cooperation among different inference engines on different machines have been considered. Further on, we were concerned with the problem of how a knowledge base can be partitioned to facilitate parallel inference and to make most efficient use of the resources of the network (multiple processors, storage). The concepts described in the paper should be applicable to any knowledge base management system using objects and inheritance.

1. Introduction

The aim to develop realistic models of the way humans interact in order to solve problems (ranging from the cooperation of individual experts, to complex interdependencies among different department of a company, every one consisting itself of a network of smaller entities) naturally leads to modeling of distributed systems ([5],[6],[16]). From a more technical point of view distribution of knowledge and control offers advantages related to reliability, efficiency and increased autonomy

---

[*] EPSILON is the acronym for ESPRIT I Project 530, "Advance KBMS Based on the Integration of Logic Programming and Databases"

of the processing nodes. Distributed problem solving is a basic method of attack for solving problems which are too complex to be solved by a single processor ([2]) where each specialist contains its own knowledge base and corresponding inference mechanism. Certain applications might naturally be modelled in terms of a collection of personal knowledge bases, in which each knowledge base must he allowed to have (partial) access to the other knowledge bases. Also, for a very large knowledge base, one single machine possibly is not performant enough to handle one global application efficiently. Distributing different parts of such an application on different nodes in a network so that the subapplications are allocated to the nodes where they are used most frequently, decreases the load of the involved processing nodes. Thus the requests of most users may be processed more quickly, although additional communication is needed in cases when subapplications have to interact.

These topics have been addressed in the work on the specification, design and implementation of the distributed EPSILON prototype ([IS], [20], [21], [22], [23]). The distributed prototype is essentially based on the EPSILON single user prototype, which will be described first.

2. The EPSILON Single User Prototype

The single user EPSILON KBMS prototype is built on top of a commercial Prolog and an available database management system (DBMS), running on standard UNIX environments. The main concepts underlying the EPSILON approach are ([4], [13]):
- the extension of Prolog with *theories* (objects I multiple worlds),
- the specification of relationships between theories using *links*,
- the definition of a transparent interface from Prolog to relational *DBMSs*,
- the use of *metaprogramming* as a basic technique to define new inference engines and tools,
- the use of *partial evaluation* as a systematic method to "compile" metaprograms.

The basic component of an EPSILON knowledge base is a *theory* ([1]). Theories are similar to worlds in MULTILOG ([9],[10]) and to unit worlds and instances in MANDALA ([7]). They are considered as objects ([17]) and a knowledge base (KB) is a collection of theories together with relations on theories defined by *links*. Each theory is accompanied by an inference engine (*theory processor*), which can itself be a meta-level theory, providing operations to query, update and search a class of theories. The knowledge component in a theory can be described by different knowledge representation techniques, e.g. database tuples, relational algebra expressions, various logic languages. The logic languages are extensions or restrictions of Prolog. A given logic language is defined by a theory processor that implements the corresponding knowledge representation features and the inference control mechanism.

Theories have two important attributes: their 'class' and their 'type'. All theory processors are of type *engine*, while all other theories are of type *object*. Every theory processor defines a *class* of theories. All theories of one class have the same theory processor. The definition of the theory processor also includes the declaration of a set of tools available for all theories of this class. Using existing classes and links, the user may compose new classes without modifying the existing ones. Inference engines defined by metaprograms in turn need another inference engine. The inference engine of the first theories in the hierarchy of meta-interpreters is Prolog extended with structuring of the knowledge base (theory handling primitives), which corresponds to the primitive *class kernel*. The *class database* and the *class deductive database* are two more primitive classes.

Relationships among theories are specified by the definition of different types of *links*. The treatment of links is embedded in the theory processors. Some predefined types of links are handled by the system. The user can define his own types of links by writing corresponding theory processors. Examples of how links can be defined and handled inside theory processors are the predefined *classes inheritance* and *constraints*. A *constraint link* between theories T1 and T2 defines in T1 the integrity constraints of T2. The four *inheritance links* (*closed_is_a*, *open-is_a*, *closed_consultance*, *open_consultance*) define different kinds of inheritance between theories. An *is_a* link defines inheritance of clauses between two theories belonging to the same class. A *consultance* link between theories T1 and T2 means that the theory processor of T2 is consulted for solving a goal of T1. The two theories can belong to different classes. An inheritance link between two theories T1 and T2 is *closed* if T2 is involved in the solution of a goal only when the queried predicate is undefined in T 1. An inheritance link between two theories T1 and T2 is *open* if each goal is first solved in T1 only, then using T2, too.

If there exists a *dict link* between the theories Tl and T2, Tl is the data dictionary of T2. In the case of a *version link*, T2 is the result of the application of a tool (e.g. partial evaluator) to the theory T1.

Information about the structure of the KB (theories, their classes and types, links, tools) are stored in the *knowledge base dictionary* (KBD). The operations for handling the KBD and the primitive classes (kernel, database, deductive database) are included in the *EPSILON Kernel*.

In summary, the EPSILON architecture is essentially object-oriented. Its main feature is uniformity. All objects in the architecture (inference engines, tools, object level theories in different logic languages, data bases, integrity constraints, results of optimization) are represented in a uniform way by the theory feature. The link feature allows multiple possibilities of defining relationships between theories, including the definition of new inference engines in terms of existing ones (composition of interpreters).

3. The distributed environment

The aim of the distributed EPSILON KSMS approach is not the distribution of the management system, but distribution of the knowledge base: an integrated application, corresponding to one knowledge base defined as a collection of theories and links between theories, is decomposed into a set of sub-knowledge bases, which are allocated to different nodes in a network. By cooperation of EPSILON instances on different nodes, every node has access to all the knowledge stored an every other node, including access to different databases (fig. 1).

The actual allocation of sub-knowledge bases and databases is fully transparent to users. They only see one integrated application and do not have to care about details of knowledge base management or distributed request evaluation. These tasks are carried out by the system, using information, stored in the distributed KBD (DKBD), about locations of theories in the network and about links defined between theories. For safety and efficiency reasons the DKBD is redundantly stored on every node, avoiding high network communication overhead, caused otherwise by frequent remote DKBD access.

The system can handle theories, that are stored redundantly on different nodes, as well as theories existing only once in the network. We assume, that metalevel theories (implementing theory processors) are redundantly stored on every node where theories of the corresponding class are stored, and that object level theories are stored on a single node each. Redundant allocation of theory processors speeds up request execution, without involving much communication overhead for updates, since inference engines are not likely to be changed frequently ([15], [21], [22]).

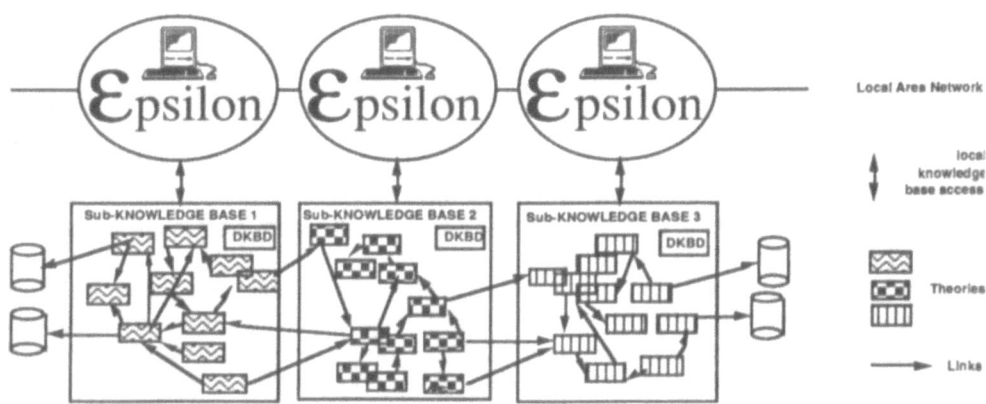

Figure 1. Distributed EPSILON Environment

Distributed problem solving is regarded as the cooperative solution of requests to the knowledge base by a decentralized and loosely coupled collection of knowledge sources, located on distinct nodes in a LAN. Cooperation is required because no node has sufficient information to solve an entire problem. loosely coupled means that individual knowledge sources spend the greatest percentage of their time in computation rather than in communication. Decentralization is related to the design decision that both control and knowledge are distributed.

In order to let problem solvers cooperate an intranode and an internode communication facility, based on a network independent communication protocol, is needed. Particular attention has been paid to control of internode communication In order to avoid reliability and efficiency problems, no node plays a central role by providing special processing capabilities (e.g. DB-server) and there is no global control of network communication or distributed request evaluation. Based on the redundant allocation of the DICBD, control of communication and processing is distributed within the network, each node beeing able to execute requests, as well as to send requests to other nodes.

Distribution of the knowledge base allows parallel processing on different nodes to take place. The overall performance of the distributed EPSILON depends on mechanisms to maximize parallelism (parallel inferencing, parallel fact retrieval) and to minimize communication. Maximizing parallelism may lead to an excessive communication among processors, which results in a saturation of the available communication channels, so that the nodes are forced to remain idle while messages are transmitted. Minimizing communication may lead to minimizing parallel processing. Therefore, we had to compromise on these two extremes.

When talking about parallel processing, we have to distinguish between two levels of parallelism:
- parallel execution of several *independent requests* originated by different users on different nodes and
- execution of one single request by parallel execution of *subrequests* (AND/OR-parallelism).

These two different levels of parallelism are treated at different levels of the system: The management of different users is done by creating one problem solver for every user (cf 5.), while the treatment of AND/OR-parallelism corresponds to an extension of the evaluation strategy of the problem solvers. The current implementation exploits parallelism by executing different independent requests from different users on different nodes concurrently. An optimization could be achieved in providing parallelism related to the execution of one request. To do this, while preserving transparency of distribution, we need
- a compiler, able to identify parallel executable subrequests, to judge if the speedup resulting from their parallel evaluation is likely to outweigh additional communication costs and to generate an efficient evaluation plan and
- a multi-threaded runtime system, able to synchronize the evaluation of subrequests and the integration of results into running evaluations.

Design and implementation of an optimizing compiler and a multi-threaded prolog runtime system, is beyond the scope of the project and is left to further work on distributed and parallel problem solving. One basis could be the extention of EPSILON with concepts from [8] and [13].

Exploitation of potential parallelism is restricted by the allocation of knowledge to processing nodes of the network. The theory and link concept in EPSILON is well suited for a distributed environment, providing a knowledge base structuring mechanism which has been exploited for decomposition and distribution of knowledge bases. Strategies have been developed, to find allocations of sub-knowledge bases on different nodes, which use efficiently the resources of the network (multiple processors, storage). Especially the tradeoff between communication and parallelism has been considered. The decomposition and allocation strategies are implemented as an additional tool, the Knowledge Base Distribution Manager (KBDM), described in [23].

4. Overall Architecture of Distributed EPSILON

We will illustrate the development of the distributed EPSILON prototype (DEP) architecture (fig. 2.b.) as an extension of the single user prototype (SUP) architecture (fig. I.a.).

The SUP is running on a UNIX workstation. The User Interface (UI) is defined on a Macintosh. It provides knowledge editing facilities and takes care of graphical operations without affecting the performance of the host UNIX machine. The Knowledge Base Manager (KBM) handles the interaction between the UI and the KS. The Knowledge Base Dictionary (KBD) contains information about the structure of the KS (theories and links). The Kernel contains the operations for handling the KSD and the primitive classes, including access to relational databases ([19]) via the Data Base Interface (DBIF).

In the distributed environment the KBD has been extended to the *distributed KDB* (DKBD) by adding meta-knowledge about the location of theories in the network and about the possible interaction of theories, as specified by links. For efficient distribution of knowledge bases and for optimization of query evaluation in the distributed environment, we extended the predefined inheritance links by *import-lists*. The import-list of a link from theory T1 to theory T2 contains those predicates of T2 which may be used by T1. It allows to determine, without (remote) access to T2, whether T2 really contains a predicate definition, that can be used to solve a subgoal in T1. The import-list can also be regarded as the view of theory T1 to theory T2.

The Knowledge Base Manager of the SUP has been replaced by
- the Knowledge Base Distribution Manager (KBDM),
  which allows to decompose an application knowledge base (KB) into a set of sub-KBs, to allocate them on different nodes and to update accordingly the DKBD on every node,
- the Process Communication and Request Manager (PCRM),
  which controls the communication between local and remote problem solvers and determines which one is in charge of evaluating a request (cf 8.),
- the Local Area Network Interface (LANIF),
  which provides network independent communication by message passing between PCRMs on different nodes.

The basic inference engine of the SUP, the KERNEL, has been extended by distributed processing facilities, yealding *KERNEL* +, the kernel of distributed EPSILON (cf 6., 7.).

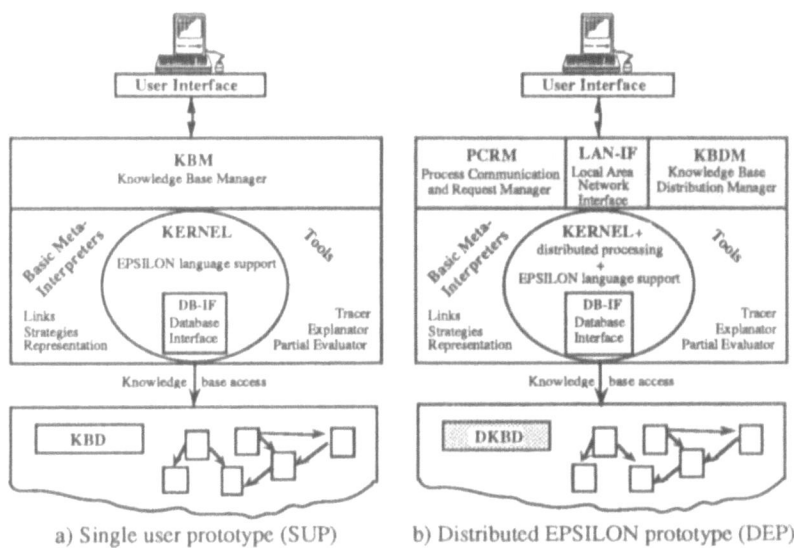

a) Single user prototype (SUP)   b) Distributed EPSILON prototype (DEP)

Figure 2. Architecture of SUP and DEP

5. Process Structure of Distributed EPSILON

Knowing how the distributed EPSILON looks at the level of system components, we still had to decide how these components should be mapped to UNIX processes. While efficiency considerations suggest to have as few processes as possible, the system is more reliable, if the overall control is separated from the problem solving facilities. If a problem solver loops or breaks down, the rest of the system (on the same node and on other nodes) can still continue its work, as long as the global control resides in another process. Therefore our approach distinguishes between two different classes of processes: *problem solving processes (PSPs)*, which produce answers to requests and *communication processes (CPs)*, responsible for passing requests and answers among the problem solving processes (fig. 3).

Communication processes provide the basis for the cooperation of problem solving processes. Their tasks cover all aspects of interaction among different local and remote processes, including the interaction with the user (e.g. initialization of the system, creation and deletion of problem solving processes, message passing). The EPSILON communication processes are: the EPSILON-Demon process, the User Interface process and the PCRM-process.

On each EPSILON node an *EPSILON-Demon process*, is permanently running. Its task is to initialize the system, when the first request is received (from a local user or from another node). Initialisation of the system is done by starting the local *PCRM-Process*[*], containing the PCRM, the LAN Interface and the KBDM. The

---

[*] In the following we will simply say PCRM instead of PCRM-process, if the distinction is not relevant.

PCRM is responsible for the management of local PSI's, for the decision which PSP is in charge of evaluating a request and for the control of communication between local and remote processes. In the current state of implementation, the *User Interface Process* just simulates some functions of the SUP user interface. The complete integration of the advanced graphical user interface of the SUP into the distributed environment will be one of the next working steps.

A *Problem Solving Process* consists of KERNEL+ together with the defined metainterpreters and tools. On every node accessed during the evaluation of a user request, there is one PSP assigned to the corresponding user. Every request and every remote subrequest of the same user is mapped by the PCRM to the same problem solving process ('one PSP per user approach'). For different users, different problem solving processes are created by the PCRM. The assignment of different users to different PSPs is mainly due to reliability considerations. It also solves the problem of fair timesharing for users, since timesharing of different processes is done by the operating system.

The process structure on one EPSILON node is illustrated by fig. 3, which shows how the system components (fig 2.b) are mapped to processes and how the processes interact The thick arrows indicate (local and remote) process communication via message passing, while the dashed arrows represent (local or remote) connection requests to the demon process, for starting the PCRM.

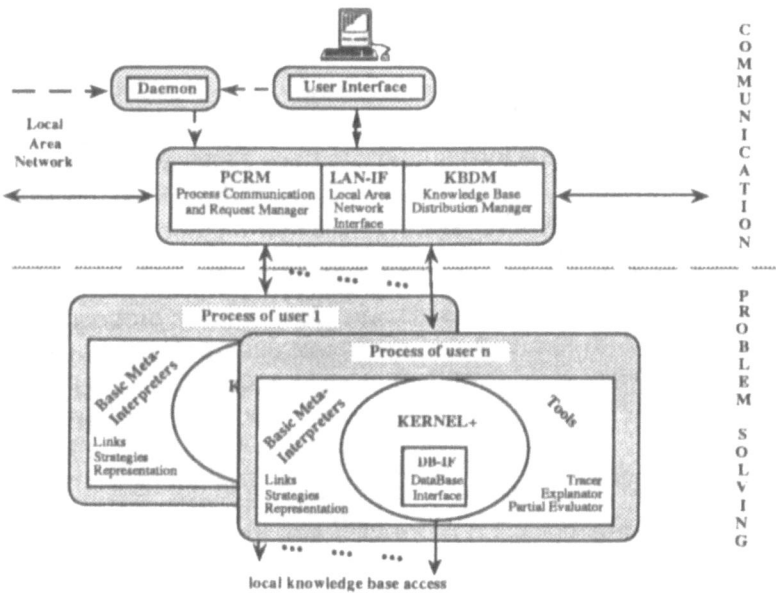

Figure 3. Processes in EPSILON

6. Cooperative Distributed Problem Solving

Summarizing our main aims, we wanted to devise a concept for cooperative distributed problem solving
- where no restriction of the allocation of knowledge in the network is required,
- every node has full access to the distributed knowledge and
- control is distributed among the processing nodes.

In order to discuss the problems arising from these aims in the design of the distributed EPSILON, we first need to introduce some notions used in this context.

A problem solving process may receive request or result messages. A request message may be a request from a user ('user_call') or a request from a user asking for more results to its previous request ('redo') or a request from another process ('process_call'). For a request which has been executed successfully a 'return' message is sent, if result values are available, and 'succeed' otherwise. If execution of a request failed a 'fail' message is sent.

A problem solver that has no request to evaluate is *idle*. A request received by an idle problem solver is said to be the initial request. An *initial request* may be a 'user_call' or a 'process_call'. If, during the evaluation of a request, a subrequest was sent to another node and no result has been received yet, the current request is *pending* and the problem solver is *waiting*. A request received by a waiting problem solver is said to be a *new request*. Because every problem solving process is responsible only for one user, a new request is always a subrequest of a pending request As a consequence, a new request must always be solved first; otherwise no results that allow the continuation of the pending request will be produced, and the process will block. A new request is received by a process P, if the evaluation of one of its remote subrequests can not be carried out completely at the addressed node, because it requires access to knowledge stored at the node of process P (fig. 4.a). Therefore new requests can only be 'process_call's'.

We consider an EPSILON knowledge base as a directed graph, with nodes representing theories and edges representing links. If, after distribution of the KS on different machines in a network, there is no path between two (not necessarily distinct) theories stored on the same node, which traverses at least one theory stored on another node, we say that the KB is *hierarchically distributed*. As illustrated by the example in fig. 4.b., new requests cannot occur in a hierarchically distributed knowledge base.

Figure 4.a. shows the same KS, but with theory 4 allocated to node I. In this case a remote subrequest from theory 1 to theory 2 may well lead to a new request to node I.

Requiring that the KS should be Hierarchically distributed ([14]) is a strong restriction, which we do not impose. Allowing non-hierarchical distribution implied the necessity to enable the execution of a new request while waiting for results to a previous subrequest.

In order to avoid blocking when a new request is received, every problem solver must be able to represent and manipulate 'computation states' of requests. Before

186

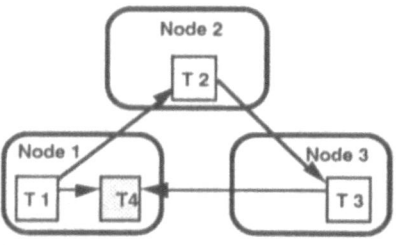
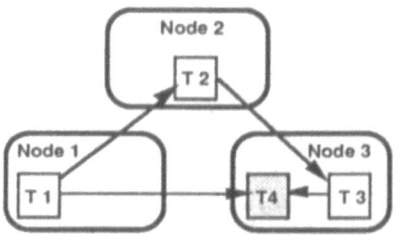

Figs 4.a Non-hierarchical allocation   Fig 4.b. Hierarchical allocation

starting the evaluation of the new request, the computation state of the current pending request must be saved. After processing of the new request, the evaluation may only be continued, if it is possible to restore the computation state of the pending request

As EPSILON is based on prolog, the most general solution would have been the implementation of a multi-threaded prolog system. But we had neither the source code of an existing prolog, to which to add new features, nor enough manpower to implement a completely new prolog system. Therefore we took another approach, which has the advantage of making the distributed EPSILON independent from special prolog implementations.

Our approach exploits the representation of computation states of requests that is implicitly offered by every prolog system. The representation of an SLD derivation in the internal stacks of prolog corresponds to the representation of the computation state of a request at the level of EPSILON. Since prolog is able to evaluate subgoals without affecting the computation state of parent goals, it implicitly offers a restricted 'state saving and restoring capability'. Restricted in this context means, that it is only possible to continue the evaluation of a pending request whose 'state' is accessible as the top of the prolog stack. To discuss the implication of this restriction, let us have a look at another basic design problem, the treatment of backtracking in the distributed environment

We talk about *distributed backtracking* if 'redo' messages are used explicitly at the level of network communication in order to generate further results for a remote subrequest. In the case of distributed backtracking, the process in charge of evaluating the request (the server process), must maintain the computation state of this request, until the process which has sent the request (the client process) does not need further solutions.

If we rely on the state saving mechanism of prolog, the combination of new requests with distributed backtracking can again lead to blocking of processes. Consider the example from fig. 4.a. m which an initial request to theory I can lead to a new request to theory 4. Suppose that the state of the new request must be maintained after returning a result. This result will enable theory 3 and theory 2 to also produce results. When the result from theory 2 arrives, it can not be used to continue the pending evaluation of theory 1, whose state is hidden in the prolog stack below the state of theory 4.

Therefore we decided to avoid distributed backtracking, by generating all solutions for a remote request (like in [20]) and transfering all results in one message. As long as we work sequentially, using the one PSP per user approach and all solution computations the results which are received by a waiting process are always results to the subrequest which is accessible at stack top, so the pending evaluation can be continued. Received results are inserted into a queue of the client process and, if needed, backtracking is done locally by using the results from the queue.

As we will describe in the next chapter, our approach can easily be implemented on top of any commercially available prolog system. The additional runtime overhead is minimal: our measurements showed, that in the case of local evaluations, the distributed prototype runs as efficiently as the single user prototype.

7. Implementation of Distributed Problem Solving

Cooperative distributed problem solving is the main feature of a distributed knowledge base system. In EPSILON it has been implemented as an extension of the kernel, resulting in *KERNEL +*, the kernel of distributed EPSILON. + consists of the EPSILON single user kernel[*], which has been slightly extended in order to be integrated with two completely new components, the internal communication manager (ICM) and the distributed processing interpreter (DIP), the main component of KERNEL + (fig. 5).

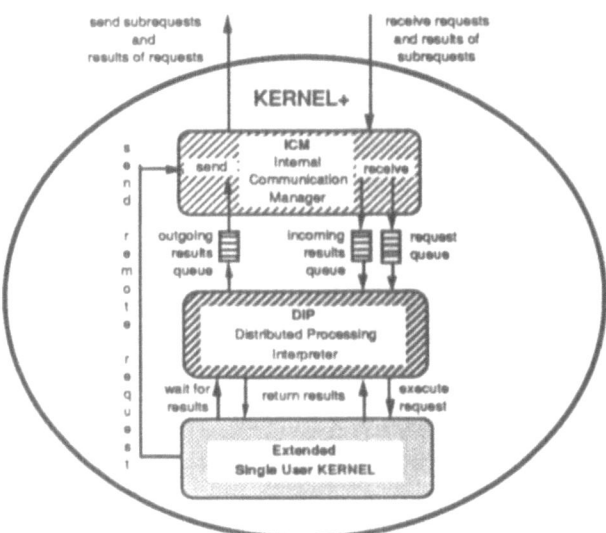

Figure 5. Components of KERNEL+ and their interaction

---

[*] In the sequel "kernel" and "extended kernel" always refer to the single user kernel, while "KERNEL+" is the kernel of the distributed prototype.

The ICM is the component that connects the + process to the PCRM and is responsible for decoding and for transforming received messages into a representation suitable for the DIP interpreter. It distinguishes requests and results and inserts them into different queues ('request queue' and 'incoming results queue'). When sending messages, the ICM encodes the internal representation into a string which is sent to the PCRM. Values to be sent in result messages are taken from the 'outgoing results queue', where they have been inserted before by the DIP.

For the management of messages in the distributed environment the ICM uses the identifier of the node on which',the process is running, the identifier of the process itself and a subrequest number, as an internal identifier for (sub)requests sent to other processes. The node identifier, together with the process identifier and the subrequest number form *message identifiers* which are unique in the network (as opposed to subrequest numbers, which are unique only within one process). A message identifier is created for every subrequest message. The message identifier of other messages (e.g. result messages) is always the identifier of the corresponding request message. Message identifiers are also used as indices for the internal queues. Every entry in a queue is associated with the corresponding message identifier. By their common message identifier, all entries in a queue, which correspond to one message, are grouped together (e.g. all results of one request).

The *DIP interpreter* is responsible for the non-blocking execution of requests from users or from problem solving processes and for the correct integration of results into pending evaluations. DIP interpretes received messages by initiating and controlling actions of the extended single user kernel. When the kernel interpreter finishes its work (failing or generating one result), the DIP inserts a corresponding entry into the 'outgoing results queue'. If needed, the DIP can force the kernel to backtrack and to generate further results. Then DIP calls the ICM to encode the contents of the 'outgoing results queue' (one or more entries) as a message and to send it (via the to the request originating process. Afterwards the DIP interpreter returns to its initial state in which it can receive and treat further initial requests.

When a KERNEL+ process receives a request from a User Interface process, it generates only one answer at a time. Only if the user wants more results (a 'redo' message is then sent by the UI), then the DIP initiates the continuation of the current request evaluation by forcing the extended kernel to backtrack. If a request from a (remote) problem solving process has to be executed, all results are produced immediately and returned in one message.

Generation of all results for *initial* 'process_calls' is not mandatory (in opposite to process_calls' that are received when a KERNEL+ process is waiting for results). It has been included in the current implementation in order to minimize the interprocess communication overhead. By changing the setting of a flag the user can enforce a tuple-oriented evaluation of all types of initial requests. We used this flag to measure on various distributed examples the influence of communication costs versus processing costs. In the case of tuple-oriented processing no processing time is spent generating useless results, but the number of messages is two times the number of needed results ('call'/'return', 'redo'/'return',...,'redo'/'fail'). When computing all solutions at once, only two messages ('call'/'return') are necessary irrespective of the number of results, but possibly some of the results are not needed. Our measurements showed, that the almost constant communication costs of the all-solutions approach outweigh the computation of a lot of possibly useless results, making the all-solutions approach very efficient in most cases. The only open problem is the treatment of goals which lead to non-terminating refutations, if all solutions are requested and the SLD-(Prolog-)strategy with a fixed (left-to-right) computation rule is used.

In order to manage requests to a theory in the distributed EPSILON, the *extended kernel* has to analyze if the addressed theory is local or remote. This is possible using the information of the DKBD (about the allocation of theories) and dynamic information about the identity of the local node. If the theory is local, the extended kernel will go on as in the SUP. If it is remote, the ICM is called to encode the request and send it to the CPRM, addressing the appropriate node in the network. Then the extended kernel calls the DIP, to receive and treat further messages.

When a result message is received, one or more entries, corresponding to the generated result(s), are inserted in the 'incoming results queue'. The queue is accessed by the DIP and one result is given (via unification) to the kernel interpreter, which may now continue the pending evaluation. When the kernel interpreter backtracks while continuing the evaluation, the DIP delivers further results from the queue. If no more results are available or if the only result was 'fail', the DIP initiates backtracking of the kernel interpreter beyond the point where the remote subrequest was sent.

When a new request is received, it has to be evaluated completely, in order to avoid blocking. Therefore, all results are produced immediately and returned to the requesting process in one single message. Then DIP calls itself recursively, to continue waiting for the expected results.

8. Communication

In a distributed EPSILON environment processes have to communicate in order to be able to cooperate. We implemented local and network communication by using stream sockets based on the TCP/IP protocol. Stream sockets guarantee reliable and efficient message passing between the communicating processes. They offer a uniform interface to local communication as well as to network communication and provide the most flexible process communication facilities in the distributed environment, allowing communication between several unrelated processes. Having the communication based on sockets has the advantage that the process structure can easily be changed /extended.

Communication management in the distributed environment is done by the PCRM, which is responsible for the management of local processes, for decisions which process is in charge of evaluating a request and for the control of communication between local and remote processes. In order to implement the 'one problem

solving process per user approach', the PCRM must
- generate a new problem solving process (PSP) for each user,
- guarantee a correct assignment of request messages to the PSP of the related user,
- guarantee a correct assignment of results to the request originating process.

For the assignment of PSPs to users and of requests to PSPs, the PCRM stores in a table the identifiers of the local PSPs, the corresponding user identifiers and the identifiers of the requests addressed to the PSPs. In this table each PSP is identified by its 'process_id'. Users are identified by the 'process_id' of the related user interface process and the 'node_id' of the node where the UI is running. Requests are identified by a unique 'request_id', which is always assigned to a request by the client process. Every result generated by the server process is accompanied by the related request identifier, allowing the PCRM a correct assignment of incoming results to the waiting client process. A request identifier is composed of the identifier of the client process, the machine where the process is running and the internal number of the request (cf. 7. ICM).

In the following example, the local problem solving process
- 'PSP1' works for the user 'user1' on node 'node1' and evaluates his first request.
- 'PSP2' evaluates a request of a remote PSP (from node 'node3'); both processes work for the user 'user1' on node 'node2'.
- 'PSP3' works for the user 'Use2' on node 'node3'. The evaluation of the initial request (the second request of the user) is pending, while a new request (the fifth request of process 'PSP2' on node 'node2') is executed.

Process	User	List of requests (current req, ..., old req)
psp1	(node1,user1)	(node1,user1,1)
psp2	(node2,user1)	(node3,psp1,5)
psp3	(node3, use2)	(node2, psp2,5), (node3, use2, 2)

Figure 6. Internal table of a PCRM process.

When a request is sent to another node, the PCPM extends the message by the user identifier of the sending process. The receiving PCPM can decide whether to generate a new PSP, by searching the internal table for a process with the same user identifier. If such a process exists, the request is sent to the process. Otherwise a new process is first created and assigned to the user whose identifier is contained in the message.

Detailed descriptions of the design and implementation of the PCPM can be found in [22] and [23].

CONCLUSIONS

We have presented an architecture for the extension of an advanced knowledge base management system by a transparent interface to distributed knowledge bases, allowing decentralized cooperation of different inference engines within a local area network. The architecture makes a clear distinction between cooperative distributed problem solving and the process communication facilities, needed for its support. The separation of problem solving from communication management increases the reliability of the system and allows to use a flexible and easily extensible process communication model. By extensions of the PCRM process, it will be possible to integrate also non-EPSILON problem solving processes into our framework, corresponding to the integration of existing data base and expert systems with heterogeneous representation formalisms.

Further we described how cooperative distributed problem solving, can be easily implemented on top of a commercially available prolog system. Our solution avoids the problems of distributed backtracking, although it does not depend on special requirements like redundant ([18]) or hierarchical ([14]) allocation of knowledge within the network

The current prototype is a solid basis for the investigation of further problems arising in distributed environments. Here are just some open questions: Is it possible to reduce the number of processes and/or to simplify the communication model, without restricting the functionality of the system? What has to be done, in order to efficiently support consistent updates of redundantly stored knowledge? Can a general purpose protocol be defined for the integration of any other system into our communication framework? What is a good model for exploiting (large grain) parallelism and how can queries generating infinite sets of answers be treated in a distributed environment?

ACKNOWLEDGEMENTS

The results of the EPSILON project are very much a team effort and include work investigated by all partners (Systems & Management, University of Pisa, C.R.I.S.S., University C. Bernard of Lyon, Bense KG, University of Dortmund), presented in greater detail in many other project documents. We thank all our co-workers for their efforts which have made this a fruitful project. Special thanks are due to all the members of the EPSILON team who contributed to the specification, design and implementation of the distributed EPSILON environment, especially to Mechthild Rohen, Harald Hönig and Ralf Schlüter.

REFERENCES

[I] Bowen, K.A.; Kowalski, R.A. (1982), 'Amalgamating Language and Metalanguage in Logic Programming' in Clark,K.L.; Tärnlund,S.A. (eds.), *Logic Programming*, Academic Press, pp. 153-172

[2] Chandrasekaran, B. (1983), 'Expert systems: Matching techniques to tasks', New York

192

University Symposium on AI Applications for Business.

[4] Coscia, P.; Djennaoui, S.; Franceschi, P.; Kouloumdjian, J.; Levi, G.; Lei, L.; Moll, G.H.; De Saint Victor, I.; Sardu, G.; Simonelli, C.; Torre, L. (1988), 'The EPSILON KBMS: Architecture and DB access optimization', Workshop on Integration of Logic Programming and Data Bases, Venice.

[5] Cromarthy, A.S. (1986), 'Control of Process by Communication over Ports as a Paradigm for Distributed Knowledge-Based System Design', in Kershberg, L. (ed.), *Proc. of First Conference on Expert Data Base Systems*, Charleston, South Carolina, pp. 47-59.

[6] Durfee, E.H.; Lesser, V.R.; Corkill, D.D.(1987), 'Coherent Cooperation among Communicating Problem Solvers', *IEEE Transact. on Computer*, Vol. C 36, No.II, pp. 1275-1291.

[7] Furakawa, K.; Takeuchi, A.; Kunifuji, 5.; Yasukawa, H.; Ohki, M. ;Ueda, K. (1984), 'MANDALA: A logic based knowledge programming system', in *Proceedings of International Conf. on 5th Generation Computer Systems, pp. 613-622.*

[8] Hönig, H. (1988), 'Parallel request evaluation in a distributed KBMS', Master's thesis at the University of Dortmund (in German).

[9] Kauffmann, H.; Grumbach, A (1986), 'Representing and Manipulating Knowledge within Worlds', in Kershberg, L. (ed.), Proc. of First Conference on Expert Data Base Systems, Charleston, South Carolina, pp. 61-73.

[10] Kauffmann, H.; Grumbach, A (1986), 'MULTILOG: Multiple Worlds in Logic Programming', Report 1986.

[11] Kniesel, G. (1988), 'Compilation of logic programs for parallel set-oriented evaluation In a distributed KBMS', Master's thesis at the University of Dortmund (in German).

[12] Lambird, B.A.; Lavine, D.; Kanal, L.N. (1984), 'Distributed architecture and parallel non-directional search for knowledge based cartographic feature extraction systems', in Coombs, M.I.(ed.), *Developments in expert systems*, Academic Press, pp. 221-234.

[13] Levi, G.; Modesti, M.; Kouloumdjian, J. (1987), 'Status and Evolution of the EPSILON System', in *Proc. of the 4th Annual ESPRIT Conference*, North Holland, pp. 593-610.

[14] Li, Y.P. (1986,1987), 'DKM - A Distributed Knowledge Representation Framework', in Kershberg, L. (ed.), *Proc. of First Conference on Expert Data Base Systems*, 1986, Charleston, South Carolina, pp.143-I52 and in *Journal of Logic Programming*, Volume 5,1987.

[15] Rohen, M.; Kniesel, G.; Cremers, A.B.; Bense, H.(1988), 'Specification of an Architecture for Distributed Problem Solving in EPSILON', in: *Proceedings of the 5th annual ESPRIT Conference 1988*, North Holland, pp. 659-673.

[16] Smith, R.G.(1980) 'The Contract Net Protocol: High-Level Communication and Control in a Distributed Problem Solver', in *IEEE Transactions on Comp.*, Vol. C-29, No. 12, pp. 1104-1113.

[17] Sterling, L.(1984), 'Expert System=Knowledge+Meta-Interpreter', Techn. Rep. CS84-17, Weizman Institute of Science, Rehovot, Israel.

[18] Warren, D.S.; Ahamad, M.; Debray, S.K.; Kalè, L.V. (1984), 'Executing Distributed Prolog Programs on a Broadcast Network',
in *IEEE Transactions on Computers*, Vol. 12, No. 3, pp. 12-21.

[19] Kouloumdjian, J.; De Saint Victor, I.; Lei, L.; Djennaoui, 5.; Moll, G.H.(1986) 'The

Communication Processor: Kernel - DBMS Communication', ESPRIT project 530, Report IS (initial phase).

[20] Kniesel, G.; Lemke, T.; Hönig, H.; Schöfer, P.; Moll, E.:(1987) 'Preliminary Study on Architectural choices for a Distributed KBMS based on the EPSILON Theory concept' in ESPRIT project 5306 Month Report, May 1987.

[21] Rohen, M.; Kniesel, G. (1987), 'Specification of the Distributed EPSILON Architecture', ESPRIT project 530 Report 13 (initial phase).

[22] Rohen, M.; Kniesel, G.; Schlüter, R.; Hönig, H.; Moll, E. (1988) 'Design of the Distributed EPSILON KBMS', ESPRIT project 530 Report 2 (continuation phase).

[23] Kniesel, G.; Rohen, M.; Höffgen, K.-U.; Schlüter, R.; Waschkowski, R. (1990) 'Implementation of the Distributed EPSILON Prototype', ESPRIT project 530 Report 10 (continuation phase).

Project No. 865

Producing Process Plans out of CAD Files through AI Techniques

Norberto Iudica
Battelle Institut e.V.
Am Römerhof 35
D-6000 Frankfurt am Main 90, FRG

Silvia Ansaldi
Italcad tecnologie e sistemi S.p.A.
Via Ravasco, 10
I-16128 Genova, Italy

Bruno Tranchero
Aeritalia GAD
Corso Marche 41
I-10146 Torino, Italy

Luisa Boato
Elsag S.p.A.
Via G. Puccini, 2
I-16154 Genova-Sestri, Italy

Abstract

MUMP (MUlti-Methods Planner) is a tool developed in ESPRIT P865 for building expert process planning systems. MUMP allows the combination of variant and generative approaches to process planning and recognizes the form features on the workpiece out of its description in a CAD system. The experience made with the application of the system to the case of aircraft manufacturing has influenced the choice and functionality of the techniques and tools made available in MUMP. We give a general view of the system and expalin the rationale behind it and report on the experience gained in building the current experimental version.

1. Introduction

The goal of a process planning system is to generate automatically lists of individual manufacturing operations with all associated information necessary to produce a part in a given manufacturing facility.

During the last ten years, several computer-aided process planning prototypes have been developed. These systems can be classified according to two different approaches: *group technology* or *variant planning* and *generative planning* [1]. The first one uses a data retrieval system based on the similarity between parts to retrieve existing process plans on which modifications to satisfy peculiar requirements are performed. While in the generative process planning approach, process plans are generated automatically for every new part without referring to existing plans.

Advanced Information Processing (AIP) techniques, and in particular, knowledge based systems, offer a solution to this kind of problems since they allow to express the heuristics commonly used by the experts. Furthemore, the structure of knowledge-based systems allows for easy updating of the planning procedures when changes in manufacturing facilihes and processes arise.

Many of the existent computer-aided process planning systems have been developed as autonomous packages in which knowledge of manufacturing is captured and encoded into efficient software [2, 3, 4, 5]. Though these systems have offered good solutions to specific process planning problems, they rarely foresee functional integration with other automation packages. Moreover, whether the input data are required by the system during an interactive session or are specified into an input file, the input definition phase is often time-expensive and tedious.

The integration of Advanced Information Processing and Computer-Aided Design (CAD) techniques can eliminate this drawback, but several factors limit the successful use of CAD models for process plan generation.

One of the major limitations is the inadequacy of the part description in 3-D CAD models for direct application to process planning. In fact some information necessary to manufacture a part is not explicitly stored in a solid modeler's data base (e.g. form features, dimensions). On the other hand, the planners express their knowledge making reference to the form features present in the part, to their shape and technological characteristics, which have no explicit counterpart in the topological and geometric data stored in a CAD model.

MUMP (MUlti-Methods Planner), the system we present here, is able to recognize form features from a CAD solid model and to automatically produce different process plans for machining a part. Knowledge processing techniques are involved in the interpretation of the model and in the automatic production of real-life process plans.

The approach followed in our application is generative, but some interesting ideas from variant planning are used. The application environment is a complex aeronautic production plant.

2. Motivation

Process plan definition is one of the most important activities in the factories that operate in the mechanical field and is often considered a bottleneck in automated manufacturing. A process plan is defined by appropriate working processes, the involved machines, the necessary operations, depending on the morphological features of the part, on tolerances, and other information like lot size and frequency.

The process planning domain is dynamic and the world model to which it refers is quite complex. The expert involved in process planning definition has to take into account:

– Part form features
– Factory situation:
    - Machine characteristics
    - Machine availability
    - Kind of material of the part
    - Tools to be employed

- Available fixtures
- Manufacturing methodologies and techniques used in the factory.

Starting from these considerations it is easy to understand why Aeritalia is interested to realise a tool able to automatically define the process plan.
The fact of having such tool available would mean:

- to perform automatically a time consuming activity
- to maintain knowledge that is present in the factory but that is scattered and subject to be lost
- to have a tool that could be used also for training.

The main Aeritalia activities in this project are :

- Providing the know-how for the application field
- Representing the knowledge base in the MUMP formalisms using the most suitable techniques
- Defining the objects (part, features, machines, etc.) involved in the process planning.

Another important task is to test the performance of the available development tools, if necessary, criticize them and suggest improvments in order to make the system more flexible, powerful and user friendly.

3. CAD Interface

Solid models are the central element in Computer Aided Design systems. They are representations which belong to the class of complete, volumetric methods based on the specification of a 3D object using either surface patches, sweeping rules or volume primitives. But "pure" solid modelers cannot be used to drive applications such as process planning or manufacturability evaluation, because some information required for these tasks (e.g. dimensions and tolerances, form features, materials) is totally absent from the solid modeler data base. In addition, the entities on which process planning is based, such as form features, require a higher level of abstraction than that which is available [6].

3.1. Form Features

The importance of feature recognition starts from the fact that each feature can be associated with knowledge about how the feature can be manufactured, and this information can be collectively used to generate a process plan.
The definition of feature depends on the context. Features that are required for machining may differ considerabily from those required for forging and so on. Features of a geometric model are relevant within a given context and highly dependent on what the model is used for.

We have adopted the following definition: "a feature is a group of geometric and topological entities with functional meaning in the process planning context" [6]. Whereas by feature recognition we mean the recognition of expected patterns of geometry, corresponding to particular engineering functionality in object parts. In fact, a form feature can be expressed on the basis of two characteristics: one depending on the application and the other on the model. Since a feature is a part of the model, its shape can be completely described by its topological and geometric information. Futhermore, each feature has a specific meaning, which depends on a set of rules characterizing the functionality of the feature itself.

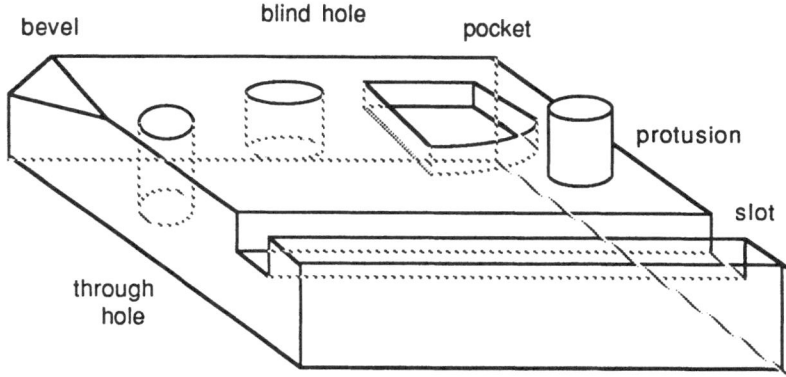

Figure 3-1 : Examples of features

Some of the information can be extracted automatically from the geometric model of the object, such as the raw material to be removed from the starting stock, the machined faces of the object, the kind of the features present and the precedence between features based on their geometry.

So, our first approach to classify form feature was based on the examination of a list of features typical of aircraft manufacturing on the basis of their geometry, topology and functionality. The features, we consider in our work, are grouped into the following classes: plates, bevels, pockets, fillets, holes, slots, blind holes, protusions. Each of them is characterized by a sequence of similar machining operations [9]. Some of these features are shown in figure 3-1.

In order to fill the gap between solid modeler possibilities and application needs we have defined a new kind of solid object representation which allows to explicitly represent feature information [10]. This representation consists of a hierarchycal structure that can model arbitrary solid objects to different level of detail and can provide abstract shape properties for matching purposes.

198

3.2. Form Feature Recognition

Our system is able to recognize features, such as pockets and holes, and to extract their technologically relevant. The form feature recognition works on a hybrid (CSG and boundary) representation of the workpiece.

The scheme illustrated in figure 3-2 shows the architecture of the CAD interface and the data processed.

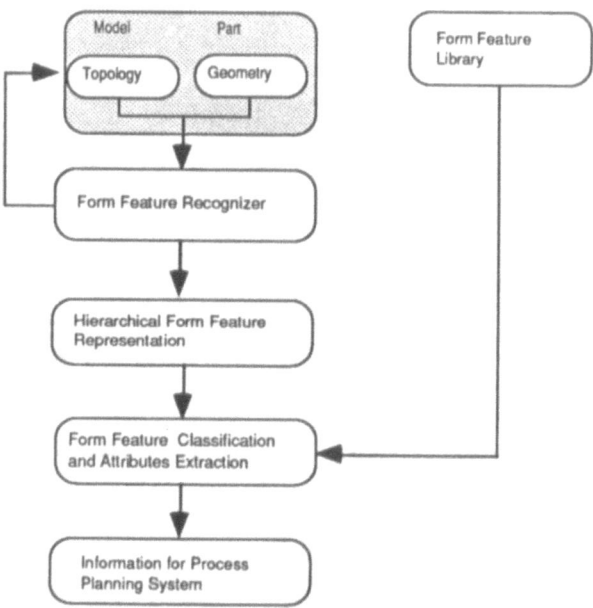

Figure 3-2: CAD interface architecture

The data structures used in the CAD interface are:

- *model data structure*, produced by a CAD system and providing explicit description of a model (i.e. geometry and topology);

- *form feature library*, containing form feature descriptions in terms of their shapes (topology and geometry), attributes (if they are admissible), geometric constrains, functionality;

- *form feature data structure*, representing the features and their possible relationships in a structured form, and also mantaining necessary links to the topology and geometry.

4. AIP in Process Planning

The complexity of the application environment, an aeronautic production plant, has allowed a complete analysis of all the possible aspects of process planning for machined parts and supplied a clear point of reference for the choice of AIP techniques proper to the problem.

Three phases can be identified in the process planning of mechanical parts. The first one is the *dressing phase* which includes preparatory operations such as the cutting of the raw material to a certain dimension, the preparations of tools for NC machines, or the drilling of holes for clamping purposes.

The second is the *machining phase* in which the part is reaches its final shape. In this phase the representation of the part in terms of form features is required.

The third is the *finishing phase* which involves operations such as stamping, quality-control processes, and surface treatment.

4.1. Planning techniques

In general, no single problem solving paradigm suffices to solve together not only all problems encounted in process planning, but not even the different aspects of a single one. Thus we came to the idea of building a system that includes multiple problem solving methods cooperating through an adequate control mechanism.

The approach we adopted implies the description of process planning as a set of "activities" to be executed. These are in our system, knowledge representation models to refer to partial descriptions of the problem. Activities can contain the direct activation of AI techniques, (for example, rule interpretation or action processing), as well as calls to standard routines and also refer to the description of other activities (activity decomposition).

Both, resolution methods and activity decomposition can be conditioned in their activation by the contents of the dynamic data base. In this way the problem solver can perform different actions according to the deduced facts.

The activities of the planning system contribute in producing information and in inserting it into the final plan. Through them it is possible to specify both how to get useful information about the application domain (through the activation of resolution methods and decompositions) and how to update consequently the dynamic representation of the plan.

For instance the activity at highest level in process plan specifies:

```
activity:process_plan_generation
execute:[read_data,
top_plan(process_plan),
fw_interpreter(raw_material_determination),
plan_expansion(process_plan, [dressing_plan,machining_plan,finishing_plan]),
decompose_in ([machining,dressing,finishing])]
```

This activity calls a procedure (read_data) to read data provided by the CAD interface, updates the plan representation using some system-defined routines (top_plan and plan_expansion), activates the forward rule interpreter on a set of rules belonging to the context raw_material_determination, and decides to decompose the problem into the three simpler activites: machining, dressing and finishing.

The decomposition of an activity into simpler sub-activities makes it possible to clearly express the sequence of planning steps needed to solve the problem and to perform high-level planning techniques, such as hierarchical planning and pseudo-reduction [11,12].

Hierarchical planning consists in structuring the problem in such a way that important and difficult goals are tried to be achieved first, while details are solved only once the important planning steps are determined.

Pseudo-reduction consists in initially ignoring the ordering of the planning steps, producing a plan to reach independently single goals and finally merge the generated plans into a single one, without destroying their important effects.

Hierarchical planning and pseudo-reduction use basic AI techniques, essentially a frame representation of the objects of the domain, rule-based reasoning, default reasoning and truth maintenance.

The use of activities allows to introduce a hierarchy among the three phases we have identified (dressing, machining, finishing). in such a way that a plan for the most complex phase is developed first. The goal of determining the machining operations is achieved first, then the goal concerning dressing and finally the one concerning finishing.

During the often very complicated machining phase, in which raw material is transformed into a machined part, a natural decomposition of the problem suggests to split the global goal into simpler subgoals that are identified with the machining of single form features.

Since it is difficult to define a manufacturing hierarchy among the form features, pseudo-reduction is used: each form feature is planned as if it were independent from the rest and then the partial results are merged. The partial plan for each feature is built and a partial ordering among them is determined.

The resources employed are selected from an examination of the partial plans of all the features. Precedence relations involving different features are determined according to the resources.

Finally the detailed plan of the machining phase is created taking into account the partial plans of all the features and the precedence relations based on resources.

Hierarchical planning and pseudo reduction produce process plans following the generative approach. Our system allows also to combine the generative and the variant approaches, where the latter is based on skeletal planning techniques [13].

Skeletal planning uses structures called *skeletal plans*, these are abstract plans containing the basic steps necessary to solve a specific problem, *in the rigth sequence*. A skeletal planner has to identify the current situation, select a suitable plan from a library, and refine single steps using rules or procedures.

We use this technique in the finishing phase in which the sequence of operations is roughly known in advance and the criteria for selecting the right plan can be easily defined. The finishing plan is then refined using rule-based reasoning to decide on the details of every operation.

4.2. Generation of Alternative Plans

An interesting feature of our system is the possibility of dealing with problems where a set of alternative solutions may be fruitfully explored in parallel. This is often true in process planning, where alternatives for a decision can not be precisely evaluated. It is then necessary to carry on different lines of reasoning and then examine competing alternative solutions.

This feature is provided by an Assumption-based Truth Maintenance System (ATMS) that allows to introduce and manage the consequences derived of hypothetical data called 'assumptions' [14]. The ATMS takes care that consequences are automatically derived out of data and assumptions and recorded with the correct dependency information.

The main use of alternatives is for the choice of the main machine, i.e. the one that is in better position to machine most of the part. This choice is to be made at the beginning of the planning process, because the paths that are followed in one or in another case are mostly different. However, the information that could definitively discriminate the best alternative is not available at that point, because it involves costs of resources (tools, fixtures) and operation changes that in turn depend on the selected machine type.

The solution given to this problem is to consider all the competing alternatives introducing different machine types as assumptions and exploring the different possibilities in parallel.

To avoid considering too many alternatives, they are a priori reduced through an approximate evaluation of their cost that involves another AI technique called "reasoning with uncertainty".

In fact, though the reasons that lead to prefer a type of machine are often dependent of many different aspects and are not easily translated into precise and certain rules, these techniques allow to define a ranking of the machines. So we can consider as possible assumptions only machines that are on the top of the ranking.

The use of the control based on "activities" has been of conclusive importance to combine uncertain and assumption based reasoning.

The capability of dealing with different solutions in parallel can be used in order to select the best solution from a comparison among the final plans, but also to reject an alternative plan as soon as it prooves to be worse than the others. In addition, the production of alternative process plans allows to avoid considering of such significant information as the work-load of the machines or the type of production (prototype or series). Inthis way, process planning can be carried out

202

only once for a given part, and when needed, the more suitable plan is selected according to the state of the shop-floor and to technical directives.

5. Know-how

The process engineer, who usually writes the process plans, uses an expertise acquired through practice during many years of work. This expertise, which is difficult to translate into a formal model, includes various kinds of knowledge, such as, for instance, machines and tools available in the shop-floor, machines corresponding to certain operations, constraints on the order of machining operations, and so on.

The human expert is also able to understand the design specifications represented by the part drawing.

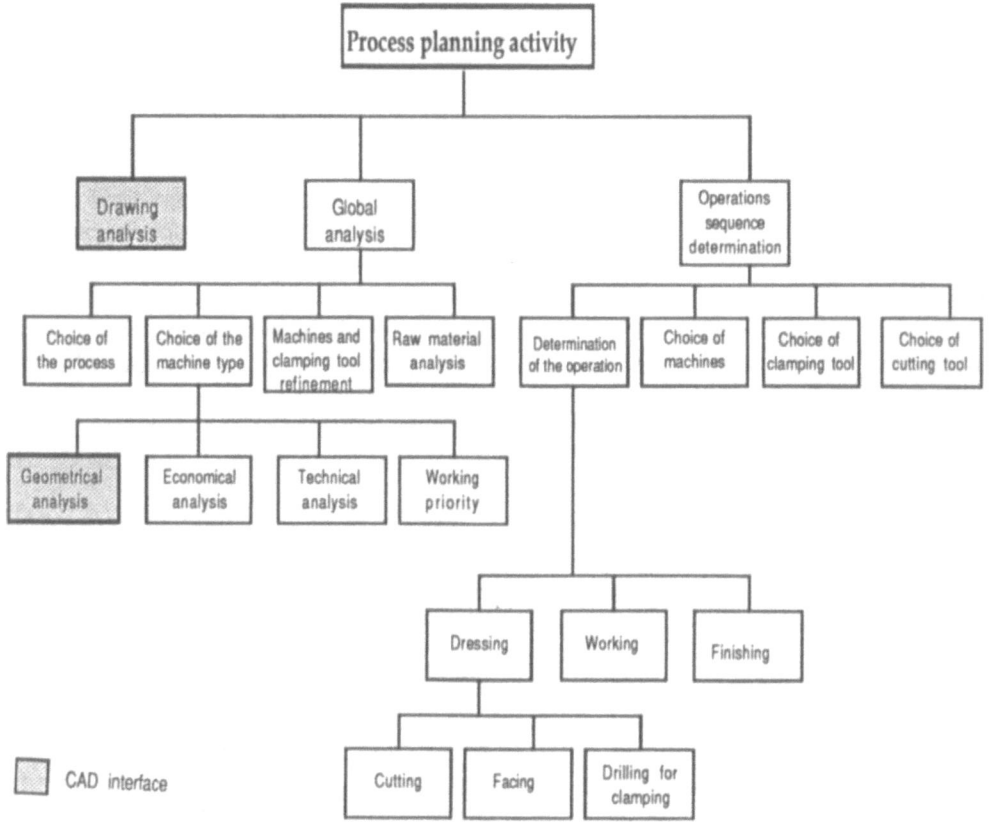

Figure 5-1: Analysis of the process planning activity

The starting point of the work is the analysis (done with the SADT methodology) of the Process Planning activity, breaking it down into subactivities and pointing out their inputs, outputs and constraints.

As process planning is an activity realised with a high degree of subjectivity, there are several representation models, each of them with its logical consistency; the one used in MUMP considers Process Planning as a hierarchically organised structure of activities . A graphic representation of this hierarchical model is shown in fig. 5-1.

6. Knowledge Representation

Knowledge acquisition concerns the process of eliciting and structuring the domain knowledge from human experts. The following steps concern organization and representation of the domain knowledge.

Figure 6-2: Main objects in the process planning system

The most useful among the knowledge representation formalisms offered by MUMP were selected and used. The "declarative" knowledge was formalized using frames, whereas the "operative" one was represented by rules.

6.1. Frames

The main aim of the frames is to represent all the objects involved in the process planning activity and their relationships.

In fig. 6-2 we represent the major classes of objects used. In each class we described the objects that MUMP has to know to be able to correctly tackle process planning problems.

The objects are hierarchically structured as it can be seen in fig. 6-3, which contains the list of the machines considered in our project. These machines are grouped on the basis of similar characteristics, the terminal instances represent a subset of the machines used in Aeritalia.

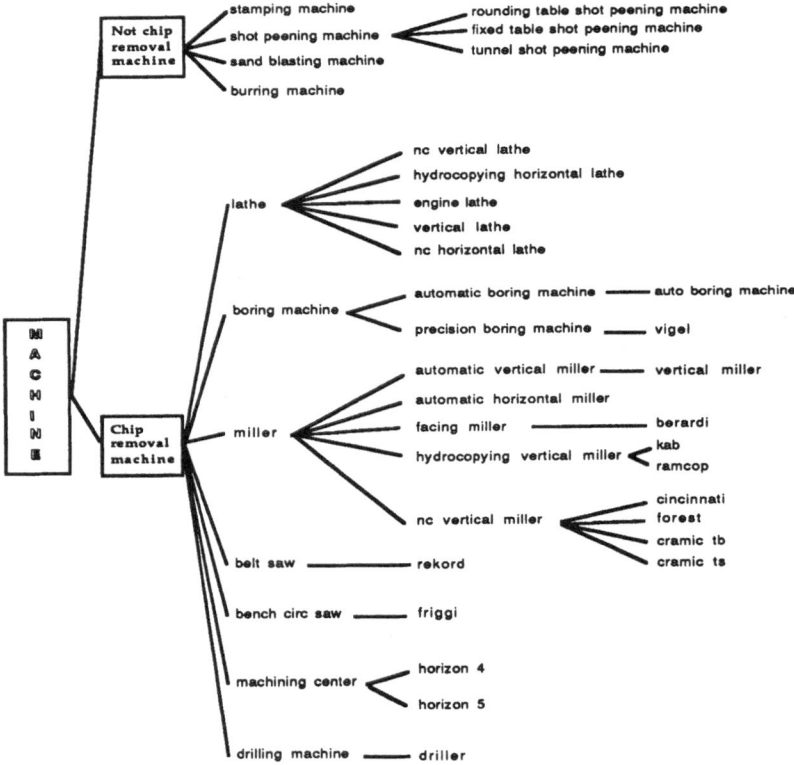

Figure 6-3: The object hierarchy of machines

Each object is described by some attributes (slots), containing information to perform process planning activity. These attributes are inherited from superclasses, which have the role to define them (fig. 6-4).

Frames are a very good formalism to represent actions too, i.e. the processes involved in the manufacturing of the part, with the associated information (type of machine, fixture, etc.). Frames are also useful to describe the relationships between objects. For instance the similarity relationship between pockets.

At present (May '89) the system contains about two hundred frames.

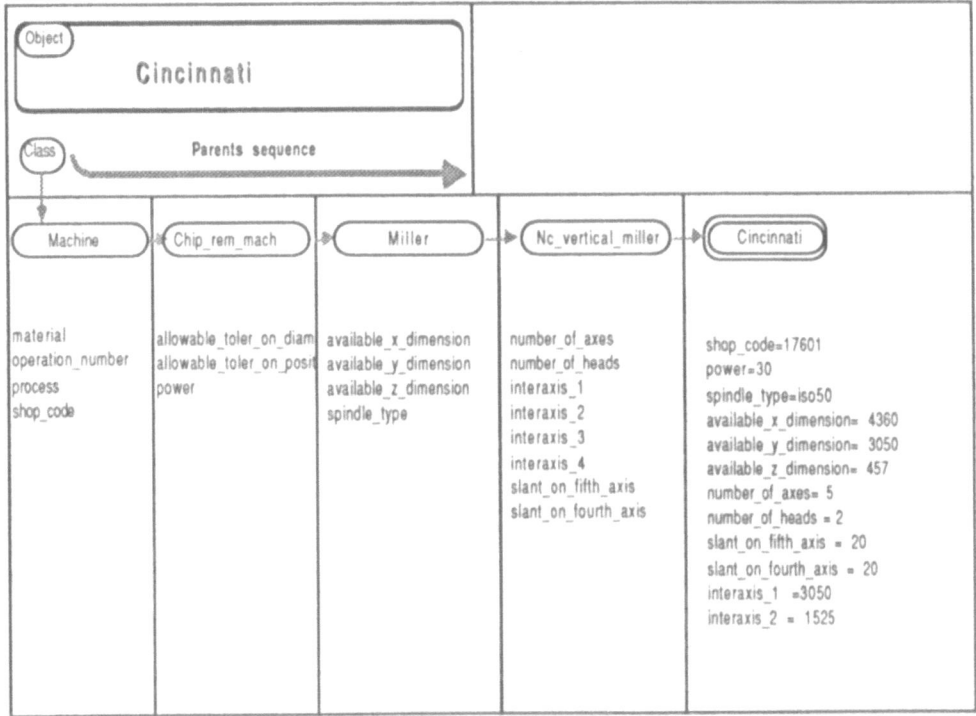

Figure 6-4: Frame-based representation of a machine

6.2. Rules

While frames have been used to represent the "descriptive" part of the expert knowledge (namely the machines, the manufacturing processes and the work-piece), the rules describe the "operating" knowledge and the heuristics (that is to say the criteria used for making decisions).

The rules used in MUMP have the following aspect:

```
srule(identifier, [context], (author,date),0,0,
      [not askable conditions],
      [askable conditions],
      [assumptions],
      [actions],
      [conclusions] ) .
```
Where:

identifier	number of the rule
context	list of names of knowledge base partitions the rule belongs to
nonaskable conditions	a list of facts comparisons or tests
askable conditions	a list of facts that will be considered as goals and thus can be satisfied in backward chaining and eventually asked to the user.
assumptions	a list of assumptions (used with ATMS
actions	a list of actions
conclusions	a list of facts that become true when the conditions are satisfied.

In fig. 6-5 we give an example of a rule for instantiating a drilling process for a through hole with diameter between 10 and 15 mm, without tolerance.

```
srule3(1005,[ process_selection] ,(fb,mar89),0,0,
        [superclasses\HOLE..through_hole,
        equal_to\HOLE..0,
        diameter\HOLE..DIA,
        DIA > 10,
        DIA < 15,
        diameter_tolerance\HOLE..0],
        [],[],
        [PD_DIA = DIA-7],
        [superclasses\DRILLING..drilling,
        feature\DRILLING..[HOLE],
        through_hole\DRILLING..yes,
        drilling_diameter\DRILLING..PD_DIA,
        superclasses\WD..widening,
        feature\WD..[HOLE],
        through_hole\WD..yes,
        drilling_diameter\WD..DIA,
        to_be_imm_followed\DRILLING..WD]).
```

Figure 6-5: A rule for a drilling operation

The rules make use of the knowledge stored in frame slots to draw conclusions in the form of values for slots of further frames. The rules are grouped in contexts, i.e. subsets referred to the solution of particular aspects of the problem at hand. At present the system contains over three hundreds and fifty rules. The contexts in our application are listed in fig. 6-6.

Process selection	Machine sel
Cutting tool sel	General ops
Precedence	Stamping ops
Nc clamping	Shot peening ops
Nc axes	Shot peening mach
Nc heads	Painting ops
Nc heads bw	Inspection ops
Nc mach refine	Leader process
Mcenter mach refine	Cutting
Economical analysis	Facing
Face analysis	Drilling
Technical analysis	Raw material size
Evaluation	
Fw working priority	
Bw working priority	

Figure 6-6: The contexts (partitions of the rule base)

The partition in contexts strictly reflects the structure of the SADT analysis. The partition of the set of rules eases the modification and extension of the knowledge base and allows a more efficient processing by limiting the number of rules to be considerd at any given moment.

The sequence in which the contexts are going to be explored as well as the strategy for rule application in each of them is determined by the activities (see the following example).

process plan,
        [read_data,top_plan_insertion(process_plan),
        decompose_in([global_analysis,operation_sequence_determination])]

        global_analysis
                [decompose_in([leader_process_generation,preferred_machine_type_choice,
                preferred_machine_tool_refinement,raw_material_analysis])]

                leader_process_generation
                        [bw_interpreter (leader_process\global _info..milling,leader_processes),
                        bw_interpreter(leader_process\global_ info..drilling,leader_processes)]

                preferred_machine_type_choice
                        [decompose_in([economical_analysis,technical_analysis,working_priority])]

        . . . .

7. Part representation model.

One of the main jobs since the start of the project has been the definition of the descriptive model of the machined parts in Aeritalia 's industrial practice.

On the ground of such model, the form feature recognizer analyzes the CAD drawings and produces the part representation needed by MUMP. The remaining part of the input data consists of information of technological nature usually reported on the part list. Data such as surface treatments, kind of raw material, etc., are at present asked to the user, but an automatic acquisition could be provided.

The actual model has been obtained through successive approximations, and is giving satisfctory results. The main characteristic of the model is that it evidentiates the morphological and geometrical features of the workpiece using the frame formalism and a hierárchical description. It allows for the definition of the features and of the relations between features and between features and part.

The model of the raw part consists of a solid and each feature is described through a set of surfaces. The features are associated to faces, of the basic solid or of other features. In this way the descriptive model develops itself hierarchically, allowing the representation of the existing relations between entities.

We show in fig 7-1 an excerpt of the representation of a part and its features, in the form of a set of frames describing the part in fig. 7-2. This is the kind of representation that is produced by the Form Feature Recognizer and then passed to the planning system to be processed in order to obtain the process plan.

The parts treated are of a reasonably high level of complexity, the same found in the Aeritalia production reality. The present version can recognize and treat several pockets and holes on a number of sides of the part, grouped and nested features, profiled walls, radii and bevels.

The process plan contains complete information about pockets and holes to be machined as well as the choice of plates (raw material), of their dimensions and the number of parts per plate to be produced.

```
superclasses\global_info..global_information
quantity\global_info..2
lot_size\global_info..15
production_phase\global_info..serie
design_number\global\info..xxxx
superclasses\examined_part..part
shape_type\examined_part. rib
grain_direction\examined_part..yes
max_x_dimension\examined_part..0.700E+03
max_y_dimension\examined_part..0.100E+03
max_z_dimension\examined_part..0.050E+03
faces\examined_part..[f1,f2,f3,f4,f5,f6]
raw_material_type\examined_part..plate
superclasses\f1..face
```

feature\f1..[p1 ,p2,p3,p4,bevel_1]
slant\f 1 ..no opposite face\f1..f2
....
superclasses\p1..pocket
bottom_thickness\p1..0.030E+02
depth\p1 ..0.235E+02
corners\p1..[0.900E+02, 0.900E+02, 0.930E+02, 0.870E+02]
fillet_radius_between_wall\p1..0.8E+0l
fillet_radius_walls_bottom\p1.. 0.4E+0l
angles_with_bottom\p1..[0.3E+01,0.000E+00,0.000E+00,000E+00]
number_of_faces\p1..6
faces\p1..[f11,f12,f13,f14,f15,f16]
constant_slant\p1..yes
reference_face\p1.f1
......

Figure 7-1: An excerpt of the part representation for the part in fig. 7-2

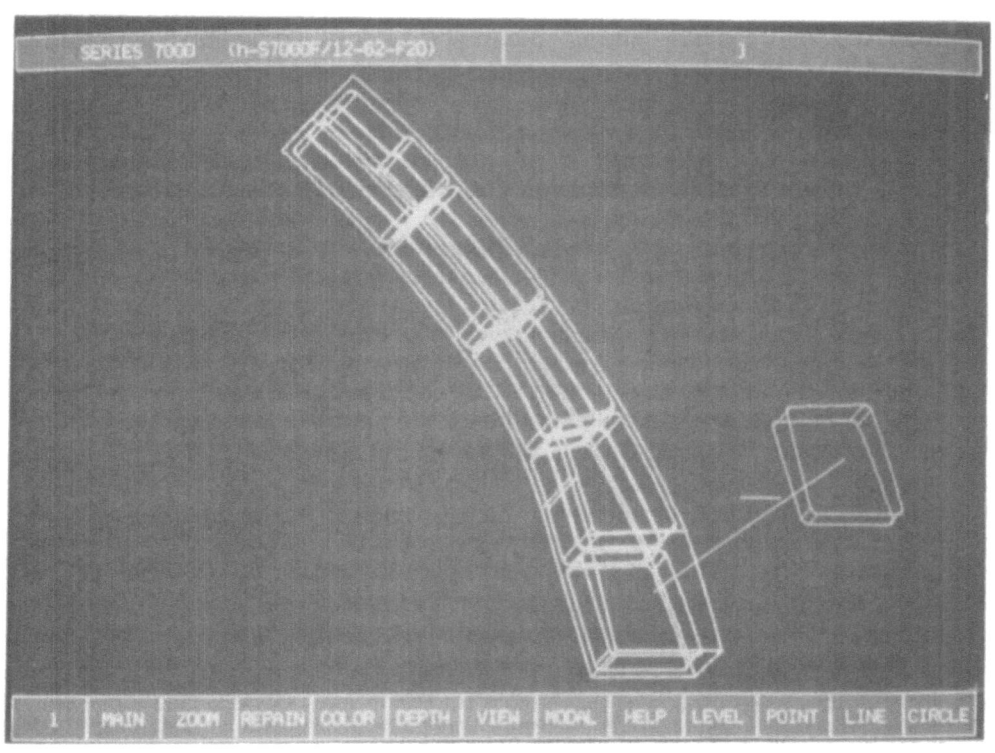

Figure 7-2: A typical aircraft part

8. Final Remarks

In agreement with the domain experts participating in the project, the rules are periodically checked in order to verify their consistency. In order to make the knowledge base as general and complete as possible, every single rule results from a mediation of the points of view of several experts. The experts also participate actively in drafting the documentation.

At this point is possible to say that the value and the flexibility of the generative approach using AI techniques has been heavily confirmed.

So far, the results are quite exciting, both from the point of view of the project, for the existing cooperation between the partners, and from the point of view of Aeritalia, for the quality of the work done and its good prospects.

Particularly good results come from the parallel involvement in process planning problems of experts in Artificial Intelligence planning techniques and knowledge engineers.

At the moment, the knowledge introduced in MUMP regards especially operations which have to be performed on NC machines, even if operations and processes performed on other machines were considered. In the future the main objectives will be the AI techniques optimization and the introduction of the knowledge able to cover the most important part of the Aeritalia production environment (NC, machining centers, hydrocopying and conventional machines).

9. Bibliography

[1] Chang, T.C. and Wysk, R.A., "An Introduction to Automated Process Planning Systems", Prentice-Hall Inc., 1985.

[2] Descotte, Y. and Latombe J.C., "GARI: a Problem Solver that Plans How to Machine Mechanical parts", in Proceedings IJCAI-81, Vancouver, Canada (1981).

[3] Del Canto, M. et alii, "An Expert System for Computer Aided Process Planning", 6th. European Conference on Artificial Intelligence, ECAI-84.

[4] Freedman, R.S. and Frail, R.P., "OPGEN: The Evolution of an Expert System for Process Planning", AI Magazine 7(5):58-70 Winter-1986.

[5] Hayes, C., "Using Goal Interactions to Guide Planning", Proceedings AAAI-87, Suttle, U.S.A. 1987.

[6] Falcidieno, B., Giannini, F. "Automatic Recognition and Representation of Shape Features in a Geometric Modeling System", Computer Vision,Graphics, and Image Processing (to appear)

[7] Joshi, S. and Chang T.C. "Graph-based heuristics for recognition of machined features from a 3D solid model", Computer-Aided-Design, volume 20 number 2 march 1988

[8] Shah, J., Sreevelson P., Rogers M., Billo R., Mathew A., "Current status of Feature Tecnology", CAM-I Report R-88-GM-04 1988

[9] ESPRIT Project Nr. 1220 (865) " Non-Monotonic Reasoning Techniques for Industrial Planning Applications", System Definition Document, November 1986

[10] Ansaldi, S., Falcidieno B. "Form Feature Representation and Recognition in a Structured Boundary Model", North Holland, Encarnacao, Mc. Laughlin Eds, Wozny, 1987

[11] Sacerdoti, E.D., "The non-linear nature of plans", Proc. International Joint Conference on Artificial Intelligence 1975, IJCAI-75

[12] Sacerdoti, E.D., "Problem Solving Tactics", Proc. International Joint Conference on Artificial Intelligence 1979, IJCAI-79

[13] Friedland, P.E., "Knowledge-based experiment design in molecular genetics", Rep. No.79-771, Computer Science Dept., Stanford University (Doctoral dissertation).

[14] deKleer,J. "An Assumption-based TMS", Artificial Intelligence, 28 (2).

Project No. 881

ESPRIT PROJECT # 881 – FORFUN
FORMAL DESCRIPTION OF DIGITAL AND ANALOG SYSTEMS BY MEANS
OF FUNCTIONAL LANGUAGES

R. BOUTE
University of Nijmegen
Toernooiveld 1
6525 ED Nijmegen
The Netherlands

ABSTRACT. The origin of this project is the observation that systems and circuits are not computational processes and that, therefore, programming language concepts are inadequate for their description. The function notation commonly used in engineering for dealing with signals and systems constitutes a better linguistic basis, especially w.r.t. clarity in description and convenience in (transformational) reasoning and system design.

The goals (and results) of the project are the design and implementation of a prototype description language (*Glass*) for analog and digital circuits, together with a supporting environment (*Glue*). The chosen approach was system semantics, which allows to associate with a single language (based on function notation) a diversity of interpretations, expressing relevant system properties such as structure, behaviour, performance (e.g. timing), cost.

Conclusions of the project are that the combination of functional notation and system semantics yields the expected advantages over traditional programming language constructs, and that the principles embodied by *Glass* will prove very useful in the design future system description languages such as the restandardization of VHDL scheduled for 1992. The implementation of the Glass language and its environment together with extensive documentation are available at reproduction cost from any of the project partners.

0 . Introduction

0.0. RATIONALE, PRINCIPLES, GOALS AND RESULTS OF THE PROJECT

The rationale for this project, as explained in section 1, is that most physical and technical systems are not computational processes and hence cannot be described adequately by algorithms. Therefore, the syntactic and semantics concepts of programming languages on which most current hardware description languages are based are inadequate for system description in the sense that they can express only a restricted simulation-oriented view. Formalisms derived from the function notation commonly used in applied mathematics and in the engineering disciplines dealing with signals and systems constitute a more suitable linguistic basis.

In section 2 we will further show how the principle of system semantics allows to associate with a single system description language (based on this function notation) a diversity of interpretations expressing relevant system properties such as structure, behaviour, performance, cost.

As elaborated in section 3, the goals of the ESPRIT # 881 project are: (1) developing a prototype language for the description of digital and analog systems, based on the aforementioned rationale and principle, (2) investigating the suitability of such a language as a notational aid for transformational reasoning in digital and analog system design, and (3) specifying and implementing a prototype system description environment supporting this language and the definition of so-called semantic functions expressing the various interpretations of the language.

The project consortium consisted of the University of Nijmegen (NL) as the prime contractor, Sagantec (Eindhoven, NL), Delft University of Technology (NL) and Alcatel Bell Telephone (Antwerp, B).

The corresponding results of this project, further discussed in section 4, are: (1) the prototype system description language *Glass* (*G*eneral *l*anguage supporting *s*ystem *s*emantics), (2) various ways of using the Glass concept in transformational reasoning and as a basis for the evaluation, the usage and the improvement of existing system description languages, in particular VHDL, (3) a prototype user environment *Glue* (*Gl*ass *u*ser *e*nvironment) supporting Glass and the definition of semantic functions in either Miranda (functional), Pascal or C (imperative). Additional results are: a set of environment generator programs to assist in software development and a general-purpose parameterized system for generating and accessing internal representations.

0.1. INTENDED USAGE OF THE RESULTS

The aforementioned prototype implementations are in the public domain and available from any of the project partners at reproduction cost (copying + media) to people wishing to experiment with or evaluate various principles and styles of system description. This is of interest to designers and implementors of system description languages and CAD environments as well as designers of electronic systems and circuits, who in fact constitute the end user community for such languages.

It will be clear from the above that the Glass/Glue system is intended for experimentation, not for usage in actual production CAD. The latter was not a goal of the project (which mainly originated from university research interests), nor would such a goal have been realistic, because

a) history shows that system description languages can gain acceptance in production CAD only through large-scale standardization (as is the case e.g. for VHDL);

b) the implementation of a production-quality environment (robust, easily maintenable software) would require a manpower investment at least three times the 20 man-years reserved for this project, using an experienced programming team rather than researchers.

It must also be emphasized that Glass describes (concrete) system and circuit *realizations*. With respect to (abstract) specification, it can be used for *verification* (compliance of the behavioural interpretation of a system description with the abstract specification). In the reverse direction (from specification to realization), the usage of Glass depends on the degree of design automation. In fully automated design (the ultimate role of silicon compilation), Glass can assume only the secondary role of internal representation. If the design is not (or cannot be) fully automated, Glass can greatly facilitate human intervention by providing easily readable realization descriptions.

Obviously, this line of research does not end with this project and with Glass. Future work will concentrate mainly on two issues:

(a) Applying the principles of and the insights gained from this project to the improvement of currently used system description languages. In particular, the IEEE standard for VHDL, i.e. VHSIC (Very High Scale Integrated Circuit) Hardware Description Language, is due for restandardization in 1992, and we are preparing a graded set of language change proposals ranging from local but significant improvements to overall redesign.

(b) Developing a next-generation system description language tentatively called Funmath (*Fun*ctional *math*ematics), which is meant to be a wide-spectrum language for expressing abstract specifications, algorithmic realizations and various kinds of circuit realizations (analog, digital) in a uniform notational framework that is very well-suited towards transformational reasoning and design. The way in which circuit realizations are described will depend to a large extent on the experience gained with Glass.

1. The FORFUN View on System Description

1.0. THE INADEQUACY OF ALGORITHMIC AND PROGRAMMING CONCEPTS

Most existing system description languages are based on syntactic and semantic principles derived from traditional (imperative) *programming* languages. This is due to a number of historical factors, viz.

a) the theory and the design of formal languages was originally developed in the area of programming and hence constituted a readily available basis;

b) the idea of formal system description was mainly restricted to simulation, with little awareness of other important issues such as formal (e.g. transformational) reasoning and symbolic manipulation (by human and by machine).

These factors jointly create a vicious cercle, because the syntactic and semantic principles of progamming languages are suitable only for expressing *algorithms*, and hence their exclusive usage in system description prevents the other issues from being addressed effectively.

The inadequacy of the algorithmic paradigm as a basis for system and circuit description can be inferred from the various definitions in the literature where the algorithm concept is carefully analyzed, e.g. by Knuth [11]. By definition, an algorithm is a set of instructions (or recipe) describing the (serial or parallel) steps in a computation. Hence an algorithm constitutes a (concrete) *realization* by means of a *computational process*. As a consequence, algorithms are *not* suitable for expressing

– (abstract) *specifications*, i.e. black-box descriptions of the input/output behaviour of programs or circuits without prejudicing particular realizations;

– (concrete) *realizations* by means that cannot be properly characterized as computational processes (in the sense elaborated rigorously in the theory of computation).

To elucidate the fact that electronic circuits and systems indeed belong to the class of systems whose behaviour cannot be properly characterized as computational processes, we present an informal epistemological argument and a formal mathematical example.

– Most natural and technical systems are essentially non-algorithmic. Firstly, their *behaviour* is governed by the laws of nature, which are better modelled mathematically by functions and equations than by algorithms. For instance, a falling object does not execute "Newton's gravitional algorithm" but behaves according to the equations of motion in mechanics. A similar observation holds for electronic circuits, both digital and analog. Superficially, synchronous digital circuits appear to be amenable to algorithmic descriptions, but further reflection reveals these to be mere simulations of one particular view of behaviour in which both the time space and the signal space are discretized. Hence this situation constitutes a degenerate particular case. Secondly, the *structure* of systems, being static in nature, is even less well-represented by algorithmic descriptions.

– The following example clearly illustrates that circuit behaviour is "equational" rather than "algorithmic". Slightly running ahead of the more carefully designed notation of Glass, consider the following definition:

$$sys\ x = box(x,\ sys\ x) \tag{0}$$

In a rather straightforward intuïtive *structural* interpretation, x is the internal name for the system input wire (the name is externally irrelevant as can be seen by substituting y for x), and the occurrence of *sys* at the right-hand side suggests feedback. Hence the structural interpretation of the complete definition (0) can be represented schematically as in Fig. 1.0.

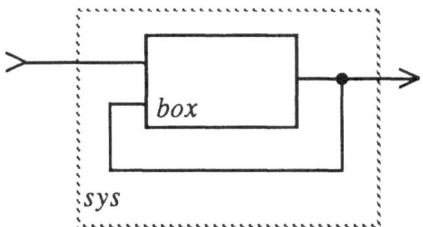

Figure 1.0 Structural interpretation of definition (0).

More interestingly, consider the *behavioural* interpretation in the very simple case where *box* is a differential amplifier with gain A, that is: $box(x, y) = A \cdot (x - y)$. Substituting this definition of *box* in the definition of *sys* yields

$$sys\ x = A \cdot (x - sys\ x) \tag{1}$$

The *syntax* (textual form) maintains the close correspondence with structure. If interpreted as a (functional) *program* (a *program* is an algorithm expressed in a formal language), definition (1) describes a nonterminating computation: "to compute *sys* x, subtract the result of computing *sys* x from x and multiply the result of the subtraction by A". The reason for nontermination is that this program is not recursive (as in $fac\ 0 = 1$, $fac\ n = n \cdot fac(n - 1)$, where the argument n is decremented with every call, until the value 0 is reached).

However, the behaviour can be correctly obtained by viewing definition (1) as a mathematical *equation*, whose closed-form solution is

$$sys\ x = A \cdot (1 + A)^{-1} \cdot x$$

In a very real sense, the circuit actually *solves* the equation in a way that cannot be properly described as algorithmic (it is even not directly obvious how to approximate this process by simulation).

1.1. A DIFFERENT PARADIGM AND ITS SUPPORTING NOTATION

We have seen that the algorithmic paradigm is not adequate for the description of general systems and circuits.This observation extends to the supporting notation, viz. constructs of traditional programming languages.

On the other hand, the *dynamical system* paradigm that has proven its value over the centuries in physics and engineering, and in recent decades especially in systems theory, electronics and communications, provides the necessary generality, even including algorithms as a particular case. The formalisms used in these disciplines have evolved in close correspondence with mathematics, in particular algebra and analysis. The most powerful concept in this respect is that of a mathematical *function*, together with the commonly used notation attributed to Euler. The advantages of functional formalisms over the concepts and notations from programming languages are the following:

– *Expressive power*:
• semantically not restricted to operational descriptions or to discrete models; hence suitable also for e.g. specification purposes as well;
• syntactically better matched to the wide variety of system classes and views of systems to be expressed (in particular also a more direct mapping to hardware realizations);
• closer to the notations commonly used in systems theory, electronics and related disciplines for the development of theory and for design.

– *Manipulative power*:
• the function notation and the equational style of expressions are eminently suitable for transformational reasoning;
• concepts and theorems from mathematical disciplines (algebra, analysis, etc.) relevant to systems engineering are more directly applicable.

Moreover, the developments in functional programming languages during the past decades, e.g. Backus [1], Bird [2], Burge [3], Burstall [8], Landin [12], Meertens [13], Turner [17], show that the same advantages are gradually being realized in the area of programming (algorithmics) itself. In particular, reasoning about algorithms expressed in functional notation follows the algebraic/transformational style based on equalities and substitutions rather than the cumbersome logic/deductive style based on pre- and postconditions.

This is why in this project the notion of *function* has been chosen as the basis for system description. The project acronym *FORFUN* stands for "*For*mal system description by means of *fun*ctional languages".

As shown next, the notation for mathematical functions that is commonly used in systems theory, electronics and related disciplines, can be given various interpretations that are of interest in formal system description. The principle of *system semantics* as proposed by Boute [3, 4] is to distinguish between these interpretations explicitly by means of semantic functions. This makes it possible to use a single formal language, rather than a collection of different languages, to express various system aspects such as structure, behaviour, performance, cost.

2. System Description with Function Notation

2.0. MATHEMATICAL FUNCTION DEFINITIONS – A BRIEF REMINDER

We recall that a *function* (or *mapping*) from (set) A to (set) B is a mathematical object that associates with every element of A a single element of B.

We write $A \rightarrow B$ for the set of all functions from A (the *domain*) to B (the *codomain*). An expression of the form $f \in A \rightarrow B$ is called a *type definition* for f. If $x \in A$ and $f \in A \rightarrow B$, then

$f x$ denotes the corresponding element in B (called the *image* of x under f). A function is completely defined by its type definition and its image definition (i.e. the image of every element in the domain). Therefore we write a function definition as follows:

def *type definition* **with** *image definition*

The two main styles for an image definition are the following:
– *Extensional definition*: the image is specified individually for every value in the domain, for instance by means of a table. Of course, this is practical only for small (and, a fortiori, finite) domains. An example is the following (where $\mathbb{B}$ denotes the set of binary values, i.e. $\{0, 1\}$):

$\quad$ **def** $sel \in \mathbb{B}^3 \to \mathbb{B}$ **with**

$\quad\quad sel\,\langle 0, 0, 0\rangle = 0 \quad\quad sel\,\langle 0, 0, 1\rangle = 1 \quad\quad sel\,\langle 0, 1, 0\rangle = 0 \quad\quad sel\,\langle 0, 1, 1\rangle = 1$

$\quad\quad sel\,\langle 1, 0, 0\rangle = 0 \quad\quad sel\,\langle 1, 0, 1\rangle = 0 \quad\quad sel\,\langle 1, 1, 0\rangle = 1 \quad\quad sel\,\langle 1, 1, 1\rangle = 1$

– *Intensional definition*: by means of an expression for the image of an arbitrary value in the domain (designated by variables), for instance:

$\quad$ **def** $sel \in \mathbb{B}^3 \to \mathbb{B}$ **with** $sel\langle s, x, y\rangle = s\ ?\ x,\, y$

Here the notation $c\ ?\ a;\ b$ abbreviates **if** c **then** a **else** b.

2.1. FORMAL FUNCTION DEFINITIONS AND THEIR SYSTEM SEMANTICS

In the above discussion, we have interpreted the notation used according to its *standard mathematical interpretation*, namely as the definition of mathematical functions (mappings). The notation itself can be given many other useful interpretations. In order to "make room" for these interpretations, we must start by considering the definitions as mere pieces of text, so-called *formal function definitions* (ffds), without any a priori or privileged interpretation, and express the desired *interpretation* (or *model*) explicitly by means of a *semantic* (or *meaning* function).

A *meaning function* is defined in a general way on the syntax of the language. Its *domain* is the set of all possible sentences (in this case ffds) in the language and its *codomain* is the space of all possible values (in the widest sense) that a certain system property or aspect may assume. For instance, the codomain for the meaning function *struct* describing the structural model is the space of all possible system structures (for the considered class of systems).

The most elegant method for describing meaning functions was developed in the context of programming languages by Knuth [10]. How to extend this method to formal system description languages is shown by Boute [3, 4]. In the present paper, however, we will not give the definition of the semantic functions themselves, but only illustrate their usage by showing the image of specific example sentences under the considered semantic function. Moreover, these images will be expressed informally, viz.
– using schematics for the structural model;
– using mathematical functions for the behavioural model.

Finally, when the meaning function is clear from the context, we designate its application by underscoring the argument, e.g., when dealing with the structural model, *system* abbreviates **struct** *system*. This reduces notational clutter and saves space. Also, in order to keep the technical part of this paper accessible to the nonspecialist, we discuss the structural interpretation only, and refer to [4] for the more advanced behavourial interpretations.

218

2.2. EXAMPLES FOR DIGITAL AND ANALOG SYSTEMS

Preliminary remark: not all expressions one encounters in mathematics can be seen as ffds having a system-oriented interpretation. In fact, only a special set of primitive functions and function composition rules can be given meaningful structural interpretations: for instance, for the class of synchronous digital circuits, the primitive functions are those corresponding to gates and memory cells (D-flipflops, JK-flipflops etc.), and the syntactic composition rules are only *function application* (representing input/output connection in the structural model) and the **where-construct** (representing fanout and/or feedback in the structural model). The *Glass* language is designed in such a way that only these combinations are possible.

2.2.0. A combinational circuit example (a data selector). Consider the following two ffds:

def $hsel \in B^3 \Rightarrow B$ **with** $hsel \langle s, x, y \rangle = (s \land x) \lor (\neg s \land y)$
def $ssel \in B^3 \Rightarrow B$ **with** $ssel \langle s, x, y \rangle = (s \land x) \lor (\neg s \land y) \lor (x \land y)$

Here B designates a so-called *base type* (still uninterpreted!).

– In the *structural model*,

struct $(B^3 \Rightarrow B)$ = the set of 3-input, 1-output digital systems;
struct $hsel$ = the circuit structure represented schematically in Fig. 2.0;

and similarly for *ssel*. Clearly **struct** $hsel \neq$ **struct** $ssel$.

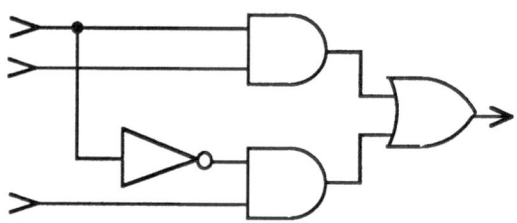

Fig. 2.0   Schematic representation of **struct** *hsel*.

– In the *simplex behavioural model* (with **simp** $B = \mathbb{B}$), the operators $\lor, \land, \neg$ are interpreted according to their standard mathematical interpretation (logical *or, and, not*). Hence

simp $(B^3 \Rightarrow B) = \mathbb{B}^3 \to \mathbb{B}$ (recall: $\mathbb{B} = \{0, 1\}$)
simp $hsel = sel$ **simp** $ssel = sel$.

where *sel* is the *mathematical* function defined earlier. Clearly **simp** $hsel =$ **simp** $ssel$.

– The *Eichelberger behavioural model* can reflect static hazards by using the ternary algebra $\mathbb{E} = \{0, 1, \bot\}$, where $\bot$ stands for "undefined" (in this case modelling a signal in transition). The primitive operators are defined appropriately. This results in **eich** $ssel \neq$ **eich** $hsel$, reflecting the fact that *hsel* is hazardous and *ssel* is safe with respect to static hazards.

2.2.1. *A sequential circuit example* (a parity checker). Consider the ffd

$\quad$ **def** *par* $\in B \Rightarrow B$ **with** *par* $x = y$ **where** $y = D(x \oplus y)$

– The *structural model* yields
$\qquad$ **struct** *par* = the circuit structure represented schematically in Fig. 2.1.

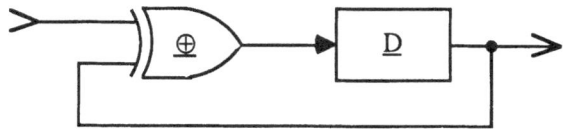

Fig. 2.1 $\quad$ Schematic representation of **struct** *par*.

– The *simplex behavioural model* is mathematically trivial and practically meaningless for systems with memory (such as this example).

– The *multiplex behavioural model* describes synchronous sequential behaviour in terms of finite sequences of binary values or infinite sequences (functions of type $\mathbb{N} \to \mathbb{B}$); see [4].

2.2.2. *An analog control system example.* The ffd

$\quad$ **def** *sys* $\in B \Rightarrow B$ **with** *sys* $x = y$ **where** $y = box\langle x, y\rangle$

has already been introduced in section 1 in a preliminary form, together with its structural interpretation and a static (steady-state, DC) behavioural interpretation. A dynamic behavioural interpretation would be of the form

$\quad$ $\underline{sys}\, x = y$ **where** $y = \underline{box}\langle x, y\rangle$

where the variables x and y designate signals in the time domain or frequency domain representation, and $\underline{sys}$, $\underline{box}$ are suitably defined functionals over signals or their Fourier transforms.

2.2.3. *Language syntax considerations.* From the above examples, it becomes apparent that, for the considered class of systems (characterized more precisely later on), as few as *two* language constructs are sufficient to make a fully-fledged system description language, namely *function application* (with structural interpretation as a signal flow description) and the **where**-*expression* (with structural interpretation as a description for fanout, i.e. output sharing, and for feedback). This illustrates the fact that *syntax design* for system description languages is essentially a matter of *variable scoping and binding* rules.

$\quad$ For systems with large repetitive structures, the usage of *macro instantiations* may avoid tedious textual repetitions. The *macro definitions* may be *parameterized* to cater for minor variations. Strictly speaking, macro definition is not a language design issue, in the sense that it does not depend on the semantics and is similar for "all" languages. Yet it is good practice to make macro definitions syntactically identical to formal function definitions, except for the fact

220

that they may contain free variables to reflect bindings in the context of instantiation. The following formal definition of a synchronous circuit *counter* may serve as an illustration:

def *counter* $\in D^4 \times D^4 \Rightarrow D^4 \times D$ **with**
$\qquad$ *counter* $\langle d, \langle nload, nclear, enap, enat \rangle \rangle$
$\qquad\qquad = \langle q, carry \rangle$
$\qquad\qquad\qquad$ **where** $q_0 = stage \langle d_0, enat \rangle$
$\qquad\qquad\qquad\qquad q_1 = stage \langle d_1, and \langle enat, q_0 \rangle \rangle$
$\qquad\qquad\qquad\qquad q_2 = stage \langle d_2, and \langle enat, q_0, q_1 \rangle \rangle$
$\qquad\qquad\qquad\qquad q_3 = stage \langle d_3, and \langle enat, q_0, q_1, q_2 \rangle \rangle$
$\qquad\qquad\qquad\qquad carry = and \langle enat, q_0, q_1, q_2, q_3 \rangle$
$\qquad\qquad\qquad\qquad$ **mac** $stage \langle d, toggle \rangle = jkff \langle and \langle j, e \rangle, and \langle k, e \rangle \rangle$
$\qquad\qquad\qquad\qquad\qquad\qquad\qquad$ **where** $j = nand \langle k, preset \rangle$
$\qquad\qquad\qquad\qquad\qquad\qquad\qquad\qquad k = nand \langle d, nclr, preset \rangle$
$\qquad\qquad\qquad\qquad\qquad\qquad\qquad\qquad e = or \langle toggle, preset \rangle$

$\qquad\qquad\qquad\qquad nclr = buf\ nclear$
$\qquad\qquad\qquad\qquad preset = omot \langle nload, nclr \rangle$
$\qquad\qquad\qquad\qquad en = and \langle enap, enat \rangle$

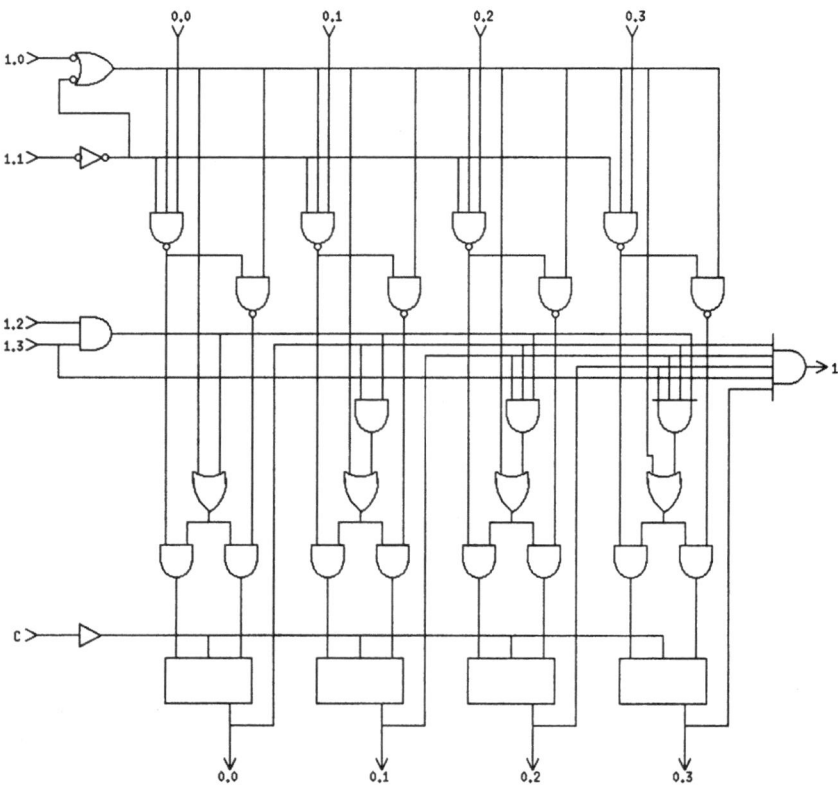

Fig. 2.2. Schematic representation of *struct counter*.

The *strutural* interpretation is represented schematically in fig. 2.2. Notice that, in the macro definition, only *d* and *toggle* occur as formal arguments because they correspond to inputs of *stage* that are different for each of the four instantiations. By contrast, the inputs *nclr* and *preset* are distributed over all four instantiations and therefore are represented by free variables which become bound for each instantiation in the context of the **where**-expression.

2.3. DESCRIPTION OF ADIRECTIONAL SYSTEMS

The examples given thus far show that formal function definitions constitute an ideal syntax for so-called *(uni)directional systems* i.e. systems, or conceptual views of systems, for which it is meaningful to attribute a *direction* to the signal or information flow inside and between the components, and hence to distinguish between *input* and *output*. Moreover, the notation is close to the standard mathematical notation for describing behaviour. We have seen that, for the examples given, it suffices to replace the formal primitive functions by the mathematical functions expressing their behaviour. Hence formal function definitions may be subject to transformations as if they were mathematical expressions, and the behavioural interpretation remains unaffected. This results in a high manipulative power for transformational design.

However, it has been shown by Boute [4] that for certain classes systems, even undirectional ones, this attractive property may be invalidated by loading effects. Moreover, formal function definitions do not have any known interpretation that makes them useful as a syntactic basis for describing adirectional systems.

For adirectional systems, it is therefore more advantageous to introduce the syntax, variable binding and scoping rules of so-called sigma terms, as proposed by Boute [6]. Here we only present a small example. Consider the following formal definition of the circuit *star*:

$$\textbf{def } star \in [\text{B}^3] \textbf{ with } star\langle x, y, z \rangle = \{A\langle x, c \rangle, B\langle y, c \rangle, C\langle z, c \rangle\}$$

where square brackets designate the type of a sigma term, which in the structural interpretation for *star* is a system with 3 terminals. The structural interpretation of *star* can be represented schematically as in Fig. 2.3.

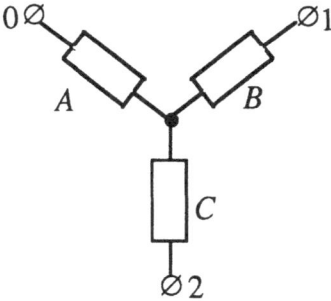

Fig. 2.3 Schematic representation of **struct** *star*.

3. Origin and Goals of the ESPRIT # 881 FORFUN Project

3.0. PROJECT GOALS

The general purpose of the FORFUN project was to evaluate the possibility of using the principle of system semantics as the basis for a uniform and coherent system description environment. System description in such an environment would involve, for a given class of systems,
 – a single system description language;
 – various system views expressed by semantic functions (predefined and user-defined).
Evaluating the possibilities of system semantics entails
  a) the design of a language syntax on which the definition of multiple semantic functions is not just possible but also convenient;
  b) exploring the style of expression and transformational reasoning in system design when such an environment is used, also taking into account the specific "flavour" of the design paradigms associated with the considered class of systems;
  c) designing and building an actual prototype implementation of both the language and the environment, including facilities for defining semantic functions.
The original project proposal specifies the elaboration of (a) and (b) for two representative system classes where the design paradigms are sufficiently different, viz. analog systems and digital systems, and the elaboration of (c) by integrating both languages into a single system description environment. The corresponding tasks are designated by A (analog), D (digital) and E (environment).
The reason why originally two languages were envisaged (for analog and digital systems respectively) is that the sigma calculus, which ultimately led to a unified language, was developed in the course of the project itself. The unification of language concepts merges the tasks A and D for what concerns item a (language design), but for obvious epistemological reasons does not extend to the design paradigms, and hence for item b tasks A and D remain essentially different.

3.1. PROJECT PARTNERSHIP

The Computer Science department of the University of Nijmegen was the originator and prime contractor for this project. Since this university does not have a School of Engineering and hence no research or teaching programme in Electronics, it was considered essential that a least one of the other project partners be actively involved in analog electronics research. The Electronics Department of the Technical University Delft has an outstanding record in this respect, in particular regarding the theory and design of high-performance (in terms of linearity, dynamic range and noise characteristics) analog circuits such as amplifiers, oscillators and mixers. Its research activities are characterized by attention to systematic design methods and to the methodological aspects of CAD for analog circuits.
In order to maintain continuous awareness of the practical applicability of the research results in industries or institutions that develop or use CAD systems, the companies Sagantec (Eindhoven) and Bell (Antwerpen) were also included in the partnership.
The task distribution was as follows:

University of Nijmegen: as the prime contractor: the overall project management, in particular maintaining the "conceptual unity", and as a project partner: the implementation of the system description environment (in cooperation with Sagantec).

Sagantec (Eindhoven): supporting the prime contractor in organizational affairs, implementation of the system description environment, monitoring the relevance w.r.t. industrial CAD needs, and integration of the software aids developed in this project, also with CAD systems or programs developed elsewhere.

Technical University Delft, Electronics department: elaboration of the language concepts w.r.t. analog electronic systems, embedding and usage of these concepts in the design process, formalization of parts of the design process, transformational methods, definition of semantic functions expressing various aspects (mathematical models) of analog systems.

Bell (Antwerpen): a task similar to that of T.U. Delft, but oriented towards digital systems.

There was also an informal cooperation with IMEC (Interuniversity Micro-Electronics Centre in Leuven, Belgium) designated as the BIN (Bell, IMEC, Nijmegen) project.

4. Project Results

4.0. LANGUAGE CONCEPTS

At the start of the project, system semantics and language concepts had evolved to the point described in section 2.2. In particular, (uni)directional systems were described using lambda terms (a formalization of function notation), whereas for adirectional circuits a formalism with combinators had been developed for two-port circuits [3]. In contrast with the K and S combinators in the theory of computation, the combinators for non-computational objects have no counterparts in the lambda calculus. Moreover, the partners in Delft have demonstrated that no finite set of combinators can be complete in the sense of being able to describe all compositions of two-port circuits. Therefore a formalism with parameterized combinators as well as a system with variables, the so-called *sigma calculus* (see section 2.3) were developed. The latter turned out the more suitable of the two.

Although, originally, explicit conversion rules were proposed between sigma terms applied to the particular case of directional systems (whereas general sigma terms cover adirectional systems as well) and lambda terms (exclusively for directional systems), it soon became apparent that these results can be obviated by suitable conventions for the syntax of the terms and for the type definitions. As a consequence, lambda terms and sigma terms can be used as subterms of each other, provided directionality is respected. This resulted in a significant unification, such that for *lumped* electronic circuits, both digital and analog, one description language suffices.

4.1. LANGUAGE DESIGN

Based on the concepts described in section 2.2 and 2.3, a prototype version of a system description language called *Glass* was designed in two phases: a first version halfway the project and a final version at the end. Its syntax has been illustrated by the preceding examples, written in the so-called *reference language* (i.e. ignoring the ASCII limitations; the *implementation language* uses the ASCII character set at the lexical level). Its semantics is *open*, in the sense that meaning has to be expressed by explicit semantic functions. The environment provides a data structure format for the abstract syntax on which such semantic functions can be defined in a convenient way, and a package of predefined semantic functions serving as examples to the user.

Moreover, semantic functions have to be defined only for a so-called *kernel language*, which is relatively small and simple (containing essentially only function application and the **where**-construct). The *full language* extends the kernel language by parameterization and macro-definitions. Conceptually as well as in the implementation, sentences in the full language are expanded according to the macro definitions, yielding sentences in the kernel language, before the semantic functions are actually applied.

A more thorough discussion of the laboratory version of Glass is given by Seutter [16].

4.2. ENVIRONMENT DESIGN AND IMPLEMENTATION

The system description environment is intended to support the language Glass and the definition of semantic functions.

In the laboratory version, the abstract parse trees (generated by the language parser from the source text consisting of system descriptions) are represented as Miranda algebraic data types (ADTs) as proposed by Turner [17]. In addition, the environment offers facilities for accessing these parse trees without having to deal with the details of this representation.

With the set of abstract parse trees being the domain, various semantic functions can be defined either in functional style using Miranda or in imperative style using Pascal or C.

Semantic functions need not always be defined to their full extent. One class of semantic functions are just transformations of the abstract syntax trees into input data for

 – numerical circuit analysis and simulation systems such as SPICE;
 – symbolic formula manipulation systems such as MACSYMA, Reduce, Maple;
 – automatic VLSI layout generation systems.

Hence existing CAD systems can be integrated into this environment.

To make the language implementation flexible w.r.t. changes during the project, and since the environment depends critically on the language definition, it was found necessary to automate part of the environment implementation process itself. A rather thorough evaluation of existing environment generators unfortunately revealed that none of these were suitable for our purpose. Hence environment generation programs had to be developed in the context of the project itself, namely TM for the C environment and COIL for the Pascal environment. Although developed by different groups (Delft and Sagantec) and slightly different in design, these generators build programs that are fully compatible through the Miranda ADT interface.

4.3. APPLICATION TO SYSTEMATIC DESIGN AND TRANSFORMATIONAL REASONING

The usage of Glass as a notational support in system design is discussed by van Reeuwijk [14] for the analog system case, by de Man [9] for the digital circuit case, and also in the chapter by Vanslembrouck in [16].

4.4. DISSEMINATION AND USAGE OF THE RESULTS

As mentioned in the introduction, Glass and Glue are not international HDL standards and hence in their present form are unlikely candidates for actual usage in production CAD environments. However, the ideas incorporated in them constitute very valuable background for two groups of professionals, namely

 – CAD language designers and implementors: in view of the advantages of functional language concepts and the principle of system semantics in the design of future system description languages;

– for electronic circuit designers using CAD languages: in order to evaluate new languages in a more critical and knowledgeable way, and as a background for using existing CAD more sytematically and effectively.

For this purpose, a public domain release package consisting of
– the implementation of the language Glass and its environment Glue (on magnetic tape),
– documentation in the 2-volume set *Books of Forfun* by van Reeuwijk [15] and Seutter [16] is available at reproduction cost to interested parties.

Furthermore, the present standard for hardware description languages, namely VHDL, is scheduled for restandardization by the IEEE in 1992. The concepts of Glass and system semantics will prove very useful input to this effort. Through the appropriate channel, the VASG (VHDL Analysis and Standardization Group), a collection of language change proposals will be submitted ranging from local but significant improvements to a more orthogonal overall language structure. Titles of the projected change proposals are:
– Separation of syntax and semantics for purely structural description in VHDL.
– Function notation and implicit names for describing interconnections of directed components in VHDL.
– Unification of typing and component instantiation in VHDL.
– An analog system and circuit description subset for VHDL.
– Enhancing the VHDL type system by simplification.
– The introduction of higher-order functions in VHDL.

Conclusion

This project has demonstrated that a very small (non-baroque) language, provided it is based on a coherent set of basic principles, can cover a wide diversity of systems and system aspects. As with any development in language design, its contribution to actual production can only be indirect. Appropriate action to enhance this transfer process has been taken.

References

[1]   J. Backus, "Can programming be liberated from the von Neumann style? A functional style and its algebra of programs". *Comm. ACM 21*, 8, pp. 613-641 (Aug. 1978)

[2]   R.S. Bird, P. Wadler, *Introduction to Functional Programming*. Prentice-Hall (UK), Hemel Hempstead (1988)

[3]   R. Boute, "System semantics and formal circuit description". *IEEE Trans. Circ. and Syst.*, *CAS-33*, 12, pp. 1219-1231 (Dec. 1986)

[4]   R. Boute, "System semantics: principles, applications and implementation". *ACM Trans. Prog. Lang. and Syst. 10*, 1, pp. 118-155 (Jan. 1988)

[5]   R. Boute, "Syntactic and semantic aspects of formal system description", *Microprocessing and Microprogramming 27*, 1-5, pp. 155-162 (Sept. 1989)

[6]   R. Boute, "On the formal description of non-computational objects", in: G. David, R. Boute, B. Shriver, eds., *Declarative Systems*, pp. 99-123. Elsevier Science Publishers B.V. (North-Holland) (1990)

[7]   W.H. Burge, *Recursive programming techniques*. Addison-Wesley, Reading, Mass. (1975)

[8]   J.R. Burstall et al., "HOPE: an experimental applicative language". *Report CSR-62-80*, University of Edinburgh (Feb. 1981)

[9]   J. De Man, "Transformational design". *Workshop on analog and digital circuit design.* Sagantec, Eindhoven (May 1989)

[10] D.E. Knuth, "Semantics of context-free languages", *Mathematical Systems Theory 2*, 2, pp. 127-145 (1967)

[11] D.E. Knuth, *The Art of Computer Programming, Vol. 1: Fundamental Algorithms* (2nd edition). Addison-Wesley, Reading, Mass. (1973)

[12] P.J. Landin, "The next 700 programming languages", *Comm. ACM 9*, 3, pp. 157-166 (March 1966)

[13] L.G.T. Meertens, "Algorithmics — towards programming as a mathematical activity", in: J.W. de Bakker et al., eds., *Mathematics and Computer Science*, CWI Monographs, Vol. 1, pp. 289-334. North-Holland (1986)

[14] K. van Reeuwijk, "Design of analog circuits with Glass". *Workshop on analog and digital circuit design.* Sagantec, Eindhoven (May 1989)

[15] K. van Reeuwijk, ed., *Glass environment implementation and maintenance*, Version 1. University of Nijmegen, Department of Computer Science (9 May 1990)

[16] M. Seutter, ed., *Glass: a system description language and its environment*, Version 1. University of Nijmegen, Department of Computer Science (9 May 1990)

[17] D.A. Turner, "Miranda: a non-strict functional language with polymorphic types", in: J.P. Jouannaud, ed., *Functional programming languages and computer architecture*, pp. 1-16. Springer LNCS 201. Springer-Verlag, Berlin (1985)

Project No 1219(967)
PADMAVATI

PARALLEL ASSOCIATIVE DEVELOPMENT MACHINE AS A VEHICLE FOR ARTIFICIAL INTELLIGENCE.

Patricia GUICHARD-JARY
THOMSON-CSF Division CIMSA-SINTRA
160 boulevard de VALMY
B.P 82
92704 COLOMBES -FRANCE
tel (33) 1.47.60.3637

Summary

This ESPRIT Project, started in 1986, is a cooperative effort between THOMSON-CSF (prime contractor - France), GEC (United Kingdom), and FIRST (Greece). PROLOGIA (France) and NSL (France) are also involved as sub-contractors.

Its objective is to develop a machine for symbolic and time critical applications, such as image recognition, natural language, and parallel expert systems. PADMA-VATI is a MIMD machine connected to a SUN host. It is composed of 16 Inmos T800 transputers, with a 16Mbyte DRAM. A Content Addressable Memory with a capacity of 48kbytes (up to 680kbytes) is also attached to the memory bus of each transputer and is especially usefull for symbolic languages. A dynamic Network and a static ring connecting the 16 transputers provide non local communications. Symbolic languages, such as Lisp, Prolog, extended with parallelism and associativity mechanisms, have been ported on this architecture.

In spite of the fact that PADMAVATI is a transputer based architecture, it has innovative features that no other current transputer machine has, some of which will be incorporated in the next generation transputer architecture which INMOS plan to release in two years time. Thus, this project can considered as an early prototype for the next generation.

1. Introduction

The ESPRIT I programme favours research in the domain of computer architecture, and supports development of new multi-processor systems and new processors on which these architectures are based. PADMAVATI (Parallel Associative Development Machine As a Vehicule for ArTificial Intelligence) -ESPRIT Project 1219(967) - belongs to the group of distributed multiprocessor machines.

PADMAVATI focuses simultaneously on the hardware level and the parallel computational model.

The objectives are to build a parallel machine, dedicated to artificial intelligence applications. Our main interest in this project was to do away with some of the limitations of transputer architectures: communications are supponed efficiently only between directly linked processors and only synchronous communications are allowed. The **PADMAVATI** architecture offers a more general parallel programming model.

In order to define this novel architecture, we concentrated our research on the following important points:

- Processor performance:
  We chose as processor the INMOS T800. It has good performance: 10MIPS and 1.5MFOPS with a 20MHz clock (15MIPS, 2.25MFLOPS with a 30MHz clock). Furthermore, multi-processing is possible, with high performance context switching (less than $1\mu s$).
- Efficient information management: (memory)
  Languages for artificial intelligence often get information from the memory in an associative way: for example, in Prolog to solve a goal, the candidate clause is choosen according to the value of its predicate and even the value of its arguments; in Lisp, we can manage association lists (A-Lists), composed of pairs "(attribute, value)" and accessed by giving the attribute value. These two examples suggest that associative or "content addressable" memory (CAM) [11] is interesting, in addition to a conventional (position addressed) memory. Associative access methods in a RAM such as perfect hashing techniques have often been used to compensate for the lack of associative memory [10] [9]. In PADMAVATI, an associative memory has been added to each transputer.
- Fast and flexible communications:
  The transputer has four links, that allow it to be connected to four other transputers at the most. The speed of the links is 5, 10, or 20Mbits/s. This fast and multiple communication, in addition to its computing performance, makes the transputer the best processor with which to build distributed architectures. Different well-known topologies of processors network can be built with transputers, such as torus, hypercube (with different dimension) ... [12]. In such topologies, distance between some processors is sometimes large, and tends towards to reduce communication performance. For example, in a three dimensional hypercube, the distance between any two nodes varies from one to three. Because of this, we developed a dynamic network, being able to interconnect all the transputers together, and to keep the initial links speed. This dynamic network allows to consider at any time all the transputers being equidistant from the others, and then guarantees a constant communication time (this time only depends on data length). This equidistance property allows the user to place processes of his application on any processor of the machine.
- Extensibility, modularity:
  The dynamic network is composed of "8x8 basic elements" (8 inputs, 8 outputs), that can interconnect 8 transputers. Its DELTA-like topology has been chosen in order to make it extensible, to interconnect up to 256 transputers.
  As far as the associative memory is concerned, the memory bus extension

interface has been developed in order to be able to cascade up to 16 CAM boards, and then extend the CAM capacity up to 16x10656 words ($\approx$160Kwords)

All these features have then been integrated into the computational model defined for this machine, and used in Prolog and Lisp languages at the high level.

In the following parts, a technical description of the machine and the software implemented on it will be given including description of:
- the different components of the machine and their characteristics,
- the communciation system allowing processes to exchange information with any other process located on any transputer,
- high level languages, extended to exploit as much as possible the features of the underlying architecture.

In the conclusion, we reiterate the interesting and original features of the project, and present it as a prototype for the next generation of transputer machines.

2. Architecture

The PADMAVATI prototype is composed of 16 processing nodes (PN) connected in a ring (for system needs such as application code loading or debugging), and also connected to the dynamic network. On each processing node, a system (see section 5) is responsible of the communications between remote processes through the network. The prototype is then connected to an UNIX host machine (SUN in this case), via a communication processor (CP). This communication processor is also a transputer, one of its links being connected to the VME bus, and two others being linked to two PADMAVATI processing nodes. The host machine is in charge of managing and answering i/o requests coming from processes located on PADMAVATI.

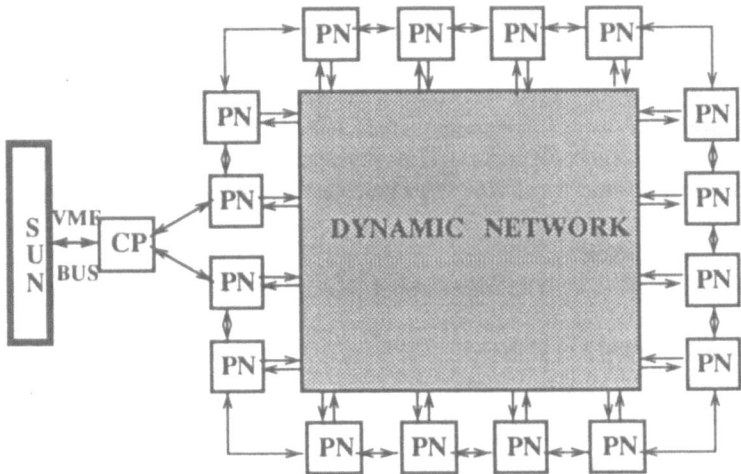

Figure 1: PADMAVATI Architecture

230

3. Processing Nodes

The basic elements composing a processing node are the following:

– Processing Element (PE), composed of:
- transputer
- RAM

– CAM: optional

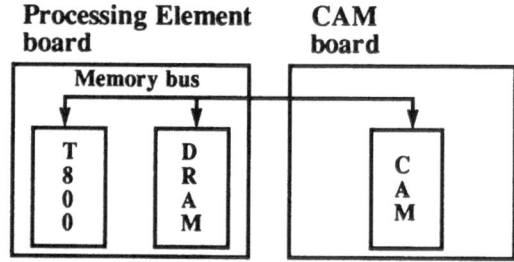

Figure 2: PADMAVATI Processing Nodes

The choice of memory capacity (RAM or CAM) per processor is left to the user, allowing this latter to build the machine corresponding to his needs. The memory capacity can then be different from one node to another.

3.1 Processing Element.

Choseen processor is a transputer T800 D-G20S, with a 20MHz clock (cycle time is 50ns). Its computing performances are 10 MIPS and 1.5 MFLOPS [7]

Four serial "full duplex" links, whose speed is 20Mbits/s, connect this transputer to its two neighbouring transputers, and to the dynamic network. Up to 16 Mbytes of DRAM are available on each transputer. This DRAM is composed of 1Mx9bits SIPS (1 bits is used for parity control). The transputer memory bus has been extended, in order to connect the CAM. The developed interface has the following features:

– demultiplexed address-data
– transformation of the synchronous feature of transputer interface into an asynchronous one
– daisy chaining of multiple CAM board
– interrupt capabilities with interrupt servicing register
– no multi-master capability

3.2 Associative Memory

Content Addressable Memory (C.A.M) or Associative memory [11] not only stores data, but can also access that data in parallel. Each bit of storage can compare the value it contains with a broadcast value (the search key). Bits are arranged in words and if the whole word matches (all its bits match the corresponding bits in the search key) then a flag bit associated with the word is set. The flags can then be used to select which words take part in subsequent operations. For instance it is possible to write the same value to all flagged words simultaneously.

Thus, CAM is accessed by specifying a value instead of a location. This would be of limited use if it was only possible to tell whether any data matched the search key or not, but it is also possible to retrieve data associated with matching data. This can be done with a partially specified search key where some bits are masked or "don't care". Alternatively, associated data can be stored in consecutive words thus allowing access by a combination of matching and relative location.

PADMAVATI CAM is organised hierarchically. A CAM array consists of a number (1-16) of cascaded CAM boards. Each CAM board contains a number (1-6) of CAM modules, and each CAM module contains twelve CAM chips. Each CAM chip contains 148 words, and so the maximum size of a CAM array is 170,496 words [6]. CAM capacity may be different on each node. The CAM is accessed by the transputer as a memory mapped peropheral. The upper address bits select the CAM array base address and lower address bits carry the encoded CAM operation (read, write, search etc ...). Other address bits are used to select which modules will take part in the current operation. The cycle time for CAM operations will be approximately 400ns. Although this is slower than a RAM cycle, mostly due to the CAM's perforrnance derives more from its inherent parallelism than from its cycle time.

A CAM chip is composed by the following elements:

- CAM cell array: each chip contains an array of 5328 static CAM bit cells arranged as 148 36-bit words. The bits in a word are grouped as four eight-bit bytes and four extra data bits. The number of words is determined by VLSI processing constraints and the width of the word was chosen to match the transputer data bus. The extra data bits can be driven by the processor's address lines.
- Flag bit: each word w has a flag bit, F[w], which stores the result of a search and can be used to select wether that word takes part in subsequent operations, for example to write simultaneously the same value in all words which are flagged, or to get the value contained in some of these flagged words.
- Exact bit: each word w has also an exact bit, E[w], which is used during a search operation, to determine the way the word matches with a search key. When $E[w] = 1$ all bits must match exactly, when $E[w] = 0$ the stored word may contain "don't care" bytes, indicated by having their top bit clear.
- Internal priority tree: this is used to determine the first requesting chip or to propagate the results of a search operation done on the whole CAM.

232

- Mask register: this allows to mask some bits of the search key during a search operation
- Write-Enable register: this is a register whose contents determine which bits of CAM words are updated in a Write operation.
  Both registers are 36 bits wide: 32 ordinary data bits, and four extra data bits.

The CAM chip is a fully-custom CMOS VLSI device developed in the PADMAVATI project. Early simulation results showed a speed-up of from three to eight times for Prolog clause selection using the CAM and much greater speed-ups should be optainable depending on the application.

3.3 Configuration

In the PADMAVATI prototype, a processing node is composed of two extended double-europe boards (233x280mm2), with two DIN41612 96 points connectors (P1, P2):

- first board (named "mother board") contains:
  - T800 transputer
  - 8 Mbytes DRAM
  - buffers (TTL and differential) for signals on the 4 links
  - 4 sockets for 28 pins JEDEC memory chips (SRAM, EPROM)
  - memory bus extension, used by the "daughter board" (see below)
  - 7 segments display to monitor the processor
  - "daughter board", plugged in the mother board, that contains:
    - 8 Mbytes DRAM
    - extended memory bus interface, in order to connect the CAM board to the processor board (via P2 connector)
- second board contains the associative memory

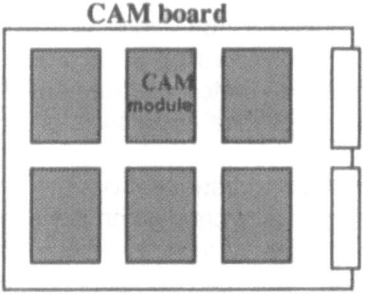 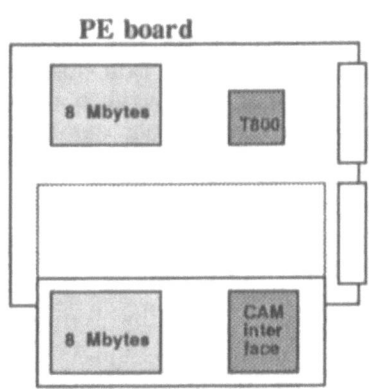

Figure 3: PADMAVAVATI boards

4. Dynamic interconnection network

4.1 Introduction

The dynamic interconnection network developed in PADMAVATI belongs to the DELTA network family. The network allows each node to establish a connection at run time to any other node, just using the number of the destination node. The connection is then like any other transputer link connection, the network being transparent after path establishment.

PADMAVATI dynamic network has the following features:

– Inputs and outputs of the network are directly connected to the transputers links
– message routing into the network is self-arbitrated, using a path switching protocol. Path establishment will not disturb existing paths (in case that there is no conflict). In that sense, the network is totally dynamic.
– basic component of the network is a VLSI device developed for the PADMAVATI machine with eight inputs and eight outputs. Up to 256 transputers can be interconnected by cascading several chips.
– The chip can be used for different network topologies, such as delta (the one used in PADMAVATI), torus or hypercube.

4.2 Network operation

For homogeniety and modularity reasons, we use two links to connect a transputer to the network, each link used only unidirectionally (-transputer links are bidirectional-). Message are sent one one link and recived on the other. This means that the network chip need only handle unidirectional paths from eight inputs to eight outputs, expandable to 256 inputs and 256 outputs.

When a transputer wants to send a message to another transputer, it sends on its output link one byte containing the address of the destination transputer (transputer address = transputer number). Two cases may occur:

– the corresponding path is available: the internal arbiter of the network establishes the path, and transmits the address byte through the rest of the network. When this byte reaches the destination transputer, a hardware acknowledge is sent back to the sending transputer. This latter can then send the whole message to the destination transputer: the network is now totally transparent!
– the corresponding path is not available: the request byte (containing destination-transputer address) is registered inside the network and waits for availability (the sender is of course blocked!)

When the whole message has been received, the destination transputer sends an acknoledgebyte back on the (otherwise unused) return half of its input link. This byte (which is the only one going that way on an input link) is considered, by the network, as an "end of message" byte, and is used to close the path.

4.3 Network VLSI

As the network VLSI is a 8x8 switching element, we only need one VLSI to dynamically interconnect 8 processors. In the delta topology, the VLSIs are cascaded according to the "n Log 2n" scheme: Logn levels, each containing n VLSIs, where n is number_of_processors/8 (8 = number of entries of a network VLSI). Thus, PADMAVATI prototype (having 16 transputers) network is composed of 4 VLSIs.

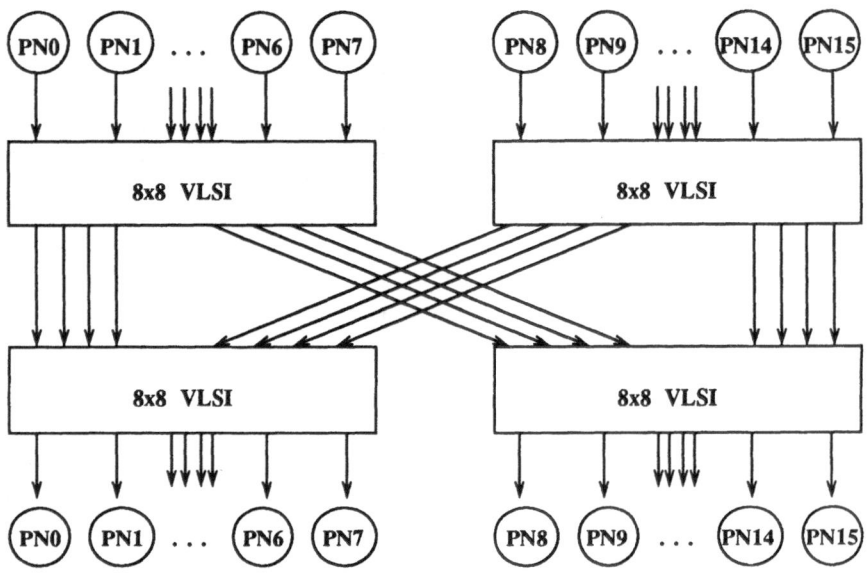

Figure 4: PADMAVATI network

An internal 40MHz clock generates two 20MHz phases, for a basic cycle of 50 nanoseconds.

Inputs and outputs of the network use the INMOS link standard. The chip has 16 link interfaces for messages and a 17th which is used to intialise.

4.4 Network board

The network board of the PADMAVATI prototype is composed of:
- 4 network VLSIs (to interconnect 16 transputers)
- 1 clock generator (40MHz)
- links buffers as on the processor board

4.5 Transfer speed

Each link speed is 20Mbits/s, this gives us a unidirectional rate of 1.2 Mbyte/s. As there are two links per transputer connected to the network (one for sending, one for receiving), the possible maximal rate of transfer is 2.4 Mbytes/s for each node. This rate is not altered by the path establishement time, which is only 1 μs. These figures give us a very good idea of the advantages of this kind of network using a path switching protocol.

For an optimal use of this network, long messages may be split into packets. The optimal size of packets must be defined as a good compromise between the time a path is blocked due to conflict and the overhead due to address transfer and establishing the path. A simulation by discrete events of the network behaviour has been done. The varying parameters were:
− size of the message (number of bytes)
− size of the machine (number of processors)
− distribution of requests.

The results for the PADMAVATI prototype (where there are 16 processors, and distribution follows a uniform law) were the following:
− optimal size of packets = 256 bytes
− throughput of the network = 55%, when
  - size of packets = 256 bytes
  - 16 processors
  - uniform law, saturating the network

5. Software Architecture

Our objective is to provide the user with a system which can run parallel applications with parts written in different languages and loaded on different nodes that can be transputers or the host(see next section).

An interprocesses communication system (called RTS - Run Time System -) has been implemented on PADMAVATI. It exploits all the particular features of the network (dynamicity, transfer speed), and provides, through high level languages, tools to the user for interprocess and interprocessor communications. These tools make the network and transputer physical addressing transparent to the user.

Our system is developed on top of synchronous and point-to-point communications provided by transputer, extends this protocol and provides multi-point, asynchronous communications.

Our model is based on communication by message passing. We define a message as: data (with variable size) + a priority + a tag (+ destination). So, a message reception can be selective, depending on the given priority, and/or the given tag.

The destination of a message is a port. A port is identified by the user with a name, and by the system with a (node number, local address) pair.

A port is like a mailbox, to which some processes can send messages, and from which others can read the messages. Thus, processes send and receive messages on named ports, and don't have to know the physical location of their communicating processes.

Following operations on ports are defined:

- port_id = create_port(port_name)

- register_port(port_name, port_id)

- port_id = find_port(port_name)

- kill_port(port_id)

- send (port_id, message)

- receive (port_id, &message)

A centralised "Name Server" records names of ports and their location on PADMAVATI nodes (= port identifier). RTS (distributed on each node) manages creation of ports, sending and reception of messages.

This model allows non-local as well as local communications in an homogeneous way.

The two most important aspects of our system are, as said before, extension of the transputer communication protocol, and performance. Extension has been described above. Performance must be balanced against flexibility - the more sophiticated the message handling system, the greater the overheads-. We found a good compromise, and provide a system where the overhead is less than $400\mu s$.

6. Languages for A.I.

Languages for Artificial Intelligence developped in this project are Prolog and Lisp. These languages are studied according to the hardware components, the architecture, and the system of the PADMAVATI machine:

- transputer
- associative memory (Content Addressable Memory)
- multi-processor machine with distributed memory
- message-passing
- asynchronous communication
- dynamic process management

For each language, an interpreter and a compiler are implemented, running on the transputer node: interpreter to get an interactive interface, compiler producing transputer code to run applications fast.

The host been considered as a PADMAVATI processing node, particular implementations (semicompiler, interpreter or simulator) for Lisp and Prolog run on the SUN.

First, the PADMAVATI Prolog system will be presented, and then the Lisp one.

6.1 Prolog

We propose to the user a Prolog system, where we have followed the BSI proposed standard for syntax and built-in predicates [3] [8].
Different ways to optimize the Prolog system on PADMAVATI have been studied:

6.1.1 Compilation

Objective is to take advantage as much as possible of the transputer by compiling high level languages into transputer code.
PADMAVATI's Prolog compiler produces Warren Abstract Machine (WAM) code [14] [15], which is then either transformed into transputer code to run on PADMAVATI (whole compilation), or interpreted to run on the host (semi-compilation).
Performance of the first version of the sequential Prolog compiler running on transputer is 15 KLIPS. We are still working on the compiler to complete it and improve it, and our objectives in terms of performance are at least 30 KLIPS (still for the sequential Prolog).

6.1.2 Parallelism

The second way to accelearte Prolog is to exploit the **PADMAVATI** architecture.
Parallelism in Prolog has been investigated through two systems: one is based on message passing (coarse grain parallelism), the other one uses a fine grain parallelism.
Coarse grain parallelism is defined as follows: an application is divided into big tasks, that can run simultaneously, and exchange messages - data, or goals to be solved -. This model is built on top of the RTS model. So, built-in predicates have been added to handle ports and message passing.
A fine grain parallellism operates within a Prolog clause rather than at task level [1]. We have only implemented AND-parallelism, denoted by "&&" and sharing of variable, e.g.

P(...) :- Q(...) && R(...).) .

First measurements of a parallel FIBONACCI program, running on four transputers showed that the speed-up is almost linear with the number of processors:

- 2 processors: 1.3 times faster than sequential version
- 3 processors :1.7 times faster
- 4 processors: 2 times faster

These two systems are not incompatible and may constitute only one system at the end of the project. Concurrent Prolog tasks may also use fine grain parallelism within their resolution.

6.1.3 Acceleration of clause selection

In this way of optimizing Prolog, we especially use the associative features of PADMAVATI. Efficient clause selection can avoid useless unifications and some choice points creations. Two methods have been studied, one uses a purely software method - Hash-coding to select clauses [10] -, the other one uses the Content Addressable Memory.

The clause selection mechanism is divided into two parts: one part is static (during clause analysis) and consists of computing all clauses heads of a Prolog program in order to put them in memory, and the second part is dynamic (during resolution) where candidate clauses are found for the resolution of the current gaol.

Measurements made on several Prolog programs running on a mono-transputer Prolog interpreter, show that the speed-up gained by using software associative clause selection is between 3 and 17. We hope that hardware selection (with CAM) will provide a better speed-up.

6.1.4 Intelligent Back-tracking[1][2]

This study is an enhancement of the Prolog language, and consequently does not exploit the underlying architecture.

The "intelligent back-tracking" method allows back-tracking to a choice point which is the most likely reason for failure (this reason for failure is determined by using a "dependence graph", which records dependences between variables and clauses in which these variables have been bound), instead of taking the last choice point as it is usually done. The resulting speed-up is important, but Prolog semantic change a little because of side effects.

6.1.5 C language interface

It is sometimes difficult, and not efficient, to write predicates in Prolog. We therefore allow the programmer to define the programmer to define predicates and functions in C and then use them in the same way as built-in predicates.

6.2 LISP

LE_LISP, from INRIA [4] [5], has been choosen to be implemented on transputer, and extended with parallel primitives.

Domains especially investigated in LE_LISP are:

6.2.1 Compilation

The LE_LISP Compiler produces an intermediate virtual machine code (LLM3), which is then transformed into transputer code to run on PADMAVATI.

6.2.2 Parallelism

The computational model is based on the RTS model, and uses a coarse grain parallelism with message passing. A LE_LISP application is then divided into concurrent LE_LISP processes, statically allocated on the processing nodes. LE_LISP has been extended with primitives to handle ports and message passing.

6.2.3 Use of CAM

Different ways to use the CAM in LE_LISP have been envisaged:

- storage of association lists ((attribute1.value1) ... (attributen.valuen))
- storage of property lists ((symbol1 value1 indicator1) ...)
- memo-functions - results of a function are cached and retrieved if the function is called again with the same arguments.
- direct user access via special commands.

Currently only the last idea has been implemented. Obviouly, we cannot give any results on the gain.

6.3 Communication between languages

C has not yet been discussed, but it also runs on the machine. As message passing primitives are provided by C libraries it is easy to write parallel C programs for PADMAVATI.

The parallel computational models of Prolog, LE_LISP, and C are the same, based on message passing using ports. So, these three languages can communicate. Typed information (e.g. lists, numbers) is exchanged between languages in an intermediate form, partly to overcome differences in internal representation and partly for efficiency (minimum message length).

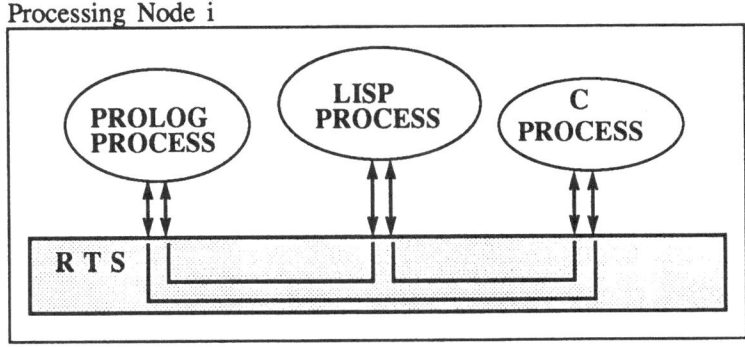

Figure 5: Cooperation between different languages

7. Conclusion

The objective of the project was to develop a high-performance parallel machine, based on transputers, but without the limitations of conventional transputer architecture. This led us to define PADMAVATI as it has been presented above, and to develop specific components such as the dynamic interconnection network, and the CAM.

Four keywords characterize the novel features of this machine:

- Dynamicity (in communications):
  Avantages for the users are obvious:
    - network is dynamically self-configurable ⇒ implementation of parallel algorithms is easier (because it doesn't depend on a static topology)
    - ports are dynamically created ⇒ the number of communication paths between processes can vary at run-time with no static limits.
    - Processors are equidistant (owing to the network)
      ⇒communication time between two any processors of the machine is constant
      ⇒placement of tasks on processors is simplified (because it doesn't depend on the parameter "distance between processors")
    - dynamic communication extends the range of parallel algorithms which can be implemented efficiently (some parallel algorithm cannot be implemented -or badly- with static communications
- Associativity (in data access)
  Associative and parallel access to data leads to better performance of symbolic languages, because it avoids combinatory explosion of the search tree.
- Heterogeneity (in cooperating languages)
  Cooperation between Lisp and Prolog processes is simplified by the choice of parallelism by message passing.
- Modularity
  PADMAVATI can be considered as a set of boards, allowing the user to build a machine corresponding to his needs. This leads to a large set of systems, whose size and performance may be very different.
  Size: for example, a machine having 8 processing nodes without CAM, may be composed of only 1 board containing the network and 8 transputers; and the same machine with CAM is composed of 17 boards: 1 network board, 8 transputer boards, 8 CAM boards.
  Performance: as CAM is optional, performance may change a lot. It is also possible to put T800-25MHz (or T800-30MHz) transputers on boards instead of T800-20MHz, to have better processing performance.

PADMAVATI is at the moment the first machine based on tranputers, which has such characteristics.

The network and RTS provide today the same functionnality as the virtual channels between distant transputers which will be available with the next generation of transputers - H1 + C104 - in two years [13]. Performance is obviously somewhat lower than is expected from the new chips because we must do part of the message routing in software and we only have 20 Mbits/s links as opposed to the H1's 100 Mbits/s. This project therefore proposes a step on the upgrade path for those who wish use current transputers but yearn for the promised functionality of the next generation.

The associative memory system offers the potential of vastly increased performance for problems involving searching and matching and applications where it can be used as an SIMD processor array. Initial results concerning acceleration of Prolog are also encouraging.

Acknowledgements

The author wishes to thank here all participants of ESPRIT project P1219(967), particularly Denis HOWE (from GEC) for all his suggestions.

References

[1] Bodeveix J.P, "LOGARITHM: Un modèle de Prolog Parallèle. Son implémentation sur transputers", Thèse Docteur en Sciences. Université Paris-Sud. Janvier 1989

[2] Bruynooghe M., Pereira L.M ,"Deduction Revision by Intelligent Backtracking" Implementation of Prolog, ed. Campbell, 1984

[3] "Prolog syntax: draft 5.0 and comments", Report BSI PS/239 (or ISO N9), march 1988

[4] Chailloux J.,Devin M., Hullot J. "LeLisp a portable and efficient Lisp system" Proc. of the 1984 ACM symposium on Lisp and Functional Programming, Austin Taxas, August 1984

[5] Chailloux J., "Manuel de Reference LeLisp 15.2" ,3ème Edition, Novembre 1986

[6] Howe D., "Padmavati CAM chip functional specification", Internal report REP- 14, July 1989

[7] INMOS, The Transputer Data Book, First edition 1989

[8] "Prolog built in predicates", Report ISO N28

[9] Jary P., De Joybert X., "Selection de Clauses en Prolog". Actes du 8ème Séminaire Programmation en Logique, Trégastel,24-26 mai 1989

[10] Knott G.D., "Hashing Functions", The computer Journal, vol 18, no 1, pages 38-44, 1975

[11] Kohonen T., "Content Addressable Memories", Springer series in Information sciences, 1980

[12] Nicole D., Lloyd E., Ward J. - "Switching Networks for Transputer Links". 1989

[13] Pountain D. - Virtual Channels: The Next Generation of Transputers - BYTE, April 1990

[14] Warren D.H, "Implementing Prolog", D.A.I research reports no 39, 40, May 1977

[15] Warren D.H, "An abstract Prolog Instruction Set", Technical Note no309, SRI International, Menlo Park, 1983

Project No. 1158

ATES: AN INTEGRATED SYSTEM FOR SOFTWARE DEVELOPMENT AND VERIFICATION

P. COUTURIER, A. PUCCETTI
C.I.S.I. Ingenierie
3 rue Le Corbusier, SILIC 232
F-94528 RUNGIS Cedex
FRANCE
Tel: + 33149 79 46 05

Abstract.

In this paper we present a programming system prototype, based on a high level, abstract language, able to express the specifications necessary to develop reliable software, in a program-to-proof approach. Within this approach, we want to conceive a program and introduce the clements necessary for its proof, at the same time. (Those formal proof elements consist in logical assertions expressing mathematically what an algorithm does and logical properties of the function realized by the algorithm). Those proof elements will be used by the system, to verify the correctness of an algorithm, guided by an interactive proof checker.

1. Introduction

ATES is the number 1222 Esprit project, which has started in 1985. The following european partners are involved: CISI Ingenierie (France), Philips Research Laboratory Brussels, University of Liege (Belgium), University of Twente (The Netherlands), Université de Paris VII (France), and Commissariat à l'Energie Atomique (France). The objective of this project is to establish the potential of the integration of a number of advanced techniques in the field of computer programming of software systems for scientific and technical applications and to provide an integrated environment to conceive and implement reliable software (from here the name of the project ATES, as an acronym of Advanced Techniques integration into Efficient scientific Software).

In recent years particular advances have been made in the areas of complete abstraction of data types and operators, formal specification and proof. Now the technology for proving formally the correctness of programs seems to be mature for real use on medium scale sequential programs. In the present paper we will focus on the main ideas involved in the ATES method for program proof. This interactive method will make use of tactics and strategics, we shall describe in details.

Using VDM ([11]), the development of a program consists of several steps of refinement and at each stage verifications should be carried out. The verifications

242

are conducted within the framework provided by the proof rules, either formally or rigorously (i.e. by appealing to intuitive arguments) according to the cost-effectiveness of either approach within the context of the particular application. It is this approach to the entire development process, and the flexibility allowed, that make VDM a most attractive development method. However, the overheads involved in carrying out these various levels of refinement are considerable and any savings in effort would be beneficial. We suggest an alternative development scenario, which is based on a methodology leading from a program to its proof: in a first step some algorithm is written, then its formal specifications, and finally the total correctness of this algorithm is proved, guided by the interactive proof checker of the ATES system. Compared with other programming environments, like MENTOR, ALPHARD, System B and LCF, we intend to realize a best balance between algorithms programming, formal specifications and proofs. In fact, systems like MENTOR ([6]) are too much oriented toward syntactical program transformations. Others, like ALPHARD ([16]), are mainly interested into formal algebraic specifications and transformations. Lastly, systems like LCF ([7]) and B ([1]) are mostly concentrated on proofs.

This led us to create a new programming language, able to express in a structured way the algorithmic realization and formal specification of a program. The originality of the ATES system lies mostly in the interactive use of general purpose strategies and tactics, and this point is essential when realizing in a stepwise manner a correct program. Indeed, most proof systems do behave poorly when fed with incomplete or incorrect data, and the link between the failure and the source code joined with its specification often cannot be easily established. Thus the ATES system suggests a method in which a close link between all the different data is kept during the entire process of program development, enabling the user to keep in touch with the algorithm when performing the proof. In a first part we introduce the basic programming elements of the ATES language, the description of its main properties, its environment and the guidelines for specifying mathematically algorithms, in such a way that the system can do their proof. Then, is explained how the ATES system establishes the correctness of an algorithm, given some formal specifications. A first experience with a theorem prover will be described. Then, a simplification tool proper to the ATES system and based on natural deduction, will be explained. In a third part, we describe the state of the ATES prototype, and mention some already written experiments carried through. Finally, some concluding remarks will be made, about the industrial interest of this system.

2. Short Description of the ATES Language

The general programming language used in this system (see [2] and [3]), offers possibilities to express in a structured way the following kinds of components:

244

- The abstract headings of elementary ATES objects: operators and abstract data types. They give an external view of the elementary objects, without any relation to their internal representation, and also some syntactical notation for them.

    Example: The heading of a function that solves a linear system of equations, of the form "A*x = B" where A is a squared lower triangular matrix (of dimension 100) and b is a vector (of the same dimension), can be expressed in the following way:

    command gauss (var x: vector; const a: matrix; const b: vector)
        syntax gauss: a, b, x

- Implementation descriptions for operators and data types, using a Pascal-like syntax, named algorithms and conerete data types.

Example:

    algorithm gauss
    var i, j, n: integer; inte : real
    begin
            for i : = 1 to 100
            do
            inte : = 0.0;
            for j : = 1 to i - 1
            do
                    *inte : = inte + a[i, j] * x[j]*
            od;
            x[i] : = (b[i] - inte) / a[i, i]
            od
    end

- Modules, describing the implementation choice for each operator and abstract data type by an algorithm or concrete data type. They generate executable code.

Example:

    module gauss
            for operator gauss
            using algorithm gauss

- The axiomatic relative to the problem treated, introducing new mathematical objects: functions, sets and logical axioms. Functions modelize operators and sets modelize data types.
    Example: The problem of solving the triangular system "A*x = B" can rise the following axiomatic, where "sum(j,i,a,x)" designates the mathematical quantity

"$\Sigma_{1 \leq k \leq j}\ a_i * x_k$" which represents the product of row number i of the matrix a by the vector b:

axioms gauss
       using axioms integer, real
var i, j: Int; a: Matrix; x: Vector
const sum: (Int;Int;Matrix;Vector) → Real   *< < function definition > >*
axiom sum_0 = = sum(0, i, a, x) = 0.0   *< < empty sum > >*
axiom sum_j = = sum(j - 1, i, a, x) + a(i, j) * x(j) = sum(j, i, a, x)
                                 < <recursive definition of the sum > >
end

- The pre and post-conditions of any operator (called its specifications), express-ing the following property: if the pre-condition is satisfied before executing the operator in an algorithm, then the post-condition will be satisfied afterwards.
Example: Concerning the operator previously introduced, we can have the following specifications:

specif gauss
pre diagonal_diff_0 = = for k: { 1 ..100 } then a(k, k) / = 0.0
                 < < The matrix must have non zero diagonal elements > >
post control_X = = for k: { 1 .. 100 } then sum(k, k, a, x) = b(k)
                 < < x is a solution of the system > >

- Some additional information coneerning the loop statements of an algorithm must be provided to do its proof. For each loop, we need a pre-condition, a post condition, an invariance property and a monotonie decreasing function (these are named "proof elements" of the algorithm). In fact, the proof of an algorithm will be cut into smaller proof steps, by considering each loop of the algorithm to be proved as a separate operator, with its own pre-condition and post-condition. Then, once a loop is proved, it can be considered as an operator call included into some algorithm or other loop. In this way, the system builds a separate proof for each loop. This feature of the ATES system is essential, because of the necessity to keep the proofs manageable and understandable. The decreasing function will be used to prove the termination of the loop.
Example: A possible set of proof elements for our example can be written:

proof gauss
       do 1 pre_loop_L1: for ii: { 1 .. 100 } then a(ii, ii) / = 0.0
                       < < pre-condition of internal loop > >
       do 1 post_loop_L1:
           a = a' and b = b'
           and (for k: { 1 .. 100 } then b(k) = sum(k, k, a, x))
                      < < post-condition for internal loop > >
       do 1 inv_loop_L1:

$i <\ = 101$ and $a = a'$ and $b = b'$
and (for k: { 1 .. i - 1 } then $b(k) = sum(k, k, a, x)$)
and (for ii: { 1 .. 100 } then $a(ii, ii) /= 0.0$)
<< invariant for the same loop >>

<u>do</u> 3 pre_loop_L2:
inte $= 0.0$ and $(0 < i$ and $i <\ = 100)$

<u>do</u> 3 post_loop_L2:
$a = a'$ and $x = x'$ and $i = i'$ and inte $= sum(i - 1, i, a, x)$

<u>do</u> 3 inv_loop_L2:
$i = i'$ and $a = a'$ and $x = x'$ and $(0 < i$ and $i <\ = 100)$
and $j <\ = i$ and inte $= sum(j - 1, i, a, x)$

<u>end</u>

where "1" (respectively "3") is the index of internal (resp. external) loop statement in the preceeding algorithm, and a' (resp. b') designate the values of a (resp. b) at the beginning of the loop statement.

To resolve a given problem with the ATES language, it is necessary to write objects of each class (operators, algorithms, axioms, specifications and proof elements). According to their class and their range of application, these ones will be grouped into chapters. (For example, operations on matrices of reals, trigonometric calculus,... each one being expressed by a list of five chapters). Further on, a collection of chapters around a certain theme will be grouped into larger entities, named "books". This structuring corresponds to the splitting of a given problem into sub-problems, and is left to the user. These chapters may use some external objects located in other chapters, which are imported by the "using" link:

For example, the previous proof chapter needs to know the definitions, pre and post-conditions of the operators involved; these can be found in axiom chapters or specification chapters of the present book or of some external book. The whole chapter will be written in the following way:

<u>proofs</u> gauss
    <u>using axioms</u> boolean, integer, real, gauss
        <u>specifs</u> boolean, integer, real, gauss
    <u>proof</u> gauss
        <u>do</u> 1 pre_loop_L1: for ii: { 1 .. 100 } then $a(ii, ii) /= 0.0$
                << pre-condition of internal loop >>
        <u>do</u> 1 post_loop_L1:
                $a = a'$ <u>and</u> $b = b'$
                <u>and</u> (<u>for</u> k: 1 .. 100 } <u>then</u> $b(k) = sum(k, k, a, x)$)
                << post-condition for internal loop >>
        <u>do</u> 1 inv_loop_L1:
                $i <\ = 101$ and $a = a'$ and $b = b'$
                and (for k: { 1 .. i - 1 } then $b(k) = sum(k, k, a, x)$)
                and (for ii: { 1 .. 100 } then $a(ii, ii) /= 0.0$)

$$<< \text{ invariant for the same loop } >>$$

<u>do</u> 3 pre_loop_L2:
   $inte = 0.0 \text{ and } (0 < i \text{ and } i < = 100)$
<u>do</u> 3 post_loop_L2:
   $a = a^{\prime} \text{ and } x = x^{\prime} \text{ and } i = i^{\prime} \text{ and } inte = sum(i - 1, i, a, x)$
<u>do</u> 3 inv_loop_L2:
   $i = i^{\prime} \text{ and } a = a^{\prime} \text{ and } x = x^{\prime} \text{ and } (0 < i \text{ and } i < = 100)$
   $\text{and } j < = i \text{ and } inte = sum(j - 1, i, a, x)$

   <u>end</u>
<u>end</u>

The correctness proof of an algorithm (often called the partial correctness, in the literature) is done by the system, when introducing an associated proof chapter. In such a proof chapter, axioms and ATES-specifications are selected. In a first step the system will create an annotated algorithm according to Hoare's method, using the WP (weakest precondition) rules (see [5]) extended to the ATES features and the specifications of the operators used in this algorithm. Formally, the WP of some statement S and some predicate Q, denoted by "WP(S,Q)", is the weakest predicate P (in the sense of logical implication), verifying the property: if P is true before executing S, then Q will be true afterwards. In fact several separate annotations are generated: one for each loop, starting by the inner ones, and one annotation for the whole algorithm, loops excluded.

3.1. Annotating an Algorithm

This step consists in inserting between the elementary statements of an algorithm A, some predicates describing the successive states of the variables, during the computation, and their relations. To do this, the post-condition of A is placed at its end, then, moving upward through A with the WP rules (one for each kind of statements), the intermediate predicates are computed. The annotated algorithm represents in some way, the flow of informations through the instructions of A. The main WP rules we introduced are the following:

– WP of an assignment statement: $WP(x := e, P) = P[x/e]$,

(where [x/e] denotes the substitution of every occurrence of "x" by the exression "e"),

– WP of a conditional statement:

   $WP(\text{ if } L \text{ then } S1 \text{ else } S2 \text{ fi}, P) = \text{ if } L \text{ then } WP(S1, P) \text{ else } WP(S2, P)$

– WP of an operator call: let O be an operator with formal input (resp. output) parameter "x" (resp. "y"), and "O(a,b)" be a call to this operator where "a" is an expression and "b" a variable name; let "pre_O" (resp."post_O") be the precondition (resp. post-condition) of O.

If "post_O" can be written "y = f(x')", where "f" designates some function and "x'" represents the initial value of "x", then

$$WP(O(a,b), P) = P [b/f(a)] \text{ and } pre\_O [a/x,b/y].$$

Otherwise, if O is not an explicit operator, then

$$WP(O(a,b)) = pre\_O [x/a,y/b] \text{ and } (post\_O(x,y\$) = > P(x,y\$))$$

where y$ designates a newly introduced free variable.

– WP of a loop statement: Let S be the loop statement "while L do S1 od" and pre S (resp. post_S, inv_S) be its associated pre-condition (resp. post-condition, invariant). Let "In" be the set of variables used by the statement S but not modified by S, and "Out" the set of variables modified by S. The loop is considered as an operator call "O(In,Out)", and the operator O has pre and post-conditions respectively equal to pre_S and post_S. Then, if the following four conditions are satisfied,

$$pre\_S \text{ and } L = > inv\_S$$
$$pre\_S \text{ and not } L = > post\_S$$
$$inv\_S \text{ and } L = > WP(S1, inv\_S)$$
$$inv\_S \text{ and not } L = > post\_S$$

we get

$$WP(S, P) = WP (O(In,Out), P).$$

A complete description of the WP rules can be found in [3].

3.2. Example

The preceding example leads to the following annotation of the algorithm, where the loop statements are excluded and the top predicate (marked by "(1)"), is a verification condition:

```
        algorithm  gauss
          var  i, j : integer;
               inte : real

{ (for k : { 1 .. 100 }
    then a(k, k) /= 0.0) => (for ii : { 1 .. 100 } then a(ii, ii) /= 0.0)
                and (a$26 = a and b$27 = b                                    (1)
                 and (for k : { 1 .. 100 }
                          then b$27(k) = sum(k, k, a$26, x$28))
                    => (for k : { 1 .. 100 }
                           then sum(k, k, a$26, x$28) = b$27(k))) }
          begin

{ (for ii : { 1 .. 100 } then a(ii, ii) /= 0.0)
   and (a$26 = a and b$27 = b
    and (for k : { 1 .. 100 }
          then b$27(k) = sum(k, k, a$26, x$28)) =>
         (for k : { 1 .. 100 } then sum(k, k, a$26,x$28) = b$27(k))) }

   i1:   for i := 1 to 100
            do
   i2:      inte := 0.0;
   i3:      for j := 1 to i - 1
               do
   i4:           inte := inte + a[i, j] * x[j]
            od
   i5:      x[i] := (b[i] - inte) / a[i, i]
         od

{ for k : { 1 .. 100 }
    then  sum(k, k, a, x) = b(k) }

         end
```

As the algorithm only contains two loop statements, the following annotation is generated for the innermost one, by placing at the bottom of the loop its invariant and computing upward the WPs until the top is reached (predicatcs markcd by "(2)", "(3)" and "(4)" are verification conditions, placed on top of the loop):

```
    algorithm gauss
      var i, j : integer; inte : real
      begin
i1:   for i := 1 to 100 do
i2:       inte := 0.0;
```

$\{ \textit{inte} = 0.0 \text{ and } (0 < i \text{ and } i <= 100) =>$
$\quad (1 <= i - 1 => ((0 < i \text{ and } i <= 100) \text{ and } 1 <= i$
$\qquad\qquad\qquad\qquad \text{and } \textit{inte} = \textit{sum}(1 - 1, i, a, x))) \qquad\qquad (2)$
$\qquad\quad \text{and } (1 > i - 1 => \textit{inte} = \textit{sum}(i - 1, i, a, x)) \}$

$\{ j >= 1$
$\quad \text{and } (i = i` \text{ and } a = a` \text{ and } x = x`$
$\quad \text{and } (0 < i \text{ and } i <= 100) \text{ and } j <= i$
$\qquad \text{and } \textit{inte} = \textit{sum}(j - 1, i, a, x))$
$\qquad \text{and } j > i - 1 => (a = a` \text{ and } x = x` \text{ and } i = i` \qquad\qquad (3)$
$\qquad\qquad\qquad\qquad \text{and } \textit{inte} = \textit{sum}(i - 1, i, a, x)) \}$

$\{ j >= 1$
$\quad \text{and } (i = i` \text{ and } a = a` \text{ and } x = x`$
$\quad \text{and } (0 < i \text{ and } i <= 100) \text{ and } j <= i \text{ and } \textit{inte} = \textit{sum}(j - 1, i, a, x))$
$\text{and } j <= i - 1 =>$
$\qquad\qquad j > 0 \text{ and } j <= 100$
$\qquad\qquad \text{and } (i > 0 \text{ and } i <= 100 \qquad\qquad\qquad\qquad (4)$
$\qquad\qquad \text{and } (j + 1 >= 1$
$\qquad\qquad\quad \text{and } (i = i` \text{ and } a = a` \text{ and } x = x` \text{ and } j + 1 <= i$
$\qquad\qquad\qquad \text{and } \textit{inte} + a(i, j) * x(j) = \textit{sum}(j + 1 - 1, i, a, x)))) \}$

```
i3:       for j := 1 to i - 1 do
```

$\{ j > 0$
$\quad \text{and } j <= 100$
$\quad \text{and } (i > 0 \text{ and } i <= 100$
$\quad \text{and } (j > 0 \text{ and } j <= 100)$
$\quad \text{and } (j + 1 >= 1$
$\quad \text{and } (i = i` \text{ and } a = a` \text{ and } x = x` \text{ and } j + 1 <= i$
$\quad\quad \text{and } \textit{inte} + a(i, j) * x(j) = \textit{sum}(j + 1 - 1, i, a, x)))) \}$

```
i4:               inte := inte + a[i, j] * x[j]
```

$\{ j + 1 >= 1$
$\quad \text{and } (i = i` \text{ and } a = a` \text{ and } x = x`$
$\quad \text{and } (0 < i \text{ and } i <= 100) \text{ and } j + 1 <= i$
$\qquad \text{and } \textit{inte} = \textit{sum}(j + 1 - 1, i, a, x)) \}$

```
          od
i5:       x[i] := (b[i] - inte) / a[i, i]
      od
    end
```

3.3. The Verification Conditions

The adequacy of the algorithm with its specification elements has now to be verified. This second task of the correction proof consists into proving formally the following correction property: *If the pre-condition is true before executing the algorithm and if this algorithm terminates, then the post-condition is true at the end.* This is equivalent to prove the two following assertions, given some algorithm A:

– The weakest precondition of the whole algorithm, with respect to its post-condition, must be deduced from its pre-condition. Formally this can be written:

$$pre\_A => WP(A, post\_A)$$

– For each loop statement S of A, the associated invariant Inv_S (from the proof chapter) must be verified at each turn of the loop and must be true at the beginning and at the end of its execution, given some pre and post-conditions Pre_S and Post_S. Formally, in the case of some while-statement "while B do S od", these conditions can be expressed by the following assertions:

$$Pre\_S => ((B = Inv\_S) \text{ and } (not B => Post\_S))$$
$$Inv\_S \text{ and } B => WP(S, Inv\_S)$$
$$Inv\_S \text{ and not } B => Post\_S$$

(Similar conditions can be written for other kinds of loop statements.)

Concerning our example, we need to prove the predicates (1) to (4); these are inserted in the annotated algorithm or loop too. Now, these predicates have to be verified using some adequate proof tool.

4. The First Experience

In a first experiment, it has been decided to orient the choice toward mechanical provers because in the last resort, the prover is to be integrated in a whole programming environment where the end user is to be, as much as possible, unaware of its existence. In any case, the final user will not converse directly with it because his prime worry is program proving and not theorem proving.

So the task assigned to one partner of the ATES project was to look for a theorem prover, adequate with the ATES language and proof requirements (see [15]). There is an overabundance of theorem provers, each with its own features, designed to answer specific requirements. However, they can be subdivided into two main classes: the interactive and the mechanical theorem provers. Interactive theorem provers act as proof checkers: a user writes the proor with or without the aid of tactics provided by the system. To this category belong LCF ([7]) and System B ([1]). Mechanical theorem provers, onee a user has set some flags, apply their own strategies. To this category belong systems such as Plaisted's theorem prover

([13]), Boyer Moore's theorem prover ([4]), the Illinois Theorem Prover ([8]) and REVE ([12]). The chosen prover was a general theorem prover, called the Illinois Theorem Prover (I.T.P.), written in Franzlisp by S. Greenbaum from the University of Illinois (see [8]). This prover has been integrated in the ATES environment, to perform the proof of the kinds of assertions previously mentioned.

The I.T.P. is based on unsorted first-order logic with equality, and whose basic mechanism is resolution. This consists in proving a given predicate by refutation, searching for a contradiction in the following assertion:

Axiom1 and Axiom2 and ... AxiomN and (not Predicate)

(where "Axiom 1 " ,. . .,"AxiomN" are the axioms collected in the proof chapter, and "Predicate" is a verification condition). The resolution method is applied to the new assertion, beforehand transformed into a clausal form (close to Horn clauses).

Finally, the algorithm A is correct if the predicates (1) and (4) are all proved to be true.

But in concrete cases, the I.T.P. lacks at the following points:

- definition of proof methodology, because of the numerous heuristics embedded in the prover,
- addition of some strategies solving particular kinds of problems, handling of real size predicates, because of the high number of clauses produced,
- feedback of the prover in the case of an erronerous predicate (indeed, running the I.T.P. on a lot of predicates resulted in an infinite loop),
- informations about the failure to prove some verification conditions, which seem correct,
- performance slow down of the prover, when an incomplete axiomatization has been input.

Concerning our example, the proof of the predicates (1) to (4) can be performed by the system using the axioms about integers and reals. These axioms can be found in some basic axiom chapters called "integer" and "real", linked to the proof chapter. But this task reveals to be tricky, because we have to choose the right axioms for the theorem prover and need to proceed tentatively without any guidelines.

5. The Interactive Proof System

Experience has shown that, with the previous version of the system, the proof of some algorithm is as much tedious and error prone as the programming task. Indeed, specifications, axioms and proof elements introduced into some book may reveal to be wrong, inadequate or even incomplete in the course of the proof. These problems have to be solved in the classical way, by modifying the source code and recompiling it, but these kinds of "errors" are difficult to find; contrarily to the algorithmic parts of a book, where the running of the executable code produced by the modules can help to find out errors, we need to perform the whole proof, to be convinced of the specification's correctness. In the previous version of the proof system, the algorithms were automatically annotated and the predicates to be proved were sent to the I.T.P. Thus the user had no control on the WP computations nor on the proof itself within the prover. We thought that a better understanding of the WP generation process and the on-line simplification of the predicates created were an efficient and important tool to find out the lacks and errors made in the formal specification elements. The availability of a simplification tool has another advantage: it reduces significantly the size of predicates, even in small programs (e.g. with arrays, pointer variables), making in turn predicates easier to understand.

In the present paragraph, we introduce the main ideas of the ATES interactive simplifier and its user interface (see [3]).

For the proof or simplification of predicates, the proof system contains two programming languages: the first one, internal to the prover, called "tactics language", enables us to describe the elementary proof steps of a proof; the second one, called "strategies language", is used to describe the chaining of the different steps into subproofs or proofs.

The informal idea of a tactic is to make the proof of a predicate equivalent to the (easier) proof of a list of sub-goals. A proof or a simplification being realized in a given context (i.e. with a list of axioms and hypotheses), the subgoals are created combining parts of the contexts with the predicate and/or modifying the context.

The informal idea of a strategy is to chain up the elementary proof steps (realized by tactics or other strategies), or to modify the context (for example, by restricting the axioms to a minimal set). The strategies serve to describe in a natural deductive way the proof of a predicate, using axioms and logical rules.

5. 1. Tactics

Axioms, along with pattern matching are used in a stepwise manner to reduce some goal to be proved into various sub-goals. The tool provided to a user by the ATES system provides a set of correct predicate transformation rules, based on rewriting.

From a given predicate, called the "current goal", a tactic generates a list of simpler sub-goals, using the following sub-goals expression:

sub-goal P [Q # R # S]

where P, Q, R, and S are the sub-goals. The proof of the current goal amounts first to prove P and, according to this result, to prove one of the subsequent sub-goals: if P is reduced to "true", then Q is tried; if P is reduced to "false", then try to prove R; otherwise prove S. Given an initial predicate, the result of a tactic will be the predicate obtained by the proof of the last sub-goal, joined with some error flag. The latter comes from the filtering of the input predicates by the goal pattern of the tactic. To the input goal of a tactic is associated a context, called a "theory": it consists of a list of axioms and a list of hypotheses. So, the sub-goals also possess an associated context, which is the input context, possibly modified by appending a new hypothesis or by enabling or disabling some axioms or hypotheses. Indeed, the elements of such a context own an activation state to restrict the search space of rewriting tactics (described later).

The following example shows a tactic handling conjunctive goals. It reduces a formula of the form "P and Q" into two separate sub-goals in the following manner: first, the left hand side of the input goal, P, is tried out; if its proof succeeds, the right hand side, Q, is tried out in turn, assuming the first one; if the proof of P fails, the entire predicate fails to be proved; otherwise Q is still tried out because this proof may fail, allowing the conclude by a failure of the input goal. When the proof of the input goal is undecidable within a tactic, it is returned unchanged; this is denoted by a ?-sign.

> *tactic conjunction*
> *var P,Q: formula; T: theory*
> *goal T (P and Q)*
> *subgoal T: P [T & hyp P: Q # T: false # T: Q [T: ? # T: false # T: ?]]*

The ATES proof system owns a set of predefined tactics, formalizing rules of first order logic. The main rules along with their respective tactic are the following:

– deduction rule:

> *tactic implication goal*
> *var P, Q: formula; T: theory*
> *goal T:P = > Q*
> *subgoal T & hyp P : Q*

A goal of the form $P = > Q$ can be replaced by the formula Q, provided that P is entered as a temporary axiom in the hypotheses list. P remains there until the produced goal is reduced.

– and-elimination, or-elimination and not-elimination rules:

> *tactic conjunction* (see above)

> *tactic disjunction*
> *var P,Q: formula; T: theory*
> *goal T: (P or Q)*
> *subgoal T: P [T: true # T & hyp (P = false): Q # T: Q / T: true # T: ? # T: ?]]*

> *tactic negation*
> *var P: formula; T: theory*

goal *T: (not P)*
subgoal *T: P [T: false # T: true # T: ?]*

– generalization rule:

> **tactic** *generalization*
> **var** *P: formula; x: variable; T: theory*
> **goal** *T: (for x then P) with FREE (x,P)*
> **subgoal** *T: P*

If x is a free variable of some formula P, then the truth of the formula "for x then P" (universal quantification of x over P), can be infered from the truth of P.

– substitution rule:

> **tactic** *axiom_subst*
> **var** *T: theory; P, Q: formula*
> **goal** *T(Q): P(Q)*
> **subgoal** *T: P(true)*

If a formula P contains instances of somc axiom, then these ones can be replaced by "true". Note: T(Q) denotes the subtheory of T restricted to the axioms matching with Q.

– equality rule:

> **tactic** *substitution_1_2*
> **var** *P: formula; T: theory; t1,t2: term*
> **goal** *T (t1 = t2): P(t1)*
> **subgoal** *T: P(t2J*

If some formula P contains occurences of the left hand side of an equality axiom, say "e1 = e2", then it can be reduced to P(e2); the system will use such axioms as rewriting rules, assuming that the axioms are oriented from left to right by the user,

– modus ponens:

> **tactic** *modus_ponens* **var** *P, Q: formula; T: theory*
> **goal** *T (P = > Q): Q*
> **subgoal** *T: P*

If some goal P is an instance of the right hand side of some axiom of the form "A = > B", then P can be replaced by the substituted left hand side of the axiom.
All these rules (tactics) are stored in the kernel of the proof checker.

5.2. Strategies

The strategy language is a language in which it must be possible to give statements, explaining what should be done next in order to come closer to the goal. Questions that a strategy has to answer are: which tactic should be called next, which axioms should be enabled (added to the goal's context), what strategy should be called to perform a certain transformation The basic commands are the tactic call, the enabling/disabling call, the strategy call. the sequence, the alternative, the repetition, and the negation. As tactics do fail or succeed, this information is used to decide what to do next in a strategy. Therefore the three basic commands are functions which deliver a boolean result too.

A strategy consists of a name and a body, which is a strategy command. In the following, we will briefly describe the semantics of all these constructions, along with their syntax, using some BNF-like syntax.

5.2.1. The tactic call

Syntax: command :: = tactic { tactic_name } +,

The tactics are called sequentialy, using the left to right order, until one of them succeeds. When a tactic succeeds, the others are skipped and the strategy is said to succeed; otherwise the tactic call fails.

5.2.2. The Enable/Disable call.

Syntax: command :: = enable [ax { axiom_name} +,] hyp { hyp_numbers} +,] / disable [ax { axiom_name} +,] hyp { hyp_numbers} +,]

Axioms and hypotheses are either enabled or disabled in the prescribed order. Notice that hypotheses have no name, so they are referred by they index in the concerned list. Such an enable/disable call always succeeds.

5.2.3. The strategy call.
A strategy is called by indicating its name, and the result of the call is equal to the one of the called strategy.

5.2.4. The sequence

Syntax: command :: = (command and { command } + and)

A sequence consists of a list of commands, separated by the keyword "and". This represents just the normal sequence; the strategies are executed successively and the result is the one of the last strategy executed.

5.2.5. The alternative

Syntax: command :: = (command or { command } + or)

This construct works like the tactic call, but concerning strategics.

5.2.6. The repetition

Syntax: *command :: = (do command od)*

The internal command is executed until it fails, and the result of this construction is always "fail".

5.2.7. The negation.

Syntax: *command :: = not command*

The result of the internal command is simply negated, i.e."fail" becomes "succeed" and vice-versa. This construction is necessary to express statements like "if strategy A succeeds then execute strategy B" or some other kinds of loop statements analogous to the ones found in classical programming languages. For example, we can introduce the following extra language construction:

(if strategy1 then strategy2) = = not (not strategy1 or not strategy2)

In the light of this language we have expressed a list of useful strategies for the simplification of intermediate predicates. They can be divided into four classes: the logical simplification strategies, the non-logical simplification strategies, the special simplification strategies and finally the general simplification strategies.

– The first class contains all strategies about logicals, and uses especially rewriting rules with the axioms located in the axioms chapter named "boolean" .
First, one strategy is devoted to the elimination of logical constants, using appropriate axioms; for some given formula, it proceeds in a left to right traversal and performs rewritings and axiom substitutions. Formally, it can be written:

```
strategy discard_logical_constants = =
( disable ax all hyp all
a n d
enable   ax
          and_true, and_false, or_false,
          or_true, not_true, not_false,
          true_imp, imp_true, false_imp,
          imp_false, if_true, if_false,
          then_true, then_false, else_true,
```

> *else_false, for_true*
> *a n d*
> *(do tactic axiom_subst,substitution_1_2 od))*

A second strategy performs structural transformations whose result is not a constant value. Typical examples of axioms used are: "(A and A) = A", "not (not A) = A", "(not (A) and not (B)) = (not (A or B))". This strategy, called "structural_logical_simplifications", is written and works in the same way as the preceeding one.

These two strategies are chained into a third more general one performing structural transformations whenever some constants have been extracted. It is written:

> *strategy logical_simplifications = =*
> *(do (if discard_logical_constants then*
> *structural_logical_simplifications)*
> *od)*

– The second class contains strategies mainly devoted to arithmetic simplifications.

First there is some strategy to eliminate expressions containing only integer or real constants. This strategy works by calling a special tactic doing this job without any axiom nor hypothesis. Another strategy is used for rewriting with axioms on integer or reals; once the "dangerous" axioms disabled, it runs in the same way as the strategy "discard logical constants". Indeed, some axioms may turn the system into an infinite loop, e.g. the ones expressing commutativity properties.

Next, one important strategy treats formulas of the form "(e = f) = > P" where e and f are non-logical expressions and P is some formula. It puts the premiss into the hypotheses list and rewrites P with it, as much as possible. Formally, it can be written:

> *strategy treat_implication = =*
> *(disable ax all hyp all*
> *a n d*
> *(if tactic implication_goal then*
> *(enable hyp* N < < *Enables the last hypothesis* > >
> *a n d*
> *(do tactic substitution_1_2 od))))*

– Special simplification strategies exist in the system to manage lambda expressions and other special operators present in our predicates (e.g the override operator, acting on functions, and the let operator, analogous to the ones of VDM).
– General simplifications strategies are integrated in the system to perform the simplifications belonging to the previous classes simultaneously.

A first strategy performs rewritings in an ordered manner: first, logical transformations are done; then arithmetical simplifications are made; finally the special simplifications are done. Formally, this strategy is written:

> **strategy** *simplify*
> (**do** (
>
>   *logical _implifications* **and**
>   *simplify_with_equality* **and**
>   *special_simplifications*)
>
> **od**)

A second strategy does structural simplifications: based on the elimination rules of the paragraph 5.1, it cuts out a formula into smaller subgoals and tries to simplify them by previously mentionned strategies.

Other strategies have been found by testing the rewriting strategies. So, users noticed that some transformations were often realized, and should be grouped into strategies. This concerns particularly the elimination of some particular, frequently used functions: strategy "normalize inequalities" suppresses all relational operators of the set { <, >, > = } and strategy "discard_binary_minus" does the same job to eliminate the concerned binary operator. Furthermore, some strategies exist for doing limited forward chaining, to generate new hypotheses .

5.3. The User Interface

Interaction with the proof checker is realized through a hierarchy of commands organized within a tree, with a menu at each node. Having successfully compiled a given book, the user enters the proof checker to perform the proofs. Successive menus show the lists of proof chapters, algorithms and elementary proofs (there is an elementary proof for the whole algorithm and for each loop contained int it: each one consists in computing some intermediate predicates and proving the verification conditions associated, according to paragraph 3.3). At each step one has to choose an item before entering the next menu. Selecting an elementary proof makes the system display the annotated algorithm. Depending upon the kind of elementary proof, the post-condition of the algorithm or of some loop has been placed in it. The system put a mark in front of the post-condition and is now ready to perform progressive simplifications and WP computations (this mark will always indicate some location in the algorithm containing the predicate we are allowed to access). At this point, one of the following commands can be entered:

- Compute intermediate predicates up to some other location, starting from the current location.
- Delete intermediate predicates between the current location and some previous one.
- Simplify the predicate currently pointed out.
- Return to the previous menu.

Using these commands, we can annotate in a stepwise manner some algorithm, by simplifying at each step the newly generated formula, reducing significantly the size of the intermediate predicates. When typing the previously mentionned simplification command, a new menu appears to handle the currently designated predicate. This menu offers the possibility to select a sub-predicate of the initial one, to perform simplifications on the desired part without altering the whole formula. A simplification can then be carried out: first, the command "simplify" is chosen; the system displays then new menus to select the strategy to be applied and, possibly, some axiom used by the latter; finally, the strategy is executed, and the result is displayed. At each moment, the user may return to the previous menu, with the possibility to insert the result predicate in the algorithm. At each level of the command tree, the system reminds us of the state of the proofs, algorithms, proof chapters or book.

Among the several other facilities existing in the user interface, the handling of "journal" files should be underlined. These functions, necessary to most interactive software, permit to create, close and read some kind of logbook during the proof task.

6. The ATES Prototype and some Experiments realized with it

Since the project has progressed for four years, a prototype running on a DEC-VAX 8350 under VMS is available, whose main subtools are the following ones:

– A compiler for ATES books (the book being the largest unit of source code) performing three basic steps:
  - syntactical and semantical analysis of all kinds of chapters; the different kinds of chapters are treated differently from now on,
  - the module chapters joined to some main module, are compiled into an executable file, the intermediate language being Fortran,
  - the formal chapters (containing axioms, specifications and proof elements), produce annotated algorithms, stored in an adequate format for realizing proofs.
– A management system adapted to a file, called the "master file", associated to the book, which contains all the information about its compilation:
  - the source code and the result of its syntactical and semantical analysis,
  - a special structure for managing the book in an efficient way, allowing selective, separate or system oriented compilation.
– A tool ordering the different compilation steps, leading the user to follow a program-to-proof methodology, in order to obtain a successfully compiled book. It keeps the book in a consistent state and reminds the compilation state of every chapter. This tool can also decide to recompile certain parts of a book and permits us to modify chapters during the compilation task.
– A run-time environment, consisting of basic libraries, a memory manager and an exception handler.

– A interactive proof environment: as described previously, the proof checker provides all the neccessary features to perform the correctness proof of an algorithm. Within such a system, the user can easily follow the different proof steps and backtrack when neccessary, loading in a comprehensive way to the proof.

A real-size application has been chosen and implemented, to validate the ATES system: this application deals with a finite element code applied to solve heat transfer equations.

A 2D and 3D version of this code (linear and steady-state case) has been written (about 50000 lines of source code). The major neccessary tasks were the following: computation on finite elements, solvers of linear equations systems, interfaces pre and post from and to a structure mechanical code (SAMCEF). This code has been extended to handle 3D cases, and an automatic 3D mesh generator has been realized, which notably increases the size of the code. The 1D version of the application (consisting of about 1000 lines of source code) has been formalized in terms of axioms, specifications and proof chapters and has been proved to be correct within 600 Men x Hours of work. Currently, the 2D version of this application is tackled.

Concerning the quality of the Fortran code generated by the ATES compiler, the following conclusions can be drawn: for the applications mentioned above, the CPU ratio between the generated code and the hand written Fortran code is about 1.12, and the execution time of ATES generated code for certain algorithms may vary in the ratio of one to ten, depending on the compiler options enabled.

7. Concluding Remarks

From a larger cleariness inducing a better understanding of the specification steps, and with the help of the tools involved, ATES contributes to produce more reliable software.

The ATES system proposes an integrated approach for the conception and production of medium and large-scale software, offering a safe, modular and progressive validation method and allowing to gain time by using a high-level programming language during the programming task.

This system is still in evolution, so that further improvements are considered: first, the development of more general strategies is foreseen; a possibility to write strategy chapters, formalizing proof methods proper to some area is being studied; finally, the integration of a certain number of useful strategies, found by running the system on medium-size examples, will be realized. The ATES language being purely sequential in its current form, it seems easy to integrate parallelism notions (at the algorithms level) by increasing the power of the language with primitives whose semantic is clearly defined, and then to extend the proof method to this langage.

Finally, the industrial interest of such a system is to build up from this prototype a tool for the programming of reliable software. Indeed most industries involved in safety critical software are highly interested in such a tool. A first step toward the industrial use of such a system would be to add a front end translator to the system, such that ATES accepts programs written in classical programming languages (as C, Fortran or Pascal) too. The translator should use the ATES language as a common description language. In this way, the system could be immediately employed for existing software in an industrial environment.

6. References

[1] ABRIAL, J.R. (1988) 'The B tool' Proceedings of the conference "VDM 88: VDM the way ahead", Lecture Notes in Computer Science, 328, Springer Verlag.

[2] ATES Project (1987) 'Specifications of the programming language (revised version)', Report of the ESPRIT Project ATES 1222(1158), C.I.S.I. Ingenierie, FRANCE.

[3] ATES Project (1989) 'Proof user's manual' Report of the ESPRIT Project ATES 1222(1158), C.I.S.I. Ingenierie, FRANCE.

[4] BOYER, R.S. STROTHER-MOORE, J.(1979) 'A theorem prover for recursive functions: a user's manual', Report no. CSL-91, Computer Science Laboratory S.R.I. International, Menlo Park, California.

[5] DIJKSTRA, E.W. (1976) 'A discipline of Programming', Prentice Hall Series in Automatic Computation, Englewood Cliffs, N.J.

[6] DONZEAU-GOUGE, V. HUET, G. KAHN, G. LANG, B. (1980) 'Programming environments based on structured editors: the Mentor experience', I.N.R.I.A. Report no. 26, FRANCE.

[7] GORDON, M.J. MILNER, R. WADSWORTH, C. (1979) 'Edinburgh LCF', Lecture Notes in Computer Science, no. 78, Springer Verlag.

[8] GREENBAUM, S. (1986) 'Input transformations and resolution implementation techniques for theorem proving in first-order logic', PhD. thesis in Computer Science, University of Illinois at Urbana Champaign, USA.

[9] HASCOET, L. (1987) 'Un constructeur d'arbre de preuves dirigé par des tactiques', I.N.R.I.A. Report no. 770, FRANCE.

[10] HOARE, C.A.R. (1969) 'An axiomatic basis for computer programming', C.A.C.M. 12(10), pp. 576-583.

[11] JONES, C.B. (1980) 'Software development : A rigorous approach', Prentice Hall .

[12] LESCANNE, P. (1983) 'Computer experiments with the REVE term Rewriting system generator', POPL Conference, Austin, Texas, USA.

[13] PLAISTED, D.A. (1981) 'Theorem proving with abstraction', Artificial Intelligence, 16, North Holland Publishing Company, pp. 47-108.

[14] PUCCETTI, A.P. (1987) 'Preuve de propriétés de fatalité temps-réel de programmes ADA, Sémantiques opérationnelle et axiomatiques associées', Thèse de 3ème cycle, Institut National Polytechnique de Lorraine, FRANCE.

[15] VANGEERSDAEL, J. (1988) 'A guided tour through theorem provers',. Report of the ESPRIT Project ATES 1222(1158), Philips Research Laboratory Brussels, BELGIUM.

[16] WULF. W.A. LONDON, R.L. SHAW, M. (1976) 'An introduction to the construction and verification of Alphard programs', IEEE Transactions of Software Engineering, vol. SE-2, no. 4.

SAMSON: AN EXPERT SYSTEM DEVOTED TO THE HELP OF SOFTWARE ASSESSMENT

Olivier SIRVENT
CRIL
Voie n°2 Imm Arizona Bât B
BP 05 31312 Labège cedex
France
Tel: 61 39 10 33
Fax: 61 39 23 73

Catherine DUPONT
CRIL
12, avenue Henri Freville
35200 Rennes
France
Tel: 99 41 74 44
Fax: 99 50 03 65

Summary

The size and the complexity of software are continuously increasing, and it is a truism that software quality measurement is necessary. However, most of quality metrics users encounter problems in selecting them and interpreting their results; this knowledge is mastered by few experts in the world. To try to overcome these problems, the MUSE esprit project (p1257) has developed a knowledge base system, SAMSON, to help software quality assessment. This paper first relates SAMSON conception and development which was based on the synergy between MUSE German and French partners, it also presents its industrial impact. The main part concerns SAMSON description, it is divided according to the two models used to design the system. The first one explains how to select the metrics with the characteristics of the assessed software and how to interpret the results. There is a lack of explanations in this model, thus the second model helps the user to understand his errors, to correct them and gives recommendations to obtain the required quality. It is important to note that the main part of the knowledge in SAMSON is standardized and that efforts have been made on the user interface, in order to give SAMSON a professional skeleton.

Introduction

Since the emergence of so called "critical systems", which in case of fault can cause human death, software production requires as much care as that of critical hardware systems like cars, nuclear stations, aeroplanes etc. The software product is built in different phases (life cycle phases). To ensure its final quality, the product is checked at the end of each phase, much as an aeroplane is along the production line. If the product assessment concludes it is adequate, the next phase can begin, otherwise the product has to be repaired and be re-evaluated.

Unlike the assessment of an aeroplane, the assessment of software is not easy (or not well known), and only rare experts are able to do it. Thus, it was useful to formalize this process and build a tool able to reproduce the work of the expert. An expert system has been developed: SAMSON.

This paper first presents briefly the MUSE project and focuses on the teamwork, the synergy and the technology transfer that existed during SAMSON's realization. It also shows some of the industrial exploitation in Europe which is proposed or which is already in process.

Next, it describes the different models used, the main features of SAMSON and also the way the different data of SAMSON are updated and displayed .

1. Project Presentation

1.1. MUSE PRESENTATION

The Esprit project MUSE p1257 (Metrics Used in Software Engineering) is a three-year project, which started in January 1987. There are four partners: BRA-MEUR LTD (software house in England), RW-TUEV (certification organization in Germany), EBO (arms producer in Greece) and CRIL (software house in France). the MUSE project follows two complementary approaches which consist of :
- selecting metrics, applying them, analysing the results and giving recommendations for different types of software: safety critical systems, clerical systems, management systems, and artificial intelligence systems.
- development tools: a quality environment devoted to maintainability and reliability and an expert system to help the selection of metrics and the interpretation of results.

1.2. Samson Conception and Development: a Real Synergy and Technology Transfer.

Quality assessment depends on the particularities of each piece of software, of the life cycle phase, and of the assessor himself. That is why this process cannot be standardized but relies on expert experience. SAMSON involved RW-TUEV experts and CRIL knowledge engineers. The first step of the knowledge acquisition was to select a field of software application. RW-TUEV is experienced in the assessment of safety critical systems and clerical systems [PHI 87], [PHI 88], [RWT 88]. A major part of the theoritical and pratical work in measurement has been performed for safety-critical systems because of the risks involved. This led us to select this type of software for our study. As soon as we had obtained formalized elements, forms were drawn up in order to check that the formalization really reflected the assessment process and save time by reducing the number of meetings between the expert and the knowledge engineer. The formalization of the software assessment process enabled us to automate it. The synergy and the technology transfer within the team during this phase are well described in [DUP 90]. This paper also shows that the knowledge acquisition process gave the expert considerable insight into his work method concerning software assessment and because of his interest in the system he learned how expert systems work. On the other hand the knowledge engineer profited from learning more about quality assessment.

Thus we developed a knowledge-based prototype called SAM (Software Advisor on Metrics) which implemented the assessment of safety critical system during the detailed design phase [DUP 88]. This prototype has been tested by all the partners (RW-TUEV, BRAMEUR LTD and EBO). A brother of SAM was born across the channel, specialized in clerical software.

The critique of RW-TUEV experts demonstrated that in SAM the explanations were insufficient and the interpretation of metrics could be improved on. Thus this prototype fathered a second one called SAMSON [SIR 90]. The knowledge added to this last system has two charasteristics. Its origin is in the standards IEEE and DIN ([IEEE 83], [DIN 88]), which will not surprise the user. But the way it is suggested and the way it is linked with the previous knowledge, is typical of the expert.

1.3. Industrial Exploitation and Impact

Industrial circles have given an enthusiastic reception to this system. In France, It has laid the foundations of another expert system which helps in quality audits. In Germany the same field has also shown interest on further developments, as well as in France.

2. Samson Description

SAMSON has been carried out around two models. The first one is devoted to the selection of the metrics and the interpretation of the results (paragraph 2.1). In order to better use this model the whole software being assessed and the set of metrics is partitioned as explained in paragraph 2.2. The objective of the second model is to explain how to try to correct the errors in the software detected by the failed metrics (paragraph 2.3). Finally, special attention has been paid to the display of these results in order to have a more professional and evolutionist system; this is shown in paragraph 2.4.

2.1. First Model Used

This is a Mc Call extended model [DUP 88], described in the figure 1. The assessment of a general software program must , in the first place, take its type into account. After that, the assessed life cycle phase has to be chosen. From then on we have a classical Mc Call model, [MCC 77], related to each couple (software type, life cycle phase).

2.1.1. The Software Type.

There are three different types :
– Safety critical systems which can endanger life and limb if they are of poor quality.
– Clerical systems which are used in operational tasks such as office data proce-
  dures.

– Management systems which are used directly by managers for decision making purposes.

2.1.2. The Software Life Cycle Phase.

The software life cycle used by the expert is composed of nine phases that are the users requirements, the specifications, the rough design, the detailed design, the coding, the unit testing, the functional testing, the validation and the maintenance. The selection of metrics is different for each phase.

The type and the life cycle phase of the software establish a set of metrics to be applied, it is possible to make a more precise choice with the Mc Call model.

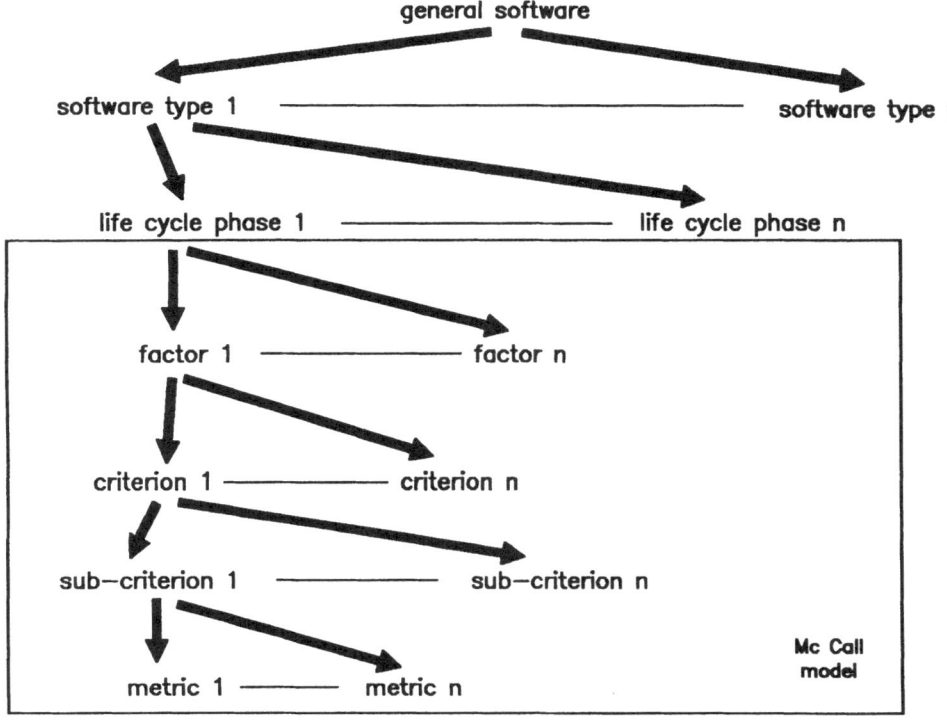

Fig 1 Extended model used

2.1.3. The Mc Call Model.

Obviously, the set of metrics to be applied is also dependent on the quality to expect in the software. To formalize this information a Mc Call model has been appended to each leaf of the previous tree.

The set of factors, criteria, sub-criteria and metrics does not stick to Mc Call's but has been extended by the expert. However, we still use the classical representation.

A factor is a quality charasteristic, whose presence or absence may be detected by users of the software product (including people who purchase the software or contract out its development and evolution). To judge and obtain a factor, qualities perceptible to computer professionals are needed, they are called criteria. To be more precise, Mc Call divided criteria in sub-criteria. Finally, to assess criteria or sub-criteria, the product is measured with metrics.

For example the user wishes to test a program to ensure it performs its intended function. This is the factor "testability". Those attributes of the software that provide the implementation of functions in the most understandable manner represent the "simplicity" (one of the criterion which is relied to the previous factor). If the modules have low complexity then a sub-criterion is satisfied. To verify this last assertion we could check if every module has only one entry-point and one exit. This is a metric.

2.2. The Partitions

SAMSON's scope is concentrated on a restriction of the extended model with the couple (safety critical system, detailed design) and a Mc Call tree limited to four factors (correctness, reliability, testabiliy, efficiency). We theretofore supposed that the different parts of the software and all metrics to measure them have the same importance. In fact that is obviously false. That is why partitions of the software and of the set of metrics are imposed.

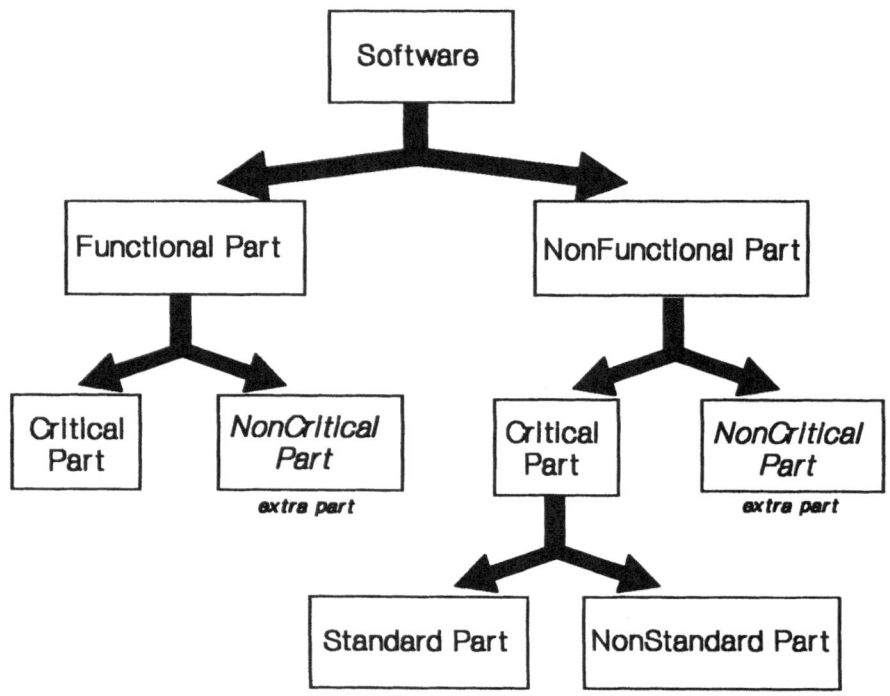

Fig 2: The software parts

2.2.1. The Software Parts.

The safety critical system is made up of different parts which have a different influence on the software risk criticallity (fig 2).

Each of these parts has a different criticallity which varies from 1 to 3 (1 is the most critical). To compute it, we must know the whole software criticallity. It is computed by using the German classification (1 to 8). If the user does not know the criticallity of his software, it is computed with run-time questions such as: the importance of the effects on human beings, the possible avoidance of risks through other means, the presence of people in the area of risk ...

All the metrics can be selected now, even if some of them are selected occasionally and need run-time information (for example numeric measures need a tool ...); for more information refer to [DUP 88]. But the interpretation can not be the same inside these different parts.

2.2.2. The Ways Of Interpretation.

Because all metrics applied to measure a product do not necessarily have the same weight, three different means of interpretation are used. Six sets of metrics are then derived with the set of metrics to be applied: the set of metrics of criticality 1 and interpretation way 1, the one of criticallity 2 and interpretation 1, the one of criticality 2 and interpretation 2 and the last ones of criticality 3 and interpretation 1, 2 and 3.

– Interpretation 1: if one measurement is not acceptable, the measured product is rejected.
– Interpretation 2: if more than 10% of the total measurements are not acceptable, the product is rejected; otherwise all defects must be corrected during the following phase.
– Interpretation 3: if the sum of non-acceptable measurements in the previous phase and in the current phase represents more than 10% of the total measurements, then modifications must be carried out in order to have less than 10% of errors; otherwise the defaults are not to be corrected immediately.

This interpretation is a little questionable, that is why another model has been created to overcome this problem and give information to try to correct the errors.

2.3. The Explanations Model

In this part two concepts are introduced in order to try to eliminate the failed measurements: "features" and "recommendations". The information described in this model (fig 3) is for the most part, standardized [IEEE 83], [DIN 88]. Only such recommendations as are necessary to complete the different links between "features" and "recommendations" have been added by the expert.

270

2.3.1. The "Features".

A "feature" is a well known and standard method or means used to get quality in your software. Each feature can be made up of sub-features. Each sub-feature is accompanied by short examples. Thus, a metric is always linked with at least one sub-feature; it can be linked with sub-features of different features; it can also be linked with all the sub-features of a feature, in this case we say that it is linked with the entire feature. When a metric result is wrong, the associated feature(s) or sub-feature(s) are not met. The glossary (fig 3) is the IEEE definition of the feature [IEEE 83]. Each feature is also linked with what we call "recommendations".

2.3.2. The "Recommendations".

A recommendation is a precise and short sentence which describes an action that must be followed to achieve a specific quality in your software. These standard recommendations are, for the most part, included in the document [DIN 88]. Each recommendation is made up of sub-recommendations. Nevertheless, a feature can be linked with one or more sub-recommendations of different recommendations (there are the same links between metrics and features). A recommendation (or a sub-recommendation) is not simply a sheet of text, it is accompanied with by up to two pieces of information: what the recommendation helps to achieve and what it helps to avoid.

2.3.3. An Example.

We suppose that the user has one metric failed in interpretation 1, his software is rejected. In the explanations, the name and definition of the metric are displayed as well as those of the correlated criterion and correlated factor. Then it is possible to use the feature and follow the recommendations to try solve the problem detected by the metric (fig 4).

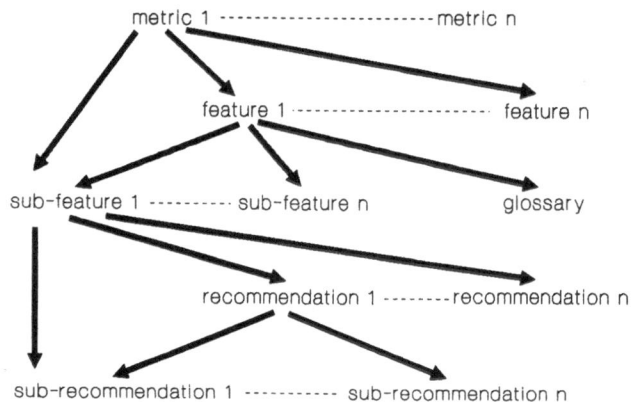

Fig 3 The explanation model

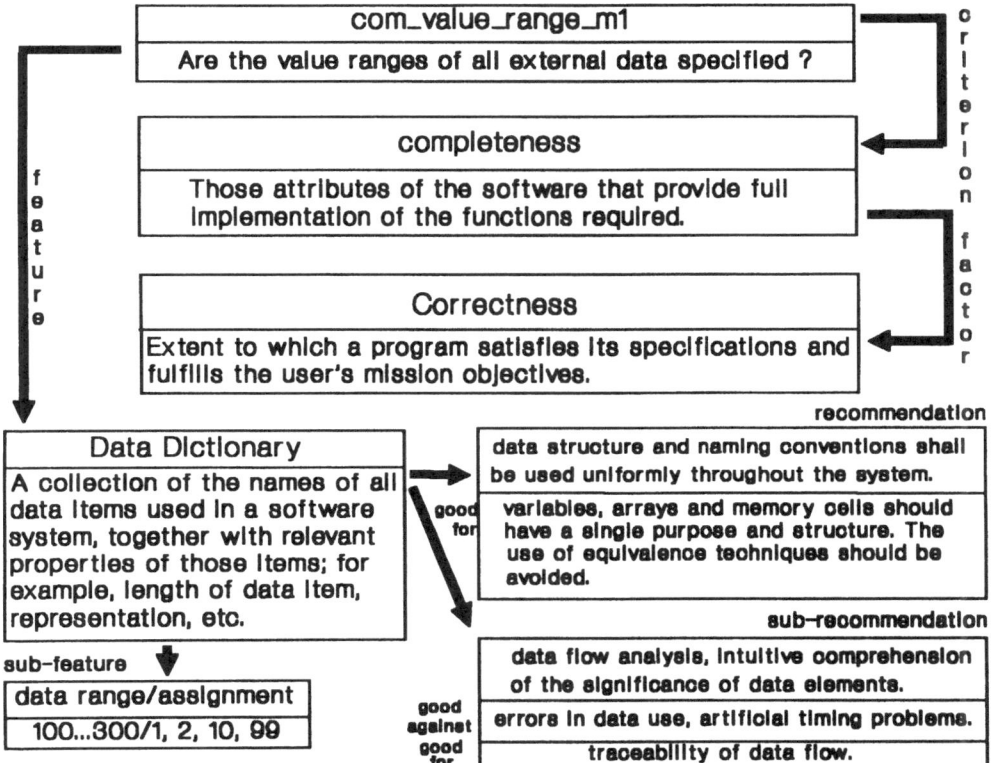

Fig 4: Example of explanations after a failed metric

2.4. Updating and Display

The expert system tool used to build SAMSON turned out to be inefficient in the display of the metrics, criteria, factors, features and recommendations (not professional and time costly). In order to improve it, external programs have been written with an object oriented representation (fig 5). Another program has been written to update all these items which are represented as objects. These objects are created or extracted from dictionaries previously loaded from the disk. They are then updated or removed and the dictionaries are stored. In this way it is possible to update the data base without modifications in the program.

272

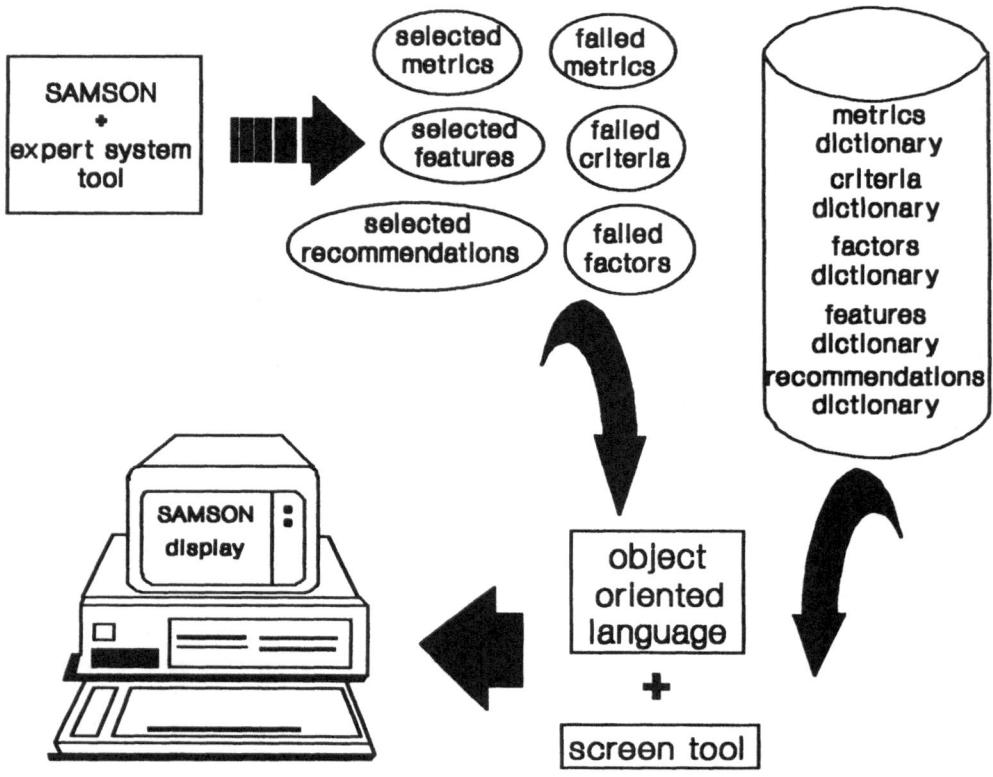

Fig 5: Items display

Conclusion

SAMSON is a prototype which covers only a part of the model described above (detailed design of critical systems software). Nevertheless, all the useful concepts needed to build a software system devoted to the help of software assessment have been highlighted, automated and tested. The process of knowledge acquisition is formalized in this case, it can be performed with forms, therefore it will be speeded up. The knowledge domain is bounded and the explanation part is standardized. It has been tested by the different partners to find improvements and, for this, the synergy inside the MUSE consortium was beneficial. Thus to make SAMSON marketable is just a question of "filling in", even if this is not a short process. That is why SAMSON is welcomed by industrial circles in Germany and in France and has fathered yet more systems.

Bibliography

[DIN 88]: Entwurf DIN IEC 45A(CO)88.

[DUP 88]: C.Dupont: "Specifications of an Expert System for Help to Software Quality

Assessment", ESPRIT project 1257 MUSE Q/R, February 1988

[DUP 90]: C.Dupont, W.Philipp: "Formalization of the Assessment of Software Quality Process and Results", Second European Conference on Software Quality Assurance, Oslo May,June 1990

[IEEE 83]: IEEE "Standard Glossary of Software Engineering Terminology", ANSI/IEEE std 729-1983

[MCC 77]: J.A.Mc Call, P.K. Richards, G.F. Walters: "Factors in Software Quality", July 1977, RADC-TR-77-369

[PHI 87]: W. PHILIPP, "Practical Experience with Certification and Testing of IT-Products", Computer Standards and Interfaces, Volume 6, Number 2, 1987

[PHI 88]: W. PHILIPP, "Verification and Validation of the Software for the Local Core Monitoring System of the Fast Breeder Reactor in Kalfar (SNR 300)", First European Software Conference, London, December 1988

[RWT 88]: B. HELLING, U. HEITKOETTER (RW-TUEV) "A Case Study: Experience and Lessons learned from Data Collection in a Safety Critical Project", ESPRIT project MUSE report, September 1988

[SIR 90]: O. SIRVENT "SAMSON REQUIREMENTS" ESPRIT project 1257 MUSE, D10.4.1, April 9th 1990

Project No. 2025

EDS - COLLABORATING FOR A HIGH PERFORMANCE PARALLEL RELATIONAL DATABASE

Frederic Andres (BULL), Bjom Bergsten (BULL), Pascale Borla-Salamet (BULL), Phil Broughton (ICL), Carla Chachaty (BULL), Michel Couprie (ESIEE), Beatrice Finance (INFOSYS), Georges Gardarin (INFOSYS), Kevin Glynn (ICL), Brian Hart (BULL), Steve Kellett (ICL), Steve Leunig (ICL), Mauricio Lopez (BULL),Mike Ward (ICL), Patrick Valduriez (INRIA), Mikal Ziane (INRIA)

Abstract.

A major goal of the EDS project is the production of a high performance, highly parallel database server, which will also offer extended capability to the programmer. The development of this system is a major task for which the companies involved find collaboration to be essential. This paper shows the extent and advanced nature of the work and demonstrates how effectively this collaboration is succeeding.

1. Introduction

1.1. EDS Database Server - Market Demands

EDS is a high performance, highly parallel system which is being designed and prototyped by a team consisting of Bull, ICL, Siemens and ECRC designers, with significant contributions from other associates. A major focus of the collaboration is the production of a Relational Database (ESQL) Server within the EDS which is capable of meeting the rapidly increasing transaction processing demands of the 90s.

There are rapidly increasing demands for both performance and system capability, whilst strictly maintaining the ability to mount existing relational databases and applications unchanged.

There are thus three major and equal thrusts for the EDS Database Server:

- support for existing standards (SQL, Unix),
- a quantum leap in performance by exploiting parallelism and large stable main memories,
- adding functionality for use by application programs and improving programmer productivity.

1.2 EDS Supporting Standards

The query language SQL is an important standard used by many existing applications. Extensions to this standard are being proposed (e.g. SQL2) and a DoD database security standard is also expected.

The EDS database server supports all the functionality of SQL and SQL2 and is designed to be extensible to cover other emerging database standards. In particular our own implementation of extensions in ESQL is putting in place a design capable of supporting the actual standards as they emerge.

1.3 EDS Meeting Increasing Performance Demands

The normally sourced transaction rate has been forecast to grow by a factor of 10 over the next 5 years - twice the current trend in mainframe growth. In addition there is already an increasing number of transactions sourced by computer systems in business-to-business interworking and by intelligent terminals in the home, office and factory.

The profile of the transaction load is also changing as management support queries and queries over knowledge bases including rules are added to the existing, largely clerical workloads. Very complex queries, generated by skilled staff working, for example, in the area of knowledge management, will contribute little to through-put demands. However, medium complexity queries such as those often macro-generated by current decision support systems or system generated as in production control will increase to demand significant throughput.

The EDS database server is designed to support all these types of query and will be optimized towards medium and low complexity queries.

The EDS database server will deliver up to 10 times the performance of conventional mainframes expected in the mid-1990's. This is expected to be 12,000 transactions per second as measured by the TPC-B benchmark; these transactions being simple lines of business enquiries.

1.4 EDS Meeting Greater Programming Demands

SQL has many desirable high level constructs, but lacks some important functions which will be needed if relational database technology is to be applied to a widening range of applications. The primary requirements of these applications are: objects with complex structures, multi-statement queries, user defined data types with associated access methods, triggering automatic actions after specified operations, and deductive capability from rules held in the database.

The EDS database server is designed to support all these features by incorporating extensions to SQL (ESQL) which have a minimal impact to the SQL syntax. Implementation priority will be given to those features required by medium and low complexity queries.

It is important to maintain and improve user productivity as the complexity of queries increases, retaining the declarative aspects of SQL, whilst avoiding the need for programmer involvement in parallelisation or optimization of query evaluation.

The EDS database server will provide extensions to SQL to facilitate a high level data definition and manipulation interface, and will ask only guidance of the user to optimize its use of parallelism.

1.5 Main Design Thrusts

The two main thrusts for obtaining high performance by exploiting parallelism and for extending the functionality of SQL lead to several areas of important design:

- the inclusion in ESQL of data definition and manipulation languages to subsume the object-oriented features within an extended relational system.
- the use of a single intermediate language (based on the extended relational algebra, LERA) at each stage of compilation and execution,
- the design of a staged compile and optimize route which fully exploits the parallelism available in the EDS share-nothing architecture whilst remaining sufficiently general to have potential for early exploitation on shared-memory systems.
- a query execution stage to manage the scheduling of work onto the light-weight process mechanisms of the share nothing parallel system,and the streaming of data to operations.
- a distributed, reliable object store which will exploit the storage of data in secure RAM and provide locking mechanisms which will not destroy the parallelism of the system.

These design areas are presented in more detail in subsequent sections of this paper. The paper concludes with a summary of how the collaboration and design is succeeding.

2. Query Language Extensions

ESQL is an SQL upward-compatible database language that integrates in a uniform and clean way the essential concepts of relational, object-oriented and deductive databases. ESQL is intended for traditional data processing applications as well as more complex applications such as large expert systems. Therefore, ESQL's salient features are:

- a rich and extendible type system based on abstract data types (ADTs) implemented in various programming languages,
- complex objects with object sharing by combining generic ADTs and object identity,
- the capability of querying and updating relations containing simple or complex objects using SQL-compatible syntax and semantics,

– a DATALOG-like deductive capability provided as an extension of the SQL view mechanism.

ESQL's functional semantics enables uniform manipulation of ESQL data and should facilitate the implementation of the ESQL compiler.

2.1 ADT Support

The support of ADTs provides a powerful typing capability [Stonebraker 86, Gardarin 90]. It makes the fixed set of system-defined types extendible by the users to accommodate the specific requirements of their application. An ADT is a type together with methods applicable to data of that type [Guttag 77]. The value of any ADT (e.g. map) can be stored in the database system and manipulated using the associated methods (e.g. map intersection).

ESQL supports this ADT capability by extending the notion of domain traditionally supported by relational database systems. The goal is to allow ADTs to be implemented in various languages such as C, C++, LISP and PROLOG. For each ADT implementation language supported, the database system must provide routines to convert between the language data structures and a supporting database system type.

2.2 Complex Objects

ESQL's support for complex objects [Carey 88] includes object identity, which enables objects to be referentially shared, and permits the construction of complex structures (e.g. hierarchy, graph). Complex objects such as office automation objects or CAD design objects can be modelled and manipulated in a natural way, while giving an ESQL compiler opportunities to optimize the access to such objects. The relational data model only supports values, imposing the use of key values to identify objects. ESQL supports both values and objects. A value is an instance of an ADT while an object has a unique identifier with a value bound to it. Data not declared as objects are values by default as in the relational model, which means that they have no system identifier. Therefore, ESQL data is divided between objects and values, and only objects may be referentially shared using object identity. The relational data model supports flat values using the tuple and set constructors at one level.

2.3 Deductive Capability

A deductive capability enables one to abstract in a rule base the common knowledge traditionally embedded with redundancy in application programs [Ceri 90]. The rule base provides centralized control of knowledge, and is primarily useful to infer new facts from the facts stored in the database. ESQL provides this deductive capability as an extension of the SQL view mechanism. This gives the ESQL user the power of the DATALOG logic-based language using statements already available in SQL.

2.4 Data Definition and Manipulation

The ESQL data definition language (DDL) augments the SQL DDL in three major ways. First, it includes a type language to create and manage ADTs with related methods, and generic ADTs. Second, it extends the table creation statement to deal with nested relations and objects. Third, it adds a statement to create derived relations defined by general rules, including recursive rules.

The ESQL data manipulation language (DML) generalizes the SQL DML in several ways. The most significant advantages are the possibility of ADT operations in SQL statements, the manipulation of shared objects by extending the dot notation, the manipulation of nested objects through nested statements and the possibility of deductive queries. Data manipulation in ESQL is more regular than SQL, much in the way of SQL2 [Melton 89].

2.5 Example

Figure 1 illustrates an ESQL database schema for a portfolio database example. The top level objects are relations, which are set of tuples. The SECTOR relation describes the different investment sectors. For each sector, an icon is given as defined in the FIGURE data type (i.e. a list of points). The shares belonging to a sector are given in a multi-valued attribute organized as a set. Shares are described in the SHARES relation. For each share, the price and the highest and lowest price of the last quotation day are given using the MONEY data type. A log list of the last 50 quotations is also given. The MOVEMENT relation describes the activity for the last 50 quotation days of each share. The DIVIDEND relation simply gives the share dividends with their types. The PORTFOLIO relation describes the portfolio of the given investors. Note that investor is an object containing the full description of a person.

Typical queries are given in figure 2. Note that a functional notation is used to invoke methods as well as nested attributes. We assume that a DRAW function has been defined to encapsulate the ICON data type. The first query draws the icon of the computer sector. The second query retrieves the name with the icon and the mean price of all shares that belong to the computer sector. The last query retrieves the total profit of each investor.

2.6 Design Progress

Further enhancements to ESQL are planned in the near future. These include the addition of conditional and loop statements to program multistatement procedures and perform explicit loops on set elements; another planned extension is triggers to manage an active database, and good support for integrity constraints, both on tables and ADTs.

RELATIONS

SECTOR: (name : memo, icon : figure, shares : set of share)
SHARES (share-id share, price : money , high : money, low : money, price-log list of (value : money,
vdate : date))
MOVEMENT (share-id : share, list of tuple
(mdate : date, buy : money, sell : money))
DIVIDEND (share-id : share, dividend : money, ddate : date, dtype : int)
PORTFOLIO (portf-id int, investor : person, portf-date : date, content : set of tuple (share-id :
share, amount : int, buy-date : date, buy-price : money))

TYPES

PERSON object tuple (lastname char(20), firstname char(20), address tuple of (number i2, street
char(20), zip i4, city char(20), country char (20)))
SHARE object tuple (title memo, icon figure)
MEMO char(30)
FIGURE list of point
POINT tuple of (x float, y float)
MONEY long integer
INT short integer

Figure 1: An example of an EDS database schema

(1) SELECT DRAW (ICON)
FROM SECTOR
WHERE NAME = "Computer"

(2) SELECT MEMO(SHARE), DRAW(ICON(SHARE)),AVG(VALUE(PRICE-LOG)) FROM
SHARES, SECTOR S
WHERE S. NAME = "Computer" AND CONTAINS (S. SHARES, SHARE-ID)

(3) SELECT INVESTOR, (SUM(AMOUNT*BUY-PRICE) -SUM(AMOUNT*S PRICE))
FROM (SELECT INVESTOR,
SHARE-ID,
AMOUNT,
BUY-PRICE
FROM PORTFOLIO
GROUP BY SHARE-ID) I,
SHARES S
WHERE I.SHARE-ID = S.SHARE-ID
GROUP BY INVESTOR

Figure 2: Typical ESQL Queries

3. Architectural Overview

The EDS database system aims above all else to exploit the power of highly parallel systems and in particular that of the EDS parallel machine.

The architecture of the EDS parallel machine is based on two main conclusions derived from a study of the trends in current hardware technologies. Firstly, the bandwidth requirements of micro-processors will outstrip that of store, and secondly the cost/performance of semi-conductor store is declining faster than that of discs and will cross over in the mid-1990s.

The design of the EDS system architecture takes account of these trends in two ways. First, the EDS machine will consist of Processor-store Elements (called PEs), where the processor is closely coupled to the local store, and the PEs are loosely connected by a network. Second, the design of the Database Server will support a two level stable store where the top level is RAM with battery back-up and the second level is disc storage. The design will be optimized for the situation where the active part of the database is held in the RAM.

The Database Server based on this architecture will exemplify the use of parallel database technology for improving performance and data reliability and availability. The architecture has thus to be such that its modules can be distributed and/or replicated over the available processors to obtain the most suitable configuration. As a first step towards achieving this objective the server is divided into three main functional components: the Request Manager, the Data Manager and the Session Manager. Furthermore, besides being adapted to the share-nothing architecture of the EDS system, the database server components are designed in such a way that a version for shared store architectures can be easily derived for early exploitation on this kind of system.

3.1 Request Manager

The role of the Request Manager is basically to compile and optimize ESQL queries to programs for execution by the Data Manager. The compilation and optimization process advances through a number of phases until the query program is produced. These phases communicate by an intermediate language called LERA (Language for Extended Relational Algebra) described in section 4.

The Analyser-Translator performs syntactic and semantic analysis of an ESQL statement and generates an equivalent LERA program. The Logical Optimizer essentially applies simplification and heuristic transformations to the LERA program to obtain a sequence of fewer operations. In particular, it handles recursive queries by applying the Alexander method and by introducing fixpoint operators. The role of the Physical Optimizer is to determine the strategy for executing the LERA program on the Data Manager that minimizes an objective cost function expressed in terms of the parallel execution environment characteristics. Both the Logical Optimizer and the Physical optimizer are designed using a rule-based and extendible approach so that different optimization algorithms and access methods can be easily accommodated (see section 5 and section 6).

The Paralleliser transforms a sequential LERA program into a parallel LERA program by applying the distributed execution strategy determined by the Physical Optimiser. The parallelisation process is divided into two phases in order to better handle data flow and control flow issues (see section 5). Finally, the Code Generator produces an object code module which contains calls to the extended relational operations and runtime primitives provided by the Data Manager.

3.2 Data Manager

The role of the Data Manager is to maintain the storage of data in the database, and to execute the parallel LERA programs generated by the Request Manager. This is provided by the Relational Execution Model, the Relation Access Manager,the Basic Relational Execution Model and the Object Manager.

The Relational Execution Model provides the local (to a PE) LERA operators required for the execution of parallel LERA programs. The Basic Relational Execution Model provides and an efficient parallel programme runtime environment, access to the low level (operating system dependent) runtime primitives and to the Object Manager services. The Relation Access Manager together with its associated Access Methods provides the data structures used to hold relations and the methods used to access them. The Object Manager provides a shared and reliable data store, including support for transactions, concurrency control and recovery facilities (see section 9).

While the long term objective of the project is to deliver a highly extensible share-nothing architecture, a shared store data manager is also being developed to ensure our ability to exploit this architecture in the shorter term. The two versions of the Data Manager being developed mainly differ in the primitives supported by the Relational Execution Model and the Basic Relational Execution Model modules. Sections 7 and 8 provide a description of the execution model for each one of the architectures considered.

3.3 Session Manager

The role of the Session Manager is the high level management of database sessions. In particular it will process requests from application programs for connection to a database session and create the initial resources required for the sessions.

A database session will consist of an instance of the Request Manager and an instance of the Data Manager. The Data Manager instance will consist of an instance of a local Data Manager for each processing element (PE) on which it is necessary to process base data. Additional local Data Manager instances may also be created for load balancing reasons. Although the Object Manager is logically considered to be a component of the Data Manager, its instances management differs from this structure in that an instance of the object manager is created for each PE in the system when the system is initialized. Only one Object Manager instance exists in each PE at any time and it is shared by all Data Manager instances running in the same PE in a given moment.

4. Intermediate Language (LERA)

LERA [Kellett 89a] is an extended form of relational algebra and has been chosen for its power of abstraction from the physical implementation.

The basic operators of the algebra are slight extensions of Codd's algebra [Gardarin 90]: 'Filter', which produces a relation of the same schema from a complex relation, and whose tuples satisfy a possibly complex condition, 'Project', which produces a new relation from a given relation by computing an expression over source attributes as target attributes, and 'Join', which may be defined as a Cartesian product of two relations followed by a filter. We also include traditional set operators on relations: union, difference and intersection.

A LERA graph is the internal representation of the user query for both logical and physical optimization, and parallelisation. Sometimes a sequence of basic operators can be implemented efficiently in a particular order. Consequently, we define in LERA Log (see later) a macro algebra with compound operations including n-ary union, n-ary join, and a compound operation ("search") of both projection, restriction and n-ary join. These compound operators are close to tuple calculus expressions, and provide the system with the necessary degrees of freedom to logically, and later physically optimize the queries.

Recursive views are expressed using a fixpoint operator. This operator produces the saturation of a relation computed recursively by an algebraic expression. Support for complex objects is provided in LERA by the generalization of the relational operators, restriction, projection, and join. For example, built-in and user defined function symbols can now appear in the criterion of a restriction, of a join and in the attribute projection lists. In addition, the standard nest and unnest operators are added to LERA to transform columns into collections, and vice-versa.

4.1 LERA Levels

We showed earlier that LERA is used at a number of different levels during the compilation process; these levels are now described.

4.1.1 Semantically Analysed LERA (LERA Sem).

LERA Sem is at the first level of ESQL compilation following parsing and semantic analysis. It is used to represent ESQL queries in an extended relational algebra, with source names replaced by internal identifiers. LERA Sem is not concerned with the way base or intemmediate relations are held but works on abstract relations.

4.1.2 logically Optimized LERA (LERA Log).

LERA Sem is processed by the Logical Optimizer which performs machine independent optimization to ease the work of the Physical Optimizer. LERA Log is the output of the Logical Optimizer, which is expressed in the same language as LERA Sem. LERA Log is the canonical form of the query using the compound operators, for example, by the collapsing of "search" operators in order to reduce the number of intermediate relations in the query.

4.1.3 Physically Optimized LERA (LERA Phy).

LERA Phy is the output of the Physical Optimizer. In terms of LERA operations, the essential difference from LERA Log is that operations are split into operations which represent basic relational operators like the join or projection. Moreover a LERA Phy program indicates the execution strategy (by using annotations) which has been chosen by the Physical Optimizer using a cost model (see section 6).

We distinguish several types of annotations:

- the localization annotation which specifies in which logical PEs (known as a "home") an operation has to be executed,
- the global method annotation which specifies the parallel algorithm which has to be applied to perform an optimal execution of the operation,
- the local method annotation which specifies the local algorithm which has to be applied,
- the consuming mode annotation which specifies for each operand the way it has to be consumed (stream or materialized) or what index to use.

4.2 Parallel Lera (LERA Par)

LERA Par [Kellett 89b] implements parallel execution strategies generated by the Physical Optimizer in LERA Phy to make programs (after a code generation step) directly executable by the Data Manager. No single centralized master process is needed to coordinate the execution, as all operations are distributed across homes, and all the control compiled into the program. The introduction of control operators provides a complete separation of data and control processing allowing a desired execution plan to be obtained by just combining the relevant control operators with a given set of data operators.

In addition to the relational operations introduced at the previous LERA levels, new operators are introduced in LERA Par such as data transmission operations (which transmit and store sets of data between homes), and control operators (which manage the synchronisation of the different parts of the request and detect data operation termination). Moreover, constructors ("local-op" for localized operator and "assoc" for association) are introduced to group operations into execution entities, localized on one home.

The local-op constructor is used to group data operators. It receives either a remote or local stream of data and processes it, or is triggered by a control operation, and consumes a locally stored relation. This execution entity is replicated on all the elements of the home. The assoc constructor is used to group local-op operators communicating locally inside each element of one home, together with associated localized control operations. These localized control operations detect and propagate the termination of the localized data operations and can also perform local synchronisation. Centralized control operators within an associate based on a single element perform the global termination detection and the synchronisation of the distributed groups of operations.

5. Optimization and Parallelisation

5.1 Logical Optimizer

The first phase of query optimization, known as query rewriting, is performed by the logical optimizer. Its main function is to transform queries into equivalent simplified ones in canonical form. Query rewriting includes syntactic transformations such as query modification with views [Stonebraker 76], redundant sub-query elimination, selection migration in fixpoint using the Alexander Method [Rohmer 86], and operations grouping. The latter transformation provides the physical optimizer with more optimization opportunities. Query rewriting also includes semantic trans-formations such as query simplification using integrity constraints[Shenoy87] and ADT operator properties.

To meet the extensibility requirements the main design decision is to generate the logical optimizer from rules which define the legal transformations of the query [Finance90]. This design provides easy extension of the optimizer as well as the handling of different optimization strategies each of them relying on a rewriting rules execution strategy. Thus the Logical optimizer architecture is centred around the strategy module which is responsible for applying the pertinent rules execution strategy.

5.2 Physical Optimizer

The physical optimizer takes as input a LERA Log program which is expressed using conceptual relations. Its role is to produce an optimal execution plan ex-pressed in LERA Phy. This execution plan consists of simple operations for which both operation localisation and the global and local algorithms used are expressed by annotations on the LERA. These algorithms have been chosen taking into account the existing access paths.

The main requirements which we want to meet in our design are support for parallel execution and extensibility. Extensibility is desirable to adapt to new or changing access methods as well as different target architectures, and to control the search strategy which should depend on the query type (decision support or repetitive). We have chosen a modular and object-oriented approach in our design to provide extensibility. Basically the optimizer's architecture consists of a language module, various optimizer experts each specialising in one type of optimization (e.g. search, fixpoint) and a parameterised cost model (see section 6). Each expert is implemented as a collection of methods attached to the operation to which it is devoted.

Support for parallel execution relies upon the direct storage model [Valduriez 86] used to distribute the relations. A relation is stored as a direct relation with secondary indexes. Direct relations are placed by declustering based on a hash function applied to one or more attributes. The set of processing elements where a relation is stored is called its home.

The parameterised cost model provides for the handling of shared memory as well as distributed memory architectures. The association of a processor to a memory module is "real" for distributed architecture whereas it is "virtual" for shared ones. The main difference is that, in the latter case, a processor can access any memory module with equal access time. However, from an optimization point of view, these two architectures are quite similar since only the cost functions for data access are different.

This optimizer consists of four layers performing successively program, control structure (e.g. fixpoint), complex operation (n- ary search) and simple operation optimization.

Program optimization is performed by the language module which has the global understanding of the program. The control structure optimization layer as well as the complex operation layer optimises their associated operations using lower layers. Simple operations have direct implementation in the target execution system and have associated access methods. The corresponding expert is responsible for access method selection; it includes a global optimizer to choose the parallel algorithm and a local optimizer to choose the local algorithm.

Isolation of units which can be independently optimised, except for their own nested units, reduces the size of the search space. However, some decisions are better taken at the global level. For example, some data transfers can be avoided if operation homes are chosen at a global level. To incorporate such global decisions, a post-optimization phase can revisit some local decisions.

5.3 The Paralleliser

The paralleliser is responsible for making explicit the execution strategy fixed by the physical optimizer through annotations in the LERA Phy programs. It transforms a LERA Phy program into a LERA Par program which is a set of communicating operations which are executed on logical homes. Some of these operations are control operations which are added by the paralleliser. The introduction of these control operations leads to the fact that the whole control of the execution is really embedded in a LERA Par program. Moreover this design decision makes the data operations completely independent of any control features.

The paralleliser supports three levels of parallelism within a single program:

- the first level directly relies on the parallel architecture. Several instances of the same operation are executed on several processing elements.
- the second one is related to operations without data and control dependencies - these operations may be executed in parallel.
- the third one is related to operations with data dependencies which may be executed in pipeline.

The paralleliser performs two main functions: abstract parallelisation which deduces the sequencing of the operations and control parallelisation which, based on this sequencing, explicitly sets the required control. The abstract paralleliser uses the LERA Phy annotations concerning the use of a global algorithm and the way the operands are consumed to integrate the communications operations in the program. Having extended the LERA Phy program in this way, the sequencing is made explicit. The control paralleliser then associates with each data operation the control operations which are needed for correct sequencing.The sequencing of a LERA program is the complete description of how the operations behave relatively to the others: if they are executed in parallel, in pipeline or if they are synchronised. To execute a program with respect to the desired sequencing, we must be able to perform the synchronisation of operations and to detect the local and global termination of an operation. Before adding this control, data operations are grouped into execution entities: the local operators. The control operations are then really associated with the local operators.

6. Designing a Cost Model

Query optimization is a growing area of interest in the database community, as it is necessary to handle specific and complex applications such as: Office Information Systems (OIS), Stock Trading Databases, Computer Aided Design (CAD), etc. The optimization of query processing [Valduriez 90] has to derive an efficient execution plan to produce the information requested by the user. This plan specifies all the information (e.g. access methods, operation graph) to compute a query.

To choose the plan to be executed, the physical optimizer needs to evaluate a number of possibilities with the help of the cost model [Andres 90a]. As well as this, the cost model can solve other problems in several fields such as data placement or performance prediction.

In the following section, we explain the main choices we made for the cost model design. Then we introduce the abstract knowledge levels exploited by the cost model.

6.1 Cost Model Design

The cost model is initially designed as a tool for the optimization of query processing. The design is based on an abstract approach based on structuring the environmental knowledge (computer architecture, target application, target system) which is stored in specific libraries. Such libraries provide extensibility and adaptability in the cost model without compromising its efficiency.

Extensibility is necessary to incorporate future language extensions (e.g. Fixpoint, Nest, Unnest). Adaptability allows us to use the cost model for several purposes. It is easy to switch between several libraries according to the local environment (single user system, transactional system, distributed memory or shared memory architecture). Furthermore, the flexibility of the design allows the development of a cost model generator. Such software will be able to quickly produce specific cost models adapted to the target applications and target computers.

6.2 The Abstract Knowledge Levels of the Cost Model

The knowledge about the environment (e.g. architecture, database, algorithms, access methods and system) is described in a language called the Environment Specification Language (or ESL). ESL is class-orientated to decompose the global knowledge of the environment into specific environment classes. Two class types are supported: the Complex class, and the Basic class. A complex class is defined by a grammar determining which environment is supported (e.g. a specific computer, a specific system). Such a grammar calls Basic classes defined as a set of attributes.

Four complex classes are supported in ESL:

- the Architecture Class (AC),
- the System Class (SC),
- the Algorithm and Access Method Class (AAMC), and
- the Database Profile Class (DBPC).

Each class is associated with a specific library which can be invoked during the cost estimation process. This knowledge allows the cost evaluator to estimate the execution time of logically optimized queries for specific configurations of architecture and application profile.

6.3 Validation

The cost model design provides a flexible framework to help the DBMS designers and the performance evaluators. The formal approach used to isolate the knowledge bases should be helpful for the verification of the design. Anyway, it will be necessary to validate the predictions of the cost model for a large number of cases by comparing these predictions to real measures.

Future work will include experimentation with the cost evaluator prototype on various target computers and operating systems. The cost model will be validated on the EDS machine (distributed memory architecture) [Lopez 89].

7. Query Execution on the Share Nothing Architecture

The Data Manager component of the Data Server software is responsible for the execution of queries. It provides a general purpose parallel programming environment (the BREM [Glynn 90]), a library of routines for maintaining and processing relations (the REM), management of relations (RAM), and a set of access methods.

A Session consists of Request Manager Module and a set of local Data Manager Modules, one for each processing element used by the session. A Session can be re-used by many queries. Initially, Sessions are created with one master Data Manager Module containing the Data Manager and query code. As the processing spreads to other processing elements local Data Manager Modules are created as required.

Code and run-time parameters are copied to the other Data Manager Modules lazily by using a virtual memory technique similar to that used by conventional systems. However, instead of the code being copied from disc, the code is copied from the master Data Manager Module across the delta network.

The tuples of a relation are horizontally partitioned amongst the processing elements by hashing upon part of the tuple contents, usually the primary key. We minimize the movement of tuples by performing work upon the processing element that contains the tuples. So, if we wish to perform a filter of relation R, and R is distributed over thirty processing elements, then each processing element performs the filter on its subset of the relation. If R is evenly distributed then the filter should take only a thirtieth of the time, plus some small overhead for synchronisation of the parallel filters.

7.1 Basic Relational Execution Model

The BREM can be incorporated into most modern imperative languages, (so far we have C and C + + versions). It provides very lightweight threads of computation, *instances,* and cheap mechanisms for performing the necessary synchronisation of these instances. It is responsible for ensuring that the lazy creation of Data Manager Modules is invisible to the programmer. The BREM has been designed to make efficient use of the EDS kernel, EMEX, a multi-processor kernel similar to Chorus and Mach.

The lightweight threads are especially useful within a processing element. Locking of relations in the EDS Data Server is performed at a tuple level and, to improve throughput, we can create an instance per tuple. Only the instance attempting access to a tuple is blocked, all other instances can continue independently.

Some relational operators, such as a join of two relations, cannot always be performed entirely locally. In these cases the smaller relation is first re-distributed to the processing elements containing the other relation. If the join is on the key attribute(s) then we make use of the hashed distribution of relations to just send a tuple to that processing element that has tuples with the same key. BREM provides special templates for instances, called *generics.* When re-distributing for a join we stream the tuples to the generics on the target processing elements. For each tuple received an instance with a reference to the tuple is spawned from the generic template. This instance can perform the local join with the local subset of the relation.

BREM allows generic nodes to be created lazily, thus they are only created on the processing elements that receive tuples at run-time. This allows us to get good performance for transactions such as Debit-Credit where at run-time we only access a small proportion of the database but at compile-time we do not know what this subset will be.

Instances have a priority at which they run with respect to other instances. This provides some control over the producer-consumer parallelism of the Data Manager algorithms.

All BREM inter-processor communications is asynchronous, message failures are routed to BREM exception handlers by the EDS kernel. The BREM run time system handles these failures invisibly to the components using it. We allow for both ordered and unordered tuple streams. Unordered streams are implemented more efficiently by the EDS kernel and are used for the vast majority of messages. Ordered streams need to be used where the order of the tuples needs to be maintained, eg. when implementing a merge sort of two or more streams.

The remainder of the Data Manager components are specific to supporting the execution of database queries. They are all built on top of the BREM and Object Manager interfaces (see next section for a description of the Object Manager).

7.2 Relational Execution Model

The REM provides a library of routines (the LERA Run-Time Library) to support the execution of relational queries. The back end of the the Request Manager (the code generator) produces C + + code that makes calls into this library for the LERA Par Data and Control Operators. This level of interface completely hides the BREM mechanisms.

7.3 Relation Access Manager and Access Methods

The RAM (Relation Access Manager) controls access to the global portion of base relations. In particular, it is responsible for ensuring that updates to a relation cause all affected primary and secondary indexes to be updated also.

The Access Methods provide access to the tuples of the local portion of the relation. The interface is declarative in style. We provide interfaces to apply a function to all the tuples that satisfy certain conditions. This contrasts with the more conventional Access Method interface of 'get_first', 'process tuple', 'get_next', 'process tuple', 'get_next', 'process tuple', etc. The declarative style allows us to provide higher throughput where some of the tuple accesses may block due to clashes in the Object Manager.

A number of different access methods will be supported. They all support the same interface but have different internal structures, such as lists, B+ trees, hash arrays etc. This allows optimum performance for relations with different access characteristics.

7.4 Design Status

We have defined the BREM interface using C++. We have implemented this interface on top of an EDS kernel simulator. An initial Access Method has been written using B+ trees and a subset of the LERA Run-Time Library has been defined.

We plan to have a prototype Data Manager running on a 16 element shared-nothing multi-processor by the end of 1990.

8. Query Execution on a Shared Memory Architecture

Although the EDS Project is concentrating on the share nothing architecture, a shared memory test bed will be developed to test various aspects of the request manager, especially those regarding parallelism, e.g., the parallel execution strategies. Note that we refer to testing the high level strategies for parallelism, which are implemented by the request manager, rather than the functionality, of the data manager itself. This test bed will allow a continuum of parallel architectures, including those of the shared memory and shared nothing to be explored. The shared memory is but one point on this continuum - the shared nothing is another.

In fact, neither the shared memory test bed nor the share nothing prototype (see section 7) are strictly shared memory or shared nothing, respectively. The shared memory test bed, in addition to testing shared nothing by emulating shared nothing strictly, will largely be shared memory, but in fact, will emulate shared nothing in horizontally partitioning most relations. The shared nothing prototype is largely shared nothing, but in fact, will emulate shared memory in the architecture to distribute code and parameters. Furthermore, exploring the continuum, as well as shared memory explicitly and shared nothing explicitly, moderates our reliance on one or the other.

9. The EDS Object Manager

The Object Manager is the lowest layer of the Database Server and its purpose is to provide a reliable object store for the objects manipulated by the LERA machine. The reliability and integrity of the data in a DBMS is a key issue in the design of the DBMS. The functions of the Object Manager are:

- to isolate from the DBMS the problems of providing a reliable shared data store,
- to raise the level of abstraction of the objects manipulated by the LERA machine.

These two functional areas split into two layers in the Object Manager. The upper layer, called the Abstract Object Support layer, provides high level objects to support the LERA machine. Examples of these "high level objects" are arrays, B+ trees, sets, rule sets, and user defined data structures. This layer is built on the lower layer which provides the reliable storage of simple, unstructured objects. This layer is called the Reliable Object Store or ROS.

9.1 Reliable Object Store

The objects provided by the ROS are entities with only the properties of unique identity, state, and persistence. That is, an object has an unique identifier which can only be used for that object during the lifetime of the system. An object has associated with it a value which is held as a sequence of bytes, and this value may be updated. Lastly, the value of an object will persist across system breaks. The type and form of the value held with an object is determined by high level software, in particular the compilers used for manipulating the values.

Thus, in summary, the function of the Reliable Object Store is to provide a shared, reliable and consistent store for unstructured data items. This abstracts the concept of a variable found in programming languages to provide a concept more suitable for shared store. The reliability of the system is supported by introducing the concept of a transaction.

The EDS Database Server must solve two crucial problems. Firstly it must allow independent programs to access the database concurrently, and secondly it must provide an acceptable level of reliability. In order to build reliable database systems the concept of a transaction has been developed. Informally a transaction is the execution of a program that accesses a shared data store such that

– each transaction accesses the shared data store without interfering with other transactions executing concurrently.
– each transaction either terminates successfully, in which case all changes to the data store are made permanent, or the transaction terminates abnormally, in which case it has no effect on the data store.

9.2 Design Direction

The level of the interface of the EDS Reliable Object Store is roughly equivalent to interfaces of such object managers as GEODE [Bellosta 90], Camelot [Spector 87], Arjuna [Shrivastava 88], and the Bubba object manager [Clay 89]. The implementation of the EDS Object Manager will be built on the ideas developed in these object managers. For example the facilities for supporting "user" written concurrency control, logging and recovery mechanisms will be used to develop advanced concurrency control mechanisms for the EDS system. Similarly the external pager technology developed in Camelot will be used to implement efficient buffering and scheduling strategies.

The second direction in which the Reliable Object Store is being developed is to take advantage of the features of the EDS system; in particular the large stable RAM memory on the processing elements.

10. Successfully Managing the Design Work

A recognition by Bull, ICL, and Siemens of the need for a powerful database server, together with common yet independent interests in both parallel architectures and relational databases led naturally to the present collaboration in EDS. In addition the design of a database server containing much that is novel, particularly in the area of exploitation of parallel machine architecture demands a commitment that any one company cannot afford.

10.1 The Teams

The involvement of Infosys, Inria, Heriot-Watt University and ESIEE has added a wealth of practical and theoretical experience in areas such as rule based extensions, performance modeling and prototype parallel systems.

During 1989, our teams have forged friendships and working relationships so that we now work effectively as one team; albeit geographically distributed. At the start the background of all these teams appeared well aligned, but many diverging ideas of detail have had to be merged or good compromises reached during the project definition phase (1989). The support and encouragement of the team at Esprit and sometimes pressure resulting from the need to present good, cohesive results to them, has played an important part in our successful team building.

With a considerable geographic distribution of the design team, it has been necessary to partition the design into clear functional pieces each of which could be worked on with a reasonable degree of independence. Tackling first the national boundaries for which traveling time and cost are major factors and within which use of native languages makes design easier, we made an early decision that the French teams should take charge of the largely compilation phase (Request Manager) and the UK teams that of the execution phase (Data Manager). The German teams decided to concentrate their efforts outside the database server, on the Kernel and hardware components of EDS. Of course these divisions are not absolute, and at the regular overall design meetings discussions range across all these boundaries.

10.2 Design Integration

Geographically partitioned design needs special care if it is to fit together as one system. We have addressed this problem from the outset in a number of ways:

Firstly we recognized LERA as a key interface, not just between the Data Manager and Request Manager but also between the stages of the compiler which are shared between four teams in Inria, Bull and Infosys. The decision to ask ICL to lead and document the joint activity to define LERA has greatly benefited our common understanding of optimizing and running queries on the parallel EDS architecture.

The joint activity to define and document the other major interface, ESQL is led by Infosys.

Another important integration initiative by the joint team has been agreement to give regular demonstrations of the evolving database system and its pilots and to make these demonstrations an important and visible commitment to our CEC partners. Quite apart from showing off our work, these are aimed very carefully to cause even the earliest prototypes of all the components to be proved as a system, though in the first place with only very simple queries. Working towards these has given our early design work a good focus.

A common exemplar database and set of queries is used by all the team for design examples, walk throughs and testing.

10.3 Management

Like all the other major EDS technical areas which are each co-ordinated and led across all sites by one nominated person, the management of the database server is primarily led by Bull, involving managers from all of the sites. A single plan covering the work for the next year on all parts of the database server and their integration has been put together by this group and is discussed regularly.

In conclusion, we have achieved a great deal in building a unified yet geographically spread team which has already proved itself with designs for LERA, ESQL and many of the components. A prototype system capable of compiling simple queries and then executing them on parallel hardware test beds is well advanced and should be working this year.

11. Glossary of Terms

AAL	Abstract Architecture Language.
ADL	Abstract Data Language.
ADT	Abstract Data Type.
AMDL	Access Method Definition Language.
BREM	Basic Relational Execution Model.
CAD	Computer Aided Design.
Chorus	Operating system developed by Chorus Systems (France).
DBMS	DataBase Management System.
DoD	Department of Defence.
EDS	European Declarative System.
EMEX	EDS Kernel.
ESQL	Extended Structured Query Language.
HOME	(of relation) The net of P.E's on which relation is stored.
I/O	Input/Output.
LAZY	(execution/creation) Execution and Creation when required, and not before.
LERA	Language for Extended Relational Algebra.

LISP	LISt Processing.
Mach	Operating system developed by Carnegie Mellon University.
OIS	Office Information Systems.
PE	Processing Element.
PROLOG	PROgramming in LOGic.
RAM	Random Access Memory.
REM	Relational Execution Model.
RISC	Reduced Instruction Set Chip.
ROS	Reliable Object Store.
TPC-B	Transaction Processing Council (Benchmark B).

12. References

[Andres 90a] F. Andres "Cost Model for Database System in a Distributed Memory Architecture" EDS.DD.11b.3901, May 1990.

[Bellosta 90] M.J.Bellosta, A.Bessede, C.Darrieumerlou, O.Gruber, P.Pucheral, J.M.Theverin, H.Steffen. GEODE: Concepts and Facilities. Submit to publication, March 90.

[Boral 90] H. Boral, W. Alexander, L. Clay, G. Copeland, S. Danforth, M. Franklin, B. Hart, M. Smith, and P. Valduriez. 1990. Prototyping Bubba, a highly parallel database system. IEEE Transactions on Knowledge and Data Engineering. 2(1).

[Carey 88] M.J. Carey, D.J. DeWitt, S.L. Vandenberg, "A Data Model and Query Language for EXODUS", SIGMOD Int. Conf., Chicago, Illinois, June 1988.

[Ceri 90] S. Ceri, G. Gottlob, L. Tanca, "Logic Programming and Database", Springer-Verlag, Surveys in Computer Science Serie, 1990.

[Clay 89] L.Clay, G.Copeland, M.Franklin. Operating System Support for an Advanced Database System. MCC Technical Report ACT-ST-140-89. April 89.

[Finan 90] B.Finance, G.Gardarin, "An Extensible Query Rewriter for Databases with Objects and Rules" Journes Bases de Donnes Avances, Montpellier France,Sept. 90

[Gardarin 89] G. Gardarin et al., "Managing Complex Objects in an Extended Relational DBMS", Int. Conf. on VLDB, Amsterdam, August 1989.

[Gardarin 90] G.Gardarin, P.Valduriez ESQL: An Extended SQL withObject and Deductive Capabilities Int. Conf. on Database and Expert System Applications, Vienna, Austria, August 1990.

[Glynn 90] K. Glynn. BREM: A Parallel Model of Computation. ICL EDS.DD.11I.0075, July 1990.

[Guttag 77] J. Guttag, "Abstract Data Types and the Development of Data Structures", Comm. of ACM, Vol. 20, No. 6, June 1977.

[Kellett 89a] S.M.Kellett. LERA 1 - Language for Extended Relational Algebra. EDS.DD. 11 I.000 1 .

[Kellett 89b] S.M.Kellett, C. Chachaty, B. Bergsten. LERA Par - Parallel Lera Definition. EDS.DD. 11I.0002

[Lopez 89] M. Lopez "The EDS Database server" EDS Working Report, EDS.D901.B 89.

[Melton 89] J. Melton, "Database Language SQL2 and SQL3", ISO-ANSI working draft, ISO DBL CAN-2b, May, 1989.

[Pucheral 90] P. Pucheral, J-M Thevenin, P. Valduriez, "Efficient Main Memory Data Management Using the DBgraph Storage Model", Int. Conf. on VLDB, Brisbane, Australia, August 1990.

[Rohmer 86] J.Rohmer, R.Lescoeur, J.M.Kerisit, "The Alexander Method, a Technique for Processing of Recursive Axioms in Deductive Databases", New Generation Computing, 4, pp.273-285, Springer Verlag Ed.,1986.

[Shenoy 87] S.Shenoy, Z.Ozsoyoglu, "A System for Semantic Query Optimization" ACM SIGMOD, 1987.

[Shrivastava 88] S.K. Shrivastava, G.N. Dixon, F. Hedayati, G.D. Parrington, S.M. Wheater. An Overview of Arjuna: A system for reliable distributed computing. University of Newcastle upon Tyne, 1988.

[Spector 87] A.Z.Spector, D.Thompson, R.F.Pausch, et.al. Camelot: A Distributed Transaction Facility for Mach and the Internet - An Interim Report. Carnegie Mellon University, June 1987.

[Stonebraker 76] M.Stonebraker, E.Wong, P.Kreps, G.Held, "The Design and Implementation of INGRES" ACM Transaction on Database Systems, Vol 1-3, Sept. 1976

[Stonebraker 86] M. Stonebraker, L.A. Rowe, "The Design of POSTGRES", ACM SIGMOD Int. Conf., Washington, D.C., May 1986.

[Valduriez 86] P.Valduriez, S.Khoshafian, G.Copeland, "Implementation Techniques of Complex Objects" Int. Conf. on VLDB, Kyoto, August 86

[Valduriez 90] P. Valduriez, M. Ziane "Optimizing a Database Programming Language for Parallel Execution" Submitted for publication, July 1990.

The MERMAID Approach to software cost estimation

P.A.M. Kok
VOLMAC
Catherinesingel 30-33
P.O.Box 2575
3500 GN Utrecht
The Netherlands

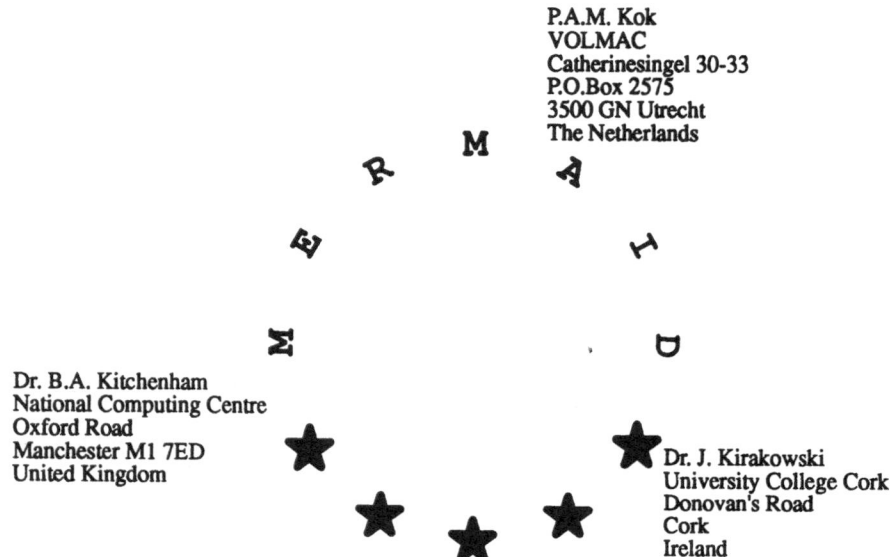

Dr. B.A. Kitchenham
National Computing Centre
Oxford Road
Manchester M1 7ED
United Kingdom

Dr. J. Kirakowski
University College Cork
Donovan's Road
Cork
Ireland

Summary

Despite the large supply of methods and tools for cost estimation, estimating the costs of a software development project remains a non-trivial activity. Research [Heemstra 1989, Mermaid 1989] has shown that the accuracy of such tools is low. It has also been shown that only a limited group of organizations uses systematic methods for drawing up a cost estimate for project-based software development. For many years, various lines of industry (for example the building industry and the catering industry) have used experience figures when drawing up cost estimates for projects. Cost estimates for software projects often lack such a basis, partly because there are no adequate tools for recording and analysing historical project data. This paper presents an analysis of the problems in the field of software cost models and describes the MERMAID approach to cost estimation. The MERMAID approach makes intensive use of local historical project data and is applicable in all sectors where project data can be collected.

1. Introduction

Over the years, software engineering has acquired a bad reputation with respect to cost estimation for software development projects. Everybody knows striking examples of projects which took considerably more time than was planned. These developments have had a harmful effect on the industry. Many companies can look back on some financial disaster or other resulting from an inadequate cost estimate before the start of the project. The importance of accurate estimates is gradually increasing for several reasons. Firstly, most organisations have already automated their most important business processes (e.g monetary exchange via the banks and the administration of complex social laws). The software to be developed over the next years will be used to start up new services and to improve existing services. Since there is no urgent need to automate these processes, the costs will play a more important role in the decision to computerise or not.

Secondly, the software development industry is growing up. Clients now reject a large number of things they would still have accepted five years ago. This affects not only the field of cost estimation, but it relates to all aspects of software engineering. The market for software products is now facing a gradual shift from demand to supply. A market characterized by supply means increasing competition. This results in a growing number of fixed-price contracts. A correct cost estimate is of the utmost importance to the software vendor. In addition there is a growing interest in the software development process, which is shown by the use of software metrics, project management tools, etcetera.

2. The MERMAID Project

The ESPRIT II project MERMAID was launched at the end of October 1988 with a budget of 45 man years. The project will end in October 1992. The goal of the MERMAID project is to improve support for estimators and project managers in the area of software sizing, effort estimation, risk analysis, resource modelling and progress monitoring. This will be achieved by the delivery of two increments of the MERMAID toolset. At the moment, the MARK 1 tools are being constructed and will be delivered to the CEC at the end of this year. Among other things, the MARK 1 toolset will be tested in commercial organizations. This evaluation combined with the results of the research program forms the basis for the MARK 2 toolset which will be delivered at the end of the project. Besides the basic concepts from the MARK 1 prototype, the MARK 2 toolset will consist of additional cost estimation tools (resource modelling and risk analysis).

Both prototypes will be generic and can be tailored to almost any environment. The flexibility of the tools gives the project manager the opportunity to optimise the use of local knowledge and data, which will lead to more accurate predictions. The prototypes will support the decision-making process, resulting in better decisions at the milestones of the project and in more effective monitoring. The MERMAID toolset is an implementation of the MERMAID approach. This approach is based on an assessment of the most significant problems in the area of cost estimation and it is a straightforward synthesis of DeMarco's concept [DeMarco 1982] of phase-based measurements and the Bailey-Basili meta-model approach [Bailey 1981].

The MERMAID consortium attaches great value to the industrial exploitation and impact of the results. The research is based on our assessment of the current and future needs of the (European) market, and the industrial partners who are primarily concerned with MERMAID tool development are, nonetheless, closely involved with the review of preliminary research results. Our research hypotheses will be tested on project data from several organizations from various European countries. The organizations operate in varying fields. In addition, both generations of the toolset will undergo industrial tests in order to asses their practical value.

The MERMAID consortium has already started technology transfer. In our opinion, acceptance of the method is the most important factor to success for the development of a new approach to cost estimation, next to the practical value of the method. From the start of the project, MERMAID has been active in the field of information dissemination by writing papers, making presentations and giving demonstrations. In addition, the MERMAID concepts are already being exploited in the form of consultancy and university course material.

3. Background

To enable the reader to view the MERMAID achievements in the right perspective, we will now give a short overview of the domain and our assessment of the major problems.

The MERMAID project is performing research in the area of project management and its supporting metrication tools as described in the ESPRIT II work plan (section II.1.11 and II.1.12) [CEC 1987]. Within this area, MERMAID focuses on support for cost estimation. Cost estimation here means estimating the staff effort and duration of a software development project.

In the past, a number of parametric models were developed which yield an effort forecast based on a limited number of factors. Most of these models assume a simple relationship between effort and project costs. They use the metric of effort, since forecasts expressed in monetary terms present problems. The problems concern the definitions (do the costs include overhead, sick-leave, profit) and the consistency of the predictions (exchange rates, inflation). Effort estimates usually refer to the software engineering effort which is necessary for software development. In other words, it usually does not include maintenance, clients' efforts to review documents, management, etcetera.

For the prediction of duration and staff levels, an extra dimension (that of time) is added to the model. In principle, effort, duration and average staff level can be deduced from each other (if 2 out of 3 are known, the remaining figure can be deduced). As every project manager knows, however, this is a nonlinear relationship. Employing additional staff will not yield a proportionate decrease in duration. In fact, it may even increase the duration in some situations.

In practice, parametric models are only to a limited extent used. Many projects are carried out without any cost estimation, or with an estimate based on informal analogy. Analogy means comparing the new project to projects carried out in the past, which had more or less the same size, in order to estimate the necessary effort, staff level and duration for the new project. The expert opinion method is also used, where one or more experts predict the effort on the basis of the data and specifications known at the beginning of the project.

In 1981, Barry Boehm introduced the COnstructive COst MOdel (COCOMO) [Boehm 1981]. This model was published with all its particulars. It is widely used, especially in the United States. Boehm developed the model on the basis of a study of some 63 projects which were completed between 1964 and 1979. One of Boehm's conclusions was that a nonlinear relationship exists between size (lines-of-code) and effort. There are three versions of the COCOMO model (basic, intermediate and detailed). These versions differ with respect to detail and complexity of the cost drivers. When using the detailed model on the COCOMO database (63 projects), an acceptable accuracy is reached (70% of the projects within 20% of the forecasts). Many commercial tools are based on this model [Mermaid 1989]. Research has shown that this level of accuracy is seldom reached in reality. There are two major causes for this.

- Firstly, at the start of the project most cost drivers (for example lines-of-code) are difficult to quantify, as opposed to the post-hoc study by Boehm, in which all factors were known.

- Secondly, the COCOMO model must be calibrated to the local environment (the weight factors and exponents must be adjusted). This is a non-trivial process and the results are often disappointing [Kitchenham 1985].

DeMarco published the BANG concept in 1982. He emphasizes the significance of measurements (You cannot control what you cannot measure) and he introduces a number of new (size) metrics, such as System Bang, the net usable function from the user's point of view) [DeMarco, 1982]. Besides a number of organizational arrangements, he stresses the importance of data collection and measurable metrics. DeMarco uses a clear phase-based approach. Each phase has its own metrics which are used to predict the next phase. DeMarco's approach offers a number of interesting solutions for the problem of quantification, but calibration is still necessary. Unlike COCOMO and Function Point Analysis (see below), DeMarco's method is not often used in real life.

Function Point Analysis was originally developed within IBM to provide a technology-independent size metric for productivity assessment. The method was worked out in further detail later on [Albrecht 1983, Symons 1988]. The functional description of the system is used to determine the number of Function Points. Because of this, Function Point Analysis can be used much earlier in the lifecycle than for instance COCOMO. The productivity, expressed in function points per day, indicates the relationship between the size of the system (multiplied by a number of correction factors) and the effort. Function Points are a synthetic metric. A complexity matrix is used to determine the number of Function Points from a number of basic product features. Various authors have drawn up such a matrix through trial and error, by means of a specific data set or local expertise. For this method, the matrix and the productivity must be calibrated to the local environment. The FPA method is often used in cost estimation tools (ESTIMACS, BIS Estimator, PROMPT, FPACS) [Rook 1990]. It is also very popular and widespread in IBM-based MIS departments. There have been a number of criticisms on the FPA method [Verner 1989], but the principle of early size metrics based on counts of user visible features seems a sound principle.

4. MERMAID achievements

In this section we focus on the achievements with respect to the development of MARK 1. We expect that the concepts and architecture of the MARK 1 will be reused in the MARK 2 toolset.

4.1 The MERMAID Approach

The MERMAID approach is based on an assessment of the problems found with existing methods and tools (particularly the lack of accuracy). The present cost estimation models (for example COCOMO, SLIM) describe the relation between product/process attributes (for example size, team experience) and effort. They are based either on data from a specific environment or on data from a number of different environments. Sometimes, the data on which the models are based are quite old (mid 1970s to 1980).

However in practice relationships are not consistent in different environments, and will change with time within a single environment. In many environments, data are highly variable and they may be inconsistent, there is no way to ensure that the metric values have been obtained using the same counting rules. In order to improve the accuracy of the models, calibration to the local environment is required. This is a non-trivial activity and involves tailoring the model to the local data definitions, the lifecycle and the local productivity rates. MERMAID takes the view that using statistical methods to generate a computer-based local model is more straightforward than manual calibration. Moreover, it allows for direct feedback from the environment to be built into the tools (in other words, the model generated by the statistical techniques will alter as the local project database grows). This is a mechanism to overcome the accuracy improvement problem. It leads to dynamic relations between attributes and effort, instead of the existing static equations.

The current cost estimation models are supposed to be applicable in any application environment. In a specific application, however, the models do not give an indication of the expected predictive accuracy of the model. The MERMAID tools will include procedures to assess the accuracy of the models used to generated cost estimates.

Many models still use estimates as input variables (for example predictions of product size in Lines-Of-Code, LOC). In order to use the models early in the development process, LOC must be predicted. There is no proof that people are better at predicting LOC than they are at predicting the effort. MERMAID will use phase-based equations which use input variables based on early available measurements, not on estimates.

In other words, we can use specification-based measurements at early stages in the process, followed by measures of size derived from design representations and finally measures derived from code. The models will use locally defined metrics that use the Function Point and BANG principles, without imposing the restrictions of these particular synthetic metrics.

Most organisations have problems fitting their work methods to the cost estimation model, as many models presuppose a specific type of lifecycle and specific metrics definitions. Unfortunately, there is no generally accepted lifecycle in the real world. The MERMAID toolset can be calibrated to the local working practices of any organisation.

Many existing models include a large number of adjustment factors. These appear to be necessary in order to deal with the variance observed from the nonhomogeneous databases that are used to develop parametric models. However, they are likely to result in unstable and inaccurate predictive equations (because correlated effects are treated as independent).

The MERMAID approach of basing models on local data should provide a more homogeneous database, requiring less adjustment factors. Statistical techniques will be used to restrict the number of factors included in the local models. The number of projects in the database will impose a basic restriction on the number of variables that could be included in a model. We will, however, use step-wise regression to try and produce a stable, unbiased predictive model. Many models do not make adequate use of the data which becomes available during the lifecycle. MERMAID believes that the concept of phase-based statistically derived models with user-defined metrics offers a good chance of improved cost estimates for organizations which have a reasonable amount of project data and a stable development process. This will also include the use of progress actuals to improve the preliminary estimates.

However, the approach presents some problems:

– many organizations have no historical project data at their disposal;

– many organizations do not have a stable development process;

– each project has its own characteristics which make it different from even a homogeneous database.

For these reasons, we have included estimation by analogy in the MERMAID approach, as an alternative method of cost estimation which will overcome some of these problems. At present, estimation by analogy is usually only available in size estimation tools (predicting LOC). It usually either involves some form of relative ranking of components/projects of known and unknown size, or the selection of similar components from a large component/project database. MERMAID will provide in MARK 2 methods of estimating effort and duration by analogy.

4.2 General Architecture

The functional architecture of MARK 1 has been based on the approach as described above. With MARK 1 a project manager has the facilities to make a forecast of effort and duration of both the next phase and the entire project. Additional support is offered in the form of product size forecasts, assessment of the accuracy of the forecast, feedback on the quality of the forecasts and progress monitoring facilities.

The results obtained can be used at various stages of the project. First, the results can be used when the project proposals are drawn up. Later on, the forecast can function as a baseline and a target for the project management. The tools can also support the project manager with replanning as a result of delay and changes in the product specifications.

The MARK 1 toolset is aimed at organizations which are developing software in a relatively stable environment and according to a lifecycle, which makes it possible to define local metrics for each milestone (i.e. a well-defined project breakpoint such as a phase end). One must also be prepared to collect data to be included in a historical project base.

The selection of the functionality of the MARK 1 is the result of the project constraints (effort and time) and the available expertise and knowledge in the consortium. MARK 1 is an intermediate prototype, and the functionality which was not selected to become part of MARK 1 is being transferred to MARK 2. Basically, the MARK 1 toolset contains models and methods based on limited research in the MERMAID project. The topics that require intensive research will be part of the end product.

The core of the MARK 1 toolset is the Estimation base. This central repository for metric definitions and estimation data is one of the mechanisms to integrate the various processes into one toolset. The structure of the estimation base is shown in figure 1. (See next page)

The Estimation base consists of one or more environments, depending on the number of software development lifecycles used within the organization. An environment consists of a set of definitions and the data of several projects. The definitions describe a certain lifecycle with its associated milestones and the attributes which will be collected in that particular environment. The data consists of the measurements collected at the milestones and the estimation data (forecasts, assumptions etcetera) generated by the estimation processes. The project data in an environment is collected according to the local environment definitions. The structure and contents of the environments in the Estimation base are defined by the user during the configuration process. They will therefore differ at each site.

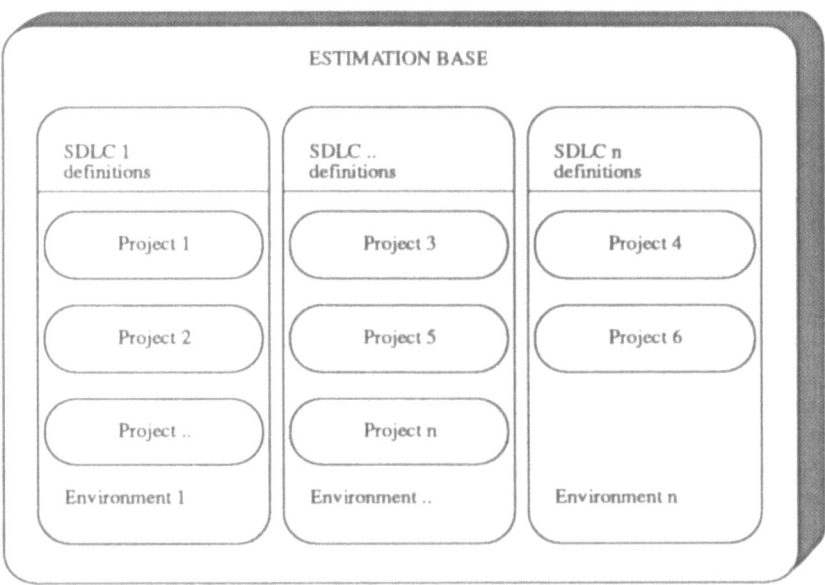

Figure 1 Structure of the Estimation base.

Figure 2 is a schematic representation of the architecture of the MARK 1 toolset. It shows that all the tools use the definitions and project data which have been centrally stored in the Estimation base.

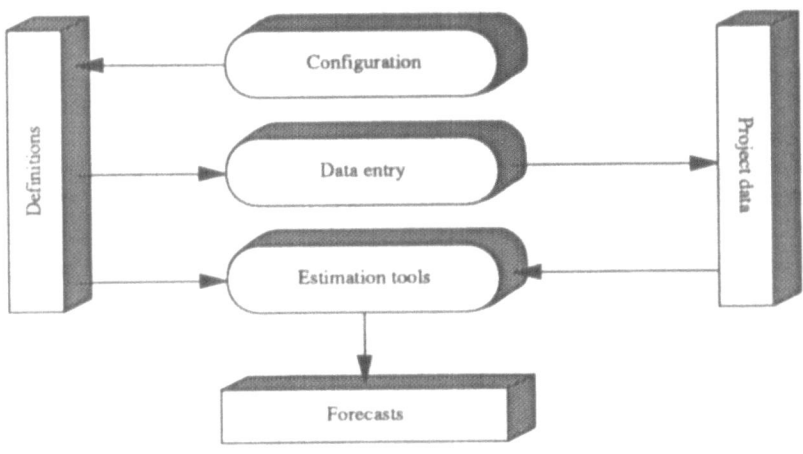

Figure 2 Architecture of the MARK 1 toolset

The environment definitions are provided by the user. They are added to the Estimation base by means of the Configuration tool. The Estimation base must contain at least one set of definitions, because the definitions control the data collection process. When the local lifecycle, the milestones and the attribute and metrics definitions have been defined, this tool is only used in case a change occurs in the metrics that are collected during the data entry process. A radical change (for example a change in the software development lifecycle) usually results in a new environment definition.

For an effective use of the configuration tool, it is necessary to understand the development lifecycle(s) and appreciate the measurement principles. Within a particular local environment we would expect senior project managers or quality assurance staff to coordinate the configuration activities.

The data collection tool within MARK 1 works interactively with the user. Automatic data collection from, for instance, data dictionaries or software libraries is not possible yet. When a project is initialized, a lifecycle must be selected. Next, the data collection process is performed at each milestone. To support the user during the collection process, the tool uses the concept of "data sources". A data source is an object in the real world (e.g. a timesheet or a design report) and contains one or more attributes which must be measured. Each data source is associated with a certain milestone.

For the development of the MARK 1 toolset, the MERMAID consortium used a SUN 3 workstation, PCTE and an object-oriented extension of C. During the development, a great deal of attention has been paid to user interaction and the definition of the various interface protocols. The user friendliness of a tool is one of the most important factors for acceptance, and therefore important for success, assuming that: SUCCESS = QUALITY * ACCEPTANCE

The project uses interface layers to make the models less dependent on the environment. The database interface makes it possible to access the estimation base as objects instead of entities and relations. The present interface supports the PCTE implementation, but it can be adapted to another DBMS with a minimum of effort. The MERMAID user interface functions as a shell between the models and methods and the way in which the interaction with the user takes place. In principle, the present (Objective C) user interface can be adapted to de-facto industrial standards. Further research is needed to determine whether a de-facto interface meets the requirements made by the MARK 1 toolset. This aspect will be playing an important role in MARK 2.

5. Functionality MARK 1 toolset

In this section we describe the statistically-based estimation methods in some detail, because they illustrate in a clear way the MERMAID approach. The other major methods, estimation by analogy and product size prediction, are described globally.

5.1 Self-calibrating Effort Estimation Function

In this section, we describe the self-calibrating effort estimation function in terms of

i)   its advances over current tools,
ii) the scope of the functionality intended for **MARK** 1,
iii) the assumptions implicit in the function,
iv) the technical approach underlying the function, and
v)   known limitations with the function.

Advances over current tools

The MERMAID approach to cost estimation is based using environment specific measurements and forecasting models, since we believe that this is the best approach to improving the accuracy of effort and duration forecasts.

The self-calibrating effort estimation functionality uses statistical analysis of past project data from a particular environment to generate effort forecasting equations. [An effort forecasting/estimating equation is a function relating effort (which is currently unknown) to other product or project attributes such as size which can be measured directly or easily assessed.]

The statistical generation of models should provide two major improvements over current tools:

– improved accuracy, as a result of i) using local data, ii) encouraging re-estimation throughout the lifecycle, and iii) increasing the chance of generating stable forecasting equations;

– assessment of model and forecast accuracy, as a result of i) applying basic tests of significance to the equations, ii) providing confidence limits on the forecasts, and iii) providing additional goodness of fit statistics.

Statistically-derived equations are by definition calibrated to the local environment and can use local metrics and metric definitions with impunity. (I.e. it is not necessary to reformulate an equation because the local definition of person hours per week differs from the definition of person hours per week built into a pre-defined model.)

In addition, the same statistical approach can be used at different stages in the development process when more information and different attributes (i.e. "metrics") are capable of being measured. The statistically-derived equations are intended to include directly measurable attributes and avoid the use of attributes which have, themselves, to be estimated.

Also, from a technical viewpoint, the use of statistically-derived equations have a number of advantages:

– they should lead to more stable forecasting equations;

– they permit an assessment of the validity of the equation.

We expect more stable equations because the step-wise regression algorithm should minimise the number of attributes included into the equations. In addition the use of statistically-derived equations provides some objective assessment of the likely accuracy of forecasts. Simple tests of significance can confirm that the attributes included in the equation have in the past influenced project effort. In addition, each prediction can be accompanied by its 95% confidence limits.

There are also a number of statistics which indicate how well an equation fits the dataset from which it was derived. These include the multiple regression coefficient and the proportion of back estimates within 25% of actual values. (A back estimate is obtained by applying a effort estimation equation to each project in the dataset which generated it and comparing the effort estimate obtained with the actual effort value.) These statistics were suggested by Conte, Dunsmore and Shen [Conte 1986].

Scope of the functionality

The self-calibrating effort estimation function will provide forecasts of total project effort, effort to next milestone, and remaining project effort. It is assumed that new forecasts of total effort and effort to next milestone will be made at the point at which a milestone is reached.

The first two forecasts will be accompanied by 95% confidence limits and information about the equation which generated the forecasts (i.e. the particular attributes included in the equation and various goodness of fit statistics). The third forecast is derived by subtracting effort to date (i.e. to the current milestone) from the total project effort forecast, so no additional information is provided.

The user will be given the opportunity of investigating the past project data in order to identify and eliminate outliers (i.e. projects which are unusual with respect to the particular environment). If the user does not exercise this option, all the projects which have the same lifecycle (i.e. set of milestones) as the current project will be used to derive the forecasting equations.

In addition, the user will be allowed to identify explicitly, the attributes ("metrics") which are to be included in the stepwise regression analysis which is to be used to generate the forecasting equations. If the user does not exercise this option, all the attributes measured at the current milestone will be included. Whether or not the user selects specific attributes, the stepwise regression algorithm will only include those which (jointly) significantly affect effort.

Assumptions

The most important assumptions underlying the statistically-derived effort estimation function are that:

– it is possible to collect data on a sufficient number of projects in a particular environment to be able to use statistical methods;

– effort and attributes are related by a formula which is well approximated by the general linear formula:

$effort = a + \Sigma f_i X_i$, where f_i is a function of X_i and X_i is an assessment of a project attribute.

The first assumption places restrictions on the potential users of the functionality. It implies that the function is only likely to be useful in an environment where a reasonably large number of projects are undertaken using the same basic lifecycle.

The second assumption is the basis of the MARK 1 **implementation** of the functional form of the effort forecasting equation. The functional form which will be used in MARK 1 assumes a multivariate, linear relationship between measures of product size (and other factors which influence effort such as staff experience) and development effort, in which first order interactions are assumed to be negligible.

The assumption that effort is influence by size and certain other factors is common to all software cost estimation models and tools, as is the assumption that first order interactions are negligible, but the assumption that the relationship is linear is not usual. The justification for this assumption is that the exponential term associated with size in effort-size equations is usually close to 1. A value greater than 1 implies diseconomies of scale, a value less than 1 implies economies of scale, but within MARK 1 we assume that the effect is likely to be negligible within a specific environment (where the range of product sizes is not likely to be great).

As part of MERMAID research, we have itemised and prepared plans to test each assumptions built into the MARK 1 tools. Any assumptions which we conclude are invalid will result lead to changes in functionality or implementation in the MARK 2 tools.

Technical approach

The function relies on using a variety of statistical techniques. For outlier detection, we provide the user with the ability to obtain box plots of various project measurements, and to view scatterplots in order to identify outliers in the past project database (figure 3). This is the procedure for outlier detection recommended by the ESPRIT 1 project REQUEST [Kitchenham and Linkman, 1990].

However, leaving this procedure under the control of the user presents a potential problem. Removal of outliers improves the chance of correctly identifying underlying trends in the data, however, if it is carried to extremes, it can cause the inherent variability of the data to be seriously underestimated and the significance of equations and the accuracy of predictions to be overestimated.

The effort forecasting equation is derived using a stepwise multiple regression algorithm to ensure both that the minimum number of variables are included in the predictive equation, and that we can perform basic tests of significance on the equations and provide confidence limits for the forecasts.

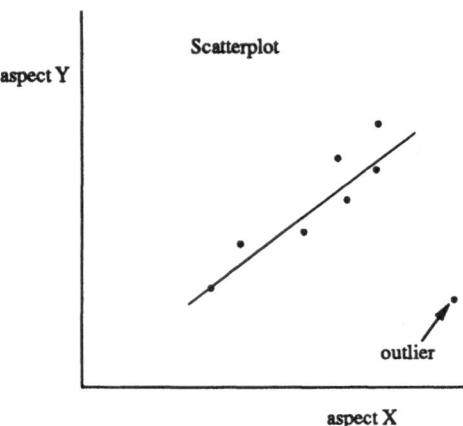

Figure 3 An example of an scatter plot to detect outliers

Limitations

In addition, to the assumptions mentioned previously, there are a number of limitations in the MARK 1 functionality.

The goodness of fit statistics advocated by Conte et al. [Conte 1986] have serious defects, and fail to distinguish properly between how well a model fits the data from which it was generated and how good a prediction from the model will be. The procedures for assessing the goodness of fit and accuracy of predictions are topics of research in MERMAID which should lead to improvements in the MARK 2 tools.

The level of granularity at which forecasts are provided is very coarse. We hope that in the MARK 2 toolset we will be able to produce predictions at component level and task level. In addition, we currently use the statistical approach only for effort forecasting, whereas using non-linear regression, it might be possible to extend the approach to predicting project duration.

5.2 Other Mark 1 Functions

The MARK 1 tools include a number of other functions aimed at improving cost estimation. These provide:

- an analogy based start-up to parametric models;
- size prediction;
- milestone-based status reports;
- DeMarco's estimating quality factor metric (EQF)
- recording ad-hoc estimates.

Analogy-based start-up to parametric models

A severe limitation upon the use of statistically-based forecasts is that they cannot be used until a reasonably large database of past projects has been assembled. In addition, they assume a stable lifecycle and will not provide accurate estimates when an organisation changes its lifecycle or development methods. Also, we do not advocate using parametric models when there is no measurable product size attribute to use as an input parameter. Therefore, in order to provide useful cost estimation tools, we recognise the need to provide methods of generating estimates under conditions when statistically-based methods will not work.

We believe that estimation methods which exploit the expertise of cost estimation staff, such as estimation by analogy, are the best means to produce estimates under conditions when statistically-based procedures will not work. The functionality provided by the analogy based start-up to parametric models provides a limited example of the use of analogy.

The function allows an existing parametric model (such as COCOMO) to be calibrated by reference to a single "similar" past project. In effect, any deviation between the parametric model's prediction of effort and/or duration for the past project and the actual effort or duration is assumed to hold for the new project. So the prediction for the new project using the same parametric model is adjusted accordingly.

The function also allows a user to "map" an existing distribution of effort or duration (used to partition a total project effort or duration forecast across the lifecycle stages) to a new distribution. This provides a means of coping with a changing lifecycle.

Size prediction

Apart from the difficulty of calibrating existing parametric models, many of them are based on input size metrics based on lines of code which are not available early in the lifecycle or function points which are oriented to 3rd generation data processing applications.

The size prediction function provides estimates of total product size in user defined scales (e.g. lines of code or function points) based on a set of predictive models which are used to predict component size for specified types of functional component. This means that there are different equations to predict the size of components which set up windows, and to predict the size of components which formulate reports, etc.

The input values used in the equations can be determined locally using the expertise of the development staff. The size prediction equations can be generated using regression analysis of previous project data. This function will use linear and non-linear regression and will select the best fitting function.

Currently, this functionality is intended to support the analogy based start-up function by providing improved size estimates for input into existing parametric models. However, in MARK 2 it may be possible to integrate this function with the statistically-based effort estimation function to provide component-based effort estimation.

Milestone-based status reporting

The project status report provides a milestone-based report on the project in terms of the current values of various product/project attributes, including actual effort and timescale information. Such information provides feedback to the estimator about the actual status of the project. It can indicate special circumstances such as a large number of outstanding fault reports, which would imply the need for adjustments to any effort and duration predictions.

Estimating Quality Factor

The MERMAID approach is to encourage re-estimation throughout the lifecycle when more information about the project/product becomes available. This is intended to contribute to the overall aim of improving the accuracy of cost estimates. MERMAID also recognises the need to provide feedback to estimators about the effectiveness of the estimation process. The Estimating Quality Factor (EQF) provides an assessment of a cost estimation process by measuring how well a series of total project effort forecasts converges on the actual total project effort. This metric was proposed by DeMarco [DeMarco 1982], but we are not aware of any existing cost estimation tools which provide this information.

Ad-hoc estimate adjustments

The MERMAID approach recognises the importance of local expertise in the production of cost estimates. We also recognise that the particular circumstances which may affect a project are too varied to be incorporated into any parametric model.

We, therefore, provide an option for a user of the MARK 1 tool to adjust any estimate made by the analogy based start-up function or the statistically-derived effort estimation function, and to record the adjusted estimate, together with any textual information required to explain the reason for the adjustment.

Future Directions in MERMAID

In phase 2 of the MERMAID project, the major focus will be the development of the MARK 2 tool, which includes further research to support this development. We anticipate that MARK 2 will rely strongly on knowledge elicitation. There are at least three strong advantages for knowledge elicitation in cost estimation which lessen dramatically the risks typically associated with knowledge elicitation projects. These can be stated as follows.

Firstly, the cost estimation area is relatively well defined already, with many tools and models already existing from which much can be learnt. Secondly, most concepts are easy to define in concrete terms, or at least, the pitfalls of imprecise definition are known if not understood. Thirdly, performance criteria for good practice exist. Indeed, there are many experts in cost estimation who routinely do a fair job in relatively adverse circumstances. It is the aim of the MERMAID consortium to harness this knowledge in a tool that will allow local expert cost estimators to do an excellent job in a much more secure working environment.

It must be clearly understood that MARK 2 is going to be a decision support system, growing out of the basic architecture of the MARK 1 prototype in the directions characterised above. We do not envisage a break in development strategy that would make MARK 2 a "conventional" knowledge- based system. For instance, MARK 2 will not be delivered on the platform of a KBS shell.

The specific areas in MARK 2 which will benefit from such knowledge input are as follows:

1. risk assessment
2. resource modelling
3. estimation by analogy
4. development of local parametric models

In the area of risk assessment, there are two main approaches, one is to produce a checklist of risk factors and assess project risk on the basis of the number of risk factors which apply. The other approach (which may be supported by checklists) is to attempt to quantify the probability of a particular event occurring (e.g. exceeding required timescales), and the impact (i.e. cost) of that event in order to identify the "risk exposure" [Boehm, 1989].

In both cases, we believe that a tool which incorporated (or responded to) human expertise would provide a major improvement over current tools. For example, this would allow local risk factors to be identified and incorporated with general risk factors and would provide support for the process of determining subjective probabilities of occurrence and assessments of costs.

Resource modelling concerns the interaction between various cost parameters (i.e. effort, staffing levels, timescales). Current tools usually include a single parametric model to describe that relation. However, different tools include completely incompatible models. For example, some models imply that reducing timescales increases effort and extending timescales decreases effort [Putnam, 1978], other models imply that reducing and extending timescales both increase effort [Boehm, 1981].

Jeffery [1987a] demonstrated that in practice **both** effects happened, as did the effect of reducing timescales resulting in decreased effort! He suggests that it is staffing level that determines whether timescale compression reduces productivity or not [Jeffery 1987b], but he, also, notes that the influence of factors such as personnel attitudes and experience could override staffing level effects.

It thus seems less and less plausible to believe that a single parametric model could describe the relationships among cost parameters. MERMAID will investigate methods which permit local resource modelling strategies to be incorporated into locally-derived parametric models.

The process of cost estimation by analogy seems to fall into three fairly clearly defined stages:

1. Discover one or more projects on which detailed records exist that are similar to the present project in terms of function.

2. Develop a "ball-park" estimate by appropriate scaling factors applied to the cost drivers of the analogous projects and estimated or already measured values of cost drivers of the present project.

3. Refine this estimate by applying corrections to the "ball-park" estimate based on relationships between cost drivers considered to be crucial in the present case and such cost drivers in other projects in the database, not limiting the search to those projects selected in stage (1).

The design of the analogy functionality depends on being able to identify the sorts of information the user needs at each of the three above stages, and to find the most appropriate method of delivering this information when requested.

Parametric modelling is the process of deriving a prediction equation. In MER-MAID this involves using a multiple regression methodology to an existing database of actual project costs and observed cost drivers. A professional statistician knows not to rely absolutely on the output of regression packages. Very often, the process of finding the best prediction equation is tempered by considerations beyond the merely statistical.

Expertise is necessary to assure the quality of the developing regression equations. There are three important areas in which expertise for parametric modelling must be focused:

1. To ensure that the set of cost drivers to be collected into the database is tolerably small and includes only cost drivers which are considered to be highly relevant to the local circumstances of the tool user; and to ensure that the cost drivers to be collected are independent in a statistical sense: there is no use in collecting both a number of observed cost drivers and a synthetic metric calculated from these. Multicollinearity will inevitably result.

2. To ensure that the data admitted to the database is "clean" and that error variation in the determination of each cost driver is kept to minimum possible. For instance, measures based on actual observations are infinitely preferable to estimates of the same cost drivers. Although "subjective" data is quite allowable, care must be taken that each user means the same thing by the same subjectively derived numbers. In practice this means developing clear and unambiguous anchor points for each subjective scale.

3. To ensure that when the final formula is being assembled during regression analysis, that cost drivers which are of marginal importance statistically should be cross-checked against the intuitions of the local expert. Statistically marginal cost drivers may have a great pragmatic importance, in which case they should definitely be included; alternatively they may be seen to be essentially trivial, resulting from historical facts of which the local expert is only too well aware.

Providing expert advice for general parametric modelling is impractical in many disciplines, for instance, in social science research. In such a discipline, the range of problems submitted to the analyst is so large that there is little alternative to judging each case on its own merits, and relying on extremely generic, often ill-defined knowledge. In cost estimation, on the other hand, we are presented with a much more constrained situation in which the resemblance between cost estimation problems are by contrast quite high. Developing facilities to advise local users therefore relies on much more knowledge, and advice can be tailored explicitly to the problem of deriving formulae for estimating cost.

In summary, the MARK 1 toolset is a synthesis of the best current approaches to cost estimation and, as such, presents an advance over existing tools. MARK 2 will incorporate a much wider range of functionality and will extend the MERMAID approach by integrating data intensive analysis methods with individual expertise. MARK 2 should, therefore, provide a sound basis for a new generation of project estimating and decision-making tools.

7. Acknowledgments

The authors acknowledge the contribution to the paper from all the members of the MERMAID consortium, Volmac (the Netherlands), City University London (United Kingdom), National Computer Centre (United Kingdom), University College Cork (Ireland) and Data Management (Italy). The MERMAID project is partially funded by the Commission of the European Countries as part of the ESPRIT II programme (ref. P2046).

8. References

Albrecht, A.J. and Gaffney, J.E. (1983) "Software Function, Source Lines of Code, and Development Effort Prediction: A Software Science Validation". IEEE Transactions on Software Engineering SE-9(6), pp. 639-648.

Bailey, J.W. and Basili, V.R. (1981) "A Meta-Model for Software Development Resource Expenditures". Proceedings of the 5th International Conference on Software Engineering, IEEE/ACM/NBS, March.

Boehm, B.W. (1981) "Software Engineering Economics". Prentice Hall, New Jersey.

Boehm, B.W. (1989) "Tutorial on Software Risk Management". 11th Conf on Software Engineering.

CEC, (1987) "ESPRIT workprogramme", Brussels.

314

Conte, S.D., Dunsmore, D.E. and Shen, V.Y. (1986) "Software Engineering Metrics and Models". Benjamin/Cummings, Menlo Park, California.

DeMarco, T. (1982) "Controlling Software Projects: management measurement and estimation". Yourdon, New York.

Heemstra, F.J. (1989) "Hoe duur is programmatuur". Kluwer, Deventer.

Jeffery, D.R. (1987a) "Time-sensitive cost models in commercial MIS environments". IEEE Trans of Software Engineering, SE-13, 7.

Jeffery, D.R. (1987b) "The relationship between team size, experience, and attitudes and software development productivity". IEEE COMPSAC '87.

Kitchenham, B.A. and Taylor, N.R. (1985) "Software project development cost estimation". The Journal of Systems and Software, no.5.

MERMAID (1989) "Comparative evaluation of existing cost estimation tools". ESPRIT II project P2046 MERMAID. Available from: VOLMAC, P.O. Box 2575, 3500 GN Utrecht, The Netherlands.

Putnam, L.H., "A general empirical solution to the macro sizing and estimating problem." IEEE Trans on Software Engineering, SE-4, 1978.

Rook, P., "Survey of Current Cost Estimation Packages". Proceedings of the European COCOMO User Group meeting 1990, 1990.

Symons, C.R., "Function point analysis difficulties and improvements". IEEE Transactions on software engineering, volume 14, no. 1, pp. 2-11, January 1988.

Verner, J.M., Tate, G., Jackson, B. and Hayward, R.G., "Technology dependence in Function Point Analysis: A case study and critical review". ICSE89, 1989.

Project No. 2059
PYGMALION

Neural Network Programming & Applications

Bernard ANGENIOL
Thomson-CSF Division Systemes Electroniques

Philip TRELEAVEN
University College London

Heidi Hackbarth
SEL Alcatel, Stuttgart

Abstract

ESPRIT II is currently funding two major neural computing projects: Project 2059: PYGMALION and Project 2092: ANNIE.

The objectives of PYGMALION are: firstly to demonstrate to European industry the potential of neural networks for various applications; secondly to develop a European neural network programming system, language and algorithm library; and thirdly to promote exchange of neural computing information in the European research community.

This paper introduces the PYGMALION project, reviews the neural network applications in image & speech processing, and then describes the neural network programming environment.

1. Pygmalion Project

In the last three years there has been a veritable explosion of interest in neural computing, covering neural applications[1], neural network models[2], neural programming environments and neurocomputers[1].

The ESPRIT II PYGMALION and ANNIE projects are intended to provide a focus for neural computing research within the European Community. The PYGMALION project, as shown by Figure 1, brings together many of the leading neural computing research groups from European industry, research institutes and universities.

315

Partner	Laboratory	Role
Thomson-CSF	Division Systemes Electroniques, Paris	Prime Contractor
		image processing
		acoustic signal processing
		high level language N
		VLSI neurocomputer chips
CSELT	Centro Studi e Lab. Telecomunicazioni, Torino	isolated word recognition(IWR)
Philips	Lab. d'Electronique et de Physique appliquée, Paris	image processing
SEL	SEL Research Centre, Stuttgart	speaker adaptive IWR
CTI	Computer Technology Institute, Patras	cellular automata tools
		3D pattern recognition
ENS	Ecole Normale Superieure, Paris	pattern recognition on images
INESC	Instituto de Engenharia Sistemas	algorithm library
	e Computadores, Lisboa	
IRIAC	Universite Paris Sud, Orsay	algorithm library
		low-level speech processing
UCL	University College London, London	graphic monitor
		intermediate level language nC
UPM	Universidad Politecnica de Madrid,Madrid	speech processing

Figure 1: Partners in the PYGMALION Project

The PYGMALION project aims to promote the application of neural networks by European industry, and to develop European "standard" computational tools for programming and simulation of neural networks. PYGMALION applications span image processing, speech processing and acoustic signal classification. Tools for neural networks centre on a programming environment comprising five major parts: (i) the Graphic Monitor for controlling and monitoring a network simulation, (ii) the Algorithm Library of common neural networks, (iii) the high level neural programming language N, (iv) the intermediate level network specification language nC, and (v) the compilers to the target machines.

Lastly, to develop a European general-purpose highly parallel neurocomputer, and to prepare the production of application dedicated neuro-chips, future funding will be requested. Nevertheless, the implementation of neural algorithms in WSI (Wafer Scale Integration) is already addressed by PYGMALION. A VLSI demonstrator for such a technology will be presented at the end of this project.

2. Neural Applications

Key real-world applications in image processing and speech processing, and a small application in acoustic signal classification, have been chosen to demonstrate the potential of neural networks to various industrial problems.

In image processing the two important applications domains being investigated are remote data sensing and factory inspection. Remote sensing covers pattern recognition and interpretation of Spot images on the earth's surface, such as road traffic, fields, and various kinds of grounds. Factory inspection covers the recognition and classification of workpieces in a factory automation context, handling normal problems relating to position, overlap and orientation of workpieces, in different lighting conditions.

In speech processing the aim is to lay the foundations for an automatic speech recognition system by developing efficient learning algorithms for the basic building blocks.

These basics include: (i) isolated word recognition for small and medium-sized vocabulary, (ii) speaker independence and speaker adaptivity, (iii) speech preprocessing, including noise reduction, (iv) isolated word recognition in noisy environments, and (v) subword-unit recognition and coarticulation.

Acoustic signal classification of underwater natural sounds will apply neural computing in two different ways:
(i) use of a processed version of the signal, obtained through classical preprocessed algorithms, as input to the neural classifier, or (ii) directly apply neural classification to the raw signal.

3. Image Processing

For image processing, we consider the first two years of this PYGMALION project as an initial exploration of neural networks. We are dealing with two distinct image data bases corresponding to two rather different kinds of applications:

Remote sensing data (namely Spot images), representing an earth area (with for instance: road traffic, fields, urban area, fluvial area...). Remote sensing has several domains of applications: geology, agriculture, meteorology, hydrology, cartography...

Workpieces in a factory automation context. The neural system constructed should be able to recognize and classify a set of workpieces, possibly overlapping, independently of their position and orientation and of the different lighting conditions.

The different tasks to be investigated may roughly be divided into low-level image processing, performed on Spot images and high-level, performed on workpieces images, after segmentation by classical techniques. Low-level processing includes Compression using learning by Back propagation on a neural network in auto-association and Segmentation into homogeneous regions via the combined approach of edge and region detection. Only supervised segmentation has been considered and the use of the Back propagation algorithm to produce edges or to classify different textual regions, after extraction of local features, is proving very efficient[3].

High-level processing mainly concerns <u>Classification</u> and <u>Pattern Recognition</u> in two and three dimensions. For 2D, two different approaches are considered:

– A description of the object in terms of statistical or syntactical features will be given as input to a neural network used as a classifier. During the learning process, the features will be evaluated by the network and the most relevant ones will be selected. Several types of neural networks will be compared (i.e. by example Hopfield's network, the multi-layer perceptron etc.).

– The image (not preprocessed) is directly used to feed the neural network. The approach is based on a constrained multi-layer perceptron: the first layer is used to preprocess the data; the synaptical coefficients are forced to be invariant with respect to translation and/or rotation of the objects presented to the network; and several preprocessings can be done using different areas of the first layer. Then, the training algorithm (e.g. Back propagation) will define the proper preprocessing(s) and perform the classification.

For 3D, we start with several images corresponding to a view of the same 3D objects from different angles. We want to be able to recognize automatically a 3D workpiece from a set of 3D objects, independently of their orientation or position on a plane. An associative memory algorithm[4] will be used to perform this task.

Finally, these image processing neural classifiers will be compared to more classic techniques such as Statistical methods and Syntactical pattern recognition techniques, for which extensive data exists.

4. Speech Processing

Many heuristic and even sophisticated methods have been tried for automatic speech recognition during the past. After initial optimism, researchers have more and more become aware of the many difficulties to be mastered. Considerable hope is being placed in the application of artificial neural networks in order to overcome current limitations and increase recognition system performance.

Main problem areas in current speech recognition are noise interference, speaker insensitivity and vocabulary size. In traditional systems, features have to be traded in for one another in order to maintain a certain level of performance, e.g. vocabulary size versus speaker independence. The characteristics of neural networks as hitherto identified, in contrast, led us to anticipate them to provide such system features at the same time, thus offering an enlarged capability profile with at least comparable recognition rates.

The objective of applying neural techniques to speech processing in the framework of the PYGMALION project is to develop and to investigate a variety of artificial neural network structures for individual tasks, such as:

– isolated word recognition for small and medium-sized vocabulary in quiet and noisy environments,

- speaker-independent and speaker-adaptive word classification,
- speech signal preprocessing, including noise reduction,
- multilingual recognition from speech data with different European languages.

Finally, the potential of artificial neural networks performing these tasks has to be evaluated in comparative tests with conventional techniques.

Five partners from industry and university are involved in the Speech Processing Task. The Research Center of SEL Alcatel in Stuttgart as Task Manager, the Centro Studi e Laboratori Telecommunicazioni SpA (CSELT) in Torino, the Politecnico di Torino, the Institut de Recherche en Intelligence Artificielle (IRIAC) in Paris and the Technical University of Madrid.

Activities at the SEL Alcatel Research Center in Stuttgart are focussing on speaker-adaptive isolated word recognition from a medium-sized vocabulary in an office environment. The recognition of whole words is considered appropriate in the scope of this project, where an immediate transfer of research results to products for the office and telecompmunication area is intended.

In initial experiments, the suitability of the neural-network approach was confirmed for speaker-dependent recognition of up to 60 words, with more than 99% being correctly classified. Regarding real environments, the network-based system must comply with high expectations with respect to noise tolerance. As demonstrated during the 1989 Technical Week in Brussels, recognition performance did not significantly drop in the noisy exhibition hall. Under such adverse conditions, neural networks can outperform the conventional recognizers, e.g. based on dynamic time warping.

The potential of network structures to differentiate among items from medium-sized vocabularies of about hundred words was furthermore studied. The particular type of so-called "scaly" neural networks (Krause and Backbarth 1989) proved to also approach perfect classification for this larger word set.

Speaker adaptivity represents a topic not investigated by neural formalisms formerly. The performance of an automatic speech recognition system can be considerably enhanced if a new speaker or one speaker in different situations (e.g. background noise, cough) is tuning the system before using it. Here, the learning capability of neural networks is exploited with the objective to adapt a recognizer to a new speaker or environment during only a brief enrollment phase. Several novel algorithms were developed, either using word input or speech feature vectors for adaptation. By such means, recognition rates can be increased by more than 10% as compared to speaker-independent recognition.

Researchers at CSELT, in cooperation with the Politecnico di Torino, are investigating the viability of neural networks for speaker-independent recognition of isolated words in a telecom environment. This is an important applicative problem not yet satisfactorily solved by conventional techniques. Practical solutions are fervidly sought and are industrially appealing, because, even if a very small vocabulary is used (e.g. the ten digits plus some command words), many useful voice-based services could be offered to the large population of the public telephone network subscribers (e.g. identification of a subscriber starting from his telephone number, statement of the account of a credit card given the number, etc.).

Although good results have already been obtained with classical methods (template matching, Hidden Markov Models) in controlled environments, there is still no solution in the typically uncontrolled telecom environment with background noise , microphone noise, naive and non-cooperative users and mispronunciations. The state-of-the-art recognition rate of approximately 95% is not sufficiently high to allow the realization of commercial applications which need 99%. The use of neural networks could contribute to overcome these difficulties due to their attractive capabilities.

Present results are encouraging: among the variety of different neural techniques investigated, even very simple neural networks of the multi-layer perceptron type, trained to classify the ten digits with a large telephone-collected voice data base,exhibit a recognition accuracy of more than 95%, showing the potential of the connectionist approach. The final goal is to develop a word classifier module based on neural techniques to be integrated with an acoustic-processing front-end module, realized with a specialized hardware and conventional techniques.

Studies at IRIAC in Paris concentrate on the preprocessing of speech signals with respect to feature extraction and noise reduction and on speaker-independent isolated word recognition from a limited vocabulary. The choice of speech signal representation for use as input to neural networks describes an important issue. The goal is to extract relevant information, such as the timing of phonetic phenomena, the intensity contour and the fundamental frequency. Results are reflected in speaker-independent and speaker-pooled isolated word recognition.

At the Politechnical University of Madrid, comparative studies on different network types are conducted. The PYGMALION library, which contains simulation software for a variety of connectionist structures, is being applied to a multi-lingual speech data base, consisting of a small vocabulary, recorded in different European languages. Thus, the generic European aspect within the PYGMALION approach to Speech Processing is coped for. This will also provide insight into particularities inherent in language structures.

Results hitherto achieved in the PYGMALION project show that artificial neural networks do in fact represent competitive methods for speech recognition as opposed to conventional techniques. They may even outperform the latter in particular applications, e.g. in noisy environments. It therefore appears to be realistic to predict the economical viability of corresponding implementations in the medium ter. (Reference [8])

5. Programming Environment

A major goal of PYGMALION is to ensure the widest usage of the neural programming environment, by making it as flexible and portable as possible. The environment comprises 5 major parts:

- Graphic Monitor, the graphical software environment for controlling the execution and monitoring of a neural network application simulation. This includes a simulation command language for setting up a simulation, monitoring its execution, interactively changing values, and saving a trained network.

- Algorithm Library, the parameterised library of common neural networks, written in the high level language and providing the user with a number of validated modules for constructing applications.

- High Level Language N, the object-oriented programming language for defining, in conjunction with the algorithm library, a neural network algorithm and application, by describing the network topology and its dynamics.

- Intermediate Level Language nC, the low level machine independent network specification language for representing the partially trained or trained neural network applications, a format analogous to P-code for PASCAL systems.

- Compilers to the target UNIX-based workstations and parallel Transputer-based machines.

To ensure uniformity and consistency, a common interface is supported by the Graphic Monitor, Algorithm Library plus the N and nC languages. The interface has the following properties. Firstly, all components present the view of a neural system based on a hierarchical structure of networks, layers, clusters, neurons and synapses. Secondly, all algorithms are parameterised. Thirdly, all algorithms and applications share a common repertoire of data structures, function names and system variables. Lastly, the specification of an algoristhm from an application are separated. The resulting, uniform, interface allows generic applications and generic algorithms; a single application to use many algorithms , or a single parameterised algorithm to be configured for many applications.

6. Graphic Monitor

Control of a neural network simulation is provided by the Graphic Monitor. The view is that the Graphic Monitor will execute on a Host computer and generate a specific net simulator/emulator to be executed on a Target computer. Monitoring of the simulation/emulation is done from the Host through X-windows graphic tools and the command language. The windows environment provides pull-down menus to select and change (i) I/O format, (ii) network architecture, (iii) network learning algorithm, (iv) network training and execution, and (v) displays of activations, weights etc.

A neural network is initially specified using the high level language N utilising modules from the Algorithm Library. It is then translated into the intermediate level language nC. Once configured and specified in nC, the neural network can then be trained or used via the Graphic Monitor. The Graphic Monitor can be used to

322

dynamically modify the network, to translate it to a particular Target machine or to save the trained or partially trained network for later usage, possibly on a different Target computer.

7. Algorithm Library

The Algorithm Library contains the classic neural network algorithms in a parameterised specification that can be configured for a specific user application. The popular algorithms provided are: (i) Gradient Back propagation with feedback, (ii) Hopfield, (iii) Boltzmann machine, (iv) Simulated Annealing, and (v) Competitive Learning etc. In these networks the interconnection geometry and the transfer equations are already specified. However, the number of PEs, their initial state and weight values, learning rates and time constants, are all user selectable.
The Library divides into five main parts:

- the Algorithm Independent Part contains the network computation subroutines, and support routines for memory and error management etc.

- the Tools Library routines for data file I/O subroutines, network architecture specification and performance measurement

- the Algorithm Modules, specifying the parameterised algorithms

- the Algorithm Evaluation Programs for testing the Algorithm Modules

- the Application Library provides the user-taylored application modules

The Algorithm Library is being constructed in two stages. Initially a _C version of the library is being implemented, for use by the image and speech processing applications in the Project. Then the full library will be implemented using N.

8. High Level Language - N

N is a high level neural network programming language for both expert and naive users. It allow development of neural algorithms and usage of operational algorithms in applications. Its syntax is a subset of C + + with additional neural oriented features. It allows the description of algorithms by defining specific types having their own data and behaviour in analogy to a class in C + + and, by assembling them in a modular tree hierarchy.

A library containing predefined types and parameterised algorithms (such as Hopfield, Back propagation ...) is available. The library may also contain unprotected programmers objects: this part of the library is used in a read-write mode and is managed by proogrammers. The source, code of both kinds of library objects is stored and thus may be easily consulted and integrated in any N program.

A typical N program consists of a list of type definitions. One type may be defined from previously defined ones. A type definition has the following structure:

```
new type xxx ( parameter-list ) parameter declaration
     {          composite types declaration
                internal variables declaration
                internal function declaration

                above:
                inherited variables declarations

                plugin:*
                communication fields (inputs)

                plugout:
                communication fields (outputs)

                connection:
                connections between communication fields

                public:
                inheritance variables declaration
                activation methods of the type beeing defined }
```

Thanks to the choice made for its syntax, any program in N can be easily translated into a C++ program and consequently algorithms can be simulated on a sequential computer such as a SUN or APOLLO. Furthermore, any neural network structure and algorithm in N can be translated into an equivalent nC structure, thus generating the nC version of any N program.

The N programming language will provide three types of facilities. Firstly, to support the conversion of an N model (i.e. algorithm) into an abstract representation which will allow the semantic analysis and the link editing together of several algorithms in the same application. Secondly, to allow the re-use of previously defined algorithms. Thirdly, to provide tools and criteria to translate N applications into nC network specifications.

9. Intermediate Level Language - Nc

The nC language acts as an intermediate level, machine-independent, representation for neural networks. A network, specified in the intermediate level language, can then be translated on to a variety of computers for training or use. After usage (ie. training) the intermediate level specification of the network is updated, so the trained network can be stored in a library, filed for later use or mapped on to a different computer.

Machine-independence is considered the major feature of the PYGMALION intermediate level language, and to enhance this we have made the language a small subset of the draft ISO standard C[6]. It is essentially a fourth generation [7] language with standard C data structures and system calls.

The nC intermediate level representation divides the neural network information into 4 different domains: (i) the network topology; (ii) the data of the system, including neuron status and synaptic weights; (iii) the functions defining the processing in the network; and (iv) the control of the network activities. An example of a framework for a neural network description is shown below.

The topology information is basically described by defining the system variables (such as NETS, LAYERS, CYCLES, LEARNING_RATE, etc), using #define commands, and by completing the system and config structures. The system structure defines the central hierarchical structure in terms of nets, layers, cluster inside layers, etc, and the config structure specifies the number of elements inside the hierarchy, like number of layers inside each network, number on neurons inside each cluster, etc. The system structure also stores data information, in terms of neuron's state and weights, as well as functional information, in the form of rules . These rules are related to the functions that should be performed by a neuron, such as weight summation and weight update.

The control information is provided by some system functions definitions, which configure, initialise and train the networks, and by a list of calls to the system functions, that gives the central control of the whole system.

10. Status

To conclude, the PYGMALION project has been running since January 1989. Already major progress has been made both on the applications and the environment.

The current status, for example with image processing, stemming from the results of the different tasks, is that the Back propagation algorithm implemented after adequate preprocessing, appears to be a good classifier in most of the cases. Further work will concern implementation of unsupervised classification algorithms and more precise performance evaluation tests.

For environment, the current status is that the Graphic Monitor, Algorithm Library, plus high level N and intermediate level nC languages are fully specified. Regarding implementation, a rudimentary Graphic Monitor (operating on an nC specification of a neural application) and a prototype Algorithm Library are both available for experimentation.

By the end of 1990 the complete PYGMALION environment will be finished.

References

1. New Science Associates Inc., "Neural Networks: Prospects for Commercial Use," New Sciences Associates, South Norwalk, CT , Spring 1988.

2. R.P. Lippmann, "An Introduction to Computing with Neural Nets," IEEE ASSP Magazine, pp. 4-22, April 1987.

3. D.E. RUMELHART, G.E. HINTON, and R.J. WILLIAMS, "323:533-536, (1986)," in Neuro-computing, Foundations of Research ", pp. 675-700.

4. T. Kohonen, "Self-Organisation and Associative Memory," Springer-Verlag Berlin, 1984.

5. B. Gold and R.P. Lippmann, "A Neural Network for isolated word recognition," IEEE ICASSP , pp. 44-47, 1988.

6. B.W. Kernighan and D.M Ritchie, The C Programming Language Second Edition, Prentice Hall, 1988.

7. J. Martin, "Fourth-Generation Languages",Prentice Hall, 1985.

8. Krause A, Hackbarth H (1989) Scaly artificial neural networks for speaker-independent recognition of isolated words. IEEE ICASSP (S1): 21-24

Project No. 2154

A Comparative Study of the Representation Languages Used in the Machine Learning Toolbox.

K. CAUSSE
Intellisoft,
28 rue Georges Clémenceau,
91400 Orsay,
France

P. SIMS
Sowerby Research Centre,
British Aerospace Plc,
FPC 267, Filton, Bristol
England

K. MORIK
Gesellscaft fur Mathematik
und Datenverarbeitung
Postfach 1240, Konrad Adenauer Str.,
Schloss Birlinghoven,
5205 Sankt Augustin 1,
W. Germany

C. ROUVEIROL
Laboratoire de Recherche en Infor-
matique
Université de Paris-Sud
Bat. 490, 91405 Orsay
France

Abstract

This paper presents some early results from the Machine Learning Toolbox (MLT) project. The MLT will be a system that recommends and implements one of several machine learning algorithms or systems for an application. The learning algorithms are being contributed by various members of the consortium, and as such have been developed with their own internal knowledge representation languages. In order for the user to supply application data in a form which can be understood by more than one algorithm, and in order for any algorithm to be capable of passing its results to any other algorithm, a Common Knowledge Representation Language (CKRL) has to be developed. The first stage in this task has been to investigate the different knowledge representation languages of the tools, with the aim of emphasising their commonalities and differences. The results of this comparison are currently being used as a basis for forming the first version of a CKRL. We also discuss the possible roles for the CKRL within the MLT, and select that of an interface language between the different sub-components of the MLT as being the most flexible. The CKRL aims to solve the problem of mapping entities of the epistemic level into the logic level (and vice versa) in a pragmatic way, but it will not attempt to solve the problems of the different expressive powers of each of the current algorithm's formalisms, or to evaluate the suitability of different languages for learning.

1. Introduction

Machine learning has now reached a level of maturity that has resulted in the commercial exploitation of some techniques. A number of simple derivates of the

ID3 algorithm (Quinlan 83) are already on the market (e.g. SD Rules). The time has come to try and make a wider range of techniques available to users wishing to benefit from the assistance machine learning can offer. Each individual learning tool or algorithm has been developed on a limited spectrum of applications. But, integrating a number of different algorithms under a common control structure will allow for a broader range of machine learning techniques to be applied to real world problems.

It is anticipated that the Machine Learning Toolbox (MLT) system will consist of a diversity of algorithms co-existing and communicating on a single host machine within a uniform environment as seen by the user (Ludwig 1989). The algorithms, contributed both by Industrialists and Academics, range from clustering approaches (for example DMP), induction algorithms (such as ID3, LASH, KBG, MAKEY), through to those algorithms which use more knowledge intensive approaches, such as MOBAL and APT (Ludwig 1989). As well as the algorithms themselves, the first version of the MLT will provide an HCI which will give a common "look and feel" to the use of any algorithm, and a module called the Consultant (Sleeman et al 1989) which will advise the user as to which learning tool is best suited to a particular problem (Figure 1).

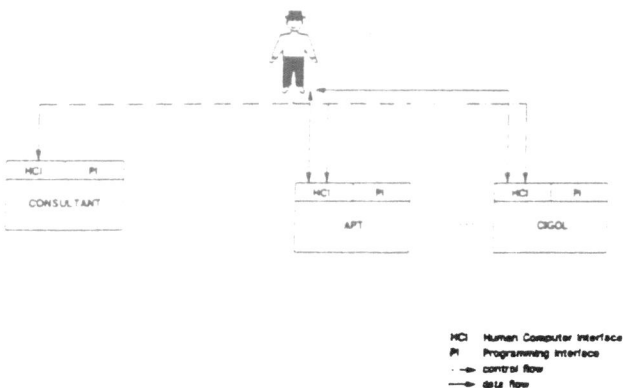

Figure 1: Basic Structure of the First Version of the MLT

Each algorithm will naturally have a different set of internal Knonledge Representation schemes from any other. Any prospective user of the MLT cannot be expected to learn all the representation languages used in the MLT. Hence, the need for a Common Knowledge Representation Language (CKRL) which is designed to be a single means by which the user may feed data to any of the learning algorithms. CKRL will also be used to pass the learning results from any learning system back to the user. This second use of the CKRL may result in algorithms being able to pass data directly to each other (via CKRL and under the guidance of the Consultant), but this will not be considered until the later stages of the MLT project. Seen in this way, the CKRL contributes towards the uniformity of the system as seen by user, the Consultant and the HCI completing this responsibility.

The remainder of this paper addresses three facets of the MLT consortium's work on the CKRL: a discussion of the possible roles that the CKRL could play in the MLT, a review of the problems of describing a number of representation formalisms under a single scheme, and excerpts of the comparison of the different representation languages used by the learning algorithms to be used in the MLT.

2. The Role of the CKRL in the MLT

The MLT consortium has discussed several possible roles for the role to be played by CKRL in the eventual system. These are presented in detail in Deliverable 2.1 (Morik et al 1989), from which we shall consider the two main options which were considered:

The first approach to CKRL for the MLT is very ambitious: CKRL is defined as the representation language for the input data, the results - including intermediate results if they are interpretable by a certain algorithm - and the background knowledge of each learning algorithm i.e it is used to represent all inputs, outputs and background knowledge required by the MLT. In addition to this, since some learning algorithms use inferences and truth maintenance, CKRL has to provide an inference engine making these activities possible on the CKRL form of any data. The CKRL would also provide a knowledge-base management system, including consistency maintenance and error recovery (in case learning for a given system leads to incorrect results), etc. The system architecture that accompanies this view shows a separate module that takes care of all these complex tasks (Figure 2).

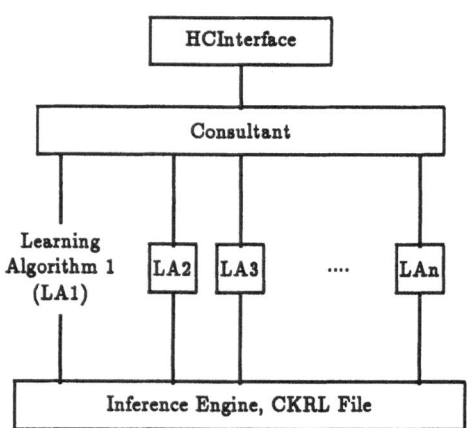

Figure 2: Architecture for First View of CKRL

Note, that this role of CKRL would mean all the algorithm and the Consultant having to use CKRL format, and each component of the MLT transforming their input and output as well as their intermediate results into a very powerful representation language. Any new form of data, or change in inferential technique used by an algorithm would require some change in CKRL if it did not have the representational power to express the new structure. In effect, the CKRL would be eventually have to represent and perform appropriate reasoning techniques with all the data structures and operations that could occur in machine learning.

System design considerations also show the disadvantages of this ambitious approach. The system would only allow for one way of input to the algorithms, namely reading data from the inference engine. However, further uses of the MLT system could include applying learning to information contained in pre-existing databases. Interfacing with a database system could be easy for some algorithms whereas it is more difficult for others (e.g. if the knowledge representation used internally by an algorithm is close to the database format). In this case, there would be a large overhead to first transforming the database representation into CKRL and then (back) into the representation used internally by the learning algorithm. The system would also rely heavily on the inference engine. Every input and output to the algorithms would use this bottleneck. Thus, not reaching the goal to create the universal knowledge representation language and inference engine for machine learning would condemn the overall system to failure. Therefore, the reasons for rejecting this role of CKRL are:

– Learning algorithms and CKRL are too closely intertwined, so that changes on one algorithm can require changes on CKRL which then have to be propagated to all other algorithms.

– CKRL has too much overhead, so that it would eventually become inefficient and the overall system becomes monolithic and inflexible.

– Success of the MLT would depend on the success of a very risky task, namely creatingthe common knowledge representation for machine learning.

The second approach to CKRL (Figure 3) views it as an interface language for assisting the communication between the user and the learning algorithms, consultant and learning algorithms, and learning algorithms with each other[1].

All the data that has to be communicated will have to be represented in CKRL. This is similar to the first approach, but it goes no further, i.e. no inference facilities are provided, and there is no absolute requirement to use the CKRL. Since CKRL will just be a communication language, it will be up to each learning algorithm developer to write translators from their existing input and output formats to appropriate CKRL items (with guidance being provided from the CKRL design team).

---

1 The passing of CKRL items to and from the algorithms is not planned to take place through the HCI which is primarily for control. The point of contact between algorithms and CKRL items called the Programming Interface (PI).

The communication between modules can be regarded as a system of message passing under a central manager. The messages are interpreted by the individual modules, in particular, any requirements for inferences and truth maintenance are performed by the learning algorithms themselves, although the CKRL will be able to express some basic operations related to its role in message passing[2]. Of course, developing a format for representing all the information that might be sent to and from the various learning algorithms is no trivial task, but it is not as risky as developing the CKRL as proposed under the first approach.

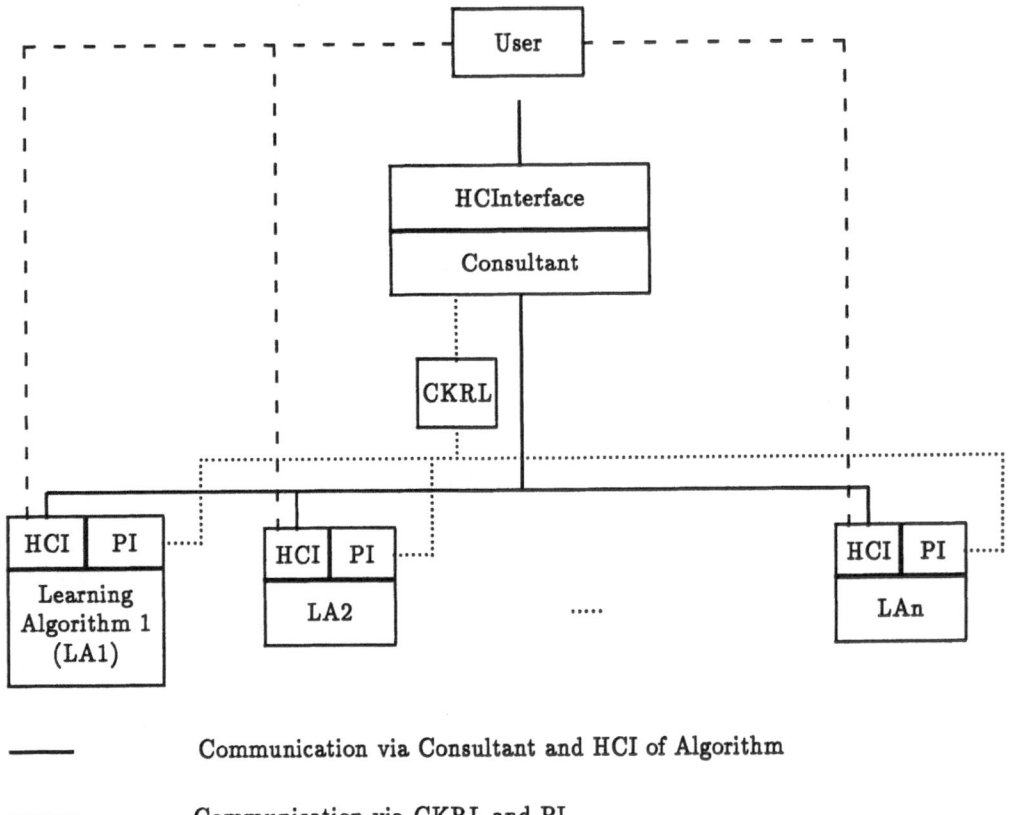

——— Communication via Consultant and HCI of Algorithm

············ Communication via CKRL and PI

- - - Communication direct to Algorithm's HCI

Figure 3: Architecture for Second Approach to CKRL

---

2 This aspect of CKRL is not addressed in any detail here, we focus on CKRL's need to express the data and information to be passed to and from the algorithms

Another advantage of this approach is that it allows for an incremental construction of the MLT in parallel with the development of the CKRL: no work has to be suspended until CKRL is accomplished. Thus, the approach is similar to the first one but allows for more flexibility and is less dependent on the CKRL to achieve milestones (e.g. the functionality of any learning algorithm can be shown without the need for CKRL under the second approach). Therefore. we propose this approach to using CKRL as a communication language.

3. The Problem of Describing Representation Formalisms

As stated above, the CKRL has to provide a format for representing all the inputs and outputs of the learning algorithms in the MLT, but not the internal data structures of a particular algorithm. Moreover, since CKRL is aimed at representing information from diverse systems in a uniform manner, it is not sufficient to simply provide a unique item to describe every entity occurring in a particular algorithm, otherwise CKRL will become a collection of all the specific items that are used in the internal representation languages of each system. For example, the AQ learning algorithm produces an output of the form:

$is\_class(Es, bmw):- has\_att(colour = red, Ex), has\_att(nationality = german, Ex).$ - (i)

Whereas MOBAL represents nearly the same information as:

$red\ (X)\ \&\ german\ (X) -> bmw(X).$ - (ii)

where "red","german", and "bmw" can be defined by other rules and declared by the sort taxonomy. The two internal representations will have to be made to correspond to each other via a canonical form (CKRL). Developing a canonical form is in itself a difficult task which consists of the following:

− what to describe in CKRL
  - find the entities of representation that are to be described
  - find out the semantics of the representation entities, i.e. their interpretation in the respective algorithms

− how to describe items in CKRL
  - find the terms which are needed to describe the input/output structures properly
  - determine the semantics of these terms and communicate it to the algorithm.

The first point is to find out which entities of representation the algorithms use. The same expression, e.g. (i), can be said to have a whole range of different entities e.g. predicates, constants, and variables; classes, objects and properties; cars, nationalities, colours and letters; or even signs like = , :,-. There are several levels of description for a given representation formalism. This follows Brachman (1979) who first pointed at the fact that it is not only a matter of granularity but also of different qualities when we talk about distinctive levels of the same expression. The second point above is, how do we describe these entities? This is now discussed for a number of representation levels.

Logic Level

On one level we could look for well-known representation schemas in algorithms, such as links and nodes in a network, columns and rows in a table, slots in a frame, or predicates, terms and clauses. However, in order to talk of underlying units of representation it is necessary to avoid treating columns and rows, links and nodes. frames and slots as all being necessarily different. For this purpose we propose to map all representations onto logic. Logic contains a set of known enitities which can be looked for in a particular domain i.e. constants, terms, variables, predicates, and quantification etc.. We call this set of entities (constant,...) the logical entities, and this level of representation the **logic level of representation.** For example, the links of a semantic network can be mapped onto 2-place predicates, and a table can be mapped into a list of 1- 2- and n-place predicates. It can be indicated which kind of logic the algorithm's representation formalism uses (e.g. propositional logic, first order predicate logic, higher-order logic), and in which way the logic is restricted (e.g. no negation allowed).

For example, on this level we may describe the representations of (I) and (II) as a Horn Clause:

$$bmw(X) :\text{-} red(X) \text{ \& } german(X) \quad \textbf{(iii)}$$

Using logic for the description has the advantage that for each construct (e.g., variable, term, conjunction, formula) there exists a well defined semantics.

Epistemic Level

The logic level is not sufficient to express all the characteristics of the entities to be represented. For example, consider 'properties' and values which are fundamental entities of representation which are important for machine learning. While these entities can be represented in logic, the problem occurs in deciding precisely which of the different logic constructs one chooses. A one-place predicate can be used to describe the value (red) of the property (colour) of an object (X) (i.e. red(X)). Alternatively, a two-place predicate (e.g. colour(X,red)) can express the same information by making the property being referred to explicit.

Conversely, structures from different algorithms looking very much the same (e.g., both being two-place predicates) can be completely different because of different interpretations. For example, consider the two expressions colour(X,red) and redder(X, Y). The two arguments of the two-place predicates map into different epistemic entities of the domain model: for one predicate the arguments are object and value, for the other they are two objects. A purely syntactic transformation of representations does not guarantee that the same domain model (or at least a part of it) is realised by corresponding expressions of different representation languages.

Finding the set of epistemic entities and determining their semantics is the task of developing an ontology or epistemology - a major concern of AI research for decades. To generate an epistemology capable of representing the semantics of all possible expressions is beyond the scope of the MLT. The approach taken to date has been pragmatic. Inspecting the learning algorithms to be used in the MLT, we proposed a small set of basic epistemic entities and described their semantics (i.e. their meaning in a given domain model) informally. We discussed the informal definitions with the partners, and then revised them iteratively and finally arrived at detailed descriptions of the entities represented in the different algorithms (the list of entities identified to date are: type, concept, class, property, function, fact value, object, relation, inheritance, subsumption,). These descriptions can be found in Morik et al 1989.

Conceptual Level

It is important to distinguish between a domain model and a domain. Whereas a domain exists in the real world, a domain model is a conceptualisation (Genesereth and Nilsson 1988) of that domain. For example, in the real world there are no such things as classes, properties or relations. These are entities which belong in domain models. (For a more subtle investigation of models and representation see Freksa et al. 1984). A learning algorithm represents the domain model of the user as best it can by means of its own formal representation language. For example, in a logic oriented representation language, an object of the domain model (e.g., **a red bmw licensed to a taxi driver, registration number THB 289Y parks at the station car space 34**) can be represented by a constant (e.g. *bmw_1)*, the object's property values (e.g., *has_att(colour = red, bmw_1))* and relations between objects (e.g., *parks (bm_1, place_1)* by predicates, and object membership can be represented as predicate (e.g., *is_class (bmw, bmw_1)* etc.

Machine learning research has shown that the choice of representation language is strongly related to both what can be learned and also how efficiently it can be learned (representation bias). The terms used in a representation language form the conceptual level of a representation. This conceptual level can have a different granularity for different domain models. For example, if a language offers several means to describe some facts, selecting one of them gives it a particular meaning which is different from a description in a language where no other way of describing it is available: in the first case, the meaning is more precise than in the second case. As a simple example, the verb *drive* exists in English, whereas the french language has both *rouler* and *conduire.* To put it into machine learning terms, the coverage of terms can be very different depending on the language. We do not aim at solving this problem. For the user of the MLT system it will be an interesting experience to see different results because of different representation languages. Using the most appropriate representation language for a particular learning algorithm remains a task of the user which can be supported by the consultant module of MLT but not automated in the lifetime of this project. It is clear however that future work would need to look closely at supporting the user in overcoming conceptual ambiguity.

The Problem of Semantics

The problem of semantics is one which encompasses all three levels of representation, and is illustrated as follows. We want the transformation (or translation) from the same piece of CKRL into internal structures of different algorithms such that the internal structures of different algorithms have nearly the same semantics. For instance, the "bmw" of (i) and the "bmw" of (ii) should be recognised as the same entity although in (i) it is an argument and in (ii) it is a predicate. We view the "bmw" to be the same entity in both representations, (i) and (ii), because the domain model to which the expressions refer has the one class "bmw". Thus, we have a criterion for a successful translation from one algorithm's representation into the one of another via CKRL:

- if two entities of different systems refer to the same entity of the domain model, they are considered the same entity and should be represented by a single entity of CKRL.

Alternatively, we can state what we do not want:

- an entity e1 of one system (e.g., "bmw" of AQ) which maps into an entity e2 of the domain model (e.g., the class of bmws) and is transformed into an entity of CKRL, must not be transformed into an entity e3 of another system (e.g. "X" Of MOBAL) which in turn would map into an entity e2+ of the domain model (e.g., a particular bmw of the domain model, bmw_1) where e2 and e2+ are not identical.

Work in machine translation has shown that translations from one language into another are never one to one except when (or if) the languages are the same, i.e. they are only notational variants of the same language. Representation formalisms of different expressive power cannot express the same domain model. The hope is, however, that epistemic entities expressed in one formalism and then transformed into CKRL correspond to the same CKRL expression produced by another formalism.

4. The Methodology of the Comparative Study

On the basis of the verbal descriptions of the algorithms' representations presented at the first MLT-Workshop we proposed a terminology and developed a questionnaire. The terminology was supposed to clarify the terms which are needed to describe the entities to be represented at all levels. This terminology was used as the basis for designing the questionnaire which requested information from each of the learning algorithm developers about their algorithm-specific representations. The first version of terminology and questionnaire was discussed and led to developing a revised one. This was redistributed to the algorithm developers. Their feedback led to a further revision, with the CKRL developers filling in the questionnaires, and then checking with the algorithm developers as to whether their understanding was correct. The terminology and questionnaire reflect the current state of discussing machine learning and knowledge representation terminology (Deliverable 2.1). Such a discussion is fruitful in its own right, it may lead to a more uniform presentation of machine learning algorithms as far as their representation formalisms is concerned. The terminology and filled-in questionnaires already point at a canonical form for inputs and outputs of machine learning systems.

5. Synopsis of Questionnaire Replies

This section presents a set of tables, summarising some of the questionnaire answers obtained for each of the algorithms of the MLT. Further details for a number of the answers are to be found in the Deliverable 2.1[3].

One of the aims of producing a synopsis of the questionnaire results is to provide a clearer indication of the commonalities and differences between the different algorithms of the MLT. For example, the complex systems MOBAL and APT have a large number of answers in common (e.g. their use of subsumption and inheritance with consistency), and the ID3 based variants of AQ, ID3 and CN2 were virtually indistinguishable in terms of the representation entities used in the questionnaire.

By referring to specific tables in the context of an application domain, some initial recommendations can be made about the suitability of appropriate algorithms. For example, if it was anticipated that there was a need to negate object or concept features, then from Question VIII.2, MOBAL, MAKEY, CIGOL and APT are the only algorithms which currently offer this alternative. Some questions will obviously be more useful for this type of discrimination than others, and brief examples are cited where applicable.

INPUT/OUTPUT/BACKGROUND KNOWLEDGE

Some algorithms use the same representation format for their input, output, and background knowledge. Others take their input in a format dissimilar to their output. For instance, ID3 takes a vector of values as input and produces a decision tree as output. ID3 is unable to interpret its own output. Also AQ, CN2, SICLA, LASH, and MAKEY have dissimilar input and output. By contrast, DMP, MOBAL, APT, and CIGOL use one uniform format for input, output, and background knowledge.

---

3 Further details of each of the algorithms can be found in the Deliverables 4.0 produced under the MLT project: Clark 1989, Feng 1989, INRIA 1989, Intellisoft 1989, Intellisoft/LRI 1989, Lebbe and Vignes 1989, Niblett 1989, Parsons 1989, Sims 1989, Wrobel 1989

Although KBG uses a different form of knowledge representation for its inputs and outputs, it is still able to interpret its own output. Closer comparisons of the data structures used by DMP, MOBAL, APT, KBG and CIGOL will produce guidelines for some of the basic forms which could be used by the CKRL.

Another important question is what a learning tool produces as output. Most of the algorithms learn concepts. Whether these concepts are represented as rules or not was asked in the questionnaire. Under rules learned are listed those algorithms which (additionally) produce rules covering regularities in the domain, dependencies among properties, relations between values, etc. Some algorithms are able to produce a learning result by introducing new terms to create an enhanced representation language.

Concept Learned		Rules Learned	New Terms Introduced	Explicit Learning Bias
One	Several			
APT	MOBAL	LASH (Regularities)	MOBAL	MOBAL
CIGOL	MAKEY	MOBAL (Several Forms)	APT	KBG
	SICLA	APT	KBG	
	KBG	KBG (Regularities)	CIGOL	
	AQ	CIGOL (Regularities)		
	CN2	SICLA		
	ID3			

OBSERVATIONS- EXAMPLES(+ /-)

All the algorithms in the MLT, except DMP, are able to learn from examples. The choice for learning from observations is clearly more limited (DMP, MOBAL, SICLA, KBG and CIGOL). The information concerning the Observations-Examples questions contains general algorithm characteristics which could be used as one basis for mapping applications to an algorithm. For example, if a learning from examples problem requires that examples are given from several classes, without multiple class membership allowed, then KBG, CIGOL,, AQ and CN2 would be appropriate (although it may be possible to use other algorithms if the problem structure is changed). Thus, results of the questionnaire concerning the representation formats of learning algorithms could be used to derive a decision tree for choosing a a particular algorithm.

Algorithm	Relation of Example to Class	Pos-Neg/ Several Classes	Multiple Class Membership
LASH	File Names	Pos-Neg	No
MOBAL	Marked as Sets	Pos-Neg	No
MAKEY	Prop of Ex.	Several	No
SICLA	Prop of Ex.	Several	No
APT		Pos-Neg	Yes
KBG	Prop of Ex.	Several	Yes
AQ	Prop of Ex.	Several	Yes
CIGOL	Prop of Ex.	Several	Yes
CN2	Prop of Ex.	Several	Yes
ID3	Prop of Ex.	Several	No

CONCEPTS

There are various ways to represent concepts. They can be intensionally defined by necessary and sufficient conditions or given extensionally by listing all instances. Also a typical instance may serve as a concept representation. If a concept is not defined but only described, the defining properties and relations are not (yet) separated from the attributive properties or relations. For example, a tomato can be defined by it being a fruit and being round and being red. An example for an attributive property is its weight. The weight is different for all the instances and cannot be used to distinguish tomatoes from other things. For another concept, of course, weight might be a defining property. For instance, the specific weight is a defining property of chemical elements.

Extensionally	Intensionally	Prototypically	Descriptionally
DMP SICLA CIGOL	MOBAL APT MAKEY SICLA KBG AQ CIGOL CN2 ID3	SICLA	MAKEY SICLA KBG

PROPERTY

Properties of Instances			Properties of Concepts		
Input	Output	B.Know.	Input	Output	B.Know.
DMP LASH MOBAL	MOBAL	MOBAL	DMP MOBAL MAKEY	MOBAL MAKEY	MOBAL
SICLA APT KBG	APT	APT	APT	APT	APT
AQ CIGOL CN2 ID3	CIGOL	CIGOL	CIGOL	AQ CIGOL CN2 ID3	CIGOL

This table provides one of the clearest indications of the nature of the AQ, CN2, ID3 group of algorithms, which all accept the properties of objects as input, and produce the properties of concepts as output.

Relations

Relations Between Objects			Relations Between Values		
Input	Output	B.Know.	Input	Output	B.Know.
DMP	DMP		DMP	DMP	
LASH					
MOBAL	MOBAL	MOBAL	MOBAL	MOBAL	MOBAL
APT	APT	APT			
KBG					KBG
CIGOL	CIGOL	CIGOL	CIGOL	CIGOL	CIGOL

Relations Between Concepts		
Input	Output	B.Know.
	DMP	
MOBAL	MOBAL	MOBAL
APT	APT	APT
		KBG
CIGOL	CIGOL	CIGOL

About half of the algorithms in the MLT have the capability to represent and manipulate relations. Of those that do, MOBAL, CIGOL and APT have the most extensive implementation. MOBAL, CIGOL, APT, and DMP deal with relations between instances and relations between concepts. In addition, MOBAL, CIGOL, and DMP handle relations between values.

EXPRESSION

There are two levels of expressive power of representation schemes used within the MLT: propositional and first order predicate logic. The tools using propositional logic are: DMP, MAKEY, SICLA, AQ, CN2, ID3. The tools using a restricted first order predicate logic are: LASH, MOBAL, CIGOL, APT, KBG.

Representation Languages of the MLT

The results from the questionnaires have allowed us to distinguish three main families of representation languages:

attribute-value formalism This formalism corresponds to 0th order logic. It is used by ID3, AQ, CN2 and SICLA algorithms.

The two following formalisms are both of first order logic:

object oriented formalism This formalism usually assumes the structuring of knowledge along the relations ISA (instantiation) and KINDOF (inheritance). Knowledge is centered on objects. This is the case in APT.

logic oriented formalism This formalism is less constrained than the previous one. The knowledge is structured using rules that may express all kinds of relations. Representation is centered on predicates (properties of objects). This is typically the case in MOBAL and KBG.

The last two formalisms are quite equivalent in the sense that knowledge can be translated from one to the other without essential loss of information. This is not the case with the first representation which is not as powerful. It is obvious that the different MLT tools can not work with the same knowledge. Some are able to use high level descriptions and background knowledge whereas others can only work on attribute-value descriptions. Since CKRL must encompass all that is representable in the MLT tools, it must be in first order logic. Therefore, all algorithms of the MLT will not be able to use all the information provided by a CKRL description of the problem. When translating CKRL in its internal representation, algorithms must try to take into account all the information provided. But a certain amount of loss can not be avoided. Therefore, when translating its outputs in CKRL, the algorithm will not be able to restore the lost knowledge.

Although not all results of the comparative study could be presented in this report, some points hopefully became clear. The distinctions of different levels and the careful investigation of semantic problems when acquiring representation formalisms of diverse systems have been presented. An example of The terminology in which the formalisms could be described with respect to a domain model was shown. On this basis a questionnaire was developed. Given the answers to this questionnaire, some classes of machine learning tools can be determined. The algorithms which use a restricted first order predicate logic (MOBAL, APT, CIGOL, KBG, LASH) have several features in common. Also the tools using an attribute-value representation (DMP, SICLA, MAKEY, AQ, CN2, ID3) share some features. Another classification occurs if the goal of learning is inspected. If in an application the data are grouped together into cases, SICLA, KBG, AQ, CN2, ID3, LASH, and MAKEY are well suited. They are different with respect to the relation between cases and classes. If all cases are in one class, KBG is applicable. If the cases are in two classes (positive and negative examples) SICLA, AQ, CN2, ID3, LASH are applicable. If each case is a class, only MAKEY is applicable, i.e. MAKEY enhances a class description. If the classes are not given by the user/data, only SICLA is applicable. If the data are not yet grouped, MOBAL, CIGOL, DMP, and APT may detect regularities in the domain (MOBAL, CIGOL, APT) or form classes of objects (MOBAL, DMP). The classifications of machine learning tools with respect to their representations form a basis for selecting an appropriate tool for an application if the representations are characterized with respect to entities of a domain model.

6. Conclusions

The results from the questionnaire have been used to stimulate discussion on the direction to be taken in specifying a first version of the CKRL. One of the main issues in MachineLearning is the representation of concepts. Definitions of what is a concept and ways of representing concepts are numerous even inside the MLT. Therefore CKRL must offer sufficient means of representation to cover all those interpretations. A concept in CKRL is a very powerful entity of representation. It authorises at the same time very complex and very incomplete descriptions of concepts.

On the other hand, we must ensure that CKRL entities are semantically well interpreted by the different MLT algorithms. The logical formalism allows too much freedom when mapping a domain to a representation (see section on the logic level). For a correct functioning of the Tools on the same CKRL representations, the distinction between what are examples, what is background knowledge and what are definitions of descriptors must be clear. Therefore, we opted for a representation of examples through a class-instance formalism ensuring a certain rigidity in their description.

Thus, CKRL should offer structuring facilities such as the class-subclass and classinstance relations when well organized knowledge is necessary or available, and high-level flexible facilities such as concepts for expressing more unorganized and incomplete knowledge.

When building the syntax of the Common Knowledge Representation Language for the MLT, we decided on the following design choices:

- The syntax of the language should be uniform.
- It should be easily extensible.
- It should produce a language that is easy to translate in and out of the MLT tools so that the algorithms developers will not spend too much time on writing the parsers
- It should offer some facilities for helping the user when writing CKRL data.

As usual, the above design choices are not always compatible together and we have had to come up with what we believe to be the best compromise to date. The current syntax will define only what is needed presently, while making sure that the it is general enough so that stretching it will not have consequences on the translators (future extensions of CKRL will be upward compatible with the previous versions). Ensuring complete coverage of every algorithm's methods of representation has resulted in some degree of redundancy, each algorithm developer will have to learn about some aspects of CKRL which are not directly relevant to their representation in order to understand the language as a whole.

We have currently completed a prototype of the CKRL syntax. We will now write the translators between this version of CKRL and the internal representations of each MLT tools. This work, as well as experimentation with real applications will provide material for further refinements.

7. References

Brachman, R. 1979: *On the Epistemological Status of Semantic Networks,* in: Findler (ed): Associate Networks - Representation and Use of Knowledge by Computers, Academic Press

Clark, P., 1989 : *Deliverable 4.0 - Functional Specification of CN and AQ*

Feng, C., 1989 : *Deliverable 4.0 - Functional Specification of CIGOL*

Freksa, C., Furbach, U., Dirlich, G. 1984: *Cognition and Representation - An Overview of Knowledge Representation Issues in Cognitive Science,* in: Laubsch (ed.) Procs. of the German Workshop on AI, GWAI84, Springer

Genesereth, M.R., Nilsson, N.J. 1988 (2nd ed.): *Logical Foundations of Artificial Intelligence,* Morgan Kaufmann

INRIA, 1989: *Deliverable 4.0 - Description of SICLA*

Intellisoft, 1989: *Deliverable 4.0 - Description of KBG*

Intellisoft/LRI, 1989: *Deliverable 4.0 - Description of APT*

Lebbe, J., Vignes, R., 1989: *Deliverable 4.0 - Functional Specification of MAKEY*

Ludwig, A. 1989: *Deliverable 1.1.1 - Specification of the Overall Architecture of the MLT*

Michalski, R., 1983: *A Theory and Methodology of Inductive Learning,* in: Michalski, Carbonell, Mitchell (eds) : Machine Learning - An Artificial Intelligence Approach, Volume 1, Tioga Press

Morik, K., Rouveirol, C., Sims, P. 1989: *Deliverable 2.1 Comparative Study of the Representation Languages Used by the Systems of the MLT*

Niblett, T., 1989: *Functional Specification if RealID*

Quinlan, J.R., 1983: *Learning Efficient Classification Procedures and their Application to Chess End Games,* in: Michalski, Carbonell, Mitchell (eds): Machine Learning - An Artificial Intelligence Approach, Volume 1, Tioga Press

Parsons, T., 1989: *Deliverable 4.0 - The DMP, Description and Status*

Ralambondrainy, H., 1989 *How to deal with categorical data using clustering methods* in : R. Coppi and S.Bolasco (eds): Multiway Data Analysis, North Holland

Sims, P., 1989: *Deliverable 4.0 - LASH Algorithm Description*

Sleeman, D., Oehlmann, R., Davidge, R. 1989: *Deliverable 5.0 - Specification of Consultant-0*

Wrobel, S., 1989: *Deliverable 4.0 - Description of MOBAL*

Acknowledgements

The Workpackage 2 partners (LRI, GMD, Intellisoft, INRIA, University of Dauphine, B.Ae) would like to acknowledge the assistance given by the algorithm developers who replied to, and helped improve the questionnaire, and the contribution of Aberdeen University to a number of discussions leading to this document.

Computerised Simulation Tools for the Design of an Oral Dialogue System

F.Andry, E.Bilange, F.Charpentier, K.Choukri, M.Ponamale, S.Soudoplatoff
Cap Gemini Innovation
118, rue de Tocqueville
75017 Paris
e-mail: charpenf@csinn.capsogeti.fr

Abstract

An important issue in the design of an oral dialogue system consists of identifying the domain-dependent knowledge needed to specify the different system levels (acoustic, linguistic, dialogue). This paper describes a original approach for the obtention of such domain-dependent material, based on a computerised version of the Wizard of Oz method, generally used for the simulation of Human-to-Machine dialogues. With this computerised environment, the accomplice managing the simulation is constrained to adopt a deterministic machine-like behaviour, so that we finally obtain a better simulation of the target system. Simulation results are presented, confirming the efficiency of the method.

1. Introduction

The goal of the SUNDIAL project (This project is partially funded by the Commission for the European Communities ESPRIT program, as project 2218. The partners in the project are CAP GEMINI INNOVATION, CNET, CSELT, DAIMLER-BENZ, ERLANGEN University, INFOVOX, IRISA, LOGICA, POLITECHNICO DI TORINO, SARIN, SIEMENS, SURREY University) is to develop cooperative oral dialogue systems for database information access in four different languages. The application chosen for the French language demonstrator is flight reservation. The intended users are people who frequently do this task, for instance secretaries that need to book airplane tickets for their company staff.

In order to ensure a good design of the different system components (Fig.1), it is necessary to collect domain-dependent knowledge on the various aspects of the system (linguistic, acoustic, task model, user model, dialogue model and ergonomy). Several methods have already been proposed to obtain such data.

In the Wizard of Oz approach, a human accomplice replaces the machine by simulating its functionalities. The Wizard of Oz method (Its etymology is explained in [WP6 D1 90]: In the children's novel "The wizard of Oz", the wizard turns out to be no more than a device operated by a man behind a curtain.) has already been advocated by many researchers [Richards 84] [Morel 85] [Guyomard 87] [Newell

87]. It involves two partners: the subjects of the simulation who believe that they are really using the final system and the accomplice (the wizard) who is simulating it. The subjects, on one hand, are given scenarios of a reservation task (flight reservation) and they are asked to call the system to achieve this task. On the other hand, the accomplice has to manage the dialogue with the callers while adopting a deterministic machine-like behaviour. The voice of the accomplice is generally made to sound artificial by using a distorting device, such as a vocoder, so that the users believe that they are actually talking to a real machine.

In the SUNDIAL project, we have chosen to adopt the Wizard of Oz approach, but we decided to bring improvements to the experimental setup so as to ensure a simulation protocol as rigorous as possible. The originality of our approach lies in the computerised tools we have designed to help the accomplice to achieve his task more efficiently. In this paper, we present the details of this computer-aided environment, and we summarise the simulation results to be used in the design of the target SUNDIAL system.

2. Design of the SUNDIAL system

The overall architecture of the target system is shown in Fig.1. An acoustic recognition stage processes the speech signal and provides a set of word hypotheses to the linguistic stage. The linguistic stage extracts the meaning of the user's utterance, given the syntactical and semantical constraints of the application. The Dialogue management module monitors the dialogue exchanges, by combining rules for the achievement of the task goal (flight reservation), and rules for the managing the system's and user's initiatives. The message generation component translates the intended meaning of the system responses into text, which is eventually converted to speech by the text-to-speech module.

The goal of the simulation experiment described in this paper is to provide relevant information for the design of the system components, especially for the three critical modules: speech recognition, understanding and dialogue management.

The data relevant to the speech recogniser is the lexicon in a phonetic form, including all phonetic variants of a given word. Significant mono-syllabic words should be identified, since they are more difficult to recognise, and possibly combined with their context into hyperwords.

The relevant information for the linguistic processing module consists, firstly of the lexicon, with its syntactic and semantic categories, and secondly of a grammar of the users' utterances, with respect to the current dialogue states.

The basic knowledge required for the design of the dialogue manager consists of a list of relevant dialogue acts and given this list, of rules governing the dialogue unfolding. In parallel, the task model adopted for the simulation should be validated, rejected or augmented. In order to avoid rejection, this model was inferred from a preliminary analysis of Human-to-Human flight reservation dialogues. A number of dialogue phenomena should be examined: turn-taking, overlapping talk, system cooperativeness, dialogue failure and repair, dialogue control (repetition or confirmation requests). The simulation should also help to build a user model (user's knowledge, reactions and believes). Finally, the ergonomy of the system must be evaluated, so as to get the users' reactions and feelings when faced with such a machine.

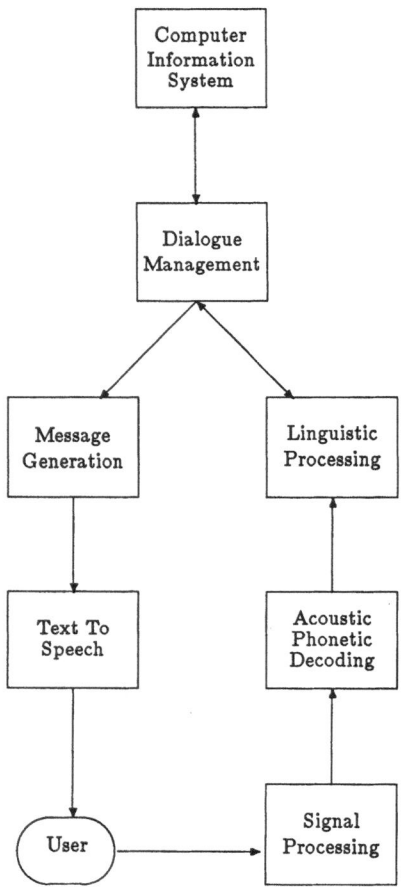

Figure 1: Overall Architecture of the SUNDIAL System

3. Choice of a simulation method

To obtain a dialogue corpus and collect such data, a straightforward approach consists of simply observing some Human-to-Human dialogues between an operator (e.g. in a travel agency) and customers [Morel 88]. Several dialogue systems are based on such observations [Roussanaly 86]. This may provide a bootstrap for the design of a system but it is far from sufficient for its complete specification, because people adopt specific behaviours when faced with a machine.

In fact, the differences between Human-to-Human and Human-to-Machine dialogues were reported in a previous work [Morel 85]. This experiment involved a free Human-to-Human dialogue phase and two Human-to-Machine dialogue phases, in the context of a railway company services enquiries. In the two Human-to-Machine phases, the machine was simulated by an accomplice whose voice was distorted by a vocoder (Wizard of Oz approach), while the difference between the two phases laid in the degree of constraints imposed to the accomplice's expression and understanding. The conclusions were that machine-like voice and behaviour do have some influence on the linguistic behaviour and the prosody of the users. The users' requests, especially the initial ones in the dialogue sessions, were more standardised, exhibiting limited linguistic complexity. In the case of high dialogue constraints, the users tended even to imitate the machine speech, using a slow and regular speaking rate, and drastically reducing their use of phatic markers (e.g. *Ok, I see, Hmm*).

In the light of this experiment, we chose to adopt the Wizard of Oz approach along with tight constraints imposed to the accomplice. In a preliminary phase, a brief study of Human-to-Human flight reservation dialogues [Morel 88] was carried out to bootstrap the design of our simulation system, primarily to design the task model and the set of system messages. The dialogue task was decomposed into several dialogue phases, from the problem formulation to the connection end, as in [Amalberti 84]. These phases included in particular: the problem formulation, the selection of a suitable flight from short or long lists, the proposal of alternative solutions if the request cannot be satisfied, the reservation transaction. The dialogue phases were then combined to form a complete dialogue automaton, to be used in guiding the accomplice through the simulation task.

In fact, the simulation task is a complicated one for the accomplice to perform, since it implies simultaneously:

– having a regular language behaviour;

– reacting promptly after the user's turn, especially in case of a non expected dialogue act;

– adopting reactions that are predictable by a trained user;

– providing plausible information.

In the following, we describe the simulation protocol adopted for our experiment, and in particular the adaptations that were brought to the Wizard of Oz framework in order to ensure a higher degree of dialogue constraints than those achievable with the usual approach.

4. Simulation protocol

4.1 System overview

The experimental setup is described in Fig.2. The subjects of the experimentation dialed the system through the telephone network. The accomplice listened to the subjects via a loudspeaker, and managed the dialogue using the computerised tools provided through a menu-based windowing environment, running on a SUN workstation. Instead of using the vocoder-distorted speech of the accomplice, a French text-to-speech (TTS) system [Stella 83] [Emorine 88] was used to deliver the machine utterances. The TTS system ran on a PC connected both to the telephone network and to the SUN workstation that sent the messages to be synthesised. Using a TTS system not only ensures a machine-like voice quality but also a more artificial prosody. Moreover, it offers the advantage of freeing the accomplice from having to speak the messages himself. Finally, the experimental setup also includes the means to keep track of the simulations sessions. Each system utterance and the dialogue phase associated with it, were automatically saved by the SUN in a specific log file. In parallel, the dialogues sessions were recorded on a tape recorder.

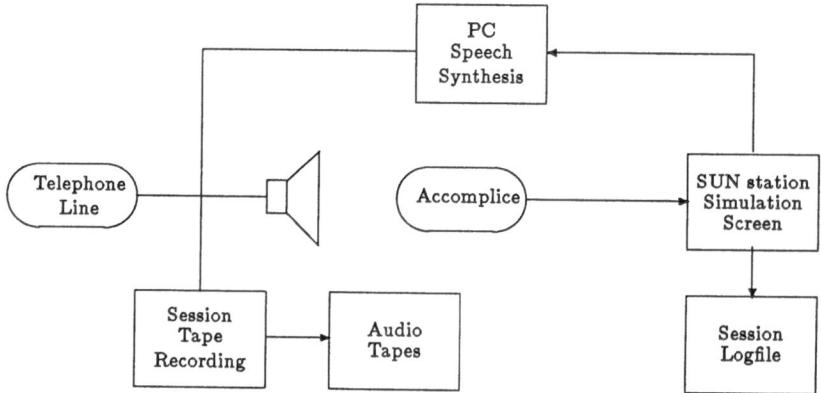

Figure 2: Experimental setup for the dialogue simulation

4.2 The simulation screen

In our simulation experiment, the accomplice replaces completely the recognition and understanding modules of the target system (Fig.1). The computerised tools help him to simulate the tasks of most other modules, including the dialogue manager, the database access and the message generator. The accomplice sits in front of the simulation control screen (Fig.3), composed of the following different windows:

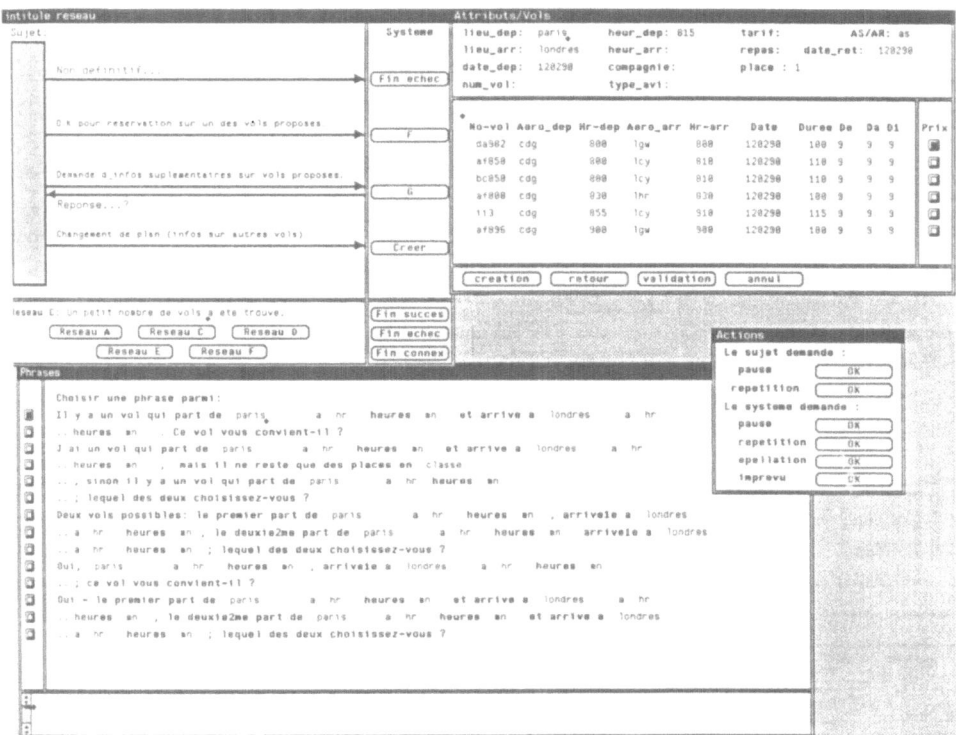

Fig.3 : The simulation screen
The accomplice uses the simulation screen to operate the simulation tools and manage
the dialogue between the user and the system. The screen is displayed on a SUN workstation and is composed of four windows: the dialogue network window (at the top
left), the machine utterance window (at the bottom), the attributes/flight window (at the
top right), and finally the smaller actions windows (on the right side).

– The *dialogue network window* provides him with a schematic view the dialogue
automaton. It represents the current dialogue phase with its possible relations to
the other dialogue phases of the automaton. There are two kinds of relevant
information corresponding to the current dialogue phase: firstly, the set of
possible reactions from the machine, and secondly, the set of possible dialogue
acts expected from the user. The first information is displayed in the machine
utterance window, described below. The dialogue network window presents the

second kind of information in the form of a transition network, using labelled arcs to represent all possible exchanges between the subject and the system. These exchanges trigger all possible system transitions to the next dialogue states. In the example of Fig.3, the network displayed represents the dialogue state (labelled C) activated when selecting a flight from a short list of proposals. The other dialogue states, labelled by letters A to F, are represented within buttons. The selection of the next dialogue state is obtained by clicking on either of them.

- The *machine utterance window* corresponds to all possible reactions from the system in the current dialogue state. The contents of this window is tied to the former window, and is updated when any new dialogue state is selected. The possible system reactions are presented as a list of sentence patterns with variable slots (canned sentences). The accomplice has to find the one which is the most appropriate, fill up the variable slots with the help of a scrolling-menu according to relevant the task parameters, and finally activate the TTS system.

- The *attributes/flights window* is separated into two sub-areas, displaying respectively the flight parameters given by the users and the cooperative responses provided by the database access module. This module relaxes constraints whenever a user parameter does not fit (for instance, a non existing airport in the destination city). When a constraint cannot be respected, the database access module proposes another close solution (closest schedule, closest airport). It also takes into account a presupposed user model for the firm staff who travel for business: for instance, if there are very few flights in the morning, the system will propose flights for the previous evening or in a more expensive class. As the database access module provides large and cooperative responses, the seat availability is managed directly by the accomplice. The database used is a toy database, which contains at the very least all the flights required by the scenarios and some other close ones.

- The *actions window* permits the management of repetition requests both from the user's or the system's side, as well as the handling of unforeseen events such as missing information in the database.

- The *guidelines window* is displayed at the opening of each new dialogue to inform the accomplice whenever he has to simulate acoustic misrecognitions. These guidelines depend upon the scenario to be developed. For instance, in scenarios including Rome as the destination city, the guideline are: *Understand Bonn instead of Rome*.

4.3 The accomplice

Prior to the experiments, the accomplice was trained to use the tools as efficiently as possible. During a dialogue session, the accomplice listened to the subject's utterance, interpreted it and selected accordingly the next dialogue state. The activation of this state triggered the display of a new network of exchanges and a new set of canned sentences. The accomplice selected the sentence which fitted best the current dialogue context, and sent it to the synthesiser. Unexpected events were dealt with by two different manners: either through the actions window, or else by switching on a specific dialogue state represented in the dialogue network window, even though it was not allowed by the dialogue automaton. After the sentence was synthesized, the accomplice waited for the next user's utterance. If the user did not take the turn, he would relaunch the dialogue by synthesizing the last sentence again. Also, in some cases, after the user had formulated his request at the beginning of a session, the accomplice was to simulate errors or omissions when entering the parameters (as indicated by the guidelines). This was also the case when he judged the user's request too complex or too long.

The total duration of a dialogue session was between 5 and 10 minutes, and contained an average of 12 dialogue system turns. The time between two successive system turns was relatively long during the phase when the accomplice was searching the right answer in the database. Generally, the system reactions were judged slow by the users, although not unbearable.

4.4 The subjects

The subjects of the experimentation were chosen in the range of the users intended for the final system. Twenty-six secretaries took part in this simulation during two months yielding some 300 dialogues. They were informed that they had been selected to evaluate a prototype of an oral flight reservation system, for the purpose of improving it. Every fortnight, they received scenarios of flight reservation requests for the next two weeks. They had regular appointments (twice a week) to call and complete the scenarios. For each scenario, they had to write down specific details such as the fare or the arrival schedule of the booked flight. This obliged them to ask explicitly for such information and to verify it through confirmation or repetition requests. After the end of the call, they had to fill up a written questionnaire to indicate their feelings about the dialogue, the synthesized voice, the efficiency of the service in comparison with that of a travel agency's human agent.

These scenarios have been decomposed into three types. The simpler ones have been used at the beginning of the simulation to avoid discouragement of the subjects. They concerned reservations which could easily be dealt with (for instance, no problem with the availability for the flight requested). The complicated ones were introduced later, and required some relaxations of constraints from the database access. The last ones were very complicated and could generate failures in the dialogues.

The choice of using scenarios was guided by pragmatic reasons, although we were aware that this methodology induced some biases, such as a lack of motivation on the subjects' side, or the tentation for them to read out the scenario at their first turn in the dialogue. Different actions were taken to keep frequent contacts with the subjects and to motivate them during all of the simulation period.

To get information about their gradual adaptation to the system, it was essential to experiment a long time with the same sample.

5. Simulation results

5.1 The corpus

After the simulation experiments were completed, the recorded material (audio tapes and log files) were used to achieve an accurate transcription of the dialogue sessions. The log files were filled up with the transcriptions of the user's utterances (interweaved with the system's utterances), according to oral transcription norms [Leroy 85, Jefferson 83] agreed upon with the project partners. The final corpus consisted of some 300 transcribed dialogues, involving 3736 utterances from the users, with an average of 12 utterances per dialogue session.

5.2 Analysis of user utterances.

From a purely acoustic point of view, variations in the users' speaking rate were often observed: they often slowed down their speaking rate when they had to repeat some information, or when they had to enter critical parameters (numbers, names).

The lexicon corresponding to all possibles user's utterances was extracted from the transcribed corpus. It consists of some 800 words including all flexional forms and some phonological variants (vowel elisions). In fact, the raw corpus also contained off-line discussions between the users and other persons in their office. Such material was tagged as off-subject and was discarded from the lexical count, since it was not relevant to the reservation task.

The user syntactical-semantical properties of the users' utterances were analysed with respect to the current dialogue state and to the previous system utterance. To do this, the subsets of the corpus tied to the specific dialogue phases were extracted, and tagged according to both dialogue acts (e.g. *wh-question*, *ask-for-repetition*) and task constituents (e.g. *departure place*, *airplane company*). A dialogic-semantic grammar has thus been built on a phase by phase basis, which contains at the highest level a combination of dialogue acts. From this analysis, it appeared that the initial user's request exhibited a too high linguistic complexity, given the expected performances of the recogniser. Part of this was due to a bias of the experiment, whereby the user heavily relied on the written booking scenario they were given, reading it out in some cases. However, as they got used to the system, the complexity of these requests tended to diminish.

To illustrate our dialogic-semantical grammar, let's consider the user's utterance when the system is in phase C (see Fig.3). In the following tagged dialogue fragment, the system presents one available flight and then requests the user's approval.

S:	ce vol vous convient-il ?	DIALOGUE ACT: posi-check
	(is this flight convenient?)	PURPOSE: ?accept(user, flight(X))
U:	oui, c'est parfait (fine)	DIALOGUE ACT: posi
	est-ce que vous pouvez	
	me donner le:: Numéro du vol	DIALOGUE ACT: wh--question
	(can you give me the flight number)	PURPOSE: flight-number(flight(X))
	s'il vous plait (please)	DIALOGUE ACT: phpolitness

The special dialogue acts *posi, posi-check*, are based on Bunt's work [Bunt 86]. When asking its question the system knows that the proposed flight fits the user desiderata and thus it expects a positive answer. The PURPOSE field addresses task specific goals such as obtain/give an information on a parameter (factual information) or obtain an agreement on a proposed solution. From a strict dialogue point of view, four main classes of dialogue acts have been used, namely *initiatives (or indirect/direct requests), reactions, evaluations* and *phatics*. Thus our work was quite similar to [Cohen 84, Whittacker 88, Walker 90]. Each main class has been decomposed according to dialogue specificities as, for example, the work done by [Bunt 86, Wachtel 86, Miclet 89], for more information about the dialogue analysis and the formal justification of these choices see [WP6 D1 90].

In the example above, the labelling is presented in a predicate structure-like formalism, but the complete dialogic-semantic grammar is described by rules written in a BNF format.

From the dialogic point of view, statistics have also been collected: for each dialogue acts uttered by the system, we describe all the possible user's responses as a list of possible dialogue acts associated with frequencies. For example, when the system prompts "en quelle classe voulez-vous voyager?" (*in which class do you want to travel*), the possible user's reponses can be classified as follows:

- the caller simply gives the desired class (thus providing an informative answer) in 93% of the cases;

- the caller gives the answer plus a complement of answer (e.g. concerning the return) in 3% of the cases;

- the caller gives the answer and asks for more information about the flight (e.g. the price), in 2% of the cases;

- the caller requests for repetition in 2% of the cases.

5.3 Criticism on the system utterances and underlying models

The task model (database requests, constraint relaxation algorithm, parameter coherence checking) adopted for the simulation appeared to be appropriate, even though a dialogue phase seemed to be missing. Such a phase, possibly named *ask for information*, would have been useful, because a lot of subjects did not take the opportunity to ask for such or such information before the system's closing message. Thus the system should have been able to ask questions of the form *Do you need more information*.

From the message generation point of view, the messages delivered by the system appeared to need improvement since they were not always well adapted to the dialogue situation. Specific messages were missing that would give the system better control over the dialogue thread. Also it seemed desirable to introduce flexibility in the formulation of the messages, allowing the system to switch from a concise formulation mode (short sentences) to a more explicit one. More precisely, the system should utilise the concise mode the first time it delivers a given speech act, and when the understanding scores of the user's utterances are high enough.

As for the quality of the speech output, the questionnaires filled by the subjects indicated that its understandability was not sufficient for delivering critical information such as schedules, prices and flight numbers. We expect to improve this both by using better techniques [Charpentier 89] and by slowing down or spelling out alphanumeric data.

6. Concluding remarks

Given the current state of art of continuous speech recognition techniques, simulation experiments of the Wizard of Oz kind are particularly useful in designing speech interfaces suitable for task-oriented applications. We have presented here an improved version of the Wizard of Oz approach for the design of such an oral dialogue system. In this approach, a computerised simulation environment is used to help the accomplice in simulating the reactions of a truly automatic system while enforcing a rigorous simulation protocol. The original features of our approach include:

- constraining the dialogue through a dialogue automaton;

- restricting the system messages to a predefined set of utterances;

- using a text-to-speech device for deliveraing the system answers.

Using this approach in the SUNDIAL project, we have obtained a large corpus of oral dialogue sessions, thus providing the relevant material for the design of the French SUNDIAL prototype. This approach was also adopted by other partners involved in the SUNDIAL project, especially by the Italian team [WP3 D1 90].

Finally, we should mention that the simulation tools developed for this work are relatively easy to adapt to the simulation of other information-seeking applications. For instance, such an adaptation was done rapidly when we transformed the cooperative free-speech reservation application simulated in the most part of our experimentation (and corresponding to 240 dialogues), into a modified application, consisting of an initial menu-driven phase, providing different menu-based functionalities such as schedule and price enquiry, and that could be chained to the original free-speech reservation phase. The simulation of this modified system was performed successfully and provided some 60 dialogues. An interesting conclusion of this experiment, run in collaboration with one of project partners (CNET), was that filtering the calls through a menu-driven phase does substantially reduce the linguistic complexity of the user's utterance, when she/he returns to a free-speech dialogue mode.

Aknowledgements

We would like to thank M.-L. Montel, J.-C. Lemasson, G.Tabuteau, H. Terrades for their contribution to SUNDIAL simulation and the Sundial partners for the fruitful discussions. In particular, we would like to thank our Sundial partners from CNET, A. Cozannet and C. Sorin, for their collaboration on the simulation of menu-driven dialogues.

References

[Amalberti 84] R. Amalberti, N. Carbonell, P. Falzon, "Strategies de controle du dialogue en situation d'interrogation telephonique", GALF-GRECO Communication orale Homme-Machine, Nancy, France, 1984.

[Bonnet 80] A. Bonnet, "Les grammaires sémantiques, outils puissants pour interroger les bases de données en langage naturel", RAIRO Informatique, vol. 14-2, pp 137-148, 1980.

[Bunt 86] H. Bunt, "Information dialogues as communicative action in relation to partner modelling and information processing", pre-acts of Structure of Multimodal Dialogues Including Voice, NATO RSG10/RSG12 Workshop, Venaco, France, 1-5 sept. 1986.

[Charpentier 89] F. Charpentier, E. Moulines "Pitch-synchronous waveform processing techniques for text-to-speech synthesis using diphones", EUROSPEECH Int. Conf., pp.13-19, Paris, 1989.

[Cohen 84] P.R. Cohen, "The Pragmatics of Referring and the Modality of Communication", Computational Linguistics, vol.10, N. 2, pp. 97--146, April--June 1984.

[WP3 D1 90] "Simulation studies", K. Choukri Ed., SUNDIAL Deliverable WP3 D1, June 90.

[WP6 D1 90] "Functional specifications for the dialogue manager", N. Gilbert Ed., SUNDIAL Deliverable WP6 D1, June 90.

[Emorine 88] O. Emorine, P. Martin, " The MULTIVOC text-to-speech system", ACL, Austin, Texas, pp 115--120, 1988.

[Fraser 89] N. Fraser, "Simulating Speech Systems", Esprit Project Sundial technical working paper, personal communication, University of Surrey, 1989.

[Guyomard 86] M. Guyomard, J. Siroux, "Phase 1 Experimental Protocol". Esprit project PALABRE, deliverable CNET/TSS/RCP WP4 TASK3, 1986.

[Guyomard 87] M. Guyomard, J. Siroux, "Experimentation in the specification of oral dialogue". NATO-ASI Recent Advances in Speech Understanding and Dialogue Systems, 1987.

354

[Jefferson 83] G. Jefferson, "Another failed hypothesis: pitch/loudness to overlap resolution", Tilburg Papers, Language and Literature 38, Tilburg, 1983.

[Leroy 85] C. Leroy, "La notation de l'oral", Langue francaise n.65, Larousse, pp 6--17, 1985.

[Luzzati 89] D. Luzzati, F. Néel, "Dialogue behaviour induced by the machine", EUROSPEECH Int. Conf., pp.601-604, Paris, 1989.

[Miclet 89] L. Miclet, J. Siroux, M. Guyomard, G. Mercier, "Compréhension automatique de la parole et intelligence artificielle", Annales des Télécommunications, vol. 44, no5-6, pp 283-300.

[Morel 85] M-A. Morel et al. "Analyse linguistique d'un corpus d'oral finalisé", rapport final d'étude, CNRS-GRECO Communication parlée, Nancy, 1985.

[Morel 88] M-A. Morel, "Corpus de dialogues avec un agent Air France. Etude des reformulations", Internal report, University PARIS III, 1988.

[Newell 87] A.F. Newell et al, "A full speed simulation of speech recognition machines", Proceedings of the European Conference on Speech Technology, Edinburgh, pp.410-413, 1987.

[Proctor 89] C. Proctor, S. Young, "Dialog control in conversational speech interfaces", in The structure of multimodal dialogue, M.M. Taylor, F. Néel, D.G. Bouwhuis Eds., Elsevier Science Publishers, 1989

[Richards 84] M.A. Richards and K.M. Underwood, "How should people and computers speak to each other?", Proceedings of Interact 84, pp.33-36, 1984.

[Roussanaly 86] A. Roussanaly, P. Mousel, N. Carbonell, B. Mangeaol, J.M. Pierrel, "Réalisation d'un corpus de dialogues oraux : application aux renseignements administratifs", Technical Report CRIN 86-R-083, Nancy, 1986.

[Stella 83] M. Stella. "Speech Synthesis". In Computer Speech Processing '83, eds Fallside Woods, Prentice Hall, 421-480.

[Wachtel 86] T. Wachtel, "Pragmatic sensitivity in NL interfaces and the structure of conversation", Coling'86, Bonn, pp. 35--41.

[Walker 90] M. Walker and S. Whittaker, "Mixed Initiative in Dialogue: An Investigation into Discourse Segmentation", to appear in Proceedings of the Association of Computational Linguistics, 1990.

[Whittaker 88] S. Whittaker and P. Stenton, "Cues in Expert-Client Dialogues", Proceedings of Association of Computational Linguistics, 1988.

[Zue 89] V. Zue, N. Daly et al, "The collection and preliminary analysis of a spontaneous speech database", Proceedings of DARPA Speech and Natural Language Workshop, Harwichport, 1989.

Project no 2252

INTRUSION-TOLERANT SECURITY SERVERS FOR DELTA-4

Laurent Blain*, Yves Deswarte**
*LAAS-CNRS
**LAAS-CNRS & INRIA
7, Avenue du Colonel Roche
31077 TOULOUSE Cedex
France

Abstract

This paper describes a new approach for security in open distributed systems. This approach is currently developed in the framework of the Delta4 project. After a few reminders about two existing distributed security architectures, the proposed "intrusion-tolerant" approach is specified. It is based on a fragmentation-scattering technique applied to a security server running on several security sites. These sites are such that intrusions into a number of sites less than a given threshold have no consequence on the global security. The different security services provided are then presented.

Introduction: Security for an Open Distributed System

The security approach presented here is developed within the framework of the ESPRIT Delta4 project. Delta-4 means Definition and Design of an **open** Dependable **Distributed** architecture. Notice the two important characteristics: openness and distribution.

Openness means that security can be implemented on any machines. The security part of an open system must not add "too much" specific hardware and software. Furthermore, security must not be so much of a constraint such that it becomes difficult to add new hosts to the system. These two considerations involve important differences with approaches such as the Red Book criteria [NCSC 87a] which impose rigid uses of software and hardware.

Distribution involves the choice of a distributed model for the security architecture. A classical one is the client-server model. An application running on one or several hosts provides services to clients on the same or other hosts. The client (the server) does not need to know on which host the server (the client) resides. This corresponds to transparent distribution. Security can follow the same model.

These two characteristics (openness and distribution) led us to give a particular approach in order to apply our security techniques.

Firstly, to keep the openness and the flexibility of the system, the intrusion-tolerant security techniques will not be integrated into the basic system design, but will be provided only as tools and applications.

Secondly, each host is considered to be under the local control of either the local administrator (time-sharing system) or the user (workstation). Local security is ensured by these persons and only global security for remote access can be ensured by specific security components. No additional security components need to be added to every new host. If and only if this host needs to do remote protected accesses or needs to be protected from remote accesses, new software and hardware components will be incorporated.

Thirdly, the client-server model must be applied to security too. The objects to be protected are servers and so the requests/accesses from clients to server must be verified. Two solutions can be adopted in order to secure the servers. The first one is to provide tools permitting construction of a secured server with authentication, authorization, audit.... The problem of this solution is that of the cost of building a new protection system for each server and all the interfaces with clients. The second solution is to provide a set of security services available for all servers which need security and which become clients of these security servers. These services are also available for clients which want to access secured servers. The advantage of this solution is that the part which is implemented on the secured servers is relatively small. The major part of security is ensured by the security servers. This is equivalent to the Kerberos approach [MILL 87]. Kerberos provides only identification, authentication and audit, whereas our approach includes sensitive data management and recovery.

The security services which are presented concern only distributed applications for which the location of application entities is transparent. In this sense, the security services can be viewed as complementary to network security services [ISO 7498-2] which are focused on communication confidentiality and integrity, rather than authorization to access application servers.

1. Distributed Security Architectures

Our intrusion tolerant security service is a new concept with regard to existing architectures such as those implied by Orange Book [DoD 85], Red Book [NCSC 87a] and Kerberos [MILL 87, STEIN 88].

1.1. The Orange Book and Red Book Architectures

The Trusted Computer System Evaluation Criteria (TCSEC), often called Orange Book, describes a multilevel mandatory policy and Bell-LaPadula model [BELL 74] and their implementation requirements. This architecture is based on two components: the **Trusted Computing Base (TCB)** and the **Security Kernel**.

- The TCB consists of all the of protection mechanisms within a computer system -- including hardware, firmware and software -- the combination of which is responsible for enforcing an authorization policy. The ability of a TCB to correctly enforce a authorization policy depends solely on the mechanisms within the TCB

and on the correct input by system administrative personnel of parameters related to the authorization policy.
- The Security Kernel is the set of hardware, firmware and software elements of the TCB that implement the reference monitor concept. It is *trusted, tamperproof, always invoked* and *verifiable as correct*.

The Orange Book provides only for a centralized system. The National Computer Security Centre also provided a Trusted Network Interpretation (Red book) of the TCSEC; the same requirements are then applied in the context of distributed system. The Red Book extends the TCB and Security Kernel concepts to distributed configurations. The idea is to connect distributed components together so as to form a global secure network. The security functionalities may be distributed.

There exists a **Network Trusted Computing Base (NTCB)** for the network. The NTCB is partitioned in such a way that the set of the partitions constitutes a global TCB. But all local accesses must be mediated locally by local mechanisms. For components which are not hosts, functionalities can be reduced since such components are not concerned by all the rules of the network security policy. This is not so for a host. The local security mechanisms must perform local security policy, because there exist subjects and objects within the host.

Thus each host possesses (Fig 1.1):

- subjects and objects which are local on the host (i.e. not distributed),
- a TCB as defined above,
- a NTCB partition which is the local part of the global TCB and which mediates all accesses from remote subjects to local objects or from local subjects to remote objects (the NTCB partition and the TCB are thus often merged),
- certain functionalities of the NTCB partition which can be used by other hosts,
- a Security Kernel which mediates *all* accesses, be they remote or local,
- a **Trusted Interface Unit (TIU)** managing communication security between hosts.

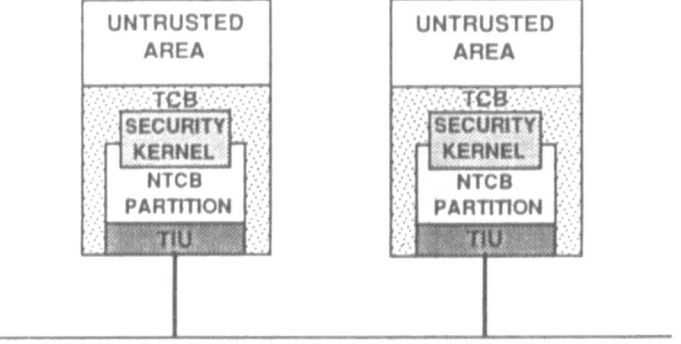

Fig. 1.1. TNI Architecture

358

This approach requires large trusted parts on all computers on the network. It is not a very open approach since it is difficult to add new computers to an existing network. The computer has to be evaluated in stand-alone mode and then in the network. Moreover, "trusted" is a very subjective property that is hard to verify and implies important software and hardware protections.

1.2. The Kerberos Architecture

Kerberos was developed at MIT [STEIN 88, MILL 87] within the framework of the Athena project. The architecture of the Kerberos is based on the Client-Server model. Kerberos provides authentication and authorization services. In the Athena project, there exist different kinds of sites: unprotected Public workstations, private workstations, protected servers... Kerberos must authenticate legitimate users who want to use remote secured services. The workstations are not considered like time-sharing systems. They are totally under the control of the user, and local security services can be easily by-passed by the local user. Consequently, the network administration cannot trust any authentication carried out on a workstation. While a user accesses only local services, such local authentication is sufficient. But when the user wants to access remote services, the authentication must be carried out by a trusted service. This is the role of the Kerberos service, split in several security servers.

The network is seen as in Fig. 1.2. Servers are trusted and so they are physically protected. Clients are not trusted. However anything occuring on the client host is under the responsibility of the user. And the server is under the responsibility of its administrator.

When a client logs in and wants to access a protected server:

- He asks the Key Distribution Server (KDS) for a session key.
- The KDS authenticates the client and gives him the session key.
- The client now asks the Ticket-Granting Server (TGS) for a ticket to access the protected server.
- The TGS provides a ticket to communicate with the server.
- The client uses this ticket to open a session with the server.
- It is the server itself which finally authorizes access.

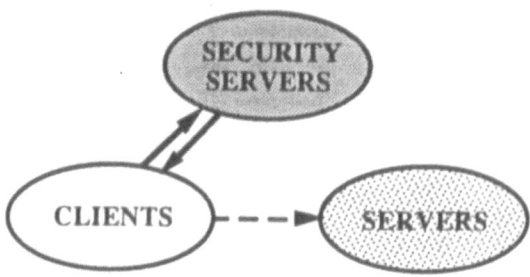

Fig. 1.2 Kerberos Architecture.

The NTCB is in part on the Kerberos server (authentication) and in part on the secured server (authorization).

An essential characteristic of the Kerberos architecture is the centralization of servers. Indeed, the security services are not distributed. The Kerberos master server is replicated on passive slaves, which can replace the master when it fails. This replication is carried out by dumping the master Kerberos database every hour. This architecture has several potential weaknesses.

- The security administrator who manages the master server can misuse his priviledges to perform unauthorized actions. The security administrator must be trusted. Furthermore, if an intruder succeeds in penetrating the master server, global security is no longer ensured.
- The slaves can also give enough information to intruders who can read them to use it later in order to act as authorized users. Each slave must be very well protected.
- If Kerberos server fails, the last database changes are lost.

There exists a single point of failure, from the viewpoint of both accidental faults and intrusions.

The openness of the Kerberos architecture induces another drawback with respect to the Red Book proposals; it is not possible to enforce a mandatory access control policy by the Kerberos servers; nothing is provided to prevent "covert" channels which can be easily implemented by communications between workstations or by a memory channel within a workstation.

2. The Delta-4 Intrusion Tolerant Approach

An **intrusion** can be defined as a *deliberate interaction fault*. The definition of fault is given by [LAP 90]. Intrusions can be treated with the same means as for other faults (fault-avoidance and fault-tolerance). The means used to provide security in architectures described above is *intrusion-avoidance*. On the contrary, the means used in our approach is *intrusion-tolerance*. It is based on the fragmentation-scattering technique [FRAG 85, FRAY 86] in order to implement an intrusion-tolerant archive server and an intrusion-tolerant security server.

2.1. The Fragmented-Scattered Archive Service

In Delta-4, fragmentation-scattering was first used to implement a secure distributed archive service [RAN 88]. Data security is provided by intrusion-tolerance and more precisely by geographical fragmentation-scattering. The basis of the fragmentation and scattering technique in this case is to cut every sensitive file into several fragments in such a way that one or more fragments (but not all) are insufficient to reconstitute the file. These fragments are then stored in geographically distributed archive sites. An intruder who could access some sites cannot obtain all the fragments of the same file unless he has almost overall control of the complete distributed system. On the other hand, in order to ensure availability, several copies of each fragment are stored on different archive sites. This service thus answers all three security requirements: confidentiality, integrity and availability. Confidentiality and integrity are directly provided by fragmentation-scattering while availability is obtained by replication of fragmented data.

The archiving steps can be described as follows:

- Enciphering of the file with the fragmentation key (private to file's owner).
- Splitting of the file into fixed-length pages.
- Fragmentation of these pages, and fragment naming using the fragmentation key.
- Fragment replication.
- Transmission of fragment replicates to the archive sites.
- Agreement between sites according random parameters in order to decide where fragment replicates will be stored.

The intrusion-tolerant security server described in this paper provides the services that are needed to complete this secure file archive service; namely, user authentication, access-rights verification, protection of fragmentation keys, etc. The same basic services can of course be used in a more general context.

2.2. Requirements and Assumptions for an Intrusion-Tolerant Security Server

The objectives of the intrusion tolerant approach are:

- openness and compatibility (for Delta-4 requirements), no specific hardware.
- reduction of TCB (by intrusion tolerance).
- modular security requirements.

The Security view of a Delta-4 network defines three kinds of sites (Fig. 2.1):

- **User sites** are untrusted computers where users can log in. The local security is ensured by users.
- **Security sites** are computers providing security services: registration, identification-authentication, authorization, sensitive information management, audit and recovery service. Each security site is managed by a different security administrator.
- **Particular servers** whose access needs to be secured. In the current system, this is the case of the Archive service located on Archive sites.

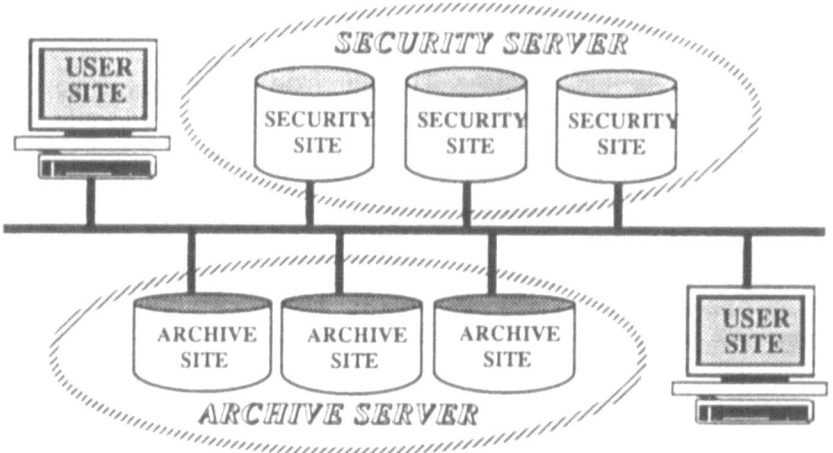

Fig. 2.1. The different types of sites of the network

The fault assumptions for security sites are:

- The probability of more than one intrusion before detection and recovery is small.
- The probability of an intrusion into a site is independent of the previous intrusion(s) in other site(s).

The characteristics of the intrusion-tolerant security approach are:

- An intrusion or a misuse on one security site is immediately masked and has no consequence on the service and on its properties.
- If errors occur on some security sites before recovery, the number which will be masked depends on the services and their properties (confidentiality, integrity, availability). The service performances can be degraded.

2.3. Intrusion Tolerance for a Security Server

The different types of intrusion depend on who makes an intrusion.

- It can be somebody outside the system who tries to access it. This is the most well known kind of intruder, but not the most important. In this case the intruder has to *by-pass* physical, procedural and logical protections.
- The second kind of intruder is a user of the system who tries to access information or services without access rights. The intruder tries to *extend his priviledges*. This the most common intrusion. The intruder has "only" to by-pass the logical protection.
- The third - and the most dangerous - type of intruder is a security administrator who *uses his rights* to perform illegitimate actions. In this later case, the

administrator has enough access-rights to do these actions but, according to the security policy, is not supposed to do them.

The architectures described in section I are intrusion avoidance architectures. A user/intruder is **trusted not to misuse his rights**. All the protection mechanisms must prevent the unauthorized actions. If an intruder succeeds in by-passing these protections, the security of the system is no longer ensured. If a security administrator decides to carry out illegal actions, there is no logical protection to prevent him from so doing. He could only be detected by using an intrusion-detection model [DENN 86] which is able to detect intrusions by monitoring system's audit records for abnormal patterns of system usage.

The principles of the intrusion tolerance are different. The system tolerates a bounded number of misuses. If one or more intruders by-pass the protection mechanisms and if the number of misuses they do is less than a given threshold, the security properties of the system (confidentiality, integrity and availability) are always ensured.

Three types of intrusion tolerance can be formulated:

- **for confidentiality**: read access to a subset of confidential data gives no information about the data.
- **for integrity**: the change of a subset of data does not change the data perceived by legitimate users.
- for **availability**: the change or deletion of a subset of data or of a server does not produce a denial of service to legitimate users.

For each property, a tolerance threshold is defined. If the reading, modification or destruction is done on a part D' of data/server D such that $|D'|$ (size of D') is less than the threshold, the properties are always verified (Fig. 2.2).

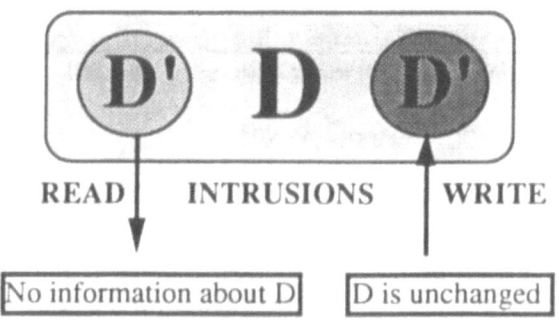

Fig. 2.2. Intrusion tolerance.

This concept is implemented on the security sites. The security sites are particular sites in that they collectively offer security services in a way such that security is always ensured in spite of intrusions on a bounded number of sites. The security sites also tolerate intrusions of malicious security administrators intrusions as they are described above. A security administrator manages only one security site. The security can thus be ensured in spite of a collusion between a minority of security administrators.

The security services are *distributed* and *intrusion tolerant*. The implementation has the essential property of requiring no local TCB.

2.3.1. Intrusion Tolerance by Distribution

Intrusion tolerance for data is relatively easy to implement. Data intrusion-tolerance techniques have existed for a long time. Confidentiality can be ensured by cryptographic tools like threshold scheme [SHA 79, BLAK 79]. The data is shared in "shadows", each shadow being stored on one security site. To build the data you only need a sufficient number of shadows called the threshold. If you do not have enough shadows, you cannot build the initial data. The same scheme can ensure availability and integrity.

Some data is not confidential, so it is also possible to replicate such data on each security site. Data is *shared* or *replicated* according to its confidentiality.

The most important point and the hardest one is the prevention of denial of service. In this case, it is not just data but a service that has to be protected. The server thus has to be replicated on each site in order to prevent denial of service in case of a security site unavailability. However the different sites cannot take certain decisions independently. They must agree by communicating data and local decisions. To ensure the last property, the servers must obey to two implementation principles: *replication* and *agreement* (Fig. 2.3).

The distribution of the security service permits a geographic distribution of the security sites. This makes the intruder's task more difficult since even if he succeeds in accessing one site, it will be more difficult for him to access other sites if they are not in the same place. It would indeed be a pity to distribute programs and data to perform logical intrusion tolerance and to not provide geographic distribution to assist physical intrusion tolerance.

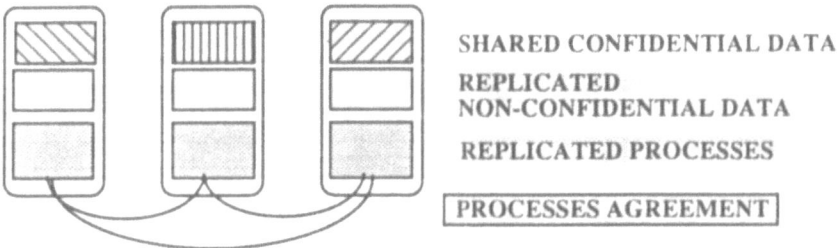

SHARED CONFIDENTIAL DATA

REPLICATED
NON-CONFIDENTIAL DATA

REPLICATED PROCESSES

PROCESSES AGREEMENT

Fig. 2.3. Distributed Security Services on three security sites.

2.3.2. Distributed TCB

In the Red Book architecture, there exists on each computer a part called the Trusted Computing Base including the NTCB partition (cf. 1.1.). The TIU assumes only communications security whereas the NTCB partition implements the local part of the network authorization policy. This NTCB partition must have a very high physical and logical protection. Moreover all sites have to trust one another. On each computer, accesses are mediated by the local TCB which is firstly the implementation of the **local** authorization policy. The **global** authorization policy is implemented in the set of NTCB partitions. When all computers are connected by a network and a subject wants to access a remote object, the NTCB partitions communicate with each other. In this case, a subject on a given site must trust all sites he wants to access. If one site has been penetrated by an intruder, the security of the network cannot be ensured.

In the Kerberos architecture, the most important part of the TCB is within the security server. There also exists on each server an important TCB which has to carry out authorization operations. However if an intruder succeeds in penetrating a server, he cannot access the other ones. The consequences of an intrusion are then limited. The only site you really have to trust to ensure global security is the security server site. However, in this case there exists a single point of failure.

In the intrusion tolerant approach, there is no local "Trusted" Computing Base on the security sites. Only the set of security sites is globally trusted (Fig. 2.4). There is also a small local TCB on the user sites and the secured servers. The servers themselves, for instance the Archive Service, can have a distributed TCB. In this case, the only single point of failure is on the user site. This can be minimized if the user site is considered as a one-user computer when a user accesses a secured service. If this were not so, there would always exist trapdoors on the local protections between users.

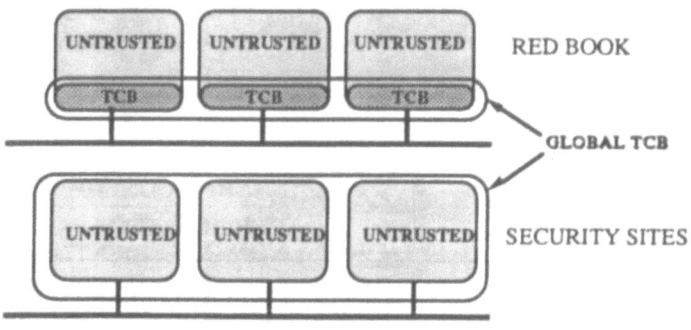

Fig. 2.4 Red Book versus Security Sites TCB.

The security server is trusted, but not the different computers. The security service is considered as one global server. If an intrusion in one computer is successful in a classical architecture with a local TCB, the security of the full system is no longer verified. On the contrary, if protections of one security site in the intrusion tolerant approach are by-passed, the security of the global system is maintained. These differences come from three different points of view about security and networks:

- In the Red Book, a network is seen as only a communication channel (low layers) between centralized systems. A subject, or an object, is located on one site and cannot be shared between several sites. In our architecture, the network is a support for distributed applications. A subject or an object can be shared on several sites.
- In the Red Book, sites are time-sharing systems with several users working on the same host. These computers are controlled by a system administrator. In Kerberos, sites are workstations under the control of one local user. In our system, whenever possible, the use of one host by several users during access to a secured service will be prohibited.
- The Red Book architecture is a support for DoD policy, a multilevel mandatory policy where confidentiality is the most important property to be ensured. All sites thus need to have an important trusted part. It is not possible to implement this policy in our system. Trusted paths between all user sites would be needed.

2.4. The Provided Services

The different services which must be provided by the security sites are registration, authentication, authorization, sensitive data management, audit and recovery service.

2.4.1. The Registration Service

The registration service permits a user to be registered by the system for future access to secured services. This operation must be carried out independently on each security site to prevent a single site from using information to impersonate the user. The operation is done under control of the security administrator of each site.

2.4.2. The Identification-Authentication Service

The role of this service is to verify the claimed identity of a subject. When a user, or a process acting for a user wants to access a secured service, he must first be authenticated. The authentication service verifies that the subject is really who he claims to be. To do this, both logical or physical techniques are available. These can be based on passwords, zero-knowledge authentication [FIAT 86, GUIL 88], smartcards or chip-cards. All these techniques use the same protocol principle: the subject must prove its identity to the authentication server by showing that he possesses a secret information, its authenticator.

In a distributed system with several authentication servers, each server must independently authenticate the subject (Fig. 2.5). Indeed, the security sites are untrusted and one site could try to use authentication information given by the user and the data stored on the site to impersonate him in another authentication phase. The independence of the authentications on each site must be complete (different authentication objects for each site or zero-knowledge authentication). The second phase of the protocol is the agreement. The servers communicate their decision to each other and they take a global decision. If a majority of servers succeed in authenticating the subject, the subject can be considered as authenticated by all the servers.

The servers will send some specific data to the subject (session key, identificator, ...) with a time-out. If the time-out expires, the subject must repeat the authentication phase.

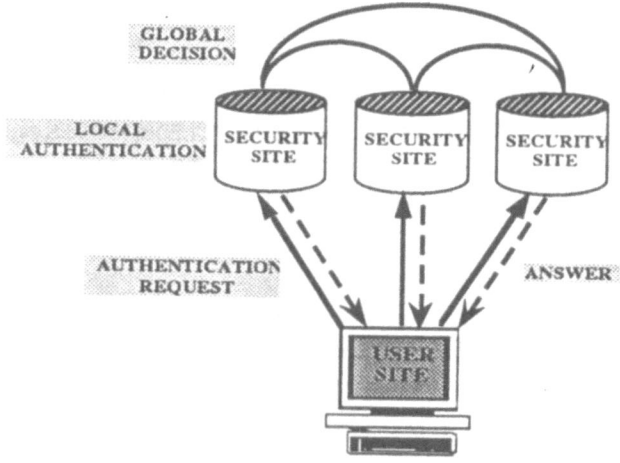

Fig.2.5. The Authentication protocol.

2.4.3. The Authorization Service

The role of this service is to check that the access to a secured service by a subject is authorized according to its access-rights. In Kerberos, this service is provided by the servers themselves because of the large size of the managed database and of the difficulty of modelling different accessrights for different services. However, these two problems can be solved.

It is possible to implement access-rights for different services using standard UNIX rights. In UNIX, all objects are viewed as files. The read, write and execute rights are thus applied to all objects of the system. The accesses to the services can be authorized using the same access-rights model. In this case, access-rights can be implemented for all kinds of services thus leading to a reduction in the cost of the access-rights database management.

The authorization service is made intrusion tolerant when it is implemented on security servers. The rights must be changed by all security administrators. One of them cannot access service if he is not authorized by others. An intruder cannot modify access rights as easily as he could do if they were located on the server. But it is obvious that if an intruder controls the service, he can do what he wants. The only solution is to build an intrusion tolerant service such as the archive service, but this is not always possible (e.g. consider a printer service).

The different phases of authorization are (Fig. 2.6):

- The subject asks the security servers for permissions to access a secured service by sending its identificator (received in the authentication phase) (1).
- The access-rights stored on the security sites enable the latter to verify that the subject is authorized to access the requested service.
- The security sites vote to decide if the access is authorized using the same protocol as that defined for authentication (2).
- If the sites agree to permit access, they send a ticket to the subject and another ticket to the secured server (3).
- With the ticket, the subject can open a session with the server (4).

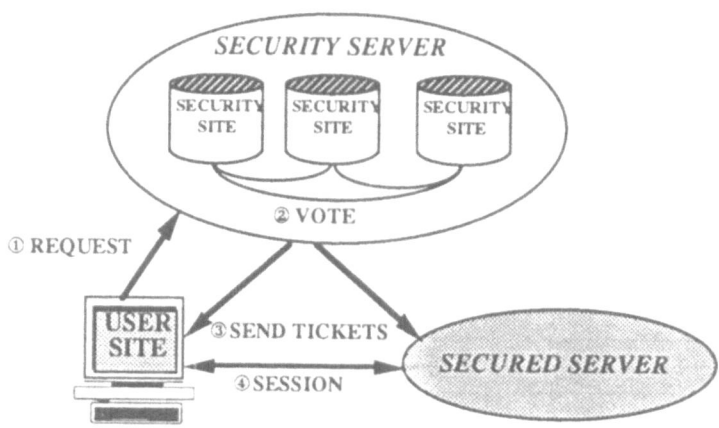

Fig.2.6. The Authorization protocol.

A subject may also want to access secured data stored on security sites, like access-rights or authentication keys. In this case, the access control protocol is the same as the one described above for the phases 1 and 2 and then, if and, when the security sites accept the access, they perform it on each site and send the information or the affirmative response to the user site.

2.4.4. The Sensitive Data Management Service

The role of this service is to store, manage and retrieve the sensitive information on the security servers so that their protection verifies the hypothesis made in §2.2. This information consists of short data items needed to achieve security services.

The data management service must enforce the three main security properties (confidentiality, integrity and availability). The integrity property is provided by modification detection mechanism such as cryptographic signatures. According to the sensitivity of the security data, it can be important to preserve both the confidentiality and availability of this data, or only the availability. For this, two storage techniques can be applied: replication (for availability) or threshold schemes (for confidentiality and availability). In function of this, the security administrators (data for a service access) or a subject (data for authentication) will store data with one of the two algorithms (Fig. 2.7).

If a data item is replicated on N security sites, it is assumed:

– with respect to availability, that N-1 replicates can be lost (modified or destroyed),
– with respect to confidentiality, that one replicate is sufficient to observe data.

If one data is shared on N security sites (in this case we speak of shadows) using a threshold T, it is assumed:

– with respect to availability, that N-T shadows can be lost,
– with respect to confidentiality, that T shadows are necessary and sufficient to observe data (less than T shadows gives **no information** about the data).

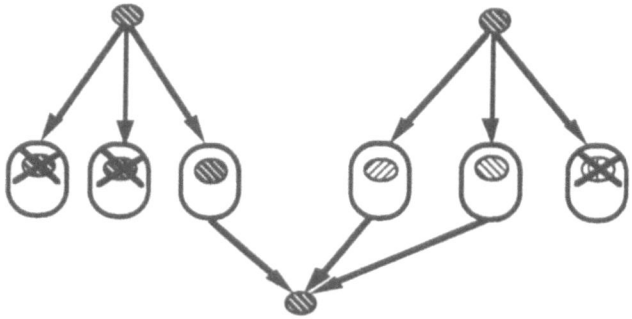

Data is replicated   Data is shared with a threshold of 2

Fig.2.7. Replication and Threshold Scheme.

2.4.5. The Audit Service

The role of this service is to record all information related to security. Such information is sent by the services defined above. There exists two kinds of information, authorized operations performed by authorized users (registration, access, rights change, ...) and attempted or successful intrusions or misuses. It is not the role of the services to determine what is an intrusion or misuse by an authorized user. This is the function of an audit trail analysis.

The audit information is sent not only by security sites but also by secured servers and user sites. For the former, it will be access-requests, and for the latter it will be, for instance, information about correct or incorrect shared data sent by security sites (bad shadows received from certain security sites).

The audit trails are stored on each security site. The information received on one site is not sufficient to compromise the security of the system.

The analysis of this audit information will be done off-line by security administrators. As one intrusion is masked, it is not necessary to detect intrusion on-line.

2.4.6. The Recovery Service

It acts as an error recovery mechanism to correct certain modified data (e.g. shadows of the threshold scheme). Other recovery functions can be performed manually by security administrators using audit trails.

3. Conclusion

In this paper, a new approach for security in an open distributed system has been described. It is based on a client-server model. Several security services are provided in order to secure the access to remote servers. These services are intrusion tolerant and reside on several security sites.

The intrusion tolerant technique is based on fragmentation-scattering to ensure confidentiality and integrity of data and availability of services. The security site architecture uses replication of data and processes, threshold schemes for confidential data and agreement protocols between processes. The different services built on this architecture are registration, identification, authentication, authorization, management of sensitive data, audit and recovery.

Acknowledgements

The authors wish to thank Jean-Claude Laprie and David Powell for their contributions to the main concepts presented in this paper. The Delta4 project is supported by the European ESPRIT program. This work is also partially supported by the Conseil Regional de Midi-Pyrénées and by the Programme de Recherche Concertée C^3.

References

[BELL 74] BELL D. E. and LAPADULA L. J., "Secure Computer Systems: Mathematical Foundations and Model", M74-244, MITRE Co., October 1974.

[BLAK 79] BLAKLEY G. R., "Safeguarding Cryptographic Keys", Proc. NCC, Vol. 48, AFIPS Press, Montvale N. J., 1979, pp. 313-317.

[DENN 86] DENNING D. E., "An Intrusion-Detection Model", Proc. of IEEE Symp. on Security and Privacy, Oakland, April 1986, pp. 118-132.

[DoD 85] D.o.D., "Department of Defense Trusted Computer System Evaluation Criteria", DOD 5200.28-STD,, December 1985.

[FIAT86] FIAT A., SHAMIR A., "How to Prove Yourself: Practical Solutions of Identification and Signature Problems", Advances in Cryptology - CRYPT0'86. Santa Barbara, August 1986, Lecture Notes in Computer Science Vol. 263, Springer Verlag, ISBN 0-387-18047-8, pp.186-194.

[FRAG 85] FRAGA J., POWELL D., "A Fault and Intrusion-Tolerant file System", in Computer Security: the practical issues in a troubled world, Proc. 3rd Int. Cong. on Comp. Security (IFIP/SEC'85), Dublin, Ireland, August 1985, ISBN 0-1-87801-7, pp. 203-218.

[FRAY 86] FRAY J.M., DESWARTE Y., POWELL D., "Intrusion-Tolerance using Fine-Grain Fragmentation-Scattering", Proc. on the 1986 IEEE Symp. on Security and Privacy, Oakland, April 1986, pp. 194-201.

[GUIL 88] GUILLOU L.C., QUISQUATER J.J., "A Practical Zero-knowledge Protocol Fitted to Security Microprocessor Minimizing both Transmission and Memory", Advances in Criptology - Eurocrypt 88, Davos, Switzerland, May 1988, Lecture Notes in Computer Science Vol. 330, Springer Verlag, ISBN 0-387-50251-3, pp. 123-128.

[HARR76] HARRISON M. A., RUZZO W. L. and ULLMAN J. D., "Protection in Operating Systems", Comm. of ACM, Vol. 19, no 8, August 1976, pp.461-471.

[ISO 7498-2] I.S.O., International Standard 7498-2: Information processing systems - OSI Reference model - Part 2: Security Architecture, Tech. Rept. no 2890, ISO/IEC JTCI/SC21, July 1988.

[LAP 90] LAPRIE J.C., "Dependability: Basic Concepts and Associated Terminology", in Dependability Concepts and Terminology, ESPRIT BRA PROJECT 3092 Predictably Dependable Computing Systems, First Year Report, Task A, Vol. 1, May 1990.

[MILL 87] MILLER S.P., NEUMAN B.C., SCHILLER J.I. and SALTZER J.H., "Kerberos Authentication and authorization System", MIT Proj. Athena Technical Plan, Sect. E.2.1, December 1987.

[NCSC 87a] N.C.S.C., "Trusted Network Interpretation of the Trusted Computer System Evaluation Criteria", NCSC-TG-OOS, July 1987.

[RAN 88] RANEA P.G., DESWARTE Y., FRAY J.M., POWELL D., "The Security Approach in Delta-4", Research into Networks and Distributed Applications EUTECO'88, Vienna, Austria, April 1988, ISBN 0111-70428-0, pp. 455466.

[SHA 79] SHAMIR A., "How to Share a Secret", Comm. of ACM, Vol. 22, no 11, November 1979, pp. 612-613.

[STEIN 88] STEINER J. G., NEUMAN C. and SCHILLER J.I., "Kerberos: An Authentication Service for Open Network Systems", USENIX Winter Conf., Dallas, February 1988.

OVERVIEW OF THE LOTOSPHERE DESIGN METHODOLOGY

L.FERREIRA PIRES
C.A.VISSERS
University of Twente
PO Box 217
7500 AE Enschede
the Netherlands

Abstract.

This paper presents the direction of work in the Lotosphere Design Methodology. It first examines some basic concepts in distributed systems' design and then identifies areas for further research and development. In this paper we define and analyse the needs of the design process for distributed systems, in order to determine how the use of a formal design language as LOTOS can improve this process. A top-down view of the design trajectory is used in order to introduce the design goals, and related definitions and concepts. In this context we discuss the role of abstraction as the concept that supports step-wise refinement, and the role of design languages, as the means to represent the relevant properties of a system. A more elaborated design trajectory takes into consideration the deviation cases from the straight top-down model. This analysis ends up at the level of design steps, at which design concerns are identified as the areas of interest: the design methods that comprise the methodology. The paper intends to provide a framework for the effective use of Formal Description Techniques (FDTs) in distributed system design, which is to be applied in the scope of the ESPRIT II Lotosphere Project (Project 2304).

1. Introduction

The use of FDTs in the design of complex distributed systems has gained considerable interest in the academic and industrial world. An evidence of this is the ESPRIT II Lotosphere project, which enjoys the participation of 6 universities, 6 research institutes, and 6 industrial companies. The Lotosphere project aims at the industrial exploitation of mathematically sound, i.e. FDT supported, system design methods. In this project the internationally standardized FDT "LOTOS" ([7]) has been chosen.

The area of distributed systems design presents some overlap with the classical hardware and software design areas. Nevertheless in distributed systems design some additional challenges appear from the distribution and cooperation among system components, which force coordination and synchronization. Although many of the concepts used in classical hardware and software design can be applied in our case, some new concepts or different interpretations of the "well-known" concepts must be introduced.

This paper introduces the Lotosphere design approach, which aims at providing to designers the methodological and conceptual integrity towards distributed systems' design. Therefore the design methodology presented here is an instance of system design methodology tailored for the distributed systems' design problem.

The paper has the following structure: section 2 introduces some terminology for the conventional top-down concept of design methodology on basis of a simplified view of the design trajectory, section 3 discusses the role of design languages in this design trajectory, section 4 presents a more realistic view of the design trajectory, section 5 identifies design methods to guide designers performing design steps, and conclusions are drawn in section 6.

2. The concept of Design Methodology

In the world of information technology frequently an *object*, we call it a *system*, is identified that is supposed to fulfil the needs of a set of future users. In case the system does not yet exist in the real world, i.e. if it is not yet a *real system* or a *product*, it has to be produced. The *design process* is the activity in which the user requirements are formulated and transformed in a real system. Figure 1 depicts the design process.

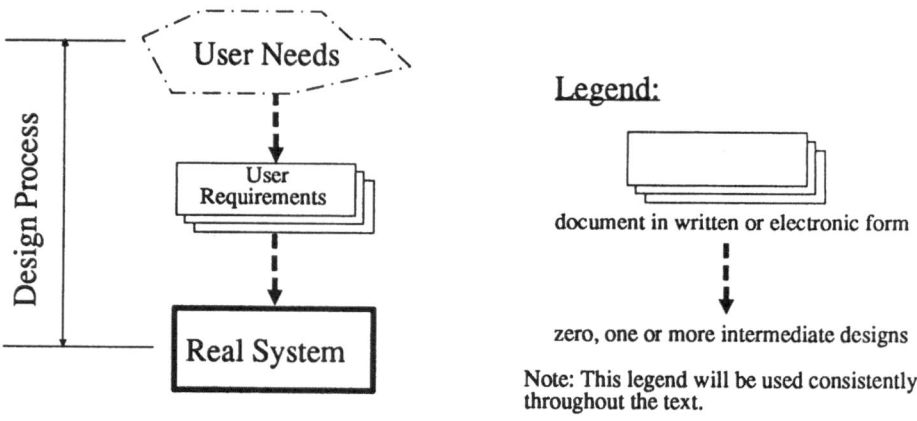

Figure 1. The Design Process.

A *design methodology* is as a set of methods that can be used to perform the design process. The prime demand for the transformation to a real system is that the user requirements must be *preserved*, however, they may be provided in various ways. We say that the real system must *conform* to the user requirements.

Handling Complexity. The Lotosphere design methodology aims at designing complex distributed systems such as communication networks and open distributed systems. Such systems are generally characterized not only by intricate user requirements but also by the involvement of many possible independent, related or conflicting *design decisions* that need to be considered during the design process. Such design decisions may have far reaching, but very difficult to antici-pate, consequences for the functionality and quality of the resulting real system. Consequently the design process of complex distributed systems is generally a large and complicated process, making it neither possible, nor desirable, to have it carried out in a single step.

Therefore the design process is split up into a sequence of *design steps*, where in each step only a limited set of design decisions are considered, evaluated, and some of them are incorporated in the design. This design acts as a starting point for a next step, in which another set of design decisions can be considered. Consequently each next design step in the sequence produces a more elaborated, we often say *refined*, version of the design that is closer to the real system. This approach towards a design process is often called *step-wise refinement* or *top-down*.

2.1. The Design Trajectory: a Simplified View

We define a *design trajectory* as a sequence of design steps, produced by step-wise refinement, which starts with the formulation of the user requirements and terminates with the production of a concrete instance of the desired system.

In the Lotosphere design methodology the design trajectory is divided into *design phases*, where each phase is characterized by different *design objectives* requiring different areas of expertise. Each phase usually consists of multiple design steps. The simplified view of the design trajectory and its design phases are depicted in figure 2.

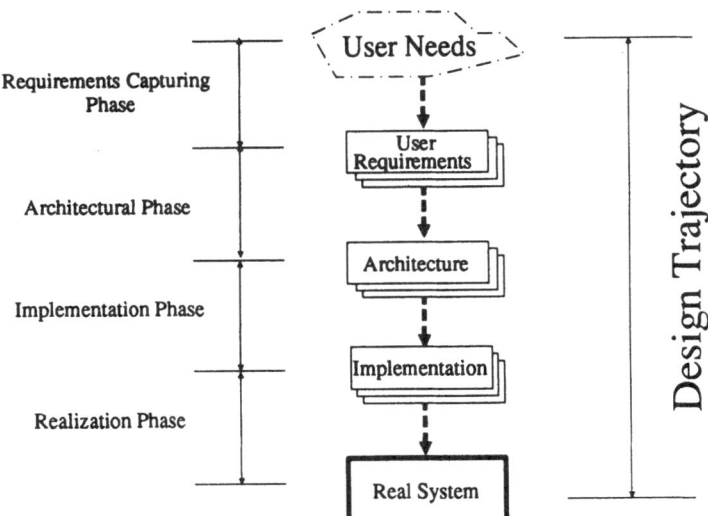

Figure 2. The Design Trajectory.

In the above design trajectory, *User Requirements*, *Architecture* and *Implemen-*

tation are consecutive designs, each one of them fulfilling the *User Needs*, but representing the real system in different ways, i.e. at different levels of abstraction.

The *User Requirements* must serve as a contract between the system user(s) and the system producer(s) with respect to the agreed expectations to the system's capabilities. The *Architecture* must express *what* properties the real system should possess, not *how* these properties are achieved. The *Implementation* must be expressed in terms of descriptive elements and combinations thereof that can be mapped directly on the physical and logical elements that will embody the real system.

2.2. Levels of Abstraction

The term *abstraction* denotes the consideration of only a limited set of characteristics of an object while ignoring its other characteristics. At the *level of abstraction* at which we want to consider the object, the limited set of characteristics that we want to consider are the *relevant characteristics*, whereas we *abstract from* the other, irrelevant, characteristics.

In the Lotosphere design methodology we relate the concept of abstraction directly to the concept of step-wise refinement. The object that we consider is the real system that we want to produce as result of the design process. The characteristics that we consider are the properties of the real system as expressed by a design in the design trajectory. Thus a design is an abstract representation of the real system, and is defined at a certain level of abstraction, the *(N)-level of abstraction*. Such a design is called *(N)-level design*. The properties of the system that we abstract from when formulating an (N)-level design ought to be addressed and incorporated in successive refinement steps, leading to more detailed designs and consequently lower levels of abstraction.

3. Role of Design Languages

We consider a *design* to be an abstraction of the real system expressed using a symbolic representation method. This symbolic representation forms the vehicle that is used for analysing and prototyping those properties that are considered relevant at a certain level of abstraction.

A symbolic representation method is only of practical use if it is based on the syntactical and semantical rules of a *design language*. This requirement is induced by the fact that there should be a consistent relationship between the properties that are expressed by a design and the language constructs expressing them.

A design language necessarily needs to be general purpose; it is impractical to reflect each different design property by a special purpose language element. The language must also allow the representation of the intended properties directly and clearly. This imposes high demands on the expressive power of the design language and its model. In many cases not all properties of a design can be expressed in the model used, forcing the designers to use multiple models and to move between models in the course of the design process, to ensure that all relevant properties of the real system are covered in the specifications at each level of abstraction.

3.1. The Spectrum of Design Languages

A design language is called a *broad spectrum design language* if it can express designs at many different abstraction levels along the design trajectory. A design language is a *narrow spectrum design language* if it can express designs at only a few different abstraction levels along the design trajectory. The use of a single broad spectrum design language throughout the design trajectory appears to be very advantageous:

- by using a single design language the designers can improve their communication, which is very important if the designer (or designers' team) that performs one design step is different than the designer (or designers' team) that performs the subsequent step(s);
- the verification of the preservation of properties described in the (N)-level design in the successive designs is also facilitated if a single design language is used;
- using a single design language may avoid re-writing of parts of an (N)-level design that are not affected by the design decisions taken in a specific design step.

3.2. LOTOS as a Broad Spectrum Design Language

Among design languages, those that are based on a formal model have a high potential to support the design process and enhance its efficiency and quality. This is because the formal model enables a rigid and unambiguous expression of designs, that subsequently can be analysed by using mathematical methods. The formality of the model also allows the development of software tools that can support the design process, and automate part of it.

In Lotosphere we base the design methodology on the Formal Description Technique (FDT) LOTOS [7], which is used as a broad spectrum design language. LOTOS was chosen because it has a high expressive power, allowing not only to make unambiguous, but also comprehensible, and concise specifications.

Comprehensibility and conciseness are favoured by a straightforward representation of the architectural concepts used in open distributed systems. An important industrial criterion is also that LOTOS is internationally standardized (IS 8807), implying that its definition is stable. Standardization also promotes the increasing availability of advanced design support tools ([5], [6]).

Properties that are expressed in a LOTOS specification ((N)-level of abstraction specification or *(N)-specification*) are called *formalized properties*, while properties that are not expressed in LOTOS specifications are called *non-formalized properties*. Formalized properties are properties that are expressed in the syntax and semantics of behaviour expressions and abstract data types; non-formalized properties are, for example, performance or cost properties of the system, still expressed in natural language.

3.3. LOTOS in the Design Trajectory

We assume that in most cases the *User Requirements* will be described in natural language. This assumption is based on the expectation that the formulation of the *User Requirements* is used for communication between users and designers, whereas users, in general, are not experts in formal design languages. Sometimes it happens that (a part of) the *User Requirements* is expressed formally. This is the case, for instance, when (a part of) the *User Requirements* imply the use of a design (e.g. a protocol) that is already expressed formally.

The formulation of the *Architecture* is the first occasion where LOTOS can be used. Experience indicates that many properties of *Architectures* can be conveniently expressed in LOTOS. However, some properties of *Architectures* cannot be formalized in LOTOS. Examples of these properties are requirements for performance, absolute time, cost, or requirements that can only be globally indicated, such as "suitable for stream oriented traffic", "fault tolerant", or "robust".

However, during the design trajectory design decisions can be taken for the (N + 1)-level design that effectively lead to the (partly) formalization of (some aspects of) non-formalized properties of the (N)-level design. After such (aspects of) non-formalized properties are (partly) incorporated in the (N + 1)-level design, they may be further ignored, or reconsidered in a different form, in successive steps. This is shown in figure 3.

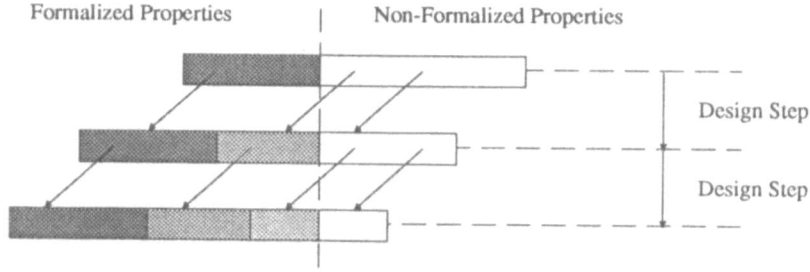

Figure 3. Formalization of Properties.

As an example we take the case in which one of the user requirements is high reliability. Due to this requirement, at some point in the implementation phase an appropriate design decision may be to duplicate some vital parts of the system, allowing to further ignore duplication in the implementation phase. The reliability requirement, however, may reappear in the realization phase, where it might invoke a decision to choose certain highly reliable hardware components.

The simplified design trajectory, thus consists of a sequence of successive transformations of properties expressed in an (N)-specification into properties expressed in the (N + 1)-specification. Some non-formalized properties in the (N)-

level design are incorporated in the (N + 1)-level and successive designs. These non-formalized properties are also transformed or simply disappear in a design step. The properties which are expressed at (N)-level design should be preserved in the (N + 1)-level design; the (N + 1)-level design should *conform* to the (N)-level design. However, the way (e.g. the structure) in which properties are specified may be rearranged in each successive design step.

In the specific case of using LOTOS to describe (parts of) designs, where both syntax and semantics elements may convey design information, proper documentation has also vital importance. Comments throughout the LOTOS text shall inform the designers that e.g. two parallel processes are supposed to be implemented as independent components, or if a gate represents a physical or logical attachment point to be identified in the implementation.

3.4. Limitations of LOTOS in the Implementation Design

From the characterization of an implementation in section 2.1 we conclude that the design language used for the implementation should support representation of concreteness and direct translation into machine executable models. Implementation languages such as conventional programming languages (e.g. PASCAL, C) and hardware design languages (e.g. VHDL, RTL) are explicitly designed for that purpose, whereas LOTOS appears to have severe limitations.

Therefore there is a point in the design trajectory where an (N)-specification expressed in LOTOS, we call it the *LOTOS implementation*, is transformed into an (N + 1)-specification, expressed in one or more implementation languages. This transformation thus implies a translation from LOTOS into those implementation languages, as shown in figure 4. We strive for this transformation to take place in the later steps of the implementation phase, because we want to make as much as possible use of the broad spectrum and formal characteristics of LOTOS.

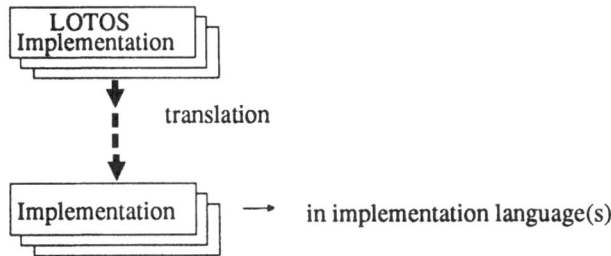

Figure 4. LOTOS in the Implementation Design.

It would be very desirable if a formal relationship could be established between the LOTOS model and the model(s) used for the implementation language(s), to allow automatic or semi-automatic translation and to ensure that the translation preserves the properties expressed in the LOTOS implementation. It appears however that such formal relationships are difficult to establish since in most cases implementation languages either have no formal model at all, or their model cannot be related to the LOTOS model.

An approach to solve this translation problem is the use of *pre-defined implementation constructs*, which is described in section 5.7.

3.5. A Design Example

In order to illustrate the concepts introduced so far we take an example design. Suppose we want to design a system, an (N)-layer service provider, which behaves according to some service definition, as for example services defined in the scope of the OSI RM. This problem appears frequently in the case of distributed systems' design, since layering is often applied in open systems architectures.

In LOTOS a service provider can be described in terms of the temporal ordering of the service primitives, using the LOTOS temporal ordering operators. Service primitives in LOTOS can be represented by atomic events, whereas the parameters of the service primitives can be represented by abstract data type definitions.

The architecture of the service provider is considered to be a *black box* (the *Service*), which does not prescribe any internal structure. In the subsequent refinement steps the service is implemented by a set of protocol entities and an underlying service. These design steps are mainly influenced by the physical distribution of the service, together with the need for an interconnection means among the protocol entities.

This more refined design represents the same service, but in a lower level of abstraction, since some internal structure is made explicit. This structure is represented again in a LOTOS specification, where each protocol entity and the underlying service is described as a LOTOS process, and the interactions between each protocol entity and the underlying service are described as "hidden" events, with the help of the LOTOS *hide* operator. This operator allows the designer to represent interactions that are not observable to the system's environment, in the example the service users.

At this level of abstraction the concept of Protocol Data Unit (PDU) can be introduced. The abstract PDUs can be represented in LOTOS by using abstract data type definitions to describe PDU parameters and their properties. Further design steps can deal with coding of PDUs into bit patterns, functional decomposition of protocol entities, refinement of interactions to real interfaces, etc. The implementation must be a sufficiently refined description, in a programming language or hardware design language or both. Figure 5 shows an example of design trajectory for a service. (See next page)

4. The Design Trajectory: A More Realistic View

Experience shows that a straightforward top-down design trajectory is unrealistic. Some reasons for that are presented below.

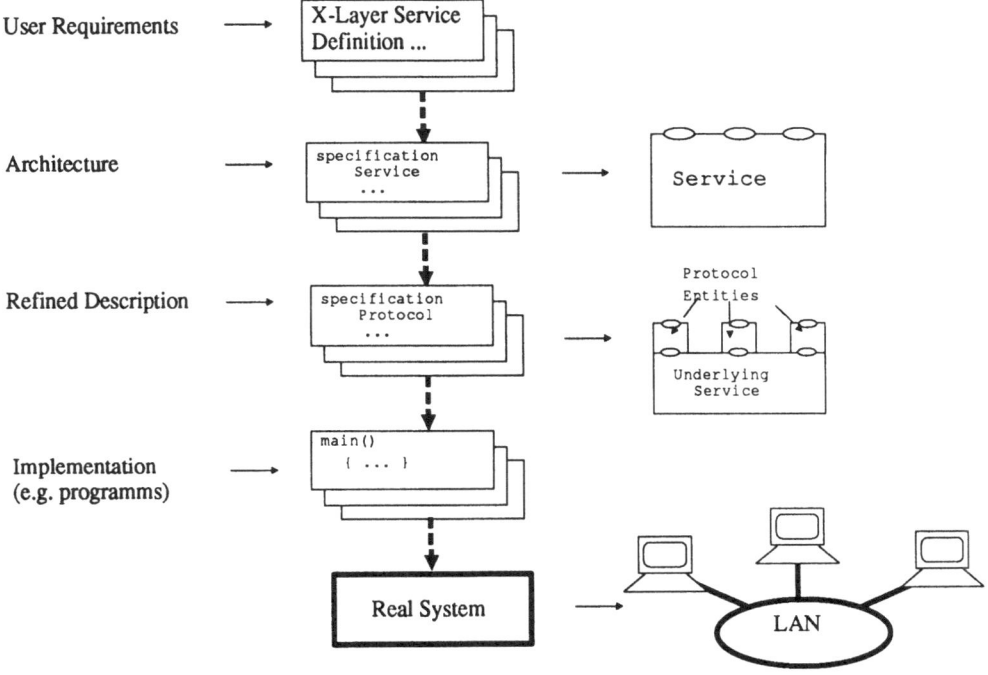

User Requirements ⟶ X-Layer Service Definition ...

Architecture ⟶ specification Service ... ⟶ Service

Refined Description ⟶ specification Protocol ... ⟶ Protocol Entities / Underlying Service

Implementation (e.g. programms) ⟶ main() { ... }

Real System ⟶ LAN

Figure 5. Design Trajectory Example.

4.1. The Design Tree

At each point in the design trajectory multiple alternative sets of design decisions can be conceived, leading to a multitude of alternative more refined designs. Therefore a design trajectory, showing alternative design choices, resembles a tree structure rather than a single sequence of steps.

Some of the alternatives in the tree may not be valid and should not be taken because they lead to designs which do not conform to the user requirements, or frustrate qualitative design criteria (see section 5.2.). Evidence of the invalidity of an alternative design choice, however, may not appear immediately, but only at later design steps. Therefore, in a more realistic design approach, the impact of design decisions along the design trajectory ought to be continuously evaluated, and incorrect or ineffective designs be pruned from the design tree. Consequently design decisions made earlier along the design trajectory must be reconsidered.

This process of continuous evaluation and reconsideration of design decisions determines iteration, resembling a tree search schema. At the end of the design process following the tree search one can reconstruct the simplified view of the design trajectory. The design tree and its eventual pruning is depicted in figure 6.

380

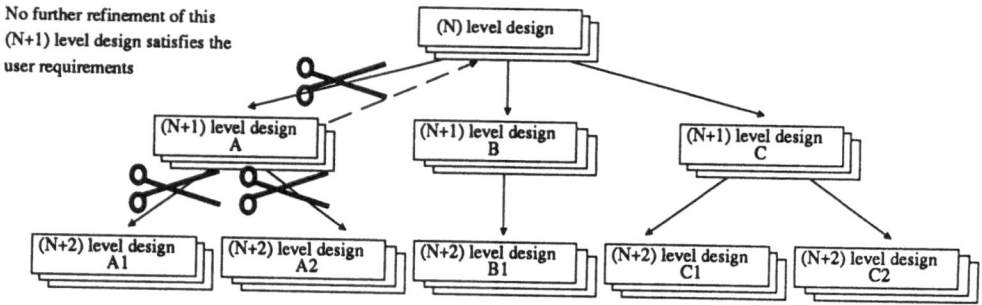

No further refinement of this
(N+1) level design satisfies the
user requirements

Figure 6. Reconsideration of Design Decisions.

4.2. Cyclic Approach

Another possible deviation of a strict top-down approach can be found appropriate by considering in a first approach towards designing the *Architecture* and the
Implementation only a limited set of the *User Requirements*, instead of all the
requirements at once. We call this a *design cycle*.

Since the real system should conform to all user requirements, design cycles
need to be repeated such that in each successive cycle more user requirements
are considered until, in the last cycle all requirements are fulfilled. This approach is
called the *cyclic approach* and is depicted in figure 7.

The main advantage of the cyclic approach is the possibility of rapid prototyping.
This rapid prototyping may be very useful to order design decisions, such that those
decisions that have the most far reaching consequences for the implementation
can be incorporated and evaluated in early cycles, whereas design decisions that
merely add details can be deferred to later cycles. From an industrial point of view
it may also be interesting to market a range of products starting with products of
limited functionality.

The cyclic approach also enforces quality requirements on designs. Extensions
made in a next cycle should preferably be done without the need for major
restructuring or re-definition of designs from later cycles. This implies a strict
obedience to generality and open-endedness design criteria when choosing the
structure of designs, influencing e.g. the specification style to be applied.

Another aspect of a realistic design methodology is the impact that decisions
taken at a high level of abstraction may have on the freedom of making decisions
at lower abstraction levels. It may very well be possible that this freedom is
unnecessarily constrained. Reversely, limitations in the possibilities of implementation and realization may have a strong influence on what should be decided at a
higher level of abstraction.

This implies that knowledge that is relevant at lower abstraction levels may be
necessary when designing at the higher levels of abstraction. We call this *bottom-up*

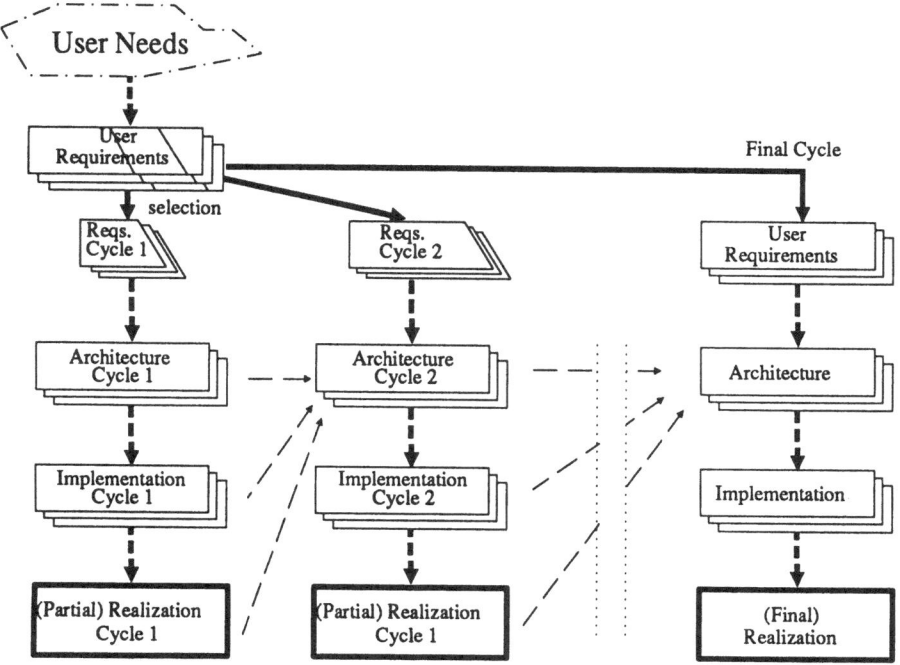

Figure 7. Cyclic Approach.

knowledge. Ignoring bottom-up knowledge at higher abstraction levels may lead to unnecessary rearrangements of designs and additional pruning in the design tree. The cyclic approach generates bottom-up knowledge in each design cycle, thus allowing this knowledge to be used in each consecutive cycle.

5. Design Steps

The design trajectory can actually be regarded as a repetition of design steps, where each step is guided by design knowledge, methods and criteria relative to its place in the total trajectory. In this section we identify some important design concerns that can guide performing each design step. In so doing we determine the areas of interest and the methods that together form the Lotosphere design methodology. In many circumstances a method can be more efficiently utilised if supported by tools. In Lotosphere we also aim at developing tools to support the most prominent methods.

Areas of interest that we consider in Lotosphere are the determination of the target (N + 1)-abstraction level, the observation of qualitative design criteria, the use of specification styles, the use of pre-defined specification constructs, correctness preserving transformation methods, testing and verification methods, and implementation strategies, among them the use of pre-defined implementation constructs. Figure 8 shows a single design step and its concerns.

382

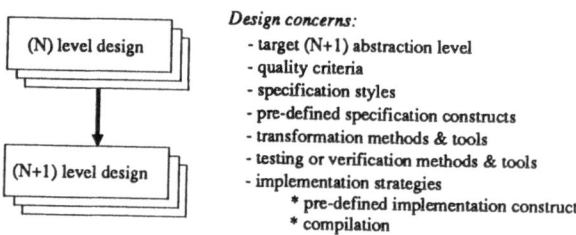

Design concerns:
- target (N+1) abstraction level
- quality criteria
- specification styles
- pre-defined specification constructs
- transformation methods & tools
- testing or verification methods & tools
- implementation strategies
  * pre-defined implementation constructs
  * compilation

Figure 8. The Design Step.

5.1. Target (N + 1)-Abstraction Level

A valuable method is the pragmatic guidance to choose the next abstraction level in a design step. This guidance shall dictate the kind of design decisions that must be considered in each design step and their relative position in the design trajectory. It shall also direct the transformation strategy to be applied, thus working as a binding element among the other concerns.

Relative evaluation of possible distinct design decisions for conformance to non-formalized properties shall be covered by this method, providing the means to designers to proceed in the design trajectory. Since this concern is quite generic, it shall relate to realistic instances of design problems, derived from the project experience.

5.2. Qualitative Design Criteria

Qualitative design criteria are principles that are used to reason about the quality of a design and at the same time can help guiding the designer during a design step concerning e.g. the structuring of the design and the consideration of design decisions. So far we have identified some concrete qualitative design criteria like parsimony, generality, orthogonality, open-endedness, propriety, completeness and consistency.

As an example we take the criterion of generality, which demands that what is inherent shall not be restricted. According to this principle inherent characteristics of an object are detached from particular ways of using it, enhancing clarity and conciseness, and allowing re-usability. Following this criterion a designer that needs to describe an instance of a FIFO queue, for example, shall first describe the general characteristics of FIFO queues in abstract data type (ADT) definitions, and then generate the instances of FIFO queues as desired (e.g. queue of characters or queue of natural numbers).

The application of qualitative design criteria in a design step can be developed by means of examples, by questioning the violation of these principles, their inter-dependences, etc. Some more refined criteria may be identified as a result of our work.

5.3. Specification Styles

Specifications can usually be written in a vast amount of different ways. This poses the question: what is a *good specification*? We assume that a good specification reflects the design properties, e.g. the functional properties and the structure of the design, in a clear way. Furthermore the specification should clearly reflect the quality of the design as mentioned in section 5.2 above.

In the Lotosphere design methodology we identify and define methods for structuring specifications that allow to pursue certain design objectives and decisions, and obey qualitative design criteria. We call these methods *specification styles*.

In the formulation of an *Architecture*, for example, the internal structuring of the specification shall not prescribe any implementation structure, since only the observable behaviour is of importance. On the other hand *Architectures* must be understood, in order to be transformed and maintained, therefore imposing some structuring. The constraint-oriented style characterized in [15] is the specification style that better suits to the description of *Architectures*. Using this style the designer describes the behaviour of the system as a composition of constraints dictated by the various concerns of the system.

In the formulation of intermediate descriptions during the design trajectory the structure of the specification in most cases reflects the structure of the implementation. The resource-oriented style is here the most appropriate, since it allows the explicit representation of structure (internal components) and communication among components (internal interfaces).

Other specification styles identified in [15] are the monolithic style and the state oriented style.

5.4. Transformation Methods

In case of a design step that starts with a LOTOS specification and results in another LOTOS specification a transformation takes place. Some properties that were described in the (N)-specification are supposed to be still represented in the (N + 1)-specification, while other properties are introduced in the design step. Therefore relevant design decisions and their impact on the specifications are investigated, leading to classes of potential standard transformations. Transformations which obey the conformance criteria are called *Correctness Preserving Transformations*.

On the level of LOTOS specifications we have a problem that can be described as: given an (N) specification, a certain correctness preserving relation, and characteristics of the (N + 1) specification, derive an (N + 1) specification with these characteristics and that is in the correctness preserving relation with the (N) specification. The formalization of these transformation problems can enable the development of tools to support the transformations. Below we present some examples of transformations:

5.4.1. *Interaction Decomposition*. Interactions in LOTOS are supposed to be synchronous and atomic, while some implementation environments do not support atomicity. This requires in some cases the atomic interactions to be decomposed into a sequence of interactions, so that atomicity properties related to the original interaction can be found back in this resulting sequence of interactions. This defines an interesting class of transformation problems, which we call *interaction decomposition*.

5.4.2. *Functionality Decomposition*. Classically we know that systems must be decomposed in order to be implemented. Functionality decomposition must meet the quality criteria already introduced above, but it is also guided by non-formalized properties of the system at a certain (N)-level of abstraction. In the case of functionality decomposition we consider that relations like bisimulation or weak-bi-simulation (Annex B of [7]) are the candidates for the correctness preserving relation, while characteristics of the (N + 1)-specification are the existence of internal hidden gates. Some attempts to define algorithms in order automatize this transformation were already made and will be further explored ([10],[11]).

5.4.3. *Resolution of Non-determinism*. The support of a partially top-down design trajectory implies that the designer needs language mechanisms to explicitly describe sets of possible implementation decisions, describing the system in terms of possible behaviour according to these pending design decisions. This characterizes *non-determinism*.

Non-determinism has to be eliminated during the design trajectory by the proper selection of the design alternatives, since implementation environments in general do not support it. This defines another interesting class of transformation problems.

Concluding, our investigation in the area of transformation aims at identifying aspects of these transformations that can be formalized, and developing theories, techniques and eventually tools for (semi-)automatic transformations. Other examples of generic design decisions for which we can envisage the corresponding standard transformations are interaction point decomposition, interaction points integration, functionality integration, functionality rearrangement, compatible representation of data and multi-way to two-way synchronization.

5.5. Testing and Verification Methods

Some properties that are described in the (N)-level design are supposed to be preserved in the (N + 1)-level design, characterizing *conformance*. Globally the real system must also conform to the user requirements.

The use of LOTOS in principle enables conformance to be verified analytically. In practice it appears that analytical verification of complex designs is still beyond the reach of pragmatic methods. Current technology allows the partial testing for conformance, reaching a reasonable degree of confidence in the correctness of the (N + 1)-level design.

5.6. Pre-defined Specification Constructs

Experience in the specification of architectural concepts like services, protocols, service access points, etc. ([8],[9]) can be used to form a useful library of architectural constructs. This library contains entries that can be customized by designers according to their specific needs. This method aims at improving the efficiency of the design steps, mainly in those cases in which the specification has still to be completely elaborated.

5.7. Pre-defined Implementation Constructs

A very promising approach towards the translation from a LOTOS implementation to an implementation is to identify and define implementation constructs expressed in (one of) the implementation language(s), which exhibit the properties that are required by constructs expressed in the LOTOS implementation. Preservation of properties can be checked e.g. by way of testing or by mathematical proofs.

This makes it possible to build a library of pairs, each pair consisting of a LOTOS construct and its corresponding translation into a correctness preserving implementation language construct. We call these pairs *pre-defined implementation constructs*. It may be possible that for each LOTOS construct there are several valid implementation constructs, since implementation languages can generally describe details that cannot be addressed using LOTOS, leading to a set of implementation choices.

The availability of pre-defined implementation constructs makes it possible to guide the implementation phase towards a LOTOS implementation that contains, or is only composed of, these pre-defined implementation constructs. Tools to support (semi-)automatic translation can also be developed.

Examples of pre-defined implementation construct are components that continuously receive inputs and deliver outputs according to some function (instance of sequential processing), schedulers, queues, stacks, and finite state machines. In its extreme form the library contains only the LOTOS language elements and their translation, in which case it effectively forms a LOTOS compiler. Care should be taken however not to apply compilation in a too early stage of a design trajectory as it may prevent the consideration of important design decisions at the appropriate abstraction level and lead to low quality implementations.

6. Conclusions

The use of a design language based on a formal model such as LOTOS may bring drastic improvements to the quality of the design process. Its mere usage, however, cannot relieve the designer of his prime design responsibilities: i.e. the actually making of design decisions. Apart from the treatment of non-formalized properties of an (N)-level design, the designer must also consider in each design step the level of abstraction of the (N + 1)-level design, qualitative design criteria, the specification style to be applied, and the availability of pre-defined specification constructs, generic transformations or pre-defined implementation constructs. Therefore the Lotosphere design methodology intends to provide a framework in which design decisions can be considered and evaluated, aiming at enhancing the quality of the design process.

7. References

[1] G.A. Blaauw and F.B. Brooks. *Computer Architecture*, Lecture Notes. Technische Hogeschool Twente - Informatica. Enschede, the Netherlands, Fall 1985.

[2] K. Bogaards. LOTOS Supported System Development. In K. J. Turner (ed.). *Formal Description Techniques*. North-Holland, 1989, pages 279-294.

[3] K. Bogaards. *A Methodology for the Architectural Design of Open Distributed Systems - PhD Thesis*. University of Twente, June 1990.

[4] P. van Eijk, C.A. Vissers, M. Diaz, *The Formal Description Technique LOTOS, Results of the ESPRIT SEDOS Project*, North-Holland, 1989.

[5] P. van Eijk. *Software Tools for the Specification Language LOTOS - PhD Thesis*. Twente University of Technology, Enschede, the Netherlands, 1988.

[6] P. van Eijk. Tools for LOTOS specification style transformation. In S.T. Voung (ed.). *The 2nd International Conference FORTE'89 on Formal Description Techniques for Distributed Systems and Communication Protocols - Participants Proceedings*. 1989, pages 54-62.

[7] International Organization for Standardization. *IS 8807 - Information Processing Systems - Open Systems Interconnection - LOTOS - A Formal Description Technique Based on the Temporal Ordering of Observational Behaviour*, 1988.

[8] ISO/IEC JTC1/SC6/WG4/Ad Hoc Group. *Formal Description of ISO 8073 (Transport Protocol) in LOTOS*, ISO/IEC JTC1/SC6/WG4/N233, August 1988.

[9] ISO/TC97/SC6/WG4/Ad Hoc Group. *Revised text of ISO/PDTR 10023 - Formal Description of IS 8072 (Transport Service) in LOTOS*. ISO/TC97/SC6/N5533, June 1989.

[10] R. Langerak. Decomposition of functionality: a correctness preserving LOTOS transformation. In *Tenth International IFIP WG6.1 Symposium on Protocol Specification, Testing and Verification*, IFIP, 1990.

[11] J. Parrow. Submodule construction as equation solving in CCS. *Theoretical Computer Science 68*, pages 175-202, 1989.

[12] J. Schot and L.F. Pires. *Pangloss - Architectural Task Final Deliverable*. Enschede, the Netherlands. December 1989.

[13] K.J. Turner. A LOTOS-based development strategy. In S.T. Voung (ed.). *The 2nd International Conference FORTE'89 on Formal Description Techniques for Distributed Systems and Communication Protocols - Participants Proceedings*. 1989, pages 157-174.

[14] C.A. Vissers et al. *The Architecture of Interaction Systems*, Lecture Notes. Enschede, the Netherlands, February 1989.

[15] C. A. Vissers, G. Scollo, M. v. Sinderen, Architecture and Specification Style in Formal Descriptions of Distributed Systems. In S.Aggarwal and K. Sabnani (eds.). *Protocol Specification, Testing, and Verification, VIII*. North-Holland, 1988, pages 189-204.

[16] C.A. Vissers. FDTs for Open Distributed Systems, a Retrospective and a Prospective

View, to appear in *Proceedings of IFIP WG6.1 Symposium on Protocol Specification, Testing and Verification X*, Ottawa, Ontario Canada, June 1990.

[17] P. Zave. The Operational Versus the Conventional Approach to Software Development. *Communications of the ACM 27*, 2 (February 1984), pages 104-118.

Project No. 2469

TEMPORA - Integrating Database Technology, Rule Based Systems and
Temporal Reasoning for Effective Software

P. Loucopoulos
Department of Computation
UMIST
P.O. Box 88
Manchester M60 1QD
U.K.

P. McBrien
Department of Computing
Imperial College
180 Queen's Gate
London SW7 2BZ
U.K.

U. Persson
SISU
P.O. Box 1250
16428 Kista
SWEDEN

F. Schumacker
Service d' Informatique
Université de Liege
Institute Montefiore
B28, B-4000 Liege
BELGIUM

P. Vasey
Logic Programming Associates
The Royal Victoria Patriotic Building
Trinity Road
London SW18 3SX
U.K.

ABSTRACT. Recent years have witnessed a growing realisation that the development of large data-intensive, transaction oriented information systems is becoming increasingly more difficult as user requirements become broader and more sophisticated. Contemporary approaches have been criticised for producing systems which are difficult to maintain and which provide little assistance in organisational developments.

The TEMPORA project advocates an approach which explicitly recognises the role of organisational policy within an information system and visibly maintains this policy throughout the software development process, from requirements specifications through to an executable implementation.

This paper introduces the philosophy and architecture of the TEMPORA paradigm and describes the conceptual models, tools and run-time environment which render such an approach a feasible undertaking.

ACKNOWLEDGEMENTS. The work reported in this paper has been partly funded by the Commission of the European Communities under the ESPRIT R&D programme. The TEMPORA project is a collaborative project between: BIM, Belgium; Hitec, Greece; Imperial College, UK; LPA, UK; SINTEF, Norway; SISU, Sweden; University of Liege, Belgium and UMIST, UK. SISU is sponsored by the National Swedish Board for Technical Development (STU), ERICSSON and Swedish Telecomm.

This paper summarises the contribution to the TEMPORA project of the following: R. Andersen, P. Bergsten, J-L Binot, J. Bubenko Jnr, G. Diakonikolaou, R. Jenssen, V. Kopanas, P. Loucopoulos, P. McBrien, M. Niezett, R. Owens, U. Persson, F. Schumacker, A-H Seltveit, G. Sindre, A. Sølvberg, U. Sundin, B. Theodoulidis, G. Tzialas, P. Vasey, R. Venken, B. Wangler, R. Wohed, P. Wolper.

1. Introduction - Philosophy and Architecture of TEMPORA

Much has been written about the benefits and problems of contemporary approaches to the development of large scale information systems [Maddison, 1983; Olle et al, 1983, 1986]. Despite the undeniable benefits that these approaches 2have brought about two challenges still remain unresolved. First, there is very little explicit correspondence between business 'processes' and information systems. This has the effect that neither the information system can be easily examined at any time to ascertain whether it is still aligned to business practises nor can the information system be effectively used in helping develop new enterprise designs. Second, the issue of maintenance and system evolution continues to be a major problem in commercial software. A large part of this problem is the way that the very same factors (mostly business policy) that are the cause of change are not explicitly maintained and even worse their representation in software is embedded together with programming code whose function is concerned with issues such as file accessing, input and output, sequencing of operations, data integrity and so on. A relatively simple example studied by Anderson et al [1986] which was concerned with 8 business policy rules for implementing a payroll system, required 3,500 lines of COBOL program to implement it. The simplicity of the problem at the conceptual level (8 business rules) was counterbalanced by complex and voluminous code much of which had little to do with the problem in hand but rather with its efficient implementation. Many of the rate determination rules were implemented by the order of the program statements. Furthermore, few people could check the correctness of the implemented policy because few people with the knowledge of the rate calculation system were able to understand the implementation. Finally, maintenance of the program was difficult, since the program described a procedure to determine rates of pay, rather than containing the policy of calculation and how and when this policy was to be activated (a problem well documented elsewhere e.g. [Fjeldstad, 1979]).

To address both these problems, TEMPORA proposes that developers must be provided with a process which assists in modelling business policy and linking this policy to the software development process. TEMPORA advocates an approach which explicitly recognises the role of business policy within an information system and visibly maintains this policy throughout the software development process, from requirements specifications through to an executable implementation. The need for such a paradigm has been recognised in the, now completed ESPRIT project RUBRIC [van Assche et al, 1988; Loucopoulos, 1989]. The TEMPORA project builds on the RUBRIC paradigm and extends this work in two directions. The first direction is concerned with the utilisation of a commercial DBMS as the underlying data management mechanism. The second direction is concerned with enhancing the paradigm with the explicit modelling of temporal aspects at both specification and application levels.

The TEMPORA architecture is shown in figure 1. An analyst/designer develops an information system specification using an interactive CASE tool environment which incorporates three tools, two of which correspond to the conceptual level in terms of the ERT (Entity-relationship-time) and process (external rule language) models and one which is concerned with the specification of application related components. Section 2 of the paper discusses the conceptual models of TEMPORA. The ERT and process part of the specification are expressed in such a way so as to enable one to reason at the business level (external level). Each component of the specification is mapped onto an execution layer (design level) which deals with data and execution mechanisms from a database schema perspective. Section 3 describes briefly the CASE shell which is used in order to develop the TEMPORA specification and analysis tools. The actual mechanics of storing data, implementing constraints and executing temporal as well as non-temporal rules is handled by the run-time environment which consists of the SYBASE DBMS and its interface to BIM_Prolog and an extension module referred to as the rule manager. The run-time environment is described in section 4.

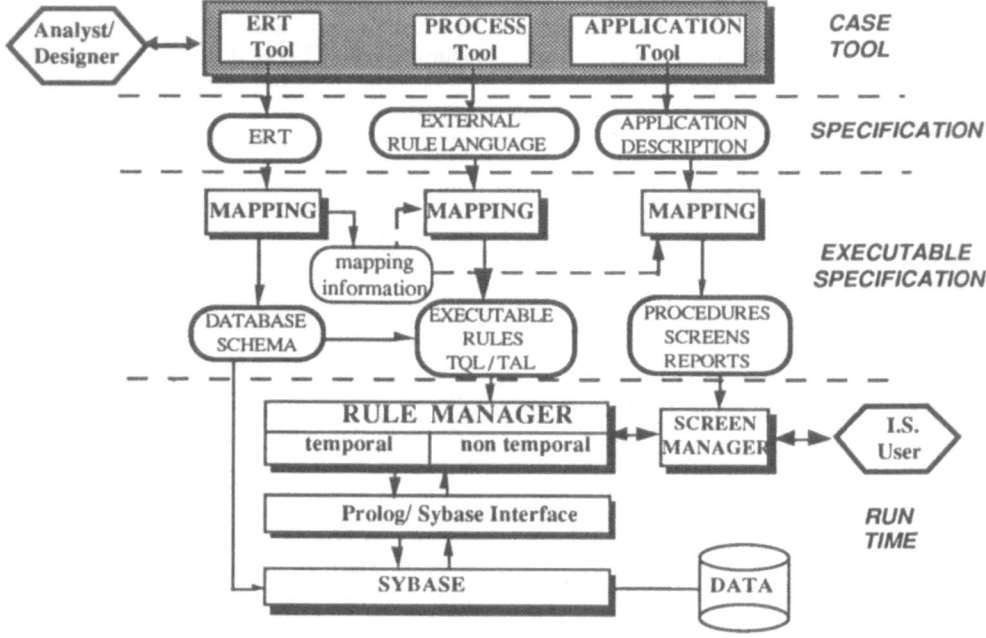

Figure 1: The TEMPORA Architecture

2. Conceptual Models

2.1 THE ENTITY-RELATIONSHIP-TIME MODEL (ERT)

2.1.1 Basic Concepts

The components of ERT are defined as follows:

- Entity is anything, concrete or abstract, uniquely identifiable and being of interest during a certain time period. Entity Class is the collection of all the entities to which a specific definition and common properties apply at a specific time period.

- Relationship is any permanent or temporary association between two entities or between an entity and a value. Relationship Class is the collection of all the relationships to which a specific definition applies at a specific time period.

- Value is a lexical object perceived individually, which is only of interest when it is associated with an entity. That is, values cannot exist in their own. Value Class is the proposition establishing a domain of values.

- Time Period is a pair of time points expressed at the same abstraction level. Time Period Class is a collection of time periods.

- Complex Object is a complex value or a complex entity. A complex entity is an abstraction (aggregation or grouping) of entities, relationships and values (complex or simple). A complex value is an abstraction (aggregation or grouping) of values (complex or simple). Complex Object Class is a collection of complex objects.

An entity or relationship can be derived. This implies that its value is not stored by default. For each such derivable component, there is a corresponding derivation rule which gives the members of this class or the values of this relationship at any time.

Time is introduced in the ERT model as a distinguished entity class. More specifically, each time-varying entity class and each time-varying relationship class is time stamped with a time period class. That is, a time period is assigned to every time-varying piece of information that exists in a schema. For example, for each entity class a time period is associated which represents the period of time during which an entity is modelled (existence period of an entity). The same argument applies also to relationships i.e., each time-varying relationship is associated with a time period which represents the period during which the relationship is valid (validity period of a relationship).

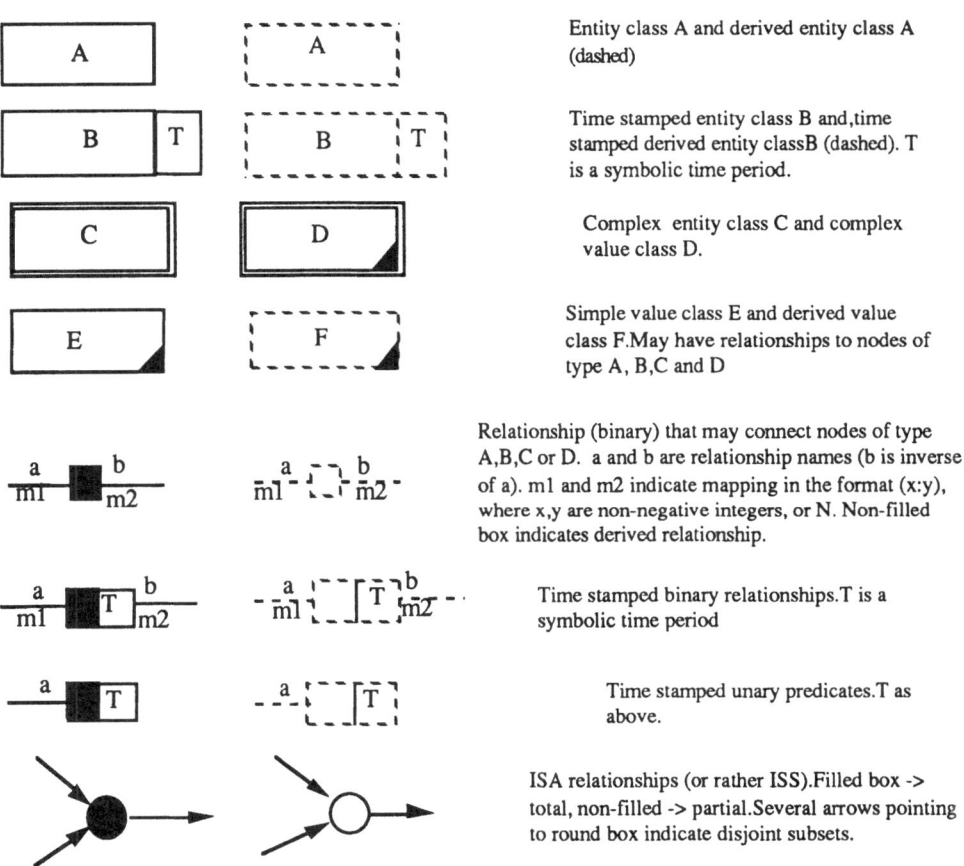

Entity class A and derived entity class A (dashed)

Time stamped entity class B and,time stamped derived entity classB (dashed). T is a symbolic time period.

Complex entity class C and complex value class D.

Simple value class E and derived value class F.May have relationships to nodes of type A, B,C and D

Relationship (binary) that may connect nodes of type A,B,C or D. a and b are relationship names (b is inverse of a). m1 and m2 indicate mapping in the format (x:y), where x,y are non-negative integers, or N. Non-filled box indicates derived relationship.

Time stamped binary relationships.T is a symbolic time period

Time stamped unary predicates.T as above.

ISA relationships (or rather ISS).Filled box -> total, non-filled -> partial.Several arrows pointing to round box indicate disjoint subsets.

Figure 2: Graphical Notation for the ERT Model

Figure 2 presents the notation for the ERT externals. Note that the graphical notation caters for the representation of some of the most common rules such as partial/total ISA relationships and cardinality constraints.

The ERT model accommodates explicitly generalisation/specialisation hierarchies. This is done through the ISA relationship which has the usual set-theoretic semantics. More specifically, it is assumed that two subclasses of the same entity class and under the same specialisation criterion are always disjoint.

Cardinality constraints may be given to all relationships (including the IS_PART_OF relationship) and also for their respective inverse relationships. Note here that there is no separate notation for the IS_PART_OF relationships. However, their corresponding cardinality constraints are interpreted in a slightly different way. This is explained in more detail in the next section.

2.1.2 Complex Objects

In general, complex objects can be viewed from at least two different perspectives [Batini, 1988]: representational and methodological. The representational perspective focuses on the way entities in the real world should be represented in a conceptual schema and on the way events in the real world are mapped onto operations on the corresponding objects. In contrast, if complex objects are not allowed, like e.g., in the relational model, then information about the object is distributed and operations on the object are transformed to a series of associated operations. The methodological perspective treats the complex object concept as a means of stepwise refinement for the schema and for hiding away details of the description. This in turn, implies that complex objects are merely treated as abbreviations that may be expanded when needed. In the context of the ERT model, complex objects are treated in terms of the methodological interpretation i.e. they serve as a convenient abstraction mechanism.

The notation of a complex object in ERT is shown in figure 3. The example ERT diagram shows a complex entity class CAR and a complex value class ADDRESS. Furthermore, the complex objects CAR and ADDRESS may be viewed at a more detailed level as shown in figure 4.

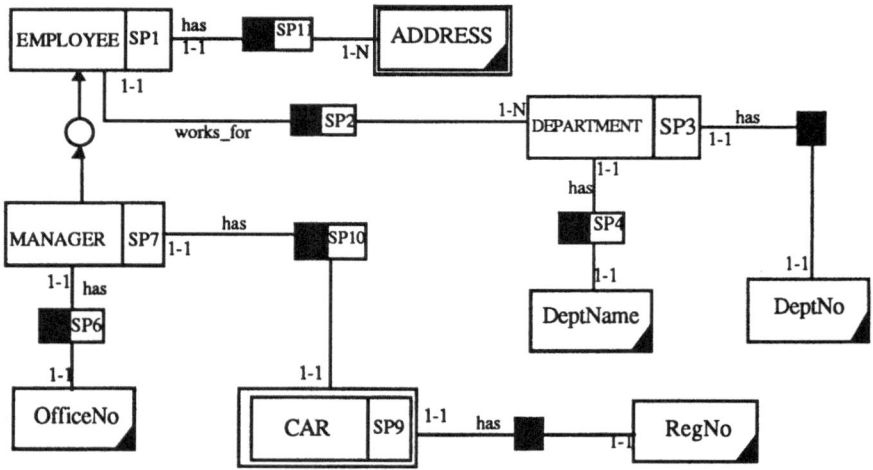

Figure 3: An ERT Diagram

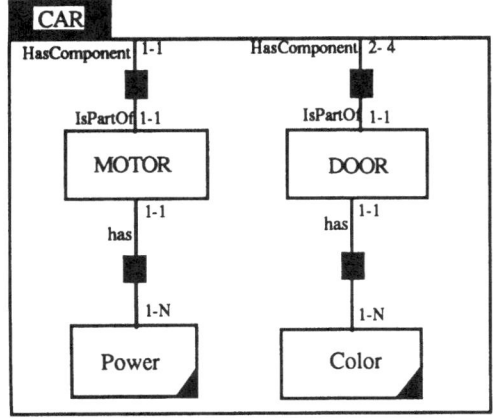 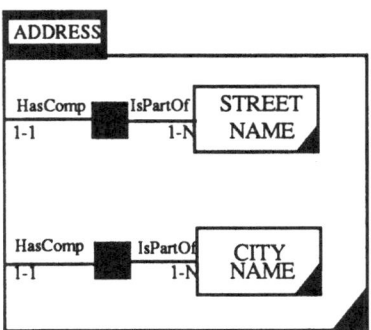

Figure 4: Components of Complex Objects CAR and ADDRESS

The components of a complex object comprise one or more hierarchically arranged substructures. Each directly subordinate component entity must be is_part_of-related to the complex object border so that the relationship between the composite object and its components will be completely defined. Whether the HasComponent relationship is one of aggregation or grouping, can be shown by means of the normal cardinality constraints. That is, if its cardinality is 0-1 or 1-1 the component is aggregate whereas if its cardinality is 0-N or 1-N the component is a set.

Most conceptual modelling formalisms which include complex objects [Kim et al, 1987; Lorie, 1983; Rabitti et al, 1988], model only physical part hierarchies i.e, hierarchies in which an object cannot be part of more than one object at the same time. In ERT, this notion is extended in order to be able to model also logical part hierarchies where the same component can be part of more than one complex object. To achieve this, four different kinds of IS_PART_OF relationships are defined according to two constraints, namely the *dependency* and *exclusiveness* constraints. The dependency constraint states that when a complex object ceases to exist, all its components also cease to exist (dependent composite reference) and the exclusiveness constraint states that a component object can be part of at most one complex object (exclusive composite reference). That is, the following kinds of IS_PART_OF variations [Kim, 1989] are accommodated:

- dependent exclusive composite reference

- independent exclusive composite reference

- dependent shared composite reference

- independent shared composite reference

Note that no specific notation is introduced for these constraints. Their interpretation comes from the cardinality constraints of the IS_PART_OF relationship. That is, assume that the cardinality of the IS_PART_OF relationship is (a,b). Then, a=0 implies non dependency, a≠0 implies dependency, b=1 implies exclusivity while b≠1 implies shareness.

Finally, the following rules concerning complex objects should be observed:

1. Complex values may only have other values as their components. In addition, the

corresponding IS_PART_OF relationship will always have dependency semantics unless it takes part in another relationship.

2. Complex entities may have both entities and values as their components. Every component entity must be IS_PART_OF-related to the complex entity.

3. Components, whether entities or values, may in turn be complex, thereby yielding a composition/decomposition hierarchy.

2.1.3 Time stamping semantics

The time period representation approach has been chosen because it satisfies the following requirements [Villain, 1982; Villain, 1986; Ladkin, 1987]:

- Period representation allows for imprecision and uncertainty of information. For example, modelling that the activity of eating precedes the activity of drinking coffee can be easily represented with the temporal relation before between the two validity periods [Allen, 1983]. If one tries, however, to model this requirement by using the line of dates then a number of problems will arise since the exact start and ending times of the two activities are not known.

- Period representation allows one to vary the grain of reasoning.

The modelling of information using time periods takes place as follows. First, each time varying object (entity or relationship) of ERT is assigned an instance of the built-in class SymbolPeriod. Instances of this class are system-generated unique identifiers of time periods e.g. SP1, SP2, etc. Members of this class can relate to each other by one of the thirteen temporal relations between periods [Allen, 1983]. The class CalendarPeriod has as instances all the conventional calendric periods e.g., 10/3/1989, 21/6/1963, etc. Members of this class are also related to each other and to symbol periods by one of the thirteen temporal relations between time periods.

Figure 5, shows graphically the definition of these concepts using the ERT notation. The symbol τ represents a temporal relationship and the symbol τi its inverse. The fact that the two classes *SymbolPeriod* and *CalendarPeriod* are disjoint is also indicated in the diagram. Note however, that for reasons of clarity, the exact definition of the calendar period units is not included. A date format like 21/6/1963 is just a shorthand notation of a calendar period.

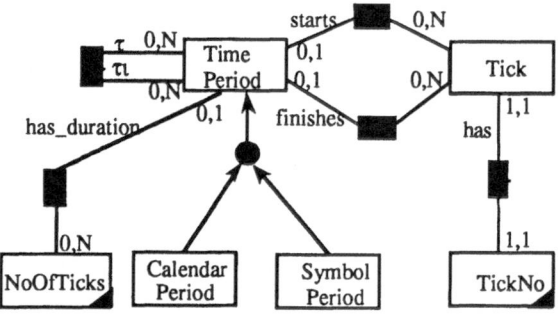

Figure 5: Time Period Metamodel

According to the above, the information in a conceptual schema is time stamped using symbol period identifiers. However, no distinction is made between time periods and time points. The fact that the abstraction level of a SymbolPeriod time stamp is say *day* can be inferred by its constraining temporal relations. The fact that an entity is time stamped only at the day abstraction level can still be represented by distinguishing between different *SymbolPeriod* subclasses according to their abstraction level i.e., SPD, SPM,..etc. (this last notation is not represented in the example of figure 3). This form of constraint is a resolution constraint which when applied to a SymbolPeriod class restricts its members to calendar periods of the same duration.

It is suggested that it would be convenient to represent directly in the conceptual schema some other notions of time such as duration and periodic time. The consequence of this is that the expressive power of the ERT external formalism is increased and so does the readability of the schema. In [Theodoulidis et al, 1990] there is a more detailed description of the time semantics of ERT.

In the ERT model, value classes and the IS_PART_OF relationships in a complex value class should always be time invariant. This is because an aggregation or grouping of values is defined through the participating value components. These assumptions affect the way that an ERT schema is mapped onto a relational schema. As discussed already, the validity period of a relationship should be a sub-period of the intersection of the existence periods of the involved entities. This does not hold for the ISA relationship since from the semantics discussed above, the existence period of the specialised entity should be a sub-period of the existence period of its generalisation and that the ISA relationship is always time invariant.

Time stamping, when applied to derived ERT components has slightly different semantics than usual. Since, the derived components are not stored by default, the interpretation of time stamps refers to their corresponding derivation formulas. That is, if a derived component is not time stamped then the derivation formula returns the value of it at all times i.e, for every valid state of the database. Alternatively, for the time stamped derived components, the derivation formula returns a value which is valid for the existence or validity period of this component. i.e. the derivation formula must refer to this period.

Finally, time stamping in a time varying IS_PART_OF relationship is translated to the following constraints. The dependency constraint in a time varying IS_PART_OF relationship means that:

The existence periods of the complex object and the component object should finish at the same time with the validity period of the IS_PART_OF relationship.

Also, the exclusiveness constraint is translated to:

If an object A is part of the complex objects B and C , then the period during which A is part of B should have an empty intersection with the period during which A is part of C.

Summarising, the above presented time semantics permit us to keep historical information for the UoD, include a strong vocabulary for expressing temporal requirements and also, model the evolution of complex objects through time in a natural way.

2.2   THE EXTERNAL RULE LANGUAGE

In using a temporal logic rule based language (TL) to control the processing of information, TEMPORA has allowed for the representation of business rules of an information system in a highly declarative manner. However, the rules inherently contain some 'design' information, by the way that they are structured as executable rules. These structural differences are necessary to reflect different procedural interpretations that can be given to a business rule. For example, consider a business rule 'the number of chairs must equal the number of tables', which leads to the following alternative procedural interpretations:

(i)   If there are x tables then make the number of chairs x.
(ii)   If there are x chairs then make the number of tables x.
(iii)   If I know that there are x tables then I can assume there are x chairs.
(iv)   If I know that there are x chairs then I can assume there are x tables.
(v)   An error has occurred if ever the number of chairs differs from the number of tables.

Although some interpretations are possibly more intuitive, one may choose any one from the list as the procedure to follow in a practical implementation, having knowledge of how other rules may affect the situation.

Business rules, and the corresponding aspect of the executable rules, may be classified into three different classes:

- **Action Rules** which imply some action must be taken if some condition holds, and will be modelled by a full TL rule of the form *condition* $\rightarrow$ *action*. The first two rules from the above list are examples of this class.

- **Derivation Rules** which express that some fact holds if some group of other facts hold, and are used during execution in evaluating the condition part of TL rules, and are represented by a rule in the condition language of the TL of the form *condition derived* $\Leftarrow$ *condition*. Rules (iii) and (iv) from the above list belong to this class.

- **Constraint Rules** which specify that some condition must not be violated. These are used during execution to validate the actions being taken. If the constraint would be violated, then the rule will *rollback*, and cause an error to be raised. The final rule of the above list belongs to this class.

At the conceptual level there is not this distinction between different classes of rule, and so we introduce the notion of an External Rule Language (ERL), which can model different classes of executable rules in a uniform manner. We leave the decision of which class a rule belongs to until the design phase, and thus allow the specification of the rules in our system to closely mirror the business rules from which they are derived. This two level approach allows users of the system to inspect the rules in a form which they comprehend, but still gives the programmer the procedural control over execution, in choosing the interpretation given to an ERL rule when translated into a TL rule. To summarise, the ERL rules closely match the business rules, and the TL rules closely match the procedural interpretation given to the business rule in the design phase.

The presence of the ERL also allows us two other important benefits:

- We may express the external rules (that the user views) as manipulating data in the ERT model, but have our executable rules manipulate data in the database model, and thus be more efficient to execute.

- We may heavily sugar the syntax of the ERL to give a semi-natural language flavour, whilst leaving the internal rule language in a more concise form that programmers would desire.

2.2.1   ERL Rules

There is one basic structure for all ERL rules, given by the following BNF definition, where the expressions in bold brackets are optional. Any free variables that appear in the rule have implicit universal quantification.

ERL_rule ::= [[WHEN <trigger_exp>] [IF <cond_exp>] THEN] <exp>

This leads to there being four valid variants of the basic ERL rule, listed here with their corresponding semantics.

- <exp>
  exp must always hold.

- IF <cond_exp> THEN <exp>
  exp must hold whenever *cond_exp* holds.

- WHEN <trigger_exp> THEN <exp>
  exp must hold when *trigger_exp* has just begun to hold.

- WHEN <trigger_exp> IF <cond_exp> THEN <exp>
  exp must hold if *cond_exp* holds, and *trigger_exp* has just begun to hold.

2.2.2 Referencing the ERT Model

To access the entities and values in the ERT model, a single general structure is used, defined by the BNF expression below, with the optional repeating sections in bold braces. Naming an entity or value class causes the access expression to hold for each instance of the class, and by enclosing a variable in parenthesis after the name to give the predicate form bindings of the variable can be obtained to each instance found. Enclosing a list of relationship names with other entities or values enables us to qualify our selection of instances by stating that the particular instance must be related to an instance of the other entity or value.

ERL_data_access ::= <entity/value name> [(<variable>)]
    [[<relationship> <ERL_data_access> { ,<relationship> <ERL_data_access> }]]

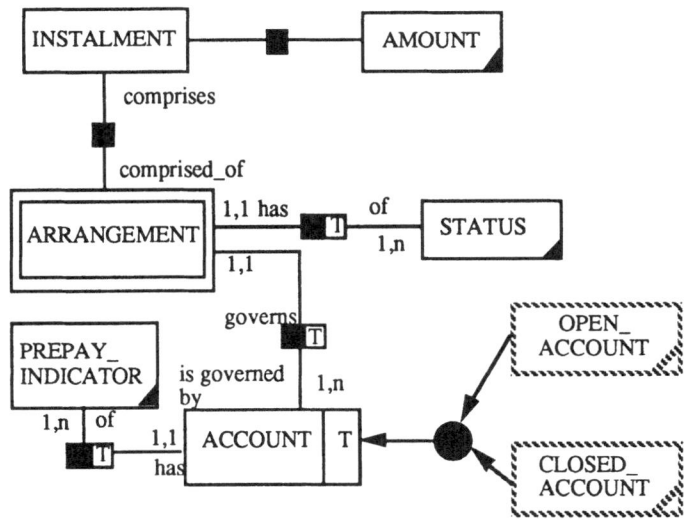

Figure 6: An Example ERT Model

As an example, consider the data access expression for the ERT of figure 6 which models an application that deals with the handling of arrangements about customers' accounts for a public utility organisation.

The following expression finds all pairs of *account* references *a* and *instalment* references *i*.

account(a) [is_governed_by arrangement [comprises instalment(i)]]

Note that we need only give variables to the entities or values we are interested in finding information about, so for the above example we were able to omit a variable for the *arrangement* entity. Also note that the enclosure of *comprises instalment(i)* in brackets is necessary to indicate that we expect the relationship *comprises* to be between *arrangement* and *instalment*, and not between account and instalment.

2.2.3 Collecting Information

A set collection construct is provided, together with a group of set operators. This allows us to group all variable substitutions for which an expression holds, and perform operations such as COUNT to count the number of instances, or INTERSECT to find common substitutions from two set expressions.

ERL_set_exp ::= { <var> { , <var> } | <exp> }

Thus we could find the number of accounts by the expression COUNT { x | account(x) }.

2.2.4 An Example of the ERL and TL

As an example of the use of ERL in interpreting business rules and of the use of the TL in executing these rules, consider the application modelled in the ERT model of figure 6.

Business Rules

An *arrangement* is an agreement between a customer and the public utility organisation for the customer to make some set of fixed payments (*instalments*) at given dates. Both the *instalments* and actual *payments* made are recorded in the database.

In the business rules listed here, the notion of a temporal database is used in order to be able to describe entities as be 'present' in the 'current' database or 'past' database, and thus phrase rules in terms of current records and old records. In fact all information is present in one database, but the TL provides a mechanism for it appearing as a series of databases or *states* over time (q.v. section on the TL).

(i)   The number of accounts is limited to 1,300,000.

(ii)   At most 25,000 accounts shall have arrangements.

(iii)   An open account is one which exists in the current records.

(iv)   A closed account is one which does not exist in the records at present, but existed at sometime in the past.

(v)   A clearby arrangement has only one instalment.

(vi)   A scheduled arrangement has more than one instalment.

(vii) If it is seven days after an instalment was due on a scheduled arrangement, and the sum of all payments made during the period of the arrangement is found to be less than the sum of all previous instalments and half of the last instalment, then the arrangement shall be terminated.

(viii) If it is seven days after an instalment was due on a clearby arrangement, and the sum of all payments made during the period of the arrangement is within £25 of the instalment amount, close the arrangement in the normal way.

ERL Rules

The business rules (i)-(viii) may be interpreted by an analyst into the following ERL rules, where all keywords of the language have been written in capitals. A short explanation of any new constructs of the language introduced follows each example rule.

(i) COUNT {a | account(a) } < 1,300,000

(ii) COUNT {a | arrangement(a) } < 25,000

(iii) IF account(a) THEN open_account(a)

(iv) IF NOT account(a) AND IN_PAST account(a)
THEN closed_account(a)

> /* NOT <exp> holds for variable substitutions for which exp does not hold.
> IN_PAST <exp> holds for each instance of exp held in the past. */

(v) IF COUNT {i | AT_ANY_TIME instalment(i) [comprises arrangement(a)] } = 1
THEN clearby(a)

> /* AT_ANY_TIME <exp> holds for each instance of exp holding at any time
> past, present or future. Note here that we derive an entity as holding in the
> present from information about entities which held in the past. */

(vi) IF COUNT {i | AT_ANY_TIME instalment(i) [comprises arrangement(a)] } > 1
THEN scheduled(a)

(vii) IF 7*DAYS AGO
(instalment [value amount(install), belongs_to arrangement(a)]
AND scheduled(a)
AND prev_instalments = SUM {old_v |
IN_PAST instalment [value amount(old_v), belongs_to arrangement(a)] }
)
AND payments = SUM { p | IN_PAST amount(p) [of account_movement
[type payment, charged_to account [governed_by arrangement(a)]]]
AND payments < prev_instalments + 0.5*install
THEN terminate_arrangement(a)

> /* <time> AGO <exp> holds for each instance of exp which held at time before
> the present. SUM <ERL_set_exp> finds the sum of all instances of the leading
> variable from the set expression variable list.*/

(viii) IF 7*DAYS AGO instalment [value amount(instal), belongs_to arrangement(a)]
AND clearby(a)

AND instal-25 < SUM { pay | IN_PAST amount(pay) [of account_movement
   [type payment, charged_to account [governed_by arrangement(a)]]]
THEN close_arrangement(a)

TL Rules

To complete the simple example of the arrangement handling system, a possible interpretation of
the rules in the TL is presented. It is during the design phase that the conceptual rules of the ERL
are translated to the more procedural rules of the TL. For the purposes of this example it has been
assumed that business rules (i) and (ii) are constraint rules, (iii) to (vi) are derivation rules, and
(vii) and (viii) are action rules. The TL rules make reference to a Relational Database (RDB)
schema and therefore a mapping is required from the ERT level to the RDB level (an automatic
generator is being developed in the project). The mapping chosen in this example has stored all the
values associated with each entity stored in a single table, with a surrogate field representing the
entity, and a table for each relationship between two entities containing the surrogates for the
entities.

(i)   declare esb_account(_,_,_,_,_) = database (cardinality : 0..1,300,000)

Constraints are not made to be part of a TL rule, but declared to the TL rule system, and used to
control the updates made as part of the actions of a rule. As part of the design phase we may also
wish to include the constraint information as part of rule conditions, to prevent the possibility of an
action attempting to violate the constraint. For instance, as part of the condition of a rule which is
intended to add a new account, we may check that the number of accounts already present allows
us to insert a new one without violating the constraint.

(ii)   declare arrangement(_,_) = database (cardinality : 0..25,000)

(iii)   open_account(Acc) ‹ esb_account(Acc,_,_,_,_,_).

A derivation rule can be stored as a clause of the Prolog-like language used to evaluate TL
conditions.

(iv)   closed_account(Acc)⇐¬esb_account(Acc,_,_,_,_,_)∧ ♦ esb_account(Acc,_,_,_,_,_).

The ♦ operator causes a search of past information to be made.

(v)   clear_by(Arr) ‹ count(• ♦ instalment(Arr,_) ∨ ◊ instalment(Arr,_),1).

The *count* predicate finds as its second argument the number of instances for which the expression
as its first argument holds. A TL formula of the form • ♦x ∨ ◊x causes all past, present and future
instances of x to be found.

(vi)   scheduled(Arr) ⇐ count(• ♦ instalment(Arr,_) ∨ ◊ instalment(Arr,_),N) ∧ N>1.

(vii)   in_past(7*days,
                    (instalment(Arr,Last_Amount)
                    ∧ arrangement(Arr,current)
                    ∧ governed_by(Esb_Acc,Arr)
                    ∧ • sum(Due,P instalment(Arr,Due),Total_Due)
)

∧sum(Payment,P(movement(Esb_Acc,payment,Payment))∧ arrangement(Arr,current)),
Total_Paid)

∧ Last_Amount*0.5+Total_Due > Total_Paid

→ O terminate_arrangement(Arr).

A full TL rule models an action rule. The in_past predicate finds if the expression given as its second argument held at a time before the present indicated by the first argument. The O operator specifies that the action takes place immediately.

(viii) in_past(7*Days,instalment(Arr,Amount))

∧ clear_by(Arr)

∧ governed_by(Esb_Acc,Arr)

∧ sum(Payment,P(movement(Esb_Acc,payment,Payment)

∧ arrangement(Arr,_)),Total_Paid)

∧ Amount-25 < Total_Paid

→ O close_arrangement(Arr)

3.   CASE Tool Support

The TEMPORA case environment is being implemented in a case-shell called RAMATIC, which is a "meta-tool" for case tool implementation developed by the Swedish Institute for Systems Development (SISU). RAMATIC includes a number of features to facilitate the implementation of a case tool for a particular, graphics oriented, method. For typical specification methods which are similar to SADT, ER-modelling, and hierarchical decomposition of data flows and processes, the time to create a case tool is in the order of weeks.

RAMATIC can be used to capture, and store in an integrated fashion, a wide range of different types of specification, be they graphical, form-oriented, tabular or pure text. This is evidenced by the current use of RAMATIC in a number of real projects in industry, where different kind of description and specification techniques are employed. It is also possible to include project control and quality control information such as design decisions, various annotations, information about designers/analysts, etc., in the specification. On the output side, various report forms can be defined as well as special checks to be performed. Various cross-reference matrices can be defined easily in order to display "where used" and "where created" information. For special analysis purposes it is simple to extend RAMATIC by analysis programs written in C or Prolog. The coupling to these extensions may be more or less tight depending on whether RAMATIC invokes them or communicates with them via files.

The core of RAMATIC is its design object data base. The design object data base consists of two parts, the conceptual data base (cdb) and the spatial data base (sdb). The cdb stores information about the incrementally developed specification in terms of design objects of a particular method, object relationships and attributes. The sdb contains information concerning how and where the various design objects are graphically represented.

A particular specification method is thus defined by design object types, object attributes, object relationship types, its syntax (textual as well as graphical), and constraints which concern completeness and consistency. The graphical syntax determines the class of graphical specifications permitted. The overall architecture of RAMATIC is given in Figure 7.

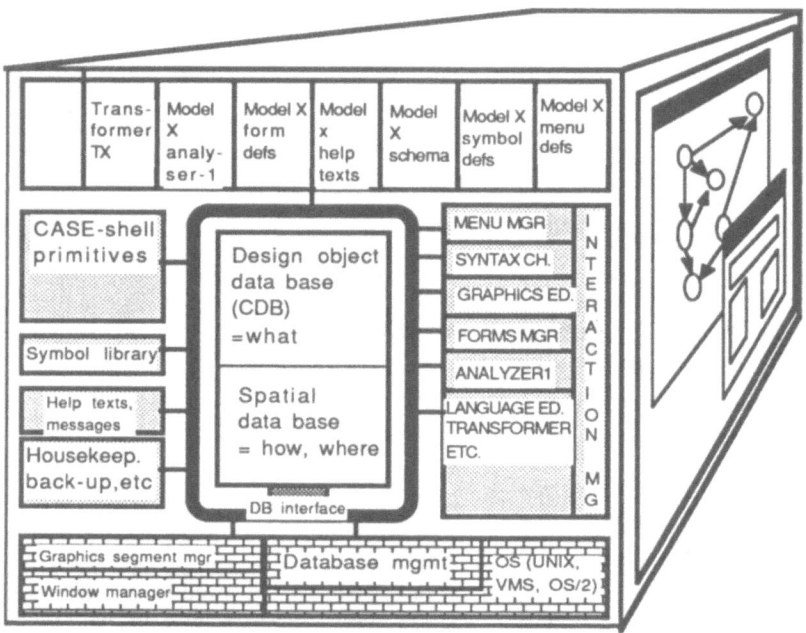

Figure 7: The RAMATIC Architecture

A methods engineer has the responsibility to supply information for the customising modules (the non-filled modules) of the architecture shown in Figure 7. For a particular method X, the method engineer has to define several components i.e. the types of design objects, attributes, and relationships needed, and the syntax constraints to be enforced. More complex (semantic and quality) constraints, or analysis of the specification requires the method engineer to write his own checking or analysis routines (in C or Prolog) by accessing the cdb contents.

The tool user, i.e. the system analyst and designer, will interact with the tool via a number of different kinds of screens for graphics input and output. The method engineer must define these kinds of screens and specify what kind of services may be obtained when using a screen of a particular type. For instance, when performing ERT modelling, the services needed are to create entity types, connect them by relationships, move them around, specify cardinality constraints, perform different kinds of changes, etc. These services are defined in the menu of that particular kind of screen.

A connection between a RAMATIC function and a menu item is defined in the menu definition. Pointing to a menu item then activates the corresponding shell function. At present, more than 100 general shell functions are available to the methods engineer. Below we illustrate a sample of such functions grouped by kinds of service they offer.

- Graphical functions, e.g Draw Line, Poly Line, Rotate etc.
- Symbol management concerns functions as, for instance, Copy Symbol, Enlarge etc.
- Groups of graphical objects may be, for instance, moved or copied.
- Picture management concerns copying and centring pictures etc.
- Window management functions provide a multiple screen environment.
- cdb management functions let us create new objects and browse old ones. It also possible

to connect two objects as equal.
- Syntactical checking according to defined rules for the actual model. This is a non interactive checking.
- Hard copy management and report management allow us to print the graphs and reports.

Another way to enter or retrieve (or "browse") design data is by the use of forms. The form facility operates on the cdb according to a form definition. Using the form definition facility, a methods engineer may define arbitrary ways of retrieving textual design information from the cdb as well as for entering textual information to the cdb. Forms may be defined for annotated descriptions of design objects such as entities, relationships, data flows, for retrieving where-used information for a particular object, etc. In addition, forms may be used for input of rules but also for verification and validation purposes.

RAMATIC is developed under Unix, but can also be run under VMS, Sun OS, Ultrix, Aix, and OS/2. It is ideally run under a windows system, i.e. X-Windows, DEC-windows, SunView, and Presentation Manager. However, it can also be run in a non-windows environment. In this case a graphics package must be available using one of the Core, GKS or Phigs "standards". In principle, RAMATIC can employ different kinds of DBMS to store and manipulate the contents of the cdb. Currently, the standard DBMS used is the binary associative DBMS cs5 of the dream/cs5 4GL application development tool. The DBMS cs5 is a separate program, which has to be adapted to the operating system separately. Communication with the DBMS is managed by a DBMS-interface, thus making RAMATIC independent of the actual DBMS used.

In TEMPORA, a requirement of the CASE tool is to provide facilities for the diagnosis and analysis of the conceptual schema. To this end, the DBMS cs5 is supplemented by the PROBE system which is an Objected-oriented layer developed on top of BIM_Prolog. Since the storage mechanisms of PROBE and RAMATIC are based on different approaches, a mapping from RAMATIC-cs5 internals to PROBE objects has been implemented. In a longer perspective RAMATIC has to be re-implemented to some extent in order to achieve a closer integration, i.e. to store the cdb/sdb directly in PROBE.

4. Run-time Environment

4.1 THE NON-TEMPORAL RULE MANAGER

The purpose of developing a non-temporal rule manager is one of run-time efficiency at the expense of expressiveness at the specification level. Where an application does not fully profit from the benefits of a temporal approach the potentially large run-time overhead involved in supporting temporal execution is unwarranted.

For an application to run in non-temporal mode certain restrictions need to be adhered to during the analysis and design phases. These restrictions relate to the ERT model and the ERL and TL rules. The restriction to ERT is that the time stamps are not considered (thus reducing the model to the ER model), whereas the restriction to ERL and TL rules is that the only temporal references concern the next time slot, which is equivalent to performing modifications only on the current database, and not on some future/past database. Furthermore, the action rules should always mention the trigger (WHEN part) to indicate the specific circumstance under which the rule should be considered. Given these restrictions at the specification level the application can benefit at run-time from a non-temporal rule manager.

The rules supported by the non-temporal rule manager are of the general form :-

WHEN <trigger_exp> IF <cond_exp> THEN <actions>

404

where the trigger specifies a modification (insert/delete/update) of a single entity in the ER diagram, and the conditions specify relationships between entities which must hold for the actions (further modifications and/or calls to foreign procedures) to take place.

The mathematical interpretation of such rules is :-

$$T \Rightarrow \forall x \, [\; C(x) \Rightarrow A(x) \;]$$

where x represents the free variables which are common to the conditions (C) and actions (A). It is the job of the rule manager to make all such interpretations true when reacting to a specific trigger (T). If any of the rules are violated for any instantiation of the variables x, then the database modification should be considered invalid and the retrieval process (e.g. a *rollback* of the database) initiated. This execution strategy can be refined to take account of active and dormant rule sets, thus enabling the dynamic rules of an application to be partitioned according to different tasks.

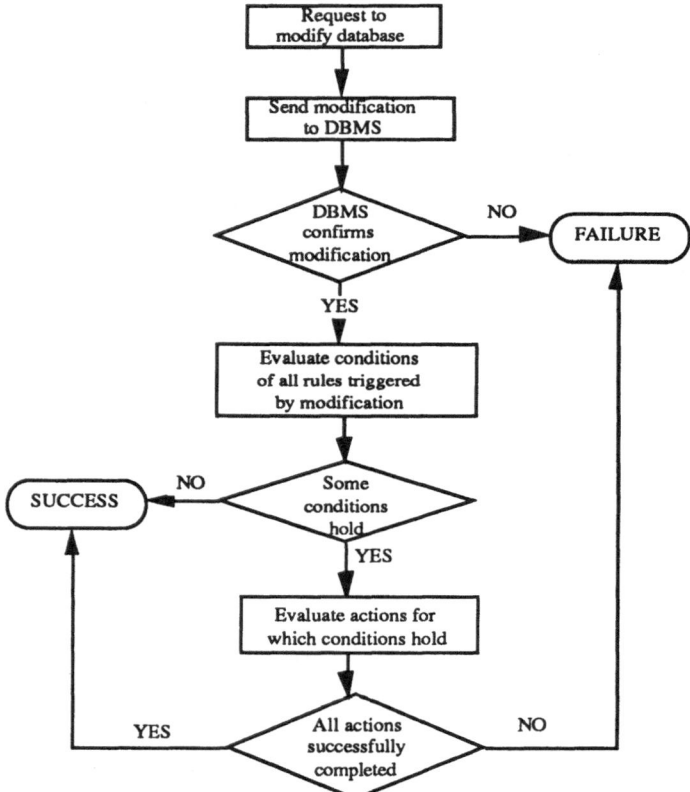

Figure 8: Reaction of Non-Temporal Rule Manager to a DB Modification

The diagram of figure 8 outlines the control flow of the non-temporal rule manager when reacting to a modification to the database. The SUCCESS label indicates that the modification is permitted, whereas the FAILURE label indicates that some recovery process is needed to restore the database back to a consistent state.

The external action rules of the specification, which make reference to the ER model are mapped directly onto Prolog data structures which refer to the database level. These data structures will be represented as Prolog clauses of the form :-

```
insert_rule( Tuple, Actions ) :-   Conditions .

delete_rule( Tuple, Actions ) :-
        Conditions .

update_rule( Tuple1, Tuple2, Actions ) :-
        Conditions .
```

As an illustration of the handling of DB updates the following demonstrates a high-level view of how this is carried out.

```
db_update( Tuple1, Tuple2 ) :-
        DBMS_update( Tuple1, Tuple2 ),
        forall(   update_rule( Tuple1, Tuple2, Actions ),
                call( Actions ) ),
        ! .

db_update( Tuple1, Tuple2 ) :-
        dbms_rollback .
```

This is a much simplified version of the rule manager which is being implemented, since there are two fundamental problems with the code given above. Firstly, the data structures representing the rules should be refined to reflect our knowledge that the Actions, just like the Conditions, are really code and not data. Secondly, the execution of Actions should be delayed until all of the Conditions of all triggered rules have been evaluated. This is to avoid the Actions of one rule affecting the evaluation of the Conditions of a later rule.

4.2   THE TEMPORAL RULE MANAGER

In TEMPORA action rules can refer to the temporal dimension of the database and thus are composed of two temporal components: the TQL (Temporal Query Language) and the TAL (Temporal Action Language). A TQL query is evaluated on the database and is concerned with the condition part of the rule. The action part is a TAL formula specifying actions to be taken.

The function of the Temporal Rule Manager (TRM) is to execute these rules is a way such that the system matches its specification. In other words, the TRM evaluates the queries and performs the actions necessary to ensure that the system actually behaves as stated in the rules.

4.2.1   A Basic Mechanism

Fundamental Assumption

As a first approach, it is assumed that the system is powerful enough to process an arbitrary large set of rules within a tick. This assumption will allow us to define a theoretical execution mechanism independently of any possible implementation constraint. Of course, in a real situation this assumption cannot be satisfied. Therefore, the basic model needs to be refined to make it practically feasible.

The Basic Execution Mechanism

Assuming that the previous hypothesis is satisfied, the basic execution mechanism can be viewed as the infinite repetition of an elementary cycle starting at each tick. The work performed during each cycle can be described as follows.

1. The Query Module (QM) evaluates the condition part of each rule.

2. For each rule R_i, the QM generates the corresponding action set[1] A^S_i.

3. The QM sends the global action set $(A^S_1 \cup A^S_2 \cup ... \cup A^S_n)$ to the Action Module (AM) and stops.

4. The AM analyses the different TAL formulae of the global action set and determines the sequence of actions to be performed directly. This operation includes several steps: consistency checking, scheduling, etc. The AM also keeps all (partial) TAL formulae describing actions to be taken at future ticks.

5. The AM sends the action sequence to the Transaction Module (TM) which performs the actions immediately.

6. Go to step 1 and start a new cycle at the next tick.

It is clear that this simple mechanism is sufficient to ensure that the system will match the specification described by the rules, because:

- the system re-evaluates all the rules at each tick and performs the necessary actions,

- the tick is the smallest time unit representable (i.e. everything is fixed in the system within a tick), and thus our execution mechanism will see all changes of the database.

If the choice of the tick as the basic cycle length appears as a sufficient condition to ensure that the system will match its specification, it will usually not be necessary to re-evaluate all the rules at every tick. This particular aspect is currently under investigation.

System Architecture

The architecture of the TRM can be represented by the diagram of Figure 9. Each sub-module has an auxiliary database attached to it, which contains all information necessary for the module to work properly.

---

[1] The action set A^S_i for a query Q_i is the set of TAL formulae $\{A^1_i,......A^k_i\}$ obtained from A_i for all instatiations of variables for which the query part is true. If the query cannot be satisfied, the action set is empty.

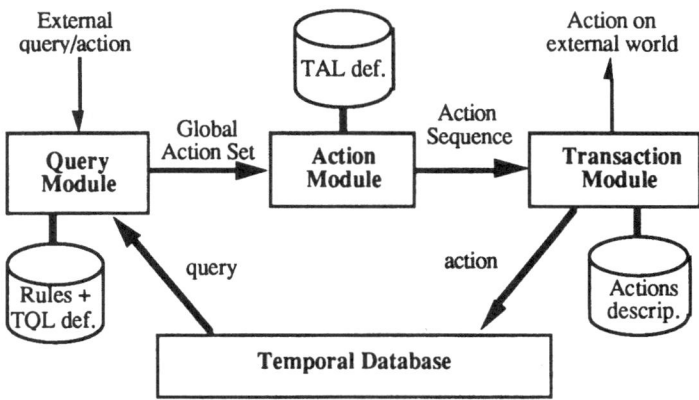

Figure 9: Temporal Rule Manager Architecture

4.2.2 Refined Execution Mechanism

Limits of the Basic Execution Mechanism

It is clear that the basic execution mechanism lacks realism. The processing capabilities of an actual computer system are always bounded, and so is the number of rules that can be processed within the duration of one tick (whatever duration we choose for the tick). Therefore, we have to modify our execution mechanism to take into account a possible delay of the temporal modules with respect to real time (i.e. the work will not be finished at the end of the tick).

The existence of a delay is specially important here, because we deal with temporal rules. For instance, if we consider the evaluation of a query, the evaluation begins in the current state of the database, which has been defined as the state that contains the current real time. It is obvious that starting the evaluation in a different state can lead to a different answer. Therefore, a query should always be evaluated at the right real-time.

The only reasonable way to solve this problem is to introduce two time references:

- the actual real time, denoted t_R, and

- a temporal module time, denoted t_{TM}, which is considered as the real time by the different temporal modules, but can actually be delayed with respect to t_R,

More precisely, t_{TM} can be defined as follows: after each evaluation cycle, t_{TM} is incremented by one tick; if t_{TM} is in advance of t_R, then the system waits until $t_R = t_{TM}$ to begin the next cycle, else the next cycle begins immediately.

When the system is (almost) idle, everything happens as if t_R was equal to t_{TM}. If the load of the system increases, the virtual clock t_{TM} slows down and t_{TM} is delayed with respect to t_R. However, if the system is not undersized, we can reasonably expect the system to catch up when the load decreases.

The Refined Execution Mechanism

The use of the virtual time reference t_{TM} for the temporal modules allows us to use our basic execution mechanism on an actual computer system without any modification, except for:

6. $t_{TM}=t_{TM}+1$ tick
    Wait while $t_{TM}>t_R$ then go to step 1.

The assumption under which the system can process an arbitrary large number of rules within a tick is verified because a tick no longer has a fixed length. Its length may vary according to the amount of work to be done.

The only impact of this modification on the behaviour of the system is that when $t_{TM} < t_R$, the system does no longer behave in "real time", i.e. at the real time t_R, the model present in the database is not up to date but corresponds to the model at time t_{TM}. Therefore, in most cases, we will have to buffer external requests until t_{TM} is greater than the time of submission, in order to make sure that these requests are evaluated on the proper model.

System architecture

Taking the architecture of Figure 9, we only need to add one input queue for every possible external input source (external query, action, signal, etc.). For example, when a query is submitted to the system, it is placed into the query queue and all the references to "NOW" are instantiated with the time (t_R) of submission. The queries of the queue are processed by the system (i.e. submitted to the QM) when t_{TM} is greater than the time of submission.

4.2.3 Implementing the Temporal Rule Manager

In the previous section, an execution mechanism was introduced which solves the problem of computing capacity by adopting a variable length for the internal tick. Even if this mechanism seems satisfactory, we must nevertheless implement the rule processing in an efficient way, because this mechanism is only usable if the average delay between t_{TM} and t_R cannot grow indefinitely.

Thus, in order to keep the average delay between t_{TM} and t_R very small, we need to reduce the amount of work to be done at every cycle as much as possible. This can be achieved in many different ways, depending on the particular temporal module considered.

The Query Module

The work of the QM is to select the rules to re-evaluate and to evaluate them. So, if we want to reduce the amount of work of the QM, we can:

1. select only the subset of rules that must be re-evaluated,

2. optimise the evaluation mechanism.

Selection of the rules

It is clear that most rules do not need to be re-evaluated at each cycle, because the value of the TQL part has not changed from one cycle to the next. In a classical database, we can trigger the evaluation of the rules with operations on the database (insert, delete, update). In TEMPORA, the problem is more complex because we have introduced a temporal dimension in our data. So, even if we do not make any modification to the database, the value of a query can change simply because time has changed.

The idea we will investigate in future work is to associate with each rule a set of "selectors" which help to determine when a rule needs to be re-evaluated. According to this mechanism, a rule will be evaluated only when one of its selectors is true .

Optimised evaluation

There are different ways to optimise the evaluation of queries.

- Incremental evaluation, i.e. use the results of previous evaluations as often as possible.

- Structuring of the rules, dependency graphs, etc.

- Global optimisation of the DB accesses.

Those aspects need further investigations.

The Action Module

The way the AM does the scheduling and the consistency checking of actions is also subject to atomisations. We will not develop this aspect here because it is highly dependent on the way the TAL will be implemented.

The Transaction Module

The TM is responsible for performing the sequences of actions issued by the AM. So far, we have supposed that the TM performs all the actions before giving control back to the QM. In fact, the evaluation of new queries must be suspended only until all operations on the database are finished. All other actions can be performed while the QM is working. This overlapping of the different sub-cycles can greatly reduce the length of the global cycle.

5.   Conclusions

Contemporary approaches to information system development, whilst attempting to improve the management of developing such systems through the use of software engineering methods and CASE, have paid little attention to the requirements for effective system evolution and for using information systems for effecting changes at the strategic level of organisations.

The process of software development can be viewed as a sequence of model-building activities. The quality of each set of models depends largely on the ability of a developer to extract and understand knowledge about an application domain which needs to be acquired form a diverse user population [Loucopoulos & Champion, 1988]. However, current technology does not provide any powerful formalisms or tools for supporting such a view. Even worse, the current fragmentation in development approaches has resulted in a situation where methodological knowledge is very difficult to obtain and use in a constructive manner.

It is also becoming obvious that current conceptual modelling formalisms are oriented towards the functional specification of the software system rather than the definition of the problem domain [Greenspan, 1984]. If improvements are to be made in the quality of software then the knowledge about the application domain must be formalised and explicitly encoded. To this end, TEMPORA follows the premise that information system development is about formalising and documenting

410

knowledge about the universe of discourse and this knowledge should be represented explicitly and independently to the way that it is implemented in data structures and algorithms thus leading to a more efficient way of developing and maintaining software.

The TEMPORA project is attempting to provide a better approach to building systems through the development of a software process and supporting tools which will explicitly accommodate those parts of a system that correspond to those elements of organisational policy which in essence impose changes to the structure and operation of information systems. Traditionally, ANSI/SPARC-style data management architectures and recent developments in HCI management tools have enabled the separation of system oriented elements from procedural code. The main initiative of the TEMPORA paradigm has been to extend this approach to include organisational policy. In particular, TEMPORA seeks to separate out and explicitly maintain throughout the software lifecycle, the notion of policy, as described by *constraint, derivation* and *action* rules.

This paper seeks to demonstrate how a set of business rules may be interpreted by analysts in terms of the objects that exist in the organisation and their structural relationships (using the ERT model) and the rules that are expressed by references to these objects (using the ERL). Furthermore, the paper demonstrates how this conceptually oriented specification can be translated into an executable specification and outlines the major components of a rule manager capable of handling transactions at the run-time application level.

References

[Allen, 1983]
 Allen J.F. *Maintaining Knowledge about Temporal Intervals*, CACM, 26(11) Nov.1983.
[Anderson et al, 1986]
 Anderson, M. et al *Report on Task A1: Research into the ability to use rules to describe the business and its activities*, E928/R2/FINAL, James Martin Associates, Brussels.
[Batini, 1988]
 Batini, C. and Di Battiste G. *A Methodology for Conceptual Documentation and Maintenance*, Information Systems, 13(3), pp.297-318, April 1988.
[Fjeldstad, 1979]
 Fjeldstad, R.K. et al *Application program maintenance*, in Parikh & Zveggintzov (1983) 'Tutorial on Software Maintenance', IEEE, pp.13-27.
[Kim et al, 1987]
 Kim W., Banerjee J., Chou H.T., Garza J.F., Woelk D. *Composite Object Support in Object-Oriented Database Systems*, in Proc. 2nd Int. Conf. on Object-Oriented Programming Systems, Languages and Applications, Orlando, Florida, Oct. 1987.
[Kim, 1989]
 Kim W., Bertino E., Garza J.F. *Composite Objects Revisited*, SIGMOD RECORD 18(2), June 1989.
[Ladkin, 1987]
 Ladkin, P. *Logical Time Pieces*, AI Expert, Aug, 1987, pp.58-67.
[Lorie, 1983]
 Lorie R., Plouffe W. *Complex Objects and Their Use in Design Transactions*, in Proc. Databases for Engineering Applications, Database Week 1983 (ACM), San Jose, Calif., May 1983.
[Loucopoulos, 1989]
 Loucopoulos, P. *The RUBRIC Project-Integrating E-R, Object and Rule-based Paradigms*, Workshop session on Design Paradigms, European Conference on Object Oriented Programming (ECOOP), 10-13 July 1989, Nottingham, U.K.
[Maddison, 1983]
 Maddison, R. *Information System Methodologies*, Wiley-Heyden.
[Olle et al, 1983]

Olle, T.W. et al (eds) (1983) *CRIS- Information System Design Methodologies: A Comparative Review*, North-Holland.

[**Olle et al, 1986**]

Olle, T.W. et al (eds) (1986) *CRIS3- Improving the Practice*, North-Holland.

[**Rabitti et al, 1988**]

Rabitti F., Woelk D., Kin W. *A Model of Authorization for Object-Oriented and Semantic Databases*, in Proc. Int. Conf. on Extending Database Technology, Venice, Italy, March 1988.

[**Theodoulidis et al, 1990**]

Theodoulidis, C., Wangler, B. and Loucopoulos, P. *Requirements Specification in TEMPORA*, To be published in the 2nd Nordic Conference on Advanced Information Systems Engineering (CAiSE90), Kista, Sweden, 1990.

[**Van Assche et al, 1988**]

Van Assche, F., Layzell, P.J., Loucopoulos, P., Speltincx, G. *Information Systems Development : A Rule-Based Approach*, Journal of Knowledge Based Systems, September, 1988, pp. 227-234.

[**Villain, 1982**]

Villain M.B. *A System for Reasoning about Time*, Proceedings of AAAI-82, Pittsburgh, Pa., Aug.1982.

[**Villain, 1986**]

Villain M.B., Kautz H. *Constraint Propagation Algorithms for Temporal Reasoning*, Proc. of AAAI-86, 1986.

Project No. 2474

ARCHITECTURE OF A MULTIMODAL DIALOGUE INTERFACE FOR KNOWLEDGE-BASED SYSTEMS.

J-L. Binot, BIM, Everberg Belgium,
P. Falzon, INRIA, Voluceau - Rocuqencourt, France
R. Perez, ISS, Barcelona, Spain
B. Peroche, Ecole des Mines, Saint-Etienne, France
N. Sheehy, University of Leeds, UK
J. Rouault, CRISS, Grenoble, France
M. Wilson, Rutheford-Appleton Laboratory, Chilton, UK

Abstract.

This paper describes the architecture and the first implementation results of a multimodal dialogue interface for knowledge based systems developed in the context of Esprit II project MMI2. The paper reviews the basic principles of the architecture of the system and the approach taken by the project with respect to user modeling issues, then describes individually each communication mode.

1. Overview of aims and basic approach

This paper outlines the first results of ESPRIT II project P2474: "MMI2: A Multi Mode Interface for Man Machine Interaction with knowledge based systems." These results were obtained through the cooperative efforts of all researchers involved in the project: Jean-Louis Binot, Fabienne Balfroid, Lieve Debille, David Sedlock and Bart Vandecapelle (BIM, Belgium), Gerard Henneron, Genevievre Lallich-Boidin, Rosalma Palermiti, Jacques Rouault, Jean-Louis Zinger (CRISS, France), Helmi Ben Hamara, Christian Bertin, Christine Jouve, Dominique Michelucci, Bernara Peroche (Ecole des Mines de Saint-Etienne, France), Bernadette Cahour, Francoise Darse, Pierre Falzon (INRIA, France), Alica. Manzanera, A. Moneta, Ricardo Perez, David Trotzig, Juan-Carlos Ruiz (ISS, Spain), Farah Arshad, N. Ghali, Mark Howes, K. Marida, Noel Sheehy (Univ. Leeds, U.K.), Helen Chappel, Graham Doe, Gordon Ringland, and Michael Wilson (Rutheford Appleton Laboratory, U.K.).

A multimodal system. The MMI2 project aims to build a man/machine interface for different kinds of users, integrating several modes of communication supported by modern workstations: natural language, command language, graphic and gesture. The interface will provide simultaneously modes suitable to support the efficiency of experienced, professional users (command languages, menus) and natural communication modes well suited to naive users, such as graphics and natural language. Natural language modules are being developed for English, French and Spanish.

Difference between modes and media. It seems first necessary to clarify the distinction that we make between the meanings of the two words medium and mode. A number of projects have already studied multi-media phenomena and in particular multi-media interfaces. The multi-media concept is present as soon as a computer system can deal with more than one type of input/output support. Multi-media communication, however, does not imply multi modal communication.

While a medium is only an information support, a mode is a means of expression and thus a means to convey information: a mode is built on a lexicon, syntax, semantics etc. Communication with a computer through graphic mode requires not only a graphic medium but the definition of one or several graphic modes using that graphic medium.

Dialogue management. Advances are aimed at in each mode separately. However the main source of improvements to interface technology will come from the integration of the different modes. To reach such an integration, one of the main aims of the project is to develop a dialogue management and mode selection system which uses knowledge of the specificities of individual modes, knowledge of the context of previous interactions, and knowledge of the application domain to interpret the input, determine the content of system output and select the most appropriate mode in which to present particular information. A user modeling module will interact with dialogue management, so that the system will react appropriately to different classes of users and individual users.

Knowledge-based backend application. On the machine side, the interface is primarily aimed to be connected to applications such as Prolog based expert systems (although we expect many of the results of this project should be usable, at least indirectly, for many other kinds of workstation application software). In order to focus on real practical problems, the interface prototype is being connected to a specific application, also developed within this project. This application, called NEST, is an expert system in computer network design. Such a system, besides having a very high intrinsic interest in its own right, given the current trends in information technology, has a great variety of potential users and offers many opportunities for multimodal interaction, including natural language and graphics.

Finally, the interface is designed to be portable across a range of potential applications of Prolog based KBS. Special emphasis is put on designing a flexible and portable architecture having well defined interconnection points with the application and on developing a set of tools for the rapid adaptation of the interface to a new application.

2. Architecture of the MMI2 system

A significant part of the work done in the first year of the project has been concerned with the definition of a clear, modular and conceptually sound architecture for the whole system. The architecture of the system is based on the notion of **"expert module "**. The name "expert" should be clearly understood. We are not proposing an architecture of "cooperating experts", or "multiple agents", which, we

414

believe, fall outside the scope of this project. What we call an expert is simply a module performing specific tasks and with its own private data structures, and which represents a sufficiently coherent set of processes to be gathered in a single module. While such a notion is clearly not new, the identification of the nature of the basic modules constituting the multimodal interface, and of the interactions between them, has been a crucial step in the project. The resulting architecture is illustrated in figure 1 below.

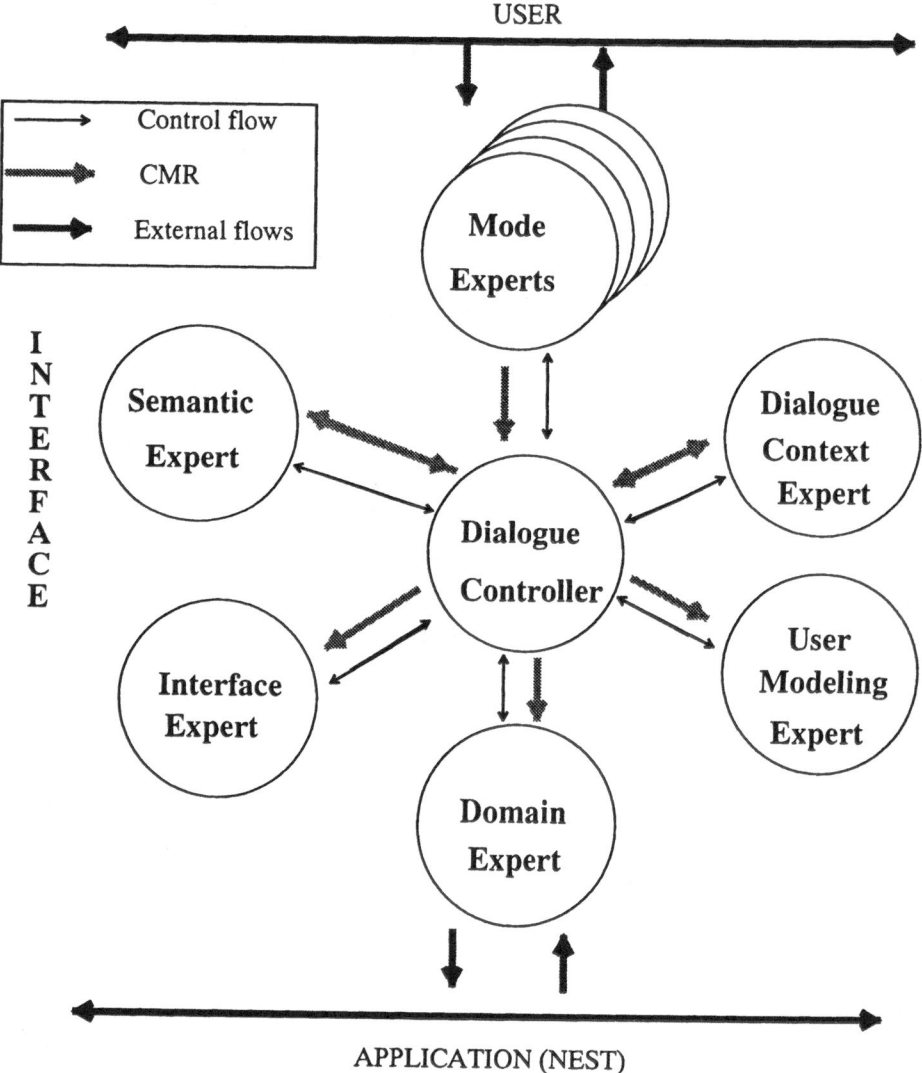

Figure 1: General architecture of the MMI2 system

The basic idea of this architecture is that every input should be cast into some suitable meaning representation, then forwarded to the dialogue controller, which will decide what to do with it. This should hold not only for input in natural language mode, but also for inputs in command and/or graphical mode (thus graphical or typed in commands, for example, should not be executed directly but go through the dialogue controller). The dialogue controller will call the dialogue context manager to update the dialogue model according to the new interaction, then reason (on communication acts, input content, state of dialogue and user models) to decide what to do with the input, how to gather results, what answers to present and in which output mode to present them.

To make such an architecture possible, there has to be a **meaning representation formalism** common to all modes, which is used as a vehicle for internal communication of the semantic content of interactions inside the interface itself and also used as a support for semantic and pragmatic reasoning. This meaning representation formalism is called the CMR and will discussed in more details in a later section.

The main fonctionalities of each of these expert modules can be described as follows:

− The *dialogue controller* deals with:
  - choosing and managing the structure of the dialogue
  - performing response determination and output mode selection
  - activating whatever experts are necessary to support dialogue interaction

− The *dialogue context* manager manages everything which has to do with:
  - identifying the dialogue structure and its various components
  - recording that structure
  - extracting relevant information from it

− The *user modeling expert* maintains and exploits the user model.

− The *domain expert* has all the expertise about the domain, including:
  - what is covered by the domain
  - how to translate the internal meaning representation into domain terms
  - how to manage "task plans" describing problems to be transmitted to the application.

− The *interface expert* has the knowledge about the interface itself, including
  - features/capabilities/configuration of the interface
  - current physical interface layout on the screen

− The *semantic expert* has all the knowledge about the general (domain-independent) properties of meaning concepts, and of semantic inferences that can be performed on them.

– The various *mode experts* perform input and output for each mode.

The modules being part of what is generally called the "dialogue management" process are the dialogue controller, dialogue context manager, semantic expert and domain expert. The following sections discuss further some basic aspects of the main modules enumerated above. But, in order to provide a practical context, we first start with some words about the application chosen as a testbed for the project.

3. Testbed application: an expert system in computer network design

The application chosen for MMI2 and developed within the project is an expert system for Network Design. It has been chosen for its own interest and for its interest in the context of a multi-mode interface development.

Indeed, information technology is evolving fast toward distributed systems. Configuring computer networks is, in this context, a crucial and difficult task both from economical and technical points of view. The development of expert system tools assisting such a task can certainly contribute significantly to the progress of European information technology. On the other hand, an expert system such as the one considered here has a great variety of users: technicians and commercials, beginners and experts. It deals with graphic and text information and must allow request and updating of data. So it supposes at least natural language, command language, graphics and gesture use. Thereby, it constitutes a credible practical test for a multi-modal interface.

Different components are involved in a design system. We decided to start the development of the application by the analysis component. This component is currently implemented. It analyzes *local area networks* using *Ethernet* technology and checks if those networks are correct or satisfying according to different evaluation criteria such as *technological validity, extensibility, client-server relation, departmentalizatior,* or *cost*. Networks which have to be analyzed are described by using an object oriented model defined in BIM_Probe, an object oriented tool built on top of BIM_Prolog.

The next step of the application will be devoted to the development of the configurator component able to compute a local area network configuration respecting some given constraints such as those specified by the customer (budget, building limitations, ...) or the technological ones already considered in the analysis tool.

4. Mode integration and dialogue context management

The MMI2 architecture is based on the fundamental assumption that

mode integration should mainly be achieved by an integrated management of a single, generalized, discourse context.

The basic idea is that any interaction, or any "discourse" between the user and the interface, in any of the modes, takes place in a common "discourse world" (which may but is not necessarily connected to the real world, or to an application). Any entity mentioned in the course of an interaction acquires, by the sole virtue of having been mentioned, an existence in this discourse world, where it shall be called a "discourse referent".

Various operations may apply to discourse referents. They may be "introduced" (or "created", or "brought into the discourse"), referred to by using a "description" of a referent, and even possibly "forgotten" ("deleted"). They may also be brought in or out of focus, and different sorts of foci can possibly be distinguished.

These notions are familiar to people involved in Natural Language Processing. What we argue is that they are common to other modes as well, and are the most natural way to perform mode integration. Let us give a few examples about each of the mode.

In natural language, new discourse referents are typically introduced by indefinite noun phrases and referred to by definite noun phrases (many special cases can be found in the literature, but will not be discussed here). Thus, in

Add a server to the network. Connect this server to....

the indefinite expression "a server" will cause the introduction of a new discourse referent in the discourse world, and this referent will receive a unique identifier. The expression "this server" can then refer to the newly introduced discourse referent, and can possibly bring it into focus.

For NL interaction, the referencing operation is done mainly through the use of descriptions which take the form of (definite) noun phrases. The basic problem is to relate such descriptions to unique referent identifiers: this process is usually known as "noun phrase resolution".

In *the graphical or gesture mode*, any basic graphical operation will have an effect on discourse referents. Creating a graphical object will obviously introduce a new referent. Selecting an object with the mouse will perform a reference to an existing referent. Moving a graphical object will bring the corresponding discourse referent into focus. Changing the graphical display (zooming, displaying another network, selecting another window where something else is displayed) would bring a different set of discourse referents into focus.

For the graphical mode the referencing operation is done by selection: the referent has a graphical description (as it could have one or several NL descriptions) and selecting the graphical description leads to the identification of the referent. The problem of identifying the referent selected is trivial, as each graphical object should have as associated property the unique identifier of the referent it represents. This association is easy to maintain: when a graphical object is created, it is either to display an existing referent (in which case the referent identifier is known) or to create a new one, in which case a new referent identifier name should be created automatically by the system.

In the **command language mode**, finally, basic operations on discourse referents are easy to identify, and bear close similitude to graphical ones. Thus a command to add an object will create a discourse referent for that object; commands performed on objects will tend to bring these objects into focus. Reference to existing discourse referents can be done by using the unique identifier of a referent as parameter in a command. The referencing operation for command language is also simple, as the referent identifier itself can serve as description of the referent.

Reasoning in terms of discourse referents and focus provides already for very interesting mode integration, which goes further than the more or less traditional graphical deixis (simply selecting something with the mouse and asking "what is this"). Thus in a sequence like:

System: < Graphical act to display new network >
User (NL)   What is the server?
System: < Graphical act to display an icon >

the first (graphical) utterance will bring a new set of discourse referent into focus. The NL query will then attempt to resolve the noun phrase "the server" against the discourse context and the current focus. The response determination module of the dialogue controller would then decide of a mode (graphics) and an appropriate graphical "communication act" (highlighting) to provide the answer.

We are thus led to study the conditions governing creation, focusing, reference and possibly destruction of discourse referents across all communication modes. A first attempt to organize the factors controlling these events is indicated in the figure below.

	Creation	Focus	Reference	Destruction
NL input	(ind.) noun phrases	sentence content & structure	(def.) noun phrases	forgetting?
Graphics	copy, paste, draw	display, zoom	click	
Command	COPY, CREATE	command content & structure	discourse referent id	

Figure 2: Operations on discourse referents across modes

5. Representing interactions - the Common Meaning Representation

A second source of integration arises from the fact that many interactions can be expressed equivalently (from the point of view of meaning, if not of ease of expression) in several modes, as the following examples illustrate:

NL: *Suppress the SUN3 connected to server Ella*
Graphics: < on icon and select delete option >

NL: *Augment performance of Ella by 100%*
Graphics: *< Select appropriate bar in a bar chart about computer performance and modify it >*

NL: *Suppress the blue servers (on color screen)*
Graphics: *< Click on icons and select delete option >*

Although expressed quite differently, all these input must have the same effect on the application (knowledge-based system) and on the dialogue context. Thus one of the results of MMI2 is to establish a taxonomy of actions across modes (e.g. the verb "suppress", a DELETE option in a a menu, a DELETE command or a gesture of crossing out something with the mouse all refer to a "deleting" operation). But integrating the representation of interactions across modes can go further than that. A second basic architectural principle of MMI2 is that

> **there is a meaning representation formalism, common to all modes, which is used as a vehicle for internal communication of the semantic content of interactions inside the interface itself and also used as a support for semantic and pragmatic reasoning.**

This meaning representation formalism is called the CMR. The purpose of the CMR is to represent the meaning of interactions between system and user, or user and system. Such interactions are called "communication actions". In MMI2 a communication action is a graphical action, a command language/gesture action, or a natural language action, which can be carried out by either the user or the system. When the communication action is expressed in a natural language, the action is an utterance.

The proposition expressed by a communication action consists of content and logical form. Following many other practical natural language processing systems, we have chosen to express the propositional content of a communication action in a language based on a first order typed predicate logic where relations are, as a general rule, reified. Specialized languages, such as frame or semantic network languages, fall short of the expressive power we expect to find used in our application. (See Chapter 2 in [Genesereth 1987]. Of course, there are extensions of these specialized languages, but these extensions end up looking like predicate logic.) On the other hand, we do not expect to need the extra expressive power that a second order language affords.

Our approach to representing meaning can thus be regarded as **logical**. However, we recognize that there are aspects to communication that are difficult, perhaps impossible, to capture on a purely logical approach. Therefore, we have decided to include extra information in CMR concerning illocutionary force, enunciation conditions and other things that we call 'annotations'. So a CMR expression contains four sorts of information: illocutionary force, propositional content, logical form, and various other annotations. A CMR expression is also part of a larger data structure that possibly includes the following additional information: processing status, mode, time of action, user presuppositions, user mistakes, and some syntactic information. The following figure illustrates one example of CMR representation. Although we shall not describe it in more details here, the full specification of the CMR language has been completed and is one of the major results of the first year of the project.

User input: What do the machines on the network cost?

```
CMR(
        [CMR_exp(
                [request,referent([var(x1)])],
                        [anno(x3,[definite_plural])],
                        (desc(the,x2,MACHINE,
                                (desc(the,x3,NETWORK,true),
                                (desc(some,x4,IS_ON,true),
                                and(
                                        [pred(SUBJECT,[var(x4),var(x2)]),
                                        pred(LOC,[var(x4),var(x3)])]))))),
                                (desc(null,x1 ,QUANTITY,true),
                                (desc(some,x5,HAS_COST,true),
                                and(
                                        [pred(PRESENT, [var(x5)]),
                                        pred(SUBJECT, [var(x5),var(x2)]),
                                        pred(OBJECT,[var(x5),var(xl)])]))))))],
        ok,
        English,
        time(1,1,1,1,1,1990),
        none,
        none,
        none)
```

Figure 3: Example of a CMR representation

A series of communication actions forms a 'dialog'. CMR is not meant to be a representation of a dialog, although it is supposed to lend itself to the construction of dialog representations. Of course, how you define actions or utterances is not so clear in general. In practice, however, each action is individuated by an illocutionary force: one force, one action, one thing to be represented in a CMR expression.

6. Dialogue flexibility and dialogue control

The application, being a knowledge-based system, has its own data and knowledge structures. A typical knowledge-based system has usually a "problem space", where the initial data of the problem to be solved are placed, a "solution space", where the expert system would build its solution, and a knowledge base containing general knowledge and expertise about the domain. What, then, should happen when the user starts to specify a problem?

Typically, for an expert system, the specification of the problem may require a set of information of different types. If the dialogue is application-driven, these information will presumably be asked to the user in some systematic order. But, in a real problem acquisition dialogue, the user may shift topics, answer questions by other questions, provide ambiguous or incomplete answers that will require sub-dialogues, request help, or even change his mind. It is obviously the job of the interface, and not of the application, to deal with such problems. If the user-provided data were sent directly to the application, they might well be in the wrong order or provide wrong values that would have to be corrected later. We have thus decided that, for reasons of flexibility, there should be a level of representation of the problem in the interface itself.

This kind of problem has started to interest researchers in dialogue management, and has been discussed notably by [Julien and all 89], who illustrate it with an example taken from financial advising expert systems:

SYSTEM: How much do you want to invest in an emergency plan
USER: Let us talk about my car loan instead!

In the face of such an answer, the interface must either enforce a strict dialogue schema, or accept a shifting of topic, which supposes that it should be able to detect it, and to remember to come back later to the "emergency plan" topic if this one is essential for the formulation of the problem.

To implement this kind of behaviour, we decided to provide the interface with three basic kinds of data structures, a "task plan", and a "communication plan", and a "discourse referent space", the two first being inspired from [Julien and all 89]. The position of these structures in the architecture is illustrated in figure 4 below:

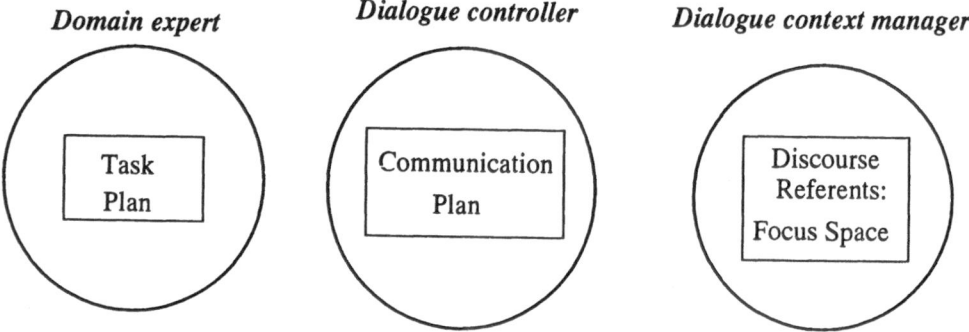

Figure 4: Key structures to support interaction with user

The basic role of the space of discourse referents has been described before.

The *task plan* is a model of the kind of information that need to be provided to the application in order to specify a well-formed problem. It can be seen as a kind of "skeleton" of a typical application problem, which will be progressively instantiated in the course of the dialogue. Task plans are of course domain dependent, and are thus managed by the domain expert. Several task plans may be available for a single domain, corresponding to the different kinds of problems one can submit to the application. The task plans will have to be provided as part of the application dependent information and will be stored in some part of the domain model. Normally, the task plan to be applied should be identified through interaction with the user.

The *communication plan* is a plan structure for guiding a dialogue with the user. Once a task plan is identified, there will be a communication plan activated to acquire the information necessary for the task plan. There will be at least a communication plan for every task plan available in the current domain; but the selection of such communication plans will also depend on the user model. There can also be communication plans for other parts or subparts of the dialogue; for example there could be one for greeting the user and ascertaining the kind of problem he wants to solve. As communication plans have to do with controlling the structure of the dialogue, they will be stored in the dialogue controller, in a "communication plan library". At any time there will be one active communication plan, which will be executed, and if needed updated by the dialogue controller.

7. User modeling

7.1. The User Modeling Expert

In MMI2, the role of the user modelling module is to increase the cooperativity of the system. The user modelling expert builds up and maintains the user model. The user model is the knowledge source in the system that contains explicit assumptions on all aspects of the user that may be relevant t the behaviour of the system. User models will be stored between sessions as User Profiles which will be downloaded for experienced user at logon.

The major uses of the information in the user model are for:

- The Dialogue Context Expert to identify speech acts, ellipsis and other pragmatic reatures on the basis of the user's goal and beliefs.
- The Mode Experts to present information using the symbols, lexical items and style the user prefers.
- The Dialogue Controller, to use the user's current goal to identify the user's current plan and thus provide co-operative responses. Responses will also be tailored on the basis of the user's beliefs according to Grice's maxims [Grice 1975] to avoid redundancy and to emphasize and explain the user's misconceptions.
- The Domain Expert will use items of user specific information in completing domain plans which will form queries to the knowledge based system.

The basic framework of the user modelling component comes from General User Modelling System - GUMS - of Finin and Drager [Finin and Drager 1986]. This provides four essential features of the user model itself:

- An interface between the user model and the rest of the system;
- A mechanism for overridable default inheritance from stereotypes with optional negation as failure;
- A truth maintenance mechanism to manage the updating of inheritance from different superordinate classes during a session as the user's membership of a class changes;
- A mechanism for storing user and class models between sessions.

Two advances on the GUMS model are required fro the underlying framework of the user modelling component in the MMI2 architecture. Firstly, the GUMS model only supports inheritasnce from a single stereotype which is limiting since users could be both a member or a class with knowledge about the domain, and another with knowledge about the interface and system itself. Inheritance from multiple stereotypes has been added to the GUMS system. Multiple inheritance brings with it the possibility of conflicts between information in different stereotypes. These are handled by searching for certain information rather than just default values; if these conflict then the certain information overrides the default. Similarly, where negation as failure is used to show the lack of knowledge or belief, and knowledge or belief is tated, then the presence of knowledge or belief is returned. If certain but contradictory knowledge is stated in different places, a temporary unsatisfactory heuristic is applied that the first version encountered in the bottom up breadth first search is returned. The second chage follows from the use of multiple stereotypes. Sparc Jones [1987] succinctly the problems with assuming that the user of the system who is entering the commands is actually the user of the system for both the interface and the reasoning components of the system. Her potent example is one of a social welfare worker entering information on behalf of a client where the view of the user act as a filter on the information obtained about the client; either approving or otherwise. The overcome this complexity, the leaf node of the user model in the present system represents the combination of the user and the client. Therefore this user model inherits from both the user and the client nodes in the lattice, each in turn from other stereotypically defined nodes.

A major requirement of the MMI2 system is to produce co-operative responses for the user. The class of co-operative responses expected follows [Kaplan 1983, Allen 1983, Carberry 1985, Pollack 1986]. Their generation require representations of the user's goals; the plans for achieving them; the user's beliefs; an evaluation of whether they are correct knowledge of misconceptions with respect to the systems view of the world which is assumed to be rue; some decision relevant information to be used in task plans for the KBS application; and various attitudes and preferences of the user for explanation style, lexical item choice, and interface style and options.

The user's beliefs are stored in the prototypes and the user's own model. These can be acquired explicitly or implicitly as suggested above. The rules for implicitly acquiring user beliefs follow those used by Kass [1988,1989]. In order to use user beliefs to interpret input to the interface or to tailor output for the user it is useful to identify the believs as true to the world, and therefore user knowledge, or false in the world, and therefore misconceptions. The test of truth inthe world is made by comparing the user's beliefs with the representations in the other experts and change their status if a judgment can be made. If no comparison is possible, the beliefs remain beliefs and are not classified as knowledge or misconception.

7.2. User Modeling Experiments

The set of classes and rules for determining if users belong to them, along with the sets of possible relevant goals, preferences, decision relevant information, knowledge and beliefs which must be included in this architecture are established for the demonstration application by the experiment described below.

The experiment included two steps: simulation of interactions between users and a system, and postverbalization supported by a record of the activity.

One expert simulated the system and different types of subjects (or 'users') requested his advice to solve two problems of physical network design that had been previously defined in collaboration with the expert: the first problem consists of designing a network for a research department and the second one consists in designing a network connecting different buildings in a university. According to the expert, these problems are typical and of average difficulty. The 10 subjects differed in computer education and in type and level of knowledge about network design.

The expert used MUSK as a tool for interaction between the subjects and the expert. MUSK is a program (designed at Rutherford Appleton Laboratory) that allows both graphic and written interaction between several users simultaneously.

Each of the ten subjects solved the two problems successively in the same order, the interaction for each problem taking about two hours. Every ten minutes the expert was given evaluation sheets on which he quickly noted the user's level and type of competence (on-line evaluation).

The second phase of the experiment consisted of a post-verbalization supported by the transcript of the dialogues: the expert was brought to look at the transcriptions of the interactions (text and graphicas) and comment onthem line by line, the on-line evaluations constituting further support for his comments. The aim of this method is to re-create the consultation situation the expert participated in. The expert's comments were focused on interlocutor modelling: he was asked to point out he clues he used for elaborating the model of the interlocutor and to stress the effects of this model on the interaction. A second expert was asked to comment on the dialogues in order to check the reliability of the evaluations.

Gathering dialogues and having the expert commenting on them have proven to be very beneficial. This method yields rich information concerning the modelling process to be gathered [Cahour 1989]. For MMI2, the analysis of these protocols allows:

- the specification of the content of the user model, i.e. the predicates and arguments that must be included, the stereotypes that categorize the different users and the inference rules that trigger them.
- the description of the elaboration of the UM, i.e. the clues in the discourse or in the drawings of the "user" on which the expert bases his elaboration of the UM.
- the identification of various effects of the modelling on the interaction, i.e. the use of the user model for identifying misconceptions, adapting the explanations (level and quantity), identifying users' plans, managing initiative distribution and disambiguations.

8. Input - Output modes

Once the architecture was defined, design and implementation work has started in parallel on all input - output modes. The following sections review briefly the aims and the first results achieved for each mode.

8.1. Graphics Mode

When designers consult with clients with problems such as our test application of computer network design, they not only talk to them but draw copiously on paper. In the expertise acquisition interviews we made, these notes contain plan and elevation views of buildings, with initially none and then reducing granularities of description of the designed network. The notes also contain lists of numbers and tables for showing the component costs, calculating composite lengths and load on the network. During a consultation both the client and the consulting expert refer to these diagrams and tables verbally, by pointing at them and by drawing signs or 'gestures' on them in 'designer shorthand' .

An analysis of this class of data from design consultations clearly shows that MMI2 must represent the plan and elevation building geography, and the designed network at different granularities. The information displayed in tabular form is sometimes used to calculate exact figures, but is often used to show relative costs, lengths or loads, so not only must tables be used, but also pie charts, histograms and graphs which more effectively convey ratios and relative values. This range of graphical tools should not only present the information to the user, but also allow it to be changed and manipulated. For example, the histogram tool should not present a set of data, but allow the user to drag a bar up or down to change the value represented; the tool representing the network must allow the system to present a designed network, but must also allow the user to state the requirements by placing devices which ought to be connected, and to modify the system's design suggestion. Therefore five graphics tools have been developed to display tables, histograms, line and scatter graphs, and building and network geography.

In order to generalise the intelligent interface to other domains, new CAD tools should be incorporable, and other removable. The graphics tools have therefore been developed to meet 3 design constraints:

1. Individual tools must be accessible for different functions
2. A Graphics Manager which provides a clear interface between all tools and the rest of the MMI2 interface must be provided. New tools can be added to this manager.
3. A window manager is used to facilitate portability. The current window manager used is SunView although their are plans to port the system to the more general X-windows manager

In operation, the graphics manager receives CMR packets from the rest of the interface, which are decomposed so a tool can be selected to display the information. The CMR is then translated into the internal representation required for that tool which opens a window and presents the information graphically. The selection of the appropriate tool is made by a rule set which assesses the structure of the information and accesses the User Modeling Expert to determine the preferences of the user. The rules determine the structure of graphs and charts draw on the methods proposed by Tukey (1977), Cleveland (1985), Beach (1985) and Mackinlay (1986). When a user makes a selection or modification this is translated into CMR by the graphics manager which passes the packet down to the Dialogue Controller. Figure 5 shows an interaction with the system illustrating the range of graphics tools.

The graphics tool to allow the presentation and manipulation of building and network geography is the most complex. This allows users to enter building geography as a free hand sketch which is then digitally sampled and adjusted to turn wavy free hand lines into straight lines and the square up angles between these lines. Linked planar maps are used to represent the building and network geography so that either may be viewed and modified separately, as well as together. This approach of allowing free hand input with a smoothing process was chosen rather

427

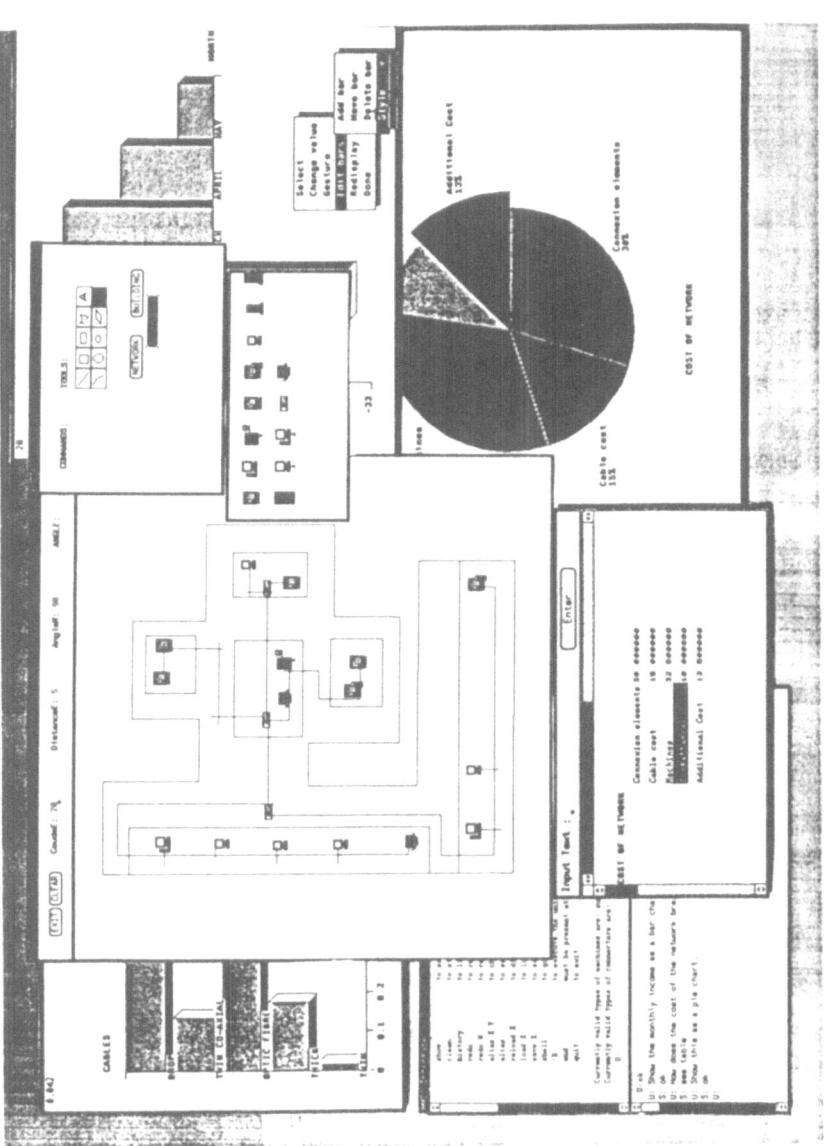

Figure 5: Example of an interaction with MMI2. The windows in the upper left and center show a tool in which the user can enter details of buildings and computer networks for those buildings, and in which MMI2 can display design solutions. The windows on the right and bottom show tools for the system to display answers as charts or tables. The window in the middle left supports command language and natural language interaction. Users can perform design gestures in any of these windows.

than the conventional use of a menu selected 'straight line' tool with handles to move it, since it fits the details of the style of the observed experts better.

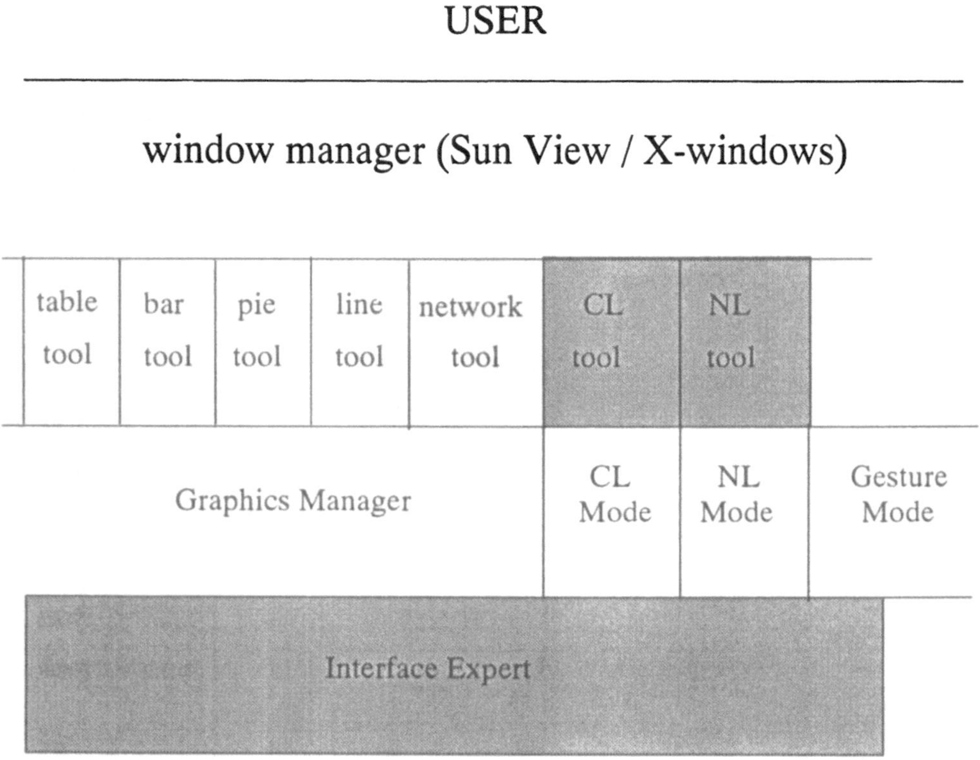

USER

---

window manager (Sun View / X-windows)

table tool	bar tool	pie tool	line tool	network tool	CL tool	NL tool	
Graphics Manager					CL Mode	NL Mode	Gesture Mode
Interface Expert							

Figure 6: Layered interaction of the graphics and text tools with the window manager. Shaded areas are part of the Interface expert.

8.2. Interface Expert

The Interface Expert serves three main roles in the MMI2 system:

Declarative Knowledge of the MMI2 Interface is stored in a knowledge base in the Interface ExpertThis is called upon by the dialogue controller when the user asks questions about the limitations, abilities, structure or components of the interface itself rather than the domain. For example, if the users asks "What natural languages can I user here ?" the knowledge that English, French and Spanish are available would be provided by this knowledge base.

Screen layout management is performed by the interface expert in as far as it provides windowing tools with a position in which to appear on the screen. This overrides the default algorithm in the window manager to position windows in task relevant positions rather than progressively across the screen. The need for this role arises since the windows within the application are overlapping rather than tiled to allow more flexibility in dialogue.

Text interaction in natural language and command language is performed in a pair of windows which are part of the interface expert. Once text is input and a carriage return typed, the string is sent to the appropriate natural or command language mode which returns a CMR packet that is sent to the Dialogue Controller. This provides a uniform image to the user which may not occur if separate interaction windows were developed for each text mode. These windows are part of the interface expert rather just another graphics tool since they include the main event handling procedures for the interface, and support the gesture mode in different windows. Consequently the graphics manager actually passes its CMR packets through the Interface Expert to the Dialogue Controller. This provides a clear interface between all modes and the dialogue controller as shown in figure 6.

8.3. Command Language

The purpose of the command language (CL) is to provide users with a language based mode but without the computational overhead associated with a full natural language. The improvement in speed is offset by a cost to the user in terms of syntactic structure: users are required to comply with the syntactic constraints of the CL. The CL comprises a syntactic part and a semantic part. The syntactic part comprises two data sets: a set of operators/actions or commands such as 'add', 'connect' etc. and a set of application objects which are common or proper nouns. A command comprises an action-object pairing. In order to determine whether the command conforms with the semantics of the application the command is coded in CMR and passed for evaluation. The Semantic Expert determines whether the pairing is legitimate and returns and evaluation.

8.4. Gesture Mode

As the complexity of graphical representations of a network increases the efficiency the efficiency of natural language modes of dialogue tends to decrease while the use of direct manipulation tends to increase. Designers routinely use a wide range of non-standard but familiar graphical annotations to their work plans in order to edit and modify them. Moreover, they usually retain a cumulative record of this part of the design process using it as an aid to quality control in a design audit trace. Thus, the use of 'design shorthand' is an integral part of the design process and the MMI2 interface permits designers to continue to use this mode of dialogue.

An algorithm has been implemented which allows designer to draw up to 19 different shapes. The shapes are recognized as commands (e.g. 'delete', 'transpose') within the application. Users can draw onto a graphical representation of a plan and make technical modifications. For example, designers can remove graphical objects by 'scribbling' over them or transpose objects by drawing the appropriate shape around the objects to be transposed. The algorithm recognizes the shape drawn by the user (the syntactic part of the process) and pairs it with the objects around which it has been drawn (the semantic part) for subsequent evaluation by the Semantic Expert. The algorithm is scale, location and style invariant. In other words the shapes can be drawn to different scales, in any location using a wide range of individual drawing styles. Improvements in the algorithm will satisfy rotational invariance, allowing users to draw the shapes in a variety of orientations. The algorithm can be used in a variety of applications where designers routinely employ graphical, gestural shorthand within the design process.

8.5. Spanish Language Mode

The Spanish System is envisaged around two components: an analysis module and a generator. The latter is to be implemented from 1992 on.

The analysis module includes a morpholexical component that does an extensive morphological analysis of Spanish - including resolution of collocations and idioms like *cable delgado* (thin cables), *de todas formas* (anyway), *tener en cuenta* (to take into account), etc.

The result of the morphological process is a graph that reveals the set of categorical ambiguities in a given input text. It is worth saying that some of these ambiguities will be solved with the aid of statistical methods. Lexical lookup is performed upon that pruned graph. This component is entirely written in C. The approach adopted is extremely lexicalist. The assumption is that lexical entries are rather huge structures containing a lot of information (selectional restrictions for arguments, semantic typing for CMR formulas, etc.)

The output information from the morphological component is the input to a syntactic parser (developed in BIM_Prolog) that yields a CMR representation by means of a compositional featurebased formalism grounded on functional dependency grammar (cf. Kaplan and Bresnan [1982], Kay [1985]). A formalism has been devised that extensively supports several types of feature manipulation: unification, constraint evaluation, and value overlapping. This parser is currently under development.

Some parts of the parser will interact with other modules of the MMI2 system, in order to solve problems like word sense and attachment ambiguities ellipsis or anaphora, that need contextual and world knowledge.

8.6. French Language Mode

The French processing system will include both the analysis of NL utterances in order to reach a CMR expression and the generation of the expert system answers. The objective of NL analysis is not to build any possible parse, but only the right ones. Concurrently, the generation system has to build a NL answer according to the user expectations.

The approach to French NL analysis has the following characteristics:

- the analyser is modular; it is made of little modules associated with limited tasks.
- the analysis is driven by the content of the NL sequences: at each step, all the accurate informations are stored, and are used for local prediction, in order to avoid to build spurious solutions.

The steps of the analysis are:

- preprocessing, which normalises the input NL sequence,
- morphological analysis, associated with a 50 000 words dictionary which tags each word with its lexical entries, grammatical categories, etc.
- disambiguation of the sequence of categories based on a Markovian filter,
- segmentation of a sentence in clauses,
- parsing of each clause through a Earley's automaton which is supervised by a linguistic expert system,
- transformation of the constituent structures in a categorical formalism. Inside this formalism, some operations like anaphora resolution, modification of the word order..., are done in order to reach a CMR expression.

The first state of the analysis, producing the constituent structures, has been implemented in C; the rest of the parser is currently under development.

8.7. English Language Mode

The English NL input mode of MMI2 will use the English parser of the LOQUI system, developed first within the context of the LOKI ESPRIT project (P 107) [Binot et al. 88], then within an internal development project called BIM_LOQUI.

LOQUI consists of a number of heavily interrelated modules, among which the following main components can be distinguished: morphology, parser and interpretation module on the analysis side and response determination module and generator on the generation side. These processing modules make use of several sources of knowledge: the lexicon, a body of morphological rules, a body of "database mapping rules", a "world model" including general semantic knowledge about word meanings as well as pragmatic knowledge of the application domain and a dialogue memory holding the discourse structure representation.

The parser for English is mainly based on the theoretical principles of Generalized Phrase Structure Grammar (GPSG) [Gazdar et al. 85] but allows mechanisms from other theories as they seem useful, the aim being to develop an implementation of English grammar which is not only theoretically sound but also computationally efficient. Prolog itself is the formal expression language of the English grammar, mainly for reasons of efficiency.

The work required to adapt the LOQUI parser to MMI2 is rather limited, due to the portable nature of the system. This work, which is currently under way, mainly includes:

432

- defining an appropriate lexicon for the domain of the application (computer network design);
- defining an appropriate conceptual model for the domain of the application; this is related to the design of the "Semantic Expert" introduced in the general architecture;
- modifying the output of the parser so that it generates CMR expressions;
- improving some aspects of the parsing itself, mainly from the point of view of extending its coverage and/or efficiency.

9. Conclusions

We have presented what should probably be considered as an ambitious architecture for a multimode interface system, including concerns such as user modeling, communication planning, multimodal meaning representation, dialogue context management, etc. We do not necessary expect to develop all above mentioned features to the same degree of completeness within the scope of this project. But we see the architecture itself, which we tried to describe in a clean and modular way, as a kind of reference framework for further research and implementation work on these topics.

The MMI2 project itself will eventually produce an interface prototype illustrating (maybe at various stages of completion) all aspects of the proposed architecture. In this respect, the current phase of the project is focused on the implementation of prototypes for the various components of the system, including the various input - output modes and on the integration of some of these prototypes in a first version of the multimodal system.

10. References

Allen J. (1983) Recognizing Intentions from Natural Language Utterances.. In M. Brady & R.C Berwick (Eds) "Computational Models of Discourse". Cambridge MA: MIT Press.

Beach, R.J. (1985), Setting tables and illustrations with style. PhD.thesis, Department of Computer Science, University of Waterloo, Canada

Binot J-L., Demoen B., Hanne K.H., Solomon L., Vassiliou Y., von Hahn W., Wachtel T., (1988), LOKI: A logic Oriented Approach to data and knowledge bases supporting Natural Language Interaction, Proceedings of the ESPRIT88 Conference, North-Holland.

Cahour B.(1989) "Competence modelling in consultation dialogues", Proceedings of the "Work With Display Units" Second Conference, Montreal, 11-14 Sept. 1989. Amsterdam: Elsevier

Canberry, M.S. (1985) "Pragmatic Modelling in Information System Interfaces." PhD. thesis, University of Delaware.

Cleveland (1985), The elements of Graphing Data, Wadsworth Advanced Books and Software: Monterey, California.

Finin T. and Drager D. (1986)."GUMS: A General User Modelling System." Technical Report MS-CIS-86-35, Department of Computer and Information Science, U. of Pennsylvania

Gazdar G., Klein E., Pullum G. and Sag I., (1985), Generalised Phrase Structure Grammar, Blackwell.

Genesereth, M.R., Nilsson, N.J. (1987), *Logical Foundations of Artificial Intelligence* (Morgan Kaufmann Pub., Los Altos, CA).

Grice, H.P. (1975) Logic and Coversation. In P. Cole and J.L. Morgan (eds.) "Suntax and Semantics", volume 3, pages 64-75, Academic Press, New York

Julien C. and Marti J-C. (1989),: Plan revision in person-machine-dialogue, Poceedings of the Fourth ACL European Chapter Conference, Manchester.

Kaplan R. and Bresnan J. (1982), Lexical-functional grammar: A formal system for grammatical representation, in J. Bresnan (ed.) The Mental representation of Grammatical Relations. The Mit Press.

Kaplan S.J. (1983) Cooperative Responses from a Portable Natural Language Database Query System. In Brady, M. and Berwick, R.C. (eds) "Computational Models of Discourse". Cambridge MA: MIT Press.

Kass, R.J. (1988) "Implicit Acquisition of User Models in Cooperative Advisory Systems" Technical Reports MS-CIS-878-05, Department of Computer and Information Science, U. of Pennsylvania

Kass, R.J. (1988) "Acquiring a Model of the User's Beliefs from a Co-operative advisory Dialogue."Technical Reports MS-CIS-88-104, Department of Computer and Information Science, U. of Pennsylvania

Kay M. (1985), Parsing in functional unification grammar, in D. Dowty, L.

Karttunen and A.Zwicky (eds.) Natural Language Parsing. Cambridge U.Press, pp.251-278.

Mackinley, J. (1986), Automating the Design of Graphical Presentations of Relational Information, ACM Transaction on Graphics, 5(2),110-141

Pollack, M.E. (1986) "Inferring Domain Plans in Question-Answering". Technical Report MS-SIC-86-40, Department of Computer and Information Science, Univ. of Pennsylvania.

Sparck Jones K. (1987) "Realism about User Modelling". Technical Report no. 111, University of Cambridge Computer Laboratory.

Tukey (1977), Exploratory Data Analysis, Addison-Wesley: Reading,Mass.

Project No 2576
ACKnowledge.

Information Pool for Knowledge Engineering Support
- The Common Information Repository

Inge Nordbø and Jorun Eggen
ELAB-RUNIT, SINTEF
N-7034 Trondheim, Norway
E-mail: inge.nordbo@elab-runit.sintef.no, jorun.eggen@elab-runit .sintef.no

Abstract

In the Knowledge Engineering Workbench (developed in the ACKnowledge project Esprit - 2576) we have replaced the traditional single knowledge base of a development tool for knowledge based system by a Common Information Repository (CIR) containing all the knowledge and information relevant for the construction of the target knowledge based system. The CIR concept represents a significant improvement of a knowledge based system development tool and enables the incorporation of new functionality; being a central pool for integration of knowledge from different knowledge acquisition tools, being a source for giving active advice and guidance in the knowledge engineering process and providing project management facilities. The Common Information Repository includes the final integrated knowledge base, which of course is the main part of the repository, but all kinds of elicited knowledge and project management information are also stored in different parts of the repository.

1. Introduction

ACKnowledge[1] (Esprit project 2576) is an acronym for ACquisition of Knowledge [9] . The objective of the project is to improve the efficiency of the knowledge acquisition process at all phases in development of knowledge based systems. This will be done by developing the *Knowledge Engineering Workbench - KEW -* an active, integrated system for development of knowledge based systems [15].

The KEW system will assist the knowledge engineer in his tasks, and partially provide automation of these tasks. The key constituents of the workbench are three kinds of knowledge acquisition tools: machine learning tools, editing tools, and knowledge elicitation tools. *Integration* of tools that support different parts of knowledge acquisition in a coherent system is the approach taken in ACKnowledge to resolve the knowledge acquisition problem. Other approaches for knowledge

---

[1] The ACKnowledge project has seven European consortium members. The partners are: Cap Gemini Innovation (Prime Contractor), Marconi Command and Control Systems - GEC-Marconi Research Center (Sub-Contractor), Telefonica - U.P. Madrid (Sub-Contractor), Computas Expert Systems - Veritas Research (Sub-Contractor), University of Amsterdam, Sintef, University of Nottingham.

acquisition support are tools for interactively modelling of the domain (e.g. Shelley [2], [3], KEATS [10], METATOOL [14]), tools taking advantage of a domain model (e.g. TEIRESIAS [4], OPAL [12]) and tools that actively interacts with the user (e.g. BLIP [6], AQUINAS [1]).

A central part of KEW is the *Common Information Repository* which contains all the knowledge that is produced during the knowledge engineering process which is relevant for the construction of a target knowledge based system. The reasons for having an information repository in the knowledge engineering workbench are firstly, to have a central pool for integration of knowledge from different knowledge acquisition tools, secondly, to have an information pool providing information about the status and history of the knowledge acquisition process for providing active advice and guidance, and thirdly, to have an information pool with background information (e.g. links to the source of the knowledge) and project information to provide project management facilities within the same environment.

The purpose of the paper is to describe how the information repository in KEW will support the knowledge engineer in developing a knowledge base. Section 2 of the paper gives a brief overview of Knowledge Engineering Workbench, section 3 describes in more detail the different components in the information repository and the sections 4 to 6 explains how the commmon information repository will be used for integration of knowledge, active help and project management, respectively.

We have tried to limit the number of abreviations, but the generally accepted KA for knowledge acquisition and KB for knowledge base are used throughout the paper.

2. The Knowledge Engineering Workbench- KEW

The ACKnowledge project is going to realize its objective of more effective knowledge acquisition for construction of knowledge based systems in two ways:

- The development of methods and techniques for integrating several knowledge acquisition techniques.

- The design and construction of an environment which integrates a number of existing automated and non-automated techniques; the Knowledge Engineering workbench - KEW.

The global architecture of the Knowledge Engineering Workbench is displayed in Figure 1.

The KEW consists of a common user interface, a set of knowledge engineering services and the Common Information Repository (CIR). The knowledge engineering services cover a number of integrated tools for giving support to the user and will be briefly outlined below. The CIR is described in more detail in the next section.

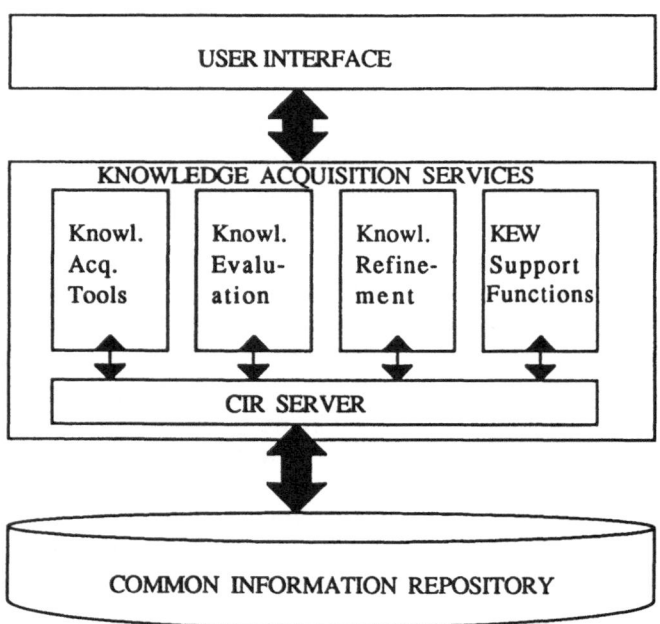

Figure 1: KEW Architecture

Knowledge Acquisition Tools consists of a selection of tools for knowledge acquisition covering tools for knowledge elicitation (card sort, laddering and repertory grid), machine learning (similarity based learning and case based learning) and modelling (protocol analyzer, knowledge structuring tool and knowledge base editors). The output from each of these tools is an Intermediate Information Structure maybe in a representation language different from the core knowledge representation language, thus functions are required to transform these Intermediate Information Structures into a the core knowledge representation language before they are integrated into the existing knowledge base.

Knowledge Evaluation consists of tools for evaluating the knowledge that is represented in the knowledge base (which is part of CIR), including tools for local checking of the knowledge, testing with cases and browsing for manual evaluation.

Knowledge Refinement consists of tools for refinement of the knowledge in the knowledge base, including tools for transformation of knowledge, integration of knowledge from different sources into the knowledge base and editing of the knowledge base.

KEW support gives the knowledge engineer advice and guidance in use of KEW covering both guidance for use of particular tools in KEW, but also guidance in selection of tools and how to proceed in the knowledge acquisition process.

The CIR server handles the communication between the four groups of services described above and the CIR which holds the knowledge base and all the information that is needed to support the knowledge acquisition process.

The main functions of CIR is to store and retrieve relevant knowledge and information that is produced by the user or by any other process in KEW. In addition to retrieval and storage CIR must support the knowledge acquisition process by additional information management functions (e.g. tracking links between related elements in CIR, version control of knowledge elements and maintenance of the project log).

3. The Common Information RepositoryCIR

Figure 2 gives an overview of the components of the Common Information Repository and the relations between them. From a conceptual point of view the information in CIR comprises three disjunctive types of information:

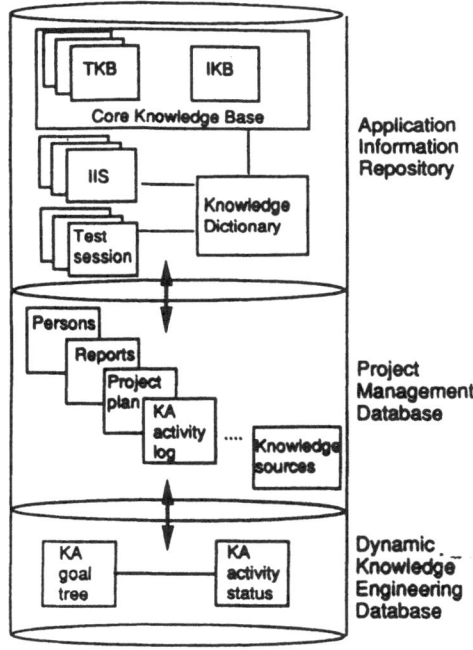

Common Information Repository

Figure 2: CIR components

Application knowledge which is the knowledge concerning the application domain. This includes all elicited knowledge about the domain that may potentially be included in the final target knowledge base. The Core KB is the part of this application knowledge that is represented in the core knowledge representation language of KEW. The application knowledge is stored in the Application Information Repository.

Project information which is the information about how the knowledge is being elicited and modelled, i.e. how the KA process within the current project is progressing. This covers the whole specter of information from the links between the pieces of knowledge and their sources to project information like how many times the repertory grid technique has been used. The project information is stored in the Project Management Data Base.

Dynamic Knowledge Engineering Information which is information about the status of the knowledge engineering process and information that is relevant to the reasoning in KEW. The Dynamic Knowledge Engineering Information is stored in the Dynamic Knowledge Engineering Database.

The most important part of the Application Information Repository is the Core Knowledge Base (CKB) as this is the knowledge that may be reasoned upon by the KEW. The Core KB comprises a set of partitions, where the Integrated Knowledge Base (IKB), which is the current view of the domain, may be regarded as a partition with a special status as this is will finally be transformed to the target knowledge base. The other partitions in the Core KB are Temporary Knowledge Bases (TKB) holding parts of the domain knowledge represented in the core knowledge representation language.

The reason for having different partitions in the Core KB is that the knowledge acquisition process is a non-linear iterative modelling process; several intermediate models may be developed, some of them as alternatives to the Integrated KB. E.g. some knowledge does not fit nicely into the current model, but could be interesting at a later stage; some knowledge is conflicting with the current view, but could not definitely be thrown out as 'wrong knowledge' at the current stage, etc.

The Application Information Repository also contains knowledge about the domain that is not represented in the core knowledge representation language. This is mainly output from the different knowledge acquisition tools, Intermediate Information Structures (IIS), including cases, repertory grids, think aloud protocols, rules, etc. In addition to this the Application Information Repository includes test sessions and a knowledge dictionary. The test sessions keeps the results when running the knowledge base in problem solving mode on a set of test cases. A test session contains pointers to the Core KB-partition that has been tested, to the test cases used as well as the results of the test (e.g. answers from KEW, answers from expert for each case) The Knowledge Dictionary is a central component that connects the different parts of the CIR. It holds the names of all the terms used in the Core KB, together with information of their origin and their relation to other elements in the CIR.

Most of the project information will be provided by the user and maintained by the system during the usage of KEW throughout the project's life cycle. The Project Management Data Base keeps project management information, like project plans, activity logs describing a knowledge acquisition activity in terms of tool used, persons involved, stage in life cycle, resulting application knowledge etc. This database also includes lists of persons, knowledge sources (experts, books, etc.) and reports.

4. CIR as a basis for integration of knowledge

The ultimate goal of the knowledge engineering process is a knowledge based system working in the target environment solving the application tasks with a satisfactory quality. The final output of the KEW will be the Integrated Knowledge Base which will have the knowledge and problem solving quality of the Target KB, but which may not be runable in the target environment and maybe lacks the required efficiency. The final generation of a Target KB from the Integrated KB will not take place until the Integrated KB satisfies the project-defined quality criteria. Thus the Integrated KB is the central goal of the user's work with KEW.

To achieve this goal, the various tools in KEW are used to produce several types and versions of *Intermediate Information Structures - IISes* - like sets of cases, repertory grids, think aloud protocols, sets of rules from induction tools etc. The Intermediate Information Structures are stored in the Application Information Repository. These structures may be represented in heterogeneous representation languages[2], and they will to some extent be acquired independent of each other due to the iterative and incremental nature of the KA-process. Thus, in order to make use of these structures to construct the Integrated KB, each of them has to be assimilated into the Integrated KB. This can be done in two ways:

The manual way: The knowledge engineer interprets the Intermediate Information Structures and by means of direct editing of the Integrated KB he creates knowledge structures like concepts and rules, and thus transforms and integrates the knowledge content of the Intermediate Information Structure into the Integrated KB.

The KEW supported way: The Intermediate Information Structure is transformed into the same representation language as the Integrated KB, and stored into a *Temporary Knowledge Base*. This Temporary KB may be integrated into the Integrated KB through interaction between the KEW and the knowledge engineer.

---

2 One of the goals of ACKnowledge is to enable integration of existing KA-tools in KEW, thus the resulting knowledge structures from these tools will be in different knowledge representation languages.

440

The former alternative comes from the fact that some of the Intermediate Information Structures have a form which is not interpretable by the KEW. E.g. think aloud protocols which are "marked" with hypertext links: there is no way the transformation function can pick out the relevant knowledge and transform it to the core representation language. Thus the user has to interpret the Intermediate Information Structure and manipulate the Integrated KB directly to integrate the knowledge[3].

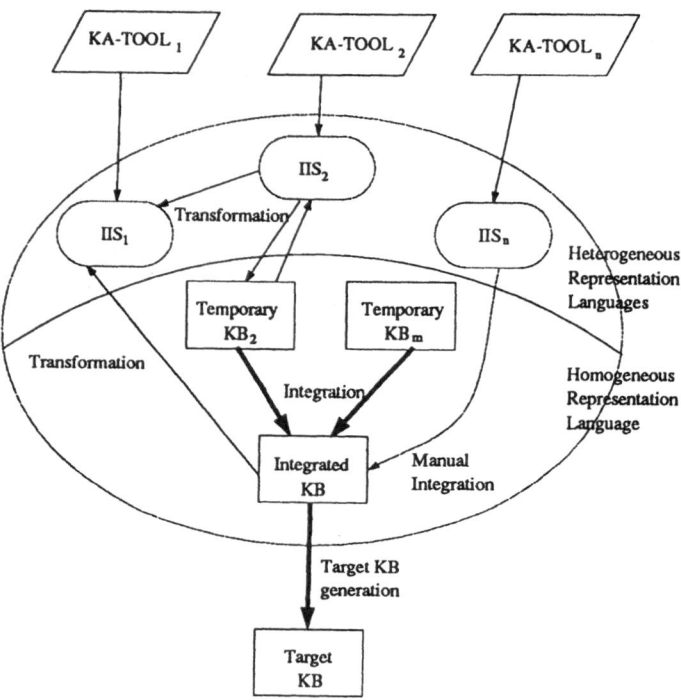

Figure 3: The assimilation process and the knowledge bases involved.

The latter alternative is applicable for the Intermediate Information Structures that have a more structured output, like a set of rules or a repertory grid.

Fig.3[4] illustrates the two alternatives of the assimilation process. The figure also shows the possibility of transformation from the Integrated KB to an Intermediate Information Structure, if the Intermediate Information Structure needs input from the existing knowledge, and the possibility of transformations between Intermediate Information Structures. The ellipse illustrates the different knowledge bases in the

3 A third alternative would be to build an interpretable Intermediate Information Structure from this one, and then go on with transformation and integration.

4 The figure is taken from [5].

Application Information Repository. The Intermediate Information Structures in the upper part may have different representation languages. The lower part is the Core KB, which has one representation language in which all the knowledge bases constituting its partitions are represented.

The constituents of the assimilation functions are:

Transformation

– converts from one formalism to another

– may be supported by domain specific background knowledge (knowledge from user, knowledge of domain, knowledge about representation etc.)

– fully automatic or semi-automatic

Integration

– input and output are in the same formalism

– highly domain knowledge intensive

– maintain (or resolve) consistency (local[5])

– semi-automatic

Studies of transformation within ACKnowledge (see [13]) discusses knowledge supported transformation, but prototyping experiments in the ACKnowledge project indicates that the transformation functions in KEW mainly will be syntax translation. This reduces the complexity and thus the transformation functions may be fully automated. The hard problem is to decide how a certain construct in one representation language should be represented in the target language. This is done once and for all when the transformation algorithm between languages A and B is written.

The consequences are that even more of the success in the approach we have taken in the ACKnowledge project relies on the integration functions.

Integration of knowledge is a process which combines knowledge from different knowledge acquisition tools which may originate from different sources. Different textbooks and different experts obviously would use slightly different terminology, have different rules-of-thumb and even slightly different theories. When a KA-process spans in time, the same problems with variations in terminology etc. may occur with a single expert as well. This of course introduces serious problems in integrating new knowledge into the existing KB. The tools may have acquired the same or different knowledge, implying that knowledge in one tool may extend or refine the

---

5 Inconsistencies may be a part of the domain, like the two theories about light - the particle theory and the wave theory - they exist in parallel and are used for different purposes. Local consistency means that each of these theories must be internally consistent.)

knowledge from another tool. Even though the same knowledge is represented in several tools, the semantics of the different tools may be different, thereby leading to possible misinterpretations. The integration must take care of the semantics in the knowledge to be sure that the knowledge piece in the Temporary KB and the Integrated KB "speaks the same language". I.e. that the meaning of the same term in the two KBs are actually equal. The knowledge in the Integrated KB must be used for this purpose.

The main objective of the integration process is to identify semantical similarities and detect possible inconsistencies and incorrect knowledge. The integration process should also reveal whether any knowledge is missing and should be supplied. This will be an interactive process, and the knowledge engineer should have a lot of support for:

- checking of constraints

- checking of completeness

- checking of correctness

- checking of consistency

The notion of integration of knowledge is a new issue in the knowledge engineering field, and very few publications exists. Murray's work [11] where he regards integration of knowledge as a knowledge rich type of machine learning, is the only closely related work we have found so far. In the ACKnowledge environment the integration of knowledge can be viewed as a **modelling activity**, where the knowledge engineer works to refine the Integrated KB - the model of the application. The integration function in KEW is a tool for supporting this activity. (See [5],-[8] for a more detailed description.)

In our approach to knowledge integration we have extended Murray's model of integration, to a computational model with the following four steps:

1. **Ordering - a search for an integration order of the knowledge pieces in the training set**
The learner[6] selects the next knowledge piece[7] to be integrated.

2. **Recognition - a search for relevant background knowledge**
The learner selects relevant knowledge structures to guide the interpretation of new information provided as training examples.

---

6 In our approach the learner is the KEW proposing and the user responding to these propositions.

7 A knowledge piece can be either a Temporary KB, a part of a Temporary KB such as the structure part of the resulting KB from INSTIL (a machine learning tool, see [7]), or a single rule or frame.

3. Elaboration - a search for consequences
The information content of the training is expanded as expectations and inferences provided by the retrieved knowledge structures are applied.

4. Adaptation - a search for accommodation
The learner's knowledge is modified to accommodate the elaborated information.

The integration process can hardly be realized without the presence of a Common Information Repository. The CIR provides common access to all Intermediate Information Structures produced by the various KA-tools in KEW, to all Temporary KBs as well as to the Integrated KB. The CIR provides linking facilities as well as searching and browsing. Integration of knowledge will release the synergy effect of using several different KA-tools within the same environment. Thus the CIR is a major contribution to this.

5. CIR as a basis for active help

As seen in fig.1 KEW support functions play a main role when using KEW. In [15] these functions are described as a knowledge based active advice and guidance module, which follows the user's operations in KEW. Because it has knowledge about the knowledge acquisition process in general and KEW in particular, it can give advice to the user on what to do next, based on the current situation in the knowledge base and the history of the current project. The information about the current state and the history comes from the CIR. The *KA-goal tree* holds the short term plan for the KA-process, the *project plan* holds the long term plan, the *project log* holds the history and the *KA activity status* holds the current status of the KA-process.

The main help capability of the advice and guidance module lies in the following two tasks:

- Select which problem to work on next.
  After evaluating the current state of the knowledge base the user must select one of the malfunctions in the knowledge base to resolve. KEW gives guidelines to the user in this selection based on the evaluation of the KB and on the current status of the KA-process. This information is provided by two parts of the CIR; Firstly by the Project Management Data Base in CIR in terms of a *project log*, a log of the activities carried out, which is linked to the *project plan* for the specific development project; Secondly by the Dynamic Knowledge Engineering Information in terms of a *goal tree* and the *KA activity status*.

- Select which tool to apply next.
  KEW holds a number of different knowledge acquisition tools for elicitation and modelling of knowledge and tools for refining the knowledge. KEW gives guidelines for selecting tools based on the selected problem and the information in

CIR about current status and history of the project (*project log, project plan, KA activity status and the goal tree*).

Without the repository of both knowledge and information about the knowledge and the actual KA-process, an active advice and guidance module would not be able to advice on short term planning, but would be limited to give general guidelines for problem and tool selection.

6. CIR as a basis for project management support

In addition to serving as a pool for knowledge integration and providing a basis for active help, the third main usage of the Common Information Repository in KEW is to support the management of the knowledge acquisition project. The project management database in CIR is the major source of information for project management.

The central part of this management information is the *project plan* with the attached *KA activity log* giving the status of the project related to the scheduled activity (e.g. application of a tool). There are also links from the activities in the activity log to the knowledge that was the result of this activity, to persons involved etc. This provides the user with valuable information about the progress of the project. Additional information for management is a description of *persons* involved in the project and a collection of *progress documents* and *annotations*.

There are two objectives for including project management facilities in the KEW:

– Some of the project management information is required as input to the active advice and guidance module.

– It will be convenient for the user to have access to more general project management functions in the same environment as used for the development. Since some of the project management information has to be provided for the sake of the advice and guidance module, a few extra functions for general project management are added to give the user a useful support for this task, although a full-scale project management tool is far from the scope of ACKnowledge.

7. Concluding Remarks

In KEW we have replaced the traditional single KB of development tool for knowledge based system by a Common Information Repository where all the knowledge and information relevant for the construction of the target knowledge based system. The CIR concept represents a significant improvement of a knowledge based system development tool and enables the incorporation of new functionality:

– The use of several different KA-tools in the KA-process will give a synergy effect leading towards a more efficient KA-process. The CIR provides a common medium for storage and retrieval of the various Intermediate Information Structures produced by the tools, and thus enables us to build functions for assimilating this knowledge into the Integrated KB. The CIR and the integration functions will amplify the synergy effect.

– The active advice and guidance module of the KEW will rely heavily on the information in the CIR during its reasoning. Based on the input from CIR such as the history and the current status of the KA-process, the long term and the short term plan, it will use its own knowledge to give advice and guidance to the user on what to do next. This could not be done on the basis of a single KB, unless through asking the user a lot of tedious questions every time he asks for help.

The development of KEW will have to face quite a few challenging issues, but our approach with the incorporation of several different KA-tools, the Common Information Repository, the assimilation functions and the knowledge based advice and guidance module, makes KEW the knowledge engineering's intelligent analogue to CASE-tools in software engineering.

Acknowledgement

The work published in this paper is part of the work done in ACKnowledge project, and thus is based on work done by other partners. In particular we like to thank Sandra Whitehouse (GEC-Marconi Research Center) and Agnar Aamodt (SINTEF) for taking part in formulation of the integration model, Bob Wielinga (University of Amsterdam), Ole Jakob Mengshoel (SINTEF) and Kjell Tangen (Computas Expert System) for taking part in developing the CIR data model and to Ingeborg Sølvberg (SINTEF) who provided valuable comments to this paper.

References

[1] J. H. Boose. Uses of repertory grid-centered knowledge acquisition tools for knowledge-based systems. *Int. J. Man-Machine Studies*, 27(1):287-310, 1988.

[2] C. Bouchet, E. Brunet, and A. Anjewierden. Shelley: An Integrated Workbench for KBS Development. In *Proceedings Expert Systems and Their Applications*, Avignon, France, June 1989.

[3] Joost Breuker and Bob Wielinga. Models of expertise in knowledge aquisition. In G. Guidaand C. Tasso, editors, *Topics in Expert System Design: Methodologies and Tools*, volume 5 of *Studies in Computer Science and Artificial Intelligence*. North-Holland, 1989.

[4] R. Davis and D. Lenat. *Knowledge-based systems in Artificial Intelligence*. Mc. Graw-Hill, 1982.

[5] J. Eggen, A.M. Lundteigen, and M. Mehus. Integration of knowledge from different knowledge acquisition tools. In B. Wielinga et al., editors, *Current Trends in Knowledge Acquisition*. IOS Press, Amsterdam, June 1990.

[6] W. Emde and K. Morik. The blip system. Report KIT-REPORT 32, Technische Universität Berlin, Berlin, West Germany, February 1986.

[7] INSTIL partners. *INSTIL, Project Summary,* January 1989.

[8] A.M. Lundteigen and M. Mehus. Integration of knowledge from different knowledge acquisition tools. Report STF40 A89225, SINTEF, ELAB-RUNIT, Trondheim, Norway, December 1989.

[9] Jean-Charles Marty, Nigel Shadbolt, and Bob Wielinga. The ACKnowledge project: Improving the knowledge acquistion process. Technical Report ACK-CSI-WID-RR-001-B, Cap Sesa Innovation, 1989.

[10] A. Motta, M. Eisenstadt, M. West, K. Pitman, and R. Evertsz. Keats: The knowledge engineer's assistant. Technical Report 20, Open University HCRL, 1986.

[11] K. S. Murray and B. W. Porter. Developing a tool for knowledge integration: Initial results. In *The third Knowledge Aquisition for Knowledge-based Systems Workshop,* Banff, Canada, November 1988.

[12] M. Musen, L.M. Fagan, D.M. Combs, and E.H. Shortliffe. Using a domain model to drive an interactive knowledge editing tool. In *Proceedings Of 1St AAAI Workshop On Knowledge Acquisition For Knowledge-Based Systems,* Banff, Canada, November 1986.

[13] Inge Nordbo, Steve Adey, Ole Jakob Mengshoel, and Peter Terpstra. Knowledge transformation. Technical Report ACK-STF-T2.1-DL-001-A, SINTEF, December 1989.

[14] 1. Solvberg. METAKREK: Methodology and tool-kit for knowledge aquisition. Report STF14 A88046, RIK-88-9, Sintef/Runit, Trondheim, Norway, December 1988.

[15] Bob Wielinga et al. Conceptualization of a knowledge engineering workbench. External deliverable ACK-UvA-T1.4-DL-010-A, University of Amsterdam, 1990.

Project No. 2615

Two Demonstrators as a Test-Bed for the Construction of Intelligent Training Systems in Industrial Environments

A. Bertin
CISE
via Reggio Emilia 39
1-20090 Segrate
Milano
Italy

E. Burguera
Iberduero SA
Gardoqui 8
48008 Bilbao
Spain

D. M. Scott
Marconi Simulation
Napier Building
Donibristle Industrial Park
Nr. Dunfermline
Fife KY11 5JZ
U.K.

Abstract

The two demonstrators under development in Esprit P2615 (Intelligent Training Systems in Industrial Environments) are described. It is argued that they are representative of ITSIEs and will provide a suitable test-bed for the construction of ITSIEs. Some information is given on the status of the domain modelling tools needed to construct them, the training requirements, the model of the trainee, and the user interfaces.

1. Introduction

This paper gives an overview of the work carried out during the first year of the ITSIE (Intelligent Training Systems in Industrial Environments) Project, Esprit P2615. Most, but not all, of this work is reported in more detail in the Deliverable "Demonstrator Requirements Specifications" [D1, 1990]. This paper follows on from "Prospects for Intelligent Training Systems in Industrial Environments" [Leitch and Horne, 1989] which puts the project in context.

In this paper an ITSIE is defined as a training system which: Adapts to the needs and abilities of the trainee; Is constructed using techniques from knowledge based systems so that the information that it contains is separated from the mechanisms that use that knowledge; Has simulation of the physical world as an integral part of the training environment. A CBT system may posses any or all of these properties to some degree: what makes an ITSIE is the presence of all of these elements in a well developed form.

The principal aims of this project are to produce a reference architecture for ITSIEs, to furnish tools with which ITSIEs with that architecture can be implemented, and to produce a methodology to be followed during the implementation of such ITSIEs. The plan for the project calls for the development of two demonstrators. This has two objectives: to produce a focus for thought whilst developing the architecture, tools, and methodology, and to provide a test-bed for all three elements on which their adequacy can be evaluated.

Both demonstrators will teach correct behaviour by means of instructional methods which use two explicit representations of domain knowledge: executable structural models (simulation), and domain expertise in the form of known associations between observables and actions. In Demonstrator A the trainee should improve his/her awareness of risk, whilst Demonstrator B will provide practice in the execution of previously taught procedures (e.g. start-up and shut-down). The demonstrators differ in that in A there is no real time scale - only sequences of events in time (the ordered actions of a procedure); conversely in B there is a time scale for events which must correspond roughly to real time (actions must be completed within certain times). Put another way, in Demonstrator A the simulation that is used is static, whilst in Demonstrator B dynamic simulation is essential. Another, less significant, feature is that in A spatial awareness is important (the trainee must know where his/her body is in relation to the electrical equipment being maintained), whilst in B spatial awareness is not so important.

The project's architecture for an ITSIE has three knowledge bases: the domain knowledge base, the training knowledge base, and the trainee knowledge base. The domain knowledge base contains information on the industrial system of concern - some of this information is in the form of executable models (simulations) of the system. The training knowledge base contains information on the various methods of teaching that are available. The trainee knowledge base contains a representation of each trainee. This representation is specific to the trainee so that the system can adapt to the individual. The separation of these knowledge bases from the mechanism that use them is crucial for the development of a generic approach to the construction of ITSIEs.

In sections 2 and 3 the demonstrators are described in outline. In section 4 domain modelling in the two demonstrators is examined, and the tools which have been adopted for this are described. Section 5 is devoted to a discussion of the training to be provided by the demonstrators. In section 6 modelling of the trainee is discussed. In section 7 the requirements for the trainee and instructor interfaces are set out. Finally, section 8 provides a conclusion.

2. An Outline of Demonstrator A

Demonstrator A will provide recurrent training to maintenance workers in the low voltage (25-400V AC, and 50-600V DC) electricity distribution network in Italy. This is an important area of training because of the number of people involved (40,000 people a year are subjected to it, each receiving 40 hours of training concentrated in a period of one month), and because of the frequency and seriousness of the accidents which occur during maintenance: at these voltages it is carried out on live equipment and in 1987 thirty accidents occurred per one million working hours resulting in 0.65 days off work per 1000 working hours. A complete ITSIE of this type would make training more effective, reduce its cost, and would provide a step towards standardization of training methods throughout Italy.

Demonstrator A will support existing training courses on the installation and maintenance of low-voltage equipment, with the emphasis on safe working practices. The analysis of existing training courses and of accident reports has shown where current training fails and why most accidents occur: workers who have been performing the same or similar maintenance tasks for a long time try new ways of working, either because they have become blase or because they wish to save time or money. In order to counter this the training will aim to foster awareness of the risk involved in carrying out a maintenance task.

The demonstrator will review information on technical matters such as the structure, behaviour, and purpose of equipment. Most importantly, the procedures for maintenance tasks will be reviewed with the emphasis on safe working practice and the evaluation of risk. The reviews of maintenance tasks will be based on simulation.

3. An Outline of Demonstrator B

Demonstrator B will provide training for operators of the Santurce II oil-fired power station in Northern Spain. Current training of operators is divided into three phases:

1) overview of plant equipment and concepts
2) operating principles and procedures
3) operation of the plant under supervision

Demonstrator B will support the third phase of training. There is a need for extra training in this phase because the Santurce power station is used only as a backup. As a consequence it is used only four or five times a year, and then for only 4-6 weeks at a time. This means that it is difficult to maintain a good progressive training programme for it is dependent upon the operation of a plant whose operation is infrequent.

Whilst simulation will be an important element of Demonstrator B it should be noted that the demonstrator will be an ITS and not merely a simulator. It will adapt to suit the needs of the trainee, and will be capable of performing expert demonstrations of standard procedures.

A simulator will obviously have benefits in providing a safe environment for training but, more importantly, it can provide a flexible environment which can be manipulated to provide abnormal and emergency conditions rarely seen at the plant. With the current training programme, carried out on the actual plant, it may take several years before the trainee operator has encountered all of even the most common states of the plant. Furthermore, in the current training programme the trainee is allowed to practise only non-critical procedures. With a simulation-based training system the trainee can be allowed to attempt critical procedures in complete safety, being limited only by the capabilities of the simulation. In sum, training can be provided irrespective of the operation of the plant and so structured to the requirement of the trainee rather that the requirements of the plant.

The training to be provided by Demonstrator B has two principal objectives. Firstly, the trainee must be able to perform standard operations on the plant, and follow normal, abnormal, and emergency procedures as laid down in the operation manuals. Secondly, the trainee must develop a conceptual (qualitative) model of the plant in order to be able to reason about the plant and so modify and create procedures to cope with abnormal and emergency situations not covered by the procedures. Gaining experience is vital to the development of these skills.

Attention will be focused on the operation of the boiler. In particular the start-up, shut-down, and certain malfunction procedures.

4. Domain Modelling

ITSIEs aim to give training in industrial domains, so ways of modelling these domains are required. Some of these models need to be executable, i.e. they must simulate some aspect of the domain.

Domain modelling is the first area in which progress has been made in the project and where the firmest conclusions have been reached. We have identified a number of tools that will be required and implementation is well in hand.

After analysing the requirements for domain modelling by the two demonstrators it was observed that a subset of the tools from the QUIC Toolkit of Esprit P820 [Leitch and Stefanini, 1989] met most of them. (These tools were developed as a result of an epistemological analysis of the domain of process control.) Consequently, those tools are being, or have been, ported to the software development environment chosen for the project, which is a CLOS (Common Lisp Object System) environment called LispWorks. The choice of this environment reflects the fact that an object-orientated approach is to be adopted throughout the project.

A common feature of the domains of the two demonstrators is that a component based description of their physical structure and behaviour is appropriate. The *Component Based Language* (CBL) from the QUIC Toolkit has been adopted to describe this structure and the equations from which a description of the behaviour of the system can be generated. The way this language is manipulated is being modified so that components can be added or removed dynamically so enabling the simulation of assembly and disassembly tasks, as required by Demonstrator A.

The CBL allows a hierarchical description of the structure of a physical system so creating structural models with varying degrees of detail [Leitch, Wiegand and Quek, 1989]. This may be coupled to various simulation mechanisms representing different abstractions (quantitative and qualitative) so producing multiple behavioural models which form a structured set:

1) Qualitative static models will be created by using a *constraint propagator* from QUIC.
2) Qualitative dynamic models will be created by using a *predictive engine* from QUIC.
3) Quantitative static models will be created by using a *linear equation solver* from QUIC.
4) Quantitative dynamic models will be created by incorporating a *numerical integrator*.

The use of a common knowledge base written in a declarative style (the CBL description) allows progressive evolution between models in a consistent manner. This is much more difficult in multiple models utilising different models of structure. Demonstrator A will provide a quantitative (numerical) static model of electrical circuits. Demonstrator B will provide both qualitative and quantitative dynamic models of a section of the plant. Static qualitative models will also be employed.

Event graphs (a generalisation of Petri nets) from QUIC are to be used for encoding information about correct procedures in both demonstrators.

Production rules will be used to represent heuristic knowledge and safety regulations in Demonstrator A.

5. Training Requirements

Training is the core aspect of this project and much effort is currently going into investigating it. A number of conclusions have been drawn but they relate mainly to what the trainer must be able to do rather than the tools that will be used to provide that functionality.

Both demonstrators are to be used in conjunction with existing training pro-grammes. As a consequence they must complement the existing programmes and fit into them easily. This means that the general approach to training must be that of the existing programme.

The training given must be adapted dynamically to the perceived needs of the trainee. The responses of the trainee should be analysed in order to identify possible misconceptions, and if misconceptions are identified remedial action should be taken.

Bloom's taxonomy of training goals in the cognitive domain [Bloom et al., 1956] has proved useful for describing our requirements. In the affective domain the training objectives of persistence and motivation have been identified. There are no training objectives in the *psychomotor* domain for either demonstrator. In the cognitive domain many objectives have been identified which can be classified into the following sub-domains:

knowledge
> Demonstrator A:
>> to reinforce and refresh knowledge of procedures, technical matters, and safety regulations.
>
> Demonstrator B:
>> to reinforce and refresh knowledge of procedures.

comprehension
> Demonstrator A:
>> none have been identified.
>
> Demonstrator B:
>> the trainee must learn to interpret the plant through the control panel of the plant.

application
> Demonstrator A:
>> the trainee must learn to apply his/her theoretical knowledge in practical situations.
>
> Demonstrator B:
>> the trainee must learn to apply the procedures learnt in phase 2 of training.

analysis
> Demonstrator A:
>> the trainee must be able to analyse equipment in accident scenarios to determine the cause of an accident.
>
> Demonstrator B:
>> the trainee must be able to analyse the current state of the plant in order to determine the source of any anomoly in behaviour.

synthesis
> Demonstrator A:
>> the trainee must be able to formulate a plan to get out of a hazardous situation.
>
> Demonstrator B:
>> once a trainee has detected an anomoly he/she must be able to devise a plan to correct it using either a known or new procedure.

evaluation
> Demonstrator A:
>> the trainee must be able to make a conscious assessment of how dangerous a given hazard is.
>
> Demonstrator B:
>> the trainee must be able to evaluate the effects of his/her actions and adjust his/her behaviour accordingly.

The provision of multiple models, whether simulations or viewpoints, is seen as an important component of training. In Demonstrator B in particular, the provision of multiple models of the plant should aid the development of similar models in the mind of the trainee.

6. The Model of the Trainee

The training system must maintain a model of the trainee (trainee representation) so that it can adapt to the needs of the individual. A separate model will be constructed for each trainee, but stereotypical models must be provided for initialisation of the system at the start of the training programme. In Demonstrator A, for example, four stereotypes of trainees have been proposed which depend on the educational background and work experience of the trainee.

Once training is in progress the model will be updated as the session proceeds. The model will include a history of the training given and the trainees performance. If any errors are detected these will be recorded, as will any identified misconceptions.

This model will of course be saved between sessions so that it can be used to initialise the next session.

The current state of the model will influence the choice of training strategy, the types of explanation used, and the choice of abstraction and resolution used in presenting models of the system.

7. The Interfaces of the Trainee and Instructor

Interfaces will have to be provided for both the trainee and the human instructor. The interface for the instructor will allow overall planning of the training programme down to the session level. It will also be possible for the instructor to influence the operation of the system, with the adjustments made between sessions. Furthermore it will allow the instructor to examine the trainee representation so that the instructor can make his/her own assessment of the performance of the trainee.

The interface for the trainee will allow the simulated performance of tasks in both demonstrators. In Demonstrator A they will be maintenance tasks, whilst in Demonstrator B they will involve operation of the control panel. In other words the system is to be a kind of simulated laboratory in which the trainee can improve and verify his/her capabilities.

New tools for constructing interfaces will not be developed by this project. The choice of tools will be made from those that are commercially available, or non-commercial tools of established good quality.

8. Conclusion

We have given an overview of the requirements for the construction of two demonstration training systems. The domains of these demonstrators, the current training methods, and the analysis of these leading to requirements specifications are described in detail in D1 [1990]. Detailed functional specification of the demonstrators is in progress.

We have shown the two demonstrators to be complementary and we are satisfied that they will prove suited to their role in the project which is to produce a focus for thought whilst developing the architecture, tools, and methodology, and to provide a test-bed for all three elements on which their adequacy can be evaluated. They are representative of ITSIEs that do not have objectives in the psychomotor domain. Psychomotor goals are currently taught through the use of full-blown simulators which are outside the scope of this project.

Acknowledgements

The authors of this paper have been chosen to represent the many people who have contributed to the work reported here. Amongst those who have made a significant contribution are: Fabio Buciol, Cristina Lanza, Emma Nicolosi, Alberto Stefanini, Mariuccia Casanova, Daniele Marini, Jose Ramon Lezameta, Jose Luis Los Arcos, Inaki Angulo, Isabel Fernandez de Castro, Arantza Diaz de Illarraza, Julian Gutierrez, Catherine Peyralbe, Erik Hollnagel, Thierry Belleli, Roy Leitch, Alan Slater, and last but not least Julie-Ann Sime.

The work reported here was partly funded by the CEC under the ESPRIT programme.

The members of the consortium are:
Marconi Simulation
CISE
Iberduero, LABEIN, University of the Basque Country
Laboratoire de Marcoussis
Axion/CRI
Heriot-Watt University.

References

Bloom B. S., Engelhart M. D., Furst E. J., Hill W. H. and Krathwohl D. R., 1956, Taxonomy of Educational Objectives- The Classification of Educational Goals. Handbook 1 Cognitive Domain, Longmans, London

D1, 1990, Demonstrator Requirements Specifications, Marconi Simulation Document 2615-1-MS-ER 002

Leitch R. and Horne K., 1989, Prospects for Intelligent Training Systems in Industrial Environments, Esprit Conference Document XIII/416/89

Leitch R. and Stefanini A., 1989, Task Dependent Tools for Intelligent Automation, AI in Engineering, Vol. 4, No. 3

Leitch R., Wiegand M. and Quek C., 1989, Coping with Complexity in Physical System Modelling, AI Communications, Vol. 3, No. 2

COMPUTER INTEGRATED MANUFACTURING

EP 932 FINAL RESULTS: CIM/AI VALUE ANALYSIS

W. Meyer
Philips Forschungslaboratorium
Vogt-Kölln-Str. 30
D-2000 Hamburg 54
now: Technical University
of Hamburg-Harburg,
Postfach 90 14 03
Process Automation Department
D-2100 Hamburg 90
FRG

Helena M.J. Walters
BICC Technolgies Ltd.
Systems Development Centre
Quantum House, Maylands Ave.
Hemel Hempstead HP2 45J
United Kingdom

1. Introduction

In this paper, we present a value analysis on the benefits of applying AI (Artificial Intelligence) techniques, especially object-oriented modelling and expert systems, to CIM (Fig. 1). As no defined methodology yet exists for establishing the value that can be placed on applying a new technology (such as AI) to a particular domain (such as CIM), the purposes of this paper are:
(1) to define the markets of CIM/AI products, thus taking a product view on the CIM/AI business and not the conceptual, technological and functional views as done in previous publications (Meyer 1986, 1987, 1988);
(2) to point out the expectations and drawbacks of appraising internal CIM/AI projects;
(3) to assess economic and strategic justification methods for CIM/AI projects; and
(4) to apply a hybrid evaluation technique based on quantitative and qualitative criteria to the CIM/AI software developed within EP 932.

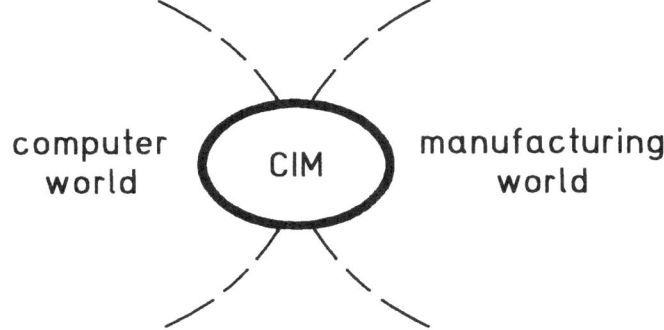

Figure 1. CIM: merging two worlds of technology and application.

2. CIM market

The worldwide CIM market, as seen from the perspectives of both factory automation vendors and information technology vendors, is shown in Table 1.

TABLE 1.   Worldwide CIM market (1990 projections): segmentation by CIM functions (left); segmentation by IT products (right).

manufacturing function (CIM)		billion $
CAD /	CAE	15
CAM/	CAQ	15
	PLC	6
	NC	4
CAL /	PMS	20
	PCS	7
	LAN	3
sum		70

information technology (IT)	billion $
mainframes and minicomputers	3
graphic workstations	5
turnkey - CAD	6
unbundled software	7
system integration, LANs	5
other IT hardware	4
sum	30

The largest market is in the US, followed by Europe and Japan (Fig. 2). 40 per cent of the 70 billion dollar factory automation market of Fig. 3 are expected to be integrated systems – whatever that means. Indeed, CIM market surveys from different sources (Macintosh, Gnostic, Arthur D. Little, Yankee Group, Bipe) largely deviate as the scopes of CIM and factory automation are seldom clearly defined. An additional complication arises from the difficult cost structure of CIM systems: implementation cost (e.g. organizational changes as a consequence of new communication structures or planning procedures) may easily surmount the actual hardware and software investments, especially in the CAD/CAL/CAQ areas (Fig 4).

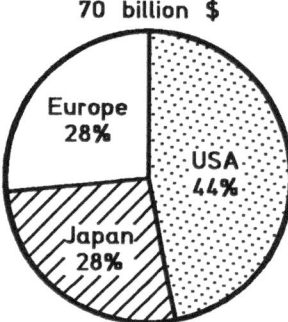

Figure 2. The worldwide factory automation market in 1990: segmentation by country.

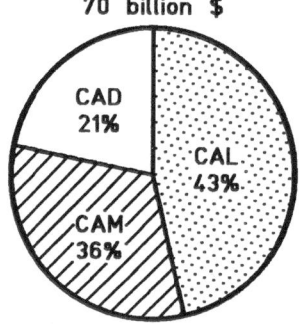

Figure 3. The worldwide factory automation market in 1990: segmentations by application area.

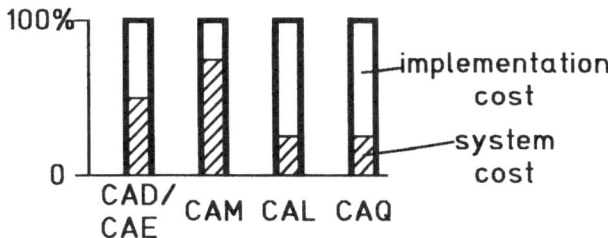

Figure 4. The cost of CIM systems are mainly organizational cost.

3. AI market

Similar to CIM with respect to factory automation, AI is a logical progression in the evolution of information technology and thus is difficult to separate from the main stream of IT industry. The integration with main stream IT is proceeding on several fronts:
- Expert systems are increasingly being developed by data processing or information departments inside companies, rather than by specialized expert system groups.
- AI and XPS tools are being seen, and sold, as part of a toolset for developing applications which will only have one AI component among others.
- The tool's potential for use in a much wider range of applications than pure AI is being recognized by suppliers and developers.

In fact this is part of an established tradition. AI research comes up against the hardest problems for computer science, and AI researchers are often the first to encounter and solve those problems. As soon as their solutions have found their way into mainstream computing, AI turns into new directions: 'If it is solved, it is not AI'. Past accomplishments include parsing techniques and virtual memory management. Recent developments include object-oriented and rule-based programming.

There are other reasons why it is difficult, if not impossible, to make estimates of the true size of the AI world. For instance, commercial sales figures as in Fig. 5 do not take into account most of the substantial inhouse investments companies are making in AI. Such investments may easily double the numbers of Fig. 5. So there may be some doubt about the absolute figures. Most industry watchers do agree, however, that the lion's share of activity in most AI segments (expert systems included) still is defence-related. Of the remaining portion, the manufacturing expert system segment is the largest, comprising about 15 per cent (Fig. 5b). From a recent (1989) inhouse investigation of a large European multinational company, Fig. 6 identifies the major industrial applications of expert systems.

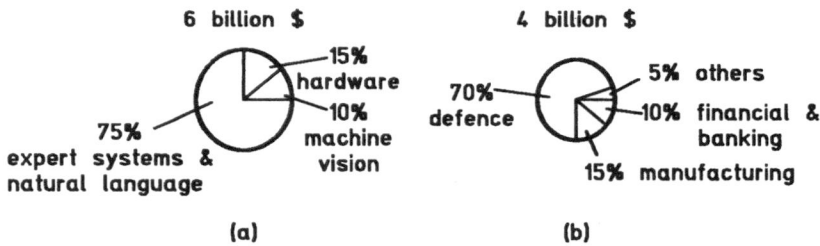

Figure 5. (a) The worldwide AI market in1990: segmentation by product type. (b) The worldwide expert system market in 1990: segmentation by application area.

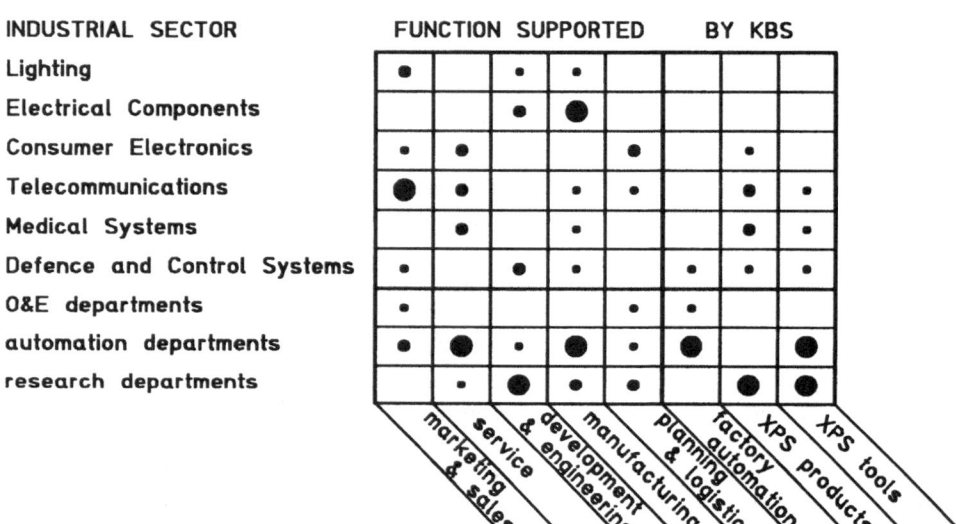

Figure 6. The importance of knowledge-based systems (KBS) per industrial sector and business function. (Dot size correlates with number of implementation.)

4. Application of AI in CIM

The following criteria can be used for assessing application areas with the highest potential for success:

– Recognized experts are available (and the expertise is concentrated in a limited number of staff).
– The experts spend a large proportion of time on routine decision making.
– A particular area of expertise is required infrequently and hence requires proportionally higher effort each time.
– Critical judgements must be made in a short time.
– A great amount of inaccurate information is available.
– The skill is routinely taught to students of the profession.

The scope for the application of AI in manufacturing industry is very wide,

encompassing most functions and disciplines, for instance:
Design
- interpretation of design guidelines
- intelligent CAD and front ends for designing programs
- component selection and product configuration.
Planning
- KBS modelling for workcell configuration and layout
- intelligent process planning
- scheduling
- project planning.
Automation of production processes
- robotics
- visual detection of parts
- intelligent sensors on machinery for monitoring
- intelligent quality systems
- diagnostic systems for maintenance
- fault prediction systems for components and machines
- intelligent energy management systems.
Commercial and management
- corporate planning aids
- product selection systems for customers
- intelligent retrieval of database information
- skills archiving
- computer-aided training.

The three most popular application areas are diagnosis of product quality and machinery, process planning and scheduling. Diagnosis is essentially a decision making process consisting of identifying the problem, determining its cause and prescribing a remedy. Process planning involves a number of evaluation and selection steps; scheduling, the allocation of resources to various product requirements within specified time constraints and a chosen time horizon. These problems involving the choice of a strategy are particularly suitable to being solved by AI techniques because solutions are equally based on logics, intuition (expressed as heuristic rules) and conventional, algorithmic techniques.

5. EP 932 result statistics

The present state of applying AI in manufacturing industry is one of prototyping and internal exploitation. Typical are the findings of ESPRIT project EP932: 'Knowledge-based realtime supervision in CIM'. This project, finished in 1990, is sufficiently large and advanced to allow for generalization of its results. These results (or products) are 55 knowledge-based systems being developed, tested and implemented for the production planning, quality and maintenance functions in a number of different manufacturing environments. The main software packages are listed in Table 2 (Meyer and Isenberg 1990). Other products address the integration

462

and supervisory control aspects. Expert system shells have been developed
specifically for the action planning and fault diagnosis domains. Further tools include
interactive modules for the acquisition of organizational knowledge and expert
domain knowledge. Several generalized methodologies for system analysis and
design have been applied and extended. The main statistics relating to the results
of the project are as follows (Schtaklef, Gilmore and Walters 1990).

TABLE 2. Main software packages developed within EP 932.

EP 932 prototypes and products	company
(1) KB Shell for Exception Handling	AEG
(2) KB Rescheduling Assistant	AEG
(3) Diagnostic XPS Industrial Prototype	ARS
(4) Maintenance Dynamic Scheduling XPS	ARS
(5) Model-based Production Planning and Scheduling Industrial Prototype MOPPS	BICC
(6) Production Flow Simulator	BICC
(7) Workcell Configurator	BICC
(8) Factory Loading Assistant	BICC
(9) Due Date Estimator	BICC
(10) Interactive Scheduler	BICC
(11) Planning and Scheduling System Design	BICC
(12) Action Planning Shell	FIAR
(13) Temporal Module	FIAR
(14) Maintenance Diagnostic Tool	FZI, IPA
(15) Unit Controller Shell	LIA, SGN, TITN, CEA
(16) Tyre Quality Assurance TQA	Pirelli, SISAV
(17) Workcell Controller FLEX	Philips
(18) Temporal Inference Component TIC	Philips
(19) Flowline Simulator FLOPAS	Philips
(20) Object Oriented Kernel COOK	Philips
(21) Blackboard Shell BOOK	Philips
(22) LISP Runtime Monitor COMO	Philips
(23) Workcell Controller - AIPLANNER	Philips
(24) Knowledge Acquisition Methodology (KAM)	Politecnico di Milano
(25) Knowledge Acquisition Expert KAE	Politecnico di Milano
(26) Quality Information System	RWTH, FZI
(27) Decision Network Acquisition Tool DNAT	SGN, CEA, LIA

Result statistics by function area

The three CIM functional areas addressed by EP932 are Production planning
(P), Quality control (Q) and machine Maintenance (M). Just under half of the results

were applications in the production planning area. Overall, about 75% of the results were specific to one of the three areas, whereas the remaining 25% were applicable to more than one area. These are shown within Fig. 7 as sectors PM and PQM. Of these, most pertain to concepts and software engineering work supporting the function area specific results.

Result statistics by organization level

The CIM reference model (section 3.5) considers the factory as a set of shops and the shop as a set of workcells. At the lowest level, the workcell consists of one or more workstation (machine) resources.

Fig. 8 shows that the majority, 58%, of the results are targeted on the workcell level. That number is double the 29% of results than can be applied to the higher levels, with shop applications in the minority. The remaining 13% of results are applicable to more than one level, shown by the WS and FWS sectors. These results consist mainly of the work carried out on concepts and software engineering methodologies.

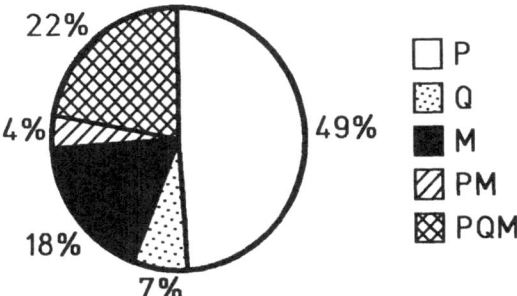

Figure 7. EP 932 result statistics by function area.
P: production; Q: quality control; M: maintenance;
PM and PQM: integrated systems.

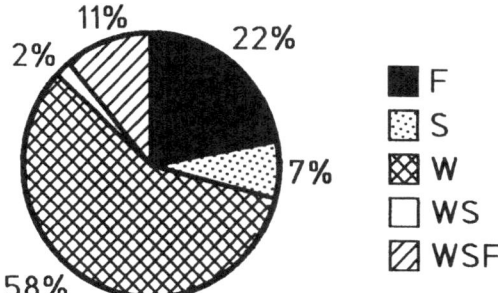

Figure 8. EP 932 result statistics by organization level.
F: factory; S: shop; W: workcell;
WS and WSF: integrated systems.

Statistics by result type

The 55 knowledge-based results of EP 932 have been categorized into one of four types which are
- standards: generally applicable methodologies
- prototypes: results demonstrated in laboratory trials
- internally exploited results: used directly by partner or associated company
- marketed products: used by external companies.

Fig. 9 shows that 89% of the results are either prototypes or internally exploited. The remaining 11% consist of five standards and one marketed product. These percentages are typical for the present state of AI in manufacturing.

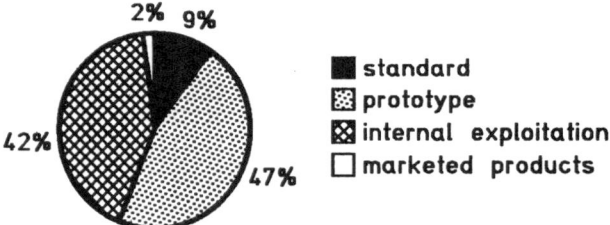

Figure 9. EP 932 statistics by result type.

6. Qualitative value analysis

The benefits emerging from the application of AI in CIM cannot always be measured in monetary terms. The cost accounting techniques in use at the particular application site determine the information that is measured for financial and management control. Information that cannot be readily measured does not form part of cost accounting. Although its effect may be directly apparent, this type of benefit can only be judged subjectively.

One example of a qualitative value is internal communications, that is, the exchange of information between the various stages of a product's life cycle, from design to distribution. No single measurable value can indicate the amount of information exchanged in the total process. Any benefits obtained through internal communications, can therefore only be estimated subjectively, not numerically.

In the course of EP 932, qualitative criteria were established that best represent the non-measurable benefits on an application site. The criteria that have been selected for the qualitative value analysis, and their explanations are as follows.

1. Labour morale   : job satisfaction due to shifts in skills.
2. Product quality   : standard of quality of the finished products.
3. Production schedule   : adaptability of scheduling to real situations and their
   flexibility   dynamic conditions.
4. Product variety   : range of products that can be made.
5. Safety   : prevention of accidents.
6. Response to change   : response to changes in the external environment,

such as market demand.

7. Company image : customers' view of the company and its reputation.

8. Internal communications : communication between departments such as engineering, accounting and production.

9. External communications : communication with the suppliers and the customers.

10. Risk : uncertainty when making high level decisions such as large investments.

11. Product innovation : supporting product development by feedback of information from production and quality control.

Since the benefits are qualitative, it is necessary to define a method for totalling up the benefit level for each criterion. The method is known as ranking method. First the benefits for each of the eleven criteria are expressed as one of four descriptive levels, namely *none, small, medium* or *large*, for each of the 55 KBS products. The total benefit per criterion is proportional to the number and magnitude of the individual benefit levels recorded for each of the criteria. Criteria with a larger number of recorded benefits are ranked higher on a scale of 1 to 11. Criteria having the same number of benefits are resolved using the magnitudes of the benefits and if this fails, they are given equal ranks. The rankings, thus arrived at, are of course as subjective as the data supplied.

The analysis of the responses to the survey are presented as a bar chart in Fig. 10. The bars are shown as cumulative totals of the contributions from P, Q and M function areas.

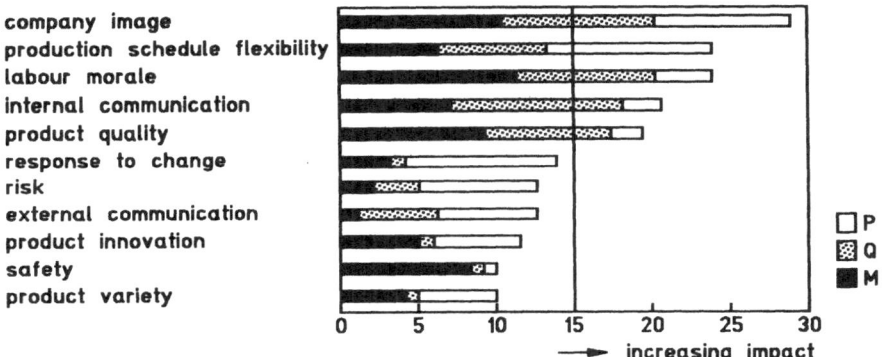

Figure 10. EP 932 qualitative benefits. (Axis shows the cumulative ranking sum which has a maximum possible value of 33.)

It is interesting to note that the relative contributions show marked differences between criteria for the different areas. Labour morale and product quality seem to benefit more from quality and maintenance applications than from production planning applications. One of the possible explanations may be that production planning is carried out at the supervisory levels, thus having minimal effects on shop floor labour morale. Production planning does not have a great impact on product quality. Production planning applications offer more benefits at the management level, providing better scheduling flexibility and better response to change. The final major benefit observed is that of improved internal communication offered by the quality applications. It is inferred that the quality applications allow better information exchange between the quality engineers and the rest of the production and design teams.

Note that the highest ranked benefit observed, improving the company image, is only an indirect consequence of KBS applications. Note also that for this criterion it is ranked equally high for each of the three function areas.

7. Quantitative value analysis

This section is concerned with the benefits of CIM/AI applications that can readily be measured in monetary terms, i.e. with the quantitative criteria. Each quantitative criterion represents a cost item contributing to the total expenditure of the application site. Any cost saving achieved by the KBS application to any of the cost items, is considered as a quantitative benefit. The sum of the individual benefits, therefore represents the reduction in total expenditure of the application site. The criteria that best represent the measurable benefits on an application site, and their explanations are as follows.

1. Labour : staffing on the shop floor, clerical and other office staff.
2. Raw materials : raw materials purchased or parts subcontracted.
3. Work-in-progress : materials, parts, components and subassemblies waiting to be processed.
4. Finished goods : stock of finished products awaiting shipment.
5. Scrap : amount of waste in the manufacturing system.
6. Reworks : parts, components and products that require remaking or repairing.
7. Productivity : utilization levels of all resources, including equipment and staff.
8. Production throughput : amount of production per unit of time.
9. Flow time : time needed to produce a finished product from the first operation to dispatch.
10. Order delay : number of orders delivered late.
11. Floor space : floor area required for production.
12. Tools : utilization of machine tools.
13. Energy consumption : total energy consumed in the factory.
14. Materials handling : equipment needed to move materials and products around the site and to customers.
15. Market share : proportion of the total market demand for products supplied.
16. Financial management : cash flow and general financial health.
17. Customer support : dealing with customer enquiries and complaints (pre- and post-sales support).
18. Training : staff training.

19. Capital   : assets held on the factory site.

20. Equipment life   : life of equipment needed to build products, depending on the type of processing.

The quantitative benefits for each of the twenty criteria are expressed in terms of two measures, an absolute monetary figure and a relative percentage figure. The absolute figure is the annual cost saving supplied as one of several ranges of values, e.g. 0 to 10 K$, 10 to 100 K$, and over 100 K$ (K$ = 1000 US dollars).

The percentage figure relates the absolute cost saving to the cost contribution of the criterion in question. The importance of this figure is demonstrated by the following example: assuming that an absolute saving of 0 to 10 K$ was achieved for the improvement in WIP level. This may not appear to be a large saving, but it could represent a 30% reduction in WIP. Hence, the percentage figure shows the actual extent of the saving relative to the part of the process being improved.

Absolute quantitative benefits

In the quantitative analysis, the worst case approach is normally adopted, where the quantitative benefits are supplied as ranges of values and the minimum values of the ranges chosen are used in the calculations. The total benefits for the absolute figures is arrived at by summing up the individual benefit levels. Each criterion is then ranked on a scale of 1 to 20, with 20 given to the largest total figure and 1 to the smallest total (or where the benefit is nonexistent for a criterion). The ranked absolute figures, grouped into the three functional areas, show marked differences between P, Q and M (Fig. 11).

Cost saved by increasing the production throughput seems to be one of the most

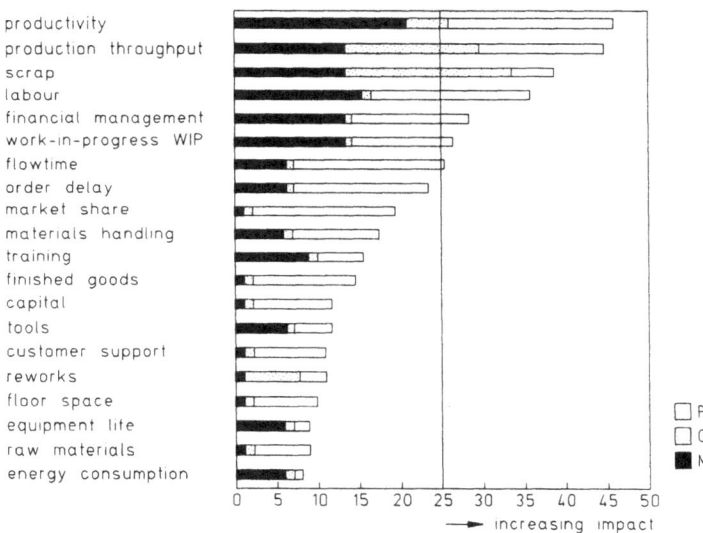

Figure 11.   EP 932 quantitative benefits. (Axis shows the cumulative ranking sum which has a maximum possible value of 60.)

important benefits in all of the three functional areas. Higher throughput is achieved by better usage of resources, equipment and labour.

Resource productivity, including labour, is seen to benefit from the production planning and maintenace applications, in turn having a positive effect on throughput. The scrap criterion indicates benefits from the quality applications, presumably leading to higher production throughput as well. Benefits in three other criteria, WIP, flow time and order delay, indicate additional production planning and maintenance applications where the effects are positive and may also effect throughput.

One criterion that offers major benefits, but is not related to throughput, is the saving in the cost of financial management offered by the production planning and maintenance applications. The remaining major benefit is the positive effect that production planning applications have on the market share. This criterion may be considered an indirect effect of improved flow times and a reduction of delayed orders.

Relative quantitative benefits

The results of the percentage benefits analysis indicate that, after ranking, the first nine criteria provide major benefits in percentage terms as were identified with the absolute value analysis of Fig. 11. In addition, three criteria that achieve minor absolute benefits show major relative benefits (Fig. 12).

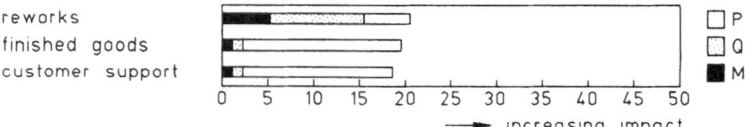

Figure 12. EP 932 quantitative criteria with major benefits in relative terms only.
(Axis shows the cumulative ranking sum which has a maximum possible value of 60.)

These three criteria indicate high benefits in relative terms but not so much in absolute terms, possibly because they form a smaller part of the total site cost than the other criteria. Lower rework levels are attributed mainly to quality control applications, while finished goods and customer support are attributed mainly to production planning applications.

8. AIPLANNER value analysis

The statistics of section 7 rest on cost-benefit analyses of each single expert system incorporated into the survey. A *cost*, or more specific, the development cost of the expert system is a measurement of the amount of resources needed required to obtain a product. Costs are expressed in quantitative dollars required. *Benefits* take the form of cost saving, cost avoidance, generation of new revenues, and intangibles, the latter focussing on improving the effectiveness of the organization. As is well known to practicioners, the main value of CIM/AI projects (and the main obstacles against them) are intangible. Is there a point, then, in doing a cost-benefit analysis if, at the outset, there seem to be unsurmountable problems in gaining accurate data? The answer is clear. Because cost-benefit analysis is the standard tool in organizations for ranking future expenditures and past performance, it must be done. A CIM/AI project cannot get far without some type of cost-benefit analysis.

Quantitative ROI analysis

IRR (internal rate of return; also called DCF: discounted cash flow) and NPV (net present value) are the primary methods for driving corporate decision making for information technology. The procedures are well known but too complex for inclusion into this basic text on CIM/AI; we, therefore, restrict ourselves to a simple ROI investigation (return on investment; also called accounting rate of return) despite its shortcomings which render it not being viable as a stand-alone justification method when competing for investment funds.

To calculate simple ROI, we use a set of three worksheets proposed by Parker et al. (1988):
(1) development worksheet (net investments)
(2) ongoing-expense worksheet (maintenance and operations cost)
(3) economic impact worksheet (cost-benefit performance).

The economic impact is based upon a straight-line relationship to calculate simple ROI of the annual net cash flow of the product or project under question (e.g., of the AIPLANNER expert system). The ROI is obtained by dividing the average annual net cash inflow by the net investment required (Table 3).

TABLE 3. Expert system economic impact worksheet (cont. overleaf).

XPS development cost worksheet			year 1
1.	Development effort		
	1.1	Incremental systems and programming (days times xxK$/day)	....
	1.2	Incremental staff support (data administration times xxK$/day)	....
2.	New hardware purchased		
	2.1	Terminals, workstations, printers	....
	2.2	Communications, others	
3.	New software purchased		
	3.1	Packaged applications software	....
	3.2	Communication software	....
4.	User training		....
5.	Other		....
	TOTAL (xx K$)		

TABLE 3. Expert system economic impact worksheet (cont.)

XPS ongoing-expense worksheet	year 1–5
1. Expert system maintenance	
1.1 Development effort (days)	....
1.2 Ratio of maintenance to development	
(based on experience, e.g. 2 to 1)	....
1.3 Resulting annual maintenance days	....
1.4 Maintenance rate (xx K$/day)	....
TOTAL maintenance (xx K$/year)	....
2. Incremental data storage required (MB)	_____
2.1 Cost (e.g., estimated MB at xx K$)	....
3. Incremental communications (lines, messages etc.)	_____
4. New software or hardware leases	....
5. Supplies	....
6. Other	....
TOTAL ongoing expenses (xx K$)	....
AIPLANNER economic impact worksheet	TOTAL
1. Amount of net investment (from development cost worksheet; xx K$)	600
2. Life of investment (years)	5
2. Expected annual net cash inflow (xx K$; cash flow can be negative)	

	year 1	year 2	year 3	year 4	year 5	
net economic benefit	50	50	50	40	30	
+ operating cost reduction	30	40	40	40	40	
= pre-tax income	80	90	90	80	70	
− ongoing expense	40	40	40	40	50	
(from worksheet)						
= net cash inflow	40	50	50	40	20	200

3. Average annual net cash inflow (xx K$)   40

4. ROI, calculated as $\dfrac{\text{average annual net cash inflow}}{\text{amount of net investment}}$ 7%

Table 3 shows the respective values for the CIM/AI shell and the AIPLANNER expert system for single and multiple application at different sites. Only the last two columns of Table 4 are realistic as they include the cost of the expert system building tool (the shell) as part of their initial investment.

TABLE 4. Economic impact of shells and expert systems.

	CIM/AI shell	AIPLANNER XPS	AIPLANNER + 0.5 CIM/AI shell	5 AIPLANNER + 0.5 CIM/AI shell
net investment (K$)	1000	100	600	700
life of investment (K$)	10	5	5	5
average annual net cash inflow(K$)	-10	50	40	200
expected ROI (%)	-1	50	7	29

This traditional approach is compatible with targets, budgets, and quotas used for business performance measurements. Refinements towards IRR or NPV methods are easily possible; for further applications and examples see Parker et al. (1988).

Quantitative risk analysis

By adding variabilities to the cash inflow estimates we arrive at Fig. 13 which quantifies the chances that the rate of return will be achieved or bettered. Clearly, this figure provides management with more information on which to base a decision than Table 4: investment decisions made only on the basis of maximum expected return are not unequivocally the best decisions. For more details on this type of risk analysis, see Hertz (1964).

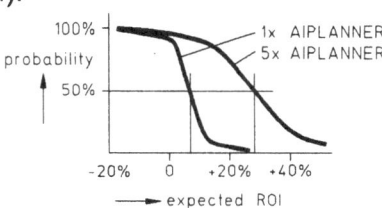

Figure 13.   Yearly expected ROI for one and five AIPLANNER applications at different factory sites.

Qualitative cost-benefit analysis

Fig. 14 shows qualitative cost and benefits arising from AIPLANNER's implementation at the car radio plant described which serves as application site for EP 932. The list reflects the view of the system users at the specific factory site; generally, any of the criteria from the previous sections could be applied.

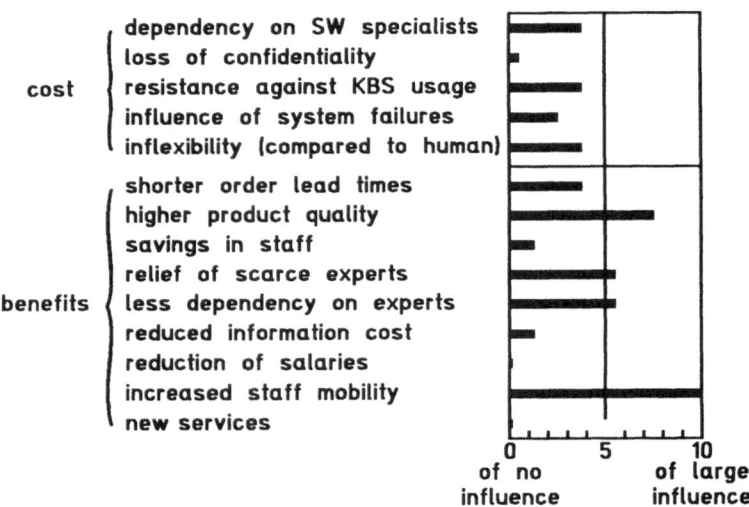

Figure 14. Qualitative cost-benefit analysis of AIPLANNER installation at car radio factory.

Goal attainment

The final proof of the expert system's worth is a field test of factory operation with and without the XPS in place. Such a comparison is difficult to achieve in practice, for obvious reasons. Moreover, AIPLANNER's implementation is part of a much wider factory improvement program aiming at CIM/JIT production. The separation of benefits as related to singular program parts is again extremely difficult (which holds for virtually all types of CIM software, either conventional or AI). For AIPLAN-NER, we had the chance to compare automatic operation with order sequencing based on heuristic set-up time optimization, and manual operation without heuristics (Isenberg 1990). Typical order patterns for the workcell were:
– planning horizon: 15 days
– number of workcell orders: 75
– number of workstation jobs: 205.

The workcell throughput diagrams for the main production goals: flowtime, lateness, productivity and WIP are shown in Fig. 15 for manual planning at the upper row, and for automatic operation under expert system support, below.

Following the advice given by the AIPLANNER expert system leads to substantial improvements as listed in Table 5. Table 5 ties the AIPLANNER results with improvements expected for general JIT implementations (Donahue 1987). Considering that such numbers normally are too general to carry any meaning, and that our data are too specific to allow for further generalization, sufficient match is obtained. At least the practical improvements obtained by AIPLANNER and similar expert systems have justified the investments already, and encourage further application of AI technologies in the modern CIM environment.

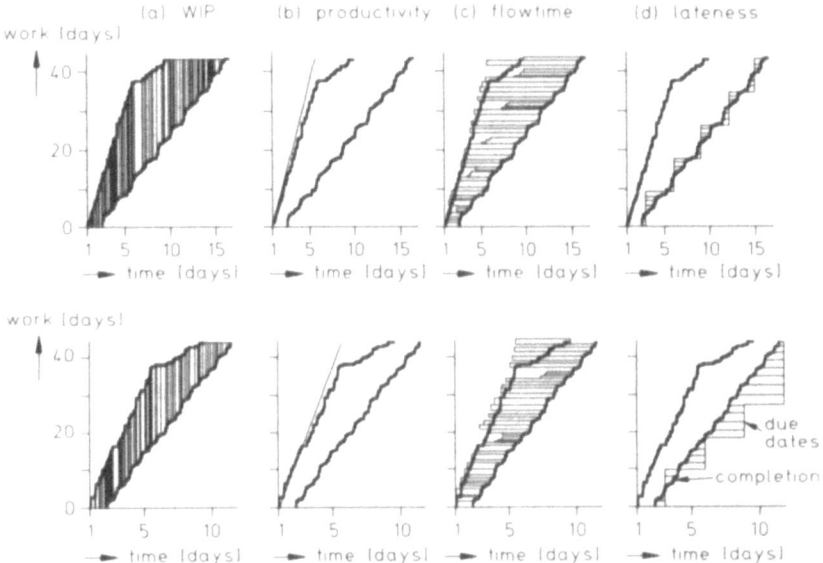

Figure 15. Improvement of workcell operation by implementing the AIPLANNER dynamic scheduling expert system: increase of productivity and decrease of WIP, flowtime and lateness (lower row) as compared to manual scheduling (upper row).

TABLE 5. Percentage improvements from general JIT implementation and specific AIPLANNER application.

benefit	JIT	AIPLANNER
flowtime reduction	80-90%	50%
set-up time reduction	75-95%	40%
inventory reduction/raw materials	35-75%	
work-in-progress reduction	30-90%	25%
productivity increase		30%
quality improvement	50-55%	
purchase price reduction	5-10%	
finished goods investment reduction	50-90%	
space reduction	40-80%	
lateness reduction		no late orders

9. Conclusion: Value of CIM/AI

CIM systems consist of all kinds of computer-oriented equipment to support the manufacturing operation in making better informed decisions at all levels of the business. As CIM in its final state integrates all levels and functions of the factory vertically as well as horizontally, the complete set of economic justification methods valid for each time horizon at all levels has to be adopted to assess cost, benefits and strategic implications of CIM investments.

CIM investments address organizational and system issues. As such, they relate to information technologies: the greater part of CIM investments, or at least the part which adds value to the system components and creates advantage over the competitition, is software. Under the heading of artificial intelligence, those software programs are classed that can solve problems, learn from experience, understand symbols or generally, behave in a way that would be considered intelligent if observed in a human being. AI programs represent the latest stage in software engineering and, therefore, will be intimately connected to CIM systems in the future.

The real aim of CIM/AI is flexibility in a diversifying marketplace. Therefore, CIM/AI is a must and a chance especially for Europe, and by far not for large companies only. The proliferation and support of CIM/AI as the logical extension of the traditionally well recognized manufacturing equipment business of Europe, is a major task for now and in the future for managers, politicians and technicians to avoid further erosion of competence compared to Japan and the US in an area which is vital for Europe in this global competition.

10. Acknowledgement

This work was partly funded by the Commission of the European Community under ESPRIT 932.

11. Appendix: EP 932 list of partners

Mr. G. Himmelstein
A E G
Goldsteinstr. 235
D-6000 Frankfurt 71
Telex: 414474
Telefax: 069/66 79 311

Dr. G. Squellati
A R S S.p.A.
Via Medici del Vascello, 26
I-20138 Milano
Telefax: 0039/2/520 29 257

Dr. H.M.J. Walters
Operational Research
BICC Systems Development Centre
Maylands Avenue
GB - Hemel Hempstead, HP2 4SJ
Telex: 051/82441
Telefax: 0044/442/21 01 01

Mr. M. Invernizzi
D.LETI - DEIN
CEN SACLAY
F-91191 Gif Sur Yvette Cedex
Telex: 042/604641
Telefax: 0033/1/69 08 83 95

Dr. R. Giorgesi
F I A R S.p.A.
Intelligent Machines Division
Via Mercantesse 3
I-20021 Baranzate di Bollate / Milano
Telex: 043/331265
Telefax: 0039/2/35 64 807

Prof. Dr.-Ing. U. Rembold
F Z I
Forschungszentrum für Informatik
an der Universität Karlsruhe
Haid-und-Neu-Strasse 10-14
D-7500 Karlsruhe 1
Telex: 17721190
Telefax: 0721/69 06 88

Mr. H.-F. Jacobi
I P A
Fraunhofer-Institut für Produktions-
technikund Automatisierung
Silberburgstr. 119 A
D-7000 Stuttgart 1
Telex: 7255166
Telefax: 0711/68 68 93 11

Prof. J.-P. Laurent
L I A
Department of Computer Science
Université de Savoie
B.P. 1104
F-73011 Chambery-Cedex
Telex: 042/309198
Telefax: 0033/79/62 79 73

Dr. W. Meyer (Project Manager)
Philips GmbH
Forschungslaboratorium Hamburg
Vogt-Koelln-Str. 30
2000 Hamburg 54
D-Telex: 21331654
Telefax: 040/54 93 210

Mr. A. Goerigk
Philips GmbH
Apparatefabrik Wetzlar AD - OT
Philipsstr. 1, Postfach 1440
D-6330 Wetzlar
D-Telex: 483847
Telefax: 06441/370 520

Dr. T. Barni
PIRELLI Coordinamento Pneumatici
Operazioni - Coordinamento
Industriale - C. 5214
Viale Sarca N. 202
I-20126 Milano
Telex: 043/323584
Telefax: 0039/2/64 42 33 00

Prof. M. Somalvico
Politecnico di Milano
Dipartimento di Elettronica
Piazza Leonardo da Vinci 32
I-20133 Milano
Telex: 043/333467
Telefax: 0039/2/23 99 35 87

Prof. Dr. H. Meyr
Lehrstuhl für
Elektrische Regelungstechnik
RWTH Aachen
Sommerfeldstr. 24
D-5100 Aachen
Telex: 832704
Telefax: 0241/80 76 31

Mr. S. Appel
S G N
1. Rue Des Herons
Montigny Le Bretonneux
F-78182 Saint Quentin en
Yvelines Cedex
Telex: 042/698316
Telefax: 0033/1/30/58 60 61

Dr. L. Bruzzi
S.I.S.AV
Via R. Vegani Marelli 1
I-20146 Milano
Telex: 043/314880
Telefax: 0039/2/412 15 47

Mr. J.-P. Cassagne
T I T N
7 Rue Louis Armand
Z.I. D'Aix en Provence BP 83
F-13762 Les Milles Cedex
Telex: 042/400221
Telefax: 0033/1/42 26 37 49 399

12. References

Doumeingts, G. (1984) Methodology to Design Computer Integrated Manufacturing and Control of Manufacturing Units, in *Methods and Tools for Computer Integrated Manufacturing*, U. Rembold & U. Dillmann (eds.), Springer: Berlin, pp. 194-255.

Hertz, D. B. (1964) Risk Analysis in Capital Investment, *Harvard Business Review*, **1**, pp. 95-106.

Huber, A. (1990) *Wissensbasierte Überwachung und Planung in der Fertigung*, Erich Schmidt: Berlin.

Hübner, M. (1988) Knowledge-Based Configuration of Flow Lines, *Int. Journal of Computer Integrated Manufacturing*, **1**, pp. 210-221.

Isenberg, R. (1987) Comparison of BB1 and KEE for Building a Production Planning Expert

476

System, in *Proceedings of the 3rd Int. Expert Systems Conference*, London, pp. 407-422.

Isenberg, R. (1988) Knowledge-Based Workcell Controller for Production Planning in the Electronics Industry, *Int. Journal of Advanced Manufacturing Technology*, **3**, pp. 67-81.

Isenberg, R. (1990) Wissensbasierte Integration von Produktionsplanungen in CIM, *Internal Report LB 778-90*, Philips Forschungslaboratorium GmbH: Hamburg.

Isenberg, R. & Hübner, M. (1989) A Workcell Controller Using a Knowledge-Based Simulation Model for Realtime Production Planning in the Electronics Industry, in *Artificial Intelligence in Scientific Computation: Towards Second Generation Systems*, R. Huber (ed.), Scientific Publishing Co.: Manchester.

Meredith, J. R. & Suresh, N.C. (1986) Justification Techniques for Advanced Manufacturing Technologies, *Int. Journal of Production Research*, **24**, pp. 1043-1057.

Meyer, W. (1987) Knowledge-Based Realtime Supervision in CIM: The Workcell Controller, in *ESPRIT '86: Results and Achievements*, Directorate General XIII (ed.), Elsevier (North Holland): Amsterdam, pp. 33-52.

Meyer, W. (1988) The Economical Aspects of CIM, in *Computer Integrated Manufacturing: Proceedings of the 4th CIM Europe Conference*, E. Puente & P. MacConaill (eds.), Springer: Berlin, pp. 185-198.

Meyer, W. (1989) Cooperating Expert Systems as CIM Modules, in *ESPRIT '89: Conference Proceedings*, Commission of the European Communities (ed.), Kluwer Academic Publ.: Dordrecht.

Parker, M. M., Benson, R. J. & Trainor, H. E. (1988) *Information Economics: Linking Business Performance to Information Technology*, Prentice-Hall: London.

Schtaklef, R. G., Gilmore, M. L. & Walters, H. M. J. (1990) Value Analysis of AI in CIM, in *Final Report EP 932 / FB 722-90*, W. Meyer (ed.), Philips Forschungs-laboratorium GmbH: Hamburg.

Schulz, H. & Bölzing, D. (1988) Lenkung von CIM Investitionen, *CIM Management*, **6**, pp. 54-60.

Stalk, G. (1988) Time – The Next Source of Competitive Advantage, *Harvard Business Review*, July/August, pp. 41-51

Wiendahl, H.-P. (1987) *Belastungsorientierte Fertigungssteuerung*, Carl Hanser: München.

ACCORD: Developing concepts towards integration of analysis and design
The Role Of ASSET In The Design Assurance Process

R.G. Parker, S.J. Denniss, N.H.W. Stobbs, A.T. Humphrey, M.A. Pearce
GEC-Marconi Research
West Hanningfield Road
Great Baddow
Chelmsford, ESSEX CM2 8HN - UK
Tel.: 44-245-73331

J.P. Patureau, A. Azarian, P. Carette, P. Excoffier, O. Vandenberghe
BERTIN et Cie
Zone Industrielle des Gâtines
BP 3 - 59, rue Pierre Curie
78373 PLAISIR Cedex - FRANCE
Tel.: 33-1-34-81-85-00

Summary

This paper gives an overview of the work accomplished in the ACCORD sub-project ASSET. ASSET's objective was to demonstrate the benefits of database management, knowledge based assistance and concurrent processing within the field of design assurance. This objective has been achieved in the implementation of the ASSET demonstrator.

The ASSET demonstrator integrates three engineering disciplines viz Reliability, Cost and Thermal Analysis, under a common management system. The analytical tools within these domains have the capability to communicate with each other through the development of a novel specification of the product hardware. Further, computational speed up through concurrent processing and new design assurance aids has been achieved.

All the software packages within the ASSET demonstrator can be used either as separate tools or in any one of a number of combined configurations.

1. Introduction

ACCORD was a 4 year ESPRIT-1 demonstrator project, which ended in August 1990. It was comprised of two sub-projects, ASSET (ACCORD Suite of System Engineering Tools) and APPEAL (ACCORD Parallel Processing Engineering Analysis Library). A broad overview of ACCORD is given by Patureau et al. [1989]. This paper concentrates on the ASSET sub-project.

paper concentrates on the ASSET sub-project.

ASSET's prime objective was to improve Computer Aided Engineering by making design assurance an integral part of design; (This is now recognised as being part of the 'Concurrent Engineering' arena). ASSET aimed to achieve this objective through the encapsulation of engineering expertise and the integration of design assurance tools. This would enhance the effectiveness and scope of product management by providing the facilities to monitor the design and design assurance activities more readily. Further benefits should accrue from providing an enlarged system engineering expertise base thereby freeing highly qualified personnel from the more mundane tasks associated with design assurance and plan preparation.

New technologies and design methods have been introduced into the ASSET system. The benefits arising from these innovations include speed up in the areas of design assurance analyses and the capability for data exchange between analytical tools. Experiments within the Reliability and Thermal domains have shown that the man effort and time required for certain tasks have been reduced by an order of magnitude.

2. Progress

The work accomplished in the ACCORD sub-project ASSET is encapsulated in the ASSET demonstrator. The ASSET demonstrator is a database application for integrated design assurance analyses (encompassing the engineering disciplines Thermal, Reliability and Cost). A major component of the demonstrator is the Knowledge based element known as ADVISE (ACCORD Design Verification Intelligent Support Environment). The ASSET demonstrator is still only a prototype system which shows the potential and can provide the basis for future exploitation. Here, its main purpose has been to investigate the feasibility of integration between the diverse range of design assurance tools and also to identify the problems associated with it.

2.1. Structure of the ASSET Demonstrator

The structure of the ASSET demonstrator, depicted in figure 1, meets the original requirements given in the functional specification. The main elements are:

- an Executive Manager,
- Domain Managers,
- libraries of technical knowledge and data,
- a selection of analysis tools,
- a product analysis database.

These elements are described in turn below.

The Executive Manager provides two separate but linked functions. The first function is concerned with project management, that is the creation and deletion of projects within the ASSET system. The second function facilitates product specification which entails the creation of an electro-mechanical breakdown of the system under analysis. The user enters the engineering domain of his choice through the analyse function.

The Domain Manager is responsible for the organisation of tasks within an engineering domain. This includes establishing the status of a project for an analysis (i.e. if sufficient data exists) and providing a common interface to the domain's tools.

Libraries of technical knowledge and data provide the user with a wealth of information at his finger tips. The libraries have been established to help the user to obtain the best predictions out of the tools which have been made available.

The selection of tools provides the analysis capability within the engineering domain.

The product analysis database manages the storage and retrieval of all project related information for each domain.

The demonstrator consists of three engineering domains, viz. Thermal, Reliability and Cost (see figure 1). Each of the domains are linked to the others at two key levels:

- at the top level, known as the Executive Manager, for project management,
- at the bottom level for product data communication through a unique database managed by a relational database management system (INGRES in our case).

2.2. Features of the ASSET Demonstrator

Common user interfaces to the database and to the knowledge-based components have been developed. This provides ASSET with an integrated user friendly environment and also has a beneficial impact on learning curves.

Extensibility has been a major design criterion. For instance, both the Cost Model library and the Thermal Template library may be easily extended. Within ADVISE, domain knowledge may be prepared by an engineer without requiring any knowledge of the underlying software. User interfaces may also be tailored to suit different users.

Another feature similar to extensibility is adaptability. ASSET provides a framework for design assurance disciplines and tools. New engineering domains and/or tools can be readily interfaced with the system.

Portability has featured largely in the development of ASSET. Consequently, an appropriate set of language and programming standards have been adopted. The system currently resides on VAX VMS but should be portable to SUNs and other

480

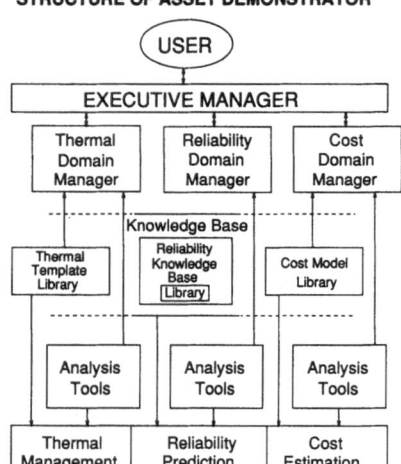

Figure 1

STRUCTURE OF ASSET DEMONSTRATOR

machines (the ADVISE sub-system already runs under SUN UNIX).

ASSET has been developed in a modular fashion. This means that packages ranging from individual tools up to the complete ASSET suite can be readily configured for future exploitation.

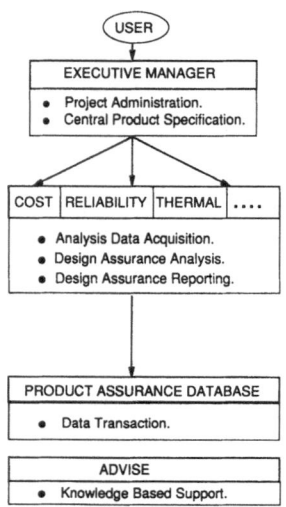

Figure 2

STRUCTURE OF ASSET'S DESIGN ASSURANCE ACTIVITIES.

3. Evaluation of ASSET'S Objectives

3.1. Review of Initial Objectives

The overall objective of ASSET was to integrate and rationalise design assurance process, concentrating on the domains of Reliability, Cost (i.e. Whole Life Cycle Costing) and Thermal analysis (Stress and Vibration analysis was also considered at an early stage, but had to be postponed due to lack of resources). This was to be addressed by the exploitation of three technologies:

Firstly, a set of existing analytical tools, and their data, would be integrated under a common environment using a Database Management System (DBMS). The adoption of a DBMS would provide centralised control, which would offer the benefits of data sharing between tools, data consistency and reduced redundancy; these in turn could be expected to yield advantages with regard to the development and integration of prototype applications. For the end user, the consistent interfaces generated with the aid of the fourth generation language would reduce learning time, while avoidance of repetitive data entry would lead to acceleration of the design process.

Secondly, a range of knowledge-based facilities for assisting with project and task planning was to be developed using software built on an Artificial Intelligence toolkit. This would exploit the expressive power of object-based and rule-based representations, together with the reasoning facilities provided by the toolkit, in order to capture and make available both theoretical and experiential expertise with regard to the domains. The use of such a toolkit would, like that of the DBMS, be highly conducive to the rapid prototyping of applications. This aspect of the project would include the development of a number of intelligent front ends to existing tools.

Thirdly, the benefits of computational speed up through parallel processing methods were to be investigated. This aspect was to be addressed through the exploitation of the results of the sister project, APPEAL.

In order to integrate the design assurance domains there was a need to establish a common view of the product specification. This product specification could then be assessed by the design assurance tools and be used as a focal point for inter-domain communication. In addition, it was recognised that if the software system produced were to be readily exploited, it would have to be user-friendly, portable and extendible.

Although the project was largely concerned with the integration of existing tools, there was little in the way of suitable software in the Cost domain. Additional goals for this area were to evaluate existing and emerging Whole Life Cycle Costing (WLCC) techniques, to incorporate a WLCC package into ASSET and to obtain recognition of the importance of WLCC in the design process (at the completion of the design phase 85% of the life cycle cost is committed).

3.2. Review of Achievements

At ESPRIT Week 89 the project was able to demonstrate the exploitation of the above technologies to enhance the design assurance process:

The Reliability, Thermal and Cost domains were successfully integrated under the INGRES DBMS; this allowed consistent user interfaces to the tools to be implemented (but see later discussion). A common representation for describing the product hardware specification, referred to as the System Breakdown Structure (SBS), was developed. The three domains are accessible through an Executive Manager, which is responsible for project management and SBS version control. An SBS is issued to the design assurance domains once it has been created in the ASSET Executive Manager, and then provides the basis for transferring information between tools (primarily within the Reliability and Cost domains).

A general environment for the knowledge-based component of ASSET (the ADVISE system), whose central feature is a basic User Interface Management System (UIMS), was developed. This was built using the AI toolkit KERIS, whose roots lie in the ESPRIT-1 PCTE project. This provided the foundation for implementing a broad range of knowledge-based design assurance support modules for the analysis domains.

Finally, a set of parallel processing software modules generated by the APPEAL project was successfully incorporated within the Thermal domain to provide significant computational speed up on complex routines.

Within the individual domains, the main advances have been the following:

Reliability

A common representation, the System Breakdown Structure (SBS), was developed using the INGRES DBMS. This provided the basis for integrating tools addressing a range of aspects of Reliability analysis.

An existing Reliability Prediction tool for determining the failure rate and mean time between failures of a system and its components was integrated into the software. This gave a more flexible environment for entering the necessary component values with the facility to build up a library of component data specific to the working environment, thereby reducing User input in the future.

A Maintainability Prediction tool which made two maintainability assessments available was integrated. The first assesses the optimum size for the maintenance team required to support the system during its working life, and the second calculates the time for which the system is likely to be unavailable due to repair and the probable time needed for preventive maintenance and repair in a given period. This uses failure rates from the same SBS module structure and adds maintainability data to the existing reliability data.

The Fault Tree Analysis tool (FTA) is one of the reliability tools developed for determining the causes of observed faults in a system using the standard SBS. It is a formalized deductive analysis tool that provides a systematic approach to investigating the possible modes of occurances of a given undesired event. The architecture of the tool encompasses the Sequencing level, the Executive level and the Data level. FTA tool has 3 main steps, viz . construction of the Fault Tree (static model), qualitative analysis and quantitative analysis. Different possibilities are offered to the user in order to select the gate events, specify the methodology and mesure the importance of the basic events.

A limited interface between the SBS (and hence the tools) and the knowledge-based modules in KERIS was developed. This allows structural models and failure rates to be passed to the knowledge-based system.

The ADVISE sub-system provided the basis for integrating a range of knowledge-based facilities for supporting Reliability Analysis. The most significant of these (all of which, like the UIMS at the heart of ADVISE, exploit fully the KERIS object representation facilities) are:

A fairly powerful graphics-oriented facility for Reliability Modelling and apportion-ment (using an overall requirement for system availability or failure rate to determine lower level goals). A variety of apportionment strategies are supported and rapid what-if analyses facilitated. Models (or sub-models) may be saved in a library and later imported at any level. The modeller has been applied to a design for a communication system containing several thousand units; the two week time-scale for analysing it from scratch suggests a productivity enhance-ment of at least one factor of magnitude over conventional tools.

A facility for generating largely complete Reliability Programme Plans for delivery to the customer. These are created automatically from a combination of project-specific data entered by the user and a set of generalised, extensible templates of text derived from standard documents. The plans thus generated may be tailored for individual customers and stored for reuse. This facility has also been successfully put to use on a project.

A suite of task scheduling facilities. This incorporates a knowledge base contain-ing information about analytical and management tasks for Reliability and/or Maintainability studies. Knowledge about the effects of various system attributes on given tasks is used to generate time-scales for each task, taking into account the characteristics of the system as specified by the user. This information (or a set of user-specified time-scales) may then be used in conjunction with knowl-edge about inter-task relationships, and constraints such as project milestones and available effort, to produce overall task schedules; these may be presented textually and/or graphically.

Cost

The cost application has achieved its initial objectives in establishing a WLCC package for ASSET. This includes a fully functional Development and Production Costing facility as well as a Whole Life Cycle Cost facility, which together are integrated into the ASSET environment.

Within the cost domain the user is advised initially to undertake a development and production cost analysis. This will predict the development engineering and prototype costs and production costs for electrical boards and electrical or mechanical modules identified within the project's System Breakdown Structure. The data requirements for these predictions include specifications of the environmental conditions and quality grades - the same information as required for the Reliability Prediction tool.

The development and production costs can be incorporated within the Whole Life Cycle Cost facility. This facility is used to predict the total cost of ownership of the system from its cradle to its grave. The model for predicting the cost of ownership is based on the concept of a Life Cycle Cost Breakdown Structure (LCCBS). The LCCBS gives an ordered breakdown of the cost defined by the project phase or function, and/or the system construction and/or the resource requirements. The cost estimates for these elements of the LCCBS are evaluated from a selection of cost models which are supplied by ASSET's Cost Model Library.

The Whole Life Cycle Cost facility currently has two tools, viz. a Cost Prediction tool and a Sensitivity Analysis tool. The Cost Prediction tool allows the user to modify the initial data prior to undertaking a fundamental WLCC analysis, whilst the Sensitivity Analysis tool evaluates the sensitivity of the LCCBS cost elements with respect to user designated parameters. The Cost Domain structure is presented in Figure 3.

Figure 3
COST DOMAIN STRUCTURE

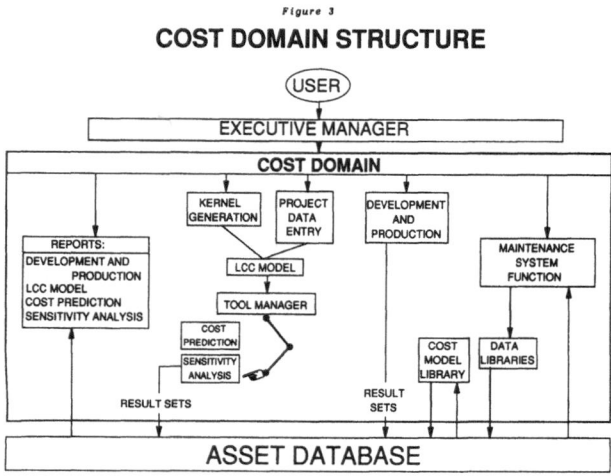

Thermal

The project aimed to develop an environment that would support thermo fluid analysis tools capable of being used in trade-off studies to optimise the enveloping packaging design. This is achieved via the provision of a prototype thermal domain package made of two components:

- the thermal domain manager: set of tools enabling the user to manage analysis data and results through a study/task reference,
- the modelling base manager that provides access through a standard interface to a range of software tools, modelling aids and also data on cooling processes including natural and forced convection. The facility also contains handbook procedures and data useful in preliminary design studies and a range of programs for assessing junction temperatures for components mounted on printed circuit boards.

It also allows the handling of complete documentation for each tool (underlying hypothesis, bibliography, user manual, sketches). Both managers use an enhanced multi-window, multi-user interface offering various access means to data, by means of keywords, catalogs ... Data transfer between the two managers is automatized, thus allowing easy storage and retrieve data and results of a previous analysis.

The thermal package includes for instance a third component that provides the interface with other analysis domains, thus allowing result data of PCB thermal analysis tool to be made available via the common standard SBS for the use of the MIL HDKB 217 tool for the prediction of failure rates. Tradeoff studies on the effectiveness of the cooling design can thus be based on actual (projected) conditions arising within the unit or enclosure including dissipation levels and distributions.

Studies at system level are completed using software tools that enable detailed assessment to be made of the airflow conditions that arise within the enclosure. The facility is based on the specification of the enclosure geometry including the internal arrangement of ducts and baffles and the location and power of any cooling fans. Thus a complete assessment can be made of alternative design options during the process of system evaluation.

Software tools/templates incorporated within the current version of ASSET can be considered in the following categories:

Thermo physical data,
Handbook cases for various fin/plate configurations,
ESATAN thermal network analyser,
ESATAN modelling templates,
FLOTHERM airflow analyser,
PCB thermal analyser.

486

Use has been made of the APPEAL parallel processing routines to enhance the computationally intensive arithmetic components of the equation solvers used within the PCB thermal analyser. The rapidity with which the data can be generated and the speed of solution permits the packaging engineer to explore a wider range of alternative layouts.

3.3. Review of Problems Encountered

In the early stages, the project suffered from incomplete and divergent views of the product specification. In consequence:

– Inter-domain views of Executive Manager and the Electro-Mechanical System Breakdown Structure differed widely. These differences were not fully recognised and consequently had not been resolved by the time the domain designs had been specified,

– There was no common Analysis Methodology. A global domain framework should have been given major priority early in the design phase to provide a common engineering approach to be applied across the ASSET system with common terminology and common views of screens and menu keys,

– Attempts to match differing representational schemas brought about integration difficulties. In particular, the intended project-wide representation of the product under analysis, proved to be of limited relevance to the thermal domain; the complex and varied schemas on which thermal modelling was based did not correspond with the tightly defined System Breakdown Structure. As a result, it was not possible to provide more than a superficial integration with the other domains, and data exchange was limited.

Although it was intended that knowledge-based techniques should be exploited, where applicable, on a project-wide basis, this objective was curtailed by budget constraints. Ideas were generated for all domains, many of which reached the design stage, but implementation of knowledge based support was restricted to the reliability domain. ESATAN, a Thermal analysis tool which is widely adopted, powerful, but difficult to use, would have benefited greatly from an Intelligent Front End (IFE), but this undertaking would have been beyond the resources available.

Some of the tools proved unsuitable for incorporation, as they were out of step with current user expectations. For instance, while the currently available FMECA tool (an old-fashioned batch-driven suite) could have benefited from an IFE, the tool itself did not offer conformity with the generally adopted standard. (In the end the tool was re-written in KERIS, with a form-based interface. While this is flexible and user friendly, it has yet to be tested on a major project). To a lesser extent, the MIL217 tool also fell into this category, being batch-driven and rather too slow for the expectations which an interactive environment generates.

Although both the INGRES-based and the KERIS-based elements of ASSET provide internally consistent interfaces, the two differ significantly. To some extent this is unavoidable, since the INGRES environment is of a traditional single-window, keyboard-oriented nature, while that offered by KERIS is aimed at the provision of multi-window, mouse-driven interfaces. This problem will be reduced when the next version of INGRES becomes available. Although the KERIS-based ADVISE user interface offers a high degree of flexibility in the interaction, it is potentially confusing to inexperienced users. ADVISE uses textual interface specifications which may be tailored for individual users; greater flexibility, however, would be offered if some of the general characteristics of user interfaces, such as the maximum permitted number of windows on the screen at any given time, were dynamically adaptable to different classes of user.

The current version of the interface between KERIS and INGRES which has been developed is too limited and slow; the fact that INGRES can communicate with a LISP (and hence a KERIS) application only via C, compounded by constraints on the degree of generalisation possible using the inbuilt LISP-C interface, tends to suggest that a large amount of low-level C programming would be needed to provide a stronger interface.

One problem arising from the above deficiency is that a module intended to provide remedial advice when analytical goals are not met is of little practical value to the analyst in its current form. At present it asks a number of questions about the results, then uses the answers to generate a list of plausible corrective actions. This, however, requires considerable work on the part of the analyst and offers only general advice. Ideally it should examine the results directly and give context-specific advice based on inferences made therefrom. Another result is that the FMECA tool is not yet robustly integrated with the Reliability Prediction tool.

4. Future Aims

Design and design assurance are recognisably different activities with the former concerned primarily with the creation of a new design concept and the latter in ensuring that the equipment system is fit for purpose when operating within the spectrum of operational environments. It is highly desirable that these activities are carried out concurrently. Design management ensures the progressive advance of the design through the various stages of the process. Close collaboration is required between the various design and development departments each with differing aims and objectives but collectively concerned with the development of an integrated engineering system. Consequently, it is ASSET's ultimate aim to provide an interface to other CAD systems, as it is shown in figure 4.

The development of an acceptable paradigm upon which to base systematic procedures is not readily achievable largely because there is no single mechanism for defining the system under consideration. Progress can be made if recognition is given to the requirement for a concurrent assembly of models each pertaining to the performance aspects of one or a number of system features. The need for a common view of the evolving product system is largely unmet at present although

488

Figure 4

POTENTIAL FEATURES OF ASSET'S EXECUTIVE MANAGER

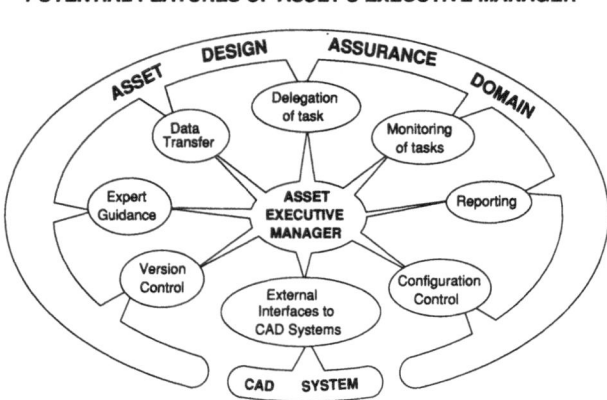

the so-called system breakdown structure goes some way to providing a universally acceptable definition of the system /sub-system /component assembly. Modelling schema pertaining to the mathematical description of some aspect aligned to system functionality provides a further acceptable description. Certain of the schema will refer to a computerised model of the hardware components (solid model) that will map closely the proposed hardware assembly. Similar links between a CAD or drawing representations and certain modelling schema have been devised by CAD vendors to enhance the capability of their software products.

The prototype ASSET system needs to address the application of security restrictions. Limited access should be provided to some functions; for example project creation and deletion in the Executive Manager should be password controlled. Also access to the domains could be monitored by means of passwords.

ASSET's Executive Manager could also benefit from enhancements to its analysis capability. Currently, what-if analysis can only be achieved at a domain level; this capability should be made available at the executive level together with a semi-automatic report generator.

Knowledge based techniques, which are currently restricted to the Reliability Domain, should be applied to support the above enhancements; in addition, Intelligent Front Ends should be provided for all of the domains, and to complex tools such as ESATAN, with a view to providing assistance with the selection of analysis modules and suitable levels of modelling refinement in defining the data, and the planning of trade-of studies.

Investigations into recent advances in technology (e.g. Object-Oriented databases) will enable ASSET to address current problem areas such as the KERIS / INGRES interface. This would allow the FMECA tool to be properly integrated with the reliability prediction tool, and would pave the way for a robust 'goal achievement expert'.

Within the existing domains, the following enhancements are envisaged:

Reliability

The Reliability and Maintainability tools should be enhanced, in particular extending the tools and methods to include an analysis of mechanical modules (development has concentrated largely upon assessing the reliability of electrical equipment up to now). The Fault-Tree Analysis tool would benefit greatly from a graphical interface.

The Reliability Modelling facility should be extended; although it provides a fast mechanism for analysis, its effective use still depends on the skill of the analyst (garbage in, garbage out). The embedding of domain knowledge, such as typical failure rate ranges for given types of unit (taking into account environmental conditions, where necessary), and information regarding different types of redundancy, would extend the range of potential users.

The Reliability Programme Plan generation facility should incorporate more robust and complete tailoring and editing facilities, while the remedial module, (or 'goal achievement expert') should provide context sensitive advice on how goals might be achieved. The latter facility will share some of the knowledge encoded for the modeller, but will also require component level information in order to draw inferences from the raw data provided by the tools.

Thermal

Further development of the thermal domain will involve enhancement of the mechanism for the characterisation of the product form primarily through the introduction of a solid modelling tool and the introduction of further software tools and modelling aids. Bearing in mind the close interaction between the hardware configuration and the mechanical performance, software tools addressing the problems arising from vibration will be introduced.

Special emphasis has been placed on providing the designer with a range of simple to use design checking tools helpful in the early stages when design concepts are still evolving. These handbook cases are to be expanded to include further data including information on materials.

The printed circuit board presents a special range of problems. The more advanced packaging concepts and component forms require improved computing algorithms and software. The programs under development by Pacific Numerix address the problems to a greater degree than those developed in the early phase of the project. The latest program has a facility for addressing the vibrational characteristics of a PCB. These programs are approaching industry standards and are having increasing usage. Advantageously they interface directly with a range of CAD systems for PCB design including placement and routing. The incorporation of this code is an immediate project objective.

The incorporation of the thermal analysis solver ESATAN and the computational fluid dynamics program FLOTHERM has provided an advanced facility for thermal analysis studies. At present both of these programs are operating essentially in stand-alone mode. The basis for an enhanced level of integration devolves upon the incorporation of a solid modelling tool. Future work will see the closer integration of these facilities.

Cost

The work already undertaken in ASSET provides a demonstratable methodology for performing Whole Life Cost studies within the prescribed framework. It is intended to enhance this facility into a marketable product. To achieve this objective three main areas of work have been identified for future development:

- The suite of prime equipment cost estimating models,
- The suite of in-service cost estimating models,
- An Optimal Repair Level Analysis Facility.

The cost estimating models will be designed to cost specific activities which together form the development, production or in-service phases of equipment. Different parts of the model will be exercised by equipment developed to meet various specification levels. It is intended that greater integration will be achieved through the selection of the cost model input parameters.

The various cost model elements will be parametric in nature and aimed primarily at personnel who have a need to perform comparative evaluations of competing equipment or system designs. The integrity of the cost model algorithms will however be such as to allow a user to perform absolute cost predictions, the accuracy of which will depend upon the time the user invests in calibrating the model elements more particularly to reflect his products and business methods. Absolute cost estimating will demand more equipment parameters to be input than purely comparative cost analyses.

As a major utility to the in-service section of the WLCC model an Optimal Repair Level Analysis tool will be developed. Although many customers specify the maintenance scenario(s) that will be employed to support their system, there are often occasions when this information is required from the equipment supplier, with documented justification for the recommendations. For these latter situations it will be necessary to evaluate various parameters against a variety of maintenance concepts. Consequently it is intended to develop a tool to assist the user in these matters.

5. Conclusions

The ASSET demonstrator shows that the encapsulation of knowledge and the integration of tools is desirable in that it does improve the efficiency of the design assurance process. This aside, in conclusion, we wish to address the main innovative segments of ASSET. Here, these are recalled as being:

i) A knowledge based interface for the Reliability domain, which lowers the skill threshold of the end user,

ii) Some newly adapted computational fluid dynamic tools which enable the packaging designer to optimise the hardware to improve the effectiveness of the cooling process. This in turn leads to improved life expectancy of the equipment,

iii) A cost estimating package which can predict the development engineering and production costs together with the whole life cycle costs for the system design under evaluation,

iv) An increase by several orders of magnitude in the speed of computationally intensive design assurance tools which has been accomplished through the incorporation of the algorithms from ASSET's sister project, APPEAL.

With the emergence of concurrent engineering, the future of ASSET is promising. It is, however, clear that (as a preliminary study) a reappraisal of the modelling methodologies and their inter-relationships is necessary for the advances in the process of design management to be realised. This should include investigations into the greater use of knowledge based interfaces applicable to the various classes of analysis tools, thereby facilitating the more effective utilisation of engineers over a wider spectrum of design disciplines. The study should also evaluate the potential for the introduction of object oriented database systems for improving communication between the various design domain interests. Nevertheless, in the course of this ESPRIT-1 project, the foundations for commercial exploitation have been firmly laid.

Appendix A: Early Exploitation of ASSET

ASSET's modular and interactive design has enabled some early exploitation to take place. However, due to the development nature of the ASSET application and the relatively recent integration of its constituent modules means that it has not been possible to date to apply the application as a complete and integrated toolset to any designs. Consequently, this section relates to some examples of the early exploitation of ASSET's elements.

The most significant example has been the application of the Reliability Modelling facility to prepare data for a proposal for a major project.

This involved the modelling and analysis of a design for a large radar network; containing several thousand units, this tested the facility to (and occasionally beyond) its limits. Nevertheless, starting from scratch the system was modelled, ans several sets of data generated in under two weeks. (This included some "on-the-fly" coding to allow certain highly complex structures to be modelled rapidly). The successful meeting of this deadline suggested a productivity advantage of at least one order of magnitude over conventional tools.

The study was valuable in demonstrating both the strengths and the limitations of the facility; as a direct result of the work, some of the weaknesses were eliminated when the demand-driven modifications were rationalised into a set of sophisticated structure copying operations. Some of these would have seemed rather esoteric, before the tool was used in earnest; yet, for any future work on a system with similar structural characteristics, the enhancements will have a major impact on efficiency.

The Reliability Programme Plan Generator has been applied internally by GEC-Marconi Research Engineers to generate largely complete Programme Plans for two major projects. The semi-automatic generation of the plan, and the ability to modify it quickly and easily to meet in changing customer and subcontractor needs, enabled a fast professional service to be provided.

In one case the work lead to the award of a substantial contract; the award of the other contract is as yet undecided, but indications are that, in view of the fast response made possible by the generator, the GEC proposal is receiving very favourable consideration.

References

Patureau, J.P, Trowbridge, W., Bryant, C., Stobbs, N.H.W., Parker, R.G. and Denniss, S.J., 1989. 'ACCORD: Developing some Concepts towards Integration of Analysis and Design.' In Proceedings of the 6th ESPRIT Conference. December.

Deliverable D7, 8, 9
ACCORD/2.2, 2.3, 2.4/GEC/DEL/005/17.1.88/SJD,
Functional Requirements, Software Architecture and Design,
February 1988, Edited by GEC.

ESPRIT PROJECT 1561 : SACODY RESULTS AND ACHIEVEMENTS.

Jean-Luc FAILLOT
BERTIN & Cie
BP 3
78373 PLAISIR CEDEX
FRANCE

Abstract.

Project SACODY focuses on the development of techniques and tools which are necessary for the implementation of advanced controls for high-speed industrial robots. Specifically, all mechanical manipulators exhibit at high speed coupled dynamic flexions and torsions which limit their performances and can be reduced or suppressed by appropriate control actions.

Issues being adressed in project SACODY deal with the development of modelling, identification and control methods and softwares, constituting a complete set of tools needed to design adapted new control actions.

The performances achieved by new control are tested on laboratory models and will be demonstrated at the end of the project on an industrial robot.

Major outputs of the project are techniques, their implementation into CAD softwares, and their programmation on a robot controller for test on an industrial robot.

This paper outlines the objectives and some of the current achievements of the project.

1. Summary of the project objectives

Reducing cycle time of robots and manipulators while maintaining or improving their global accuracy is the traditional problem to be solved by robot manufacturers, who have to satisfy ever increasing specifications in terms of operational speeds.

Up to now, it has been possible to rely only on the mechanical design of the robot to avoid vibration phenomena damaging the robot accuracy. Nevertheless, despite the use of very stiff materials, some applications, especially those involving large robots carrying heavy payloads, have their performances limited in order to avoid excitation of the robot structure.

SACODY set out to solve this problem by proposing new control actions capable of ensuring simultaneously fast motions of the robot tool and control of the structure vibrations. The expected outcomes of SACODY are a new methodology for the design of robot control, the corresponding software tools enabling:

- Modelling and simulation of the robot including deflections and vibrations of the links,
- Identification of the model parameters through experiments,
- Design of the control algorithms,
  and sensor systems

At the end of the project, to occur in 1991, this methodology will be applied on a demonstrator, involving an industrial robot controlled by a preindustrial numerical controller.

SACODY is an ESPRIT CIM project involving six european organisations namely:
- BERTIN et Cie (France, prime contractor),
- AEG AG and KUKA Roboter GmbH (F.R.Germany),
- Katholieke Universiteit Leuven and LMS International (Belgium)
- University College of Dublin (Ireland).

This paper first describes the main achievements of the project, and overviews the results obtained so far on the project demonstrator.

2. Main achievements of the project

2.1. Control Algorithms

The main aspect of SACODY project is the development and the test of active control of flexible manipulators, i.e. control which takes into account the structural dynamics of the manipulator. The main advantage of such control methods is that the control bandwidth is not limited by the lowest natural frequency of the manipulator. Hence, by using these new control methods, the natural frequencies of the manipulator structure can be lowered without affecting in a negative way the stability, bandwidth and dynamic accuracy of the control system. The weight and stiffness of the manipulator can be reduced, thereby allowing higher operational speeds and accelerations, reducing energy consumption, and maintaining or even improving the overall accuracy. Since the beginning of SACODY, innovative methods have been defined (De Schutter et al, 1988) and experimentally validated on laboratory models of flexible robots (Faillot, 1989).

Once a model representative of the behaviour of a flexible robot has been derived, it is possible to design control actions in order to obtain a proper positioning of this robot despite the elasticies lying in its structure and/or its gearings.

To the variables classicaly controlled by kinematic control, i.e. the joint degrees of freedom and their time derivatives, some variables are added in order to represent the elastic behaviour of the robot. The obtained set of controlled variables is regrouped in a vector of generalized joint coordinates for which the classical stages of robot position control (kinematic inversion, dynamic inversion, servo control) are extended.

- Inverse kinematics:
  Static deflection of each link must be taken into account when solving the inverse kinematics problem, i.e. determining the desired joint displacements and their

time variations starting from the desired values expressed in a cartesian reference frame.
- Inverse dynamics:
In order to achieve the tracking control, a feedforward action, using inverse dynamics, is computed. This consists of the calculation of the nominal actuator commands at the robot joints as a result of position, velocities and accelerations of the desired robot generalized joint coordinates.
- Servo control:
For ensuring the precise tracking of the desired joint trajectories and the active damping of vibration modes of the structure, a new servo control is implemented. It is designed on the basis of a low order model (including essential structural modes) ; an observer reconstructs the whole state of the system in minimizing the spillover effects, i.e. disturbing effects of ignored vibration modes on control stability.
In order to validate experimentally the theoretical algorithms, five laboratory models have been constructed for SACODY, representing increasing complexity in terms of control of flexible manipulators. Collectively, these models form a complete collection of testing situations enabling verification of control algorithms against such defects as configuration dependant modal frequencies, gear plays and other mechanical backlashes, various unmodelled dynamics, torque saturations, etc.
Details on the results obtained are given in (Swevers et al. 1989).

2.2. Dynamic Modelling and Simulation

2.2.1. Modelling a flexible robot.

The design of a control law satisfying the simultaneous goals of fast motions and vibrations control requires the derivation of a model of the robot, describing not only the evolution of the robot configuration, but also interaxes couplings due to inertia forces and vibrations of the robot structure.
Derivation of such a model is known to be a complex task because of non linearities caused by change of robot configuration during motion, and difficulties for modelling accurately elastic behaviour of polyarticulated flexible components. In fact, a systematic tool is necessary to compute equations of motion of such systems.
Within the framework of SACODY, a general purpose computer code developed by BERTIN, called ADAMEUS, has been adapted to the particular case of robots or manipulators with possible flexible links or joints. In its current commercial version, it enables the modelling and closed-loop simulation (i.e. simulation with a feedback control) of any mechanical polyarticulated chain, including actuators and sensors.
ADAMEUS involves an adapted method for obtaining automatically dynamical equations of flexible mechanical systems. It enables a general approach of the problem of flexible systems dynamics for complex bodies behaviours and system

496

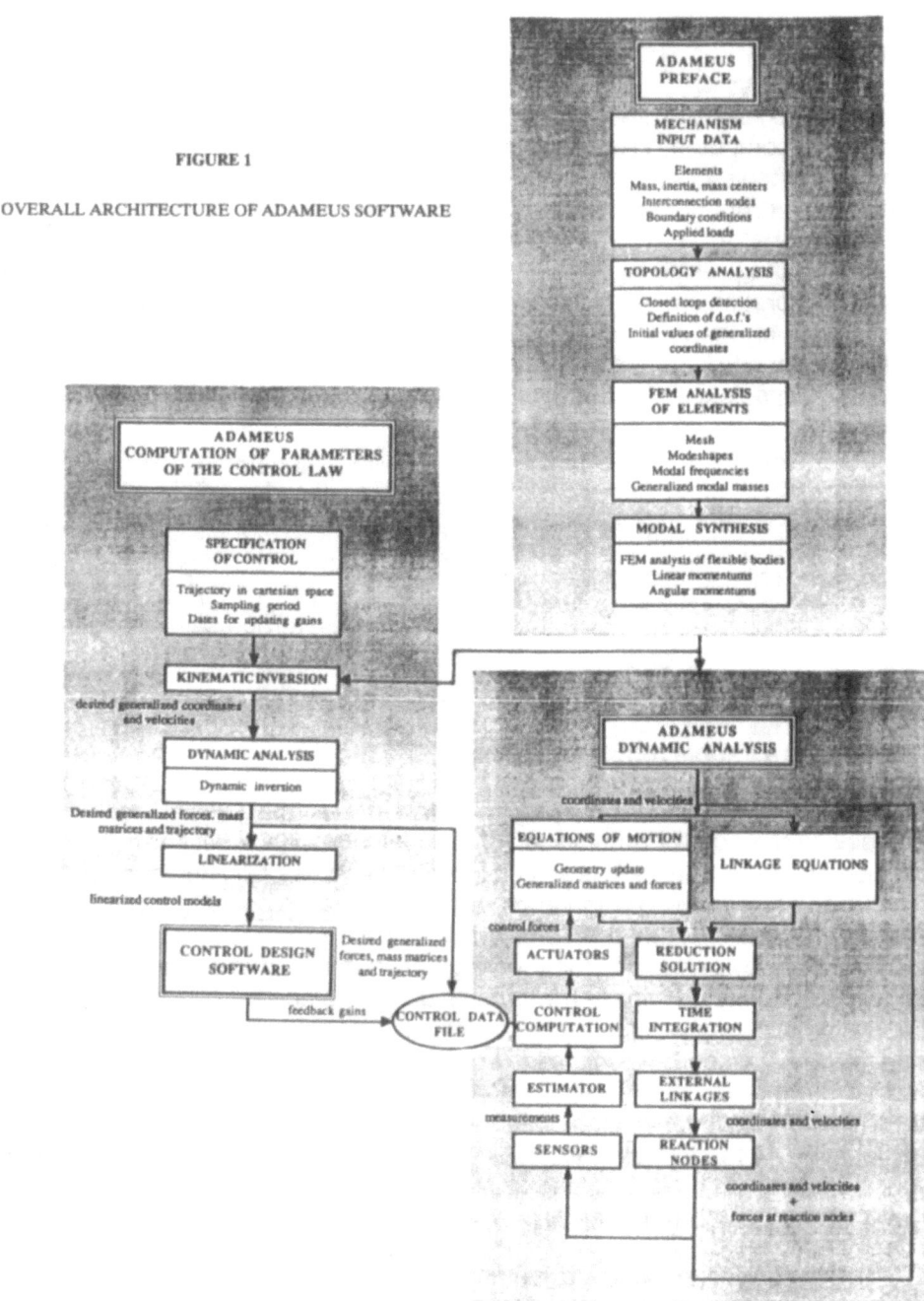

FIGURE 1

OVERALL ARCHITECTURE OF ADAMEUS SOFTWARE

topologies (open or closed chains with possible constraints). The difficulty of the coupled rigid-flexible dynamics problem lies in the difference of magnitude order of displacements induced by these two types of behaviour.

The equations are built by use of the principle of virtual work (KANE's method). Rigid body kinematics uses relative coordinates, and a modal synthesis method is associated to deal with the problem of flexibility.

The method used consists of separating the two contributions, and of choosing for the second one a modal representation of the deformation, the modes being the natural modes of vibration or more general deformation modes, as yielded by standard FEM codes such as NASTRAN or SYSTUS.

The advantage of this choice is to minimize the number of variables, in order to run numerical simulations on different computers of reasonable size. ADAMEUS therefore satisfies both criteria of generality (complex configuration topologies, 3D geometries, flexibilities), and efficiency (minimization of the number of variables, automatic elimination of non working degrees of freedom).

The architecture of ADAMEUS is described in figure 1.

2.2.2. Using modelling software for computing control.

Obviously, the robot model is needed to perform all the steps mentionned above. For this reason further developments of ADAMEUS have been undertaken, in order to provide the control designer with all the parameters needed for computing the orders to be applied at the joints.

By inclusion of the resulting new modules, the model yielded by ADAMEUS is used not only for providing open-loop and closed-loop simulations, but also to compute :
– the kinematic inversion yielding the trajectory of joint coordinates while taking static flexibility into account,
– the predictive feedforward resulting from dynamic inversion along the desired trajectory,
– linearized models about prescribed trajectory points, to be handled by control design software to obtain the feedback gains needed by tracking control stage.

All these items are loaded in a control data file, and used during closed-loop simulation for computing the control orders to be applied at the actuators.

2.3. Identification Software

The experimental assessment of the dynamic robot behaviour is required to provide the necessary information for the optimization of the robot design as well as for the design of a controller which can take these dynamic characteristics into account.

Also for simulation purposes, the experimental identification results can be used to optimize and update the theoretical system model.

In the framework of SACODY, a frequency domain identification technique deriving directly the parameters of a reduced state space model of a flexible robot has especially been developed (Lembregts et al, 1990).

This identification technique features the following characteristics :

- The state transition, input and output matrices of the state equation are readily available,
- Numerically reliable tools are used : least squares solution and eigenvalue decomposition techniques,
- The number of unknowns is limited, using only a second order linear model with matrix coefficients of dimensions equal to the number of modes. This model is valid for a reduced state. Since the output matrix is obtained from a real symmetric matrix, the usual singular value analysis is replaced by an eigenvalue decomposition, drastically reducing computer requirements and load as compared to traditional methods (e.g. ERA),
- A very important problem, namely the dimension of this state space model, is eliminated by the data measurement matrix,
- All measurement data can be analysed simultaneously, reducing the measurement noise, and yielding a consistent model for the system's response.

The integration of this frequency domain direct parameter identification (FDPI) technique into the LMS CADA-X software package for Computer Aided Testing (CAT) has been completed. This package has been set up for general dynamic signal acquisition and analysis, and contains the necessary hard- and sofware to perform for example modal analysis tests of mechanical and mechanical/servo systems. This CAT system also provides general software libraries for modern linear algebra, least squares problem solving, eigenvalue and -vector decomposition, singular value analysis, system theory, etc...

The integration of the identification algorithm in the LMS CADA-X system for computer aided dynamic analysis (CADA) and computer aided testing (CAT) makes communication with other sofware modules straightforward. The measurement data can be retrieved directly from the relational CADA-X project data base, making all necessary information available: used test equipement and circumstances, transducer sentivity and calibration values, the needed frequency response functions, etc... The obtained modal parameters can be stored to the analysis tables of the same data base, and interpreted via a graphical animation module to display the deformation patterns (mode shapes). Other interesting analysis information (such as the singular values, indicating the state space dimension) can also be stored to this data base for later use.

2.4. Sensor Systems

2.4.1. Real Time deflection sensor.

The success and the feasibility of the implementation of control algorithms for flexible robot arms are to a large extent depending on the availability of appropriate real-time end-point deflection sensors. For this reason, an important activity has been devoted at K.U.LEUVEN to the development of suchlike sensors. Two systems have been tested : one based on strain gauge measurements and the other based on laser diodes and position sensitive devices (DIOMEDES), which we address in detail in this paragraph.

The robot structural deflections to be measured by DIOMEDES (Laser DIode Optical System for MEasuring Structural DEflectionS) are the deflections at the tip of each flexible link of a robot. This measurement is crucial because the deflections affect the position and orientation of the robot end-effector. Figure 2 defines the deflections to be measured.

The sensor system shown in Figure 3 measures the structural deflections d_x, d_y, a_x, a_y and a_z. The elongation of the link, d_z can not be measured. Since d_z is usually very small compared to the other deflections, neglecting d_z causes no real error. The system is therefore able to measure the global deformation of a robot link.

The system consists of three semiconductor laser diodes mounted at one end of the link and of three PSD's of a planar type, one mounted in the focal plane of a lens, at the other end. Since the laser beams go straight in spite of the deflections of the link, the positions of the spots that the laser diodes make on the PSD's contain all information regarding the structural deflections of the link.

An experimental set-up of DIOMEDES has been built by K.U.LEUVEN in order to demonstrate the feasibility of the sensor system to determine in *real time* the spatial structural deflections of a flexible link. This set-up involved laser diodes mounted at the origin of an aluminium link and delivering a maximum output power of 1mW for a 790nm wavelength, as well as three PSD's mounted at the other end of the link. One PSD is mounted in the focal plane of a lens and has an effective sensitive area of 2x2mm. The lens is an achromatic lens with a focal length of 56mm. The PSD's without a lens have an effective sensitive area of 12x12mm. The laser diodes and the PSD equipped with a lens are easily adjustable concerning their pointing accuracy.

Analog evaluation electronic circuits have been developed to measure the coordinates of the laser spot on the PSD. The coordinates are fed to the calculation algorithm in the microcomputer using a 12bit Lab Master DMA board. The micro-computer is a PC-AT, operating at 8kHz and equipped with a floating point processor.

The software allows to calibrate the set-up and to calculate, in *real time*, the five structural deflections. A graphical simulation routine is available to show continuously the spatial deformation of the end of the flexible link.

The presented sensor system has a measuring range of 5mm, with a resolution of $3\mu m$, for the structural deflections d_x, d_y, depending on the value of the torsion angle a_z, and a measuring range of 14mrad, equivalent to 0.8 degrees, with a resolution of $7\mu rad$, for the bending angles a_x, a_y. The time to measure the six coordinates and to calculate the five structural deflections is 5msec. The bandwidth of the sensor system is over 50Hz.

The system is able to measure static as well as dynamic deformations of flexible links. Its development is being finalised, to optimise the system behaviour and to

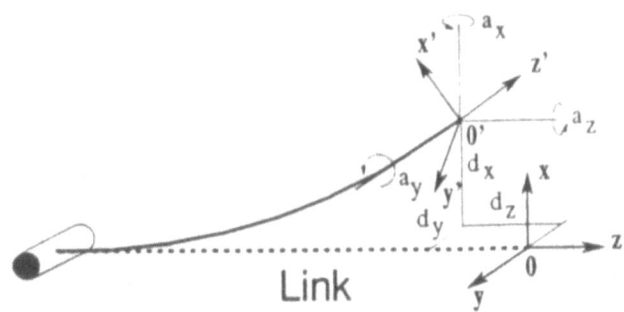

FIGURE 2

STRUCTURAL DEFLECTIONS OF A ROBOT LINK

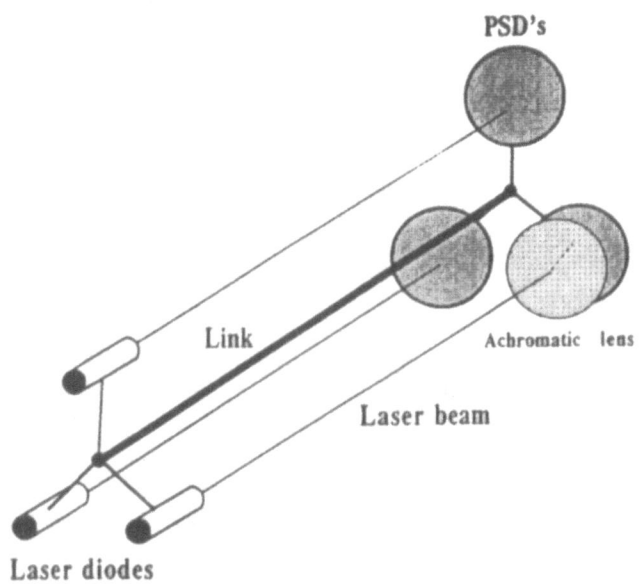

FIGURE 3

LASER DIODE OPTICAL SENSOR SYSTEM TO MEASURE

THE STRUCTURAL DEFLECTIONS OF A ROBOT LINK

apply it for real-time control purposes, first on the laboratory models and ultimately for the final demonstration on the Kuka-robot.

2.4.2. Sensor systems for robot evaluation.

A second and equally important item under the sensor development chapter of the Sacody project is to provide for means to objectively verify the improvements brought about by the improved control strategies and to compare the results with those obtained with traditional control methods. No systems were available in the market at affordable prices at the time they were needed for the project. Some commercial systems are not accurate enough (a few mm) and are extremely expensive. Others are still under development.

One simple 2D-system, based upon a digitizing tablet (RODYM) has been finalised and turned into a commercial product. It has been continuously used to assess the developed control algorithms for flexible robots.

3. Application: the project demonstrator

The last year of the project is dedicated to the application of the project outcomes on a demonstrator made of a KUKA spot welding robot controlled by a AEG robot controller prototype.

Spot welding remains the main robotized application especially in the automotive industry where the spot welding robots represent 90% of all robots at work in car body assembly operations.

Because of their large range and the heavy tool they have to carry (up to 150 kg), these robots usually exhibit elasticities in their gears, yielding low frequency vibrations which damage the welding gun positioning accuracy, especially when maximum speed is sought.

The purpose of the demonstrator is to show that the application of the SACODY control algorithms enables to get rid of these vibrations and therefore improves the robot operationnal speed as well as the repeatability of the weld spot.

The main development steps leading to the delivery of the demonstrator are described in this section.

3.1. Identification of the Robot Structure

As a first development step, a dynamic analysis has been carried out on the demonstration robot in order to determine the dynamic model for this robot in the low frequency region.

The characterization was mainly achieved in two different ways. First, experimental modal analysis was used to describe the dynamic behaviour in terms of natural frequencies, damping values and modeshapes. These were determined out of frequency response functions of acceleration over a force being externally generated with an vibration exciter. Next, frequency response functions of acceleration over motor current were measured. In this case, the excitation was generated by

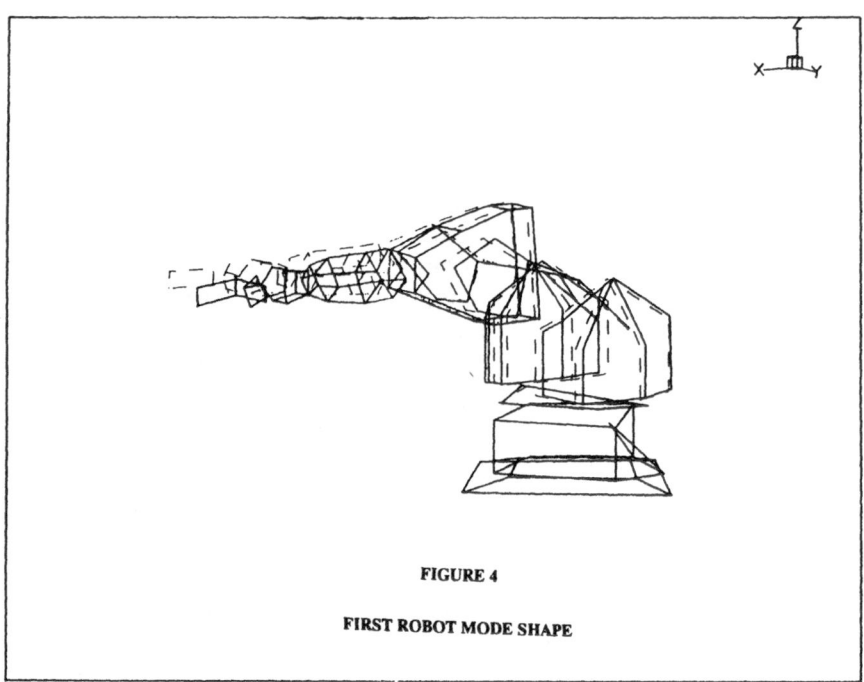

FIGURE 4

FIRST ROBOT MODE SHAPE

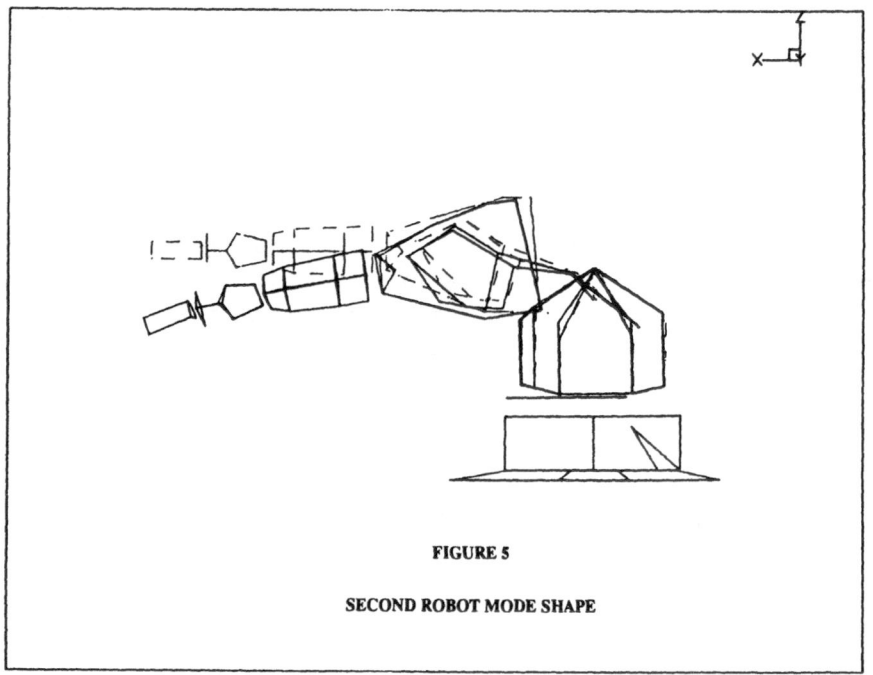

FIGURE 5

SECOND ROBOT MODE SHAPE

each of the first three motors separately.

As the dynamic properties vary with the configuration of the robot, the analysis was carried out for different position configurations. The modal analysis was carried out in three configurations. The frequency response functions with respect to the motor current were measured in ten different configurations.

Four robot modes and one foundation mode were found in the frequency region of interest. The first two robots modes are one horizontal and one vertical rotation due to flexibilities lumped in the motor gearings as shown on figures 4 and 5. The two following modes are a horizontal and vertical bending with global structure deformation.

The study has therefore enabled to focus on the presence of elasticities in the robot which can be easily explained by the difficulty for the mechanical design to stiffen the robot gears in order to suppress transient oscillations and static deflections, especially in this case where a heavy load (45 kg) has to be carried with a range of 2 meters.

The control will therefore be defined in order to prevent the oscillations induced by these elasticities and therefore lessen the time needed to position the robot tool with a prespecified accuracy.

3.2. Model and Simulation of the Robot

A base model has been established by describing all the mechanical elements (links, speed reduction stages, motors,..) of which the robot is composed and their interconnection. The model features globally 45 connection nodes, 20 gearlike links and 16 rigid links or constraints.

In accordance with the results of the identification tests described above, two gearing shaft have been defined as flexible bodies in order to model the flexibilities lying in the joints of the first two axes. By proper choice of the shaft stiffnesses introduced in the model, it has been possible to reproduce the measured modal frequencies and their evolution with the robot configuration with an accuracy of 3%.

After optimization of the number of variables and the automatic elimination of the non-working degrees of freedom, 8 variables are kept by ADAMEUS, out of the 47 declared in input, namely the six motor rotations and the two shaft torsions.

The model is now used to design the robot control algorithms and assess by simulation their performances, by applying the methods developed within the project (De Schutter et al, 1988 - Froment et al, 1989).

Figure 6 shows a graphic output of an ADAMEUS simulation.

3.3. Real Time Control

After modal identification and modelling of the demonstration robot, development of the real time control by application of the methods defined at the beginning of the project has been initiated.

A preliminary implementation of the algorithms on a laboratory multiprocessor is

504

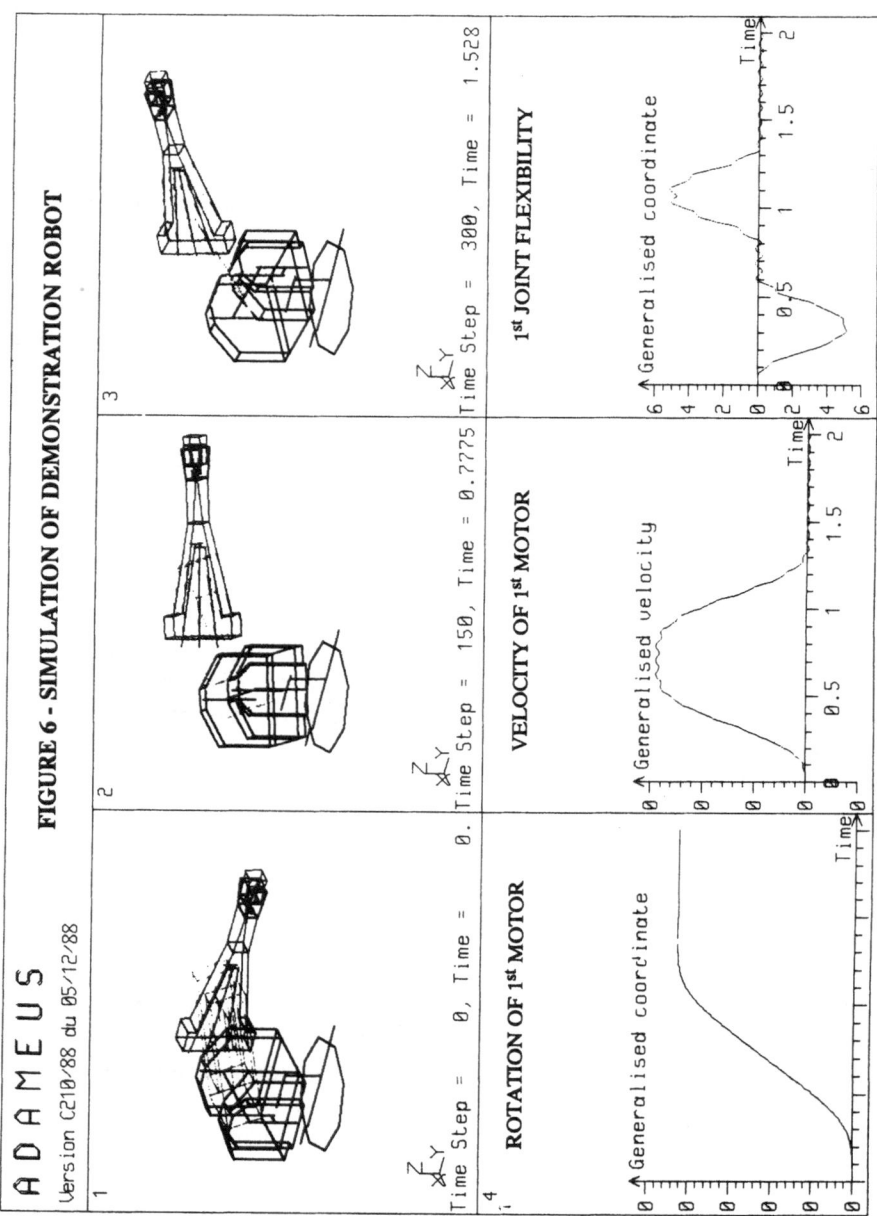

ADAMEUS

Version C210'88 du 05/12'88

FIGURE 6 - SIMULATION OF DEMONSTRATION ROBOT

ROTATION OF 1st MOTOR

VELOCITY OF 1st MOTOR

1st JOINT FLEXIBILITY

currently going on. The implementation on the target system, a prototype of industrial digital robot controller developed by AEG, will follow.

The preliminary developments have already enabled to test successfully the control of the first axis flexibility. Figure 7 compares the tool acceleration when the robot in extended configuration has its first axis submitted to a conventional control (top) and to a control involving flexibility control (bottom).

It appears clearly on this figure that the robot vibrations are almost completely suppressed, although higher acceleration rates are applied on the bottom plot. In this case, the gain in positioning time reaches 45%, thanks to the combination of the vibration control and of an improved trajectory profile enabled by the laboratory multiprocessor.

The final implementation on an industrial controller will be carried out at the end of 1990, and will be dedicated to the development of the combined control of the six robot axis and of the main flexibilities in the gearings.

Gain of performance achieved will be assessed by a series of standard robot performance tests, carried out with conventional control and with the new control.

4. Conclusion

SACODY is a multi-discipline project which aims at associating advances :
– in dynamic modelling and simulation,
– in theoretical control methods,
– in real-time numerical computer hardware and programming software,
to make practical use of advanced control methods in industrial (CIM) applications, with a view to compensate for unavoidable mechanical deficiencies of manipulators.

The software packages for identification on the one hand, and for modelling, simulating and control design on the other hand, have been completed and experimentally tested on models of flexible robots.

They are currently applied to design the control of a 6 axis spot welding robot, in order to obtain, thanks to a better knowledge of the robot structural behaviour under extreme accelerations, faster motions and, thus, reduction of its operational cycle time.

5. Acknowledgements

This paper is a representation of a collective work carried out at each of the six organizations of the consortium. The dedicated work of a large number of people, under a spirit of cooperation, is acknowledged here.

The contractors would like to acknowledge the support of the Commission of the European Communities, as represented by DG XIII, on Telecommunications, Information Industries and Innovation, and more especially its CIM Group.

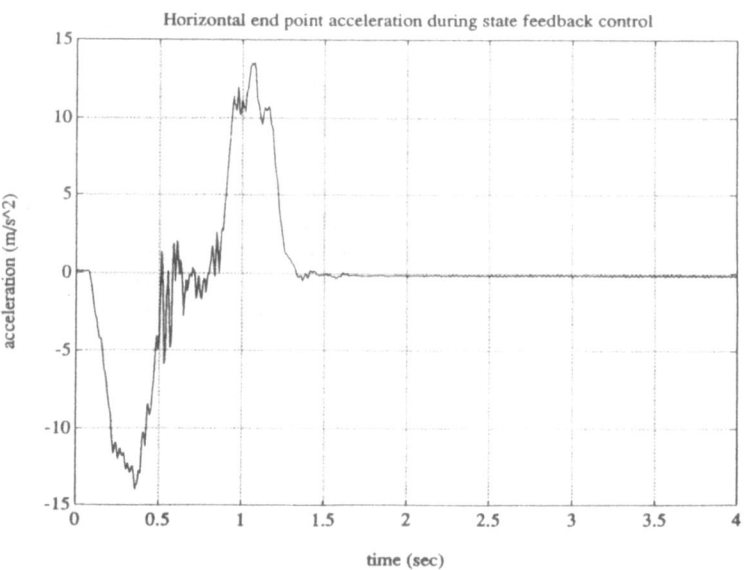

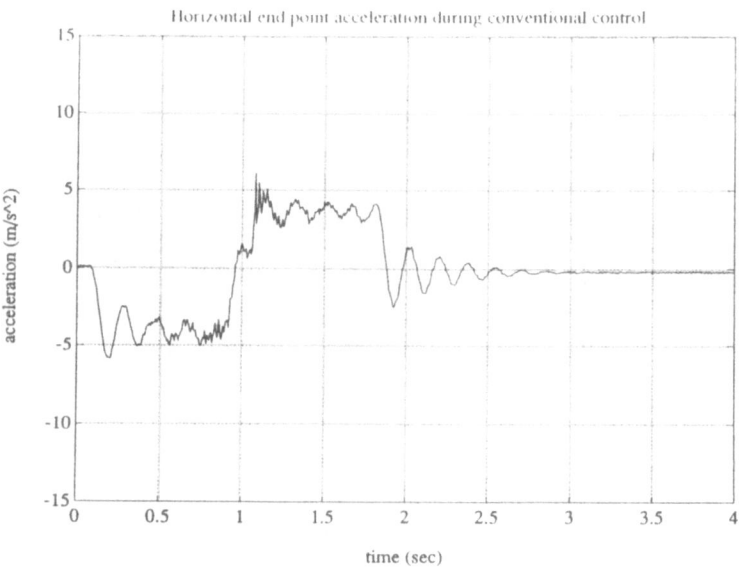

FIGURE 7

CONTROL OF FIRST AXIS ROBOT ARM EXTENDED
Comparaison of Results between conventional control (top)
and new control (bottom)

6. References

De Schutter, J., Van Brussel, H., Adams, M.,Froment, A., Faillot, J-L., Control of flexible robots using generalized non linear decompling, IFAC Symposium on robot control SYROCO, Karlsruhe, October 1988.

Faillot, J-L., Advanced active control for industrial robotics, ESPRIT '89 Conference Proceedings, pp 757-773.

Froment, A., Faillot, J-L., Gallay, G., Gerbeaux, F-X., Modélisation et commande des structures flexibles polyarticulées, Colloque SMAI "l'Automatique pour l'aéronautique et l'espace", Paris, Mars 1989.

Lembregts, F., Leuridan, J.,Van Brussel, H., Frequency domain direct parameter identification for modal analysis : State space formulation, Mechanical systems and signal processing (1990) 4(1), pp 65-75.

Swevers, J., Adams, M., De Schutter, J., Van Brussel, H., Thielemans, H.

Limitations of linear identification and control techniques for flexible robots with non linear joint friction. International symposium on Experimental Robotics, Montreal, Canada, June 19-21,1989.

Project No. 2090 EPIC : Early Process Design Integrated with Control

TOWARDS AN INTEGRATED ENVIRONMENT
FOR THE EARLY STAGES OF PROCESS DESIGN

N. KARCANIAS   *and*   G. CHAMILOTHORIS
CITY UNIVERSITY   *PLANET S.A.*
Control Engineering Centre   *L. Riencourt 64*
Northampton Square   *11523 Athens*
London EC1V 0HB   *Greece*
United Kingdom

Summary : The objective of the EPIC project is to develop an off-line Computer-Aided-Design environment, assisting engineering teams involved in the design of complex plants. These teams belong to different fields such as chemical process engineering, control systems engineering, instrumentation engineering and optimisation. Techniques from all these engineering fields are integrated into a coherent conceptual framework and embodied into a software prototype, through the application of modern software engineering tools and methods.

At the second year of the project's life, work within EPIC has produced a number of important products. These achievements concern the use and adaptation of advanced software engineering tools and the promotion of these in the CIM community and the creation of a generic database structure and of a common user interface, tailored to the requirements of process engineering activities.

EPIC has also produced foundations for an integrated approach to the design of process control systems. This led to advances in theoretical aspects of the analysis and synthesis of control systems for multivariable processes and to the specification of novel techniques for the optimisation of a plant's operating conditions. At the same time, EPIC's innovatine approach to process design, embodied in the workbench functionalities, is assessed through engagement in concrete industrial engineering case studies.

Main :

1. Introduction

In order to conceive industrial plants which are safe to operate, straightforward to command and efficient both in terms of capital and operating costs, designers of modern complex processes must mobilize a array of techniques from different engineering fields, such as chemical process engineering, control systems engineering, instrumentation engineering and optimisation.

ESPRIT II project EPIC is currenty engaged in formulating an integrated design framework bringing together advanced tools from different engineering fields, by providing a collection of methods, techniques and tools and by developing an off-line Computer-Aided Design system : the EPIC prototype.

To achieve this goal, the project reviews, assesses and, when required, augments the techniques and tools currently used in process design and integrates these into a conceptual framework for the development of a software prototype, through the application of modern software engineering tools and methods.

A brief account on the use and adaptation of software engineering techniques for the development of the EPIC workbench is given in Section 2. Section 3 discusses the building of a generic database structure and of a common user interface, tailored to the requirements of process engineering activities.

EPIC has also produced foundations for an integrated approach to the design of process control systems. As briefly indicated in Section 4, this led to advances in theoretical aspects of the analysis and synthesis of control systems for multivariable processes and to the specification of novel techniques for the optimisation of a plant's operating conditions.

Section 5 describes the project's engagement in concrete industrial design situations that will help the EPIC workbench evolve into a complete, off-line, interactive software environment which shall integrate facilities offered by existing conventional flow-sheeting programmes, as well as advanced control design and optimisation facilities. This is expected to meet the requirements of practicing engineers, by offering a wider potential for carrying out the coordinated tasks of industrial (re)design projects.

Summarised in section 6, progress of work of the EPIC project points out to promising fall-outs in the way of the ESPRIT programme objectives.

2. Use Of Software Engineering Techniques

2.1. METHODOLOGICAL APPROACH

The development of the EPIC workbench is carried out along three major methodological guidelines:

- adhesion to an object-oriented philosophy,
- adoption of a rapid prototyping strategy,
- extensive use of automated software engineering facilities.

The object-oriented character of the system allows for ease of expansion, modification and maintenance of the existing data structure and knowledge base, by splitting or renaming a class-object, adding or removing a class-object along an inheritance path, etc.

Also simplified are the construction of data models for new composite objects and dynamic generation and/or the elimination of "instance" objects, as needed by a particular reasoning strategy. This is achieved by associating the same messages (of standardised protocols) with the instances of various class-objects, even when new class-objects are introduced, removed, or replaced along an inheritance path.

In order to tackle the problem of imbalance between analysis and synthesis, several development models have been proposed : rapid prototyping, two-legged model, operational model, risk-analysis model etc, which may also be used in conjuction with the traditional life-cycle model.

The first phase of the prototyping activity concentrated on the selection and combinations of various "state-of-he art" hardware and software products (workstation, data base manegement system, software development environments etc.) with applications software (implementation of several tools and techniques) in order to demonstrate the functionality of a basic subset of the systems' facilities and features.

The second phase of the prototyping activity included a higher proportion of project-built prototype modules for process design techniques. During the successive versions of prototype development, new software modules increase the degree of functionality until a suitable representative subset of the problems applicable to the EPIC domain is

510

addressed.

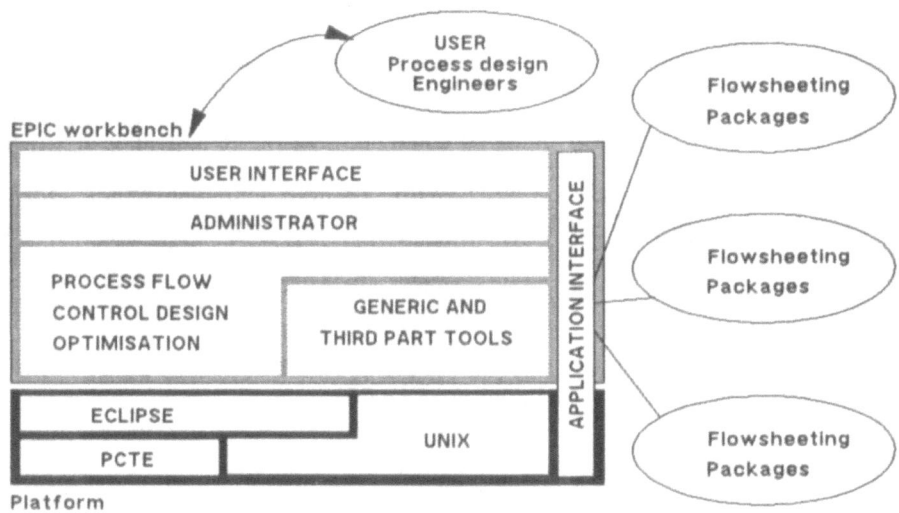

Figure 1. : Architecture of the EPIC Workbench

This activity has led to the specification of a number of technical options and to the specification of the basic architecture of the EPIC workbench (see Figure 1).

2.2. TECHNICAL OPTIONS

The necessity of having enough processing power, considerable capacity and adequate man-machine interface, while keeping to a widely used hardware platform, led to the choice of the SUN workstation.

In the context of the object-oriented approach adopted for EPIC, it was found that the PCTE covers successfully the database management requirements as well as solving a number of UNIX deficiencies. The PCTE (basis for a Portable Common Tool Interface) is the result of an ESPRIT project, offering an object-oriented database management system based on the entity-attribute-relationship approach, the capacity of transaction handling, as well as a number of distribution facilities.

Built on top of the PCTE kernel, is a layer of software : the ECLIPSE kernel, designed to provide improved support for tool integration. The ECLIPSE run-time system consists of the kernel and a number of tool components.

The tool-components take the form of libraries of functions and implement the database inteface, the application interface, the message interface and the help Interface. The last three are collectively known as the user interface. The interfaces form the Public Tools Interface (PTI) and are available with both C and Ada bindings.

On top of this kernel, a number of meta-tools have been constructed which form a tool builder's kit. This provides a powerful base from which integrated toolsets can be built rapidly.

The meta-tools enable particular kinds of tools, specifically editors for textual and graphical data, to be generated with very little coding. The meta-tools themselves utilise the ECLIPSE interfaces to ensure that toolsets will be integrated. One such meta-tool is the design editor, a computerised diagram editor providing an active system which enables neat drawings to be maintained as well as checks that a design conforms to the rules of the method.

3. Database Structure and User Interface

Computer-aided engineering must be supported by an engineering information system; EPIC takes a very broad view of CAE in the process industries including design and development of processes and products, design of plants, project engineering, engineering construction, and process engineering for operations, retrofits and upgrades.

Once the linking of engineering tasks through database support is achieved, one can begin to incorporate a host of user-friendly facilities such as powerful command languages, menus, windows or views, spreadsheets, graphic displays and natural language queries.

Nonetheless, the key issue in producing a usable and sharable software environment is to ensure an initial structuring of the information, adaptet to the needs of users and as close as possible to the physical context of entities involved in the activities of the specific engineering fields. This is primarily achieved during the requirements analysis stage (or procedure, in the prototyping approach).

3.1. DATABASE STRUCTURE

Most of the entities that engineers manipulate cannot be represented by simple data elements (such as vectors or matrices), particularly where geometry is a significant aspect. Such entities may also contain embedded operations which generate data as needed. These higher level data structures must be accomodated too, and are called complex objects.

Extensive analysis of requirements, repeatedly followed by database re-design activities, led EPIC to the specification of an initial 'skeleton' of data entities and of basic relationships among these data objects.

All types of declarative knowledge, e.g. graphical representation of unit operations, control loop elements, characterization of processing units (isothermal, adiabatic, perfect mixing, partial condenser, siphon type of reboiler, continuous stirred-tank reactor, two-stage compressor, etc.), description of chemical reactions, catalysts of solvents to enhance kinetics, sets of declarative equations for modelling processing units, equipment design, sizing or costing equations, sensors, controllers or control elements, tables and accompanying assumptions or heuristics, expert-system's rules, etc., are described by objects or attributes of objects.

This object-oriented character allows a flexible implementation of "data-directed programming", where the same programmatic procedures behave differently for different types of inputs. Modularity and polymorphism (different forms of data models) should be the two principal strengths of the EPIC system endowed by its object-oriented character.

3.2 USER INTERFACE

Engineers involved in process design activities rely heavily upon graphics, both for representing the (static) complex collections of information used, and for picturing the (dynamic) transformations of information throughout the design procedure. This fact stresses the importance of securing the qualities of the outer layer of the workbench, i.e. the user interface.

The EPIC workbench has inherited the graphical facilities of conventional flowsheeting packages, currently used in chemical plant design. These features, imposed by open-endedness and interfacing conciderations, are further enchriched with the creation of a consistent menu pattern, closely reflecting the inherent organisation of engineering tasks.

The EPIC user interface, then, offers an integrated toolset environment, the central features of which seem, at this stage of work, sufficiently stable. Supported by the ECLIPSE tools, however, the implementation of additional characteristics is straightforward. Forthcoming tasks of the project, such as the execution of case studies and the further promotion of the environment in the plant engineering sector, shall contribute to the refinement and adaptation of the graphical interface of the EPIC workbench.

4. Conceptual Framework and Integration of Tools

An efficient and operable CAD software system consists both of a number of software facilities assisting the design, and of a conceptual framework that guides the engineer throughout the design stages.

The multidisciplinary nature of the EPIC project (requiring expertise from chemical process engineering, process instrumentation and control, process optimisation and control theory and design) necessitates an integration of concepts, techniques and methods into a coherent body of useful engineering guidelines.

Within the EPIC project, work in this direction has led to the creation of new, cross-breed techniques issuing from the integration of traditional process control design tools and to the development of novel techniques in the area of control systems theory and of process optimisation.

4.1. INTEGRATING TRADITIONAL TECHNIQUES

Of crucial importance to the penetration of the EPIC system in the community of practicing engineers, the inclusion of tradional techniques for the design of control schemes has been an urgent priority since the early stages of the project.

These traditional techniques include procedures for the selection and pairing of measurements and actuations in an optimal manner, static and dynamic diagnostics of process operability based on rough plant models and procedures for initial plant decomposition into manageable sections and sub-sections.

Also incorporated are a number of standard methods for the design of single-input single-output regulators such as PID tuning procedures, feedforward control techniques or frequency responce analysis methods.

These techniques, supported by the EPIC workbench, are re-formulated in order to communicate with and rely upon an unique common information pool : the Process and Instrumentation Matrix (PIM) which contains all the data relevant to the control design activities.

The PIM data structure is organised in an measurement/actuation cross-reference pattern, provides meaningfull and confortable access to important information items and serves as a navigator throughout the design procedure.

4.2. DEVELOPING TECHNIQUES FOR CONTROL DESIGN

In the area of new techniques for control systems design, work in EPIC is concerned either with reformulating and expanding advanced theoretical methods which have not being used before in a systematic way for problems of early process design, or with investigating promising methods which are the outcome of recent research activities.

Of the latter kind, new measures for assessing the operability of a process have been investigated and partial results have been obtained. These include the calculation of different degrees of controllability, the evaluation of energy-based coupling using the input-output Grammian, the use of tracking indicators etc.

Also, work in EPIC has led to new results on the development of the Determinental Assignment Problem (DAP) method. This is a unifying mathematical context for the formulation and the solution of many control design problems : the selection of input-output structures, the synthesis of multivariable regulators as well as the design of simple coupling and low order control schemes.

Finally, a novel procedure for the derivation of mathematical models for composite systems from mathematical models of subprocesses and the process functional diagram is under development.

4.3. DEVELOPING OPTIMISATION TECHNIQUES

An important part of research activities is devoted to the development of efficient process optimising control algorithms. A large number of existing methodologies and algorithms have been evaluated and tested against various practical problems using computer simulation techniques.

The objective of the work is to provide the designer with a sophisticated tool box, based on the current practices of on-line steady state optimising control, incorporating existing techniques and algorithms and to provide methodologies, experience and software facilities for future development.

A number of a techniques based on existing methodologies have been developed, as well as novel tools such as an optimisation routine for constrained nonlinear optimization; the latter has been proved to be succesfull in a variety of examples and expected to be used as a new basic tool also in other areas of the EPIC work.

5. Application Domain and Industrial Potential

5.1. INDUSTRIAL CASE STUDIES

Industrial case studies are used to focus, refine and prove various developments during the implementation and testing phases of the EPIC project. Covering the full range of activities adressed by EPIC, the cases studies are real-life industrial situations so that the workbench is presented with all of the problems that it must overcome

Nonetheless, to ensure that the studies can be completed within the project timeframe, while meeting the general case study requirements, a single, fairly large process or process section which allows a variety of designs for the flowsheet and control system, and which lends itself to optimising control applications should be used.

An ideal example was found within the consortium : a fluid catalytic cracker unit (FCCU) as operated by Motor-Oil Hellas. Data is available from the industrial user, by consortium partners involved in process engineering and published papers, to allow modelling and to provide alternative design ideas.

The FCCU process has been reduced to a series of sub-units, each of which will form the basis for applying and testing certain EPIC workbench tools. These sub-units are : catalyst preparation, air blowers, feed preparation, reactor-regenerator, flue gas, heat recovery, main fractionator and wet gas compressor.

The first case study will be carried out, as far as possible, in the same manner as a real project would be conducted using th 'marketed' EPIC system. One partner acts as a customer, the others as design consultants and contractors. The aim of this action is twofold:

- To investigate possibilities for revamping the regulatory control system design and for applying advanced or optimising control on the existing FCCU. Small changes to the process may also be considered.
- To specify feasible alternatives for complete redesign, paying particular attention to control characteristics and optimal operation.

5.2 APPLICATION POTENTIAL

Refinement and assessement through concrete industrial design situations will help the EPIC workbench evolve into a complete, off-line, interactive software environment which shall integrate facilities offered by existing conventional flow-sheeting programmes, allowing for the bookkeeping, manipulation, selection and assesment of basic process units. These facilities are enriched with the integration of a set of tools for the analysis, evaluation and synthesis of control structures and optimisation schemes.

Whithin the EPIC environment, designers shall be allowed to consistently move among the tasks of flowsheeting , control design and optimisation design. Each task, then, is composed of a structure of sub-tasks, requires information at various levels of detail, generates information which may be prerequisite for the completion of another task and may also involve the designer's participation at various levels of interaction.

Moreover, advanced process design means maximum energy and material recovery, achieved through the introduction of more and more recycles which, in their turn, make the operation of the plant more sensitive to disturbances and possibly lead to inherently unstable behaviour. As these advanced plants are extremely complicated and the economic advantage involved is substantial, early intervention of control engineering is strongly needed.

Also, early evaluation of the operability of a plant can help discriminate between plant areas where control is easy and straightforward and plant sections where control would be critical. The latter must receive more detailed attention in subsequent design phases, before time and money is running out.

Finaly, a thorough analysis of hazard and operability aspects is required before construction of a plant is authorised. Evaluation of the control and operability characteristics of a plant is closely relevant to the elaboration of these studies.

6. Conclusions

In line with the objectives of the ESPRIT programme, project EPIC claims contributions to the dissemination of information technology and the advancement of aspects of engineering and scientific knowledge. Actual benefits concern the use, and subsequent promotion, of advanced software engineering methods and concepts within the environment of practicing process engineers and the development of a prototype workbench version featuring a well-adapted database structure and a flexible user interface.

Also, work within the EPIC project produces progress in theoretical results and gives rise to novel applications of advances in the fields of control engineering, optimisation and integrated plant design.

These achievements encourage the EPIC partners to anticipate significant impact in the field of Computer Integrated Manufacturing in as much as the project produces a useful, open-ended and flexible software product, on one hand, and compiles valuable information on the development and use of software for the needs of process designers, on the other.

Keywords: Industrial Processes, Chemical Processes, Process Design, Industrial Plant Design, Computer-Aided Design, Computer-Aided Manufacturing, Control Systems, Optimisation, Flowsheeting.

The EPIC consortium are:

INTRASOFT S.A.
PLANET S.A.
METEK S.A.
MOTOR-OIL HELLAS S.A.
ITP/TNO
SAST LTD
CITY UNIVERSITY

Project No. 2165
IMPPACT - Improvements in Integration by a Feature Approach. Product and Process Modelling move closer together.

ANDREAS MEIER
Krupp Atlas Datensysteme GmbH
Niederlassung Bremen
Am Wall 142
D-2800 Bremen 1
Tel. (+ +49) 421/14214

Abstract

The objective of IMPPACT is to develop and demonstrate a new generation of integrated modelling systems for product design and process planning including machine control data generation. The main goals are to improve software vendors efficiency by making software more easy to integrate and to improve the users efficiency by enhancing the functionality of existing CIM-modules. Limitations of integration will be overcome by a conceptual approach of using features in all stages of the manufacturing process.

A reference model as a general approach for deriving specific software development strategies builds the framework. A flexible, adaptable architecture will be used, open for the integration of future software. The integrated product model contains all information about the product and its manufacturing incl. standards and processes.

The main functional improvements are to combine solid and surface techniques for all planning stages to offer practice oriented unfolding techniques for 3-D models and to combine generic and knowledge based process planning to offer full automated NC-Part programming.

By simulation and evaluation, the capabilities of machining are used to optimize the product. Know-how gained is made available as feedback information not only by knowledge based modules cooperating with the product and production process model.

The economic objectives are being more competetive on time and cutting costs by increasing quality and productivity.

Results of IMPPACT will be demonstrated in the application areas of the sheet metal parts manufacturing (for aircraft spares) and complex shaped parts manufacturing (for ship propellers).

1. Objectives of IMPPACT

To approach the global aim of "doing it right-the first time" in order to increase efficiency in making products by being quick and keeping high quality and in this sense being able to react more flexible to the requirements of the market was the mainly initiation of the IMPPACT project. The desired information flow between design, process planning and production requires a conceptual approach that can only be based on an integrated view of design, planning and production activities.

To reach that goal, a flexible, adaptable system architecture is necessary with the capability of supporting the integration of various existing software packages of the involved partners with individual structures and environments. To increase the performance and the efficiency of the functions of the different systems, their existing structures and functionalities are combined in order to bring into line this with the generalized product and process reference model.

Knowledge-based information for both product and process modelling will be made available on the basis of production experience, material properties, machine and tool facilities. Each obtained input to production knowledge can be seen as upward feedback guiding the designer and the process planner to do the manufacturing from the beginning as efficient and complete as possible.

The architecture will be open to permit the integration of existing autonomous systems with future CIM-modules on a function oriented application level.

The objectives are, on the one hand, to develop tools which enable "easy" integration and, on the other, to enhance functionality as a benefit for the user by combining different techniques related to company-specific product model structures, as well as developing modules for evaluation and optimization of the product through simulation of machining processes at an early stage.

The capability of generating intermediate workpiece stages is regarded as an essential requirement in providing planning systems for successive manufacturing processes.

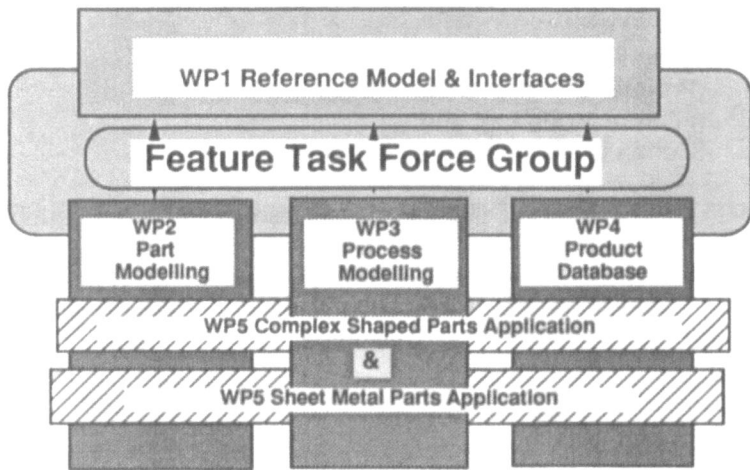

Fig. 1: IMPPACT project structure in relationship to Workpackages

The integrated product and production process model for IMPPACT will be adjusted to design tasks and will be able to involve all available information about standards, supplier parts and processes.

2. Product and Production Process Modelling: State of the Art

2.1 System Architecture

The work on Computer Integrated Manufacturing (CIM) draws upon the linking of many of the traditional areas of manufacturing automation including CAD, CAM, MRP into a complete system that will satisfy the enterprise's business strategy and objectives.

Concerning the architecture, the planners in IMPPACT take into account the results of the AMICE project (CIM-OSA, Esprit Project 688) (1). This architecture allows current systems to evolve towards possible future systems in manageable steps.

Open system architecture standardisation takes place in ISO TC184 (Industrial Automation Systems) SC5 (System Integration and Communication). IMPPACT influences the development of STEP (Standard for the Exchange of Product Model Data) within ISO TC184 SC4 (Industrial Data and Global Manufacturing Programming Languages).

2.2 Reference Models for Product and Production Process Modelling

The current developments of reference models for data exchange in an open integrated environment are strongly influenced by US DOD sponsored research programs (2, 3, 4, 5) like

ICAM   (Integrated Computer Aided Manufacturing),
PDDI   (Product Definition Data Interface),
GMAP   (Feature Modelling) and
CALS   (Computer Aided Logistic Support).

The results eyert a strong influence on standardisation efforts such as PDES (Product Data Exchange Standard) and therefore on STEP (6).

The ESPRIT projects CIM-OSA / AMICE (7) and the CAD*I (8) have some influence on the development of reference models for product and production process modelling, referring on the development of standard interfaces between CAD systems.

2.3 Data Handling

CIM components may store, access and exchange data in various ways. The PDDI project has combined the two aspects of shared databases as well as using local databases by exchanging via neutral files such as IGES, VDA-FS, SET or STEP. Furtheron, this project makes use of metafiles in order to be independent of the conceptual schema on pre-/postprocessor level. STEP applies a similar concept, based on the object oriented specification language for information modelling

EXPRESS (11) ESPRIT CAD*I (9) has developed a concept for a common database for analytical and experimental models.

2.4 Application Interface

To enable independent development of CIM application modules and to ensure software exchangeability, CAM-I's Geometric Modelling Program (GMP) has developed the neutral programming interface Applications Interface Specification (AIS) (12) which is unfortunately restricted to mainly geometrical functions with poor separation of the logical content and its physical, language-oriented representation.

Developments are needed to extend neutral programming interfaces to the reference data model that will be the basis for IMPPACT.

2.5 Part Modelling

Most of the geometric modellers with respect to solid modelling are based on the CSG (Constructive Solids Geometry) approach. Geometric volume primitives are combined with special volume operators. A second approach is boundary structure oriented using analytical defined geometry.

New developments are using the Non Uniform Rational B Splines (NURBS) representation, or, better, combine NURBS and boundary structures based on an analytical representation. A system like this one will have a strong competitive position in the market (14,15, 16, 17, 18,19).

Another trend is to use object-oriented techniques. Worldwide, most of the approaches are using feature-based and constraint modelling techniques to relate product related information to the geometric model.

Some experiments have shown the usefulness of form feature data, particular in the interface between solid modelling and process planning (20, 21, 22, 23, 24) but it will be shown that additional types of features and methods are necessary to solve integration on an application level.

2.6 Process Modelling for Parts

Process planning, being the link between design and production, is affected to a high degree by company internal and external developments and requirements. This fact is expressed in an increase in new additional planning tasks (16). An integrated overall solution is still missing. Complete integration between CAD, CAP and NC programming has only been realised in some simpler instances e.g. for rotational parts (14) in a vertical line for restricted facilities.

In the case of complex tasks, the information exchange between design and process planning is nowadays still being done conventionally. Technical drawings and parts lists are still set up by the designer manually or with the help of CAD systems. Today's existing feasibilities in changing information including neutral file (e. g. IGES) data exchange can not sufficiently consider the requirements of process planning.

New methods for information processing, e.g. generating intermediate work-piece stages based on expert systems, offer new possibilities of developing Computer Aided Process Planning systems which support heuristic activities of process planners. The contents of the planning process will change by facing of a number of influences (25, 26).

3. The IMPPACT Framework And Feature Approach

Integrated modelling systems for product design, process planning, and oper-ation planning with reference to a product and process model, will be developed.

Before starting IMPPACT, the wise decision was made to take into account such different applications like manufacturing of ship propellers on one side, and the simple appearing sheet metal parts manufacturing on the other. The imaginary integration of both broadened the horizon not only for each of the applications but also for the reference model.

The main tasks of IMPPACT are on one side to increase the functionality of CIM-modules and, on the other, to enable an "easy" integration of CIM-modules. This was taken into account for the development of part modelling and process modelling software systems and for demonstrating these for sheet metal and complex shaped parts applications and led in a natural way to the corresponding project structure (Fig. 1). In addition, it became more and more clear during the runtime of the project that the use of features would be, besides the common use of information in a shared product database, the most important integration factor. Therefore, a Task Force Group with members of all Workpackages was formed in order to harmonize the integration aspects by using features and not to predeter-mine this by only one area, e.g. part design.

3.1 Reference Model and Interfaces

To provide a general specification and concept for an open system architecture describing an integrated product and process modelling system in Workpackage 1, first of all the detailed requirements from engineering, design, process planning and shop floor production have been considered and different models have been built. From the two more general models for functions and data structures the expected reference model is derived. The feature model is also mapped to the reference model. Doing this, integration will be done on a semantic level. This will guide and permit work to be done especially for part (or product) and process modelling and for the design of a product database in the Workpackages 2, 3 and 4.

The reference model, as a result of the three models stated above, will consist of general terms and more specific extensions for the two application lines complex shaped part (CSP) modelling and sheet metal part (SMP) modelling. Attempts will be made to keep the reference model as general as possible. Workpackage 1 builds the frame for the other workpackages. The reference model has to reduce the gap between the development of the software vendors and the demand of the users. It has to make sure in this sense that developed software makes most benefit for the user. In this sense, it plays a consultant role - maybe not only within IMPPACT.

3.2 Part Modelling

Integration of part modelling with production process modelling is more effective if the CIM-Modules to be integrated already have a high standard in functionality. Therefore, the most important issue of part modelling in IMPPACT is to overcome the functional limitations in the design of parts by making several views available to the user through the combination of solid and surface techniques (Fig. 2), in addition, specific needs for the two lines CSP and SMP will be addressed.

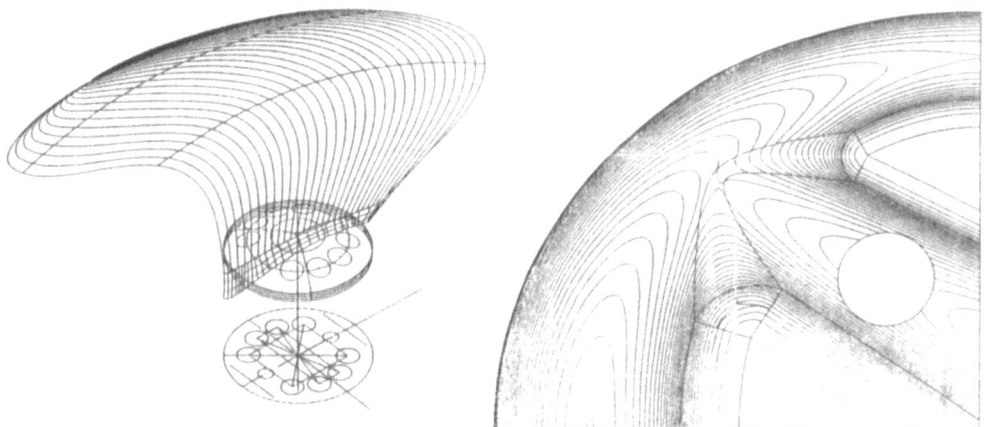

Fig. 2: Smooth transition between solid and surface regions

The second issue is of course the integration aspect itself. Both issues will reduce the effort of the designer and improve the quality. To overcome the handicap for the upstream of information for unique parts in the manufacturing process, three solutions could in principle be used:
- parametric definition of parts or
- decomposition of the parts into regions (features) which are used or applied more frequently or
- both, parametric definition of features

Relating to feature modelling, it is known that different experts in the manufacturing process have different views of the product (Fig. 3). The feature based CAM oriented design system will offer modelling by features as one possible tool for the designer to create a part. Besides this kind of design with features also the design of features will be covered.

522

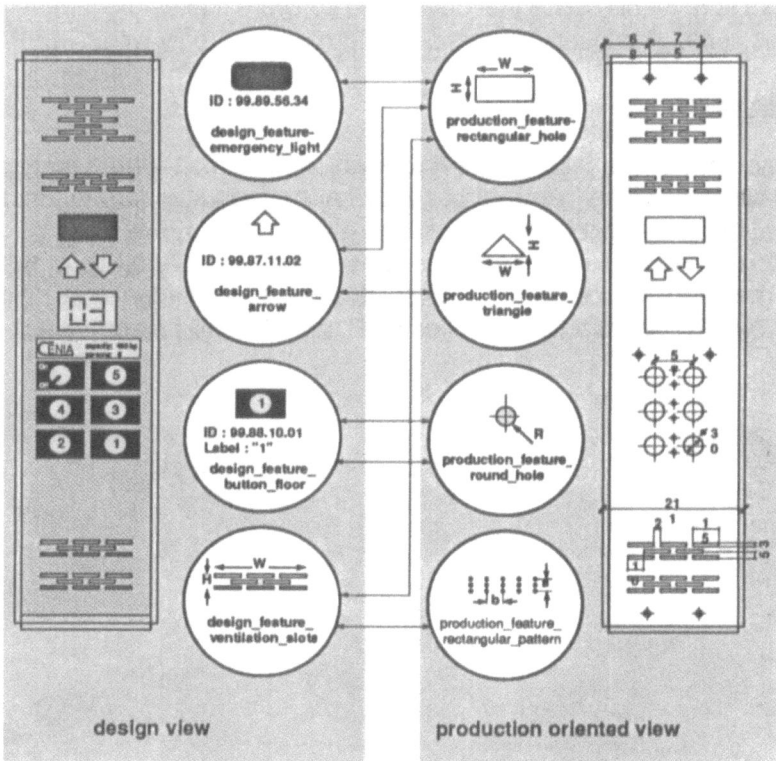

Fig. 3: Different views of design and production

Both functions can be automated for certain one-to-one relations between design and production for certain applications. Automated design of features is feature recognition which is only possible, if production knowledge is available.

Of course, the shape representation of each feature with idealized, implicit and explicit representations (Fig. 4) is one important submodul.

In addition, the analysis at the demonstrator sites leads to the definition of at least three types of features and corresponding feature models have been developed within IMPPACT.

a. Design features or functional features are defined from the viewpoint of the functionary behaviour of the product and are oriented towards the primary function of the part.

b. Analysis features play a role in different stages of the manufacturing process, from design to production. They differ from design features in the sense that they may not refer to the primary function of the feature

c. Production features. For these, one or more production processes exist. Although they can be produced, no direct selection of a specific production process is made. Whether the one-to-one relation to a specific production activity (Fig. 5) should also be regarded as a feature, has been in discussion for a long time. It is decided for IMPPACT, that this would reduce flexibility of the production process.

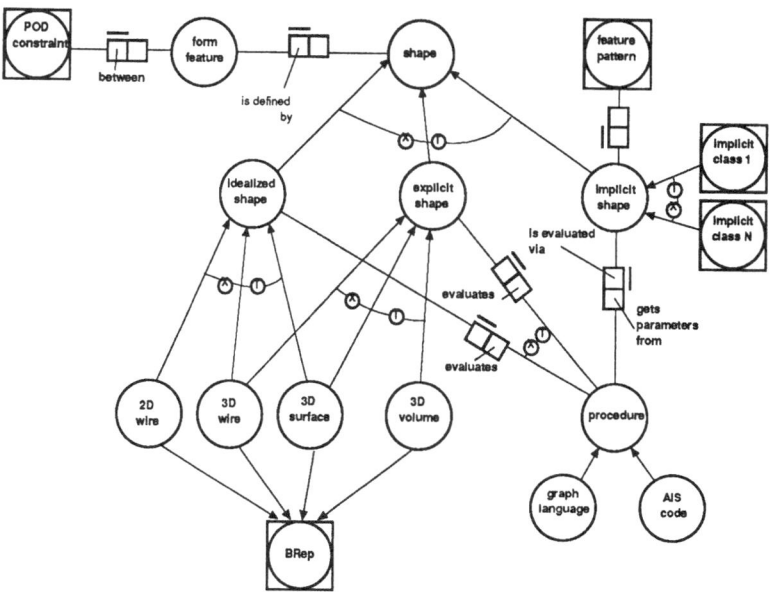

Fig. 4: The Shape Model as a Submodel to the Feature Model

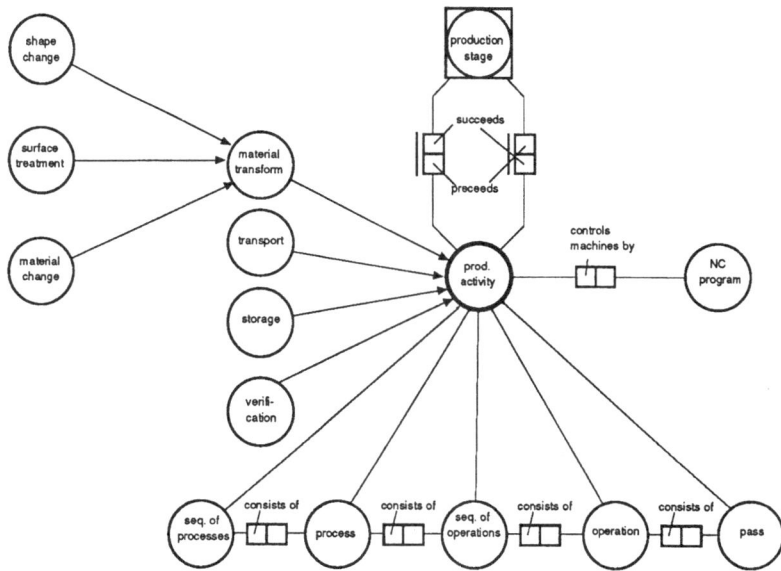

Fig. 5: Subschema of the Production Activity Model

A consistent mapping of the features and their use in different feature models must be ensured (Fig. 6). A general feature model than can be deduced (Fig. 7).

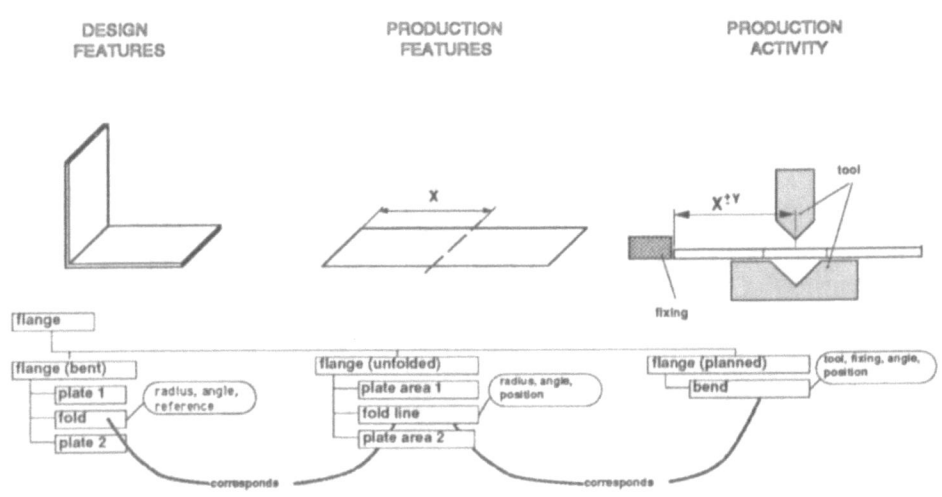

Fig.6: Mapping process between feature models

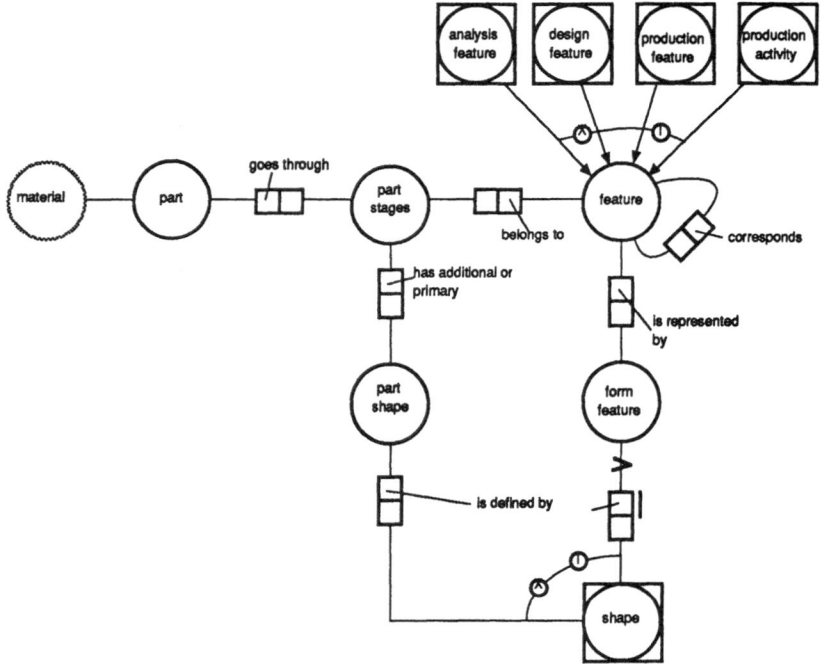

Fig. 7: Feature Model in NIAM

3.3 Process Modelling

General process modelling and operation planning is the processing, manipulation, extension and representation of general as well as application-oriented process knowledge and information based on the factory model oriented part containing available production facilities such as machines, tools and fixtures.

The access to all information and the extension of company-specific process knowledge is based on planning-oriented, feature based workpiece models containing all geometric and technological information needed for the planning task. According to each machining operation, all intermediate stages will be generated and represented.

One central part to fulfil the planning tasks in an optimised manner contains the general technological knowledge about available processes in the company, including their suppliers. It contains, for example, methods for the determination of cutting data, times and costs, machining strategies, production features and standard process plans as well as company-specific heuristic knowledge about machining processes that is experience from the production process.

The generalized process model is a complete description of how to make the part independent from particular production technologies. On detailed levels, special functions have to be developed in Workpackage 3 for each technology to prepare all the necessary instructions to human operators and machine controllers for direct manufacture.

A Process Planning Supervisor will evaluate the product design (i. e. analyse design features) and administrate submodules (both knowledge-based and generative) to make complete process plans for all types of products and processes. A challenge will be to prepare the system to take full advantage of the new techniques using features whenever necessary or at least useful.

The operation planning will of course be based on the feature oriented workpiece description. The operation planning system as developed (Fig 8) will either request an external system to generate the operation details, or it can make them internally using its own knowledge data. The system will try to run automatically with the ability to ask for user interaction if the available information is incomplete.

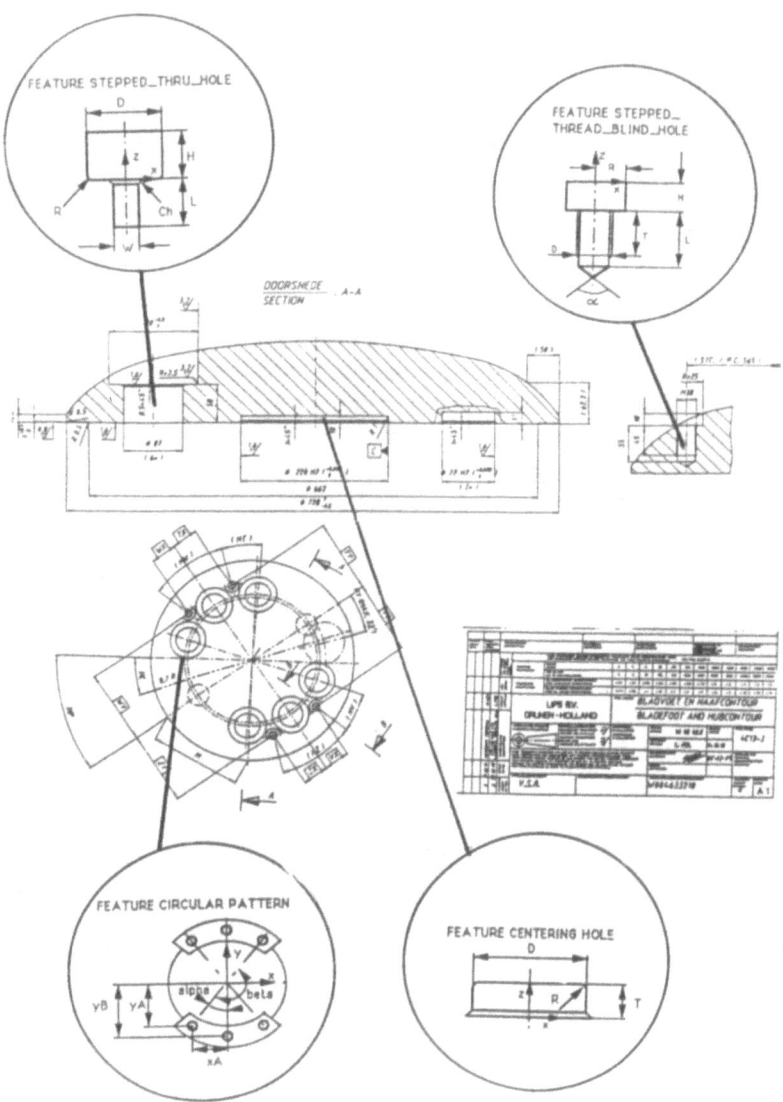

Fig. 8: Propeller Blade Foot with typical Features

3.4 Product Database

In IMPPACT, the task of providing a product database system for the integrated product and process modelling environment is done in Workpackage 4. This system is required for accessing all the information relevant to integrated product and production process modelling in a manner independent of the physical storage. The contents comprises design requirements, geometric models at various intermediate stages, process plans, NC programs, analysis results, production and quality reports.

And, in addition, the feature library has to be managed here as an archive. As features can in principle be hirarchical, their physical storage and access has to regard this topic. The access to the features has to be quick - otherwise they can not be used successfully on an simulation level. This has to be regarded in the existing product database configuration (Fig. 9).

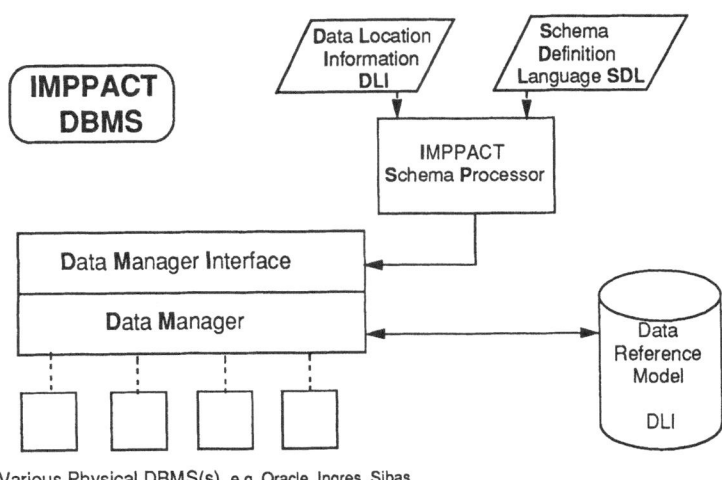

Fig. 9: IMPPACT -Product Database Configuration

4. Conclusion

The production process itself consists mainly of integration. Each action, whether performed by man or machine, is a combination of objects and information. Which items of information and how many of them are needed depends on the smoothness with which this communication process takes place. It is not enough to have a common distributed network-wide database into which all subsystems input what they know, trusting that whoever has a need (and perhaps requires help) will search in it to find the information he needs. Information should not have to be requested and searched for, but should be active and make itself available. This can be achieved by integration on application level like it is planned within IMPPACT by the feature approach. The use of features will support integration on a semantic-level. This makes the necessary communication more easy and more acceptable.

Without this communication, computers and manufacturing are not combined in a way that really increases efficiency. Total monitoring from receipt of order up to delivery is then certainly possible, but the aim of being able to react more flexibly and more quickly to the requirements of the market is not achieved. The amount of work needed for this task will be enormous. With IMPPACT, the partners will all contribute a step towards this aim, and the feature approach in all manufacturing stages will help to set an important milestone.

IMPPACT was initiated in 1987 and started in 1989 to develop a new generation of integrated systems for product design, process and operation planning in order to "doing it right the first time". The goal of reducing the time from market requirements to the availability of a product without increasing the cost of manufacturing but, on the contrary, reducing, leads to this functional investigation of the manufacturing areas of design, process planning and production.

The expected results are to improve on one hand side users efficiency by offering various integrated modelling techniques for products and processes and on the other side the efficiency of software vendors by using standard interfaces and data structures. Doing this, "easy integration" will be enabled, so that software vendors can concentrate on further improving the functionality to user's benefit. By this they beat the competition in the market by being more competetive on time and cost.

References.

(1) CIM-OSA: Reference Architecture Specification, ESPRIT-Project Nr. 688 Report, 1988

(2) Product Definition Data Interface - System Requirement Document Dept. of the Air Force document SRD5601130000, July 1984

(3) Product Definition Data Interface - Needs Analysis Document Dept. of the Air Force document NAD560130000, July 1984

(4) Product Definition Data Interface - System Specification Document, Draft Standard; Dept. of the Air Force document SS560130000, July 1984

(5) Product Definition Data Interface - System Specification Document Dept. of the Air Force document SDS560130000, July 1984

(6) Weiss, J. STEP & Project Plan - Document o. O, ISO TC184/SC4/WG1 Document Nr. 297, Aug. 1988

(7) Trippner, D. CAD*I Interface development - Experience Gained for the Benefit of Improvement of CAD/CAM Data Exchange; Presentation CAD*I Workshop, Karlsruhe, December 1986

(8) Schlechtendahl, E. (ed.); Specification of a CAD*I Neutral File for-CAD Geometry,Version 3.3 Springer 1988

(9) ANSI/X3/SPARC Database System Study Group. Reference Model for DBMS Standardization, U. S. Dept. of Commerce, National Bureau of Standards, Report NBSIR-85/3173 (1985)

((11) Altemueller, J. Mapping from EXPRESS to Physical File structure, ISO TC184/SC4/.WG1 Document N280, 1988

(12) Pratt, M. J. Solid Modelling and the interface between design and manufacturing, IEEE Computer Graphics and Applications 4, 7, page 52 - 59 (1984)

(13) Normung von Schnittstellen fur die rechnerintegrierte Produktion (CIM); DIN-Fachbericht 15, July 1987

(14) Bjorke, O. (ed), APS - Advanced Production System Tapir-Verlag, Trondheim, 1987

(15) Grabowski, H., Anderl, R., Paetzold, B., ESPRIT, Technical Week 1987

(16) Spur, G., Krause, F.-L. CAD-Technik, Carl Hanser Verlag, München, 1984

(17) A. G. Requicha, Reproduction for Rigid Solids, Theory, Methods and Systems, Computing Surveys, Vol. 12, No. 4, Dec. 1980

(18) A. G. Requicha, Solid Modelling: Current Status and Research Directions, IEEE CG & A, Oct. 1983

(19) A. G. Requicha; Growing and Shrinking Solids blending,tolerancing and other application (CAD-Kolloquium) Nov. 1986, Berlin, SFB 203

(20) Bunce, P. G., M. J. Pratt, S. Pavey, J. Pinte; Features Extraction and Process Planning; CAM-I report R-86-GM/PP-0l, Arlington 1986.

(21) Falcidieno, B. and F. Giannini. Extraction and Organisation of Form Features into a Structured Boundary Model, Proc. Eurographics 87 Conf., North Holland, (1987)

(22) Henderson, M. R., and D. C. Anderson, Computer Recognition and Extraction of Form Features: a CAD/CAM link, Computers in industry 5, page 315-325 (1984)

(23) Krause, F.-L.; Vosgerau, F. H.; Yaramanoglu, N.; Using Technical Rules and Features in Product Modelling;Proc. IFIP Working Group 5.2 Workshop on Intelligent CAD, Boston, Oct. 1987,

(24) Pavey, S. G.; Hailstone, S. R.; Pratt, M. J. An automated interface between CAD and Process Planning; Proc. International CAPE conf., Edinburgh, April 1986, Mech. Engineering, Publications Ltd., Bury St. Edmunds, England

(25) Major, F., Grottke, W. Knowledge Engineering within Integrated Process Planning Systems, Proceedings of the Int. Conference on Intelligent Manufacturing Systems, Budapest, 1986

(26) Mäntylä, M. Feature-Based Product Modelling for Process Planning

Project No. 2192
AIMBURN

MODELLING FLUID FLOW AND HEAT TRANSFER IN AN INDUSTRIAL GLASS FURNACE

M.G.. Carvalho and M. Nogueira
Instituto Superior Tecnico
Mechanical Engineering Department
Av. Rovisco Pais
1096 Lisboa Codex
Portugal

Abstract.

In the present work a three-dimensional mathematical model describing the physical phenomena occuring in a glass melting furnace is presented. This model is based on the solution of conservation equations for mass, momentum, energy and combustion related chemical species, and comprises two main submodels, for the combustion chamber and for the glass melt tank. The model incorporates among others physical modelling for the turbulent diffusion flame, soot formation and consumption and thermal radiation. The time-averaged conservation equations set was solved using a finite volume technique. The discrete transfer procedure is used to solve radiative transfer in the combustion chamber. Predicted distributions of soot, oxygen and fuel concentrations are presented as well as temperature field and the fluid flow pattern inside both the combustion chamber and the glass melting tank.

Nomenclature

f   Mixture fraction
g   Mean square of fluctuations
k   Turbulent kinetic energy (m^2/s^2)
m_{ox} Oxygen mass concentration (Kg/m^3)
m_{fu} Fuel mass concentration (Kg/m^3)
ε Dissipation rate of k (m^2/s^3)

Introduction

The resources limitations growth, on one hand, and the pollution problem awarness in the other are turning heat transfer research attention towards the importance of improving performance of industrial process furnaces, such as those used in glass, cement and steel assemblies manufacture < and power plants.

That commitment cannot be achieved without to improve control strategies, personal training, furnace design, particularly combustion equipment design.

Traditional zero-dimensional methods based on empirical knowledge, largely ignoring spatial variations are unable to assist that purpose. The physical understanding of the phenomena occuring in the furnace/boiler environment based on three-dimensional mathematical modelling is becaming an essential requirement to assist operating conditions changes and further design modifications.

A three dimensional simulation of a glass furnace combustion chamber in which the flow field and heat transfer characteristics were predicted, was presented at a first time by Gosman et al. [1].

Lockwood and Shah [2] presented an accurate and efficient technique for handling thermal radiation, the "discrete transfer method". This method is built upon the assumption that radiative transfer can be modelled by means of representative rays emitted in specified directions. The enclosure is subdivided in cells. The total cell radiative source is the sum of the contributions from all of the beams which happen to cross it. The present method combines ease of use, economy and flexibility of application which is of particular importance in geometrically applications. The discrete transfer assumptions was used in the first simulations of a full scale industrial glass furnace (Gosman et al. [1] and Carvalho [3]).

In the work of Carvalho [3] a geometry with a horseshoe flow shape and combustion of fuel oil were considered. The same end-port regenerative glass furnace was considered in the study of Carvalho and Lockwood [4] which brought

In these early works, time averaged equations for the conservation of momentum were solved as well as the equations for energy conservation and mass continuity. The "two-equation" turbulence model [5], in which equations for the kinetic energy of turbulence and its dissipation rate are solved, was considered to be appropriated.

Semião [6] has applied a three-dimensional prediction procedure to a ceramic glass smelting-kiln with oxygen-rich burning conditions. Carvalho et al. [7] extended that work and used a two-dimensional axisymmetric model to simulate the burner region, providing with these results the inlet conditions for the three-dimensional calculations of the combustion chamber. The results were extensively validated with experimental data acquired in the furnace. Measured and predicted temperature profiles corresponding to transverse paths exhibit a satisfactory agreement and quantify the ability of the present calculation method in simulating the mean scalar characteristics of a furnace combustion chamber.

The glass industry is significant user of energy resources. The mathematical model presented in this paper was applied to a glass bottles production furnace, which is part of a plant where energy consumption accounts for some 20/25% of glass production costs. The furnace accounts for 80% of the energy consumption.

The Physical Modelling

The heat transfer, fluid flow, combustion and glass melting phenomena occuring inside a glass furnace considering its combustion chamber and glass melting tank parts, may be predicted by solving the governing partial differential equations set in its steady-state time-averaged form.

The Mean Flow Equations. The time-averaged equations for the conservation of momentum were used as well as the equation for the conservation of energy. In addition to the referred equations, one must also include the equation of mass continuity.

The Turbulence Model. The "two-equation" model [5], in which equations for the kinetic energy turbulence k and its dissipation rate ε are solved, was considered appropriated to the flow in the combustion chamber.

The Combustion Model. The combustion model is based on the ideal of a single step and fast reaction between the gaseous fuel and oxidant, assumed to combine in stoichiometric proportion. Equal effective turbulent mass diffusion coefficients for the fuel and oxidant and an instantaneous reaction are also assumed (see [8, 9]).

The transport equation for the mixture fraction f was solved. Furthermore, we have assumed that the chemical kinetic rate is fast with respect to the turbulent transport rate; then fuel and oxidant coexist, so $m_{fu} = 0$ for $m_{ox} \geq 0$ and $m_{ox} = 0$ for $m_{fu} \geq 0$, and concentrations m_{fu} and m_{ox} are related linearly to the mixture fraction.

The above modelling, it may be noted, presumes a stationary thin - flame envelope. The fluctuating nature of the turbulent reaction is more usually accomodated (e.g., in the work of Spalding [10]) through a modelled equation for the variance of the mixture fraction fluctuations.

We have adopted a statistical approach to describe the temporal nature of the mixture fraction fluctuations. In the present work we have assumed the "clipped normal" probability density function [11], which is characterized by just two parameters: f and the mean square of f fluctuations $g = (f - f)^2$.

The Soot Model. The distinctive feature of oil-fired flames is their significant soot content. Soot is of concern because its presence greatly augments the radiation heat transfer and because it is a pollutant. To predict the spacial distribution of soot, a transport equation of its mass concentration was solved. To characterize soot production, a simple expression similar to that used by (Khan and Greeves [12]) was chosen. A straightforward method of estimating the rate of soot burning proposed by (magnussen and Hjertager [13]) was used in the present work.

The Radiation Model. The "discrete transfer" radiation prediction procedure of (Lockwood and Shah [1]) was utilized in the combustion chamber modelling. This method combines ease of use, economy, and flexibility of application. This last feature is of particular importance in the real world of geometrically intricate combustion chambers.

The gas absorption coefficient was calculated from the "two grey plus a clear gas" fit of Truelove [14]. Water vapor and carbon dioxide are the prime contributors to the gaseous radiation.

Glass Melting Tank Model. The model for the glass tank flow incorporates physical modelling for the laminar flow and energy in the molten glass. The radiation transfer inside the glass bulk was handled by using an effective thermal conductivity in the energy balance equation. The radiative exchanges between the glass and surrounding surfaces was handled using a standard formulation adequate for semitransparent medium which on the glass free surface ensures a quite realistic description of the penetration of the radiative fluxes. Air bubbling effect was calculated using a Langrangean formulation to predict the bubble rising flow accounting heat and momentum transfer with the glass melt. The batch melting distribution was imposed following observations performed through a vision system instaled in the studied furnace. A quadratic velocity profile was imposed at the molten glass outlet port.

Coupling Algorithm. The combustion chamber and the molten glass flow were studied by a cyclical iterative way, by separated calculations, matched by the relation between the heat flux from the flame to the glass surface and its temperature. Using the heat flux to the glass, calculated by the chamber model, as a boundary condition, the flow and the temperature distribution of the molten glass were predicted by the tank model. A similar procedure is described in [15].

Numerical Solution

The finite difference method used to solve the equations entails subdividing the combustion chamber into a number of finite volumes or "cells".

The convection terms were discretized by the hybrid central/upwind method [16]. The velocities and pressure are calculated by a variant of the SIMPLE algorithm described in [17]. The solution of the individual equations sets was obtained by a form of Gauss-Seidel line-by-line iteration.

The Studied Glass Furnace

The studied glass furnace is installed at Barbosa & Almeida Factories (Portugal) and produces glass bottles and other glass containers. The furnace which is of the endport regenerative kind, is shown in figure 1. The furnace roof and side walls are refractory lined and the roof is arched as indicated on figure 1. The fuel is admitted from 3 tube burners (see 5 on figure 1) located below a port (3) which admits preheated air from a chamber of regenerators previously heated. The air flow is directed downwards and the fuel jets have an adjustable upward inclination with a value of 8°. The flame forms a loop within the combustion chamber and waste gases leave via an exhaust port (4) heating-up a second chamber of regenerators.

The batch enters via the side "DOG HOUSE" port near the back wall (8), while the molten glass exits from the throat at location (7). The working compartment, is connected to the glass melt tank by a throat in the lower part of a separation wall (9).

Typical values, which describes the essential features and the operation pattern of the furnace, are as follows.

Fuel:
- Flow rate - 0.18 Kg/s
- Composition-thick fuel oil (ASTN D 396-N°6) aproxim. 85% C, 11% H
- Inlet temperature - 105° C

– Upward inclination of the burners-8°

Air:

– Flow rate-2.031 Kg/s
– Inlet temperature-1325°C
– Air Excess- 13% aprox.
– Downwards angle of inlet duct-8°

Results Discussion

A parametric study of the combustion chamber efficiency was performed. The varied parameter was the fuel mass flow rate. Results of this study and a description of the phenomena occuring in the combustion chamber and in the glass melting tank, through the predicted flow field, temperature and species concentrations, are presented.

Parametric Study

The above referred mathematical model was applied to the furnace described in the previous section for the typical operating conditions. One of the most important control variables is fuel inlet rate. The AIR/FUEL ratio was, in the present study, assumed constant and equal to 15.87.

The efficency was defined as a ratio of the total heat flux to the batch and molten glass divided by the inlet fuel enthalpy.

The inlet mass flow rate of fuel was varied from 0.16 Kg/s up to 0.21 Kg/s in steps of 0.01 Kg/s. The remaining operating conditions were kept constant. Figure 2 shows the variation of the efficiency and of the total heat flux to the glass with mass Flow rate of fuel. The furnace efficiency decreases with the increase of the heat flux rate to the fuel. This expected result suggests the need to search for an optimum point.

Mathematically forecasted results presented here may supply a valuable numerical information which may assist control system tunning.

Combustion Chamber Predictions

In this section, predicted distributions of temperature, fuel, soot and oxygen concentration obtained with a fuel inlet rate of 0.20 Kg/s are presented. The fluid flow pattern inside the furnace is also shown.

The flow is characterised by one main recirculating motion as shown in the figure 3 which presents the trajectories of unweighed particles imaginary seeded in the flow facing each burner. The calculated paths, which are coincident with streamlines of the time-averaged flow, show the "horse shoe" shape of the predicted main recirculation region, typical of this kind of furnace.

Figure 4 presents a predicted gas temperature field for different y - z planes. The highest temperatures level is found in the half of the furnace near to the fuel and air inlet. High temperature gradients appear in the interface between air and fuel stream denoting the presence of combustion reaction. The temperature near the outlet is quite uniform, about 1850 K, which is a value well accomodated with the few industrial temperature measurements obtained in the outlet duct. Near the glass melt surface the temperature distribution denotes the cooling effect of the batch melting process and presents an hot spot under the hottest gas region. The location of the glass surface hot spot matches quite well with the pattern expected by the furnace operators.

Figure 5 shows the fuel concentration in horizontal and vertical planes crossing the burners (figures 5a and b) and in an horizontal plane 0.06 m away the glass surface (figure 5c). The main flow recirculation from the inlet ports to the outlet imposes a displacement of the fuel distribution towards the nearest side wall. This effect, which certainly will yield a wall degradation increase in that region, can be avoided introducing changes in the furnace design. This task would be assisted by mathematical modelling.

The inlet air and fuel relative location imposes the flame shape displacement towards the glass surface. This effect, which is explained by a secondary recirculation below the burners row, produce a redutive atmosphere just above the batch melting region, even for oxigen-rich flames. For some glass compositions, this effect cannot be neglected. A control strategy able to avoid this effect, without damaging the thermal efficiency, may be studied using mathematical modelling skills.

The predicted contours of soot concentration in a vertical plane crossing the burner region are shown in figure 6. The maximum soot mass concentration occurs near the flame tip. Thereafter soot mass concentration decreases rapidly by oxidation.

In Figure 7 the fluid flow inside the furnace is characterized. The representation method used takes the paths produced by imaginary unweighted particles arbitrarily seeded in the flow. The time step referred to each plane is the time during which the trajectories are followed. In this figure four horizontal planes are presented. The main feature of the present flow is a large recirculation region which is well apparent in all planes of Figures 4. The informations contained in these predictions may be, for example, very useful in order to evaluate the residence time of the gas inside the chamber, which is a determinant parameter in the point of view of thermal NOX production levels.

Glass Tank Predictions

Free convection effect, which predominantly drives the flow when no air bubbling is being used, can be observed in the pattern shown in Figure 6. This representation uses the seeking of particles trajectory technique above referred. The imaginary particles were seeded in the melting region. A minimum residence time of eight hours was obtained. Figure 7 presents a vertical view of the flow field. High velocity values can be observed in near the outlet region and above the step, where severe corrosion effect may be expected.

The temperature field in a plane near the middle of the tank, in which the cold

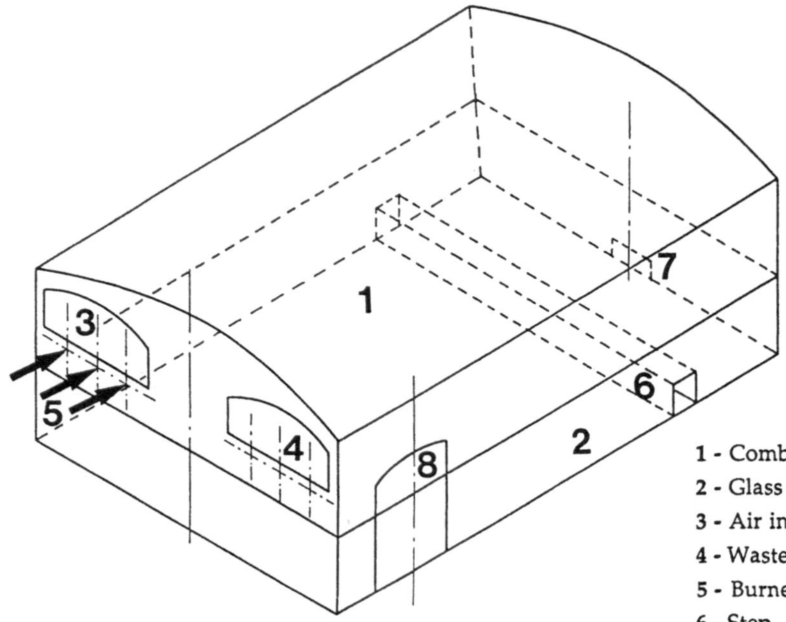

Figure 1:   Schematic diagram of the furnace.

1 - Combustion chamber.
2 - Glass melting tank.
3 - Air inlet port.
4 - Waste gases outlet port.
5 - Burners.
6 - Step.
7 - Molten glass exit.
8 - Batch inlet port.

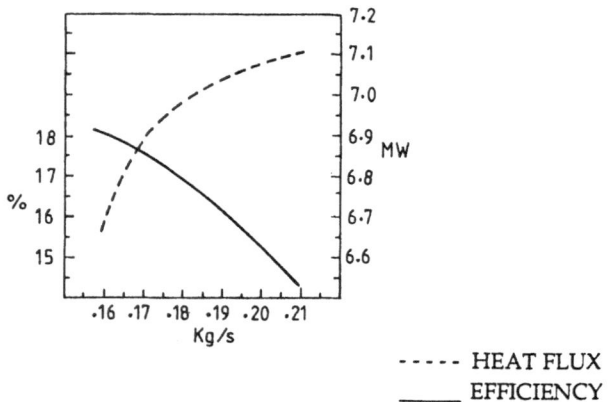

----- HEAT FLUX
_____ EFFICIENCY

Figure 2:   Variation of effciency and heat transfer to the glass on the fuel
input rate.

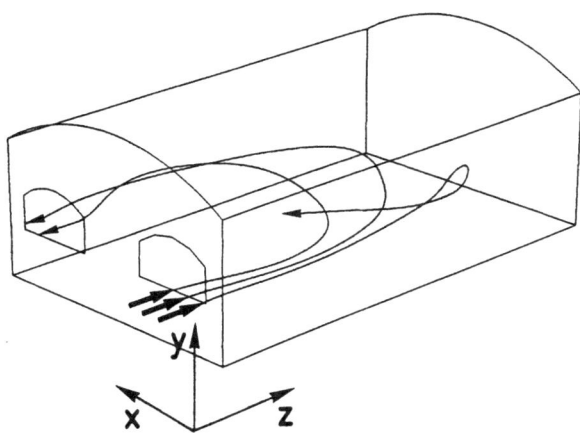

Figure 3:   Fluid flow patterns inside the chamber.

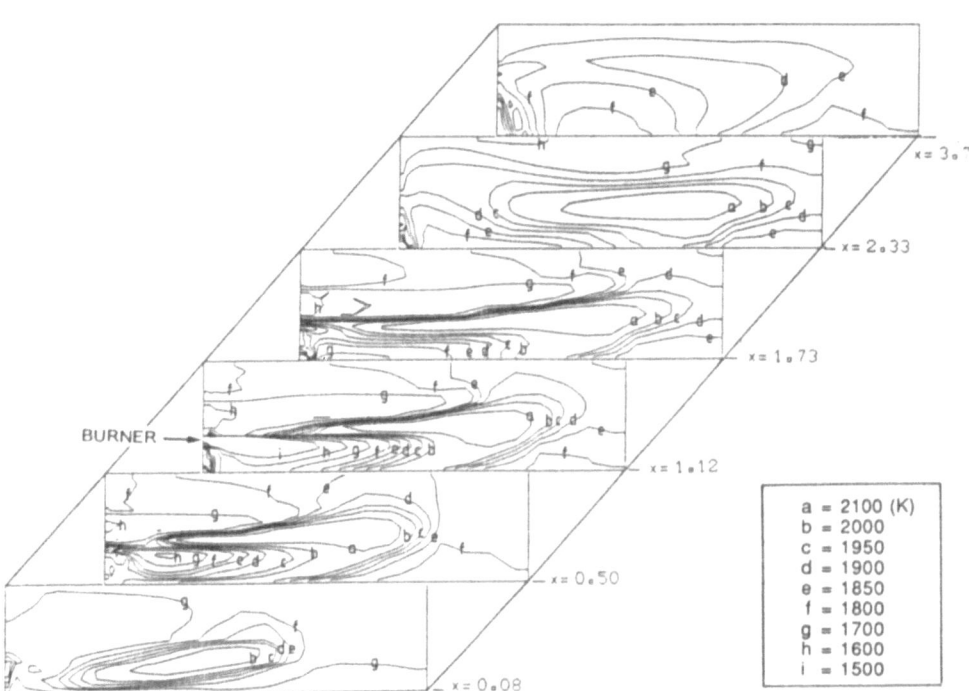

Figure 4:   Predicted temperature field.

538

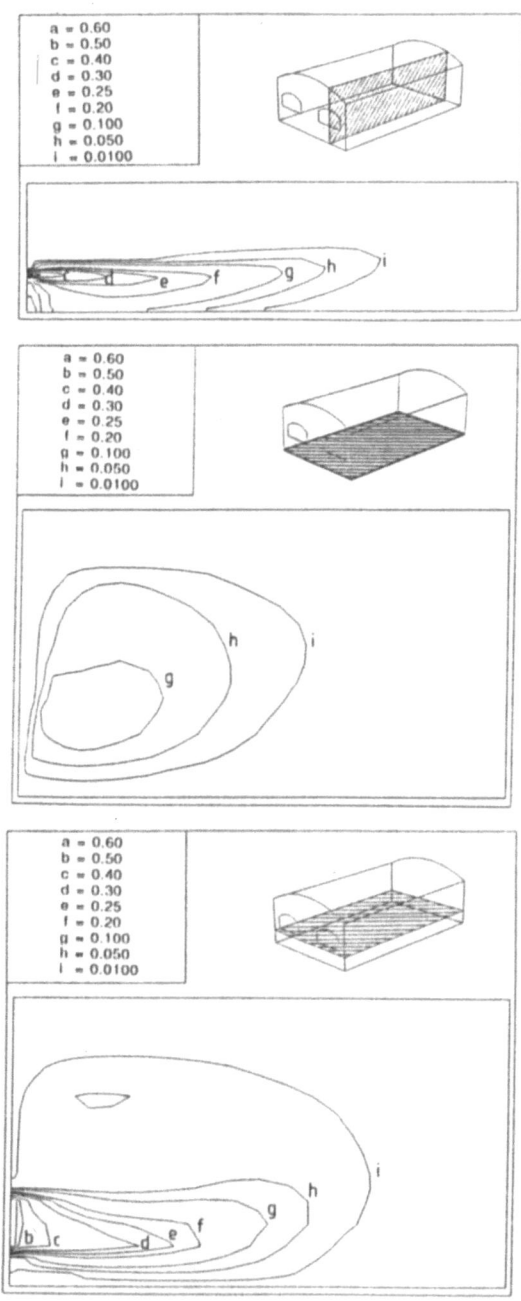

Figure 5:   Predicted fuel concentration field

539

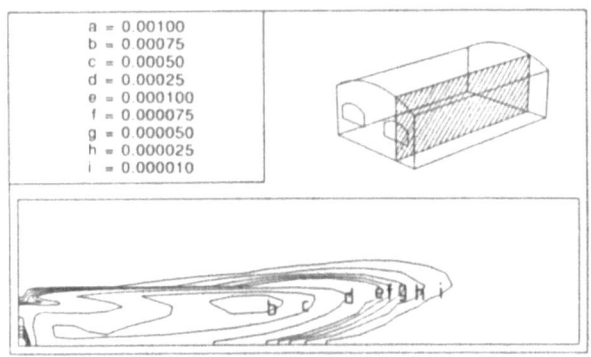

Figure 6:   Predicted soot mass concentration field.

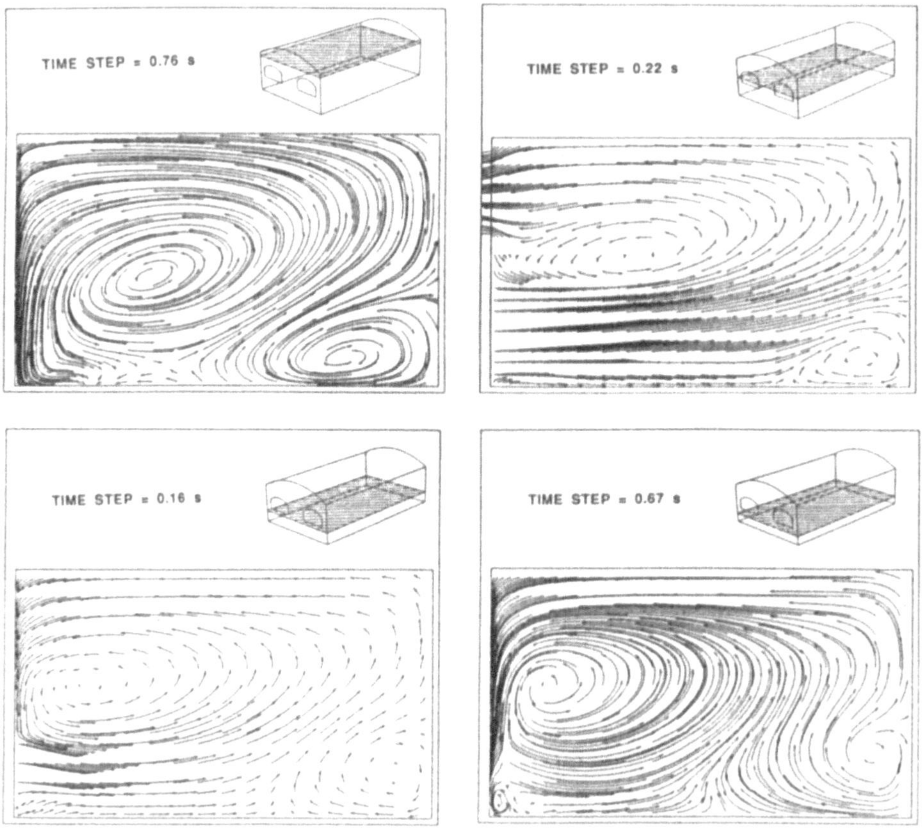

Figure 7:   Predicted fluid flow inside the chamber.

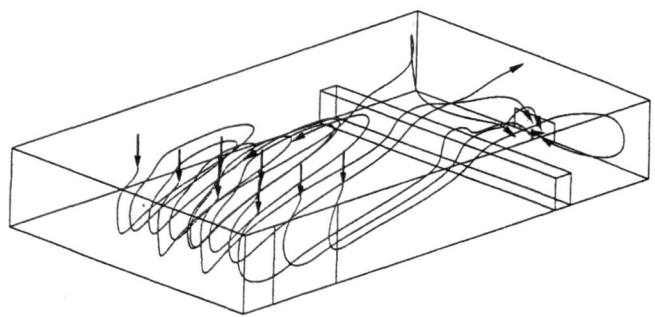

Figure 8:   Predicted flow pattern in the tank.

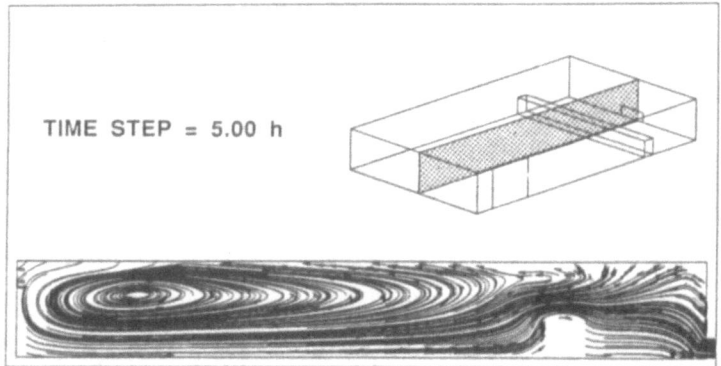

Figure 9:   Predicted flow pattern in the tank.

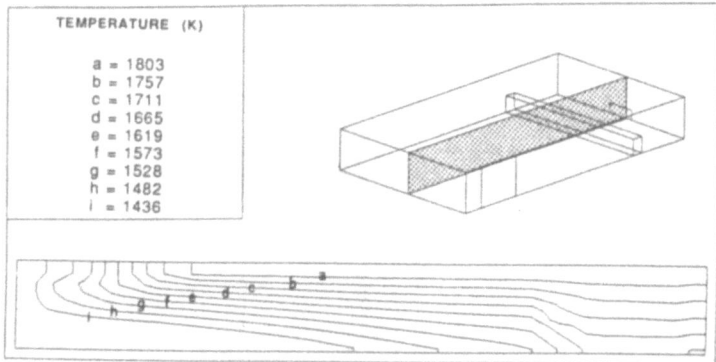

Figure 10:   Predicted temperature in the tank.

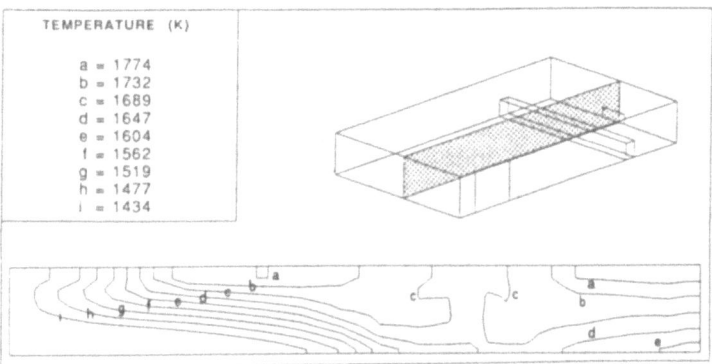

Figure 11:   Predicted flow pattern in the tank

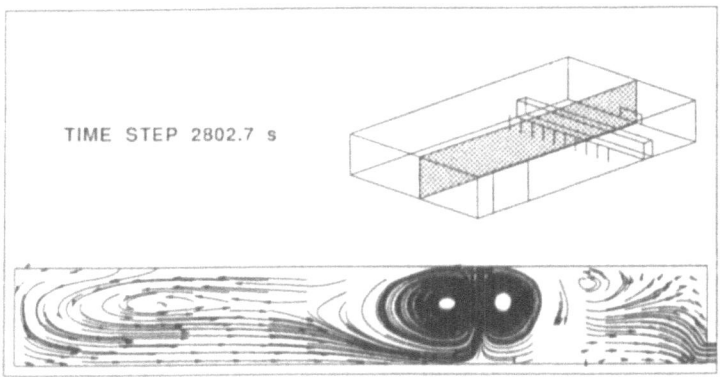

Figure 12:   Predicted flow pattern in the tank

542

regions below the batch zone are well apparent, can be seen in Figure 8. For no air bubbling conditions, a stratified temperature field is produced as shown in that figure. This effect does not help the glass refining process. However, air bubbling is used mainly to produce a barrier in the molten glass flow, preventing the motion of unmelted batch particles towards the refining zone of the furnace.

When the furnace is being operated with air - bubbling a completely different flow pattern occurs as can be seen in Figure 9, where the main flow is represented by a vertical plane in the middle of the furnace. The effect of the bubbling is determinant lo produce two strong recirculations which actually ensures the barrier effect above mentioned. These recirculations also produce a mixing in the central region of the tank yielding a homogenizing effect well characterised in the temperature field presented in Figure 10.

Concluding Remarks

This paper describes the application of an useful and general prediction procedure to real industrial furnaces. Predictions were obtained using a three-dimensional algorithm for the solution of the fluid flow, heat transfer and combustion phenomena occuring inside the combustion chamber and glass tank regions of an industrial glass furnace. The paper suggests how the modelling of relevant physical processes in industrial furnaces may supply an important basis for improvements in control strategies. A parametric study of the efficiency sensitivity on the fuel inlet rate is presented. Results of three-dimensional predictions of fluid flow and temperature was given, as well as predictions of fuel and soot concentration within the combustion chamber.

References

[1] Gosman, A.D., Lockwood, F.C., Megahed, I.E.A. and Shah, N.G. (1980). "The prediction of the Flow, Reaction and Heat Tranfer in the Combustion Chamber of a Glass Furnace"- AIAA 18th Aerospace Sciences Meeting, January 14-46, Pasadena, CA, U.S.A.

[2] Lockwood, F.C. and Shah, N.G. (1980). "A New Radiation Solution Method for Incorporation in General Combustion Prediction Procedures". 18th Symp. (Int.) on Combustion, The Combustion Institute.

[3] Carvalho, M.G. (1983). "Computer Simulation of a Class Furnace". Ph.D. Thesis, London University.

[4] Carvalho, M.G. and Lockwood, F.C. (1985). "Mathematical Simulation of an End-Port Regenerative Glass Furnace". Proc. Inst. Mech. Engrs., 199, (C2), pp.113-120

[5] Launder, B.E. and Spalding, D.B. (1972). "Mathematical Models of Turbulence". Academic Press, New York.

[6] Semião, V. (1986). "Simulacão Numérica de uma Fornalha Industrial". MSc Thesis, University of Lisbon.

[7] Carvalho, M.C., Durão, D.F.C., Heitor, M.V., Moreira, A.L.N. and Pereira, J.C.F. (1988). "The Flow and Heat Tranfer in an Oxy-Fuel Glass Furnace". Proc. 1st European Conference

on Industrial Furnaces and Boilers. March 21-24, Lisbon.

[8] Pun, W.M. and Spalding, D.B (1967). "A Procedure for Predicting the Velocity and Temperature Distributions in a Confined Steady, Turbulent, Gaseous Diffusion Flame". Proc. Int. Astronautical Federation Meeting, Belgrade.

[9] Bilger, R.W. (1980). "Turbulent Flows with Non-Premixed Reactants, Turbulent Reacting Flows". Topics in Applied Physics, Ed. Libby, P.A. and Williams, F.A., Springer-Verlag.

[10] Spalding, D.B. (1971). "Concentration Fluctuations in a Round Turbulent Free Jet". Chem. Eng. Sci., 26, p. 96.

[11] Lockwood, F.C. and Naguib, A.S. (1975). "The Prediction of the Fluctuations in the Properties of Free, Round Jet, Turbulent Diffusion Flame". Combustion and Flame, Vol. 24, No 1, pp. 109.

[12] Khan, I.M. and Greeves G. (1974). "A Method for Calculating the Formation and Combustion of Soot in Diesel Engines, Heat Transfer in Flames". Ed. Afgan and Beer, pp. 391-402.

[13] Magnussen, B.F. and Hjertager, B.H. (1976). "On Mathematical Modelling of Turbulent Combustion with Special Emphasis on Soot Formation and Combustion". 16th Symp. (Int.) on Combustion, Combustion Institute.

[14] Truelove, J.S (1974). "Mathematical Modelling of Radiant Heat Tranfer in Furnaces". Heat Tranfer & Fluid Flow Service, Chemical Engineering Division, Aere Harwell Rept., NQ. HL76/3448/KE, September.

[15] Carvalho, M.G., Oliveira, P. and Semião, V. (1988). "Mathematical Simulation of an End-Port Regenerative Glass Furnace", Journal of the Institute of Energy, pp. 143-156, Sept.

[16] Spalding, D.B. (1972). "A Novel Finite Difference Formulation for Differential Expressions Involving both First and Second Derivatives. Int. J. Num. Methods Eng., 4, 557.

[17] Carreto, L.S., Gosman, A.D., Patankar, S.V. and Spalding, D.B. (1972). "Two Calculation Procedures for Steady, Three-Dimensional Flows with Recirculation". Proc. of 3rd Int. Conf. on Numerical Methods in Fluid Dynamics. Springer-Verlag, New York, p. 60.

Project No. 2434

INTEGRATING OPERATIONAL RESEARCH AND ARTIFICIAL INTELLIGENCE IN A DISTRIBUTED APPROACH TO DYNAMIC SCHEDULING: THE B.I.S. PROJECT

Marco Guida
Pirelli Informatica S.p.A.
Via dei Valtorta 48
20127 Milano
Italy

Giorgio Basaglia
Pirelli Coordinamento Pneumatici S.p.A.
Viale Sarca 222
20100 Milano
Italy

Abstract.

We claim that Operational Research and Artificial Intelligence, while proposing different approaches to scheduling, seem to offer complementary techniques rather than antithetic ones.

Starting from this consideration, we have devised an approach to scheduling based on a framework for the description of problems and a distributed architecture for the development of systems, according to which an high degree, effective integration among O.R.-based and A.I.based techniques can be achieved.

The framework and the architecture have been both experimented in the development of an industrial real size dynamic scheduling system for the compound area of a tire factory.

1. Introduction

Operational Research (O.R.) and Artificial Intelligence (A.I.) propose different approaches to scheduling and seem to offer complementary techniques rather than antithetic ones.

Starting from this consideration, we have devised an approach to scheduling based on a framework for the description of problems and a distributed architecture for the development of systems, according to which the desired degree of integration among O.R.-based and A.I.-based techniques can be achieved.

First, the framework is used as a reference for analyzing those aspects of a scheduling problem that are relevant to further identify where and to what extent O.R. and A.I. can play a role in its solution.

Secondly, the expected role of O.R. and A.I. in the solution of the problem is identified.

Thirdly, the architecture is used as a conceptual tool to analyze and decompose the scheduling problem into sub-problems.

Finally, the architecture is used as an implementation framework, in order to solve each sub-problem with the most suitable technique and to allow subsystems to cooperate each other, achieving the desired integration of O.R. and A.I.

This paper briefly sets the background of our work and describes the main theoretical achievements, focusing on their industrial exploitation and impact in the development of BIS (Banbury Information system). BIS solves the real-size, complex problem of dynamically scheduling the compound area of a tire factory.

In particular, Section 2 sets main theoretical backgrounds of our work and defines a framework for describing scheduling problems, derived by the analysis of the role that O.R. and A.I. can play in their solution.

Section 3 describes the use of the framework in analyzing and describing the scheduling problem chosen for experimentation and in defining suitable solution techniques.

Section 4 describes the fundamentals of the architecture we have designed focusing on its use in the BIS project as both a conceptual tool for problem decomposition and a system implementation framework.

Finally, section 5 discusses, under both a theoretical and a practical perspective, the obtained results, identifying future research directions.

2. A framework for the description of scheduling problems

The general scheduling problem consists in deciding, given

- n jobs (orders), each composed by a given sequence of operations;
- m resources (machines);
- The machine order to process each job;
- A cost function to estimate the cost of each possible solution

the processing order of the operations needed to complete the jobs on the available machines, assigning the starting processing time of each job on each machine in order to minimize the cost function.

Operational Research (O.R.) and Artificial Intelligence (A.I.) propose several techniques to solve scheduling problems. The aim of our research is to define an approach to scheduling allowing the integration of those techniques to gain an advantage over approaches based on single ones.

In order to achieve this goal, we must first of all state the scope of each technique under different problem condition. Then, we have to develop a framework to describe a scheduling problem in terms of those conditions, to easily identify more promising techniques for solving it.

2.1 O.R. and A.I. Approaches to Scheduling

O.R. proposes two broad classes of techniques for schedule generation (Richel, 1988; Elleby and Grant, 1986; Smith et al., 1986):

- Techniques based on complete enumeration: a subset of legal and consistent schedules, among which we are mathematically guaranteed to find an optimal solution, is generated and the optimal one is searched by testing them against the chosen cost function. Algorithms do exist under very strict problem conditions (flow-shop, static and not stochastic) and are computationally feasible only for small size problems.

– Techniques based on implicit enumeration: the set of schedules among which the desired one should be searched is implicitly enumerated. The general methods, described below, assume that the environment is static and deterministic, a single, regular cost function is to be optimized and the constraints can be represented in terms of mathematical equations. Methods belong to four different streams (Rinnooy Kan, 1976; Rickel, 1988; Graves, 1981):

  - Linear and non-linear programming, where constraints and cost function can be represented in terms of linear or non linear equations. Unfortunately, most of the scheduling problems are integer programming problems.
  - Branch and Bound, where the problem is an integer programming one. It seems the most successful O.R. approach for both flow-shop and job-shop problems (Rinnooy Kan, 1976; Graves, 1981; Bellman et al., 1982).
  - Dynamic programming, where the problem is decomposeable into subproblems according to the mathematical structure of cost functions.
  - Heuristic search. Since the search for optimal solutions is computationally feasible only for small size problems, heuristic search techniques have been developed to generate feasible and good schedules. A prominent role is played by priority dispatching rules (e.g. SPT, EDD, ST, CR, ..), used in order to determine which operation is to be scheduled next (Graves, 1981; Rinnooy Kan, 1976; Vollmann et al., 1988).

A.I. contribution to scheduling problems solution appears particularly relevant, considering that scheduling is a knowledge intensive activity and that A.I. offers several techniques to represent and deal with knowledge, allowing to (Lepape et al., 1986; Bensana et al., 1988; Rickel, 1988):

– Define a comprehensive model of the factory environment (products, machines, constraints, experience of shop floor personnel, ...).
– Let general purpose knowledge cooperate with dedicated information provided by shop floor managers.
– Build flexible scheduling systems, easily adaptable when shop conditions change, since factory environment is not represented as a set of mathematical equations, but as a set of explicitly stated knowledge structure.
– Come in where O.R. fails, offering several approach to cope with the computational complexity of purely O.R.-based search techniques.

An important contribution of A.I. consists in having developed techniques to represent and reason with non-numerical constraints and related knowledge, as opposed to numerical ones, which pertain to O.R. techniques (Fox and Smith, 1984b; Smith et al., 1986; Fox, 1986).

While O.R. uses constraints as a mean to define legal values for the problem variables, A.I. uses them to define also preference values, distinguishing among "hard" constraints, that must be satisfied, and "soft" constraints, that should be satisfied, if possible, but can be relaxed, if needed. As a consequence, constraints can be used not only to test the feasibility of a schedule, but also and mainly to choose, among alternative decisions arising during schedule generation, the most suitable one.

Furthermore, while O.R. assumes the existence of a fixed set of constraints, A.I. offers constraints propagation techniques to dynamically change the set of relevant ones during the solution of the problem (Ow et al., 1988a).

A second contribution of A.I. to scheduling comes from problem decomposition techniques, as a mean to cope with the complexity of schedule generation in real environments. A.I. emphasizes the role of domain specific knowledge in decomposing the initial problem (Fox and Smith, 1984b; Fox, 1986; Smith et al., 1986; Lepape et al., 1986; Shaw, 1988; Keng, 1988; Ow and Smith, 1987; Ow et al, 1988b).

A third contribution of A.I. consists in having proposed hierarchical search techniques, based on abstracting from a problem by aggregating information and ignoring less important details, in order to dominate the combinatorial explosion of the search space (Elleby and Grant, 1986; Smith and Hynynen, 1987; Smith, 1987; Rickel, 1988).

In a dynamic environment, changes can occur asynchronously, during the execution of a schedule, due to the unreliability of machines or the stochastic nature of the environment. Artificial Intelligence supplies several techniques to enable opportunistic, reactive scheduling (Ow and Smith, 1987; Ow et al., 1988a; Fox and Smith, 1984a), and incremental constraint satisfaction (Elleby et al., 1988).

A major consequence of dynamic changes in shop conditions is that different techniques and approaches can be suitable, for the same problem, under different circumstances (Lepape et al., 1986). Thus, instead of using a single technique or approach to solve a problem, they can profitably be dynamically and opportunistically chosen while developing or revising a schedule.

The comparison of O.R. and A.I. techniques for solving scheduling problems can be carried out, considering them as two extremes of a spectrum of possible solution approaches, as shown in the following page.

2.2 A Framework for the Description of Scheduling Problems

In order to support the choice of the most suitable O.R. and A.I. scheduling techniques, we have developed a problem description framework.

The framework, derived from the analysis of state-of-the-art scheduling approaches described in section 2.1, divides problem characteristics into two sets:

General characteristics

They are particularly useful in identifying if and to what extent O.R.-based techniques are suitable, since most of them have been classified in the literature according to such properties (Rinnooy Kan, 1976; Bellman et al., 1982):

- Number of jobs (n);
- Number of machines (m);
- Problem type (flow-shop or job-shop);
- Scheduling goals;
- Valid assumptions on the set J of jobs;
- Valid assumptions on the set M of machines;
- Valid assumptions on job-machines relationships (such as processing times such as processing order of jobs are known and fixed);

	O.R	A.I.
PROBLEM CHARACTERISTIC	Static & Deterministic	Dynamic & Stochastic
PROBLEM STRUCTURE	Problem as a whole	Decomposition & Abstraction
CONSTRAINTS	Numeric, used as restrictions; Fixed	Non numeric, also preferences Flexible
SOLUTIONS	Optimal & Good	Good
HEURISTICS	Domain independent	Domain dependent
DECISIONS	Unique, local decisor	Unique or several global decisors

Detailed characteristics
These properties are particularly useful in identifying the role that A.I.-based approaches can play in problem solution.

The constraints, relevant for generating schedules, can be classified, from the A.I. point of view, into two main categories according to the role they play in the search process (Fox and Smith, 1984b; Fox, 1986):

- Restrictions ("hard" constraints), that can be used to define the space of admissible schedules, and allow to verify, at each step of the search process, the feasibility of the current solution.

- Preferences ("soft" constraints), that can be used as problem specific heuristics to focus the schedule generation process towards good solutions, allowing, at any decision point, to choose among different alternatives.

3. A case study: scheduling the compound area of a tyre factory

In order to evaluate the suitability of the proposed framework to solve real-size, difficult scheduling problems by integrating O.R. and A.I. approaches, our attention has been focused on one of the most critical areas in a tire factory: the compound area.

Here, the first stage in the tire production process is performed.

3.1 General Problem Description

In the compound area of a tire factory, raw rubber, chemical substances and oils are mixed to produce finished compounds.

Production process consists in a sequence of transformations of rubber, taking place in successive stages. Between two stages, a given amount of standing time should be allowed, to obtain the desired quality of the compound.

During each stage, raw materials, different chemical substances and previous stage compounds are mixed, in order to make them more and more homogeneous. Each stage requires different mixing speed to gain the desired degree of homogeneity and take place in few minutes.

Usually, about 40 different finished compound types are produced, each one requiring from 2 to 4 production stages.

The compound area consists of 4 or more machines of a unique type (namely the Banbury machines), responsible for all production stages. In principle, each machine can work each stage of each compound; actually each stage of each compound is allowed to be performed by a subset of machines, according to quality and productivity constraints.

The input for scheduling the compound area is a production plan that states:

- Finished compounds to be produced (quantities and due dates).
- Production priorities.

Usually, production plans vary day by day, but are relatively stable on a weekly basis, varying in the percentage of 20%, but involving the 80% of mix.

Production orders, thus, are defined for a whole week, but revised day by day.

The scheduling activity in the compound area usually takes place with an horizon of 1 day. During the first shift of each day, the existing schedule is revised and a new schedule for the other two shift of the day and the first shift of the next day is generated.

The compound area sets dynamic scheduling requirements.

Firstly, small revisions are needed at the beginning of each day, in order to account for small delays, due to very short stops of machines.

Secondly, more substantial revisions to the current schedule are required each time:

– A Banbury machine breakdowns and requires more than few minutes to be repaired.
– A machine in one of the further area of the factory breakdowns, with a consequent change in production priorities.

It should be noted that, since each job has, in principle, a different routing among the available machines, a delay in producing a compound on a particular machine propagates to all other products and to all other machines.

3.2 Analyzing the Problem According to the Proposed Framework

According to the framework we have defined in section 2.2, the scheduling problem described can be synthetically characterized as follows:

General characteristics
– Number of jobs = 40 (each one requiring 2 or 3 production stages);
– Number of "Banbury" machines = 4 to 9;
– Problem type = job-shop;
– Scheduling goals (in decreasing order of importance):
  - respect due dates;
  - maximize machine utilization;
  - minimize set-up times;
  - maintain intermediate stock levels within a given range;
– Violated assumptions on the set J of jobs:
  - jobs are not equally important (priorities do exists);
  - jobs are not independent each from the others (forced sequences are defined);
  - jobs can change dynamically (new orders can arrive and old one can be cancelled);
– Violated assumptions on the set M of machines;
  - machines are not independent;
  - machine are not equally important, for a given job;
  - not every machine can work every job;
  - the set of available machine can dynamically vary (due to breakdowns, maintenance activities, ...).

Detailed characteristics
– Restrictions ("hard" constraints).
  Several restrictions do exist, such as:
  - known setup and processing times;
  - known and limited machine capacities;
  - known precedences among operations;

- known alternative machines for a given job;
- minimum and maximum allowable time between different stages of the same job;
- Preferences ("soft" constraints).
  Several preferences do exists:
  - Organizational constraints, such as acceptable lateness, acceptable machine saturation levels, allowed stock levels;
  - Operational preference, such as preferred machines for a given job, preferred job sequences on the same machine, desired standing times and related allowed variability ranges, ...
  - Availability constraints, such as desired machine stops.

3.3 Identifying the Role of O.R. and A.I. for Our Problem

The analysis of our scheduling problem according to the framework allowed us to draw some conclusions about useful scheduling techniques. In particular:

- O.R.-based approaches.
  Considering problem characteristics (dynamic and stochastic, with possibly conflicting goals) several assumption that should hold for the applicability of optimization techniques actually drop.
  As a consequence, only heuristics-based approaches seem to be viable. In particular, since the primary goal is to minimize the job lateness, the Earliest Due Date (EDD, that helps minimizing the maximum lateness) and the Shortest Processing Time (SPT, that helps in minimizing the average lateness) dispatching rules seem to be useful in selecting the next job to be scheduled.
- A.I.-based approaches.
  Restrictions will be used in order to verify the feasibility of scheduling decisions. Preferences acquired from shop floor scheduler, on the other side, play the role of problem specific heuristics, i.e. criteria on the basis of which alternative decisions are to be weighted:
  - which individual operation is to be scheduled next;
  - which machine is to be selected for a given operation;
  - when a given operation is to be started on a given machine;
  Furthermore, since the environment is both dynamic and stochastic,
  A.I. techniques can be used in order to support rescheduling. In particular:
  - constraint propagation techniques can be used to propagate on the current schedule the effect of new constraints due to unexpected events;
  - domain specific heuristics (expressed as preferences) can be used both to assess if revision of the current (final or under generation) schedule is actually needed and to identify what it is better to revise;
  - backtracking techniques can be used to achieve more flexibility in the solution search process;
  - domain specific heuristics can be combined with domain independent dis-

patching rules. In particular, decisions taken on the basis of dispatching rules can be revised or refined according to problem dependent criteria.

Having identified the expected role of both O.R.-based and A.I.-based techniques, it is now worth identifying how they can actually be made cooperating to achieve a good solution of our scheduling problem. The next section will describe our architectural proposal and its specific implementation in the BIS project.

4. From an architecture to BIS

In this section, main features of the architectural model we propose for dynamic scheduling systems will be briefly outlined. Then, focusing on the BIS project, an implementation of such an architecture will be described.

4.1 The Proposed Architectural Model

The solution we have designed to integrate O.R. and A.I. techniques to cope with scheduling problems is based on the blackboard architectural model for distributed system (Erman et al., 1980; Nii, 1986). The model is defined as made up of three components:

- A set of separate and independent knowledge sources (KS), each one supplying informations of different nature, to be used to perform different steps in the problem solving process.
- A common data structure (namely the blackboard), that contains dynamic information from the solution space explored by the KSs and represents the communication mean among them.
- A control mechanism, that allows the KSs to react opportunistically to changes in the blackboard.

The problem solving process takes place according to the following loop schema:

1. A KS executes, performing changes to the blackboard and defining a new solution state.
2. The control mechanism selects, among the KSs that can contribute to the solution in the new state, the next to be executed and the objects of the blackboard on which it will act.
3. The selected KS executes.
4. If the blackboard does not contain the solution of the problem, goto 2; otherwise terminate.

As a model, the blackboard architecture does not specify how its components should be implemented, representing thus a general and flexible conceptual tool. We have specialized this model according to our goal, i.e. the integration of O.R. and A.I. techniques. In particular, the following design decisions have been taken:

- Each KS supplies the knowledge needed to perform a particular step in the schedule generation process, including a set of available, state of the art relevant techniques. Techniques included in the same KS can be both O.R. and A.I. based.
- Each KS is rule-based. Each rule (or set of rules) represents a technique that, under the conditions specified in its premise, can profitably be used to perform the schedule generation step which the KS refers to. The conditions have to be matched against the current content of the blackboard. Rules define operators of different grain size. The grain size must be chosen according to a tradeoff between problem solving efficiency and ability to revise and fine tune the generated schedule by retracting individual decisions and backtracking over them.
- The control mechanism is a specialized KS. It includes meta-rules that, according to the current content of the blackboard, identify which KS can contribute to the solution in a given state.
- The problem solving process takes place according to two inferential processes. The first one refers to the meta-rules in the control KS and is aimed at opportunistically selecting, at each step, the most suitable KS to be activated. The second one refers to the rules in the activated KS and is aimed at opportunistically selecting, at each step, the rule to be applied next. According to the current problem solving status, it can be, as mentioned before, an O.R. or an A.I. based technique.
- Dynamic scheduling is supported by the architecture by means of a specialized KS, responsible for activating an intelligent backtracking. It can be performed at the level of the control KS (leading to the decision of activating a different KSs) or at the level of individual KSs (leading to the decision of firing a different rule).

According to general blackboard model and our scheduling-oriented guidelines, we have designed and implemented, as described in the following section, a real size dynamic scheduling system.

4.2 BIS Development

BIS (Banbury Information System) is an information and dynamic scheduling system for tyre factories producing compounds for themselves and for other factories. "Banbury" is the name of the machines of the related factory area.

BIS consists of two main modules:

- A relational database, including information on machines, products, working sequences, stock levels, resource availability and orders. Particular attention has been payed to the user interface. It is window-oriented and menu-driven, allowing a friendly interaction during the updating and maintenance of the database.
- A dynamic scheduling module, to which our attention will be devoted in the following, that accessing information in the database, generate the related schedule.

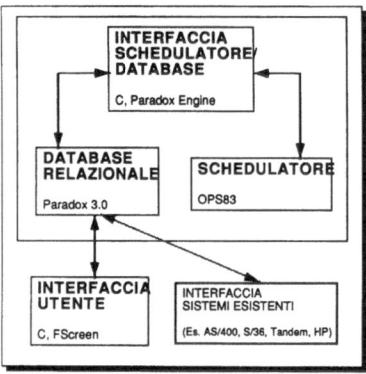

Figure 2

In order to develop the dynamic scheduling module, our first task has been to analyze the problem in the perspective of the decisions that have to be taken while developing a schedule:

- Decisions about jobs and machines:
    - Which is the next machine(s) (or job(s)) to be considered?
    - Which is the next job(s) to be allocated on the considered machine(s) (or the next machine(s) to be assigned to the selected job(s))?
    - When the job(s) are to be allocated on the machine(s)?
- Decisions about schedule performances:
    - Is the schedule complete?
    - Is the current schedule a "good" schedule?
- Decisions about the need of revising a schedule:
    - Is it convenient to revise the current schedule now?
    - Which decisions should be retracted in order to revise the current schedule?

Different decisions represent different steps within the schedule generation process. The knowledge used to take such decisions can be suitably represented in terms of KSs of our architecture. The architecture itself is used as a reference during such a decomposition. The BIS architecture, in particular, includes the following KSs:

- Job Priorizer KS. It is responsible for selecting, among the jobs to be allocated, the next to be considered. It is based on both O.R. and A.I. techniques. In particular, dispatching rules (EDD and SPT) are normally used, but domain specific rules are occasionally fired, under particular conditions, in order to modify the default ordering criteria.
- Operation Selector KS. It is responsible for selecting the next operation to be

considered for allocation, among the operations related to the last selected job and those ones considered before, but not yet assigned. It is based on domain specific heuristics.

– Machine Selector KS. It is responsible for selecting those machines, that are actually available to work the selected operation, according to current constraints. It is based on an O.R.-oriented view of constraints, used in order to define legal values for given variables.

– Operation Allocator KS. It is responsible for actually assigning the current operation to one of the selected machine, with specified starting time. If needed, the allocation can be suspended, to be reconsidered later. It is based on domain specific heuristics, to take the allocation decisions, and on A.I. constraint propagation techniques, in order to propagate the effect of decisions to the current constraint set.

– Backtracker KS. It is responsible for evaluating the quality of the current schedule and to decide if and how decisions should be retracted to explore alternative solutions. It is based on both general and domain specific search heuristics, in order to perform an intelligent, but not too computationally expensive, backtracking.

– Control KS. It is responsible for managing the sequence of activations of the other knowledge sources. It is based on domain specific heuristics, defining the conditions under which a specified knowledge source should be activated.

BIS ARCHITECTURE

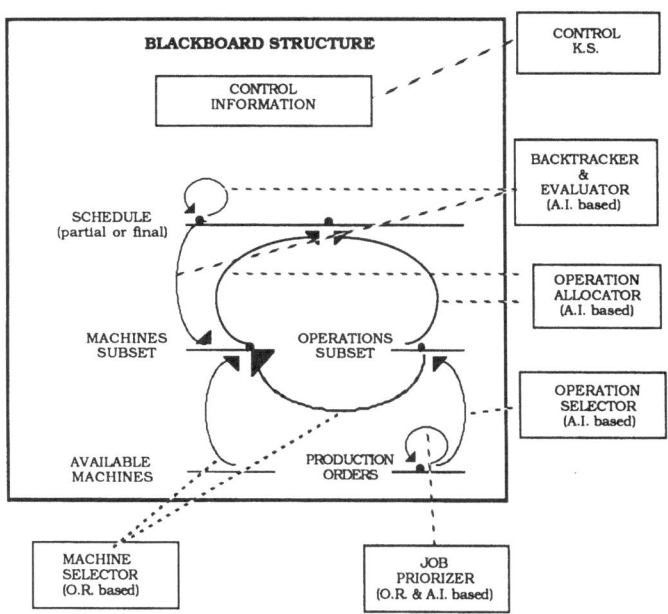

Figure 3

According to this problem decomposition and to the proposed blackboard architecture, a first prototype of the BIS system has been developed on a 386-based hardware platform, under MS/DOS operating system, using the following tools:

- Paradox, to develop the relational database.
- OPS83, to develop the scheduling architecture and the KSs.
- C language, to develop (trough Paradox Engine) the interface between the database and the scheduling modules.
- FSCREEN, a Pirelli Informatica proprietary window package, to develop the user interface.

5. Conclusions

The BIS project allowed us to verify the suitability of our architectural approach to the integration of O.R. and A.I. techniques for scheduling real-size, difficult industrial environment. Since each knowledge source can include O.R.-based operators, A.I. based ones or both, integration have been achieved at different levels:

- Within individual knowledge sources, by allowing different techniques to be opportunistically chosen in order to perform at the best extent specific steps in the schedule generation process.
- Among different knowledge sources, possibly based on different scheduling approaches, by allowing individual knowledge sources to be opportunistically activated, according to the current state of the schedule generation process.

The BIS experience and the experimentation of the same architectural approach for few other simpler problems related to tyre factories have highlighted the existence of few recurrent knowledge sources (e.g. Machine Selector. Operation Selector, Job Allocator, Control KSs).

The content of each KS depends upon the characteristics of individual scheduling problems; furthermore, specific problems possibly require other KSs to be defined. Nevertheless, the BIS experience have shown us that inference and cooperation mechanisms can be considered general enough, to be implemented in an easily reusable way.

Thus, future efforts should be aimed to further the architecture toward the development of a scheduling toolkit, allowing both to actively support problem decomposition according to the designed architecture and to simplify the implementation of scheduling systems.

6. References

Bellman, R., Esogbue, A.O., and Nabeshima, I. (1982) 'Mathematical aspects of scheduling and applications', Pergamon Press.

Bensana, E., Bel, G., Dubois, D. (1988) 'OPAL: a multi-knowledge based system for industrial job-shop scheduling', Int. J. Prod. Res., 26 (5), 795-819.

Elleby, P. and Grant, T. (1986) 'Knowledge-based scheduling', in G.Mitra (ed.) Computer aided decision making, North-Holland, The Netherlands, 175-186.

Elleby, P., Fargher, H.E., Addis, T.R. (1988) 'Reactive constraint-based job-shop scheduling' in M.D. Oliff (ed.) Expert Systems and Intelligent Manufacturing, Elsevier Science Publishing Co., 1-10.

Erman, L.D., Hayes-both, F., Lesserr, V.R., and Reddy, D.R. (1980) 'The Hessay-II speech understanding system: integrating knowledge to resolve uncertainty', Computing Surveys, 12(2).

Fox, S.M. and Smith, S. (1984a) 'The role of intelligent reactive processing in production management', Proc. of CAM-I's 13th Annual Meeting and Technical Conference, Florida, 6.13-6.17.

Fox, M.S., Smith, S.F. (1984b) 'ISIS: A knowledge-based system for factory scheduling', Expert Systems, 1 (1), 25-49.

Fox, M.S. (1986) 'Observations on the role of constraints in problem solving', Proc. of 6th Canadian Conf. on A.I., Montreal, 172-185.

Graves, S.C. (1981) 'A review of production scheduling', Operation Research, 29(4), 647-675.

Keng, N.P., Yun, D.Y.Y., and Rossi, M. (1988) 'Interaction sensitive planning system for job shop scheduling' in M.D. Oliff (ed.) Expert Systems and Intelligent Manufacturing, Elsevier, 57-69.

Lepape, C., Smith, S.F., Ow, P.S., McLaren, B., and Muscettola, N. (1986) 'Integrating multiple scheduling perspectives to generate detailed production plans', Proc. 1986 SME Conf. on AI in manufacturing, Long Beach, CA, 2.123-2.137.

Nii, H.P. (1986) 'Blackboard Systems: the blackboard model of problem solving and the evolution of blackboard architectures', The AI Magazine, Summer 1986, 38-106.

Ow, P.S. and Smith, S.F. (1987) 'Viewing scheduling as an opportunistic problem solving process' in R.G.Jeroslow (ed.) Annals of Operations Research: approaches to intelligent decision support, Beltzer Scientific Publishing Co..

Ow, P.S., Smith, S.F., and Thiriez, A. (1988a) 'Reactive plan revision', Proc. of AAAI '88, 1-6.

Ow, P.S., Smith, S.F., and Howie, R. (1988b) 'A cooperative scheduling system' in M.D. Oliff (ed.) Expert Systems and Intelligent Manufacturing, Elsevier, 43-56.

Rinnooy Kan, A.H.G. (1976) 'Machine Scheduling Problems: Classification, complexity and computation', The Hague.

Rickel, J. (1988) 'Issues in the design of scheduling systems' in M.D. Oliff (ed.) Expert Systems and Intelligent Manufacturing, Elsevier 7089.

Shaw, M.J. (1988) 'Knowledge based scheduling in flexible manufacturing system: an integration of pattern directed inference and heuristic search', Int. J. Prod. Res., 26 (5), 821-844.

Smith, S., Fox, M.S., and Ow, P.S. (1986) 'Constructing and maintaining detailed productions plans: investigations into the development of knowledge-based factory scheduling systems', AI Magazine, Fall, 45-61.

Smith, S.F. (1987) 'A constraint based framework for reactive management of factory schedules', Proc. Int. Conf. on Expert Systems and Leading Edge in Production Planning and Control, Charleston, South Carolina.

Smith, S.F. and Hynynen, J.E. (1987) 'Integrated decentralization of production management: an approach for factory scheduling', Proc. 1987 Symposium on Integrated and Intelligent Manufacturing, Boston, MA.

Vollmann, T.E., Berry, W.L., and Whybark, D.C. (1988) 'Manufacturing Planning and Control Systems', Irwin, Homewood (Illinois).

Neutral Interfaces for Robotics. Goals and First Results of ESPRIT Project 2614/5109: "NIRO"

Dr. Ing. Ingward Bey
Kernforschungszentrum Karlsruhe GmbH
Postfach 36 40
D-7500 Karlsruhe 1
Federal Republic of Germany
Phone: +49 7247 825271

Summary

The project NIRO started in December 1989 with 9 partners from 5 EC countries. Continuing the line of ESPRIT Project 322: CAD Interfaces (CAD*I) the project is aiming at the development of neutral interfaces (neutral data formats, processors, and software tools) for the exchange of geometrical, technological, and programming functions and data between CADCAM modules, simulation programmes, and robot control. Results will be introduced directly into the international standardisation process at ISO. First prototype results which cover only partially the finally intended goals will be demonstrated at the end of 1990. Further demonstrations at several user sites (automotive industry, ship building, aerospace manufacturing, and plastic parts manufacturing in 1992 will show the practical advantages of open system architectures in the industrial CIM environment.

1. Baseline of Project NIRO

The project is a logical extension of ESPRIT Project 322: CAD Interfaces (CAD*I). CAD*I has so far produced a number of off-spring projects including two which are specially devoted to information exchange via neutral interfaces: Project 2195 (CADEX) and this project.

– Project 2195 (CADEX) aims at the development of STEP processors for commercial CAD and FEM systems on the basis of CAD*I results.
– This project aims at a continuous and vendor-independent flow from CAD to robot-based manufacturing via robot programming.

The principal applicability of the CAD*I approach to this problem was already demonstrated at the ESPRIT Conference '88 exhibition where an industrial robot and a simulation system were shown to perform according to information generated in CAD systems and utilized in the robotic application. The information flow used for that demonstration was vendor-independent (neutral) and in conformance with the STEP predecessor CAD*I as far as the geometric information was concerned. For kinematic information, path definitions, and other technological information no neutral format was available at that time. This project will extend CAD*I into one specific area: robotics.

– The following results of CAD*I are used as a basis:
– The specification technique: The relevant information models are specified formally using the CAD*I-developed specification language HDSL and its subsequent extensions in the STEP specification language EXPRESS.
– The processors for transfer of geometric models can be used as such.
– For extending the capabilities of the existing processors to other kind of information (kinematics, etc.) the CAD*I-developed software tools (generators for scanners and parsers, low-level routines for writing the neutral file) will be used.
– Beyond this, the CAD*I methodology for structuring information will be applied.
– The well-established contacts to the ISO-STEP community are used to introduce NIRO results into future STEP versions.

A main target of research in the domain of computer integrated manufacturing systems comprehends the properties, conditions, possibilities, and means by which system components can be integrated and the way in which the integrated system can be modified, updated and controlled.

Integration in this context does not mean to centralize and merge all kinds of sub-systems, but integration should allow for the building of integrated manufacturing systems of a type, size, order, degree of automation and implementation speed that fits into the practical needs of a given user company - whether it is a small, medium-sized or a large company. A development like the one described above demands the standardisation of the interfaces between the individual sub-systems in a CIM environment. A standardisation like this is a prerequisite in the development of the future CIM systems, namely for replacing existing individual components in the CIM system by more effective ones as they appear on the market leading to the application of multi-vendor systems, the "open system" concept.

In a CIM-context the concept of standard interfaces will lead to a conceptual scheme as in Figure 1. Three operational system levels are identified: CAD systems, application systems like planning or simulation, and real-time production systems. Systems in adjacent levels are interfaced through two levels of neutral formats such that any system in principle can exchange pertinent data with any other system.

The key issue in the open system concept is the intelligent, neutral interface between sub-systems. This corresponds precisely to the goals of NIRO.

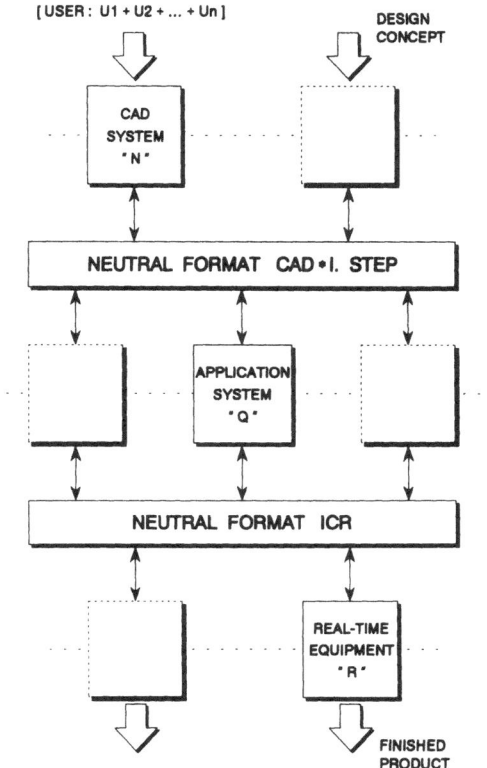

Figure 1: CIM concept with standardised interfaces in the information flow from product design to product manufacturing and delivery. The interrelation with robot programming is shown in Figure 2.

2. Areas of standardisation activities within NIRO

There are three fields of work in which the NIRO team is trying to improve standardisation. The first one is the *standardisation in CAD with respect to robotics*. Based on the CAD*I approach of transferring solid models (they are essential for robotic applications and the corresponding collision checks) the existing solutions are being extended for the transfer of kinematic information. Partners in NIRO just had developed a topological information model for kinematics to be included in the international standard proposal STEP. It is presently being evaluated on the ISO level and includes kinematic topology, geometry association, and kinematic configuration sequences and paths.

During the ISO meeting in Reston/USA in April 1990 the project plan for the STEP kinematics model was agreed upon by the ISO group (with representatives from Japan, USA, Canada and Germany). This plan will lead to a draft international standard (DIS) by the end of 1991.

It is very interesting to see that the US-driven activity in this field, PDES (Product

Data Exchange Specification) has joined the STEP line in the ISO meeting in April '90. PDES stands now for "Product Data Exchange using STEP".

The second field of action concerns *robot programming*. These activities are conducted in close relation to the international standardisation of a robot programming language started at ISO TC 184/SC2/WG4 in 1988. It is planned to have a first draft proposal of the IRL (Industrial Robot Language) during year 1990 which then will compete with the Japanese approach called SLIM (Standard Language for Industrial Manipulators), for which first implementations are just available.

The third aspect is related to standardisation for *robot control*. The definition of the neutral software interface ICR (Intermediate Code for Robotics) is discussed in ISO TC 184/SC2/WG4. Representatives of NIRO are strongly engaged in this field. The proposed ICR format as well as the IRDATA format defined in Germany (see DIN 66313, part 1) are the basis to the NIRO work. They will influence each other, similarly as the CAD*I work did with the geometry definitions in STEP. A rough overview is given in Figure 2.

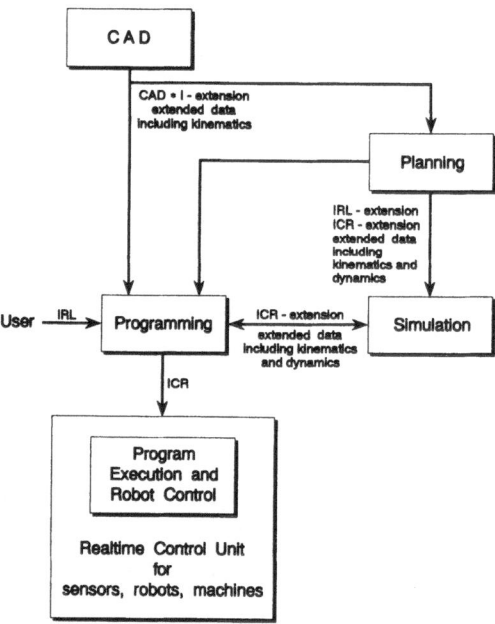

Figure 2: Robot programming and robot control in the context of CIM including planning and simulation.

Robot move instructions which describe not only the target but also a great variety of move parameters like speed, acceleration, path smoothing and the like can be given by both IRL <u>and</u> ICR. This is true also for technological parameters, program flow instructions, or arithmetic/geometric calculations, as well as for sensor data processing. The main difference between RL and ICR is that IRL is a programming language and therefore designed for human operators whereas ICR as an executable code can be implemented on a robot control more easily and cheaper than IRL and, of course, be executed faster.

Both IRL and ICR will essentially contribute to build up open systems, that means the easy realisation of multi-vendor robot programming and robot control systems in industry. The more standardised robot programming languages are, the less difficulties arise for training people when equipment is being changed. Existing application programs can easily be transferred to new generations of controls or equipment of other vendors using a standardised intermediate code as intended with ICR.

Practical work for these three aspects is carried out following the scheme:

– analysis of requirements
– critical review of existing approaches
– definition of basic scheme and reference models
– definition of interface format
– implementation of prototype processors (pre- and postprocessors)
– test
– re-design and improvement of formats and processors
– practical demonstrations in a CIM environment.

3. Development of Interfaces and Software Products

The development work follows the above main lines.

First of all neutral interfaces for kinematics and robotics CAD data are being developed. This includes documentation on initial neutral file specification, a library of text suites for the exchange of kinematics and robot models, the design and implementation of skeleton pre-and postprocessors (prototypes) for test bench facility for some commercial systems used at the partners sites. These are, among others, for CAD: BRAVO, Technovision, CATIA; for robot programming: GRASP; for robot simulation: KISMET. Software products are made accessible to all partners in the project.

Secondly, neutral interfaces for robot programming and planning are considered for both aspects: "language" and "code". Activities in this area include:

– elaboration of ICR enhancements
– design of the IRL programming system
– implementation of the IRL sub-systems
– integration of sub-systems
– prototype implementation of processors:
  between STEP and off-line programming systems
  between off-line programming and simulation
– prototype processors for on-line programming between simulation and robot control.

To explain in more detail what is happening in the above mentioned scope of tasks, a short comment hereafter on the point: elaboration of ICR enhancements. The ISO Draft Proposal ICR was reviewed by the NIRO consortium as one of the starting activities. This led to a long list of possible improvements which will be treated in the following steps of activities, i.e.

– some important facilities are missing, like dynamic allocation, variables as target addresses in jump intructions or links between modules
– missing comfort at teaching and programming level
– inclusion of advanced robot dependent facilities like interrupt possibilities, searching functions or change of parameters in servo controllers.

Thirdly, neutral interfaces for robot control systems and corresponding processors are defined and implemented. This work considers the following steps:

– specification of the integration of the neutral interface in the robot control system (design of software architecture, etc.)
– prototype of the software transformation package between the neutral interface and the robot control applied to the actually available robot systems Reis, ABB and INISEL
– development of test environment at research institutions, vendors and users sites.

Figure 3 (see next page) gives an overview on the included commercial systems, the considered interfaces and the different integration paths which will be implemented at the end of the project. Additional systems will be included during the project according to work progress and market developments and needs.

4. Demonstrations

In the difficult area of interface development and standardisation it is of vital importance to demonstrate the validity of concepts and their implementation by demonstrating them in a practical, industrial environment. This is, of course, a main goal and aspect of work within the NIRO project. The participating users are preparing applications of NIRO results in their industrial environment. Thus, prototype implementations will be shown at least at three industrial sites.

One task of demonstration which tentatively was selected is a carbon fibre filament winding manufacturing cell for fabrication of high quality parts at CASA facilities at Getafe. An ABB five-axis robot is available with CAD information coming from CATIA.

At Odense Steel Shipyard in Denmark the welding of large ship components was selected for practical demonstration. Robots mounted on gantries will be programmed using the reference scheme of the NIRO project with all the levels of interfaces described.

Reis is looking for application of NIRO results in a flexible production island for

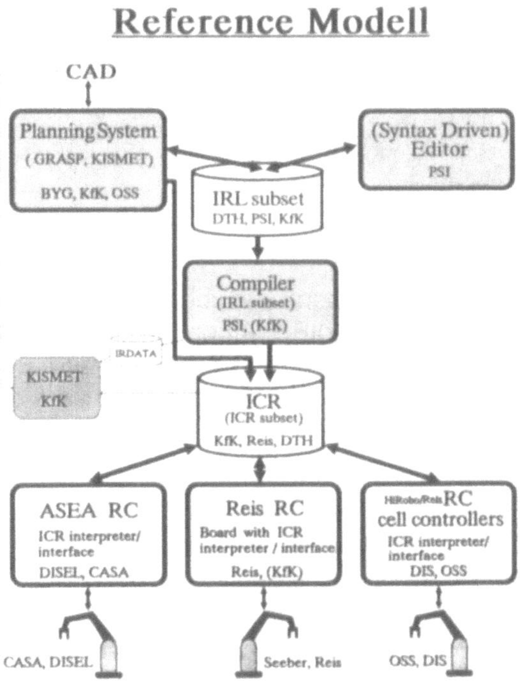

Reference Modell

Figure 3: NIRO overall scheme of relationships, commercial equipment involved, products to be developed, participating organisations and envisaged user sites.

automotive parts treatment after injection moulding at Seeber S.R.L., Italy.

The work to be done for these demonstrators follows more or less the same steps:

- definition of requirements from user point of view
- construction of additional peripheral equipment for the robot application
- installation of prototype processors to realize the full integration between CAD and shop floor
- adaptation of software to the real environment
- development of test and demonstration cases
- validation, evaluation of modules
- documentation.

During the ESPRIT Conference 1990 a simulation of one of the industrial applications will be shown in Brussels. Included are the first prototype processors covering a first subset of the industrial requirements for data transfer in the CAD-robotics area. A detailed and actual description will be available at the exhibition itself.

5. Outlook

The NIRO project can be considered as a user driven effort in Europe to substantially improve the present situation of information flow in the CAD-robotics domain. It will contribute to strengthen European influence to the upcoming international standards in this field and it helps both vendors and users to profit from an open system architecture.

6. Partners in the project

Partners from Denmark, the Federal Republic of Germany, Italy, Spain and the United Kingdom form the team for NIRO. Among them, CASA, Seeber and KfK act as users, REIS as robot vendor, PSI, DIS, DISEL and BYG as engineering, consulting and vendor companies and KfK and DTH as scientific research and development partners.

6.1 List of Partners

BYG: B.Y.G. Systems Std., Nottingham/UK
CASA: Construcciones Aeronauticas S.A., Madrid/Spain
DIS: Dansk Ingenior System as, Glostrup/Denmark
DISEL: Diseno e Ingenieria de Systemas Electronicos s.a., Madrid/Spain
DTH: Danmarks Tekniske Hojskole, Instituttet for Styreteknik, Lyngby/Denmark
FIAT: FIAT Auto, Torino/Italy
KfK: Kernforschungszentrum Karlsruhe GmbH, Karlsruhe/Federal Republic of Germany
OSS: Odense Steel Shipyard, Odense/Denmark
PSI: PSI Gesellschaft für Prozeßsteuerung- und Informationssysteme mbH, Berlin
          Federal Republic of Germany
Reis: Reis GmbH & Co Maschinenfabrik, Obernburg/ Federal Republic of Germany
Seeber: Seeber S.R.L/GmbH, Leifers/Italy

7. References

[1] E.G. Schlechtendahl (Ed.): Specification of a CAD*I Neutral File for CAD Geometry. Version 3.3. Third edition. 1988 Springer Verlag. Research report Project 322 ESPRIT, 283 pp.

[2] ISO: External Representation of Product Definition Data. (STEP), ISO DP 102030, 1989.

[3] E. Trostmann: CAD data interfaces for robot control. 2. Duisburger Kolloquium. Automation und Robotik, 15-17 Juli 1987, Universität Duisburg, Proceedings, page 73-96.

[4] C. Blume, W. Jakob: Programming Languages for Industrial Robots. Springer Verlag Berlin, Heidelberg, New York, Tokyo, 1986.

[5] DIN 66 313: IRDATA - Interface between programming system and robot control. Beuth-Verlag, Berlin, published 1989.

[6] Industrial Robot Language (IRL)- Language Description (Draft). NAM in the DIN

committee working paper N 96.2.2/ 89, 1989.

[7] ICR-Intermediate Code for Robots. ISO TC184/SC2/WG4 N20. Draft proposal. 1989

Project No 2617
Communicatiorls Networks for Manufacturing Applications

OSI-Based Network Management for Industrial Networks

Dr.W.Kiesel and K.H.Deiretsbacher
CNMA Network Management Working Group
Siemens AG, Automation Group
Department AUT E 51
P.O. Box 3220
D-8520 Erlangen
Federal Republic of Germany

Abstract

Network Management (NMT) is one of the key areas within CNMA Phase 4 and the currently planned Phase 5. Most NMT standards are still immature and their promotion has therefore highest importance.

CNMA NMT is based on the OSI Management Framework (OSI 7498-4). The NMT Applications are based on several standards. With the exception of CMIS/P (status DIS) these standards have at most Draft Proposal status.

Currently CNMA supports one Manager System and several Managed Systems. NMT is performed through communication between System Management Application Processes (SMAPs) executing on these systems. Agent SMAPs collect or modify management information under control of the Manager SMAP. Manager Applications process all collected information in order to achieve Configuration Management, Performance Management and Fault Management.

1. CNMA and Network Management

1.1. Functional Overview

CNMA Network Management provides tools for the purpose of:

- problem detection and diagnosis
- installation and checkout
- performance monitoring.

The basis of CNMA NMT is the OSI Management Framework [OSI 7498-4]. It is limited, however, to the Functional Areas (FA) of Configuration Management, Performance Management and Fault Management. Specifically, the chosen System Management Functions are:
- Object Management Function (Configuration Management)
- Confidence and Diagnostic Testing Function (Fault Management)
- Workload Monitoring Function (Performance Management)

– Management Association Control Function (common for all FAs)

Within their Functional Area these functions have reached the most advanced state in the standardisation process and have been identified as most useful.

The Network Management Application Protocol is based on several ISO standards. These standards currently have at most Draft Proposal status (exception: CMIS/P are DIS -- and meanwhile IS).

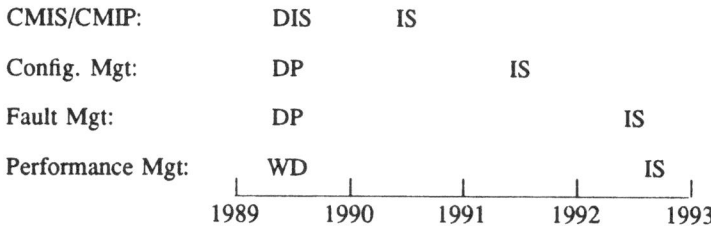

CMIS/CMIP:	DIS	IS			
Config. Mgt:	DP		IS		
Fault Mgt:	DP			IS	
Performance Mgt:	WD				IS
	1989	1990	1991	1992	1993

State of Standards.

IS:	International Standard
DIS:	Draft International Standard
DP:	Draft Proposal (new term = CD - Committee Draft)
WD:	Working Draft

1.2. MAP V3.0

CNMA specifications have also been largely influenced by MAP V3.0. MAP V3.0 is based on standards which existed approximately end of 1987. Since one main CNMA goal is the use of most recent ISO Standards, and since several concerned standards in the meantime are much more stable, there are some differences between CNMA and MAP V3.0. However, these are mostly enhancements, so that (from a general point of view) CNMA is a superset of MAP V3.0. To prove this, the major vendors of NMT systems plan to produce manager systems that are able to manage CNMA and MAP stations.

CNMA is only concerned with OSI systems and so special MAP nodes like MAP/EPA nodes or Mini MAP nodes are not addressed.

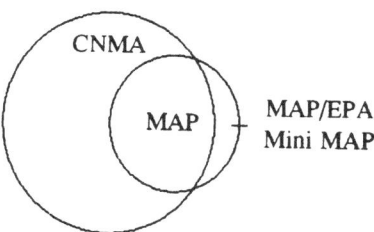

Comparison of Functional Content.

2. Implementation Guide (IG).

One of the work items within CNMA is the production of an **Implementation Guide.** This guide specifies the functional standards (profiles) and thus the requirements for Manager Systems (applications) and for Managed Systems (agents). In addition to protocol profiles, the IG specifies the needed management information and rules for the Agent behaviour. An Addendum to the IG specifies the architecture and functionality of the NMT applications.

In general, the IG content is based on the functionality identified as a requirement by user representatives.

3. Management Systems Overview

CNMA in Phase 4 supports one Manager System (the NM Console) and several managed systems.

Network Management is performed through communication between System Management Application Processes (SMAPs) residing on the various nodes. The SMAP on the manager system is called Manager-SMAP. The SMAPs on the managed systems are called Agent-SMAPs.

Agents collect or modify management information under control of the Manager. They also have the ability to send events related to predefined event conditions set by the Manager.

Communication between SMAPs is performed by use of the Common Management Information Protocol (CMIP).

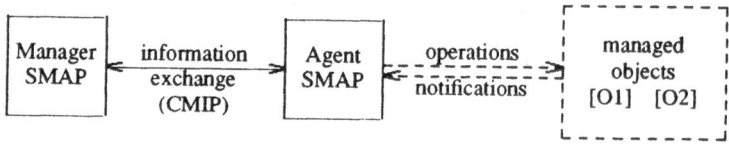

Interactions between Management SMAPs.

The Network Manager application processes all collected information in order to achieve Configuration Management, Performance Management and Fault Management.

4. Network Management Components

4.1 Management Information

Management information is defined in terms of managed objects. A managed object is the management view of a system resource (e.g., a layer entity, a computer system, a scheduling queue, or a printer). The set of managed objects in a system, controlled by a single Agent SMAP is called that system's Management Information Base (MIB).

As outlined above, a managed object is an abstracted view of a system resource. Each managed object is defined by:

- the attributes visible at its boundary, the management operations which are applied to it,
- its behaviour to management operations, and
- the notifications emitted by the object.

Within CNMA, the definition of the **Management Information** was an essential work item. Main focus was thereby on the selected "System Management Functions" and on MAP V3.0 upward compatibility. CNMA Phase 4 has selected more that 20 Object Classes (Layer Entities, Test Performer, Connections, Service Access Points) covering all important communication aspects. A total number of more than 350 attributes, actions and events has been defined half of which is mandatory for all vendors of the CNMA project.

4.2 Management Information Transfer

The interactions between management SMAPs is realized through the use of OSI's Common Management Information Services and Protocol (CMIS/CMIP). CNMA is based on the Draft International Standard (DIS).

The following CMISE (CMIS Element) services have been chosen by CNMA:

- M-INITIALISE service: to establish an association with another CMISE service user for the purpose of exchanging management information.
- M-TERMINATE and M-ABORT service: to release an association orderly or abrupt.
- M-EVENT-REPORT service: to report an event about a managed object to another CMISE service user.
- M-GET service: to request retrieval of management information (attribute values of managed objects).
- M-SET service: to request modification of management information (attribute values of managed objects).
- M-ACTION service: to request another service user to perform an action (e.g., start test, deenroll) upon a certain object.

4.3 Management Application

The Network Management Applications provide a means for the network administrator, a human, to read or alter data, control the data and access reports. It is part of a manager process that is outside the ISO standardization activities of OSI.

CNMA specifies and implements also the Management Applications. A subgroup of CNMA (the NMT Kernel Group) has defined a common architecture and has developed the components of the Management Applications in a joint effort.

<u>Decomposition of Activities</u>. The basis for the architectural structuring is the decomposition of the application into four levels of activity:

Level 0: Information display

Level 0 deals with the user interface. It contains all necessary functions to give the user access to administrative information, access to perform administrative operations, and access to signals coming from the managed objects.

Level 1: Functional Area Specific Information Processing

Level 1 contains management application services specific for the functional areas: configuration management, performance management and fault management.

Level 2: Common Information Processing

Level 2 contains services common to one or more management applications. It also manages a Manager Data Base, containing all administrative information not contained in the Management Information Base (MIB).

Level 3: Information Transfer

Level 3 contains all the standardized functions giving access to the Management Information Base (MIB).

Each level is defined to be independent of the level built on top of it. This means in particular, that Level 2 modules are independent of User Interface issues. The advantage of this separation is manyfold:

-   it is a method of modularization (parallel development, maintainability, extensibility), and
-   it makes large parts of the application independent of the user interface. This allows, for instance, to use the modules underneath by different user interface implementations (command level, semi-graphic, full graphic).
    This in particular is an important requirement for CNMA, where Level 2 modules will be ported between the vendors in order to produce a complete Network Management System (NMS).

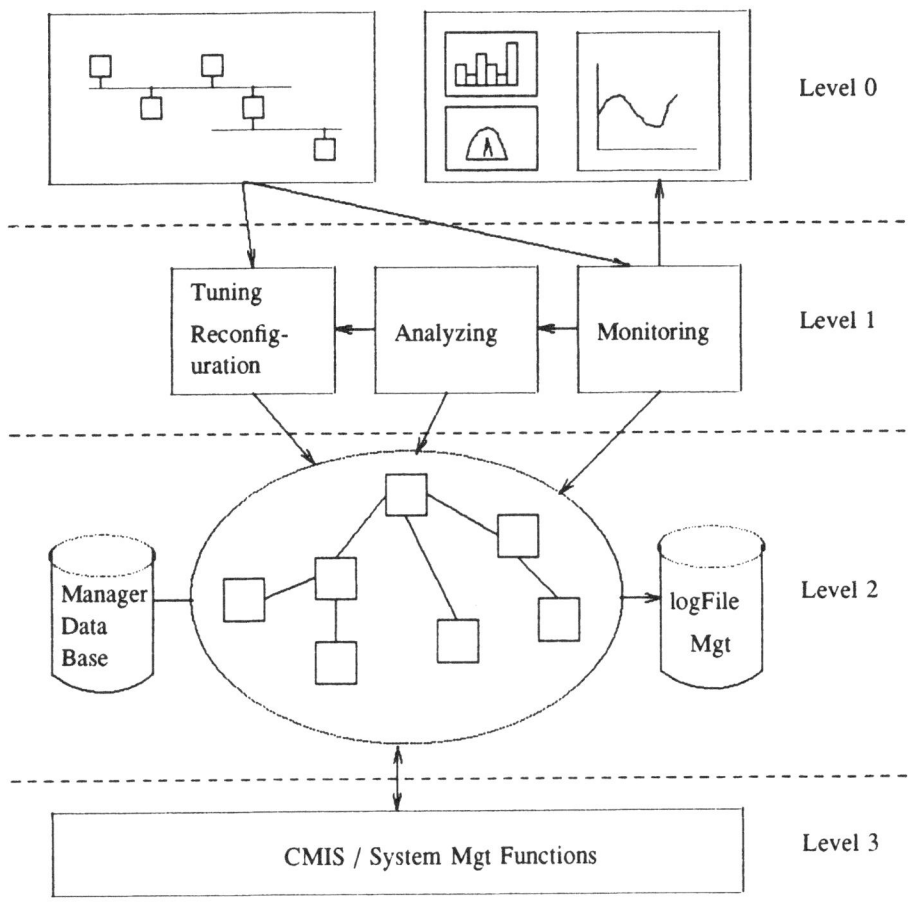

Management Activity Levels

Phase 4 Applications: For the "Manager System" Applications have been implemented, which primarily integrate the following management aspects:

- **Configuration Management** (to identify and describe the network topology and the network components, and to change and monitor configuration characteristica during installation phase and operation phase).
Two types of objects are handled by configuration management services: hardware components (numerical controllers, computers, network equipments, ...) and software components (application packages, support tools, ...). Configuration Management offers the ability to install, modify, remove and observe hardware and software components.
Installation involves services to set up the manager database as well as services to ensure that the installed component is up and running.

Observation means the monitoring of the operational status (e.g., MIB of managed systems, transport connections, associations).

- **Performance Management** is the portion of Network Management concerned with system and communication performance.
  Performance Management provides functions for the administrator to monitor the communication quality at runtime [workload, throughput, reliability], and to change certain parameters for performance improvement (tuning).
  Several jobs may run in parallel, monitoring different entities of the network. Diagrams and gauges are used to display statistics or show a system's performance. Monitoring results are usually written to logFiles. The recorded data can be interpreted for longterm statistics or trend analysis.

- **Fault Management** (to identify and localize faulty components, and to improve diagnostics and necessary corrective actions by use of knowledge based systems).
  Fault management includes a knowledge based system in order to facilitate diagnosis and to support the operator in repair and configuration tasks. These functions will be performed by the Network Diagnosis System. An inference engine carries out deductive and abductive reasoning in both forward and backward chaining.
  Static information about the network (real world) is derived from the manager database; dynamic state information is received via event notifications reported by Agent-SMAPs in various fault situations; additional data may be obtained by asking the user.
  Diagnosis concentrates on hardware faults. The diagnostic mechanism is based on associational knowledge of the meaning of symptoms and about the verification of suspicions and is supported by a structural model of the network system. The quality of the Network Diagnosis depends on the number of fault situations recognized and reported by managed systems and the amount of knowledge coded in the system's knowledge base. This expert knowledge increases when experience with operating CNMA networks has been gained, and thus is only partially available at the beginning of the pilot installation. Therefore the Fault Manager is designed as an open system with a fundamental knowledge base, which can be extended in an easy way during the operating phase of the pilot.

Wherever possible and reasonable, user interactions and presentation of monitoring results are performed in a fully graphical manner.

5. Conformance and Interoperability Tests

Before their integration in Pilots, the vendor's implementations of the communications protocols will be tested against a suite of Conformance and Interoperability Test Tools. Experience from previous phases of CNMA has shown that conformance testing is vital when installing complex systems to tight timescales. The Test

Tools for Phase 4 are being produced by a complementary ESPRIT Project, known as TT-CNMA.

6. CNMA Pilots

The profile defined in the Implementation Guide will then be validated and demonstrated by use in a number of real production cells, known as 'pilots'. The use of the communications software in these pilots ensures that the project addresses genuine CIM communication problems, and implements practical solutions.

The implementations of Network Management will be tested and demonstrated specifically at the Renault Experimental Demonstrator in Boulogne Billancourt, France.

In addition to the features mentioned above, this Pilot will provide a fault injector to input simulated faults into the network. With that the ability of Network Management to tune and reconfigure the network, to diagnose and manage fault conditions will be illustrated.

Once the Network Management Application has been proven on the Renault Experimental Demonstrator, it will be installed in the Aerospatiale Demonstrator, thereby showing its value in a real production environment.

7. Promotion

Throughout the project, the partners are promoting the use of OSI communications. CNMA will hold a workshop at the ISW Demonstrator site in September 1990. The CNMA pilots will all be used for demonstrations to visiting members of the public.

The CNMA partners are liaising with national and international standards bodies, as well as other related organizations, for example EMUG, NM Forum, X/Open, and the American NIST.

8. Outlook -- The Phase 5

The planning for CNMA Phase 5 is in progress. Phase 5 work will be based on the results of Phase 4. Newest, then available protocol standards will be used.

It is expected that additional System Management Functions (like Response Time Monitoring) will have reached a state where they are usable.

Following are the highlights on which Phase 5 is planned to focus on:

Integrated LAN Analyzers for Performance Monitoring

A LAN Analyser is a specialised equipment for monitoring a single LAN. With its specialised features (special MAC board working in promiscuous mode) it is able to offer a level of detail impossible to gather via manager to agent communication. The objective would be to have such a monitoring device per LAN.

- Provide means to program this device for analysing / filtering by the Network Management Console
- Gather the results and make the informating accessible at NMT Console (for interactive user or for expert system).

Distributed Network Managers

The CNMA Phase 5 network manager application is based upon a distributed computational model. Its functions may be distributed on CNMA devices of different types (PCS, Minis ...). This allows for:

Parallelisation:
modules of the manager application can run on multiple machines to speed up execution.

Resource usage optimisation:
modules of the manager application can run on devices targeted to answer to specific needs: high quality graphic workstations, low cost end stations, knowledge based systems.

Reliability:
modules of the manager application can be duplicated on multiple devices to increase the management reliability and the global quality of service. This leads to a fault tolerant manager application.

The distributed computational model of the CNMA network manager application is expected to be transaction oriented. The transactional model naturally enforces consistency of the results of operations carried by distributed cooperating modules of the manager application. This property is necessary in a heterogeneous open architecture to allow the growth of the manager application.

The concept of "Management Domains" currently introduced within the management related ISO-world will be adopted, exhanced, and applied to the distributed network manager. Management Domains form an appropriate construction principle to assign and coordinate management policies for the (hierarchical) subareas of CNMA - Networks according to the various functional requirements.

Cooperation with non-CNMA NMT systems

In a given plant (including a factory), several activities cooperate in order to achieve the objectives of the plant. These activities are of different nature. We can distinguish:

- Technical activities including engineering, CAD/CAM, quality control..., etc. In this activity domain Digital Equipment with DECNET, LAVC, LAT is the predominating vendor today.

- Administrative activities including Accounting, Personnel Management, Plant Management ... etc. In this activity domain IBM with SNA is the predominating vendor today.
- Manufacturing activities where SIEMENS with SINEC H1 predominates today in Europe.

With the objective of management system integration in mind, it will become quickly necessary to interconnect heterogeneous management systems in order to make them cooperating. Only such an action will allow plant telecommunication people to get a global view of the plant communication system covering all activity domains and then to master it. It will be a task of CNMA phase 5 to study and demonstrate the cooperation of NON CNMA ISO based network management systems and CNMA network manager, with the integration objective.

Security

"Systems that are used to process or handle classified or other sensitive information must be designed to guarantee correct and accurate interpretation of the security policy and must not distort the intent of that policy. Assurance must be provided that correct implementation and operations of the policy exists throughout the system's life cycle" [1].
CNMA will evaluate the current network management architecture regarding the requirements of a secure environment and a secure network administration.

Knowledge aquisition for Fault and Performance Management

Acquisition of expert knowledge gained from experience with operating CNMA networks. Because the installation and test of CNMA network at different pilot sites takes place at the end of CNMA 2617, no experience about working CNMA networks will be available before that date. This knowledge should be gathered, analysed and integrated in the knowledge based system in order to handle networking problems encountered in the field.
A second aspect of this work is the improvement of the model of the network within the knowledge based systems to make the systems able to handle greater degrees of complexity more efficiently.
Integration and evaluation of performance management related information will improve the fault diagnosis quality and enable the automatical tuning of the network's performance.

References

[IG 4.0] ESPRIT Project 2617 - CNMA Implementation Guide - Volume 2, A-Profile (Sep 89), Addendum I (March 90)

[Mgt Framework]: (ISO/IEC DIS 7498-4) Information Processing Systems - Open Systems Interconnection - Basic Reference Model Part 4: Management Framework. (Oct 1988)

[Mgt Overview]: (ISO/IEC DP 10040) Inforrnation Processing Systems - Open Systems Interconnection - System Management Overview. (Feb 89)

[CMIS]: (ISO/IEC DIS 9595-2) Information Processing Systems - Open Systems Interconnection - Management Information Service Definition Part 2: Common Management Information Service. (Dec 88)

[CMIP]: (ISO/IEC DIS 9596-2) Information Processing Systems - Open Systems Interconnection - Management Information Protocol Specification Part 2: Common Management Information Protocol. (Dec 88)

[SMI]: (ISO/IEC DP 10165) Information Processing Systems - Open Systems Interconnection - Structure of Management Information (Parts 1-4).

[Object Mgt]: (ISO/IEC DP 10164-1) Information Processing Systems - Open Systems Interconnection - Systems Management - Part 1: Object Management Function (Feb 89)

[PMFA]: (ISO/IEC JTC1/SC 21 N3313) Information Processing Systems - Open Systems Interconnection - Performance Management Working Document (Jan 89)

[FMFA]: (ISO/IEC JTC1/SC 21 N3312) Information Processing Systems - Open Systems Interconnection - Fault Management Working Document (Jan 89)

OFFICE AND BUSINESS SYSTEMS

Gil E. Gordon
Gil Gordon Associates
Monmouth Junction
New Jersey USA

SHOULD WE SAY GOODBYE TO THE OFFICE?
THE ROLE OF TELEWORK TODAY AND IN THE FUTURE

Abstract

It is becoming increasingly popular to discuss, and even to use, telework in the U.S., Japan, Germany, the United Kingdom, and other countries. Telecommuting, working from home, or the use of satellite offices are all different terms for the process of decentralizing the workplace. Among the forces behind this trend are traffic and air pollution problems, need for flexible work arrangements among today's workers, the appeal of telework's business benefits, and convincing evidence that it works.

By contrast, there are some reasons why it is not more widely used: the force of tradition, managerial resistance, mixed public policy messages, and some doubts about what will happen if telework actually does become much more popular. On balance, we can expect telework to keep growing slowly but surely. We will see more use of satellite offices, better vendor support to clear up technical confusion, limited support from unions, and better guidance and example-setting by governments. Also, each new generation of workers should be increasingly supportive.

Telework is a valid new way of working because, in an information-based economy, we no longer need to bring everyone to the workplace at the same time. It is supported by, but not dependent on, technology; it is an innovative management method that makes use of the technology to solve business problems.

Should We Say Goodbye to the Office?
The Role of Telework Today and in the Future

Imagine that you are a member of the Board of Directors of a large company. You are asked to review a proposal for a capital budget item of several million dollars, and you are told that your business cannot operate without this asset. However, you are also told that the asset will be used only 25% of the time at most, and will be idle the rest of the time.

Would you be likely to approve the spending of that money for an expensive asset that would be used so little of the time? Probably not, but this is exactly what happens now when companies decide to build or expand office buildings for today's workers. Our entire approach to creating workplaces for information workers goes back to the Industrial Revolution, when there was no choice but to have everyone working in the same place at the same time. That approach leads to the kind of wasteful investing in office space that can no longer be tolerated in the 1990s.

This realization that everyone no longer must work in the same place at the same time is one reason for the growing interest in telework, or telecommuting as it is commonly known in the United States. Telework has progressed from an oddity to a more widely accepted business practice in the U.S. and elsewhere, although it is not yet fully accepted and there is still a lot of resistance to its use. This paper will describe some of the forces moving telework ahead and those restraining it, and will also assess the future for telework development.

The "Push-pull" Forces on Telework

Let us examine why telework has progressed so far and why it has not yet become a more common work method. First, a short definition: we are concerned here only with the practice of having employees spend two to four days a week working at home or elsewhere off site, instead of going to their regular office location every day.

This is a rather small segment of the growing remote work population that includes home-based self-employed businesspeople, professionals who do increasing amounts of work in hotels, on airplanes, or even at ski resorts, and the large numbers of workers who take work home to do in the evening or on weekends

These are all interesting cases but do not represent the kind of fundamental change in work methods that telework does.

There are six main reasons why telework is growing:

1. THE ROADS DON'T WORK ANYMORE - There is hardly a city or a suburb in the U.S. today that is not struggling with having too many cars on too few roads. The "gridlock" that used to be limited to Los Angeles, New York, and other large cities has now spread to smaller cities and to the suburban areas around them. It is becoming widely accepted that we can no longer bring all the workers to the workplace, and we must use telework, car pools, staggered start times and other methods.

   There is a regulation in Southern California that requires employers to increase the "average vehicle ridership" for their employees - or face large fines. This means that employers have a direct financial incentive to promote telework (along with other traffic-reduction methods), and similar regulations are being prepared for other high-traffic, high-pollution cities.

2. THE OFFICES DON'T WORK ANYMORE - A recent survey of executives in 200 of the Fortune 1000 largest firms in the U.S. cited excessive noise as the biggest factor restricting productivity. One reason why teleworkers are more productive at home (a fact confirmed by their managers) is simply that they are away from the unproductive office environment, with all of its noise, interruptions, and distractions. The knowledge worker of today cannot perform effectively in a setting that in many ways is unchanged from the clerical work "factories" of the early 1900s'.

3. THE WORKFORCE NEEDS (OR WANTS) FLEXIBILITY - With the growth of single-parent households, dual-career couples, aging parents to be cared for, and a large installed base of personal computers in homes, today's workers are putting pressure on management to provide more flexible work arrangements. This is why we often see telework implemented in the same firms that use flextime, job-sharing, compressed work weeks, and other alternatives. Also, the declining numbers of younger workers in the U.S. means that employees have more bargaining power, and (in some cases) can exert upward pressure as a nego-tiating tactic.

4. THE TECHNOLOGY IS (USUALLY) HELPFUL - We can now make many jobs location-independent; wherever we move the person, we can move the work. However, everything does not always work as smoothly as the advertisements would suggest; telecommunications connections are not always easy, and data transfer between different computers sometimes fails. But we have made great advances in the last few years, and the vendors are starting to take the remote-work market more seriously. The two technologies that seem to encour-age telework the most are voice mail and fax; my own explanation for their popularity is simply that they allow the professional to escape from the keyboard and typing, and from all the perceived complexities of computer dial-up connec-tions.

5. TELEWORK HAS MANY DIRECT BENEFITS - The problems for which telework is a good solution continue to grow, and the benefits it offers are attractive. In addition to the ability to help reduce "gridlock" and air pollution, telework can: reduce office space costs, raise productivity, allow businesses to provide ex-tended hours of customer service, and attract and retain valuable workers. These are important benefits to consider as employers face the more competitive business environment of the 1990's and the need to reduce operating costs.

6. TELEWORK WORKS - Finally, we now have enough documented examples (plus much more anecdotal evidence) about the success of telework to show that we are past the experimental stage. We know how to select the right jobs, the right employees, and the right managers; we know how to do the needed training; we know how to protect against various risks; we know how to keep the teleworkers feeling a part of the social network of the office, and more. In short, we have demonstrated (to almost everyone) that telework is no longer a risky experiment with a high probability of failure.

Balancing against these positives are four negatives that hold telework back:

1. WE ARE VICTIMS OF THE PAST - No matter how many benefits are offered by telework, there is the undeniable fact that telework is different. Many companies talk a lot about innovating and being willing to try new alternatives, but they still approach telework with much hesitation. Also, when we get down to the level of the individual managers, many still rely on "eyeball supervision" - if they can see

their people working (whether or not they are actually being productive), they are satisfied.

2. THE "SLIPPERY SLOPE" PROBLEM - A colleague of mine (Dr. Kathleen Christensen) uses this term to describe the reaction of the typical manager who is considering offering telework or any other flexible work option to the workforce. The fear is that there will be so many employees who want to work at home that it will be uncontrollable - and the manager will feel like he/she is skiing uncontrollably down an icy mountainside. In reality, many people definitely do not want to work at home, and many of those who think they do can be persuaded to change their minds once they are given adequate information about how demanding it is.

3. PUBLIC POLICY MESSAGES ARE MIXED - In the U.S., telework has been noticeably absent from the list of solutions to traffic and air quality problems as promoted by most politicians. They are willing to spend millions of dollars for roads, trains, parking lots, sophisticated computerized traffic control systems, and even to give employers a tax credit if they provide mass transit vouchers for their employees. But as yet, there has been very little attention paid to telework as an equally suitable solution. This is slowly starting to change, and in fact, the Office of Policy Development in President Bush's White House is actively promoting telework as part of the President's Clean Air legislation.

4. FEAR OF SUCCESS - Last, I detect a small but significant form of resistance from managers who are afraid of what they THINK will happen if telework works. Here are two examples: telework implies that employees self-supervise for part of the work week; does that suggest that there will be fewer managers needed? Telework might lessen the value of the manager's job, as it is calculated in job evaluation systems that give points for size of staff, scope of work supervised, etc.; does that suggest that managers might earn less money if they permit telework? I do not believe that these are logical outcomes, nor do I believe that the better managers think they are true. However, the world is ruled neither by logic nor by the better managers.

Some Factors Affecting Telework's Future Growth

I have been involved in promoting and implementing telework since 1982, and have been both encouraged and discouraged by the prospects for the future since then. As a new way of working, telework offers great promise that will only be realized if we get beyond some of the problems that limit its usefulness.

Using the restrictive definition given earlier, it still seems likely that 5% to 10% of the office-based workforce can easily be teleworkers at least two days a week. We have about 60% of the workforce in the U.S. classified as information workers, so that is the theoretical upper limit. I do not believe we should start selling our skyscrapers and planning for the mass exodus from office buildings. I do believe that we will slowly but surely rethink what we mean by office work and where that work will be done. (There is, for example, a small group of researchers studying the practice of "officing," much in the same way that earlier researchers studied farming, manufacturing, and other commercial activities.) Here are some of the issues facing telework in the future, and an educated guess about how each will develop:

1. HOME WORK VS. REMOTE WORK - Alvin Toffler's book THE THIRD WAVE was responsible for popularizing the term "electronic cottage" in the mid-1980's. This was helpful because it raised awareness about the possibility of telework, but harmful because it planted the image of the sole worker at home (surrounded by various pieces of computer equipment) as the predominant form of telework. In reality, all telework is not high-tech, and it is certainly not restricted to the home. There is growing pressure to consider various kinds of satellite offices or neighborhood work centers as telework sites. These are more suitable than the home for certain kinds of work and/or employees, more cost-effective for certain technologies, and may turn out to be better traffic-reduction methods. One possible problem with home-based telework is that it might encourage the very kind of urban sprawl it is intended to combat; workers who have to commute to the office only two days a week may choose housing even farther from the office.
   THE PREDICTION: Look for many more examples of satellite or neighborhood offices in the next few years, especially in countries like Japan and others where homes are smaller and cultural norms about separation between work and home life are stronger.
2. THE EFFECT OF "TECHNO-CONFUSION" - It seems we are always being told that the solution to our telecommunications and computer problems is "just around the corner." Yet every time we turn that corner, there is another obstacle in the way. It is as if we live in a building with all corners and no smooth surfaces. The latest "corner" we are waiting to turn is ISDN, with its promise of digital pathways, multiple uses of single lines, high speed data transmission, and so on. I think the technology infrastructure for telework is quite well in place, although certain technologies (such as ISDN and lower-cost video conferencing) would get us beyond some barriers that limit certain kinds of telework today. The problem that remains is that there are painfully few people who seem to be able to make all the pieces work together well. The computer and telecommunications experts in the companies considering telework typically do not see their role as facilitators of telework solutions; for them, telework is often an after-thought.
   THE PREDICTION: Look for one or two aggressive vendors to fill this gap. They will not need to invent new solutions or design new hardware; they simply have to provide the overall plan for fitting existing pieces together. They will also begin packaging their existing products for telework applications, and by doing so will open the eyes of their customers.
3. THE UNIONS: PARTNERS OR PROBLEMS? - Telework has almost universally been attacked by labor unions, owing largely to the history of abusive home work arrangements in many countries. The unions generally do not make a distinction between industrial home work and telework as we know it today, and this is an unfortunate mistake in my mind.

In the few cases where unions have been partners in telework programs, the programs have worked well and the unions have fulfilled their proper role as representing the interests of the workers. Given that telework is one of the rare "win-win" programs, i.e., both management and labor can benefit, the unions are not doing their members much good by continuing to resist.

THE PREDICTION: Look for a few isolated cases where the more enlightened union leaders support telework projects; otherwise, expect continued resistance. There is the possibility that unions may exert enough influence to cause the passage of laws that restrict telework, though this almost definitely will not happen in the U.S.

4. GOVERNMENTS: LEADERS, NOT FOLLOWERS? - Governments as employers do not generally have a reputation of being early adopters of innovation. However, there are some encouraging signs that this is changing, at least in the U.S., as telework programs are seen at the local, state, and Federal levels. Most of these are in response to traffic-reduction statutes as noted earlier, while others are in response to the employees' needs for greater flexibility or for quasi-benefits to help make up for lower government pay scales.

In the U.S., there is an interesting reaction to these government efforts: some private-sector employers are shocked that government bureaucracies are actually taking a leadership role for a change. It is as if the private-sector companies feel ashamed that they do not have telework in place, if even the Federal government is doing it!

THE PREDICTION: Look for more government telework activity, though it may be slow to come and it will be weighted down with all the typical rules, regulations, and policies. Government programs can be excellent models for the private sector to follow, if only because they demonstrate that telework is feasible and practical.

5. THE NEXT GENERATION OF TELEWORKERS - What will happen when today's youth - who grew up with Commodore computers and Nintendo sets at home, and with personal computers in school and college - enter the workforce? Will they expect telework opportunities as workers and provide them as supervisors? There are some early indications that this will be true. People who entered the office workforce since 1985, approximately, are more computer-literate and have different expectations about the need for supervision.

They are fiercely independent in many ways, and want their supervisors to be coaches and helpers but not overseers. The concept of telework makes sense to them. However, we do not know yet whether their values will change as they move up into management - assuming they do - and whether they will be more accepting of new work options in general.

THE PREDICTION: Look for the widespread resistance to telework seen among middle-aged managers today to begin fading away, though not entirely, as younger workers assume management roles. I do not want to blame today's middle-aged managers nor do I want to congratulate their younger successors too early. My observation is that the younger workers will find it much easier to accept new ways of working and thus will encourage their use.

6. GOVERNMENT INCENTIVES VS. FREE MARKETS - Finally, there is a question of what role government should play to encourage the growth of telework. One side says that the societal benefits are so strong that there should be incentives: tax credits, direct investment in pilot programs and satellite offices, and so on. The other side says that if telework truly does offer employers significant benefits, the employers will not need any outside incentives to create telework programs. In the U.S., at least for now, it is clear that the free-market argument is stronger.

THE PREDICTION: Look for selected cases of government support for telework programs, especially when they can be used to foster regional economic development and job creation. Telework is a way to import jobs from the cities to the rural areas, instead of exporting workers in the other direction. Also, as traffic and air pollution problems continue to worsen, look for some kind of tax incentive or tax credit in most developed countries to encourage the movement of information, not workers.

Conclusion: New Ways of Working - and New Places to Work

Much of this workshop is directed at examining new ways of working. We must also look at new places to work as we move toward the 21st century, because it is no longer necessary or desirable to work only in the old places. Everyone will not be working in their spare bedrooms, to be sure, and office buildings will not become extinct relics.

Today's information technology frees us from the need to BE together just because we WORK together, and telework is an important tool in making this transition. But it is vitally important to remember that telework is not just about the technology. It responds to the need to reverse the costly, outdated trend of centralizing the workplace, and the need for managers to stop being paid for maintaining the old traditions and start being paid for innovating.

Telework is an innovation but not a radical or frightening one. Technology is the catalyst that should help managers see other ways to organize work and workers. When we give that technology to managers who comprehend the demands of global business in the 1990s, and we encourage them to use it sensibly, we can look forward to more use of telework as well as other management techniques that will take us into the next century.

History has shown that we can be our own worst enemies when it comes to envisioning the possibilities of technology. When the telephone was invented, the established business people said, "Why do we need it - we have messengers to carry messages." When xerography (the photocopy machine) was invented, the established business people said, "Why do we need it - we have carbon paper." Let us not make the same mistake about telework because we are more comfortable with what exists, and responding with "Why do we need it - we have all these nice office buildings."

Application tools and end-user applications:
The Object Management Architecture

Dr. Richard Mark Soley
The Object Management Group

Applications developers and end-users alike today clamor for similar features in application architectures: interoperability, portability, maintainability, reusability, distribution, heterogeneity and consistency. Although there may be multiple approaches to achieving these goals, repeatedly many different projects have chosen object-oriented programming styles to solve these problems and alleviate the expense of large-scale software development and integration.

The Object Management Architecture, under development by the Technical Committee of the Object Management Group, is approaching just such an object-oriented application integration platform. This corporate-sponsored international group has developed an application framework that categorizes and structures the problem of integration. This Object Management Architecture framework will be presented, with an overview of how it will grow over time. How integration is achieved through this approach will be examined.

Project no. 385
Human Factors in Information Technologies - HUFIT

USER CENTRED DESIGN, PROTOTYPING, AND COGNITIVE MODELLING WITH THE SANE TOOLKIT

TOM BÖSSER, ELKE-MARIA MELCHIOR
Institut für Allgemeine & Angewandte Psychologie
Westfälische Wilhelms-Universität
Spiekerhof 40/43
D-4400 Münster
Germany
Tel.: +49 251 834125 / 834108 /374208
Fax: +49 251 37223

Abstract

The SANE toolkit is a set of CASE tools based on a language for cognitive modelling of user behaviour in well defined domains such as human computer interaction. A SANE model, called 'the interaction model of the device', represents precisely the user-visible functionality of a device and the tasks performed by users. Activity charts can be refined into the considerably more precise interaction model, which can be used as a prototype. User procedures are derived by simulation and render the user knowledge model, which serves for validation of the interaction model and as a basis for constructing user support functions. Validation includes verification, the proof that the defined tasks can be performed with the device, and evaluation, a quantitative estimate of the cognitive workload involved in performing the tasks. The decisive advantages offered by SANE are the speed and efficiency of modelling, prototyping by executable specifications, specification-based evaluation of the interaction model and the automatic generation of documentation. Typical application domains are interactive systems tailored to specific applications, for example NC-machines, transaction processing, office process migration systems, telecom terminals or systems in vehicles. Further development will establish links to interface object builders, providing support for a complete top-down development method for interactive systems.

1. Introduction

Information technology offers new capabilities for implementing computer based applications and work places. The domain of application we are concerned with are interactive application systems for specific tasks, rather than user interface standards or systems for general application areas, for example office automation products. The underlying technology are modern computer systems and user interfaces, where the activity of the system is controlled in interaction with the user.

The software designer must foresee the functions the user needs for performing his tasks and make them available in an optimal form.

At new, technology-generated workplaces the worker (user) may be faced with a new tool for existing tasks, or entirely new tasks. Learning and training are costly in this situation. Previous studies (BÖSSER [1], MELCHIOR & BÖSSER [2]) have shown that for efficient text editing clerical workers require two weeks of training, training for Computer Aided Design (CAD) effectively requires three to twelve months training time (loss of work). In these and comparable cases investment into user knowledge by training and practice is more expensive than the purchase price of hard- and software. (It is important to distinguish between task- and tool-specific knowledge. Tool-specific knowledge, which we discuss here, must amortize in a much shorter period than task-specific knowledge.)

The choice of criteria for an optimal design must consider a compromise between 'ease of use' and 'easy to learn' (BÖSSER [1]). Usability is defined as 'The degree to which a user can achieve specified goals effectively, efficiently, comfortably' (BEVAN [3]), learnability is the effort to learn performing a task. Simple systems are easy to learn, but learning makes complex systems easy to use and raises the efficiency of work. Taking into account the learning curve, the criterion is efficieny, calculated as the ratio of total benefit in terms of work performed to total effort and time expended during the total duration of usage of a system. Complex tasks can not be made simple by tools, but inappropriate design adds unnecessary complexity for the user and impairs performance. User characteristics are reflected in the tasks the user has to perform with the system as a tool, in the frequency of these tasks and in the knowledge which the user brings to the task before using the device.

We consider the task, the device functionality and user knowledge as separate entities with characteristic dependencies. The SANE (Skill Acquisition NEtwork) represents the user visible device functionality as a network, the task as goal states, and user knowledge in terms of states and procedures. Design usually centers on one factor, whilst the others are fixed or constrained. If the task and the device are defined, the user's knowledge must be adapted through training. If the device and the user knowledge are fixed, the tasks which can be performed are limited. Our main concern here is the third case: Assuming that the task is given, how is the efficiency of cognitive work optimized by the design of interactive systems?

1.1 Human factors test and validation is desirable early in the design process - 'just in time'

Human factors testing and the effort for improving the human factors quality of IT products adds additional cost and time to software development. Evaluation of prototypes, the standard procedure for human factors testing, occurs late in the system development cycle and causes delays and redesign costs. Considerable benefits accrue from system evaluation concurrent with design, so that the results of evaluation become available before implementation is completed - 'just in time'. If human factors validation is based on the design specification rather than a finished product or fully functional prototype, it must be analytical and predictive rather than empirical. The desiderata for an analytical test and validation method follow from these constraints.

The effort for human factors evaluation can be accepted more easily when it is linked to activities offering other benefits, for example documentation. A tool for analytical human factors evaluation should therefore be linked to design tools - Computer Aided System Engineering (CASE) tools and User Interface Management Systems (UIMS) - and use a compatible representation of design objects.

Evaluation delivers information of predominantly negative content because its function is to detect deficiencies and shortcomings, understandably this is often not seen as helpful. Support for the designer can be given by a precise and detailed analysis and indication of the causes for shortcomings.

1.2 Prototyping, test and validation

Executable specifications offer a common basis for early prototyping and analytical and predictive human factors evaluation. This supports an evolutionary form of system development, where prototypes are built early, evaluated, and iteratively improved.

Prototyping serves different purposes in different phases of the development cycle. Early prototyping may help to elicit from experts or prospective users more detailed descriptions of the intended user tasks (requirements capture). In a later phase, prototyping helps to test, enlarge and complete specifications.

Test and evaluation of a product must relate to a specification of defined or intended properties. Validation of user interface characteristics includes verification and evaluation. Verification tests correspondence of specification and product, meaning in this context: Can the tasks for which the product is intended be performed with a product conforming to the specifications? Evaluation is a quantitative or comparative estimate of the efficiency with which these tasks can be performed with the device (workload, learning and knowledge requirements, performance). Validation of user interfaces requires precise specifications and adequate testing procedures, empirical and analytical.

1.3 Interface builders

For a number of window-based user-interface systems interface-building tools are available today, e.g. GUIDE for OPEN LOOK, or the Interface Builder for NextStep. These tools support the development of the low-level interface objects in a bottom-up fashion, which are linked by other program components to the application functions.

The SANE toolkit supports a top-down design procedure, where first the functionality (the interaction model of the device) is specified and designed, and then linked to appropriate interface objects and background functions (Figure 1).

592

2. Modelling of user knowledge

2.1 Task modelling and the design of new work environments

In the context of system design and evaluation, a task model describing the goals and constraints of use for the system is essential. This cannot be a description of existing work procedures, because a new system and new tools are expected to change the working procedures. Task descriptions as part of the requirement specification for new IT systems influence the functional specification of a product. When for the future office system the task defined is 'write a letter', a system will result which in the end can generate docu-

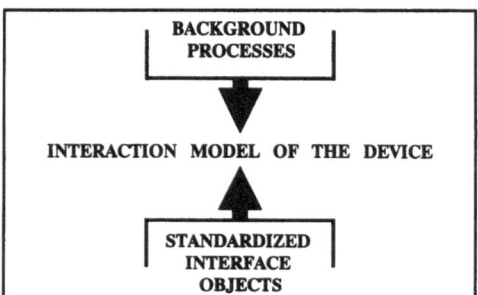

Figure 1. The Interaction Model of the Device connects Background Process and the Interface Objects

ments on paper. However, a more correct description of the task might have been 'pass a message', where electronic mail can be an adequate solution. User centred system design must be based on an adequate analysis of the requirements for the new device, which cannot rely on the analysis of ongoing work only, because this would reproduce the old style of working. The objective is to optimise systems design such that defined tasks are performed efficiently. The task model is the basis for the specification and design of a new device. The roles of task modelling, task analysis and design are ordered into an analysis / design cycle as shown in Figure 2 (see next page).

In the concrete world working procedures are observable, but not tasks. Tasks are an abstraction which in practice is generated in steps: The result of measurement and observation is a description of activities. Common forms of description are job inventories or goal action protocols, constituting a first level of abstraction. A further level of abstraction describes these activities in terms of the overall goals.

The process of design proceeds in the opposite direction, from given abstract representations of tasks, to which the functionality of the new device is adapted. As a result new workplaces are designed. In very few areas of technology, for example aerospace operations and process control, operator procedures are designed as part of the total system. In all other cases the user of the system considers his goals and preferences, and adjusts his working procedures, including possibly a new interpretation of his tasks, to the new workplace.

Whether the user will find the innovative technological device right or wrong for his purposes, or whether he uses it in the way expected, is a function of whether a correct representation of the users tasks were in the mind of the designer.

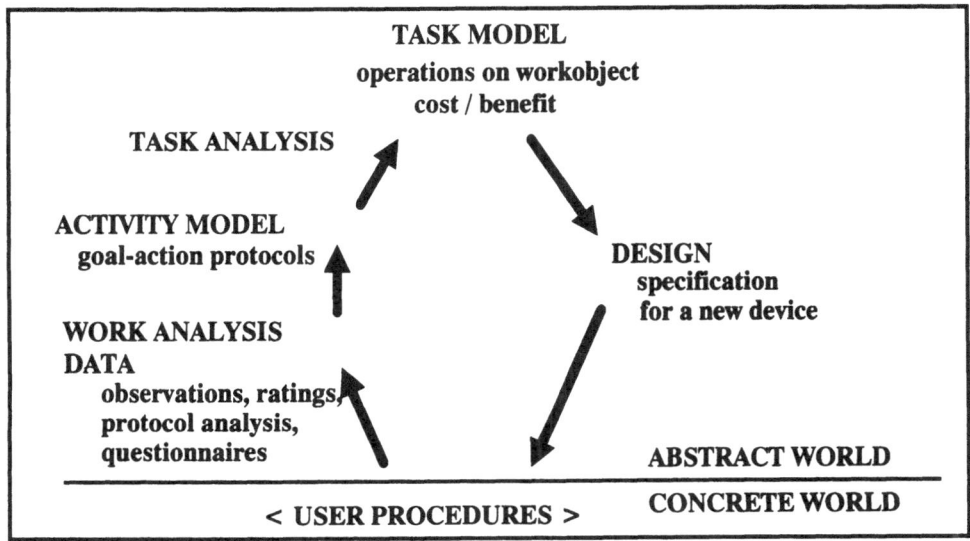

Figure 2. Task Analysis and Design Cycle

Forecasts related to man-machine interface and functionality have generated some obvious mispredictions, for example the use of electronic mail vs. fax, or the paperless office vs. more laser-printer output. Some predictions with undecided outcome are under investigation now, for example natural language dialogue, voice input/output and intelligent agents. This clearly indicates the difficulties of describing user tasks adequately, so that it can be reliably predicted, how users will adjust their working procedures to new functionality.

Task analysis starts from observations and generates descriptions or models. (Job analysis moves in the opposite direction, starting with a scheme into which observed behaviour is classified.) Task description and analysis methods differ according to the objectives and purpose of the analysis (JEFFROY [4], MAZOYER [5], PINSKY & THEUREAU [6] and JEFFROY & LAMBERT [7], JOHNSON et al. [8], [9], [10], SIOCHI & HARTSON [11]).

The representational principle we use is a state space representation in which tasks are specified as goal states, as detailed below. This approach is limited by the difficulty in representing complex goal states.

Common to most task representations is their limited generality. The least general form of task representation is the procedural representation where not the task, but the procedure for performing the task is represented. The representation of tasks as hierarchically ordered goal states does not take into account that in the real world behaviour is a compromise between conflicting goals, and the resulting course of action represents the best known solution under prevailing conditions; changing conditions make other options preferable, and this is not captured by the representation of goals as procedures.

The representation of meaningful, generally valid goals in terms of states meets with the difficulty that not precisely defined goal states, but an ordered set of many acceptable states represents real tasks. No concise and precise description is known today.

We conclude that task analysis performed in real working environments is a prerequisite to capture the requirements for the design of new systems. However, both observational and analytic tools are of limited power today, the process of task analysis is tedious and task models are largely hand crafted. Both the analysis problem and the representation problem of task analysis are in need of further research and innovation.

Task analysis and -modelling with an appropriate degree of empirical precision and detail is essential in the system design cycle but difficult to do, especially for innovative workplaces. The Skill Acquisition Network (SANE) method includes efficient task modelling tools, which will be extended in future.

2.2 Models of procedural knowledge for using computers

Computational models of human problem solving and of skilled behaviour are closely related: Problem solving occurs when procedures for solving problems cannot be retrieved ready-made from memory. Skilled behaviour develops from problem solving behaviour by learning and practice. Models of problem solving have been at the centre of artificial intelligence research, and advances have been made both in research and in applications of these methods. The modelling of skilled behaviour is a domain of empirical psychology, and computational models like 'Automatic Control of Thought' (ACT* by ANDERSON [12], [13]) or SOAR (LAIRD, NEWELL & ROSENBLOOM [14]) have been advanced, which include the capability to model skilled behaviour. KIERAS & POLSON ([15], [16]) and CARD, MORAN & NEWELL [17] have developed models and representational languages specifically to represent the procedural knowledge of users of computers.

The goal of using cognitive models in the context of system evaluation is to develop a representation and computational model which can solve tasks in the same way as the user of a device. The benefit expected from these models is that they show interesting formal properties, which can be the basis for analytic and predictive methods of evaluation and validation of user procedures.

GOMS type models (CARD, MORAN & NEWELL [17]) represent user procedures in the form of productions (containing goals, operators, methods and selection rules), which has the desirable property to offer an extensible and additive structure, very similar to how human procedural knowledge seems to be structured and emerges by adding element to element. Incomplete knowledge or incorrect knowledge may be represented as a subset or superset and limits the capability to solve tasks.

Tasks, as a model of what the user wants to do, are represented as goal hierarchies in these and similar approaches. Goals and procedural operations are combined in a rule-based representation, there is no separation of knowledge and of tasks. This is a main feature where we attempt to improve on the capabilities offered by other approaches.

A second aspect where we try to advance the methodology are the evaluation capabilities based on cognitive modelling of user procedures. Cognitive Complexity Theory (CCT) as proposed by KIERAS & POLSON [15], [16] almost exclusively relies on the common elements theory of transfer of knowledge as a basis for estimating learning requirements. The empirical validation has shown that the derived parameters vary depending on details of the model building procedure. Our approach to evaluation is based on a multiple resources theory of workload (GOPHER & DONCHIN [18], NAVON & GOPHER [19]). This view assumes that a number of resources are available for cognitive processes. Cognitive actions use these resources to a variable extent. Neither the number and nature of dimensions of this space are known, nor are standardised and precise measurement procedures available. However, in any realistic and practical situation it seems feasible to estimate relevant parameters, for example for reading or input actions. An advantage is that cognitive parameters, for example memory load and decision difficulty, can be estimated with satisfactory precision for specific tasks and application areas.

Other approaches to user modelling, only briefly to be mentioned here, are distinguished by different principles and objectives. An approach to communication between humans in natural language assumes that communicating persons need reciprocal models of their respective knowledge and intentions, both to interpret and to form correct statements of their own (WAHLSTER & KOBSA [20]). User modelling as advocated by RICH [21] aims at a classification of users. In HOPPE's task-oriented parser [22] the user is modelled implicitly in the definition of the language by which the user can express himself.

3. The SANE Model

3.1 The representation of task and device-related knowledge

Our objective is to develop a well structured representation of user knowledge, representing the prescriptive knowledge of the user - what the user must know if he is to perform correct procedures with a given device. Important features to be realized are:
- separate representation of task- and of device-related knowledge
- the representation of the device rests upon the principle of isomorphy between device functions and user knowledge.

The latter is an important aspect of computer based devices, differentiating them from mechanical tools for example. In software systems all functions which can be accessed and used by the user have to be programmed explicitely into the device. This means that user knowledge must represent device functionality.

A more structured representation of device function than, for example, a rule based one, seems to be appropriate for representation of the device and device knowledge. The user knowledge model in SANE is based on a state transition network (or similar formalism) representing the functionality of the device. Knowledge is represented as an overlay of this model, which defines an upper bound for the knowledge available to the user. This excludes some forms of incorrect knowledge, which cannot be represented in this framework. Tasks and task related knowledge are to be represented separately.

State transition networks have been used as a method for specifying interactive systems, for example by WASSERMAN [23] and JACOB [24]. State transition networks have the undesirable property of exploding complexity for detailed models. An appropriate solution is the grouping of states according to a meaningful principle, which suggest a form of augmented transition network (ATN) for the representation of a device model. An ATN model of an interactive system defines the syntax of the device, i.e. acceptable user input sequences (procedures). The semantic component is added by defining goals to be attained by procedures. We call this the interaction model of the device. It defines the state variables and operations which are user visible, i. e. define possible operations and actions of the user. Tasks are defined in a separate task space, but must be linked to operations.

Conceptual knowledge is a prerequisite both for understanding tasks and for being able to understand the displayed states of the device and messages on the user interface. A procedural representation must include a representation of conceptual knowledge, possibly in a somewhat simplified form. Conceptual knowledge is implied by the state variables in the task model and in the device model. In ACT* (ANDERSON [13]) conceptual knowledge and memory search processes are represented by semantic networks and activation processes, which are obviously needed for problem solving, but may be dispensible for models of skills.

Procedural knowledge suggests how to do things. Conceptual and procedural knowledge representations are exchangeable and translatable to a certain extent. Adequate representations of procedural knowledge are rules and rule sets, for example in production systems. In the specific case where procedural knowledge relates to the procedures for interacting with a computer-based device, the knowledge model can be limited to the device functionality.

3.2 The interaction model of a device

The Skill Acquisition NEtwork (SANE) is an interaction model of a device. It represents user knowledge and procedures on which we base analysis and evaluation methods. It's basic element is the operation (as shown in Figure 3 below), which is a transition in a state space defined by the state variables of the work object (called 'task attributes') and of the device (called 'determinators', because they are only relevant when they determine the behaviour of the device).

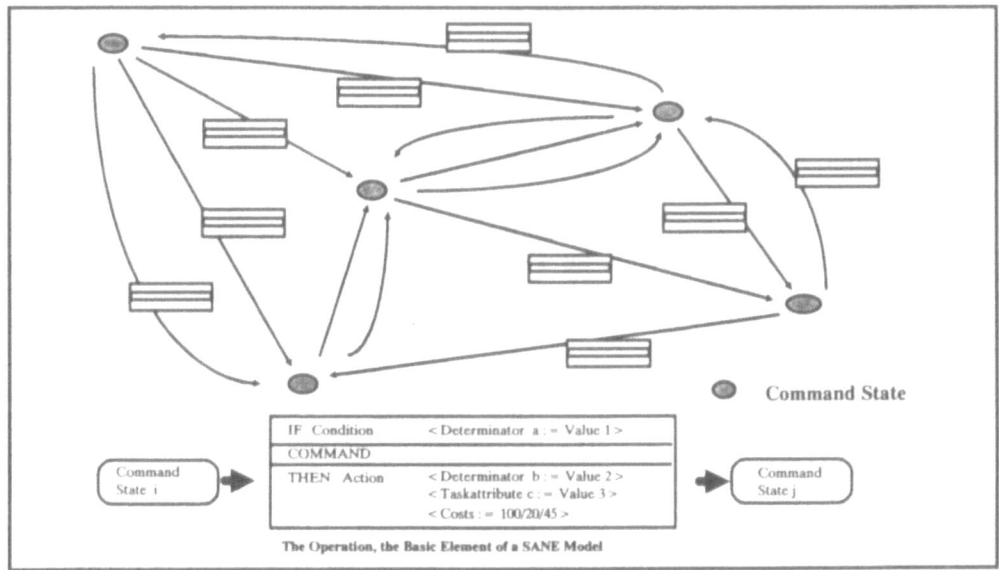

Figure 3. SANE (Skill Acquisition Network) is a Network composed of Operations

An operation is a state transition defined by
- a command state where the operation can be applied
- conditions, which limit the applicability
- a command (user input) for effecting the operation
- actions, modifying determinators and task attributes
- the next command state
- optional cost values representing workload and learning state.

Operations combine into a network (SANE), which represents skilled behaviour, including practice learning, which is not discussed further here.

The state space of the device is constructed implicitely by formulating operations. The task space is constructed implicitely by defining tasks, or explicitely by defining the dimensions of the task space (task attributes) and the values they can assume. Tasks are combined into task lists and task models.

A SANE can be considered as a representation of the syntax of the device, defining the interaction language. The task model represents the semantic aspect.

A SANE represents both the functions of the device, and the operations of the user of the device, which are isomorphic. The representation is on a functional level, excluding display states and input operations, which are not specified in early design phases. The SANE as an interaction model of the device does specify, however the state information which is represented, and the information needed by the user.

The cognitive workload associated with performing operations is represented by a multi-dimensional cost criterion. When the I/O functions are unspecified, neither is the workload associated with performing them. The workload parameters can be empirically determined and included in the model for commands with specified I/O operations.

3.3 Simulation of user procedures

User procedures (as shown in Figure 4 below) for solving tasks are generated by simulation. A task is defined as a goal, and, provided a computable solution exists, sequences of operations and associated workload estimates will be returned. A large task model, composed of many task lists, requires considerable computational effort.

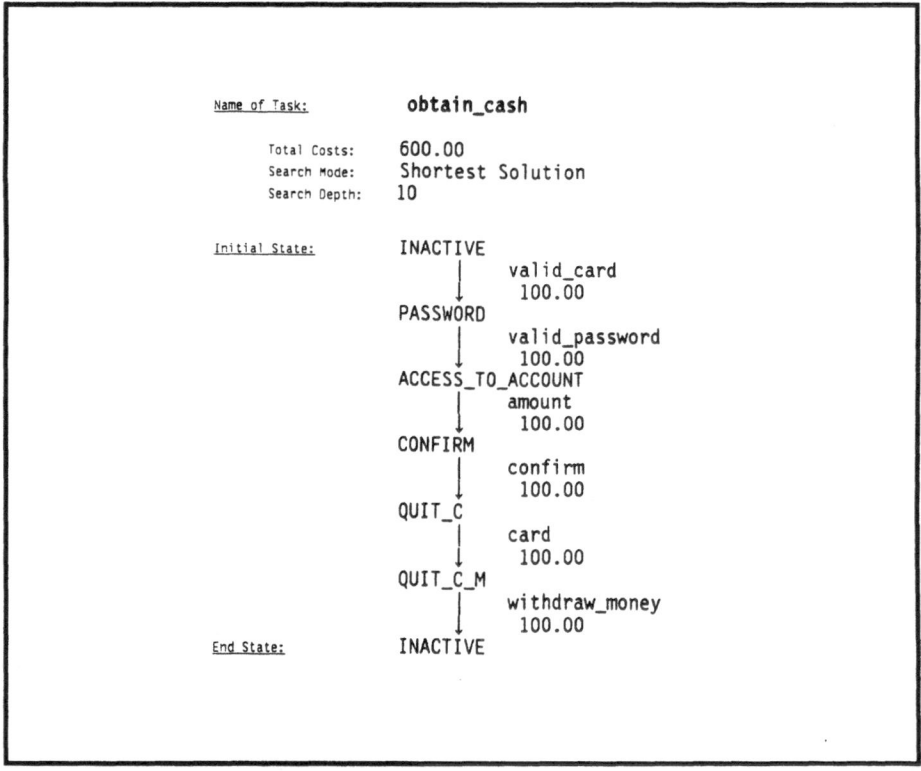

Figure 4. Simulated User Procedure to obtain Cash from Cash Dispensing Machine

3.4 Complexity of user procedures

Summary workload estimates for task models are calculated and permit assessment of the complexity of user procedures for different combinations of device models and task models, representing different user groups.

The calculated complexity of user procedures is an estimate of workload for performing the procedure in terms of time and of utilization of cognitive resources. Rough estimates of complexity are the numbers of operations or the branching structure - comparable to CCT - which indicate learning times or performance. Higher precision of cognitive workload estimates can be based on multiple resources theory (NAVON & GOPHER [19]), a principle which is widely accepted but hard to verify because parameters cannot be estimated in general. SANE provides a computational framework for modelling, but does not solve the empirical problem of workload estimation. The modelling capabilities offered by SANE invite explorations of parameters by a sensitivity analysis. Future work aimed at estimating workload parameters can benefit by concentrating effort on sensitive parameters only.

A set of complexity metrics has been defined, validation according to test-theoretical principles would have to be based on empirical tests with representative populations of users and devices, which theoretically and in practice is not feasible. The analytic measures are well defined and preliminary tests show that they indicate interesting properties of devices. Our proposal is to validate complexity metrics for precisely defined, constrained domains of application only. This can be done efficiently by using the prototyping capabilities of SANE.

3.5 User knowledge

A precise representation of the prescriptive user knowledge model is generated, comprising command states, state variables, operations, and procedures, which the user must know in order to perform the defined tasks.

3.6 Two strands of SANE: Cognitive modelling and computer-aided design of interactive applications

Sane is based on two principles, cognitive modelling on the one hand, computer-aided support for user interface test and validation on the other hand, complementing other design tools like CASE tools and UIMS.

Comparable cognitive modelling principles are offered by Cognitive Complexity Theory (POLSON & KIERAS [25]), ACT* by ANDERSON [13] or SOAR by LAIRD, NEWELL & ROSENBLOOM [14]. SANE is more domain specific than these modelling languages, and does not include a representation of conceptual knowledge, like ACT*. It does not include a representation of heuristic knowledge, as does SOAR. The strict separation of task and device related knowledge differentiates SANE from the other approaches.

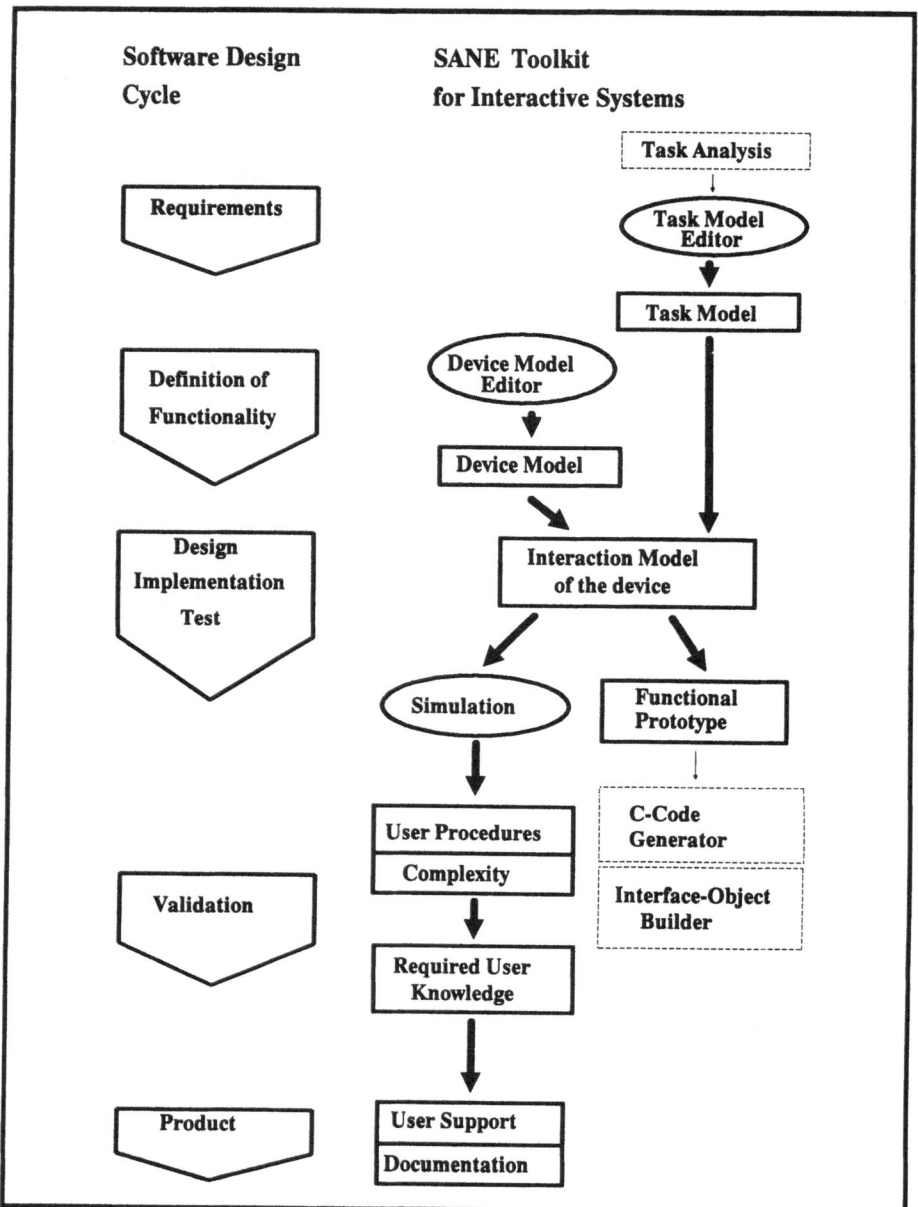

Figure 5.   The Use of the SANE Toolkit in different Phases of
the Software Design and Development Cycle

The objective to support the design of devices has prompted the representation of device functions as a basis for cognitive modelling. A specification of device functions is a starting point for building a SANE model, and assures correspondence of the design specification and the model of user interaction. The connection to CASE tools or design aids like INTUIT (RUSSELL et al. [26]) is possible at the level of Activity Diagrams. The pragmatic goals of SANE to support the design of software in a top-down fashion has lead to the implementation of efficient tools to support the method.

4. The SANE toolkit

Building SANE models requires extensive programming. It was recognised early that the efficiency of modelling depends critically upon efficient tools. A set of software-tools spanning the whole range of modelling activities was constructed, based on the principles
– graphical representation of objects
– syntax checks during editing operations
– efficient usage.
The SANE toolkit is implemented on SUN workstations in BIM-PROLOG, using SUNviews. It is extensible, and extensions for other application domains and added functions are planned for the future.

With the SANE toolkit models can be built, tested, evaluated and modified quickly and efficiently. The model components correspond to phases of the design cycle as shown in Figure 5: Task modelling is connected to requirements specification, the benefit is a precise description of the tasks to be performed with a system, which is considerably more specific, detailed and precise than requirements in terms of verbally formulated design goals and constraints. Multiple user groups can be defined by different task models. Building a device model is essentially a design process, generating an increasingly precise model of the device functionality. Of particular advantage is the fact that changes can be made quickly and easily, and tests which have been performed once can be repeated automatically.

4.1 Evaluation capabilities offered in SANE

Evaluation of the design of an interactive system can be seen as the validation of the interaction model of a device, comprising two parts, verification and evaluation.

Verification answers the question 'can the defined tasks be performed with the device ?'. This is shown by the reachability table, and proves that the defined tasks can be performed with the device. The interaction model can be used as a functional prototype on which the developer, customer, or end user can test if the intended work procedures can be performed. More efficient is the definition of a task model and formal test of reachability.

Evaluation is a quantitative or comparative evaluation of the efficiency of the user in performing the tasks, and of the cognitive workload of the user. This is represented in SANE by the complexity of user procedures and by associated workload indices, if appropriate data and parameters are available. Complexity measures include memory load, decision complexity, complexity of command states and others. Some of these complexity measures are analytic, others require empirical estimates of parameters.

Comprehensive validation of these complexity indices (CX) would require empirical tests with a large number of subjects and representative devices. This is not possible in principle, as for all comparable approaches. We therefore rely mostly on well defined analytical and predictive measures.

Table 1 summarizes some complexity indices for interaction models from various domains, which reflect the different knowledge and workload requirements which are typical for these domains.

The development of standards and metrics for user interface quality will require adequate models and measurement procedures. A form of cognitive modelling and analytic complexity metrics such as SANE will be indispensable for this purpose.

Table 1. Complexity Indices for Devices from different Application Domains

Application Domains	Measures of Complexity					
	(1)	(2)	(3)	(4)	(5)	(6)
Interactive systems for use by the general public (ticket vending machines, access systems, telecom)	< 30	< 10	< 3	< 0.0	< 5	< 5
Specific Applications (IT applications in vehicles, NC-machines, transaction processing)	< 100	< 15	< 5	< 0.1	< 10	< 15
Office Systems (Textprocessing, DBS, Spreedsheet, Graphic)	< 300	< 30	< 10	< 1.0	< 30	< 50
CAD	< 1000	< 50	< 15	< 5.0	< 50	< 200
Plane, NPS	> 1000	> 50	> 15	> 5.0	> 50	< 200

(1) Device Complexity (McCabe Cyclomatic Complexity Metric)
$e - n + 2$ (e = operations, n = command states)

Procedural Complexity:
(2) Operations / Procedure (list of tasks)
(3) Operations / Procedure (single task)
(4) Memory Load (Average number of latent determinators per operation)
(5) Command State Complexity (Average number of commands from which a command is selected)
(6) Choice Complexity (Average number of operations from which an operation is selected)

4.2 Documentation and user support functions

User support functions include error messages, help, user documentation (paper-based or software based) and tutorials. They have to be added to the system after the design is completed, and contribute considerably to the end-user's perception of the quality of a product. A SANE model represents the conceptual knowledge by the state variables and values occuring when performing tasks with the device, and procedural knowledge by precise descriptions of tasks, operations, procedures, command states and commands. A SANE model can be transformed into other useful representations, knowledge for user support functions can be extracted automatically. This formal knowledge model constitutes a full specification of the user support functions.

The knowledge model offers the advantage that to the same extent that the SANE model is correct and complete, user support functions can be correct and complete. In addition, the effects of changes in the system functionality represented in the SANE are mirrored in the user support functions. The generation of technical documentation and the maintenance of user support functions is likely to be simplified and to achieve high quality through use of SANE models.

4.3 Designing the functionality with SANE, prototyping and interface building tools

SANE modelling is effective for the functional design of application programs. This coincides with current trends in user interface technology, where efficient tools are becoming available for designing the presentation level of user interfaces. Programming of high quality user interfaces is becoming much easier with the advent of new tools (interface object builders) which are offered now as part of programming environments, for example GUIDE for OPEN LOOK or NextStep. This does not solve the problem of the design of functionality nor the evaluation of the efficiency of the system in relation to user tasks. Interface builders make prototyping and interface building easier, prototyping can be made part of the regular design process and development time is much reduced. The role of functional modelling with SANE is to evaluate the interaction model behind the user interface.

In other projects we are extending the SANE toolkit in such a way that the SANE interaction model can be cross-compiled and linked with interface objects designed in other environments, including GUIDE and OPEN LOOK, hypermedia environments, and high-quality graphical simulation environments like VAPS and DATA-VIEWS. The challenge is to define an architecture for interactive systems which achieves the modularization defined in Figure 1.

4.4 Conclusions

Use of the SANE toolkit for exploring applications in areas like NC-machines, telecom terminals, systems for use by the general public, hypermedia applications, and transaction processing has shown that designers find the capability to produce abstract, executable specifications, which can be much less detailed and specified than prototypes, most useful, especially the interactive testing of the functional prototype. Generating adequate task models proved to be difficult because it requires a certain amount of abstraction and data which very often are not available during the design phase. This indicates that there is a lack of data rather than a lack of methods. However, a conceptual task model must always underly design, and the advantage of modelling is that it is made explicit rather than remaining implicit and vague.

A further successful application area is reverse engineering of interactive systems. We have participated in some projects where interactive systems have evolved into a working system, but no appropriate documentation had been maintained. A SANE model was developed from the working system and provided documentation and specifications which were considered most helpful by the design team, and were made the basis for further product development.

Validation of a user interface model is an effort which adds additional time and effort to software development. It is a means to improve the quality of a product by tailoring the application to the precise tasks of the user.

Interactive systems, where the benefit of analysis and validation outweighs the added cost and effort, are application specific systems for transaction processing, manufacturing or similar tasks. Many systems using WIMP interfaces, as our analysis has shown, are largely composed of formfilling components, and include relatively few interactive capabilities. Systems which are tailored to a specific task characteristically have a complex interaction syntax.

4.5 Further development of SANE

An underdeveloped area of cognitive user modelling is task analysis and task modelling, generating an abstracted task model from observations and requirements specifications. More research is needed, and we consider including new functions for this purpose in the SANE toolkit.

As the work of POLSON & KIERAS [24] shows, universal parameters for estimating learning requirements and performance can not be identified in the sense that they are valid for all types of systems. However, to identify such parameters for specific application systems with specific interface objects seems to be relatively straight forward. Efficient methods to do this in development projects are needed.

A closer link to user interface management systems will be implemented. The result of SANE modelling and evaluation is a functional specification of the system, which must be extended by a specification of the interface objects to arrive at a fully developed system. Today these processes are quite separate and should be combined into a design method for interactive systems.

Interactive systems for process control in vehicles demand modelling principles which take dynamic processes into account. For this purpose the SANE modelling language will be extended into a Hybrid Operator Model.

5. References

[1] Bösser, T. (1987): Learning in Man-Computer Interaction. A Review of the Literature. Berlin, Heidelberg, New York: Springer Verlag.

[2] Melchior, E.-M., & Bösser, T. (1989): User Modeling and Validation. ESPRIT Project 385 - HUFIT. Deliverable B7.1c.

[3] Bevan, N., personal communication.

[4] Jeffroy, F. (1988): Course of Action. A Theoretical and Methodological Framework for Analysis of Users and Tasks Characteristics. ESPRIT Project 385 - HUFIT. Deliverable A4.3/4.

[5] Mazoyer, B. (1986): The Analysis of Tasks and Users for the Design of an I.T. Product. ESPRIT Project 385 - HUFIT. Deliverable A4.4a.

[6] Pinsky, L., & Theureau, J. (1987): Study of course of action and AI tools development in process control. First European Meeting on Cognitive Science Approaches to Process Control, Marcoussis (France), October 19-20.

[7] Jeffroy, F., & Lambert, I. (in press): Ergonomics framework for user activity software centered design, in: H.-J. Bullinger & B. Shackel (Eds.) Human Factors in Information Technology - From Conception to Use. Esprit Project 385. Heidelberg, Berlin, Tokyo, New York: Springer, Chapter 4.

[8] Johnson, P., Diaper, D., & Long, J. (1984): Tasks, skills and knowledge: Task analysis for knowledge based descriptions, in: Proceedings of the INTERACT'84: First IFIP Conference on 'Human-Computer Interaction' (Imperial College of Science and Technology, London, September 4-7, 1984) (Vol. 1, pp. 23-27). Amsterdam: Elsevier.

[9] Johnson, P., Diaper, D., & Long, J. (1985): Task analysis in interactive systems design and evaluation, in: G. Johannsen, G. Mancini, & L. Martensson (Eds.), Analysis, Design, and Evaluation of Man-Machine Systems (2nd IFAC/IFIP/IFORS/IEA Conference, Varese, Italy, September 10-12, 1985) (pp. 74-78). Varese: CEC-JRC Ispra.

[10] Johnson, H., & Johnson, P. (1987): Task modeling - identifying the structure and rules for modeling tasks. Technical Report No. 3. London: Department of Computer Science, Queen Mary College.

[11] Siochi, A.C., & Hartson, H.R. (1989): Task-oriented Representation of asynchronous user interfaces, in: K. Bice & C. Lewis (Eds.), Proceedings of the CHI'89 Human Factors in Computing Systems , New York: ACM, pp. 183-188.

[12] Anderson, J.R.(1982): Acquisition of cognitive skill. Psychological Review, 89, 369-406.

[13] Anderson, J.R. (1983): The Architecture of Cognition. Cambridge, MA: Harvard University Press.

[14] Laird, J.E., Newell, A., & Rosenbloom, P.S. (1987): SOAR: An architecture for general intelligence. Artificial Intelligence, 33, 1-64.

[15] Kieras, D.E., & Polson, P.G. (1983): A generalized transition network representation of interactive systems, in: A. Janda (Ed.), Proceedings of the CHI'83 Human Factors in Computing Systems (pp. 103-106). New York: ACM.

[16] Kieras, D.E., & Polson, P.G. (1985): An approach to the formal analysis of user complexity. International Journal of Man-Machine Studies, 22, 365-394.

[17] Card, S.K., Moran, T.P., & Newell, A. (1983): The Psychology of Human-Computer Interaction. Hillsdale, NJ: Lawrence Erlbaum Associates.

[18] Gopher, D., & Donchin, E. (1986): Workload - An examination of the concept, in: K.R. Boff, L. Kaufman & J.P. Thomas (Eds.), Handbook of Perception and Human Performance (vol. 2, chap. 41, pp. 1-49). New York: John Wiley and Sons.

[19] Navon, D., & Gopher, D. (1979): On the economy of the human processing system. Psychological Review, 86, 214-255.

[20] Wahlster, W., & Kobsa, A. (1986): Dialogue-based user models. Proceedings of the IEEE, 74, 948-960.

[21] Rich, E. (1979): User modeling via stereotypes. Cognitive Science, 3, 329-354.

[22] Hoppe, H.U. (1988). Task-Oriented Parsing - A Diagnostic Method to be used by Adaptive Systems. in: Proceedings of the CHI'88 Human Factors in Computing Systems (Washington, DC, May 15-19, 1988), New York: ACM, pp. 241-247.

[23] Wasserman, A.I. (1985): Extending State Transition Diagrams for the Specification of Human-Computer Interaction. IEEE Transactions on Software Engineering, 11, 699-713.

[24] Jacob, R.J.K. (1983): Using formal specifications in the design of a human-computer interface. Communications of the ACM, 6, 259-264.

[25] Polson, P.G., & Kieras, D.E. (1985): A quantitative model of the learning and performance of text editing knowledge, in: Proceedings of the CHI'85 Human Factors in Computing Systems (San Francisco, April 14-18, 1985), New York: ACM, pp. 207-212.

[26] Russell, A.J., Pettit, P., & Elder, S. (in press): INTUIT, in: H.-J. Bullinger & B. Shackel (Eds.) Human Factors in Information Technology - From Conception to Use. Esprit Project 385. Heidelberg, Berlin, Tokyo, New York: Springer, Chapter 7.

Project No. 2001

VERSIONED DOCUMENTS IN A TECHNICAL DOCUMENT MANAGEMENT SYSTEM

LUCY HEDERMAN
Dept. of Computer Science
Trinity College
Dublin 2
Ireland

HANS WEIGAND
EIT/KUB
P.O.Box 90153
5000 LE Tilburg
The Netherlands, Hans Weigand
fax: +3113 663069
tel: +3113 662688

Abstract.

A Document Management System (SPRITE) especially suited for technical documentation is described. The system includes a powerful WYSIWYG editor and scanning, archiving and printing facilities. A distinctive feature of SPRITE is its version model. SPRITE not only allows the user to maintain historical versions of documents, but also variants, or configurations, that represent slightly different versions of a certain document. The SPRITE version model is described in detail and compared with other versioning mechanisms.

1. Introduction

The commercial market of today offers a number of systems for technical documentation. They range from simple systems working with single documents up to sophisticated systems for an integral document management system, from batch-oriented systems up to state-of-the-art interactive WYSIWYG workstations (Walter, 1988). Examples are Documenter (Xerox), The Publisher (ArborText) and KEEPS (Kodak).

Technical documentation differs from normal documentation as created by average text processing systems in the following areas:

– Technical documents are often very large in size. Documents exceeding l000 pages are no exception.
– Technical documents are created by a group of authors often working concurrently.
– Technical documents have a long lifetime, following the progressive development of the product described. The development has to be supported by multiple versions of one document.
– Technical documents have to incorporate information from other sources (other documents, but also paper and remote CAD/CAM or database systems).
– Since technical documents are generally complex, there is a need for management support for such documents.

- Since technical products are often designed in series, as configurations differing on details only, the various documents describing the products also exist as close variants of each other.

To cope with these requirements, we are developing the SPRITE Document Management System as an integrated system for the production and maintenance of technical documents. We call such a system a Technical Document Management System (TDMS).

The rest of this paper is organized as follows. In section 2, we give a short overview of the functionality of the SPRITE system. In section 3, we spell out one feature, that is, how SPRITE supports document versions. The versioning mechanism is compared with other sytems, such as MINOS (a multimedia database), EXODUS, ORION and ONTOS (object-oriented database systems).

2. Overview of the SPRITE system

The document management system consists essentially of six components:

- a screen-oriented WYSIWYG document processor
- a browser and retrieval component
- a high-quality printing component
- a scanning & recognition component
- an information acquisition component
- a multimedia database (MMD)

Both the database and the document processor allow the use of text, graphic and raster data. In composing a document, an author can access other documents (by means of an *import* mechanism), or retrieve information from other systems (for example, CAD systems) by means of the information acquisition component, or extract inforlation from paper by means of the scanning & recognition component.

The system offers several functions for project management. Work on documents can be delegated to several authors with well-defined permissions. Project information can be attached to a document reflecting its lifecycle ("draft", "ready for review" etc).

Documents may exist in the system in different versions. By means of the browser component the user can easily view all versions of a document. The MMD supports efficient data sharing between different versions.

The system can also maintain links and interdependencies between documents. Such links can be used either to locate related documents with the browser (hypertext-like function), or to trigger an action in case one of a group of related documents has been updated and the other documents have to be updated as well.

Since technical documentation generally uses a large amount of data and has a long lifetime, the MMD is supplied with mass storage capabilities, which will be accomplished by integration of an *optical disk* within the MMD.

The SPRITE system also supports *multi-authoring.* It is possible for several authors to edit a document simultaneously. Provisions are taken so that concurrent access to a document does not cause inconsistencies, that authors have easy access to assigned document parts, and that a manager may easily supervise the progress of the work.

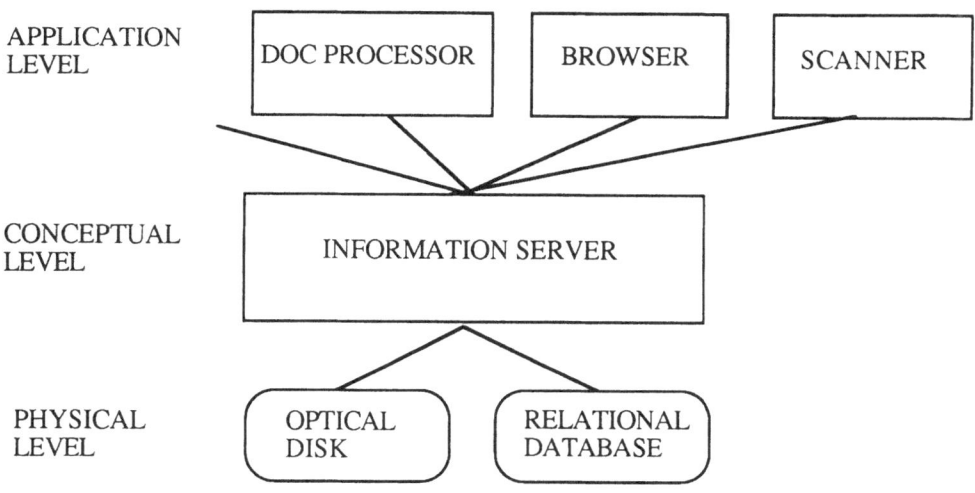

APPLICATION LEVEL

CONCEPTUAL LEVEL

PHYSICAL LEVEL

Figure 1 SPRITE system architecture

The architecture of the Document Management System is split up over three levels to enhance data-independence and extensibility (see figure 1). The bottom layer (physical layer) is responsible for the storage of content and structure of the documents. Structural information is stored in a relational database, and content on magnetic and optical disk. The middle layer (conceptual layer) comprises an object-oriented multimedia database (MMD). The MMD reflects the conceptual model of the document and the document space and supports conceptually meaningful operations such as CREATE document, INSERT document IN folder, SELECT (search condition) etc. On top of the MMD, the various applications are located, such as the document processor, the browser, and the scaning & recognition.

The MMD of the SPRITE system concentrates on all media that can be printed on paper, such as text, drawings, photographs, tables (structured information) and business graphics. Sound and video are not included. The document model used by SPRITE has been based on ODA (ISO 8631 international standard for Office Document Architecture; for an introduction, see e.g. Krönert, 1987) but for practical reasons this standard was not followed completely.

3. Version model

As stated in the introduction, a technical document management system (TDMS) should support the use of versions. Two types of versions must be distinguished:

historical versions

configurations

Historical versions correspond with either the derivational or logical history of a document. Technical documents are developed on a project basis, in consecutive steps, over a long period of time. The TDMS should support the derivation of new versions of documents.

Different configurations of a product require (slightly) different versions of the documentation. For example, a car may exist in a manual transmission or automatic configuration. The documentation system should support corresponding configurations of the car's manual. It should also support versions that differ in style or language.

The MINOS system (Christodoulakis et al, 1986) and the EXODUS system (Carey et al, 1986) also support the derivation of new versions of a document, but no distinction is made between historical versions and configurations. However, the combination of these two dimensions is not trivial, and can easily lead to chaos if no special organizational measures are taken. Distinguishing the two has the advantage of a clear conceptual picture. In several situations, it also allows a greater level of data sharing, that is, higher efficiency in storage use. A distinction between historical versions and alternatives is made in several CAD/CAM systems and in the object-oriented database system ONTOS (Andrews, 1989). For an overview of some open questions about versions, see (Kent, 1989).

3.1. Historical Versions

Technical documents exist over a long period of time and are developed in several steps. SPRITE allows the user to keep historical versions of a certain document, alternatively called *checkpoints*. This mechanism has several functions:

- recovery from mistakes. In the course of development, the author may find out that he is on the wrong track, and wants to start again from some previous version. In that case, he can use the browser to locate that old version and start editing again from there.
- data sharing. When the author wants to rewrite some existing document, he need not copy its entire content. Deriving a new version from it is sufficient and guarantees efficient content sharing.
- project management. Technical documents typically have more than one edition. The version mechanism represents the logical relationship between the subsequent editions.

The historical version mechanism in SPRITE is implemented by the following operations:

NEW_CHECKPOINT(oid): oid
FREEZE(oid)
CHECKIN(oid)
CHECKOUT(oid)
DELETE(oid)
ARCHIVE(oid)

NEW_CHECKPOINT takes an object identifier as argument and returns the object identifier of a new object. This new object (document) is initially the same as the old object; attribute values are copied and the content is shared. When the user starts editing the document, the affected components in the logical structure are automatically replaced by new versions. Replacing a component by a new version triggers the replacement of the parent component by a new version, up to the root component. The updates are performed on the new versions. In this way, the data sharing is maximal; it is essentially the same as used in the EXODUS system.

The new checkpoint is connected to the old checkpoint by means of a previous/next relationship. In this way, it is easy to go back in the derivation history of a document.

Note that new checkpoints can be derived from any existing checkpoint. The relationships between checkpoints therefore form a *tree.*

The effect of FREEZE is to make an object no longer revisable. Any attempts to update its attributes or content are blocked. However, frozen documents can be displayed, printed and used to derive new (revisable) checkpoints. At present, a FREEZE (of the old document) is triggered by NEW_CHECKPOINT, so that all internal nodes of the checkpoint tree are always frozen, and hence immutable, but FREEZE can also be done directly by the user.

Not all checkpoints are equally important in the project history. Usually, authors will work on a document for some time, and then decide to turn the last checkpoint into an edition. An *edition is* defined as a special checkpoint with a certain public relevance; it may be the checkpoint that is actually printed and shipped to the clients, or it is accepted by the author's manager. The operation CHECKIN is used to promote a checkpoint to the status of an edition. Editions have editionnumbers, so that it is easy to go through all editions of a document. Editions are always frozen, and, even more strictly, cannot be used to derive new checkpoints. An explicit CHECKOUT command is needed beforehand.

DELETE just deletes a checkpoint. Any checkpoint can be deleted, unless it has been archived (by means of the ARCHIVE command).

Figure 2 gives an example of a document history. Note the difference between the derivation history and the edition history. Checkpoint 4 is the next edition of checkpoint 8, but is derived from checkpoint 2.

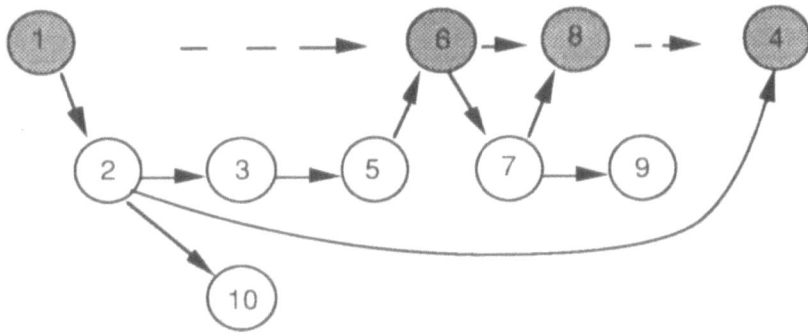

Figure 2: Checkpoint tree with editions

The history mechanism of SPRITE is especially aimed at documents, but we believe it also has a more general applicability. For example, instead of documents, we could also consider historical versions of employee objects in a personnel department. Each time the employee information is updated, for example, because of a new salary, a new version is created.

3.2. Configurations

SPRITE supports the creation and retrieval of documents describing different configurations of a product. Two perspectives are provided, one for authors or creators of configuration documents, called the aggregation view, and the other, the specialisation view, for those wishing to retrieve a configuration document.

Let us call the collection of documents for the different configurations of a product a document set. The individual documents in this set will be called final or configuration documents, to distinguish them from the objects from which they are built up, which are called building block documents. These terms will become clearer soon.

Our examples are based on documentation for a car. The car exists with automatic or manual transmission. Manual transmission cars have either four or five speeds, and come with either a sports-style gear stick or a standard one. See figure 3. Thus there are three configuration dimensions - "transmission", "speeds", and "stick", with the following domains of configuration values: {manual, automatic}, {4-speed, 5-speed}, {sports, standard}. Speeds and stick are sub-configurations of the manual transmission configuration.

All of the documents in the car manual document set will be largely the same. In some places they will be transmission specific. Similarly, the documents for manual transmission cars will have additional parts in common with each other, but will have some parts specific to the speeds dimension and other parts specific to the kind of stick.

The author's or creator's perspective of the SPRITE configuration model is of a configuration document as a complex object made up of common components and components dependent on particular configuration dimensions. The author creates all the components (chapters, text, graphics, ..) corresponding to one configuration value (e.g. the manual transmission, without any sub-configuration parts) in one building block document.

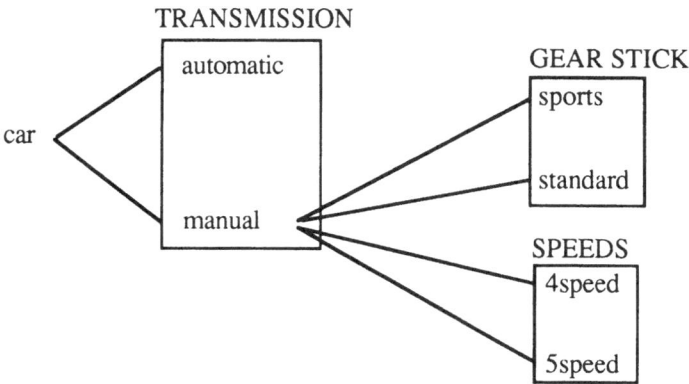

Figure 3: Configuration model example

Each building block document is a separate identifiable object in the system, with its own history (as described earlier). It can be edited and manipulated separately from the other building blocks making up the document set. Separate building block documents can be assigned to separate authors.

The other important point about this model, from the point of view of creation and update of configuration documents, is that a particular component or sequence of components is stored in only one place, even if it appears in many of the final configuration documents. So when the picture of the sports stick needs to be changed in all the final documents, it is replaced in one place - in the sports stick building block document.

As well as containing content such as text and graphics, building blocks include information about how they fit together with other building block documents to create final documents. This is explained further in the next section. As an example the sports stick building block, the 5-speed building block and the manual trans-mission building block are combined with the building block containing content common to all the car manuals to produce the final document for the sports stick, 5-speed, manual transmission car. It is this building up of final documents from building blocks which leads to the name "aggregation view" for this perspective.

The other perspective on the SPRITE configuration model is that of those retrieving configuration documents. In this case we are not concerned with the construction of the document, only with its content. In this perspective a document describing a specific configuration of a product is a specialisation of a document describing the generic product. For example the five speed manual transmission car manual is a specialisation of the manual transmission car manual which is in turn a specialisation of the ordinary car manual. Another way of looking at it is as

614

inheritance - the more specialised document inherits all the components of the more general document, (probably) adding some of its own.

3.3. The Building Block Model

In this section we explain more fully what a building block is and how it relates to other building blocks of the same document set.

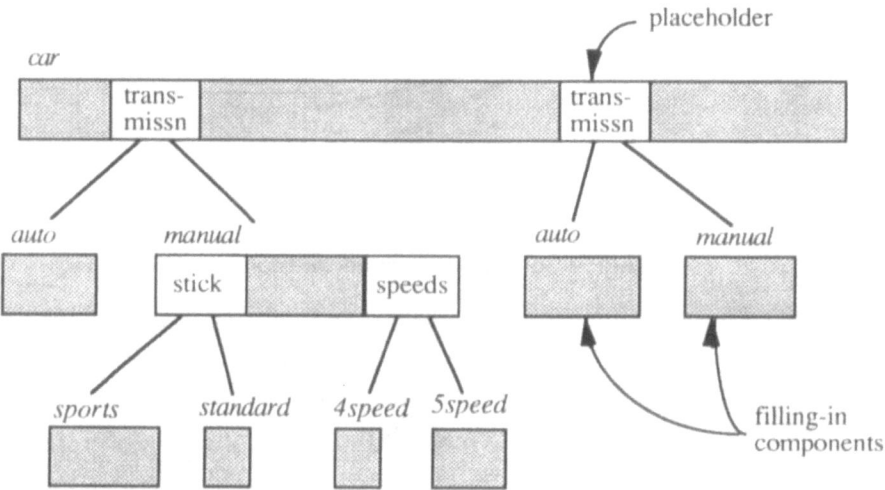

Figure 4: Placeholders and filling-in components for an example car manual

Figure 4 shows the conceptual structure of a sample manual for our car example. The top part of the diagram contains content common to all the car configuration documents. In two places final documents will contain transmission specific content. These are indicated by placeholders. Each placeholder is marked with a domain, in this case "transmission". At the next level we see the components which would replace these placeholders in final documents - one component for each place-holder for each dimension value. These are called filling-in components.

One of the manual transmission components has a speeds-specific part and a stick-specific part, each indicated by a placeholder. At the third level we see the filling-in components for these placeholders.

A building block is the collection of components for a specific configuration value. In Figure 5 we show the building blocks for the car manual document set. There are seven of them. Each one, other than the common one marked "car", is linked to a higher level building block. The links indicate which filling-in components fill which placeholders.

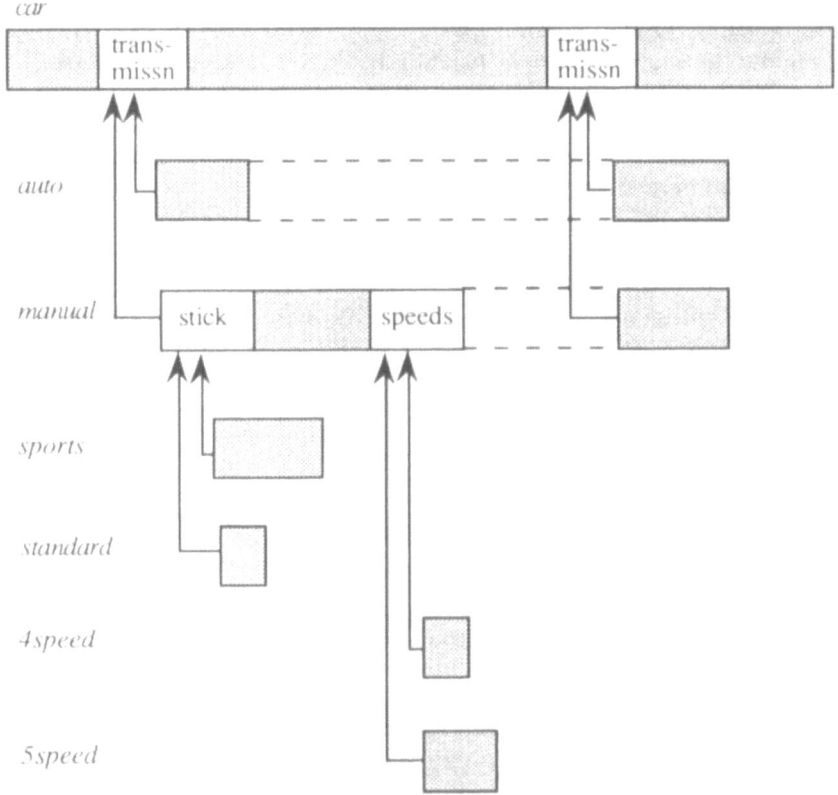

car

trans-
missn

trans-
missn

auto

manual

stick

speeds

sports

standard

4speed

5speed

Figure 5: Seven building block documents for the example of figure 4

Final documents are produced by combining the set of building block documents corresponding to the desired configurations. Each placeholder in a higher level building block document is replaced by a filling-in component from a building block of the corresponding configuration dimension.

3.4. Combining Checkpoints and Configurations

As stated above, building blocks have their own checkpoint history. If we consider all checkpoints of building blocks, the configuration tree, as exemplified in Figure 5, becomes more complicated. For example, we may have three checkpoints of the 4speed building block, and two of the manual transmission building block. Furthermore, it may be the case that only the first checkpoint of the 4speed building block is compatible with the first checkpoint of the manual transmission building block, while the second and third checkpoints are compatible with the second. "Compatible" means here that for each filling-in component there exists one placeholder component in the higher level building block. The compatibility relationship is explicitly recorded and automatically updated by the system. In this way, the user can see immediately how a certain building block can be combined. The system also ensures that for each lower level building block at least one compatible higher level building block exists. If this integrity constraint were not enforced, we could get building blocks that could never be edited or printed, since editing and printing requires a context that specifies, for each filling-in component, its position in the logical structure of the document.

Usually, the user is not interested in all checkpoints and configurations but only in one particular checkpoint of each configuration. It is possible to keep record of the "currently active configurations" in a so called "composed document". This composed document is just a set of identifiers representing compatible building blocks. This composed document can be opened by the user when starting the editor. Composed document objects can be regarded as offering a simple idea of *context*.

3.5. Comparison with Other Approaches

Distinctive features of the SPRITE version model are:

– versioning is defined at the conceptual level;
– a clean distinction is made between historical versions and configurations;
– the version model is supported by an efficient storing mechanism;

As for the first feature, SPRITE differs for example from EXODUS (Carey et al, 1986 and the general version model presented in (Klahold et al, 1986). EXODUS supports basic versioning operations on the internal level, but leaves the conceptual level open. (Klahold et al, 1986) represent versions by means of version graphs and partitions. The user has the possibility to define several version graphs, and to insert versions into these graphs. To make one object a version of another object, the user must define an edge between the two nodes. Several version graphs can coexist, so that for example our version model could be implemented by means of a combination of a historical graph and a configuration graph. Partitions are used to class nodes of a graph together; for example, our notion of editions corresponds to a partition of the historical version graph. Although this mechanism could be used to implement our version model, it does not have the attached semantics.

A difference between configurations and histories is made in the VISION object-oriented database system (Caruso & Sciore, 1988). The difference is here between TimeSeries and versions. TimeSeries are ordered and accessed on date; when an object (function) is declared as a TimeSeries, automatically all historial versions are stored (no explicit freezing is necessary). Versions are derived from objects by means of a "newVersion" operation, and versions can be frozen and reactivated. VISION also implements an *interesting context* mechanism; however, it is not a multimedia database and apparalltly has no data sharing at the internal level.

4. Conclusion

In this paper, we described the SPRITE techical documentation system and in particular its versioning mechanism. SPRITE makes a distinction between historical versions and configurations. A special feature of this versioning mechanism compared with other approaches is that versions are defined at the conceptual level. It allows for efficient data sharing,but this occurs as a result of the user's modelling the document rather than by direct manipulation on data structure level. We compared SPRITE in this respect with a couple of other multi-media and/or object-oriented databases.

One limitation of the SPRITE version mechanism is that configurations must be fitted into a hierarchy. Multiple inheritance is not possible. Taking up the example of section 3, there is no place to put content that is specific to "5 gear, sport stick". This would require a building block that is both a subconfiguration of "5 gear" and of "sport stick". We have prohibited such situations in order to keep the system simple.

One interesting point of future research is to transfer this versioning mechanism from the particular case of the document to the general level of "object". For example, for a "person" we might also keep a history record, as well as different configurations. John may be both a teacher and a researcher; as a teacher, he earns 30K and as a researcher he earns 40K. Although for these cases the implementation issue of data sharing is less relevant (because the data items are not very big), there is of course the question of how to handle these versions cleanly on the conceptual level.

Acknowledgements

This work has been carried out under ESPRIT contract 2001 (SPRITE). The authors are grateful to all participants in this project, Oce-Nederland B.V., Alcatel TITN, AEG Electrokom Gmbh, Tilburg University and Trinity College Dublin. Special acknowledgements are due to Hans Daanen and Olga de Troyer for their significant contributions to the version model. However, nothing here should be taken to commit any partner, and the opinions and any errors are our own.

References

Andrews, T., C. Harris, K. Sinkel, 1989. *The ONTOS Object Database.* Ontologic.

Carey, M. et al, 1986. Object and File Management in the EXODUS Extensible Database system. *Proc. Int. Conf on VLDB,* Kyoto.

Caruso, M., E. Sciore, 1988. Contexts and MetaMessages in Object-Oriented Programming Language Design. *Proc. ACM SIGMOD.*

Christodoulakis, S. et al, 1986. Multimedia Document Presentation, Information Extraction, and Document Formation in MINOS: A Model and a System. *ACM Trans. on Office Information Systems, 4(4).*

Kent, W., 1989. An overview of the versioning problem. Panel Contribution. *Proc. ACM SIGMOD* .

Kim, W. et al, 1989. Composite Objects revisited. *Proc. ACM SIGMOD.*

Klahold, P., G. Schlageter, W. Wilkes, 1986. A general model for version management in databases. *Proc. Int. Conf. on VLDB,* Kyoto

Krönert, G., 1987. Standardized Interchange Formats for Documents. *ESPRIT'87,* North-Holland .

Walter, M.E., 1988. *Technical Documentation Systems.* Seybold Publications.

THE APPLICATION OF MORPHO-SYNTACTIC LANGUAGE PROCESSING TO EFFECTIVE TEXT RETRIEVAL

A.F. Smeaton
School of Computer Apps.
Dublin City University
Glasnevin, Dublin 9
Ireland.

A. Voutilainen
Research Unit for
Computational Linguistics
Hallituskatu 11
University of Helsinki
SF-00100, Finland.

P. Sheridan
School of Computer Apps.
Dublin City University
Glasnevin, Dublin 9
Ireland.

Abstract

This paper describes some of the work done as part of the ESPRIT II SIMPR project. The aim of SIMPR is to research and develop methods for the effective retrieval and manipulation of textual information, using automatic natural language processing techniques. This paper describes the automatic language processing techniques developed during SIMPR and how these have been used to develop a method for retrieving and matching textual information. Input texts are processed automatically at the morpho-syntactic language analysis level and from this processing are generated tree-structured internal representations which are matched to perform retrieval. This paper describes the language analysis and how it is used in matching texts.

1. Introduction.

The SIMPR Project (Structured Information Management: Processing and Retrieval) is a 64 man-year project running from 1989 to 1992 whose broad aim is to apply automatic natural language processing techniques to the effective manipulation and retrieval of textual data. Specifically, the project aims to apply a morpho-syntactic level of language analysis to aid in text indexing, subject analysis, subject classification and interactive retrieval.

What this paper concentrates on is a description of the automatic language processing on which the SIMPR work is being based, and how this language analysis is being used to provide effective retrieval of text. In the following section we introduce the notion of using automatic natural language processing in text retrieval. We then describe the SIMPR language analysis process. This is followed by a description of how we use the language analysis to generate representations of text which are used in retrieval or text matching. We then review the processes of language processing and text matching that we have developed and briefly compare them to related approaches.

2. Natural Language Processing in Information Retrieval.

Research into text retrieval has been ongoing since the 1960s yet really effective information retrieval has still not really been achieved. In the 1960s, automatic natural language processing (NLP) techniques were seen as a desirable computational process to be included within an overall information retrieval process. However, the successful integration of NLP and information retrieval did not prove to be as easy as was initially supposed, and the hoped-for integration of the two processes did not materialise. Consequently, in the intervening decades, researchers turned their attentions to working on methods of statistically-based text retrieval.

The automatic processing of natural language has achieved significant levels of improvement in terms of robustness, accuracy, efficiency and quality of output, since the 1960s. While we cannot claim that we fully understand how to do automatic natural language processing using a computer, we can say that we have made great progress over the decades and we can apply such processes to systems for querying databases, performing text critiquing, etc. An obvious question for current research and development in information retrieval now is "What NLP techniques can we apply to text retrieval and how can we apply them ?".

In the SIMPR project it was decided to use a domain-independent morpho-syntactic level of language analysis to produce an output which is used to assist in the retrieval and manipulation of textual information. It was chosen because it was possible to produce a robust language analysis process at this level of linguistic processing, and the domain-independence aspects mean that the system we produce will not have to be tailored to new domains except perhaps to add new words and their word classes to a lexicon. This level of language analysis will, of course, produce syntactic ambiguities which would normally be handled by making recourse to higher levels of language processing, like the semantic level, in order to perform any disambiguation necessary. In our work on text retrieval we have decided that we will tolerate and handle whatever syntactic ambiguities that may arise. The language analysis processes we are using explicitly encodes ambiguities at this level of language processing, and these ambiguities are then used as an integral part of the text matching processes. Our philosophy is that instead of forcing a complete and unambiguous syntactic analysis to every single utterance, we use syntax only as far as it can be used without the risk of missing a possible analysis. Instead of forcing a top-down functional assignment to every single word, we allow at a very early stage of syntax a maximal set of analyses to every word, and then, on the basis of contextual information, discard as many analyses as can be done with full safety. We are ready to live with remaining ambiguity, if we can be positive that no correct analysis is missing

Our view of language processing as being a constrainer of interpretations rather than a complete resolver of interpretations is shared with some other researchers working in the area. For example, at the University of Pittsburgh (Metzler [Metz90]) and at the Siemens research Laboratories in Munich (Schwartz [Schw90]) similar projects have adopted a related viewpoint. We shall contrast our work with other approaches in a later section.

3. Language Processing in SIMPR.

The SIMPR project is developing a prototype system during 1990 and as newly-created documents are entered into the SIMPR system they go through a document markup process. This phase of the indexing process is used to reveal the structure of the document by marking the boundaries of sentences, paragraphs, sub-sections, sections, chapters, headings, etc. The marked up document is then passed to the morpho-syntactic language analysis phase which implicitly does spelling error detection and which is being implemented by the Research Unit for Computational Linguistics (RUCL) at the University of Helsinki.

The morphological description of language is based on Koskenniemi's Two-Level Model [KOSK83], and the basic ideas of the syntactic description are taken from Karlsson's Constraint Grammar (CG) framework [KARL89, 90a]. The whole formalism is language-independent. The most advanced application in SIMPR is the present one on English, but work on Finnish will also be started in the SIMPR project in the near future.

CG parsing is modular in nature. It consists of interactions between the Lexicon, Disambiguation (DA), Clause Boundary Determination (CBD) and Assignment of Syntactic Functions (ASSF). The whole linguistic analysis can be represented as the following sequence of operations:

– Morphological Analysis (MA).
– Elimination of contextually impossible readings (DA/CBD).
– Determination of sentence-internal clause-boundaries (CBD).
– Assignment of syntactic functions (ASSF).

A note on the treatment of ill-formed text is due. Prior to MA proper, the Lexicon update and Text correction modules are run on the text. Strings surrounded by blanks are checked against the Lexicon, and the unrecognised ones are either converted into new entries (Lexicon update), and the remainder indicate that there are spelling-errors in the text itself. The spelling-errors are then corrected manually (Text correction). Note that text correction is operated only on word-forms, not on text that might need correction on the syntactic or semantic level. Grammatical well-formedness is overall a problematic issue - it is often difficult to make a distinction between well- and ill-formed text on levels other than that of the word.

On the whole, the aim of syntactic analysis here is not to mark sentences as well-formed or ill-formed; the primary aim is to say something true and useful about all words of the text.

3.1 Morphological Analysis: The Lexicon

The Master Lexicon for English is modelled on Koskenniemi's Two-Level Model [Kosk83]. The English lexicon was compiled by Voutilainen [VOUT89a, 89b], and, to a lesser extent, by Heikkil. The basic advantages of the formalism for the present purposes are economy of representation with most of the lexical entries in the lexicon representing more than one actual word, and the possibility of attaching grammatical information to any entry. The Master Lexicon consists of many so-called mini-lexicons. One of the mini-lexicons contains the lexical entries (beginnings of words with pointers to certain other mini-lexicons containing the endings legitimate for the given lexical entry) and information about those endings. The categorization of the lexical entries is based on syntactic properties of words. The basic distinctive factor is part-of-speech: Noun, Verb, Adjective, Adverb, and a few others.

Lexemes with a membership in more than one part-of-speech are coded with several entries. For example, the word *bottle* as a noun and a verb would have two entries:

> bottle N;
> bottl Ve-0;

The minilexicon "N" contains endings for nouns of this type: nominative singular (), nominative plural (-s), genitive singular (-'s), genitive plural (-s') and pointers to some other minilexicons containing derivational suffixes (*-like, -ness, -less*, and some others). The minilexicon also contains the labels for the endings and the same arrangement holds for verb entries. In the minilexicons are the verb endings with appropriate information *-ing* (participle), *-ed* (past tense and past participle), *-able, -ably, -ability* (derivational suffixes resulting in change of part-of-speech), and some other endings.

Information specific to a particular entry can be attached to the lexical entry itself. If, for example, we want to code in the verb entry for *bottle* that it is a verb that takes both a subject and an object as arguments, the entry would look like this:

> bottl Ve-0 " = <SVO>";

In the output of MA, information attached to the lexical entry itself is represented left of the information on the endings. A morphological analysis for *bottles* is accordingly:

> bottles
>> bottle " N NOM PL "
>> bottle " <SVO> V PRES -SG3 @ +FMAINV "

The analysis line consists of two parts: the base-form of the word and the morphological analysis. The first analysis line says in effect that *bottles* is a nominative plural form of the noun "bottle"; the second analysis line says that *bottles* is a present tense third person singular form of the transitive verb "bottle". The existence of more than one analysis line for this word-form indicates that the word-form is regarded as morphologically two-ways ambiguous in the present formalism.

MA is meant to operate on words. There is a limitation, however: MA works only on strings containing no embedded blanks. However, there are word-like units that contain embedded blanks, such as idioms, complex prepositions like *in spite of*, certain adverbials, and some other categories, and nominal compounds, like *tea time*. A part of the present approach to this problem is to detect such compound expressions prior to MA, and convert them into one string corresponding to a similar entry in the lexicon. This reduces the work load of the syntax modules considerably: a number of ambiguities are resolved at this stage. During the preprocessing stage the text is scanned for instances of compound expressions listed in the preprocessing module. Thus the string *in spite of* is converted into "in = spite = of", and during the morphological analysis, the string is analysed as a single word (as a preposition in this case).

The list of compound expressions contains those of the some 6,000 compounds in the Collins Cobuild Dictionary [COLL87] that can be trusted to be compounds in all contexts, and not strings accidentally adjacent to each other, as parts of different structures. There are also some 200 phrasal prepositions etc. like *in spite of*. It may be noticed that most of the analysis of compounds is left to syntax.

The morphosyntactic description used in the English Master Lexicon is based almost entirely on [QUIR85]. The MA module assigns a part-of-speech label to every recognised word-form. In our description, they are: Adjective, ABBReviation, ADVerb, CC and CS (co- and sub-ordinating conjunction), DETerminer ("the", "any", ...), @INFMARK (infinitive marker "to"), INTERJection, Noun, NEG-PART (not), NUMeral, PCP1 and PCP2 (present and past participle - "walking", "walked"), PREPosition, PRONoun, and Verb. In the analysis of input, the part-of-speech label is immediately to the right of the base form. There is only one exception to this: the so-called "inherent features" (that are within angle brackets " < > ") are always given between the base form and the part-of-speech label. A base-form plus a part-of-speech label alone do not constitute the whole analysis line. Each part-of-speech label contains further distinctions, which are coded to the right of the part-of-speech label. Further details of feature patterns can be found in [HEIK90].

3.2 Disambiguation

MA output contains analyses, some 50% of which are contextually impossible. The task of the DA module is to eliminate as many inappropriate analyses as possible. After DA, only the appropriate analyses and the least detectable of the inappropriate ones will be left. The constraints are expressed in the CG formalism developed in [KARL89, 90a]. The main features of the formalism are summarized here.

The disambiguation and syntax modules process the morphologically analysed text in sentence-sized chunks. In outline, the DA constraints obey the following format:

("object_of_constraint" operation_type "ipret" (context_1) (context_2) ... (context_n))

Object_of_constraint can be a word or a feature.
Operation_type is one of the following:

" = !!" X is the only correct interpretation for object_of_constraint Y in context Z, and for Y, X is impossible in any other context.

" = !" X is the only correct interpretation for object_of_constraint Y in context Z

" = 0" interpret. X is impossible for object_of_constraint Y in context Z.

As can be seen, the first type of operation is the most powerful one, and the last is the weakest one; it merely discards the "ipret" as impossible.

Ipret can be a real feature, like "NOM", or it can be a set of features, like "A/PCP2", which is declared elsewhere as a list consisting of the name of the set and the features and/or words in the set; in this case, (A/PCP2 "A" "PCP2"). If ipret is a set, then a rule of type = 0 can eliminate more than one interpretation line.

Context_1 to n consists of one or more context-conditions which are predicates examining positional features of words in sentence chunks.

It is also possible to use information on relative position in context. One can e.g. express a constraint saying "do operation X if there is somewhere in the preceding part of the sentence feature X, and there is not feature Y between the most recent instance of X and position 0. We will not go into the formal details here, suffice it to give a concrete example. We might want to say that a word ambiguous between < Rel > and some other reading is < Rel >, if the word in position -1 has the feature CC (co-ordinating conjunction), and there is to the left an unambiguous < Rel >, with no intervening clause boundaries. The rule for this would be:

(@w = ! "< Rel >" (-1 CC) (*-2 REL *R) (NOT *R CLB))

"@w" denotes any word; i.e. the rule is feature-oriented. The set REL contains the feature "< Rel >"; the set CLB contains words and features that mark clause-boundaries (like features "PRON WH", "CS" "< Rel >"). All of the context-conditions have to be satisfied before any interpretations are discarded.

In the present formalism, clause boundary detection is not a goal for its own sake; clause boundaries are not syntactic categories. CBD is presupposed purely because of its usefulness for the other components of the syntax. Input with clearly marked clause boundaries is simply easier to analyse than input without these boundary markers. CBD is partly done by DA proper. Words like subordinating conjunctions, WH-words, and some others have typically as one of their functions the marking of clause boundaries. Therefore it is more useful to disambiguate words which have as one of their functions the markup of clause boundaries. CBD is not complete with DA alone. There are certain types of clause boundary where almost any word can mark the clause boundary. Relative clauses with no relative pronoun, and the endings of subordinate clauses preceding their mother clause (like "If you go the rest will too.") are fairly frequent examples.

3.3 Assignment of Syntactic Functions

ASSF aims at the assignment of unambiguous labels of a brand of functional dependency syntax [KARL90b]. The main idea is to indicate whether a word has a head or a modifier function in a clause. For all words with a modifier function, the head is also indicated. The head of a modifier is shown by indicating the direction of the head ("<" for left; ">" for right), and, depending of the type of modifier, by numerically indexing, or by telling e.g. that the closest member of category X is the head of modifier Y. Our syntax is surface-near, i.e. no structures are postulated that are more abstract than being in direct linear correspondence with the real word-forms occurring in the sentences to be described.

Next, the syntactic functions are presented [KARL90b]. All syntax labels are prefixed by "@" for recognition purposes. The present collection of codes does not exhaust all intricacies of English syntax. However, the code system should be delicate enough to capture those aspects of clause structure that are relevant for information retrieval purposes.

Assignment of syntactic functions takes as input, sentence-sized chunks of text that are morphologically disambiguated as fully as possible. A third of the words get their unambiguous syntactic label directly from the lexicon. That means that if a word of this kind is unambiguous after DA, it is also syntactically unambiguous.

What syntax has to work on is interpretation lines with no syntactic functions. ASSF is a two-part process. The first part consists of a so-called mapping from morphology to syntax. From any given morphological feature, there is a mapping to a limited number of syntactic functions. Different morphological features have different mappings to syntactic functions.

Mappings from morphology to syntax are expressed with mapping functions given as lists consisting of 1) a string of the input - usually a morphological feature or a base-form - 2) context-condition (optional), and 3) the syntactic function(s). The mapping functions are listed in order of increasing specificity. First are given mappings to a single syntactic function only, and last are given mappings to all allowable syntactic functions a word or feature can have a mapping to. An example of a specific mapping could be a mapping from N to @SUBJ in a clause-initial position:

("N" ((-2C CLB) (-1C DET/GEN) (1C VFIN)) "@SUBJ")

A non-restrictive mapping has no context-conditions. The second part of ASSF consists of the syntactic rules proper. The rules are expressed basically within the same formalism as is used in DA. These rules work on those interpretation lines containing more than one syntactic function. As an example is given a rule of type "=0" stating that a word can have the function "@I-OBJ" only if there is somewhere to the left a word with the label "<SVOO>" (marking a verb governing two objects and a subject):

(@w =s0 "@I-OBJ" (NOT *-1 I-OBJ))

(The set I-OBJ contains the label "<SVOO>".)

Let us have a sample of the output from syntax.

```
remove
    remove " <SVO> <SV> V IMP VFIN @+FMAINV "

the
    the " <Def> DET CENTRAL ART SG/PL @DN> "

fuel
    fuel " <-Indef> N NOM SG " @NN>

pump
    pump " N NOM SG " @NN>

sediment
    sediment " N NOM SG " @NN>

bowl
    bowl " N NOM SG " @OBJ

and
    and " CC @CC "

filter
    filter " N NOM SG " "OBJ
    filter " <SVO> <SV> V IMP VFIN @+FMAINV "

from
    from " PREP " @<NOM @ADVL

the
    the " <Def> DET CENTRAL ART SG/PL @DN> "

top
    top " N NOM SG " @<P

of
    of " PREP " @NOM-OF

the
    the " <Def> DET CENTRAL ART SG/PL @DN> "

pump
    pump " N NOM SG " @NN>

unit
    unit " N NOM SG " @<P
```

Syntax labels to the left of a double-quote come directly from the lexicon as unambiguous syntactic readings for the interpretation in question; the rest of the syntactic labels are assigned by the mapping declarations, and in the case of post-mapping syntactic ambiguity, "selected" by the rest of the syntax rules as the correct ones.

4. Using Natural Language Processing for Text Matching.

Once a document text has been entered into the SIMPR system, the morpho-syntactically analysed text is then passed to a module which identifies analytics or potentially rich word sequences which may indicate content. These analytics may be one word, or a sequence of words not necessarily in the same sequence as in the original text. Analytics could in fact be as large as sentence clauses. This component of the system is being researched by Strathclyde University and the research component of this is outlined in [GIBB89]. In the prototype being developed by SIMPR this year, this process will involve the author or editor of the original document in the role of validating the analytic representation, but we hope that eventually this will be an entirely automatic process. The process is being implemented by defining a set of rules for analytic identification and these rules will use the output of statistical analysis routines (frequencies and co-occurrences of words etc.) as well as morpho-syntactic information from the analysis described in the previous section.

The original marked up text, and the generated analytics mentioned above, are then passed to a subject classification expert system (SCES) which classifies input documents in terms of a classification hierarchy that it maintains. This classification hierarchy is dynamic and changing as new texts are processed by the system. Having evaluated the suitability of several classification approaches, Strathclyde University has decided to use a faceted classification technique and this work is described in [SHAR89].

When a user wants to retrieve documents from the database (s)he has a choice of retrieval strategies available. A user may retrieve by browsing through the heading hierarchy, by specifying a single keyword to start retrieval, by browsing through the subject classification hierarchy or by specifying a natural language statement of information need. Fundamental to this latter type of search, and also used in automatic document classification, is the facility to be able to match two pieces of natural language text; a query against document texts in interactive retrieval, or document texts against document texts in automatic classification. The analytics which provide input to the automatic subject classification system are also used to generate tree-structured analytics (TSAs), which yield a rich representation of the input text or query. These TSAs, their design and matching, form the research task being done by Dublin City University within the SIMPR project, and are described in more detail now.

We have defined an intermediate representation of text and a matching algorithm for that representation which is simple enough to be derived from the output of the

cannot tell us. Our approach here is to accept both interpretations as we have stated earlier, but to assign degrees of preference during matching, to some interpretations over others.

The matching algorithm we have devised for matching TSAs takes the form of a basic tree or search algorithm but with certain enhancements due to the requirements of our application. Two of these enhancements are:

1. The need for inexact or relaxed matching of nodes and of structures. For example we must be able to match the noun *"removal"* to the verb *"remove"*. This is required for matching TSAs in Figs. 1 and 2. The matching algorithm will have available all information regarding the language analysis of the word as it appeared in the text, its base form, and morphological and syntactic categories, in performing this match. Inexact matching is also required in the inexact matching of structures; i.e. matching the TSA for *"engine removal"* to the TSA for *"removal of the engine"*, *"remove the engine"* and *"removing the engine"* (or indeed *"carefully remove the big broken engine"*).

2. The manipulation of weights in the weighted matching of nodes and of structures, i.e. a measurement of the degree of "relaxation" as described above. In node matching this would entail scoring the noun *"removal"* matching the noun *"removal"* higher than scoring the noun "analysis" matching verb "analyse". In the weighted matching of inexact structures, i.e. a weight reflecting the "exactness" of a match, we would want to reflect the difference in exactness between the match of *"engine removal"* to *"engine removal"* and *"carefully remove the broken engine"*. The manipulation of weighted links is necessary to take into account the uncertainty involved with ambiguity from the language analysis. This involves suitable modification (decrease) of the match score of an analytic that has several interpretations - the final scoring must reflect the fact that the match may have occurred with some interpretation of that analytic that was not intended by the author.

In order to illustrate our match algorithm, we shall sketch it using the examples given earlier where Figs. 1 and 2 are regarded as TSAs for queries and Fig. 3 is a TSA from a piece of text.

1. Matching *"Sediment bowl removal"* with *"Remove the fuel pump sediment bowl and filter from the top of the pump unit"*

We always start our matching of TSAs by trying to find a match for the head of the query or the head of the smallest TSA in the case of a text being matched against a text. In the case above, we choose to find a match for *removal* (@NPHR) and we find that it matches *remove* (@mainpred) on the grounds of base forms of words, and a rule which scores a match between the @NPHR and @mainpred labels. We next try to match the subtree for *sediment bowl* with *bowl* being the head of this sub-tree, and we find an exact match in the TSA in Figure 3. In this case *bowl* in Fig. 3 is the leftmost node under the conjunction so there is no ambiguity associated with its relationship to *sediment*. The final score for the match between the TSAs in Fig. 1 and Fig. 3 is made up of the match score for *removal* and *remove*, the sub-tree match score for *sediment bowl* which reflect that it was a sub-tree match, and a weight to reflect the fact that the structure in Fig. 1 was a sub-structure of that in Fig. 3, and not an exact match.

2. Matching *"The fuel pump filter"* with *"Remove the fuel pump sediment bowl and filter from the top of the pump unit"*

We start with the head of the query, *filter* and we find it in Fig. 3 so we match the labels @SUBJ and @OBJ to some degree. Now we look for the query modifiers *fuel* and *pump* in Fig. 3. Both are found in Fig. 3 exactly as they occur in Fig. 2, however some uncertainty must be introduced here to reflect the fact that *filter* is on a right link under a conjunction node, so there is some ambiguity regarding the distribution of the modifiers over the conjunction. The final score for the match between the TSAs in Fig. 2 and Fig. 3 is made up of the score for exact word matches but with differing syntactic labels (@SUBJ and @OBJ), the uncertainty with the attachment of the modifiers in the TSA to the head *filter*, and the score for the structure match which reflects that the query is a sub-structure of the TSA in Fig. 3.

With regard to the representation and manipulation of weights and uncertainties, we have surveyed many different theories of uncertainty including certainties, probabilities, Dempster-Shafer theory and, in particular, the calculus used in the RUBRIC system. RUBRIC is a commercially available document management system, a commercial derivative of the research work done by Richard Tong while at Carnegie Mellon University and it is outlined in [TONG89]. In this system texts are retrieved by searching for textual evidence in the form of word patterns to indicate that a concept is being discussed in the text. This evidence is gathered using a numerical uncertainty representation and is propagated upwards when concepts are combined, using an appropriate and specially devised calculus which is what we are interested in. The RUBRIC system has also demonstrated that it is possible for a retrieval mechanism to perform complex weight manipulation like we do in SIMPR, quite efficiently. At present we are evaluating whether or not to use the RUBRIC calculus as the framework for implementing our score calculations.

5. A Review of Our Approach.

5.1 Our Approach to Language Processing.

An inherent feature of the formalism presented in section 3 is the use of lexical information at all stages of analysis. Much of the success of the whole linguistic analysis is directly dependent on how carefully the lexicon has been compiled. While it is customary in some grammar formalisms largely to neglect lexical description, every effort has been done here to ensure that all relevant information is available directly after morphological analysis. A detailed lexical description is a good basis for disambiguation. It is again fairly common for some syntax formalisms to presuppose an input with some kind of lexical description with little or no ambiguity, thus the task of disambiguation is not recognised as an integral part of syntactic description. The approach to disambiguation adopted in SIMPR is different. Disambiguation is seen as a real problem that should accordingly have a non-peripheral role in syntactic description. A rich set of categories presented at an early stage of the description makes it possible to finish a large part of syntax during disambiguation. Although there are a number of difficult problems in the syntax of English, it is still true to say that a successful disambiguation makes a large part of the assignment of functions a fairly easy task. In other words, we believe that syntax is a more manageable enterprise if the whole process can be divided into distinct subcomponents, and the subcomponents are processed one at a time. However, distinctness does not mean here full isolation. The components are interdependent in the sense that e.g. clause boundary determination would be easier if a full disambiguation were done, ASSF would benefit from disambiguation, and so on. But it is worth noticing that the subcomponents can be finished to a very high degree without constant reference to other modules.

Ambiguity in natural language is a problem that cannot always be fully resolved. If we accept this then the problem thus becomes a problem of the representation of ambiguity. There are a number of ways to react to this predicament. One is to assign no label to ambiguous words. Another would be to make a statistical guess. Still another would be to devise new labels specific to types of ambiguity. An advantage of constraint-based syntax operating on a detailed morphological analysis is that although all ambiguity cannot in every case be eliminated, remaining ambiguity need not mean a full surrender. The nature of rules of type "=0" is the idea that although we cannot tell in positive terms which of the competing interpretations is the correct one, we can at least tell that a given interpretation is impossible in a particular context. A partial resolution of ambiguity is better than no resolution, and this function is a central property of the CG formalism. A word-form with more than one interpretation may seem awkward, but if any interpretation has been discarded, the analyses of the still ambiguous word-forms approximate more the correct interpretation(s) of the word than would be the situation where no interpretation were given, i.e., where any would remain a candidate analysis. The other advantage is that the use of "=0" rules contains fewer risks of error than the use of the more powerful, assignment-type "=!" rules, hence the present formalism contains better tools for the maintenance of a low error percentage than a formalism employing only assignment rules and thus sometimes having to resort to the "guess or skip" approach.

Still another advantage of the CG formalism is that there is no need for labels for (predefined) types of ambiguity. The presence of ambiguity is signalled by the language analysis described earlier, flexible enough to allow encoding of the different types of language ambiguity that we are interested in, yet rich enough to provide a more effective retrieval mechanism than simple phrase matching. This has been achieved by defining a structured intermediate representation. In essence, our tree structured analytics (TSAs) have the following features:

– TSAs are basically binary trees with all word information stored at the leaf nodes. This includes the original word from the text and all linguistic information supplied by the language analysis (base form, lexical category, morphological and syntactic labels).

– Conjunctions identified in the language analysis are stored at the non-leaf node covering all of the conjuncts.

– Nominal compounds with modifier/head sub-trees are specially marked in the TSA (with asterisks) at non-leaf parent nodes.

– Prepositions are stored on the branches of the tree from the parent node to the sub-tree representing the prepositional complement.

Some sample TSAs includinging those examples used earlier to describe the language analysis are given in Figs. 1, 2 and 3, where we show only the words as they occur in text, and the syntactic labels at the leaf nodes, although other information is available.

Fig. 1 ("*Sediment bowl removal*") illustrates how a modifier/sub-tree relationship is represented. From the labels assigned during language analysis we know that sediment modifies bowl as sediment is the pre-modifier and bowl is the head of the clause sediment bowl. In "*The fuel pump filter*" example, the language analysis tells us that fuel modifies pump and pump modifies filter, but this level of linguistic processing cannot discern whether fuel also modifies filter or not, hence there is no asterisk at the parent node of fuel and pump. Our TSA for this in Fig. 2 encodes this ambiguity. In Fig. 3 we see an example of a conjunction stored at the non-leaf node covering bowl and filter and we also see an example of a preposition (*of*) on the branch of the sub-tree representing the prepositional compliment (pump unit).

In an earlier part of our work we carried out a comprehensive study of the different types of ambiguity that can occur in natural language, and their likely frequencies of occurrence [SHER90]. The TSAs sketched above encode only the ambiguities that we are interested in and that would possibly be of use during TSA matching for retrieval, and these are principally prepositional attachment and conjunction. However, one subtle aspect of ambiguity that arises particularly in nominal compounds and that we must address in our work is the preference of certain interpretations over others, in certain cases of ambiguity. In Fig. 3 it is certain that the input sentence is about *sediment bowl removal* but is it also about *sediment filter removal* or just *non-sediment filter removal* ? Because *bowl* and *filter* have a common ancestor node which is a conjunction we do not know and our level of language analysis

632

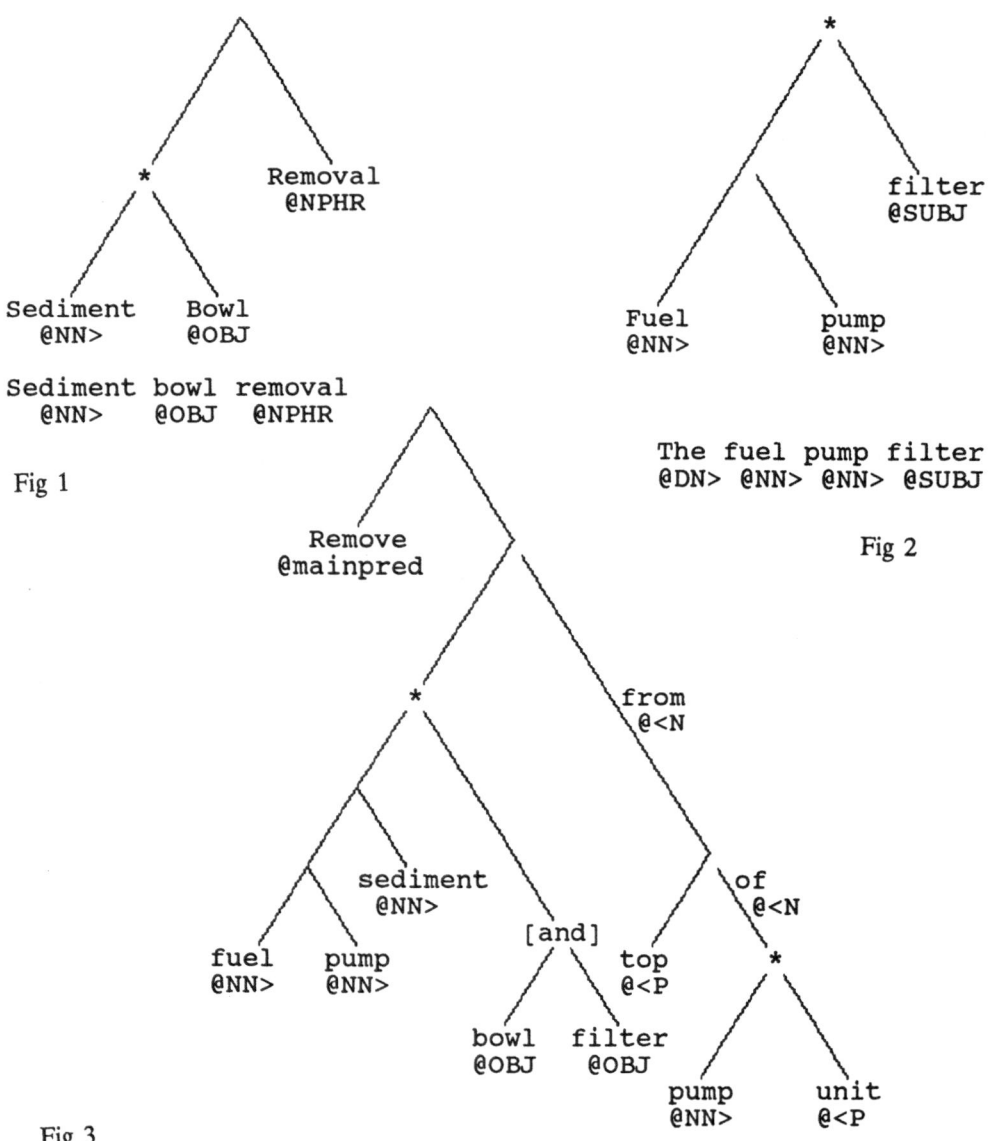

*   Removal
                     @NPHR

Sediment   Bowl
  @NN>   @OBJ

Sediment bowl removal
  @NN>   @OBJ   @NPHR

Fig 1

*

                    filter
                    @SUBJ

Fuel   pump
@NN>   @NN>

The fuel pump filter
@DN> @NN> @NN> @SUBJ

Fig 2

Remove
@mainpred

*   from
                     @<N

          sediment
          @NN>
                        of
              [and]   @<N
fuel pump
@NN> @NN>   top   *
                  @<P
             bowl filter
             @OBJ @OBJ
                        pump   unit
                        @NN>   @<P

Fig 3

Remove the fuel pump sediment bowl and filter
@mainpred @dn> @nn> @nn>   @nn>   @obj @conj @obj

from the top of the pump unit
@<n @dn> @<p @<n @dn> @nn> @<p

existence of more than one analysis line per word-form (morphological ambiguity) or the presence of more than one syntactic label per line (syntactic ambiguity). The kind of the ambiguity can in turn be determined by observing the labels themselves. As a result, kinds of ambiguity emerge; there is no need for devising a more or less complete set of ambiguity-specific labels.

5.2 Our Approach to Text Matching

Recent years have seen other researchers try to incorporate syntactic analysis into document retrieval processes. Martin Dillon developed the FASIT system [DILL83] which did grammatical tagging and searched for grammatical tag patterns as content indicators for indexing. As part of an earlier ESPRIT project (MINSTREL) the first author of this paper has done a syntactic analysis of user queries from which typed word-word dependencies were derived and piggy-backed on top of statistical retrieval [SMEA88]. Fagan at Cornell performed a similar analysis on document texts as well as user queries and used this information to index by phrases [FAGA87]. Again, Fagan used the syntactic structure and patterns of the texts as indexing criterion. The Siemens Research group in Munich has also used shallow syntactic analysis to construct dependency trees from noun phrases [SCHW90]. These trees are used in matching by fitting both keywords and dependencies together. Another significant application of syntactic analysis of text to information retrieval is being undertaken by Metzler at the University of Pittsburgh [METZ89, METZ90]. This approach is based on stressing the hierarchical structure of syntax in matching texts and operates on a hierarchical structure of dependency relationships in the Constituent Object Parser (COP). COP produces binary trees with the branch containing the head of the construct always marked as being dominant. This representation allows the matching algorithm to note ambiguities and match texts which have the same meaning but use a variety of linguistic constructs. A brief review and comparison of these approaches can be found in [SHER89].

Of the previous approaches which use syntactic analysis in information retrieval processes, it seems that those which decompose language analysis information into phrases and store the phrases, like Dillon, Smeaton and Fagan, are losing a lot of information as well as creating seemingly unnecessary problems. The decomposition process must be heuristic and complex and there must also be a complex set of normalisation rules to allow matching between syntactic variants. On the other hand, storing intermediate representations of text as complex structures and performing some kind of graph matching as the retrieval operation does seem more inherently attractive as ambiguities in the language analysis can be encoded within the structure and these ambiguities taken into consideration during retrieval. Thus the advantage of our approach to indexing and to retrieval is that we do not have to explicitly cater for syntactic variants of language which is one of the major problems in text retrieval, as our matching algorithm inherently does this for us. The computational expense of any graph matching algorithm can be quite large but we have shown in SIMPR that graph matching problems can be simplified and made efficient when dealing with a specific application [SMEA89].

6. Conclusion.

At present, the English Master Lexicon contains some 40,000 lexical entries, and its coverage of standard English running text is some 98%. The syntax module is run by a LISP interpreter [KARL90a] and a constraint file which contains some 500 disambiguation rules [VOUT91] and some syntax which is still immature. From the disambiguation module, some 95 to 98 per cent of all running-text word-forms emerge unambiguous (some 40 to 50 per cent of the input words are at least two-ways ambiguous), and the error rate is 0.3 to 0.9 per cent. The early version of the ASSF module contains some 100 rules. Their coverage is reasonably high; of all output word-forms, some 80% are syntactically unambiguous. All of the modules are still under development; a mature stage is expected to be achieved by the end of 1990.

Part of the power of SIMPR's retrieval environment is that it allows the user to specify a request that is matched to the database of texts at several different levels - word/word, phrase/phrase and structure/structure matches in addition to searching via the subject classification system. In this paper we have described a method for matching fragments of text with other fragments of text. The generation of TSAs from language analysis has also been implemented in LISP. The matching algorithm is currently under refinement. The next stage of our work is to scale up and move from matching text fragments to matching chunks of texts and sets of TSAs and addressing the normalisation of TSA match scores.

References.

[COLL87] Collins Cobuild English Language Dictionary 1987. Collins: London and Glasgow.

[DILL83] "FASIT: Fully Automatic Syntactically-Based Indexing of Text", M. Dillon and A.S. Gray, *Journal of the ASIS*, 34(2), 99-108, 1983.

[FAGA87] "Experiments in Automatic Phrase Indexing for Document Retrieval: A Comparison of Syntactic and Non-syntactic Methods", J. Fagan, *Department of Computer science, Cornell University Technical Report*, 87-868, September 1987.

[GIBB89] "Knowledge-Based Extraction of Analytics from Text", F. Gibb et al., *SIMPR Document No. SIMPR-SU-1989-4.1e*, December 1989.

[HEIK90] "Morphological Features for English", J. Heikkil, *SIMPR Document SIMPR-RUCL-1990-13.2e*, 1990.

[KARL89] "Parsing and Constraint Grammar", F. Karlsson, Unpublished paper. Research Unit for Computational Linguistics, University of Helsinki, 1990.

[KARL90a] "Constraint Grammar as a Framework for Parsing Running Text", F. Karlsson, in: *Proceedings from the XIII Conference on Computational Linguistics*, M. Karlgren (Ed), Helsinki, Vol 3, pp168-173, 1990.

[KARL90b] "The Constraint Grammar Parser CGP", F. Karlsson, Unpublished paper. Research Unit for Computational Linguistics, University of Helsinki 1990.

[KOSK83] "Two-Level Morphology: A General Computational Model for Word-Form Recognition and Production", K. Koskenniemi, *Publications of the Dept. of General Linguistics, 11. University of Helsinki,* 1983.

[METZ89] "The Constituent Object Parser: Syntactic Structure Matching for Information Retrieval", D. Metzler and S.W. Haas, *Proceedings of the ACM SIGIR Conference on Research and Development in Information Retrieval,* Boston, 1989, also in *ACM TOIS,* 7(3), 292-316, July 1989.

[METZ90] "Conjunction, Ellipses and Other Discontinuous Constituents in the Constituent Object Parser", D. Metzler et al., *Information Processing and Management,* 26(1), 53-72, 1990.

[QUIR85] "A Comprehensive Grammar of the English Language", Quirk, Greenbaum, Leech, Svartvik, *Longman: London and New York,* 1983.

[SCHW90] "Context Based Text Handling", C. Schwartz et al., *Information Processing and Management,* 26(2), 219-226, 1990.

[SHAR89] "Subject Classification Research", C. Sharif et al., *SIMPR Document No. SIMPR-SU-1989-6.1e,* December 1989.

[SHER89] "Syntactic Processing for Text Analysis: A Survey", P. Sheridan, *SIMPR Document No. SIMPR-DCU-1989-8.2e,* November, 1989.

[SHER90] "Structured Analytics: A Method for Handling Syntactic Ambiguity", P. Sheridan and A.F. Smeaton, *SIMPR Document No. SIMPR-DCU-1990-16.1e,* March 1990.

[SMEA88] "Experiments on Incorporating Syntactic Processing of User Queries into a Document Retrieval Strategy", A.F. Smeaton and C.J. van Rijsbergen, *Proceedings of ACM SIGIR Conference on Research and Development in Information Retrieval,* 31-51, Grenoble, 1988.

[SMEA89] "Searching Chemical Structures: Implications for Searching Parse Trees in SIMPR", A.F. Smeaton, *SIMPR Document No. SIMPR-DCU-1989-8.2e,* May 1989.

[TONG89] "A Knowledge Representation for Conceptual Information Retrieval", R.M. Tong et al., *International Journal of Intelligent Systems,* 4(3), 259-283, 1989

[VOUT89a] "Compilation of a Computerised Master Lexicon for English" A. Voutilainen

[VOUT89b] "Inflectional Categories in the RUCL Master Lexicon (Version 1.1) for English", A. Voutilainen

[VOUT91] "Constraint Based Disambiguation of Lexical Ambiguity, with Special Reference to English", A. Voutilainen, a tentative title for a forthcoming Ph.D thesis.

THE MAX PROJECT APPROACH TO MULTISERVICE BUSINESS COMMUNICATIONS

Angelo LUVISON
CSELT - Centro Studi e Laboratori Telecomunicazioni S.p.A.
via G. Reiss Romoli 274
10148 TORINO (ITALY)

Abstract.

Project 2100 "Metropolitan Area Communication System (MAX)" aims at an effective solution to meet the communication requirements of heterogeneous business user over large (e.g. metropolitan) areas with a high concentration of information-based organizations. Major objectives of the project regard the definition, design, prototype development and testing of an integrated communication system for the flexible provision of data, video services and applications. The paper highlights these issues, outlines the basic technical choices and emphasizes the project rationale and scope, i.e. market opportunities for viable products and services, policy towards standards and compatibility with related developments, in particular the broadband integrated services digital network (B-ISDN).

1. Introduction

The today's telecommunication environment is mainly characterized by the existence of different networks, each dedicated to different applications and/or services. Even ISDN, which offers an integrated access to the public network, foresees two transit networks based on two different switching techniques: circuit switching and packet switching. As a result, each network requires its own products, operation and maintenance.

In this scenario the advantage of a future fully integrated network, the Broadband ISDN (B-ISDN), is unquestionable. It is well known, in fact, that CCITT Study Group XVIII has recommended the Asynchronous Transfer Mode (ATM) as "the target transfer mode solution for implementing a B-ISDN" (CCITT Rec. I.121) for its characteristics of flexibility and service independence at the bearer layers. Since the ATM technology is still under development, the deployment of ATM-based equipment to offer voice, data and full-motion video services is only foreseen as a long-term goal. Therefore, to achieve a cost-effective introduction of the ATM technique, a pragmatic approach based on two phases appears more promising, i.e. a soft products/services evolution by means of new technologies (initial stage) followed by the generalized offering of products/services exploiting well consolidated technologies (second stage).

Following the guidelines stated above the reference scenario for a first-entry strategy is depicted in Figure 1 [1]. Key feature of this scenario is an ATM-based network that should not only be considered as a pure "core" network, but also as a structure capable to provide customer-to-customer connections. It is important to stress that the future evolution of switching should ensure an extremely high degree of flexibility about the provision of new services, even not yet defined, to cope with the rapidly evolving needs of the telecommunication market. For this reason the network strategy shown in Figure 1 is focused on two co-operating guidelines:

1. introduction of nodes with ATM functional capabilities; this can be considered as a centralized approach;
2. introduction of Metropolitan Area Networks (MANs) for gathering business traffic; this can be considered as a distributed approach.

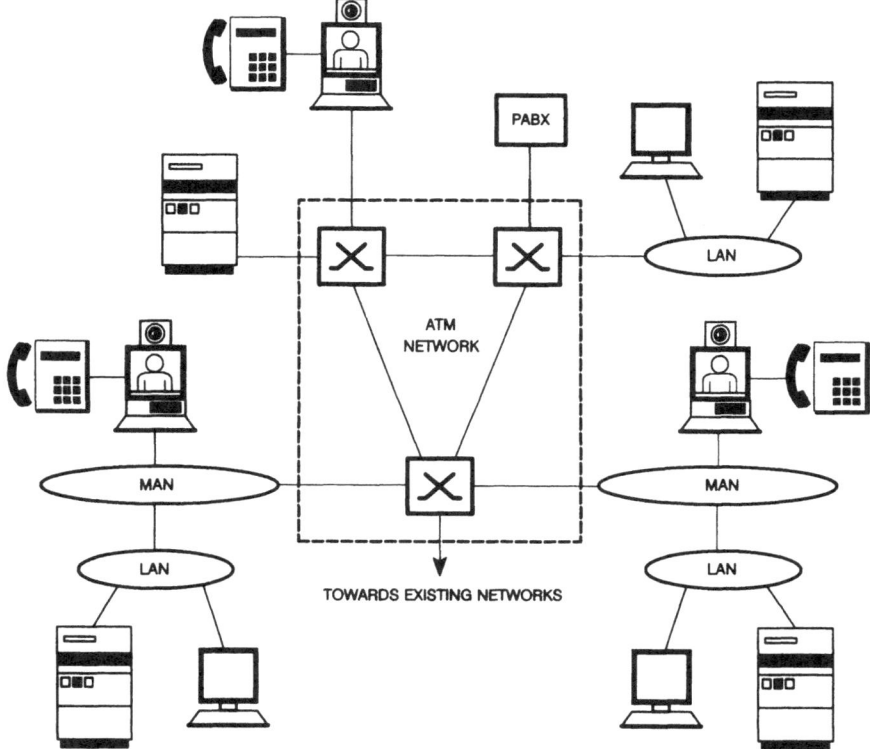

Figure 1. Target scenario for broadband communications

In this paper, emphasis is put on the medium term bearing in mind the second guideline, which suggests the implementation of distributed architectures such as MANs, for achieving the objectives of effectiveness and flexibility in the public

network, particularly when data applications are taken into account. Because of the MAN characteristics of shared access, high bit rate and multiuser environment, this concept is perceived to be of strategic value in connection with its use as gathering network towards an ATM-based transit network. MANs represent an efficient approach for interconnecting LANs with service and performance characteristics comparable with the ones available within a single LAN. Therefore they are particularly interesting for telecommunication operators, which, with the deployment of such distributed architecture, could give a timely answer to the business user requirements in terms of:
− high bit-rate information flow;
− very short response time;
− integration of heterogeneous data communication environments;
− dynamic resource allocation;
− high network reliability;
− new criteria for network management.

Section 2 outlines the MAX Project organization and objectives, discusses some key issues impacting on the project choices and summarizes the MAX policy on standards. The general system architecture, defined during the first year of work, is described in Section 3, clarifying the relationship of MAX with OSI model and with MAN architectures under definition by international standadization groups. Section 4 highlights market perspectives and opportunities in the segment of multi-service LANs and MANs; finally, conclusions are drawn in Section 5.

The work summarized in this paper reflects a joint effort that sees the cooperation from several participants, viz. CSELT (I), SIRTI (I), ALCATEL-TITN-ANWARE (F), KTAS (DK), NKT (DK), BRITISH TELECOM (UK), HEWLETT PACKARD (UK), L-CUBE (GR), DNAC (F), 3IT (F), UNIVERSITY OF PATRAS (GR).

2. MAX Project Overview

2.1 Organization and Objectives

The MAX Project is part of a comprehensive activity which develops in the time frame of five years through two subsequent phases: Phase 1, represented by the current ESPRIT Project n.2100 "MAX" and Phase 2, of which a proposal is under negoziation. The overall goal is to provide an innovative, efficient and cost-effective solution to the communication needs of heterogeneous business users distributed over large (e.g. metropolitan) areas. The system is asked to support multi-megabit applications deriving e.g. from LAN to LAN linking, mainframe to mainframe communications, rapid transfer of high resolution (CAM/CAD) images, audio and video communication and conference. The specific objectives of MAX Project (Phase 1) consist of the definition, specification, design and start development of the main MAX building blocks (including software and hardware breadboards and gate arrays) to be completed, assembled, tested and demonstrated in Phase 2 (MAXI Project).

The first year objective was the definition of the overall architecture and a preliminary functional specification of the major subsystems, after having performed the analysis of requirements and comparisons of the most promising technical solutions. Moreover, the analysis of emerging standards in the field of telecommunication network management and the definition of a suitable MAX management system architecture were carried out.

A parallel objective was to assure an early favourable market impact on future MAX products. This has been pursued through the participation in international standardization bodies (in particular the European Telecommunication Standards Institute - ETSI) not only to monitor the activity progress in the field, thus granting that MAX system incorporates the emerging concepts, but also to influence standards in directions conforming to the MAX Consortium policy.

The first year activity resulted in a positive response to the framework presented in the Technical Annex (February 1989): all the planned Tasks have been activated and major expected results achieved. The functional specifications of the overall MAX system is now available; MAX general architecture, network topology, access protocol, transmission infrastructure, internal and external interfaces, suitable technology, and the network management system have been identified and the specification has been started.

MAX has been positioned as "gathering network" (public or private) capable to provide reliable, high-speed switched connectivity across distances typical of metropolitan areas and supporting different narrowband and broadband data, video and voice services. A mapping of the MAX overall layered structure has been set in relation to OSI, IEEE and ANSI protocol reference models. For the MAX transmission infrastructure (using monomode optical fibre as shared medium) the CCITT Synchronous Digital Hierarchy (SDH) has been selected [2], whereas the Telecommunication Management Network (TMN) concept has been adopted to define the MAX management system.

For the "Medium Access Control (MAC)" sublayer, the IEEE 802.6 "DQDB" draft standard has been chosen as the basis for MAX distributed multiple access protocol [3]. The DQDB proposal has achieved general agreement, although it is not covering, in the available version, all the aspects needed for meeting the MAX requirements. MAX will then address additional issues, e.g. the provision of advanced management capabilities and of connection-oriented services to support both isochronous and non-isochronous communications.

These qualifying aspects, along with the reference transmission speed of MAX, i.e. the SDH with a STM-4 link bit rate at 622.080 Mbit/s, will position future products based on MAX at the top of the market, with respect to the announced DQDB-based products with lower speed and supporting connectionless data-only communications.

2.2 Strategic Issues

MAX Programme is addressing a new and already hot telecommunication sector dominated by a number of problem areas which are clearly emerging and are influencing both the international marketing policy and standards activity. In the following some key issues, which directly impact on the complexity of managing the choices for a project like MAX, are identified and discussed.

1. The demand for services a MAN should provide is already well established in the market. It mainly comes from a growing community of users, e.g. business users, in particular the enterprises devoted to the so-called "tertiary" activities. They are already well equipped with personal computers, hosts, workstations and LANs for internal use: now they are looking to extend their communicability towards the outside world, first of all in the metropolitan range reference, 1, [4]. On the contrary the today's offer can only provide ad-hoc solutions of limited performance, which result costly, inefficient, and lacking of flexibility and integration capability. Most of the demand still remains unsatisfied.
2. Manufacturers, system integrators and PTTs have identified that this will be one of the major opportunities in the coming years. But finding proper solutions is not a simple problem. This sector, unlike the LAN sector which addresses the private area and has grown rapidly with a plethora of proprietary solutions, is put at the boundary between the private and public area. The primary way to tackle this problem is on the ground of the promotion and early implementation of suitable standard solution; the only way to assure the compatibility needed for interworking.
3. The overseas manufacturers and operators have first identified these market opportunities and, after a number of trials with ad-hoc solutions to meet the demand, are now actively working on standards for private and public compatible solutions. In addition, another result of the USA activities has been the preparation of the "Switched Multi-megabit Data Service SMDS" Technical Advisory from Bellcore, specifying a public, packed-switched service that provides for the exchange of variable-length data units on a connectionless basis [5]. It is worthwhile noting that these solutions are not directly applicable to European public networks.
4. From a technical viewpoint, European public operators were not used to MAN technologies, which directly derive from an extension of the LAN concepts, i.e. are based on distributed access architectures, that represent a recent ground of computer manufacturers. On the other hand PTTs are concentrating their effort on a future scenario of high-speed telecommunications based on the B-ISDN concept, which, apart from the advanced technologies to be used, is inspired to more traditional architectural models. The inclusion of MANs in the overall scenario therefore requires a new outlook.
5. As a matter of fact it could be of strategic impact to include MANs in the B-ISDN scenario and this is also recognized by a number of PTTs. MAN makes use of

an assessed technology and can be put in operation before the ATM technology will be available. Therefore, the major concern of public operators is related to the compatibility of MAN with B-ISDN and a clear identification of the role of MAN in this context [6].

6. The future role of MANs is also a concern for manufacturers which need to invest in products with a guaranteed lifetime. This convergence of interests is the rationale for a common presence of both industry and PTTs in the international standardization bodies. In this particular area it seems that the role of PTTs could be of strategic impact and will also stimulate manufacturers. The direct involvement of companies participating in the ESPRIT Programme can also provide a valuable stimulus for pursuing a common European position in the field. While standardization on MANs is progressing, such a common viewpoint is a need for an early development of MAN products.

7. MAN products from overseas manufacturers are coming to be launched on the market. As a matter of fact they do not reflect the actual status of the standard neither provide the top performance claimed in the literature; this first-generation products can be seen as provisional solutions of limited impact, mainly devoted to test the technology and the suitability of technical choices. In parallel manufacturers are preparing the second generation products. The challenge of European industry should be directed towards advanced products, also covering additional features, e.g. higher speed, extended range of services, full management capability, etc.: this is the direction undertaken by MAX Programme.

2.3 MAX Policy on Standards

One of the major objectives of the Project is to foster the convergence towards a common European standard on MANs through ETSI.Currently the IEEE and ANSI standards committees are actively developing MAN standards for North America. ISO and ETSI are setting up activities to extend the scope and international coverage of the standards. It is anticipated that in the time frame of the MAX Programme the initial system specifications for MANs emanating from North America will be completed, with chips and well advanced implementations. However, the standards development programme in North America network environment was up to now aimed at supporting data service requirements. Extensions of service coverage for voice and video, together with the increase in operational speeds and management options, must be included to meet all service requirements of future MANs.

Having these general issues in mind, the MAX policy in relation with standards is now clearly assessed and can be summarized by the following statements.

1. Emerging international standards have been assumed as a basis for MAX where available; new proposals have been taken as reference where standard solutions are still lacking.

2. Active participation in ETSI-NA5 MAN Working Party is now well established, monitoring of other bodies is going on; particular attention is to be paid to

possible active contributions inside IEEE 802.6, influencing the future work with the MAX ongoing solutions.

3. CCITT Synchronous Digital Hierarchy (SDH) has been chosen for the MAX transmission infra-structure, operating at 622 Mbit/s (the so-called STM-4 level). This solution is suitable to implement unidirectional buses by cascading a number of point-to-point links (solutions are under study for geographic networks also); moreover the SDH, through the concept of "virtual container", provides a wide degree of flexibility and makes also possible the partitioning of the payload by following a number of strategies.

4. IEEE 802.6 "DQDB" is the choice for the MAC sublayer. This standard has now reached a stable basis--after a dozen of draft versions--although it is not covering all the aspects; a prototype implementation can be undertaken on the available basis--in line with the MAX workplan--without the need of waiting for further progress.

8. For the time being only the specification of the connectionless service for data is available and assumed to be supported by DQDB products, while other bearer services will be addressed in the future work of IEEE. The MAX Consortium has reached a common understanding that an additional effort should be directed towards the relevant aspects not yet covered by IEEE, with specific reference to the provision of connection-oriented services (CO) to support both non-isochronous (i.e. data) and isochronous (i.e. stream) traffic.

9. IEEE 802.6 is still lacking of reference to the network management system. The Consortium has already analyzed the impact of OSI and TMN proposals and has defined a suitable MAX network management architecture. Advantages are expected from this early position to be reflected in key management capabilities of the MAX prototypes.

3. General Architecture

3.1 Reference Model

The MAX architecture ought to achieve the maximum degree of flexibility and is aimed at defining a network structure suitable to be positioned between the "end system environment" (including local communications systems and data, voice, video equipment) and the wide area network facilities, therefore covering the access/gathering network functions.

The functional architecture of the system is shown in Figure 2. A key architectural element of MAX is the "MAX node", capable of communicating with other nodes of the same type by means of a distributed multiple access protocol over a shared medium (optical fibre).

In Figure 2 three boundaries, delimiting four macro-levels, are identified. The so-called M-boundary (i.e. MAX boundary) is the boundary between MAX system and the "outside world"; it is not a "unique" interface, on the contrary, it is only a "reference point" at which the coexistence of a plurality of different physical and logical interfaces is allowed. This approach is in line with the requirement that MAX has to gather different commercial equipment.

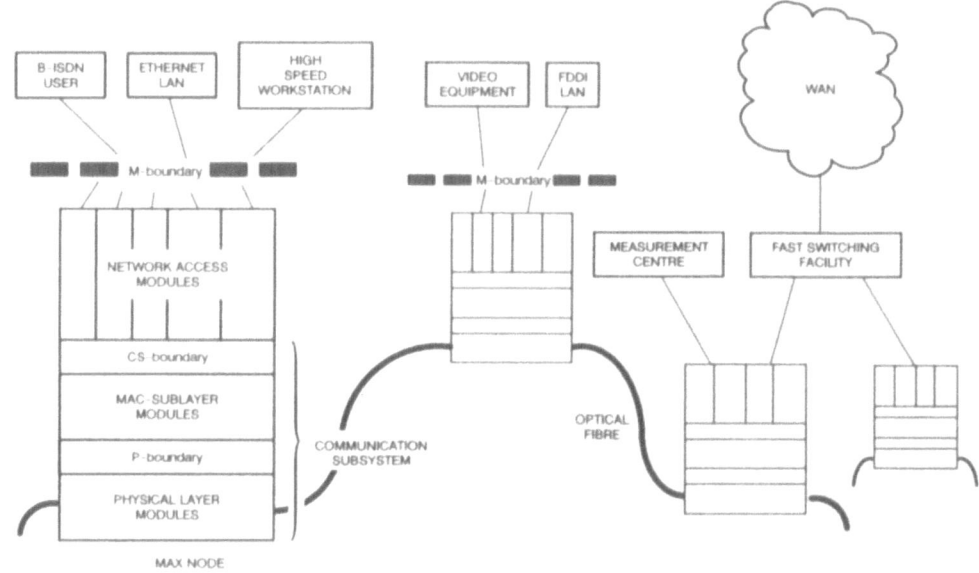

Figure 2. MAX reference architecture

Below the M-boundary, the so-called CS-boundary (i.e. communication subsystems boundary) is placed between the Communication Subsystem (CS), covering layers 1 and 2a of the OSI stack, and the Network Access Modules (NAMs), that are in charge of the transfer and adaptation functions between the devices external to MAX and the communication infrastructure developed inside MAX. Through the CS-boundary, both connection less services and isochronous services are provided. Moreover, the CS-boundary supports the exchange of both management primitives and signalling primitives between the communication subsystem and the upper part of the node.

The communication subsystem provides its services to the NAMs. For each type of device gathered by the MAX node a specific NAM is foreseen. For example, different NAMs will provide interfaces to:

A) End-user equipment as LANs, workstations, video equipment (e.g. video codec)

B) Wide area network devices as dedicated high speed links of the wide area network, fast packet switching nodes, and ISDN or B-ISDN transit nodes.

The NAMs represent the "flexibility point" of the architecture; all MAX nodes will have the same communication subsystem, but with different NAMs for each interface requirement, allowing a node to be configured for example as an Ethernet router or B-ISDN network adapter.

Inside the communication subsystem, the Physical Layer modules and the MAC sublayer modules are interfaced through the P-boundary (i.e. the Physical Layer boundary). The Medium Access Control (MAC) sublayer is based on a distributed multiple access protocol capable of supporting different kinds of traffic ("hybrid switching"). The functions of this sublayer are: to recognize the right to start a transmission, to transmit and receive data of different priorities (corresponding to the different services offered to the NAMs), to detect bit errors in received data and to perform communication subsystem start-up and recovery procedures (under the control of the management systems).

The Physical Layer modules provide the digital baseband point-to-point communication between MAX nodes and include optical transmitters and receivers with clock and data recovery circuits. Moreover this layer performs mutual clock synchronization, line encoding/decodi ng, frame synchronization, serial/parallel conversion of data; most of the electronic circuitry will be implemented on dedicated chips. It should be possible to use different appropriate standard transmission systems for the MAX Physical Layer, thanks to the P-interface handler (one of the Physical Layer modules) that performs the required adaptations and mapping of the P-boundary services onto the specific transmission system services.

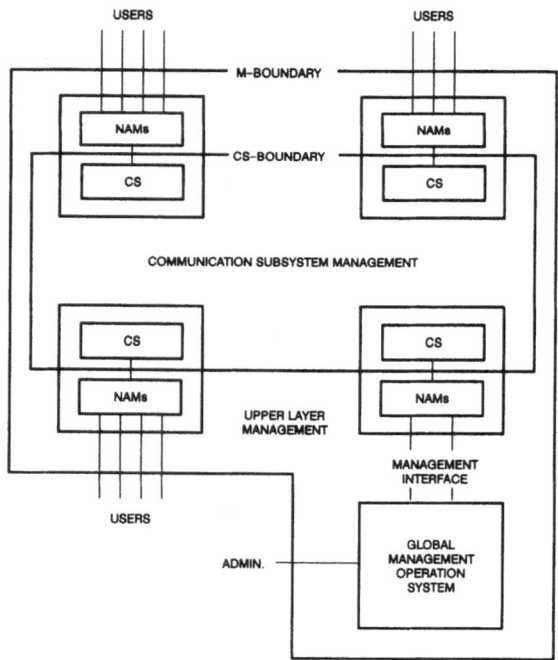

Figure 3. MAX network management model

With the increasing complexity of MANs and associated user requirements, a key element of MAX architecture is the Management System (Figure 3)[7]. Each MAX subsystem (Physical Layer modules, MAC sublayer modules, Network Access Modules, interface handlers, etc...) incorporates a management entity that performs management functions specific of the subsystems and that is controlled by the MAX management system.

3.2 Relationship with ETSI MAN Architecture

The MAN architecture under study within ETSI NA5 "MAN" Working Party [8] is shown in Figure 4. The Subscriber Premises Networks (SPNs) are connected to Access Facilities (AFs) that terminate on a MAN Switching Systems (MSS); several MSSs are interconnected through the Transit Network with the Inter-MSS Interfaces (ISSIs). For OA&M functions a Management Interface Q* is identified. At least two options exist for the AF:

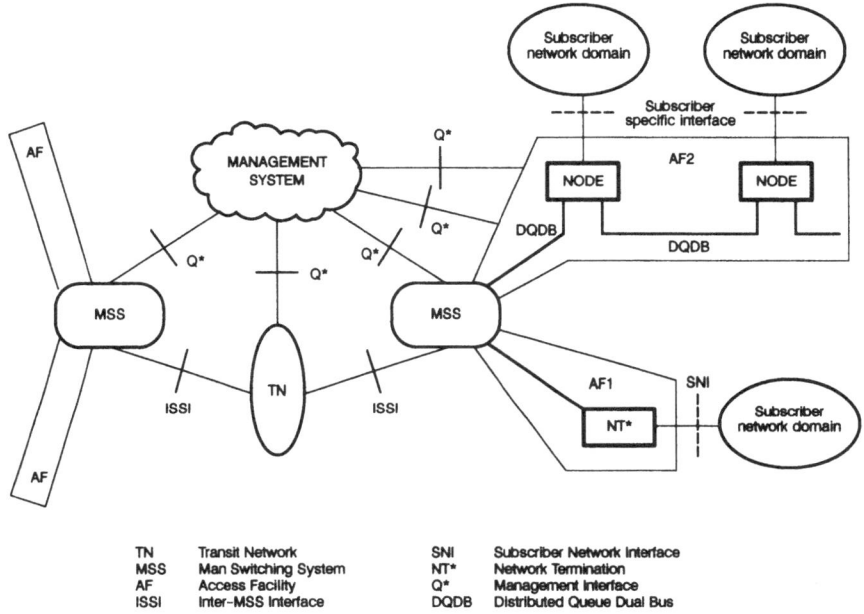

TN	Transit Network	SNI	Subscriber Network Interface
MSS	Man Switching System	NT*	Network Termination
AF	Access Facility	Q*	Management Interface
ISSI	Inter-MSS Interface	DQDB	Distributed Queue Dual Bus

Figure 4. ETSI architectoral model of MAN

1. AF1: based on a distributed multiple access protocol with shared access medium; the subscriber devices are connected to this link through the Subscriber Network Interface (SNI).
2. AF2: based on a set of MAN nodes forming a gathering network; they communicate through a distributed multiple access protocol. Each node is connected to

one SPN; as for the interface between the SPN and AF2 several options are under study.

The MSS is a multimegabit communication switching system. It can be implemented through distributed or centralize d switching technology, and will provide functions such as termination of AF, routing of communications to the destination AF and routing of communications towards other MSSs

The ISSI links remote MSSs; in this way, wide networks can be formed. Several options are envisaged for the ISSI, also depending on the time frame, e.g. dedicated links on WAN carriers, digital cross-connects, satellite links, B-ISDN, etc...

The relationships between MAX and the ETSI MAN are straightforward; this is also due to the great effort put by the companies of MAX consortium in the ETSI MAN standardization activity. The identification of the "access/gathering role" of MANs, that was a starting point of MAX programme, is now fully endorsed by ETSI.

Because of the modular structure chosen for the MAX node it is evident that by using MAX nodes as basic building blocks it is possible to develop:

– Access Facilities (in particular AF2) operating at 600 Mbit/s; in this case the "M-boundary" coincides with the boundary between the public domain and the private domain.
– Distributed MAN Switching Systems (MSSs); in this case the "M-boundary" is the MSS boundary, and the corresponding NAM should operate as a fast bridge (or router).

3.3 Relationship with OSI Model and IEEE 802.6

The Open System Interconnection (OSI) basic reference model is a fundamental concept for architectural modelling and it is the best widely accepted reference available. Nevertheless, the OSI model was conceived mainly for data services and it has been recognized not to be perfect to model completely networks supporting a multitude of services including voice and video communications.

This inadequacy was recognized in the CCITT definition and specification of ISDN; therefore, the ISDN protocol reference model (CCITT Rec. I.320) was introduced bringing together the OSI modelling principles and the ISDN requirements.

In the broadband communication scenario the need to extend the OSI model is still challenging; the B-ISDN protocol reference model (CCITT Draft Rec. I.321) introduces a layering that refers to the three "planes" used also in narrowband ISDN, i,e. U-plane for user information C-plane for control information and M-plane for management.

The same kind of problems arises in the application of OSI model to LANs and MANs supporting integrated communications. As far as the simple data-LANs are concerned, it was only necessary to use the "sub-layering" concept fully allowed by OSI: the IEEE 802 family standards identify, inside the OSI Layer 2 (Data Link) two

647

sublayers: the "Logical Link Control - LLC" and the "Medium Access Control - MAC". In the case of integrated-service LAN/MAN (e.g. FDDI-II and IEEE 802.6) it was recognized the need of introducing, on the top of the MAC sublayer, parallel "stacks", respectively for connectionless data services (LLC), for isochronous services and for connection oriented non-isochronous services (this is not different from the introduction, in B-ISDN, of "ATM Adaptation Layer protocols" service-specific on the top of the "ATM Layer").

In MAX protocol architecture the same approach of IEEE 802.6 has been adopted. In particular it can be noted that the MAX Communication Subsystem covers the functions of the Physical Layer and the MAC sublayer; the MAX Network Access Modules (NAMs) implement data protocols until the Transport Layer and the functions related to isochronous and connection ori ented non-isochronous applications.

In Figure 5 a comparison of the protocol architectures of OSI, FDDI-II/ANSI, IEEE 802 data-LANs, IEEE 802.6 MAN and MAX is shown.

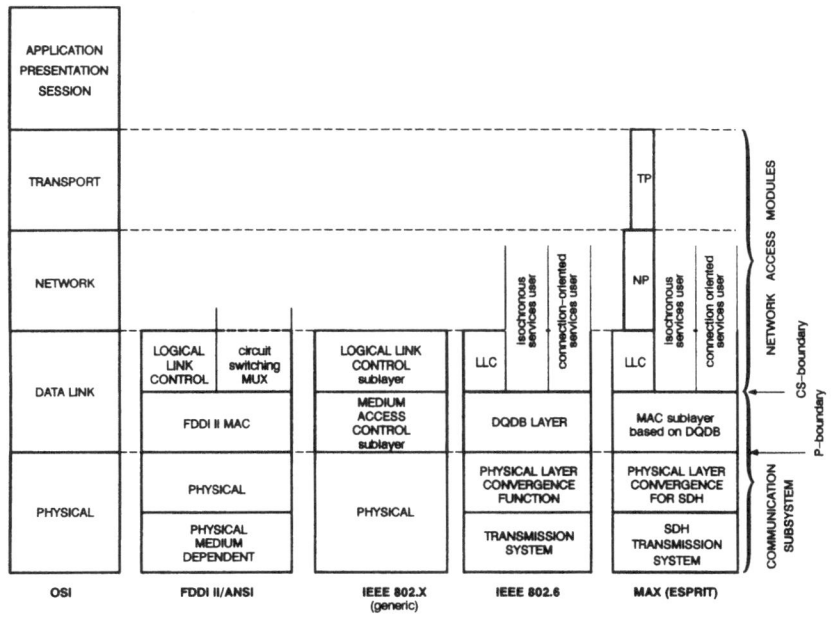

Figure 5. Comparison of protocol profiles

3.4 Extended Network Applications

Figure 6 shows various possible uses of MAX nodes, all having the same Communication Subsystem but equipped with different Network Access Modules:

- *Node A* allows the interconnection of commercial user equipment like medium-speed data LANs (e.g. Ethernet, Token ring) or video codec. In this type of configuration a MAX node can serve a few user equipment.
- *Node B* bridges a high-speed LAN "HS-LAN" (or Extended-LAN "E-LAN") that, in turn, gathers medium-speed LANs and other user equipment. The HS-LAN could cover the role of "corporate MAN", being the backbone of an intelligent area, or could represent a tributary subnetwork of the public gathering MAN. The HS-LAN can be implemented through a data-only product (e.g. a commercial QPSX working at a speed of about 45 Mbit/s, or a commercial FDDI working at 100 Mbit/s), or a multi-service communication system (e.g. a future commercial "DQDB-type" or a "LION-type" [9] network, both operating at less than 150 Mbit/s).
- *Node C* supports the interconnection to a node of the Transit-Level Network; in ETSI MAN terminology, this refers to the "MSS" function.

In the design and prototyping activity of the current MAX Project only type A and type B applications are considered (type B limited to the FDDI I case); this is in line with the market analysis that has identified the "2 Mbit/s connections" and the "LAN (low and medium speed) interconnection" as the major market sectors for MANs. The Consortium is now evaluating the opportunity to address type C applications and to increase the effort related to type B (with "LION-type" and "DQDB-type" tributary networks).

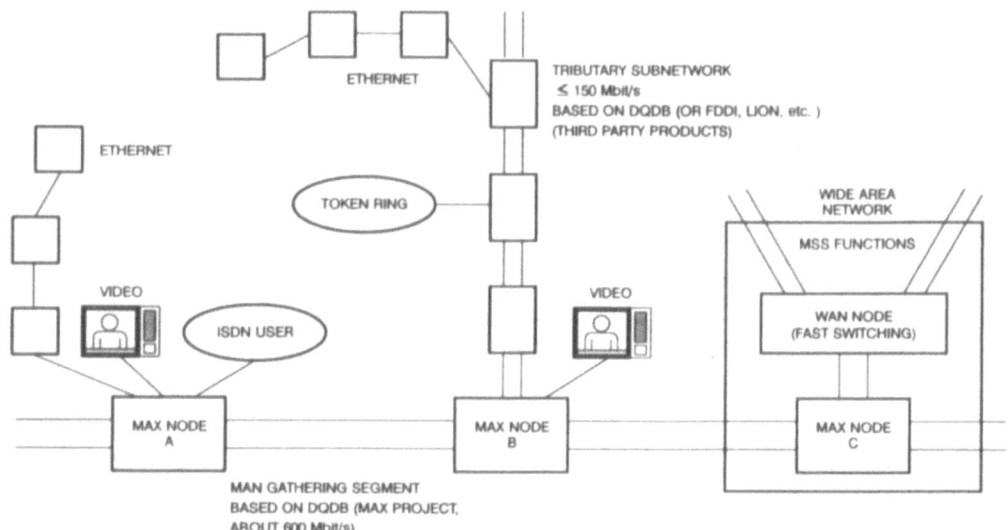

Figure 6. MAX extended network application

4. Developments from a Market Viewpoint

4.1 Evolving Market Scenario

In the telecommunication field a fast evolution is in progress, due to a growing market demand for new services coming mainly from business users. This market demand is the result of the wide diffusion of Office Automation in small and medium size companies and of the growth of distributed systems within large business organizations.

This trend is stimulating a market demand where two different segments can be identified:

- in large companies there is a growing need for high speed networks capable of providing "backbone" functions to existing Customer Premises Equipments (CPEs) and Local Area Networks (LANs) and integrating different types of traffic, ranging from that related to traditional services (i.e. telephony, low speed data, telefax) to that related to advanced services (i.e. electronic mail, high-definition images, high speed data);
- in the whole business sector there is an emerging need to interconnect heterogeneous users spread over metropolitan and wide areas.

The former case is typical of a private environment, where the use of a single integrated network could give the company advantages in terms of interconnection capability between different locations and of powerful network management tools. For these types of networks terms like Extended-LANs (E-LANs) or Corporate-MANs are currently used, depending on the geographical coverage of the network.

The latter case is strictly related to the public environment, where the public operator could provide the required interconnection facilities between different users spread over metropolitan and wide areas, ensuring a high degree of security. In this case, it is common to use the term of public MAN and the multi-user needs have a strong impact on the network management system which has to be much more advanced than the one suitable for private single-user applications.

Worldwide market perspectives and opportunities in the segment of multi-service business networks for both private and public applications may be supported by figures assessed in recent market analyses Autonum, reference, 1, Value = [10] . These studies, covering the next five years time frame, give explicit evidence to the following data:

- a 20-25% average annual growth rate for LANs;
- a 20-50% of LANs and PABXs interconnected over local and metropolitan areas;
- a growing demand for broadband services and applications, initially in the business sector (see, for example,Table I).

	NUMBER OF ESTABLISHMENTS	PERCENTAGE
VIDEOCONFERENCE	48.700	16%
VIDEOPHONY	63.100	21%
HIGH-SPEED DATA	51.200	17%
HIGH-SPEED COULOR FACSIMILE	9.800	3%
HIGH-QUALITY SOUND	1.100	-
HIGH-QUALITY TV	10.400	3%

THE BREAKDOWN OF THE DEMAND IS APPROXIMATELY:

- 70% FOR FRANCE, F.R.G., ITALY, UK
- 20% FOR THE OTHER 8 EEC COUNTRIES
- 10% FOR THE 5 EFTA COUNTRIES

SOURCE: CEC [11]

TABLE 1. Estimated demand for btoadband services in 1996 (EEC and EFTA countries): number of establishments with more than 50 employees applying for a broadband connection

Today some products suitable for single-user applications already exist on a commercial basis. These products are based on the ANSI X3T9.5 standard (FDDI), operating up to 100 Mbit/s and handling data only. It is likely that in the short run (up to 1992) a certain diffusion of such systems will occur in the private environment, while in the medium run (from 1992 to 1996) a smooth but continuous increase of the transmission rate, up to several hundred of Mbit/s, will be needed to face user requirements in terms of network throughput and service integration. In the long run a further evolution will take place with the exploitation of innovative networking technologies, based on coherent multi-channel communications and photonics.

The main evolution might occur in the public domain, where the deployment of early MANs is foreseen in 1992-93, in parallel with the introduction of SDH based transmission equipment. The first MAN applications will be dedicated to data only traffic, while later service integration will probably take place to meet user requirements in this sense. Therefore MANs could play the role of the first evolutionary step toward the future B-ISDN, that will be based on ATM technologies and will be started to be deployed not before the late Nineties. With reference to Figure 7, it is expected that MANs will develop and grow to cover the time lag between ISDN and B-ISDN.

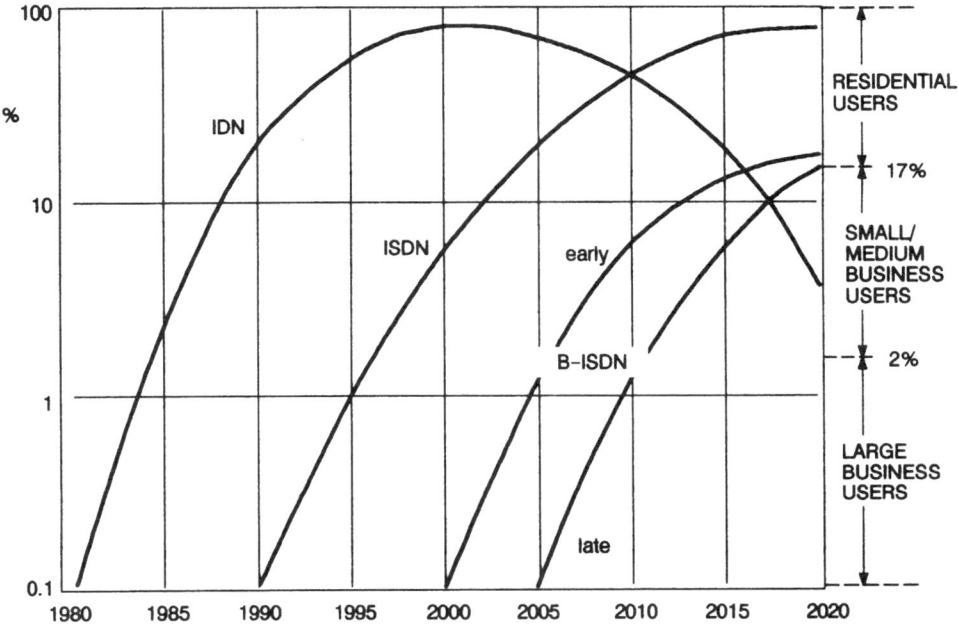

Figure 7. Network evolution (percentage of total installed main lines)

4.2 Network Requirements from the User Viewpoint

In the business market segment, the main network requirements from the user viewpoint in the short and medium run include:

- the adoption of a distributed network architecture, allowing a higher reliability, flexibility and modularity in network components;
- the provision of both bearer and tele-services, both to carry user traffic and to encompass upper OSI functions, related to security, billing, accounting and access to service providers and centralized computing facilities;
- the support of multimedia communications;
- the implementation of all-level network management functions, allowing the development of a reliable, flexible, high performance, secure and dynamic integrated network environment. This requirement comes from the user need to increase its control and management capabilities on the overall network domain;
- the implementation of security management functions, coming from the typically distributed business environment, where multi-tenant and inter-organization applications must run;
- network interoperability with other external public and private communication systems, to support the connection of a variety of CPEs with full exploitation of their capabilities.

5. Conclusions

A weak point of the European industrial position in the IT area is its historical strong dependence on the USA industry. This is more apparent in the new segment of high speed business local and metropolitan area networks, where today the majority of technologies and products available on the market comes from the USA.

The same consideration can be made for standardization activities. All the most important solutions currently under discussion in the field of LANs and MANs (i.e. FDDI, DQDB and SMDS) originated mainly in the USA, within the ANSI X3T9.5 committee for FDDI, the IEEE 802.6 committee for DQDB and Bellcore for SMDS. In Europe, an ETSI working party on MANs has been recently established, which has chosen the IEEE 802.6 DQDB protocol as the basis for European standardization.

From the industrial viewpoint, many development activities are in progress in the USA involving large telecommunication operators and manufacturers. Some FDDI-like products have recently been put on the worldwide marketplace and early field-trials based on the SMDS concept and the DQDB protocol are foreseen to be installed in the second half of this year.

In Europe PTTs and manufacturers are looking at this market and technology evolution with a strong interest, but it does not seem that at present the development works are so advanced as in the USA and Australia.

In this scenario, the ESPRIT Project 2100 MAX could play an important role. Actually, as the project choices and the results achieved so far are in line with the general technology and standard trends, MAX is to be considered for the European IT industry an important milestone towards the industrial development of a product capable of providing high performance services according to European standards and user needs.

The project goal is to achieve a proof-of-principle of a high speed multiservice MAN through the development and evaluation of a prototype network. Parallel objectives are to promote the convergence towards common European MAN standards and to explore worldwide market opportunities for MAX-based products and services. Therefore, the project results can help in reducing or eliminating the gap between the European IT industry and the USA one in the growing market segment of multi-service LANs and MANs.

References

[1] Forcina, A., Luvison, A. and Perardi, F. (April 16-19, 1990) 'A strategy for ATM introduction into public networks', in Proc. IEEE Int. Conf. on Commun. (ICC '90), Atlanta, GA, pp. 1596-1601.

[2] CCITT Recs. G. 707-708-709 on Synchronous Digital Hierarchy, Blue Books, vol. III.4, Geneva, 1988

[3] IEEE 802.6 'Distributed Queue Dual Bus', Proposed Standard P.802.6/D13, June 1990.

[4] Freschi, G., Roffinella, D. and Trinchero, C. (April 16-19, 1990) 'Requirements and prospects for multiservices business networks', in Proc. IEEE Int. Conf. on Commun. (ICC '90), Atlanta, GA, pp. 1446-1450.

[5] Bellcore TA-TSY-000772, Generic System Requirements in Support of Switched Multi-megabit Data Service, issue 3, October 1989.

[6] Biocca, A., Freschi, G., Forcina, A. and Melen, R. (May 27-June 1, 1990) 'Architectural issues in the interoperability between MANs and the ATM network', in Proc. XIII Int. Switching Symp. (ISS '90), Stockholm, Sweden, Vol. II pp. 23-28.

[7] Gavi, L., Vercelli, R., Perretti, E., Ciccardi, A. and Roffinella, D. (June 27-29, 1990) 'Architectural and management issues for Metropolitan Area Networks', in Proc. European Fibre Optic Commun. and Local Area Networks (EFOC LAN '90), Munich, Federal Republic of Germany, pp. 270-275.

[8] ETSI NA5 Working Party on MAN, Report of the meeting. Tampere, Finland, June 18-21, 1990.

[9] CEC DG XIII A5, Report of the Communications and Network Steering Committee, March 1990.

[10] Luvison, A. (January 30 - February 3, 1989) 'The LION approach to multi-service business networks', in Proc. 12th European Congress Fair for Technical Commun. (ONLINE '89), Hamburg, Federal Republic of Germany, paper IV-2.

[11] Bigi, F. and Cordaro, G. (April 16-19, 1990) 'Marketing perspectives for business broadband services in Europe', in Proc. IEEE Int. Conf. on Commun. (ICC'90), Atlanta, GA, pp. 76-80.

Project No. 2105

MULTIWORKS - MULTimedia Integrated WORKStation

ROBERTO DEL MORETTO
Technical Direction
Olivetti Systems and Networks DOR
Via Jervis 77
I- 10015 Ivrea, TO, Italy
Tel. -39-125-528913

ABSTRACT. It is a recognized trend that the workstation of the 90's will be characterized by the ability to process multimedia information. This will provide support for the development of new applications, that will take advantage of the added value provided by the combination of the "traditional" computer media and existing information systems, with digital video and sound technology. The goal of MULTIWORKS is to develop a workstation for the office that manipulates video, graphics, text,voice and sound with facilities comparable to those available today for traditional media: text and graphics. This paper presents the rationale behind the workstation's design, the technical choices made and its general architecture. It also discusses some of the key, state of the art technology that will be implemented in the workstation. It is expected that industrial partners will incorporate some of Multiworks components in their multimedia products for the 90's.

1. Introduction.

It is a recognized trend that the workstation of the 90's will be characterized by the ability to process multimedia information.

In current trade terminology, multimedia has come to mean the use of digital sound and moving pictures in a computer system, for which current estimates foresee a total European market of about 12 Billion dollars in 1994, including the home market, with 3 Billion dollars coming from software shipments alone.

Application areas include high end simulations, video post-production, authoring workstations, home PC, home entertainement and games, geographic information services, promotion,education and training.

This new kind of applications will take advantage of the added value provided by the combination of the "traditional" computer media and existing information systems, with digital video and sound technology. The existing, mostly analog video industry will be able to use digital, computer-based information technology and data processing applications will be enhanced with the use of new, very expressive and friendly media. This way, for instance, reservation systems will be able to show movies of resort areas, computer aided instruction programs will be able to include short movies or documentaries about the subject at hand, movie editors can easily include text and computer graphics in their productions.

In this context, a four year project, called Multiworks, has been sponsored by the European Community as part of the ESPRIT program. It is developed by a consortium comprising some of the

654

most prominent European computer manufacturers such as Olivetti, Bull, STL/ICL and Philips.

Multiworks is a Technology Integration Project (TIP) where each partner brings its experience in a specific field to be shared and integrated with the others' during the development of the workstation.

Multiworks' goal is to develop a workstation for the office that manipulates video, voice, sound,graphics and text with facilities comparable to those available today for traditional media: text and graphics.

This multimedia technology is just emerging and showing in some specialized products like CD-I from Philips, CD ROM XA from Sony and Philips, DVI from Intel, and Digital Video Overlay from a variety of manufacturers.

It is foreseen that a high volume market will develop in 1994. Since the results of Multiworks will be available in 1992, there will be enough time for the industrial partners to evaluate them and incorporate some of Multiworks components in their multimedia products.

The challenge faced by this new technology is to create a base of applications that use it. This can happen if software developers see a business opportunity and are enticed to invest in it. The market potential, of course, represents a strong motivation, but it must be balanced with the investment required to tackle it.

A successful multimedia offering therefore must minimize the investment required to develop new applications, which is affected by the expected life time of the software, its use with different products (portability) and the cost of development.

To accomplish this, a multimedia work station must provide:

-adherence to standards to improve portability

-high performance, high level media processing tools so that the development of the "traditional" as well as the new components of the application can be made at low cost

-fast prototyping tools to test the introduction of the new media.

2. Multimedia.

A medium may be thought of as a means of conveying information. In a computer system, information is internally processed and stored or exchanged with the user. The internal representation of the information by the computer is mainly determined by the efficiency with which it can be processed. There is not necessarily a one-to-one relationship with the way it is presented to the user. Information that has the same internal representation can be exchanged with the user by different media. Text, for instance, which is usually stored as a sequence of ASCII characters, may be presented as written words by an interactive editor, or synthesized and relayed to the human being as voice in a voice-response inquiry program. On the other hand, information with different computer internal representations may be presented using the same interaction medium. Text and numeric data, for instance, may be stored internally as ASCII characters and floating point numbers, respectively, but they can be presented together as written text by a compound document editor.

From the user's point of view, different media are characterized by the amount of information that each element can deliver to human beings. Text, for instance, is composed by a sequence of words and sentences used to represent concepts and ideas. Text can be read carefully and it can transfer precise information. Typed text, however, carries somewhat less information than handwritten text. Handwritten text also preserves the personal touch of the writer's calligraphy that may reveal hidden moods and feelings. Finally, spoken text, which still uses a sequence of words to exchange information, is of a different nature altogether, as it is less precise and it adds information about the speaker's tone of voice.

With this characterization in mind, according to the current state of the art, we can consider "traditional" computer media:

-typed text which handles both simple, single font text and text in structured, multi-page, multi-font documents

-audio which is an unstructured sound that is stored as is and played back with no recognition

-spoken text that can be stored in ASCII format and automatically synthesized into voice or stored as sound and played back to the user

-still graphics and pictures which handles 2D and 3D drawings as well as unstructured computer generated graphics and real images

-animated graphics where parts of drawings can be moved to emulate real life movements

-composite documents without handwritten characters that have come out in applications for desk top publishing.

New media are:

-music which is a structured sound that is stored as music sheets equivalent and synthesized by the system

-handwritten text which includes handwritten characters with no recognition (bitmaps) and recognized handwritten text. In Multiworks special devices are provided for handwritten input: electronic paper and document recognizer.

-moving pictures which handle real life animation of images

-all combined media.

The availability of a large number of media gives the application developer the possibility to choose the most effective one for his specific purpose. This choice is many times affected by all the media characteristics that we mentioned above: effectiveness in human communication, characteristics of the environment and efficiency of computer representation.

Availability of new media can improve traditional applications and create the opportunity to address new fields.

2.1 MULTIMEDIA PROCESSING

A media that is computer processable usually can be processed in digital form, stored in and retrieved from direct access devices, transferred in and out of external supports (tape, paper), exchanged over communication lines, accessed from programming languages and manipulated by the user sitting in front of a terminal.

General purpose computer systems have evolved from the processing of numbers, to numbers and text, to today's processing of numbers, text, graphics and images.

From the point of view of computer processing, the new media listed above have the following characteristics:

Handwritten Text.

The computer representation of handwritten text, which is composed of raster or vectorial graphical elements, is part of the traditional media and therefore can be adequately processed by most existing computer systems. What is new and challenging with this medium is:

- The recognition algorithms to improve the speed and quality of the recognition process(connected letters).

- The exploitation of this medium to improve interaction between the user and the computer. New metaphores, like an "intelligent writing pad", where the stylus is used like a pen writing on an electronic screen, need to be concieved and experimented.

Both issues will be investigated in Multiworks, in cooperation with other Esprit projects that are addressing similar technology.

Music and Audio.

Digital Audio and sound processing, in general, has been used in the musical instrument and consumer electronic industry to provide improved quality into functionally specialized components, such as digital audio players or music synthesizers. The introduction of this technology in a general purpose computer adds flexibility to the production of music systems and, at the same time, provides a basic component of the multimedia workstation. The availability of this technology, in itself, becomes an opportunity for the information processing industry to open new markets in the professional and home segments. Coupled with the other media, it is an essential component in the production of multimedia applications, like motion movie strips.

Music and audio processing pose tight architectural requirements and they introduce a new dimension to the traditional information processing framework. Uncompressed, studio quality audio may require as much as 200 KBytes of data for each second of sound. At this rate, the storage of one hour of sound will require about 700 MBytes of space. For high quality reproduction, this data rate must be sustained by the system with virtually no delay. These characteristics place this medium at the edge of the processing capabilities of current workstation design.

Audio and music are "unusual" media in the classical information processing framework. In general, they do not have any intrinsic, meaningful data that can be easily processed by the computer. Text, for instance, is composed of a sequence of characters that can be individually processed. Single characters in a text file can therefore be searched, replaced and displayed. Text files can be stored and retrieved by their contents and printed on a hard copy device. Audio files do not have any of these characteristics. They are sequences of raw data in which the computer and the user himself can hardly identify meaningful elementary components. The only exception to this is Music, when it is represented as a sequence of discrete elements like music sheets or MIDI events.

For all these reasons, audio and music require:

-specialized Digital Signal Processing hardware components capable of performing real time operations with high data throughput

-privileged routes through the system to provide the required bandwith or, alternatevely, dedicated resources (intelligent controllers)

-compression-decompression techniques to reduce space requirements

-new means to represent to the user and the system the information carried by these media.

Appropriate interfaces must be defined to store, retrieve and edit audio sequences.

In Multiworks we want to integrate music and audio in a general purpose workstation architecture with all the features necessary for their complete processing.

Moving Pictures.

Of all the new media, digital moving pictures, in the most general definition, are beyond the processing capabilities of current workstation design. A typical uncompressed video signal with normal enconding requires circa 15 MBytes/sec data rate. With this bandwith, all system components become critical:

-the storage subsystem, in terms of capacity (one hour of video requires 54 GBytes) and access time (transfer rate of 15 Mbyte/sec including possible seek operations)

-the system bus itself becomes a critical component, even with 32 bit architectures, like EISA, VME and MCA

-CPU speed (the execution of 1 instruction per byte would require a dedicated 15 MIPS CPU).

To solve these problems, three different approaches are possible:

a) reduction of quality and dimension of the image (size and resolution reduction)

b) overlay of video source on the computer screen by external, specialized hardware (Digital Video

Overlay)

c) compression/decompression of digital video.

All these methods have draw-backs and they pose restrictions on the kind of processing that can be done.

Solution a) requires a very small image to be shown, in order to reach acceptable data rates.

Solution b) allows only the play back of the video source, with practically no processing by the workstation. The use of external video devices with direct access capability (like video disk players) together with overlaying windows techniques may make this solution viable for some applications. In this case, the computer is basically used to index and control a pre-recorded, unchangeable "video-base" whose data flow proceeds in parallel with the computer route, converging only on the screen. This is the major disadvantage of this solution. Since the video stream is practically not accessible by the computer, it cannot be fully processed by it with the same facilities available for other on-line information (mailed, inserted into documents, etc.).

Solution c) has the potential to overcome all the problems mentioned above. Once the video stream is compressed down to an acceptable size it can be treated as any other on line information by the workstation. During play-back in human understandable form, it is decompressed to the original size. Specialized, high performance hardware is used for video input/output and compression/decompression. Unfortunately, at this time, there is no reasonable hardware that can perform real-time compression and decompression to an acceptable data rate with no loss of information. The current compromise is to use asymmetric algorithms that perform very time consuming compression off-line and real-time decompression during play-back. The ratio between compression and decompression time can be as much as 400; that is, it takes 400 seconds to compress 1 second of video stream. There are currently two major compression/decompression systems available on the market: DVI from Intel and CDI from Philips. DVI is a chip-set that can be incorporated into specific system designs, CDI is available as a separate box that connects externally to the computer. As DVI allows selection of the compression/decompression algorithm at run time, size and resolution reduction techniques can be used to implement symmetric compression/decompression algorithms.

In Multiworks, solution b) will be implemented during the first two years of the project. Solution c) will also be provided after careful evaluation of the existing technologies and their market potential.

2.2 WHY MULTIMEDIA?

It may be appropriate to ask ourselves why we should introduce such "exotic" media as music, audio and motion video in a general purpose computer system.

If we look back, we see that the traditional role of computers is to process (store, retrieve, manipulate and distribute) numbers. It was invented by mathematicians and it was a wonderful tool for them. The first applications of this tool was, therefore, as an aid in solving all mathematics-related problems: financial-accounting and scientific tasks. Medium and large corporations first started using this expensive tool and, in this process, they created huge amounts of sensitive and vital data stored into these computer data bases. With the advent of cheaper, smaller versions of computers, their use spread widely. We started talking about "a computer on everybody's desk".

At this point, two major evolutionary lines can be identified:

-in the traditional, number processing line, the big challenge is to provide controlled and effective access by these computing tools to the company's data base. Implementation of company-wide and inter-company communication networks is the natural evolution.

-down at the desk level, the ever more powerful computer sitting there could do more than just number processing. Text processing, to generate and exchange documents, became a normal application. With the increase in the power of the small computers, graphics and images were added to provide general desk-top-publishing facilities to every user. This is the Personal Work Station evolution line. The basic idea is to use the computer to process and control all information that is accessed or produced by the single user.

Although multimedia is mostly concerned with the improvement of personal productivity and therefore with the latter category, it can only be fully exploited when appropriate communication technology will be available to support it.

If we look at the way computers are used today as a general office tool, we find that they are merely intended to replace existing office communication media: paper and telephone. With the introduction of new media, computers can be used to improve the effectiveness of inter-personal communication by providing the facilities to chose the media that best fit the message that we want to convey. As mentioned before, different media have different characteristics which may make them more or less appropriate for the purpose. No medium is superior to any other one in all circumstances. Motion video will not replace paper communication, in the same way that TV news have not replaced printed newspapers.

As far as the computer industry is concerned, the introduction of multimedia workstations represent a new opportunity to:
- sell more sophisticated workstations and peripherals
- expand their customer base in existing market segments
- create new potential markets

In particular, if we look at the consumer and professional market segments, we see the following opportunities for expansion:

The **home computer** market could be boosted by the possibility of using cheap computers for home video production (coupled with a video camera, VTR or CD-WO), for home music composition from existing song collections and for interactive home training. A lot of these electronic gadgets are already present in many households. Most of them have draw-backs and their usage could be greatly improved by the use of computers. Let's take two examples. Most people, who have tried to use their video camera to film personal events, are familiar with the frustration felt with the quality of the resulting movie. The possibility of post-production editing to improve the image and/or the sound track would be of great help.

Video tapes on "Do it yourself ...." home training courses are becoming very popular. These movies are made to teach people how to do small home repairs or improvements without calling in the expert. They could be greatly improved by using the interactive devices and intelligent teaching techniques that can be provided by a computer.

The largest expansion that we see in the **professional market** is the use of multimedia workstations for the production and delivery of new applications for Education (Corporate training, School Education, Maintanance and Installation Training), On-Line Information Systems (Point of Information, Point of Sale, Electronic Catalogues and Visual Data Bases) and Business Presentations.

In most existing workstations, except perhaps Apple's McIntosh and Next, traditional media are handled as add-on features developed by indipendent software houses for specialized applications.

Multiworks is designed to integrate "traditional" and new media in a consistent environment where they can all be processed with comparable facilities and tools.

The following table shows how Multiworks compares to existing multimedia products:

Workstation \ Features	Next	Apple	Sony	Multiworks
Adherence to Standards	No	No	Yes	Yes
Programming Tools -O.O.	Yes	Yes	No	Yes
Integrated Digital Video	No	Yes	Yes	Yes
Multi - Processing	Yes	No	No	Yes
Distribution Functions	No	No	No	Yes
Integrated Knowledge Engineering	No	No	No	Yes
Integrated new Interaction Devices (Script, Stylus, Voice)	No	No	No	Yes

TABLE 1.   Comparison of Multiworks features with existing market products.

3. Multiworks.

If we consider the multimedia application development cycle, we can identify two different environments:

-original authoring system, where the application is concieved, studio work is eventually done and the movie, sound, text and graphical sequences are put together to solve a specific problem. An example of this is a courseware development system where an author prepares the teaching material and student's evaluation techniques.

-customization and delivery system where the application is used and, possibly some changes are made to customize it to specific requirements such as language translation for a courseware or film dubbing.

This classification affects the power, cost and degree of sophistication of the workstation used in the two environments. The former must be capable of supporting a dynamic, highly demanding environment where changes are made frequently and multiple activities are taking place at the same time. The latter is a more static system where applications are basically "played back" with little or no modifications.

In Multiworks we address both levels of requirements by providing two system configurations: a high end version, called MIW, and a low cost, delivery system called MIW L.

MIW is based on a industry standard, open 32 bit architecture with intelligent controllers to manage multimedia devices. The first instance of MIW will be based on the CP486 workstation available from Olivetti. It is expected that other partners of Multiworks will develop different versions of it with compatible architectures. MIW L is built around a highly integrated, low cost workstation based on the RISC processor developed by Acorn in England. Both versions will be able to support the same operating software and applications.

Multiworks provides:
- a peripheral subsystem of intelligent controllers that manage the new multimedia devices
- enhanced user interfaces including new user interaction devices like voice input and electronic paper
- Unix operating system enhanced with real time features to handle multimedia
- a consistent, object-oriented programming environment for the development of traditional applications that use the new media
- a set of authoring tools based on hypertext and expert systems.

A multimedia integrated workstation faces many technological challenges:
- media such as motion video and CD-quality audio require high throughput at hardware, software and communication levels and/or very sophisticated compression techniques
- real time synchronization of media coming from different sources poses strong requirements on system design and it may require specific solutions
- storage and retrieval techniques must be capable of accessing contiguous streams of data with real time constraints
- new user environment models to access and manipulate multimedia objects must be defined and designed.

The current state of the art makes the true general purpose processing of motion video at a reasonable cost not feasible.

Uncompressed video requires too much bandwith and compression algorithms require too much computer proccesing time. As a matter of fact, unless there is a significant leap in hardware technology or in compression algorithms, it is not clear when motion video will be completely manageable by the computer, which requires real time high quality compression or very high bandwith at both hardware and software levels.

Currently, the usual compromise is to play pre-recorded films in a window on the computer's screen and to use the computer as a control device to randomly access video sequences on the external device. Whereas this solution may be enough to develop some specific applications, it is far from what we consider true video processing by the computer.

In Multiworks we accept that uncompressed video processing is outside of our scope and we introduce "compressed video" as a media that can be fully managed by the workstation. Compressed video may be stored and retrieved, processed, combined with other media in documents, output to the screen or external devices in uncompressed form and sent over communication lines.

The consequence of this choice is that no real time data collection of video is possible, therefore leaving out applications, such as video conferencing, that require such a feature, unless an external real time device will be able to supply compressed video to Multiworks.

Applications developed in Multiworks will be able to use and combine video sequences recorded off-line. This is not a great restriction since, except for video conferencing, most applications do not use the computer for live video capturing. Authoring and play-back applications will use collections of video sequences recorded by professionals off-site and then made accessible to and processed by the computer.

Combination of media such as sound, animated graphics, voice and video poses difficult synchronization problems, particularly when developed on business oriented systems which are not capable of handling real time constraints and pre-determined maximum delays.

Applications like cartoons, talking heads, lip-synch, for instance, require tight coordination between the graphics software and the voice synthesizer and they may require extensions to standard software components that were not designed with this kind of requirements in mind.

Video editing, where pre-recorded video sequences are combined with text, graphics and synthesized sound, requires synchronization at the lowest system levels.

In Multiworks, since the workstation is based on standard, commercially available software, synchronization is achieved, whenever possible, by extending existing software. If synchronization is lost because of uncontrollable events, the system will be able to re-synchronize itself at pre-defined intervals, therefore avoiding unreversable degradation of the system.

Motion video and CD quality stereo sound require pre-defined maximum delays between successive blocks of information. Full motion video, for instance, requires one frame every 40 msec.

In general purpose, business oriented computer systems, storage and retrieval techniques traditionally have been designed to handle with maximum efficiency relatively small collections of data that were frequently changed. One consequence of this model is that access to contiguous streams of data may not get done at the best speed theoretically possible with the given storage device. Since each access is actually translated into a sequence of separate accesses to smaller blocks of data, not necessarily recorded contiguously onto the device, unnecessary overhead may be added. Whereas this is acceptable in managing static data, such as documents, it may not be appropriate to handle dynamic, real time information.

In Multiworks this problem has been addressed at the physical and logical access level by providing an array of disks that increases the direct access storage's capacity and throughput, and a new File Management System.

4.   Multiworks' architecture.

The key elements of the architecture are:
-the use of intelligent controllers to distribute processing load therefore increasing system performance,
-the choice of Unix as the standard operating system enhanced with real-time facilities needed to effectively handle real-time media,
-the development of a rich programming environment to allow for easy development of interactive, multimedia applications,
-the provision of multimedia authoring tools.
Multiworks' general architecture is structured in four layers: hardware, operating system services, programming and application environments.

4.1 HARDWARE

At the hardware level, the industry standard EISA bus has been chosen as the connective structure. The EISA bus has the following characteristics: speed of about 30 Mbyte/sec, Multimaster capability and PC AT bus compatibility. The choice of a standard bus allows for a modular design in which intelligence and processing load can be distributed on a number of different , independent controllers that can evolve and be replaced indipendently from the rest of the system. The Main Processing Unit itself may have different implementations with different characteristics that use the same architecture and peripheral subsystem. As mentioned earlier, two main CPU's have been chosen for the project: the high end version, MIW uses a i80486-based computer manufactured by Olivetti (CP486), which is also used as the development platform; the low cost version, MIW L, is based on the RISC computer, ARM3, developed by Acorn. The peripheral subsystem consists of a set of controllers to provide the following functionalities:

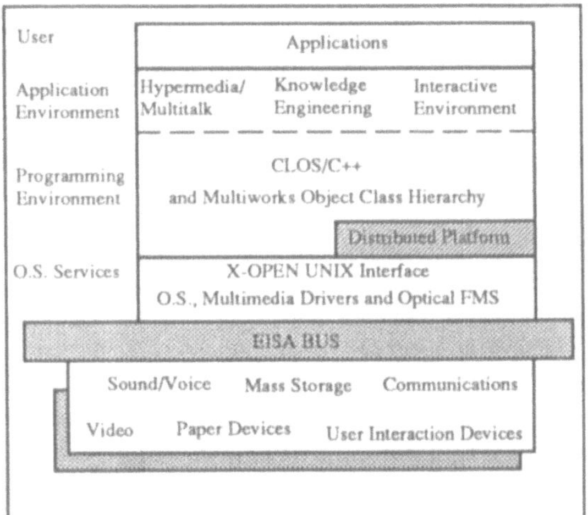

Figure 1. Multiworks' Architecture

- access to multimedia devices
- improved man-machine interaction
- high performance, high capacity mass storage to meet multimedia requirements
- access to public and local area networks

4.1.1 *Access to Multimedia Devices*. The MOVIE board allows the merging in a window of moving colour video images with the computer video output. The Movie board is placed between the usual video controller board and the computer's screen. It takes the signal coming from the video board and, when required, overlays a specific area of it with the analog (in the case of camera or VCR) or digital (in the case of DVI or CDI) movie signal. It supports the VGA and Enhanced VGA video controller standards. In the case of digital input video signal some decompression is necessary. There are currently two compression/decompression algorithms being considered for Multiworks: CD-I developed by Philips and introduced as a black-box in the cosumer market and the more open DVI from Intel. When an analog source is used as input for motion video, the computer processing of the data stream is very limited. Aside from the control commands that can be given to the external device, like start, rewind and seek, the only operation that is usually allowed to get the video data is the "single frame capture" command. This operation lets you read the current digitized frame that is stored in the board's buffer. When digital, compressed video sources are used, the data stream can be transferred over the bus to the main CPU and fully processed by it.

During the development of the movie board two VLSI chips will be designed: a YUV to RGB signal converter and a frame buffer controller.

The Voice/Audio Processing System is an intelligent controller based on a general purpose processor running all the voice and signal processing software. It includes a DSP (such as the TMS320) for elementary signal processing functions, AD/DA converters of CD quality and a MIDI interface to connect MIDI devices. It provides real time, speaker independent, connected word speech recognition for a small vocabulary of a few hundred words. A word spotting system will also be implemented. Text to speech synthesis is provided, with the ability to generate multi-speaker, multi-style speech.

This board provides the music synthesis functions of Multiworks. It generates music from some internal, symbolic representation equivalent to a music sheet. It handles mono as well as poly-phonic compositions. The choice of an intelligent, dedicated controller comes from the consideration that these are heavy, "MIPS consuming" operations that could decrease the main CPU's performance down to an unacceptable level for interactive operations.

The document recognition module is designed as a peripheral device connected to the workstation via the SCSI bus. It is capable of taking documents coming from the scanner, disk or network in raster form, it can partition them in segments and recognize in them multifont typed, printed and hadwritten characters. It consists of dedicated HW and software that implements improved existing algorithms.

4.1.2 *Improved Man-Machine Interaction Devices.* The Electronic Paper consists of a flat screen display overlaid with a transparent tablet that senses a stylus' position and movement. This device allows therefore the user to interact with the computer by writing with a pen on a dynamically changing, intelligent piece of paper: the screen. In Multiworks, an intelligent controller running the appropriate software will manage the device in two operating modes: in the local editing mode, the electronic paper is handled locally by the software running in the controller which then interacts at macro level with the application running on the CPU; in the keyboard/mouse emulation mode, the stylus' movements and the handwritten characters are recognized and transferred to the application as if they were coming from the usual keyboard/mouse devices. In the latter mode, existing applications can take advantage, with virtually no changes, of the improved man-machine interaction technique.

Voice interaction with the user is provided in Multiworks by using the previously mentioned audio controller board.

4.1.3 *High Performance, High Capacity Mass Storage Devices.* A high performance array of disks and its intelligent controller form the primary mass storage solution for the workstation. The controller uses the Acorn ARM 3 processor running Chorus kernel as the real time exexcutive. A low cost archival option based on CD Write Once as well as the Digital Audio Tape technology for backup functions are being evaluated for integration in Multiworks.

4.1.4 *Access to Public and Local Area Networks.* Access to public telephone networks is provided by the ISDN controller. It implements ISDN access protocol functions (layers 1 and 2) for network speeds of up to 144kbit/sec. It is capable of transferring voice, data, images, fax and video. Two new VLSI chips will be developed in the project.

The FDDI2 controller provides access to fiber optic local area networks with speeds of up to 100 Mbit/sec transmission rate.It conforms to FDDI 2 standard for multiservice applications using voice, video and data transmission. Isochronous, real time data, like voice, can be transmitted between FDDI nodes and to external nodes via the isochronous bus that connects the board to the ISDN controller. The board also implements configuration and error reporting functions.

4.2 O.S. SERVICES.

The operating system layer provides the first level of abstraction of the underlying hardware devices. It implements standard services to access system resources such as main memory and CPU (process and memory management), communications (ISO/OSI, TCP/IP protocols), files and multimedia devices. In Multiworks the O.S. interface is a POSIX/X-OPEN compatible Unix. Unfortunately the current Unix standard does not provide the real time features that are required to handle multimedia in Multiworks. For this reason, a small real time, event driven kernel will be provided by Chorus. This kernel runs below the standard Unix operating system and it extends its functions by providing the facilities to write time dependent processes. Above it, all standard Unix functions are available, therefore providing applications with a fully compatible environment. As mentioned earlier, the standard Unix file management system will be extended to provide the possibility to read and write contiguous streams of data to/from mass memory devices such as CD Write Once and CD ROM. These extensions will be completely transparent to application programmers as they will be accessible by the usual Unix file interface, with the obvious limitations coming from the device's physical characteristics, like the read-only possibility of CD ROM's. Further extensions to the operating system will be developed to solve the problems related to synchronize multiple media sources. In particular, extensions to the X-Window System will be required to manage motion video and animated sound graphics. These extensions will be made in close contact with the relevant standard committees to make sure that they will be compatible with future standards.

4.3 PROGRAMMING ENVIRONMENT.

The Programming environment provides the programmer with device independent facilities to develop applications that use system resources.

In Multiworks we have chosen the object oriented paradigm as a means to provide an extensible environment of re-usable software components that can be used to quickly develop new, multimedia applications. In hardware design, engineers have learned to cope with complexity by defining building blocks - Integrated Circuits - as black boxes that implement standard components with pre-defined, well documented interfaces. This has resulted in considerable increase of individual productivity and cost effectiveness of the design process. In Multiworks we propose a similar approach to software design technology by providing a basic set of building blocks as libraries of re-usable objects that can be extended by external suppliers.

Tools offered to programmers are probably the most important component of a workstation offering new technology, as application developers will be the first and unavoidable "users" of the new features.

As mentioned earlier, it is very important that the workstation provide means to protect the investment made by the software developer. One way to accomplish this is to isolate the application specific problem domain from the system's hardware and low level software characteristics that may vary with the fast pace of technology.

In an application we can distinguish three dependency factors that may affect portability and that are not related to the application problem domain:

-language dependency

-logical and physical resource model dependency

-user model dependency

The first one refers to the programming language used to write the application.

The second one refers to the target machine's architecture. In this category we include operating system dependencies, special controllers dependencies and logical object dependencies.

The third category refers to the way the application models the user interface for command and data I/O.

In Multiworks we address these portability problems by providing:

- a C-language-based programming environment extended with object oriented features: C++. C is a widely used language for complex system and application developments. For fast prototyping and as support for Artificial Intelligence applications, the Common Lisp Object System (CLOS) is also provided. The two language environments are closely interwoven, so that objects from either environment can be used in the same progam.

- a set of objects that model the logical and physical resources of the system. They encapsulate these resources' representation and implement their functions. An example of an object is an external video source. This object contains and hides from its user the actual device and the computer's internal representation of it. It also implements standard functions such as play, stop, rewind, search-frame that will always be accessible with the same commands, independently of the actual device connected to the machine. A persistent object storage system will be available to store and retrieve instances of these objects.

- a set of pre-defined graphical objects that can be used in the development of the application's user interface. The basic objects provided implement simple user interaction concepts, such as selection from a menu or command line input, that are of general use in many applications. These basic graphical objects correspond to the widgets defined in OSF Motif kit. However, complex applications may have to model higher level user interface objects, such as forms, envelopes, calendArs, etc. For this reason, a set of tools, called Interactive Environment will be provided. With these tools, the programmer-user can generate his own graphical objects interactively by using page-layout software. These user interface objects may become part of the system and incorporated in the application. They provide a good encapsulation of the user model dependencies into components that can be changed interactively at any time with great ease without affecting the rest of the software.

If we compare this approach to the model-view-controller paradigm (MVC), the former set of objects provide an application framework of interchangeable components that can be used to develop the application's model. The latter ones correspond to the view hierarchy where views incorporate controllers functions for managing cursor and keyboard actions. This way each view implements a complete user interface avoiding the need of a separate class hierarchy for controllers.

Domain specific objects may be added to these libraries to model objects commonly used by vertical applications. In Multiworks, objects for authoring and expert system applications are provided.

Whereas these are powerful facilities to build self contained, stand alone applications, they do little to solve the problems related to distribution of resources. In modern office environments, work and facilities are usually spread through the organization to specialized functional groups. This creates a natural prolif-eration of in-compatible sys-tems and distri-bution of re-sources, like specialized hardware de-vices, informa-tion bases and programs. In Multiworks we provide two different main processing units and we expect other partners to

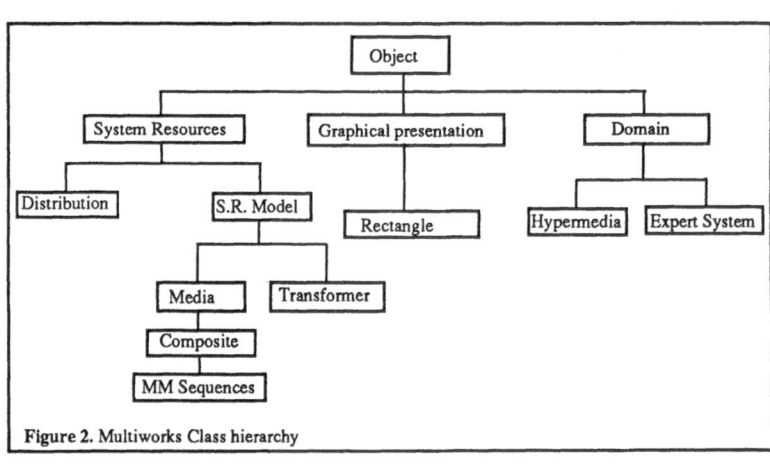

Figure 2. Multiworks Class hierarchy

provide more. All these systems must be able to communicate and share resources and multimedia devices in a network. The distribution platform in Multiworks provides the facilities to develop applications capable of using resources distributed over heterogeneous hardware and software, independently of location, communication subsystem characteristics and operating system. Resources can be re-assigned or moved from local to remote systems without any changes in the applications. This is the architecture developed by ISA, that is already working on a number of different computers. Multiworks therefore will be able to cooperate and share resources with a wide variety of systems.

Figure 2 shows a part of the class hierarchy provided in the system.

The media class of the System Resource Model provides objects for all multimedia facilities of Multiworks: Sound, Video, Graphics, Text and Electronic Paper. The Composite class defines structured documents containing a combination of text, graphics and pictures. The MMSequences object provides representation for the most general form of production. As shown in figure 3, a production in Multiworks is a stream of atomic blocks each containing a fixed number of "frames" of different media. A "frame" can be thought of as the MMSequences' clock used for synchronizing the beginning of each block.

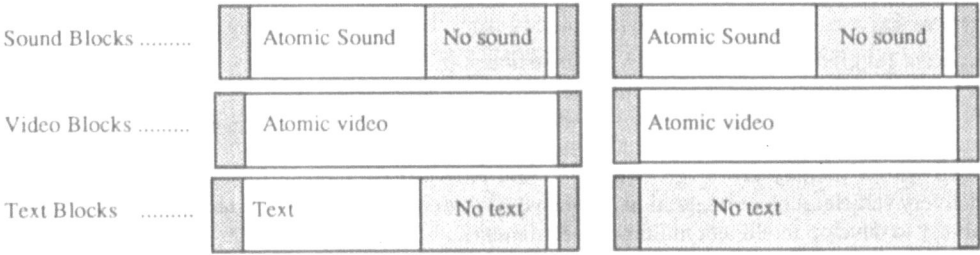

Figure 3. MMSequences. Atomic blocks are the indivisible units of media that must be treated as a whole. An example of this are compressed video blocks using inter-frame compression techniques where single video frames are reconstructed from differential information with previous ones.

During the authoring process, the single streams must be synchronized together so that the appropriate sound will start playing at the beginning of a selected video sequence and the appropriate text will be displayed at the same time. This way, for instance, a production can be interactively built by changing individual media blocks, tested out for audio visual effects and stored for later use.

These atomic blocks provide logical synchronization between the different streams as they specify when a block of media information should be played relatively to the beginning of the other blocks.

In theory, all media blocks should begin playing "at the same time". In practice, these different streams are fed into different modules that introduce different delays. The video stream, for instance, will be processed by the video decompressor that will take a different time than the sound decompressor device. Although special synchronization modules have been designed at low level into the system, it is still possible, under stress condintions, to exceed acceptable delays. As mentioned earlier, this atomic blocks approach will allow the system to recover at the beginning of a new block.

One particularly relevant aspect of the general synchronization problem deals with the use of X-Window, that Multiworks has chosen as the standard window management system. Currently the X server's architecture has no concept of real-time response and it can introduce unknown delays to a display request. This is partly due to the possibility of having multiple requests pending from different clients and partly due to the X-server's handling of mouse events. Within the Multiworks project this problem will be thoroughly investigated and, where appropriate, proposals for extensions will be made to the X standard consortium.

4.4 APPLICATION AND USER ENVIRONMENT

The application environment provides a set of tools for the development of highly interactive applications. As mentioned before, it is expected that the first applications that will take full advantage of the new media will be the ones concerning the authoring and production of promotional, educational and training material. Such markets are already addressed by all analog video movies on tape. It will be a natural step to improve these applications with the interactive and direct access facilities provided by Multiworks. For this reason, a hypermedia based authoring tool, called Multicard, will be developed in Multiworks. Multicard provides functions to create and change interactively hypermedia documents, containing all simple and combined media available in Multiworks, with associated "script programs" written in the Multitalk interpretive language. Multitalk combines the advantages of interpretation (immediate execution) with an english-like syntax and a modular design deriving directly from the events that activate each portion of the scripted program. Another common characteristics of the above mentioned applications is their knowledge transfer goal accomplished with the aid of audio-visual material. It becomes a requirement therefore to combine authoring and knowledge management tools to make available a complete authoring environment. The knowledge engineering tools in Multiworks provide an expert system shell with a friendly kowledge acquisition and processing interface, together with an efficient delivery vehicle. A close integration between the Multicard and the knowledge system provides the ability to develop intelligent multimedia documents.

5. Conclusions.

Multimedia processing workstations represent the next technological leap that the computer industry will take. They have the potential of opening huge markets as they make the professional and consumer segments come close together. For this reason, it is extremely important that European industry be prepared and ready with the technology necessary to build one ahead of the competition.

As I hope I have presented in this paper, multimedia processing presents still many challenges and some operational questions that, having not being thoroughly explored before, need some experimentation and refinements. Multiworks, within the Esprit program, attempts to solve these problems by bringing together the experience of the European industry to provide a working multimedia workstation before demand is expected to pick up. The workstation's architecture has been designed to be modular and open to smooth integration of different components (Central Processing Unit, for instance), so that individual partners can very easily draw from Multiworks technology to build products that fit their own systems line.

The European project for the development of a standard, open architecture for hospital information systems

Fabrizio Massimo Ferrara
RICHE Consortium &
GESI Gestione Sistemi per l'Informatica
Via Rodi, 32 - 00195 Rome
Tel +39 6 32.52.335
Fax +39 6 32.52.336

Abstract

RICHE - Réseau d'Information et de Comunication Hopitalier Européen

In this paper we present the main characteristics of RICHE, the European project carried out by a Consortium of hospitals, specialised IT companies and manufacturers, for the design and development of a standard, open architecture for hospital information systems.

The paper is organised as follows:

In the first section we present the fundamental RICHE's objectives and the underlying requirements of the users, in terms of standardisation in medical informatics, which the project is dealing with.

The second section presents the overall methodological considerations and framework, which has been adopted for the project, considering a formal methodological approach as one of the major success factors of a complex, international initiative like RICHE.

In section four we outline the main results of the project up to the end of April 1990, and the subsequent evolution guide-lines.

Finally, in the fourth chapter, we discuss the principal concepts of the RICHE architecture, in terms of application areas and functional modules .

1. Objectives and users requirements

An overall increase is foreseen world-wide in the investments for the medical informatics, to support political, organisational and scientific requirements, in order to rely on faster and more complete information.

In such a situation, the primary requirement for the user has been identified in the availability of a standard conceptual framework for organising, storing and retrieving the information of interest in the various application areas. It will be able to ensure the consistency and coherence of the various healthcare activities, also with respect to the necessary connections between different institutions, requiring to interchange data for medical, administrative, and statistical purpose.

From the technological point of view, the availability of a physical, open, architecture and software components, based on the industrial IT standards, is a fundamental need of the market, in order to permit a real synergy among various manufacturers and software supplier, enabling them to actually cooperate by integrating different applications into a unique homogeneous healthcare information system.

These considerations represent the rational basis, underlying the effort of the RICHE Consortium, formed by the following nine European organisations: **BAZIS** (NL), **BULL** (F), **Conseil de Filière STAF** (F, Coordinating Partner), **ICL** (UK), **IIRIAM** (F), **Irish Medical Systems** (IRL), GESI (I), **Lombardia Informatica** (I), **SIG Services** (NL), **Università Cattolica del Sacro Cuore-Policlinico A. Gemelli** (I). These organisations, from five European countries, include hospitals, specialised IT companies and manufacturers, ensuring, therefore, the maximum synergy of the different expertises, to propose advanced solutions, aligned with the actual users needs.

The project was started in January 1989, and will last three calendar years. It will involve a total effort of more than 90 man/years, for a total budgeting investment of more that 10 millions of ECU.

According to such organisation, the objectives of RICHE can be summarised in four main issues:

– A common architecture for hospital information systems, where different applications (developed also by third parties) can be integrated and effectively cooperate to support the users' activities.

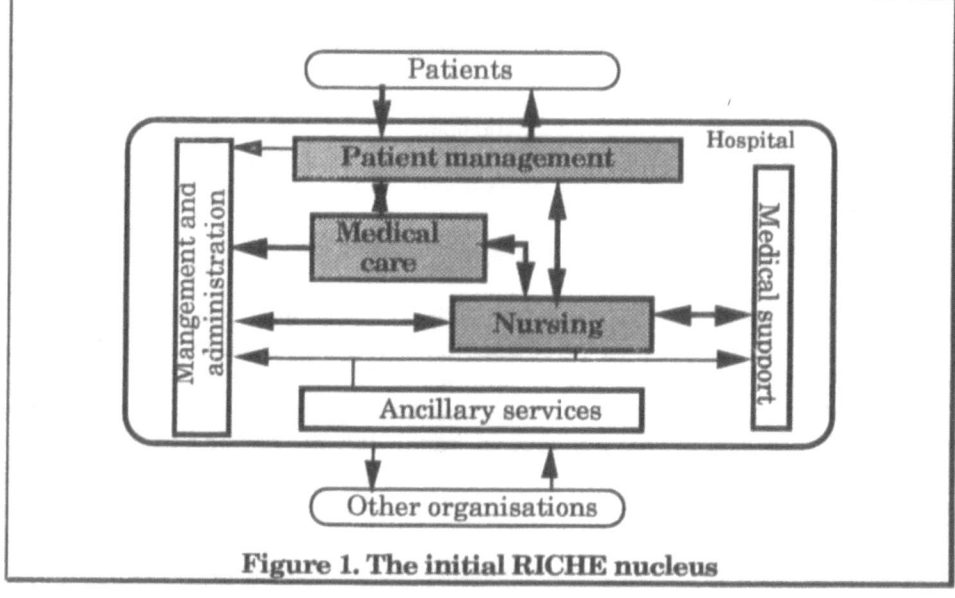

Figure 1. The initial RICHE nucleus

- Standard methodologies for the design and documentation of hospital information systems and local applications, from the initial formalisation of the users requirements and conceptual design, up to the physical implementation level.
- A common, standard conceptual model for the data and processes of a nucleus of professional hospital activities (as shown in Figure 1), applicable in all European countries and able to be customised and evolved according to local requirements, without loosing its common validity.
- A number of working, prototypal applications, esperimented in pilot units of some of the over 300 hospitals connected to the Consortium.

2. The RICHE methodological framework

Designing, developing and maintaining a hospital information system is a complex exercise, involving organisational, application and technological issues, mutually interacting. The effective synergy among the various elements has been considered a fundamental need, if -as in the RICHE case- a standardisation objective underlays the project, aiming at the identification and development of a common system, applicable in different Countries, regions and institutions, to provide a uniform basis enabling the reliable interchange of consistent healthcare information.

The adoption of formal methodologies supporting and coordinating the various activities has represented, therefore, a vital concern in order to ensure the consistency of the design and development tasks, up to the final result of an integrated system.

Moreover, it cannot be underestimated that, the healthcare scenario is (more than others) characterised by a high degree of complexity and variety. In fact, although the overall missions and objectives of hospitals are similar, each application must be able to concurrently support many types of different requirements (e.g. medical, managerial, administrative). In addition to this the various organisations are different each other for size, organisational aspects and technological constraints The definition of methodologies is therefore necessary also to enable a wider adoption of the RICHE approach (and marketable results) in the European market, through the specification of formal criteria supporting the analysis of the customisations locally requested and the integration of the common RICHE nucleus in the individual environments.

Different requirements and characteristics are specific of the various phases of the information system's life-cycle; as a consequence different methodologies are being identified, to efficiently and effectively support each individual activity, from the initial conceptual design, up to the physical implementation, in a complete and consistent *RICHE Methodological Framework* (as shown in Figure 2).

The adopted methodologies and models have been selected and refined in order to ensure their mutual consistence. As a consequence they support the continuous evolution of the various phases of the project, directly using the result of each task as input of the subsequent one, without the need of conversions or reorganisations.

672

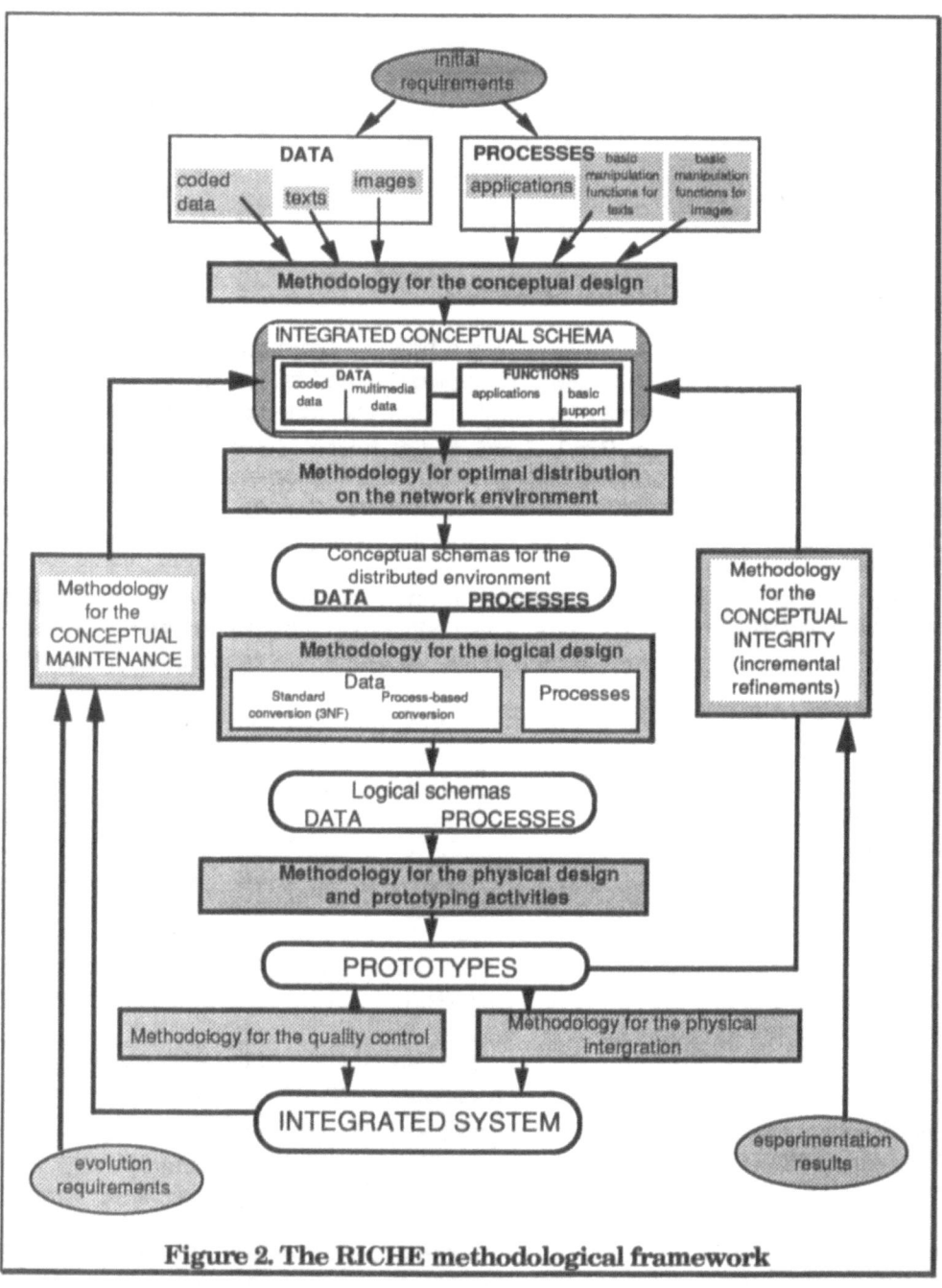

Figure 2. The RICHE methodological framework

Finally, it must be emphasized that, apart from the necessary support to the operational activities of the RICHE project, the envisaged methodological framework also has an autonomous validity, due to the possible significant contribution it can provide to the overall standardisation effort in the medical informatics. This strategic consideration has been also stated by CEN/ CENELEC, where it is presently active a Technical Committee (TC 251), in charge of the standardisation needs in medical informatics, with a basic group dealing with "Methodologies and models". In fact, defining common methodologies and models can facilitate the adoption of compatible strategies in the design of healthcare information systems, as well as the interchange of consistent information among the various organisations, due to the following benefits:

– The identification of a standard design methodology permits the comparison and integration of different projects, on the basis of a reliable working platform, able to ensure the completeness and coherence of the analysis and design activities.
– The definition of a standard conceptual model for the data ensures the formal (syntactic and semantic) compatibility of the data managed by different organisations, with the consequent reliability (and possibility) of the interchanging processes.
– The definition of a standard conceptual model of the processes ensures the complete consistency of the information managed by different organisations, due to the adoption of common procedures and criteria for manipulating the individual data.

Thanks to its overall approach, RICHE framework can provide a valuable contribution to the standardisation effort. As a consequence, also in cooperation with AIM programme, connections are being established with TC 251, which has formally requested to the RICHE Consortium to provide detailed documentation about the results of the projects.

In order to ensure the actual consistency of the RICHE applications with the users needs, actually users and experts are playing a significant and leading role up to the initial conceptual design phase, in the specification of their requirements and in the active definition of the conceptual models.

The strategic and innovative assumption, adopted by RICHE to carry out this task is that:

Users and experts must play a fundamental, leading role in the design activities

To obtain this result, it is necessary to identify a sort of common language (consisting of methods and models) between the technicians and the users, in order to permit them to cooperate in an active and effective way, without the risk of mis-understandings. In RICHE we have adopted the Entity-Relationship model to describe the data and the-Data Flow Diagrams to represent the processes of

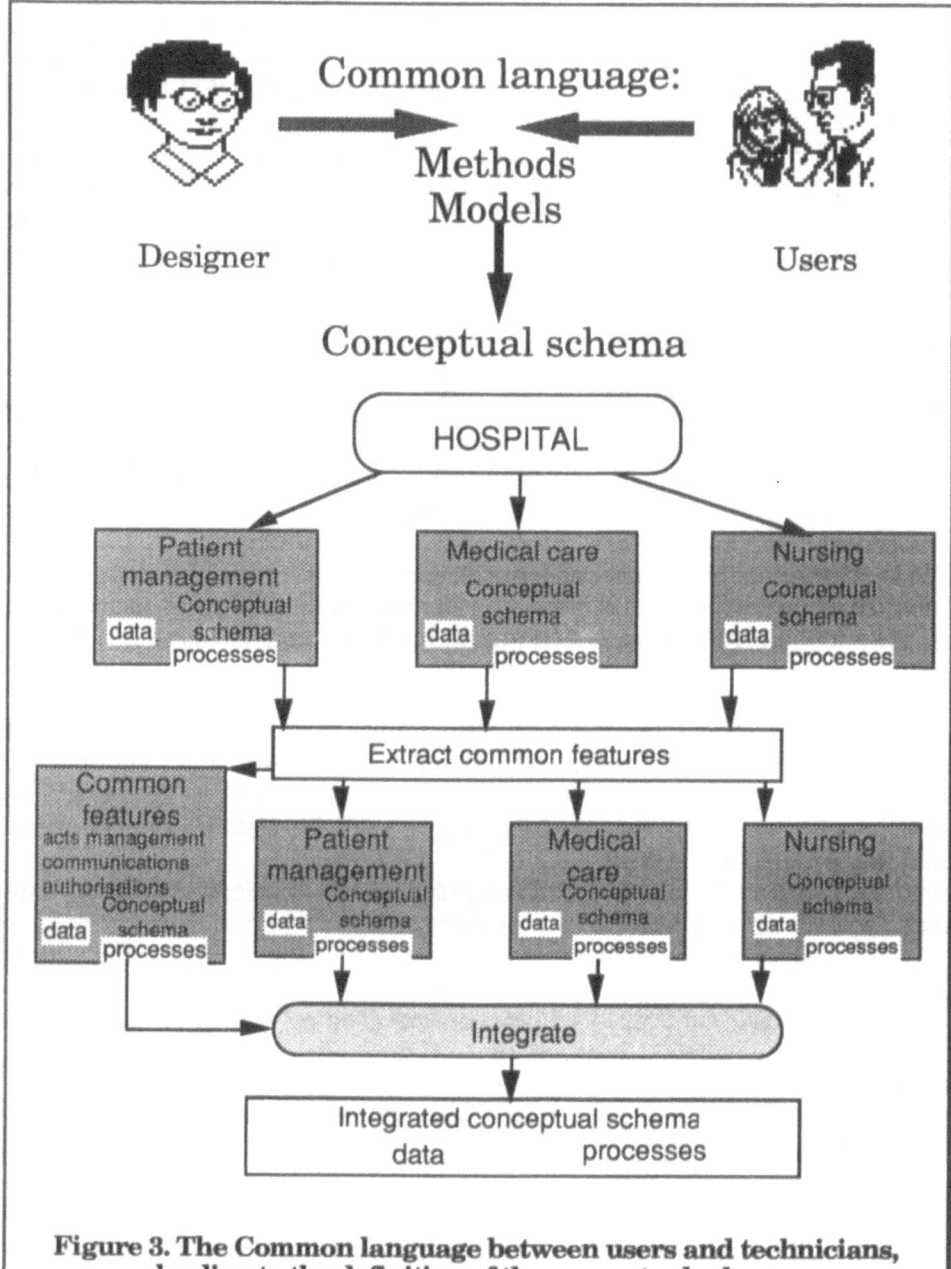

Figure 3. The Common language between users and technicians,
leading to the definition of the conceptual schema

interest. The result of this common effort is represented by the *Conceptual Schema* of the system: a formal description of the requirements, data and processes -subdivided according to the different areas of the hospital taken into account-, representing the system through (graphical) models and formal criteria not depending on any specific technological environment (as shown in Figure 3).

This working approach has demonstrated its validity; in fact the active participation of the users is being carried out both on a day-by-day basis in the design teams (i.e. nurses and physicians of Università Cattolica del Sacro Cuore-Policlinico A.Gemelli, Leiden University Hospital, Hospitals of STAF Consortium), and through panels of European experts, cooperating to discuss and validate the proposed solutions.

Also the detailed design activities of Work-Package 2, responsible of the conceptual design and architecture have been carried out according to this working methodology. The basic conceptual schemata identified by groups of (at least two) partners in cooperation with experts and users have been incrementally discussed and verified by the whole design team, during monthly meetings, and finally integrated into a unique consistent schema containing all the data of interest.

3. Present results and future evolution[*]

After one year of activities, the project has obtained significantly satisfactory results, which can be summarised in the following main items:

- A methodology for the conceptual design, starting from the user requirements.
- The detailed model (partially already validated at European level) for the data and the processes of the:
  - patient management, in the admission, discharging, transfer and patient identification areas
  - nursing area
  - medical care
  - unified management of acts

  The data model contains the detailed description of the entities and relationships of the various areas, as well as the formalisation of more than 1000 relevant attributes.
- The design of a standard communication system, to interface the external applications (already existing in the hospital)
- The overall design of the architecture, based on standard components (UNIX and relational DBMS), to obtain an open, flexible and modular system, customizable for local requirements.

MastER, the CASE system manufactured by GESI, has been adopted for the conceptual design phase, in the various phases of the complete methodology, up to the generation of the physical data structures.

---

[*] Up to April 30th, 1990

The conceptual schemata of data and processes (with Entity-Relationship and Data-Flow Diagrams) are stored in a project's database, easily accessible also during the future development activities. The availability of an automated system and of a project's database facilitate the interchanging of the standard RICHE models and their adoption in local applications.

An incremental prototyping is foreseen for the development activities. According to this approach, the live databases and software programs are iteratively developed and refined, starting from the basic conceptual specifications delivered by April 1990.

This approach leads to refine the specifications of the system, through its actual implementation. In a first cycle (approximately up to April 1991), a limited set of key functional elements will be developed, able to characterize the main services and infra-structure of the system. In the next cycles, the basic modules will be refined and additional functions can be added.

Both technical and operational advantages are widely acknowledged to such approach:

- The system may better support **the actual organisation's requirements**. An incremental development permits to quickly obtain working examples (i.e. programs, data elements, screens, reports, etc.), which can be shown and discussed with the users. This permits the designer to have an immediate feedback from the users, in terms of comments and suggestions for the developers, identifying the possible improvements and amendments.
- The **technical implementation** can more evolve in a more reliable way. In fact, developers have more time to familiarise themselves with the target technological platform and tools, identifying the best solutions to improve the technical efficiency of the system. Since formalized and rigourous development methodologies will be adopted, it will be possible to reduce critical paths in the whole project, by developing a first version of prototypes using consolidated technology (i.e. outside the integrated RIBA technical platform, jointly developed by RICHE and BANK 92 projects), and moving the system in the new environment, only when it will be ready and fully tested.

4. Main aspects of the RICHE architecture

The objective of the RICHE architecture is to represent an overall technological and information framework, where different applications can cooperate, sharing common data and services to support the users activities, as it is schematised in Figure 4.

From the technical point of view, this basic objective is achieved through the adoption of two fundamental, strategical guide-lines:

- The design of a **distributed system**, able to ensure the integrity and integration. The various software modules are designed in order to permit easy customisa-

677

tions, to optimise each RICHE installation with respect to the actual size and (logistic and operational) requirements of the Organisation.

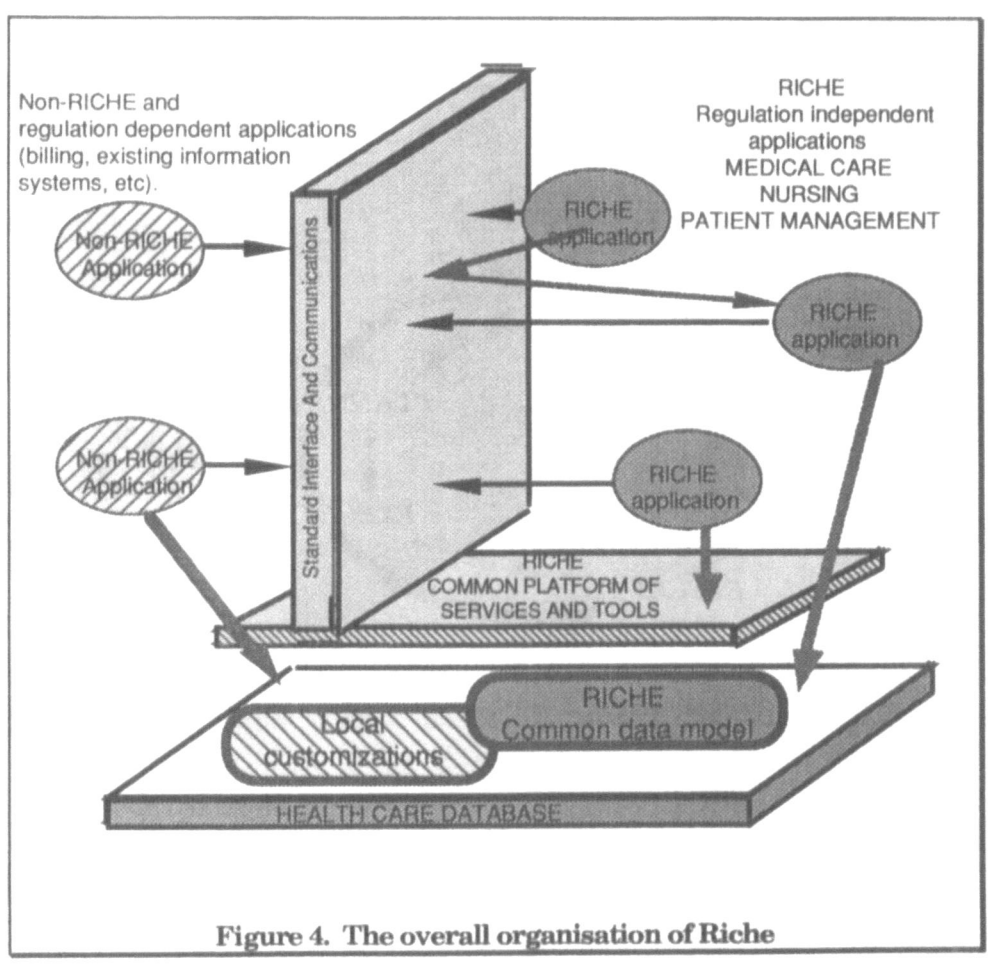

Figure 4. The overall organisation of Riche

- The adoption of **IT standards and open architectures** (UNIX and MS-DOS operating systems, as well as tools interfacing the major Data Base
- Management Systems), to ensure the integrity of different products, as well as the independence of the user from any possible (hardware or software) monopoly.

The unified management of acts

An act is every action of interest, performed in the hospital.

The life-cycle of an act

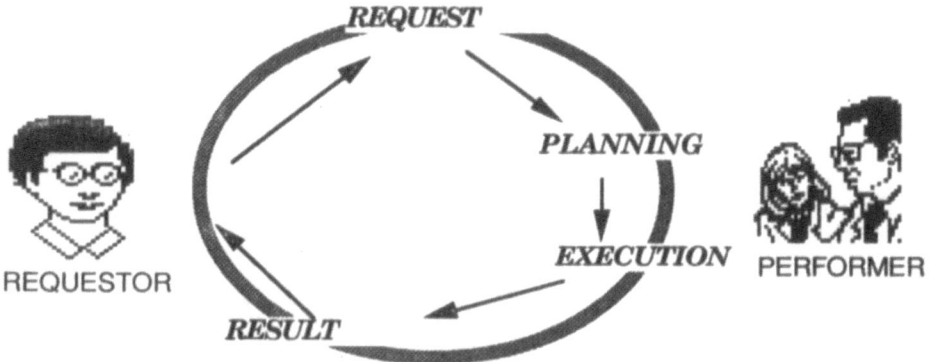

Examples of acts:

* a visit
* a consultation
* a radiography
* a lab test
* the subministration of a meal
* the subministration of a drug
* the registration of the vital signs

....

Figure 5. The unified management of acts

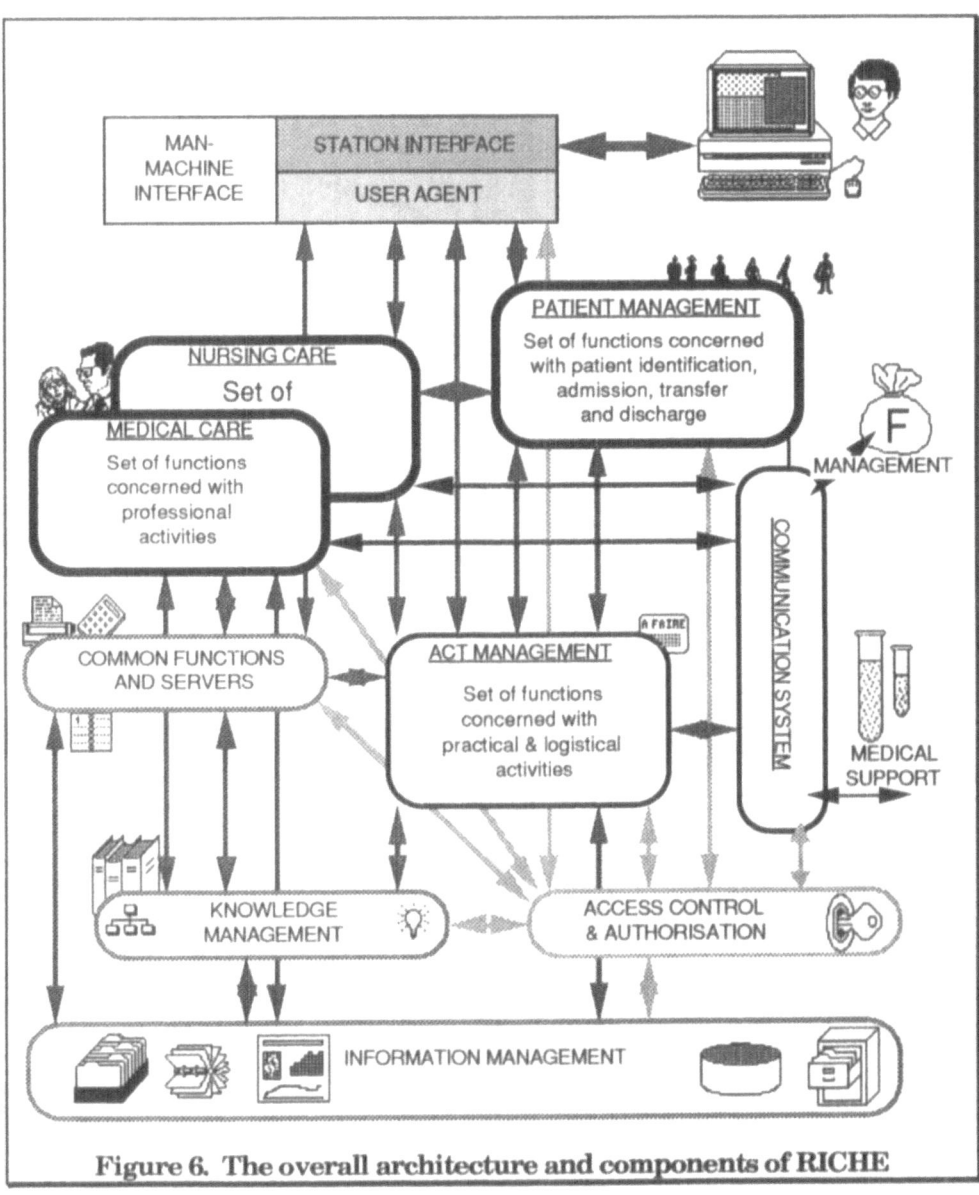

Figure 6. The overall architecture and components of RICHE

680

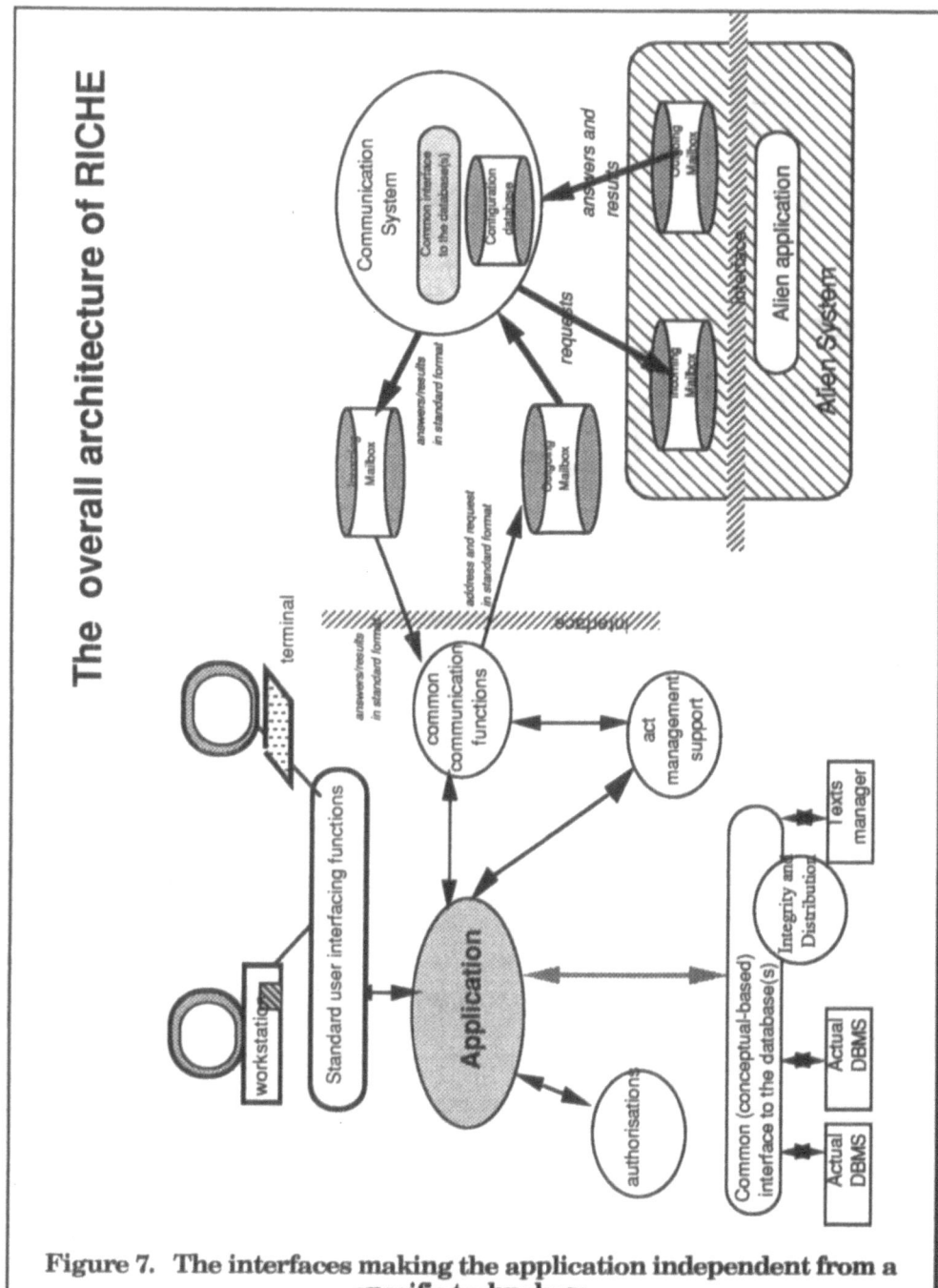

The overall architecture of RICHE

Figure 7. The interfaces making the application independent from a specific technology

On the basis of this consideration, as it has been shown in Figure 1, the project is developing the prototypes of applications aiming at the support of three professional areas of activities: patient management, medical care, nursing.

Such applications rely on two basic services: the **Management of Acts**, providing an integrated and knowledge-based support to all activities carried out in the various areas (Figure 5), and the **Communication System**, to interact and exchange information with the non-RICHE applications (either already existing or developed by third parties).

Figure 6 shows the overall architecture and mutual interactions between the various components.

From the technical point of view, it must be emphasized that all RICHE modules have been designed in order to ensure the maximum level of portability and independence from specific environments and/or proprietary products, as shown in Figure 7.

According to this approach specific products are encapsulated into objects, interacting with the other applications through interfaces defined according to the industrial standards.

Particularly, in order to make the actual organisation of the data in the specific installation independent from any particular product as well as from their actual physical organisation in the network, a standard interface is foreseen between the application and the actual DBMS, as shown in Figure 7.

This interface (the **Information Manager**) will operate according to common Data Definition and Data Manipulation Languages based on the conceptual (Entity-Relationship) description of the data. This means that the application can refer to the data simply specifying the *concepts* of interest (e.g. patient name, visit date, etc.). It does not need to know the actual physical organisation (i.e. files and fields), which can change in the different sites, due to technical and optimisation needs.

Essential Bibliography

F. M. Ferrara (editor) "The RICHE methodological framework" - RICHE document PC-65-0200-001, Version 1.1

F. M. Ferrara (editor) "The conceptual model of processes" - RICHE document PC-65-0200-003, Version 1.0

F. M. Ferrara "The conceptual data model" - RICHE document PC-650200-002, Version 2.0

ICL and BULL (editors) "The System architecture" - RICHE document PC-65-0200-005, Version 1.1

D.Whysall (Editor) "RIBA platform, Release notes for version 0.0"

P.P.Chen "The Entity-Relationship model: toward a unified view of data" ACM Transactions on data base systems, 1976

H.Tardieu - D.Nanci - P.Ascot "Conception d'un System d'Information: construction de la base de donée" G. Morin, 1979

C. Batini - M. Lenzerini - S.B. Navathe "A comparative analysis of methodologies for database integration", ACM Computing surveys, 1986

F.M. Ferrara "An integrated system for the conceptual design and documentation of data base applications", Proceedings of the 4th International Conference on the Entity-Relationship Approach. Chicago, 1985

T. Catarci - F.M. Ferrara "OPTIMER: a tool for the automatic conversion from the conceptual to the logical design of data bases", Proceedings of the 7th International Conference on the Entity-Relationship approach. Rome, 1988

C. Batini - F.M. Ferrara "An integrated architecture for CASE systems", Proceedings of the CASE '88 Conference, Boston

F.M. Ferrara "Extending databases with natural language texts", Proceedings of CASE '89 Conference, London

H.Sakai - "Entity-Relationship Approach to Logical Database Design" - in C.Davis, S.Jajoida, P.Ng, R.Yeh (ed.) Entity Relationship Approach to Software Engineering, North Holland, 1983.

M.Schkolnick, P.Sorensen - Denormalization: performance oriented data base design technique - Proc. of Congresso AICA, Bologna, Italia, 1980.

G.Schuldt - ER based access modeling - in S.Spaccapietra (ed.) Proc. of the 5th International Conference on Entity Relationship Approach, Dijon, France, 1986.

T.Teorey, J.Fry - The logical Record Access Approach to Database Design - pp. 179 - 211 in Computing Surveys, Vol. 12, 2,1980.

T.Teorey, J.Fry - A Logical Design Methodology for Relational Databases Using the Extended Entity-Relationship Model - pp. 195 - 207 in Computing Surveys, Vol. 18, 2,1986.

Project No. 2267

Progress in Distributed Systems - an Overview of the ISA Project

D. M. Eyre, D.A. Iggulden
APM Ltd
Poseidon House
Castle Park
Cambridge CB3 0RD
United Kingdom

Summary

The ISA (Integrated Systems Architecture) Project is defining an architecture for distributed systems which is designed to allow heterogenous computing systems to work together in a wide variety of application areas. The main technical orientation of the work can be described as providing a suitable infrastructure for distributed object management within a larger framework that enables ISA systems to be described and manipulated from a number of points of view. This is more than, for example, the provision of a distributed file system. The architecture is known as the Advanced Network Systems Architecture (ANSA) and is derived from the Alvey project of the same name. This paper will review briefly the main features of the architecture as well as other features of the ISA approach including the provision of a working exemplar of the architecture, the work in standards and the structure of the project team.

1. The Rationale for Open Distributed Processing

Increasingly end-users are indicating a need to move away from a single-vendor central IT system policy and towards multi-supplier distributed systems. This growing market trend offers an opportunity to turn the fragmentation of European IT industry into a strength by developing a standard generic architecture for such systems, one which will enable integration and interworking of applications from a wide variety of domains using heterogenous equipment from multiple vendors.

The existence of an architectural platform for distributed systems will create important opportunities for European suppliers. It can be exploited not only by the major product-supply companies but also by skilled niche-suppliers, enabling them to integrate their products into a wide variety of IT systems and facilitating their ability to produce integrated multi-supplier offerings. By leading in the development of the architecture, important advantages are gained, for example an early ability to produce conformant systems.

The ISA project is developing and demonstrating a generic architecture for Open Distributed Processing (ODP) in multi-vendor multi-domain heterogenous systems.

As systems become distributed, the future of IT lies in the ability to achieve integration of computing and telecommunications; the ISA consortium has been structured to include strong representation by companies in both sectors so that each may benefit from the experience and insights of the other and that the architecture may cater for the needs of both.

2.The ISA Approach

The ISA project is concerned with themes of integration and standards. Integration of systems from multiple vendors will be achieved by creating a set of common architectural constructs for distributed computing systems. These architectural components and rules for their composition will for the basis of distributed computing standards. A third theme will the practical demonstrations of application interworking, based on standards derived from the project.

In pursuing these themes the project has a four-fold objective:

- to provide software and tools oriented towards technical standards, upon which applications of distributed processing can be constructed, and to develop an underlying rigorous basis and process for distributed systems design and implementation;

- to ensure - through active support of, and participation in, national, European, and international standards bodies - that there will be a widely accepted set of open distributed processing standards, onto which all ESPRIT distributed processing projects can converge and through which they can be introduced into a world-wide multi-vendor marketplace;

- to validate the proposed infrastructure and standards by enabling and encouraging the development of practical demonstrations; and

- to encourage European industry to take a lead in exploiting the ISA Architecture and the related standards especially in fostering liaisons with international groups concerned with defining and promoting functional standardisation;

and which it realises by deliverables in five major areas:

- architecture
- manuals
- standards
- software
- technology transfer.

3.The Architecture

Underlying the development of the architecture is the objective of providing complete descriptions of distributed systems together with a sufficient set of simple concepts to achieve this. This has been done by providing a sets of models within a framework of viewpoints. The set of models within a viewpoint provide a complete view of the distributed system being described but emphasize particular aspects or concerns. This principle of the ANSA architecture ensures therefore that the sets of models are strongly connected. There are five viewpoints in the ANSA architecture Enterprise, Information, Computation, Engineering and Technology. An introduction to these concepts is given in 'An Engineer's Introduction to the Architecture (APM 1989) and a fuller description and explanation in the ANSA Reference Manual (ARM 1989).

Strategically, the architecture articulates long-term goals for a distributed object-oriented programming environment. At present there is no single distributed computing environment which could accomplish all of the functions identified in the ANSA architecture, largely because parts of the architecture demand advanced operating system and network structures not generally available.

3.1 Technical Orientation

The ANSA architecture is a logical design for a distributed computing environment comprised of two major parts: a computational model and an engineering model. Both models follow a similar object-oriented approach to systems design and implementation. More generally ANSA is discussed in terms of five related models called enterprise, information, computation, engineering and technology. While all are necessary to a complete systems design, the computational and engineering models are the ones that most distinguish object-oriented distributed systems from those of process-oriented centralised ones. The ISA project, in developing the architecture, has so far concentrated on the development of computational and engineering models. The computational models allow application and systems programmers to develop suitable descriptions of the problem abstracted from unnecessary concerns. This is reflected in the adoption of the object-oriented approach and by the development of mechanisms for handling distribution transparency.

ANSA recognises the importance of distinguishing between a distributed computing environment as seen by an applications programmer and that as seen by a systems programmer. Many environments have an *application programmer's interface* as the dividing line between the two views. In practice this style of interface turns out to be clumsy and lets too much of the system detail show through. Moreover, checks must be made at run time to ensure that application programmers are working the interface correctly. ANSA has taken a programming language view; that is to say, distributed computing concepts should be represented by extra syntactic constructs added to existing programming languages. These can then be compiled directly into calls at the systems level. The advantages of this approach are threefold:

- a simple programming model for applications programmers
- checking at compile time rather than run time
- independence of application programmer view from system view.

The first two increase the confidence that application programmers have in their programs. The third provides for separate evolution of the two views of the environment, making applications and systems more 'future-proof'.

3.2 Architecture: Computational model

The computational model defines the programming structures and program development tools that should be available to programmers, in whatever application programming language they choose to use. This model addresses the topics of:

- modularity of distributed applications
- naming and binding of module interfaces
- access transparent invocation of operations in interfaces
- parameter passing
- configuration and location transparency of interfaces
- concurrency and synchronisation constraints on interfaces
- atomicity constraints on interfaces
- replication constraints on interfaces
- extending existing languages to support distributed computing.

Maximum engineering flexibility is obtained if all computational requirements of an application are expressed declaratively. This permits tools to be applied to the specifications to generate code satisfying the declared requirements. It leads to a clean separation between application programmers (stating the requirements) and system programmers (providing tools which use the requirements to generate template code satisfying those requirements in the environment in which the applications are to operate). For example, generation of stub routines from an interface specification permits error free marshalling/unmarshalling of application data between interacting objects.

The *interface* is central to the ANSA architecture. An object supports a group of interfaces, and an interface consists of a number of operations. It is the interface which is named and traded, and qualified by administrative attributes, such as being the unit of access control. Object Interface references therefore feature prominently in the computational model; an interface is called to obtain service, and interface references are returned as results.

A *thread* is an independent execution path through the system, and so may pass through multiple objects, interfaces and operations. Many *tasks* may exist within an object to provide threads with resources to make progress. Hence threads represent potential, and tasks real concurrency. Queueing must be imposed when the number of threads exceed the number of available tasks. Where a local operating system offers only one task, it is possible to obtain the illusion of multiple tasks via a coroutine package. Threads are provided by the interpreter through the computational model, and tasks by the ANSA platform through the engineering model.

Event counts and *sequencers* (Reed 1979) were chosen as the mechanism for providing synchronisation between threads. They are particularly suited to a multiprocessor environment since they minimise the extent of mutual exclusion necessary to obtain synchronisation and ordering guarantees.

3.3 Architecture: Engineering model

The engineering model defines a set of logical compiler and operating system components that realize the computational model in heterogeneous environments, namely:

- thread and task management
- address space management
- inter-address space communication
- distributed application protocols
- network protocols
- interface locator
- interface traders
- configuration managers
- atomic operation manager
- replicated interface manager.

The engineering model shows the system designer the range of engineering trade-offs available when providing a mechanism to support a particular function defined in the computational model. By making different trade-offs the implementor may vary the quality attributes of a system in terms of its dependability (reliability, availability, performance, security, safety), performance and scalability without disturbing its function. This is an important feature of the ANSA architecture since it decouples application design from technology to a significant degree. By conforming to the computational model, a programmer is given a guarantee that his program will be able to operate in a variety of different quality environments without modification of the source. The engineering model gives the system implementor a toolbox for building an environment of the appropriate quality to the task in hand. In other words, by making this separation it is possible to identify what forms of transparency are required by a distributed application and allow choice of the most appropriate technique for providing the required transparency.

Sockets, plugs and *channels*, are engineering views of client/server object connection. A socket is associated with a server interface, a plug with a client call, and a channel with a path or association between them. They are concerned with management of the **physical** resources needed to support object interactions, such as for multiple object invocations (or interface instances), concurrency control, and thread dispatching. A concurrency limit is a resource associated with a socket. If

688

resources exist when a message arrives, the message will be attached to a thread, otherwise it will be discarded and the client expected to repeat.

A *session* provides functions for resource management and synchronisation between the interpreter and thread dispatcher, and a cache for channel information. A session will be created for a sequence of client/server messages and will remain extant until it becomes inactive. The resources may then be re-assigned.

3.4 Architecture: Overall structure

The way in which the aspects of ANSA fit together is shown in Figure 1. The host systems represent the computers and networks used to resource a computing environment composed of distributed objects. The nucleus objects take the basic resources and combine them to provide a uniform platform, or environment, for distributed application objects. The transparency objects provide additional functions to those provided by the platform. Similar objects could be written by users to encapsulate existing applications, since inevitably users will wish to retain existing software investments.

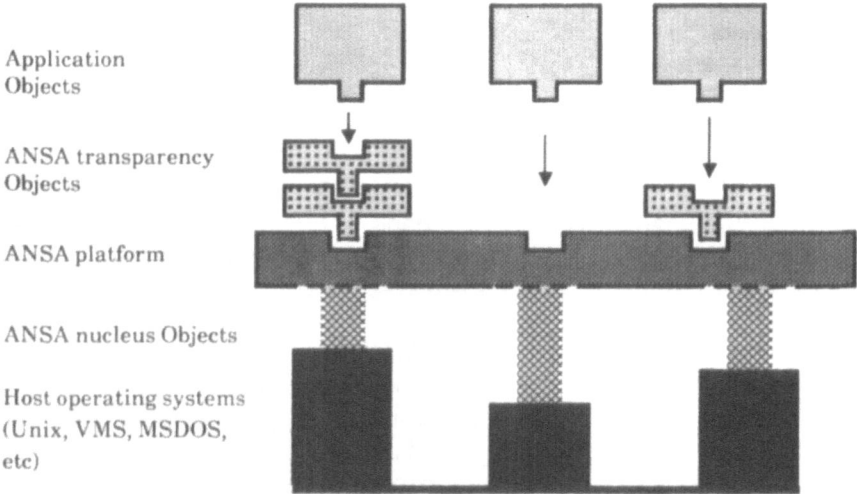

Application
Objects

ANSA transparency
Objects

ANSA platform

ANSA nucleus Objects

Host operating systems
(Unix, VMS, MSDOS,
etc)

Figure 1: An ANSA distributed object system

Distributed computing systems are seen as a way of facilitating cooperation within the communities they serve. Typically, distributed systems are constructed to permit one or more client systems to access services on one or more server systems. Trading is the process of choosing service offers such that they match service requests according to information supplied by community members. A trading service is often the crucial component in linking clients and servers in the distributed system.

The ANSA platform provides the environment for application object trading and interaction; with the ANSA Trader providing for registration of object interfaces, and the platform for subsequent dynamic binding between client and server objects.

The engineering model specifies the mechanisms needed to provide the various kinds of transparency and the protocols for interaction between nucleus components on different hosts. Application components are structured according to the computational model and the distributed computing aspects of the application are compiled into calls on the interfaces to the appropriate transparency and platform components.

It is possible to imagine many system implementations of the architecture which make different engineering trade-offs. To provide interworking between such systems it will be necessary to provide gateway functions, but this will be confined to simple interface adaptors that match the different engineering trade-offs rather than changes to the applications themselves. The engineering model can be taken as a template for implementation of nucleus, platform and transparency components, although this is not mandatory for either application portability across implementations, nor for interworking between them.

Many hosts will provide a range of functions and resources beyond those needed by the platform and will wish to contribute them to the distributed computing environment as potential application objects. This can be achieved by extending the nucleus with additional distributed computing environment interfaces that map onto the locally available functions.

3.5 Architecture: Advanced concepts

The ANSA architecture has been developed in an open-ended fashion with requirements for some advanced facilities in mind, although these are not yet present in the current software. Design specifications are either completed or partially prepared for: object groups (with object replication), transactions, protection and object migration.

4. Software functionality

The ISA project, in developing the architecture, has concentrated on the development of computational and engineering models as these are of immediate interest in providing support for product development and exploitation. The computational models allow application and systems programmers to develop suitable descriptions of the problem abstracted from unnecessary concerns. This reflected in the adoption of the object-oriented approach and by developing mechanisms for handling distribution transparency. Figure 2 shows a typical engineering model derived from the ANSA Architecture. An implementation of the engineering model has been provided as an example of the architecture which can be used to develop applications and explore the ramifications of distribution in particular instances (Figure 3). The ANSA Testbench is supported with a number of tools to support its

690

development, integration and use in typical distributed applications. A considerable number of ANSA Testbenches and supporting software tools have been distributed for teaching, research, and use in project development and demonstration.

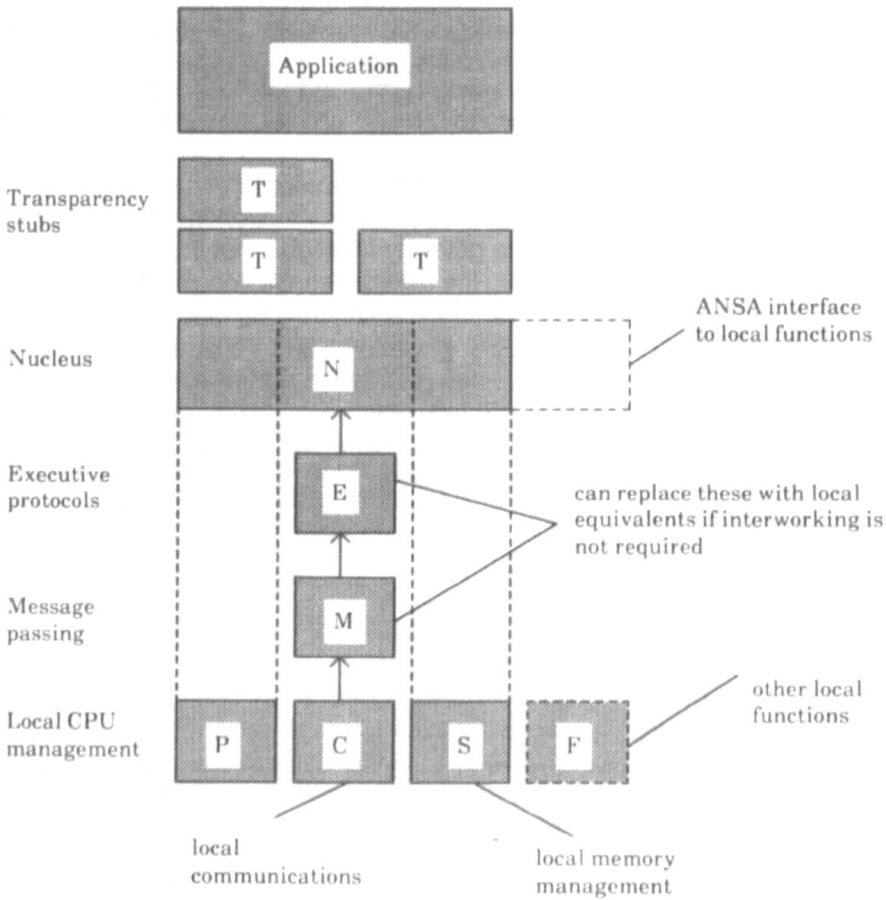

Figure 2: An ANSA capsule

The development and maintenance of this demonstration of the ANSA engineering model has now been put on a production oriented basis because of this widespread use. The documentation of the Testbench is part of the ANSA Reference Manual and is updated with releases of the Testbench.

The current version of the Testbench has been engineered to provide appropriate support for future code deliverables of the Project. The latest release (3.0) provides support for factories that allow the instantiation of multiple objects within a capsule and, within an object instance, the instantiation of multiple interfaces. A notification service is provided which allows objects to register their interest in the existence of other objects and which will inform them of any subsequent termination. A node manager has also been added which provides an architectural method for managing the services available on a particular node. Services may be provided statically or on demand. The Testbench Implementation Manual provides the necessary details. The Testbench is distributed with examples of programs using X11 and the Testbench communication facilities.

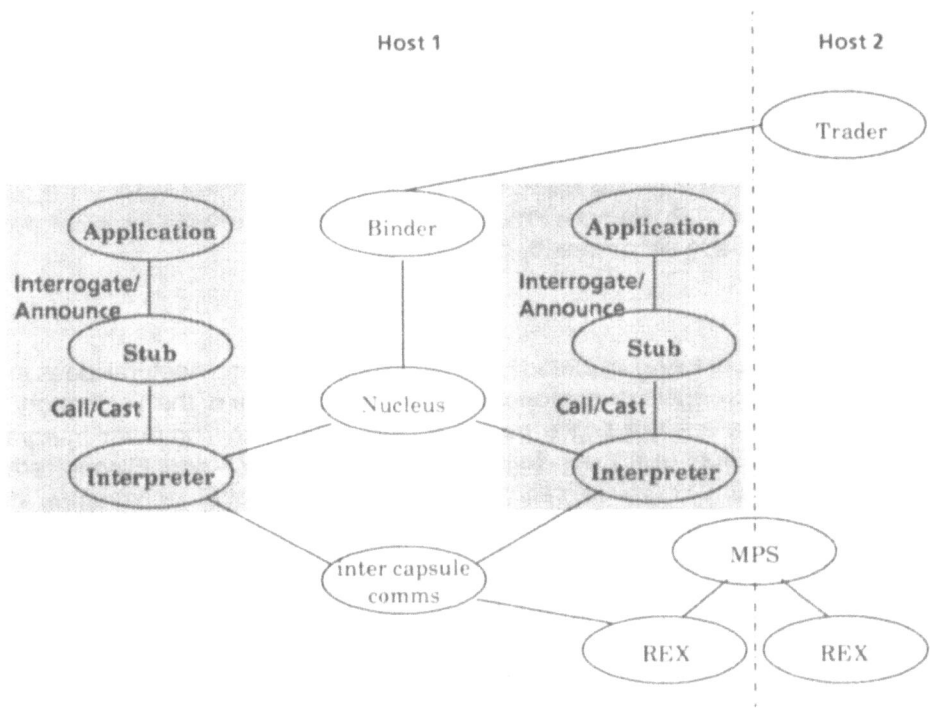

Figure 3: The engineering model

5. ISA and Standards

Standards are an essential part of the development of distributed processing systems. This was recognised early on the ANSA phase of the project and strong efforts have been made to introduce the architecture into standards work. The main activity has centred on the ISO/IEC JTC1 WG7 Open Distributed Processing project where project members are active at the national and international level. In this particular forum the ideas of the ANSA Architecture have been accepted and incorporated into the working draft of a prescriptive model of Open Distributed Processing.

There are two other standards activities where the project is active through the participation its members. The first is ECMA whose technical reports are directed to the ISO ODP work and the second is CCITT whose work on a Distributed Applications Framework has a technical orientation based more on telecommunications but which nevertheless has a strong overlap with the ISO work. This overlap shows itself in a number of project members who contribute to both activities. An agreement recently made between ISO and CCITT for joint working and this is expected to lead to joint text. Work has also started on specific standards related to the ODP framework, notably Remote Procedure Call, and plans to generate new work items, for example, on trading, are emerging as the framework activity matures. Other relevant standards activities, reflecting on the large scope of the topic, such as document architecture, dictionaries, application programming interfaces, user architectures, database reference models, upper layer architecture, etc are kept under review by the team. Contributions to the ECMA work on Support Environments for ODP, Remote Procedure Call and Open Systems Architectural Framework are made either directly or by review.

6. Validation

ISA concepts are being validated by the employment of architectural ideas in the development of products or demonstrations and by the use of the Testbench. For example Siemens are building a management system for Computer Integrated Manufacturing (CIM) using the Testbench and AEG, GPT and Ellemtel are all building a number of communications related products using architectural ideas from ANSA and in some instances employing the Testbench. Other projects that are exploiting the Testbench to develop projects and demonstrations and at the same time providing feedback into to the ISA architectural work include Riche/Bank92 led by STC, who are also developing a multi-media system, the Advance/Race project on Integrated Broadband Networks, and the Multiworks multi-media architecture project. Chorus are porting the Testbench onto their operating system for use by SEPT in an office document exchange project. The ANSA Architecture together with the Testbench is also being used in Technology Transfer and Training as part of special University courses. On a wider front the Testbench has been used in the NASA ADS - Astro-Physics Data System - which has over 10,000 users at a large number of scientific institutions spread across the United States and which encapsulates six very large databases of differing designs. Efficient implementation and scalability were two features that made the Testbench attractive for this project.

7. Liaisons

The project is establishing or has established liaisons with a number of projects, in some cases as noted above, where the ISA work can be put to immediate practical use but also with others where it is important to align, or at the very least exchange, concepts and philosophies. Among these are the Comandos project and CIM-OSA where plans have been made for joint working to enable, for example, the CIM work in Enterprise models available to ISA and for the ISA Testbench to be used for interworking between CIM components. The first phase of such work is seen to be that of an alignment of philosophies with the ultimate aim of producing harmonized architectures and concepts. A new Esprit project (ISA-DEMON) will build on results of the ISA project to add a novel graphical Design and Monitoring tool to ANSA.

Further liaisons are also envisaged with particular industry groupings with an interest in establishing industry norms or harvesting research in particular areas. This work is often alongside and complementary to the work of regular standards bodies. Of particular significance is the liaison with the Object Management Group (OMG) where both groups have agreed to work together in furthering the development of object standards. Interest has been shown in the ANSA Architecture by a number of industry groupings, including OSF, Unix/International, X/Open, the Network Management Forum, and the Open Access Group which is interested in remote database access over non-OSI communications.

8. The Team and the Partners

The project is somewhat unusual in that it has a laboratory at a single location where employees of the management company and secondees from the Collaborators work on the Architecture and the development of the Testbench. The Collaborators in addition to direct involvement with the laboratory, are concerned with validating the architecture by developing demonstrators and products. Information is disseminated throughout the project by document release and by workshops and staff rotation. The work on standards is co-ordinated through document exchange and is monitored at regular management reviews. This has proved to be suitable method for ensuring that a common view of the work of the project is presented to the appropriate standards bodies though it could be elevated into a more formal process if the current debate on standards within the CEC results in a more European focus to standards involvement.

APM (1989) 'An Engineer's Introduction to the Architecture', Architecture Projects Management Ltd.

ARM (1989) 'The ANSA Reference Manual', Architecture Projects Management Ltd.

Reed, D.P. & Kanodia, R.K. (Feb 1979) 'Synchronisation with Eventcounts and Sequencers', Communications of the ACM, 22(2), 115-123

Project No 2294
(TOBIASI)
Tools for Object-Based Integrated Administration of Systems

Managing Management
The TOBIAS Approach

V. Kolias
Intrasoft S.A.
Athens Tower
2 Adrianiou & Papada st.
GR-115 25 Athens
Greece
Tel. +30.1.6917763

I. Lefebvre
Intecs International S.A.
385 Avenue Rogier
1030 Brussels
Belgium
Tel. +32.2.7355571

L.F. Marshall
University of Newcastle upon Tyne
6 Kensington Terrace
Newcastle upon Tyne NE1 7RU
U.K
Tel. +44.91.2228267

R. Mackenzie
GIE Emeraude c/o Bull
38 Boulevard Henri Sellier
92154 Suresnes Cedex
France
Tel. +33.1.39024094

P. Moukas
Planet S.A.
Apollon Tower
64 Louise Riencourt st.
GR-115 23 Athens
Greece

Summary

Management of computer systems is an area that has had little attention paid to it by researchers or developers. So the tools and the interfaces used by system managers have not matured in the way that those provided to the ordinary user have. With the introduction of distributed systems into the workplace, administration has become significantly more complex and the inadequacy of current system management support is becoming obvious. The TOBIAS project is developing an integrated support system for managers that will help to ease these problems. To this end we have developed an object oriented model of system administration and this paper describes its key aspects. These concern the way in which management policies can be implemented and enforced in a dependable way.

Introduction

Considering how important a well managed system is to any computer installation, the tools used by system administrators have received little attention from software designers. This has not proved to be a major problem in the past when sites consisted of small numbers of large machines often with their own operators. But as the scale, power and complexity of computer systems increases so to do the problems of the system administrator. Faced with poorly designed, command driven interfaces and little protection from their mistakes, system managers find it hard to provide the service to which users have become accustomed. Standardisation work on the system managers' interface (ISO 1988, Smith, Quarterman, and Vasilatos 1989) is in progress but it will be some time before its results will be sufficiently developed to be useful in real situations.

Luckily, the improved reliability of today's hardware has meant less time spent on maintenance and recovery. Unfortunately the increase in the number of machines in systems is eroding this advantage. This increases the likelihood of hardware failures and the scheduling of regular maintenance becomes more involved. If the systems installed are different the problems are worse as the system manager must now have expert knowledge about all the platforms available. This often reinforces the traditional "guru" image of system managers which itself leads to problems. Their employer immediately loses the specialised knowledge that they have built up should they change jobs or have an accident. This lack of continuity can have a serious impact on the functioning of a business.

Software suffers from the same problems as hardware—upgrading a program across several hundred machines safely and consistently is almost impossible, even if they all have the same (or similar) architecture. Add to this the need to ensure that users can only access the systems that they require and it becomes obvious that many system managers have more responsibility than they can cope with. Unfortunately, the design of most operating systems is such that the privileges needed to carry out these administrative tasks are too dangerous to be made generally available. Thus it becomes difficult safely to spread the work load over more people. Also, only a small number of system managers can work together effectively as communications' problems start to arise when there are too many.

The most common solution to this problem is to organise system support into several independent, administrative domains. Though this reduces the number of machines that particular managers have to deal with, it introduces the possibility of conflicts and inconsistencies between domains. Often these can be ignored, but where users use facilities provided by several domains they will not expect to have to change their way of working for each system they use. For this to be possible some higher level of coordination must exist.

The aim of the TOBIAS project is to help solve some of these problems by providing a safe, easy to use interface for system administration. To do this the project team is developing a model of the system management process that will drive the implementation stage of the project. This paper describes the basic structure of this object oriented model.

Fundamental Ideas

Underlying the model are two concepts mentioned briefly above. Before looking at the TOBIAS view of system administration it is important that the reader understands these properly.

Domains

The idea of using domains as a structuring principle for organizing complex systems occurs in many areas of computer science, but it is most often met in connection with computer networks. Domains provide an excellent method for coping with the physical reality of machines, their inter-connection and their administration. The principle of grouping like with like that they embody (no matter how "like" may defined) is valuable and is particularly helpful for organising higher levels of systems. Yet, it is hard to define a domain in a way that will meet with everyone's approval and there are differences between people's definitions (ANSA 1989, Moffett 1989, Sloman 1987). In particular there is often disagreement over whether domains can overlap, though it seems that this difference is not significant as it always possible to map overlapping domains into non-overlapping ones. The TOBIAS model makes considerable use of domains and this will be described below in more detail.

Policies

Policies concern "the business objectives of the organisation operating a system and the effects of the system's use on the performance of the organisation" (ANSA 1987). They may therefore seem divorced from the day to day running of a computer installation. Still, decisions made at high levels with respect to, for example, security or availability, will frequently have extensive effects on what a system manager must do. Besides, policies are not always explicit. By making people aware of their existence, the kinds of interactions that take place in their systems (and the reasons for them) will become much clearer. Mentioned above are two good examples of policiesthe decision to split a system into administrative domains and the subsequent maintenance of consistency between them. At many sites this division will be natural and will not need to be imposed, though this does not imply that there is no policy involved. It is just that a policy adopted elsewhere in the organisation (for example, the departmental breakdown) has made itself felt indirectly. Maintenance of consistency will usually involve explicit policies, such as controls on equipment purchasing, and will require checks to be made that managers do obey the rules. The TOBIAS model does not attempt to describe policies directly but does introduce ways in which their schemes can be realised and monitored. The following section describes these.

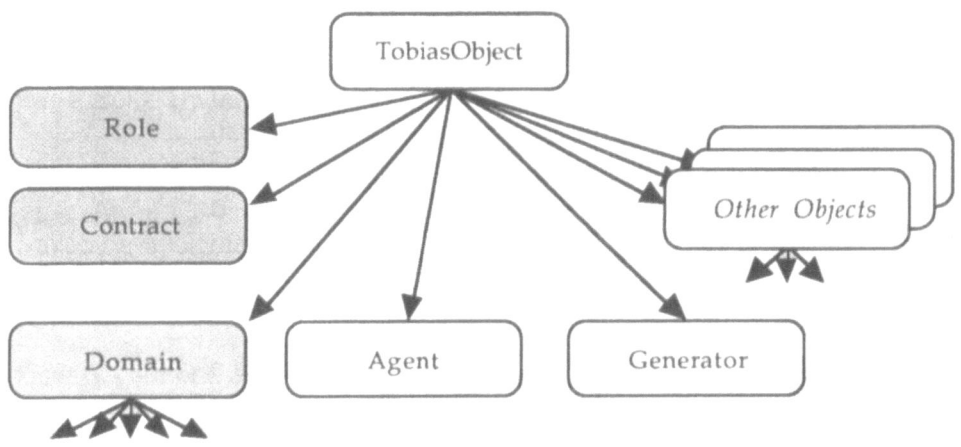

Fig. 1.

The Model

The TOBIAS model must cope with the many different kinds of entity that a manager meets when administering a system. To do this and avoid having an unmanageably large number of different types of object, it is essential to have a small set of classes that abstract out the specific properties of particular entities. These include such things as a Device class that is the basis for describing such things as printers and terminals, a User class to describe individual users of the system and a Medium class that represents tapes, discs etc. The details of these classes are not relevant here. What follows concentrates only on those classes that are central to modelling the management process. The development of the model has been driven both by the experience of the members of the TOBIAS project and by a systematic study of the system management aspects of the UNIX[1], VMS[2] and PCTE systems. All the classes in the model derive from an abstract base class and figure 1 shows this structure.

Domain

Domain is the base class of what are described as structuring objects. Its function is similar to that of the Collection class in Smalltalk (Robson 1983). That is, it implements a "bag" into which objects can be put. There are no restrictions on the types of object that can be put into a Domain nor any on whether the same object can be inserted more than once. Thus a simple tree structure can be modelled by

1 UNIX is a registered trademark of AT&T in the USA and other countries.

2 VMS is a registered trademark of Digital Equipment Corporation.

having Domain objects inserted into other Domain objects. Frequently this will be too general and specialisations of the class will be made. Thus the class UserGroup derives from Domain and can only contain User objects and other UserGroup objects. Another derivation is the Zone which is a collection of System objects, which are themselves Domains that contain many different kinds classes of objects.

The most important derivation from Domain is the class ManagementDomain. This can contain only objects of the filter classes Role and Contract and of the class Agent, all of which will be described in the following sections. The part ManagementDomains play in the TOBIAS model is described in the example given below.

Filters

There are two filtering classes in the TOBIAS modelthe Role and the Contract. Together these form the heart of the mechanisms that control system management activities. This control is achieved by enforcing the principle of "need to know" and by making sure that administrators meet the responsibilities determined by the policies currently in force. That is, managers can manipulate only those objects that they can control and no others. They are responsible for carrying out clearly defined tasks involving these objects and if responsibilities are not met the agents may incur penalties.

Role objects provide support for "need to know". Looking at the dictionary definition of role is perhaps the best place to start when trying to understand what Role objects do:

"role, rôle rol, n. a part played by an actor: a function, part played in life or any event" (Chambers 1988)

Another way of looking at a role often used in conversation is, for example, to speak of someone "wearing their hardware hat". It is this aspect of management that Role objects capture. When carrying out a particular administrative task, a manager only wants to see those objects used for the task in hand. Any others are a distraction and their presence may even be the cause of an error. Role objects provide this cleaned up view of the environment by filtering the collection of objects "visible" to a system manager. They only allow through objects whose classes match one contained in the list that characterises each Role. For example the Role "user manager" might only allow objects of the classes User, UserGroup and Domain to be seen by a manager who had adopted it. In the same way, Roles also filter the operations that managers can carry out on the visible objects. The classes of object and operations allowed through the filter are what characterise a particular Role.

To add an even finer "need to know" control and to bring in the dimension of responsibility we use Contracts. The idea of using contracts to control the interactions in a complex system is not new (Dowson 1987, Kopetz 1989, Stenning 1986), but their use seems restricted to those areas that directly involve humans. Meyer (Meyer 1988) introduces the idea of programming as a contract where pre-conditions represent obligations on a client and post-conditions obligations on a server. We wish to expand and generalise the notion and define a *contract* as

"an agreement negotiated between objects regarding the kind, quantity and quality of their interaction."

More specifically, a Contract is an agreement between a ManagementDomain and a managed object (or collection of managed objects) that certain services, defined by particular Roles, will be carried out. The collection of managed objects may be specified either directly by naming the objects, or by the provision of conditions that objects must meet before they fall into scope of the Contract.

Contracts are therefore the glue that holds the system together and which will ultimately define the level of dependability that can be achieved from a given system. Figure 2 shows how their position in the general scheme of things.

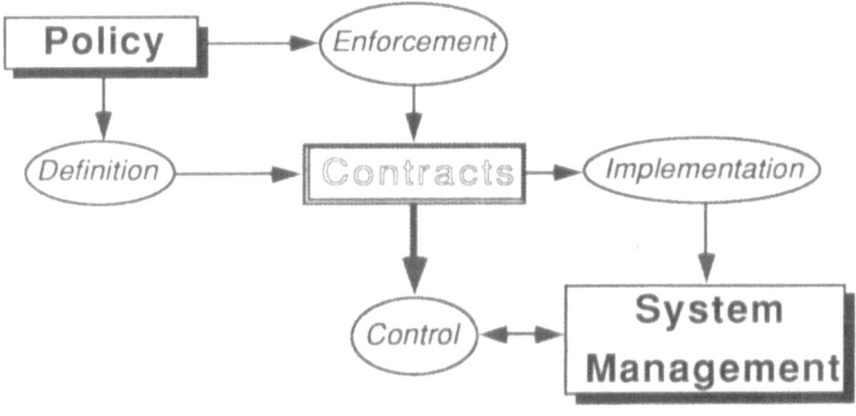

Fig. 2

In environments with a centralised management structure it is likely that the contracts between objects are defined and fixed by the policies of the parent organisation. However in more loosely structured environments there may be a considerable amount of negotiation between interested parties before contracts are settled and the process will be similar to that of *trading* (ANSA 1989).

In general, a site system manager will offer a set of services (e.g. file archiving and user management) to any computer owner who wishes to make use of them. In return for receiving these services, the computer owner must make available certain facilities and privileges to the system manager, and it is these that will be the subject of the negotiation between them. This is usually an informal process that does not result in any kind of binding agreement. However TOBIAS formalises this and encapsulates the resulting agreement so that it can be used for controlling the activities that take place when the contracted services take place.

There are two ways in which such a contract can be used in a system. First, it can be used statically as a guide for system configuration. The contract specifies exactly what the activity to which it relates requires. To carry this out the system need only include those objects involved in it. Everything else can be omitted when putting together the system. This increases its dependability considerably as now it can only perform the tasks it was specified to and no others. The second use is for dynamic control. Some contract requirements are not amenable to static control, for example cpu utilisation restrictions, and so must be monitored at run time. Selecting a particular Role within a ManagementDomain activates Contract objects. The manager will then have access to the objects specified in the Contract and no others. A contract will normally be between a System and a ManagementDomain, and there is no need to identify particular individual objects within the System as the Role mechanism will filter out those that are inappropriate. Not that a contract can never identify specific objects, only that mostly it will be unnecessary.

Agents

Agents represent the TOBIAS usersthat is, the system managers. They are distinct from User objects which stand for the users of the managed systems. Agents get their powers by being members of management domains and thus becoming subject to the responsibilities placed on them by the contracts that are also part of these domains. As mentioned above, agents may become liable to penalties if they do not meet these responsibilities. This implies that Agents as well as being "responsible for" certain tasks, are also "responsible to" other agents further up the management hierarchy. We do not envisage that the TOBIAS model will develop this aspect of management at this stage, but it presents an interesting avenue for future work.

One further point to note about Agents is that they can represent non-human system managers. They can just as well stand for processes implementing routine administrative tasks or providing the services of an "expert advisor" to other Agents. This freedom allows use of the model in describing such things as mail daemons and tasks automatically initiated at specific times of day.

Generators

Generators are not directly connected with the modelling of the management process, but exist to solve a perennial problem suffered by administrators. When asked what their biggest bugbear is, system managers often identify the work involved in adding large numbers of users to a system. This occurs, for example, in universities where there is mass registration of students at the start of each academic year. The idea behind Generator objects is that they can create a stream of new objects based on a prototype provided by an Agent. These prototypes will have "mail merge" like facilities so that parameters can be supplied to configure each object created. When copied or moved to another object the effect will be as if many individual copy/move operations had taken place. They also could provide a form of lazy evaluation, objects from them not coming into existence until some other object needs them.

System Structure

The management policies currently in force determine the domain structure of any environment. In particular, the management domains, the agents who are members of them and the roles determined for each domain must be made explicit. Figure 3 shows a management domain (Domain1) whose Agents can take on any of the three roles A, B or C. Domain2 could be any other domain the system, even another ManagementDomain. There is a contract between Domain1 and Domain2 for the former to provide the services defined by role B. Any Agent who is a member of Domain1 can do this, and it is also possible for Domain2 to check that the services requested have been carried out.

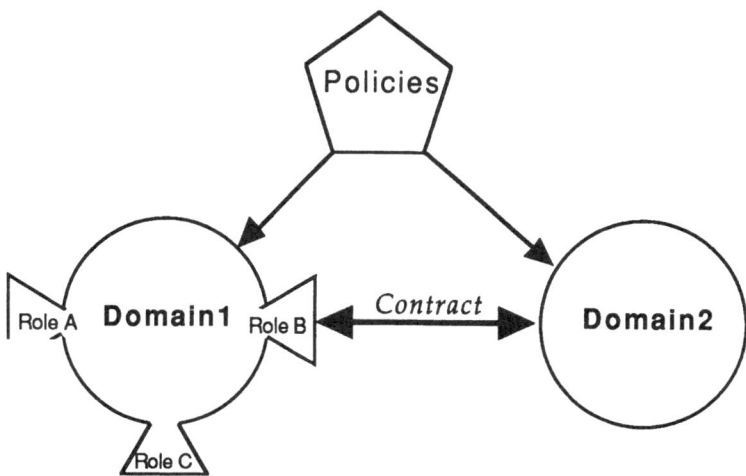

Fig. 3

An Example

In this section, we will give an example of how the model relates to real world system administration by giving an idealised description of some objects and operations on them that might take place when the Role "User Manager" is active. Throughout the example we will assume the existence of an underlying persistent object store that provides a system management knowledge base (SMKB).

First a system administrator (SA) must authenticate herself in some way to the TOBIAS system by identifying herself with a particular Agent. Once this has succeeded the system retrieves information about this agent from the SMKB. This will contain various housekeeping data about Agent preferences, but the most important information will be the set of ManagementDomains of which the Agent is a member. The Agent can then open one or more of these domains and will be presented with a variety of Roles that she can activate. Selecting a particular role, also will select those contracts involving the current ManagementDomain providing the services of the selected Role to other objects. The agent will now have the capability to navigate through those parts of the managed system that are referred to (explicitly or implicitly) in the active contracts.

Having adopted the role 'User Manager', an Agent will be able to "see" User and Group objects as well as the various structuring objects that may contain objects of those classes. The agent can carry out the following activities:

- create/delete/modify a User
- create/delete/modify a Group
- add/remove a User to/from a domain (e.g. a System)
- add/remove a Group to/from a domain.
- validate the consistency of the user and group information

A User object will have the following basic properties, some of which must be supplied on creation and some of which may be defaulted:

- real world: information about the person associated with this object (see below)
- login name: the name the user uses to login in to a system
- password: the SA can modify the password of the user, but cannot read it.
- groups: the groups to which this user belongs
- system: system specific information e.g. uid number
- quota: any quotas that may be allocated to the user
- security: information about the user's capabilities

Group objects, as mentioned above, are specialised Domains that can only hold User objects. They can therefore be treated just like other structuring objects that an Agent can access. However, it is possible for a User Manager to create and destroy Groups as well as to modify they so an Agent's access rights to them are more comprehensive than for other Domain instances such as those that describe systems. Group objects have properties similar to those of User objects though the uses to which the various pieces of information are put may be different. For instance the real world information for a Group will show someone who is responsible for all the users in the group (say a project leader or departmental manager).

The environment the Agent is operating in is completely controlled. She can see only those objects that she should and can perform only those operations determined for her Role. Thus it is perfectly safe to allow non-specialist users to control some aspects of user management. For example, the addition and deletion of users involved with a particular project could be done by the project administrator when staff join or leave rather than by the system manager at some later stage. This fine level of control achieved by the use of the Contract objects. An Agent in ManagementDomain X looking after *all* of system P's users will have a simple contract represented by the tuple < X, "UserManager", P >. That is, the contract is between the ManagementDomain X and the object P to provide all the services of the Role UserManager. A more restrictive contract, suitable for the manager of a specific project can be shown by this tuple- < Z, "UserManager", P & User("group") = "project1" <. Here the extra condition in the tuple would be applied to any User objects encountered in P and only those that met this condition would be made visible to the Agent.

Present Status

At the time of writing (August 1990) a prototype version of the system has been demonstrated that supports the TOBIAS Model. This runs on SUN computers and provides a MOTIF style interface allowing a TOBIAS user to take on the User manager role and carry out the tasks associated with it. The basic filtering activities of Roles and Contracts are implemented and these are being expanded to provide finer levels of control. Currently those parts of the SMKB that are not system specific (for instance Management Domain information) are built in to the running prototype but work is being done so that TOBIAS can interface to existing data management systems. The system specific data, such as user and group information, are collected dynamically by the TOBIAS Object Controller that runs on each managed system. Future development of the system will see the inclusion of support for backup management and for roles more geared towards system monitoring such as performance management.

Conclusions

None of the individual elements used to build the TOBIAS model are new. In fact some have the advantage of being tried and tested over many years. However the combination of these elements is new. It provides a representation of the system management process that is closer to the way that administrators work than most other models. The implementation of the management environment based on this model will provide several advantages when used with large systems.

First it will allow responsibility for administrative tasks to be spread over a much wider set of people. The use of authentication mechanisms alongside the filter structure will provide tight control over the actions an individual manager can carry out.

Secondly, and for the same reason, TOBIAS need not restrict the freedom of action of more experienced users. When surveyed by the POSIX 1007.3 committee, many system managers expressed a fear that the introduction of standards would limit their "creativity". The structure of Roles and Contracts described above is

x

a

adaptable enough to allay this fear and thus removes what could be a considerable stumbling block to the acceptance of any management workbench.

Finally, the object oriented nature of the model encourages the generalisation of management activities. This means that the interface provided to system managers can be consistent across many different operating systems and machines. The consequent flattening of the learning curve will mean that system managers can be more productive and less error prone in their every day tasks.

References

ANSA. 1987. *ANSA Reference Manual Release 00.03 (Draft)*. Cambridge, UK: Advanced Networked Systems Architecture.

ANSA. 1989. *The ANSA Reference Manua Release 01.00*. Cambridge, UK: Architecture Projects Management Ltd.

Chambers. 1988. *Chambers English Dictionary.* Edinburgh: W.R. Chambers and Cambridge University Press.

Dowson, M. 1987. An Integrated Project Support Environment. *Proc. 2nd Sigsoft / Sigplan Symp on Practical Software Development Environments* SIGPLAN Notices 22 (1) : 27 - 33. *SIGPLAN Notices* 22 (1) : 27 - 33.

ISO. 1988. *Information Processing SystemsOpen Systems InterconnectionBasic Reference ModelPart 4: Management Framework.*, ISO/DIS 7498-4.

Kopetz, H. 1989. Personal Communication. MARDS Project, Technical University of Vienna.

Meyer, Bertrand. 1988. *Object-oriented Software Construction*. Edited by C. A. R. Hoare. Prentice Hall International Series in Computer Science. Prentice Hall.

Robson, Adele Goldberg and David. 1983. *Smalltalk-80The language*. Edited by M. A. Harrison. Addison-Wesley series in Computer Science. Addison-Wesley Publishing Co.

Sloman, M. 1987. *Distributed Systems Management : A Report*. Department of Computing, Imperial College, London.

Sloman, Morris S., and Jonathan D. Moffett. 1989. *Domain Management for Distributed Systems*. Proceedings of the IFIP Integrated Network Management Symposium, Boston, May 1989

Smith, Susanne W., John S. Quarterman, and Alix Vasilatos. 1989. *White Paper on System Administration for IEEE 1003.7*.

Stenning, V. 1986. *An Introduction to ISTAR*. In *Software Engineering Environments*. Edited by I. Sommerville. 1 - 22.

Project No. 2404
(PROOF)

Primary Rate ISDN OSI Office Facilities

B. Patel et al.
3 NET Limited,
Ringway House, Bell Road,
Daneshill, Basingstoke,
Hants. RG24 0QG. U.K.

PROOF (Primary Rate ISDN OSI Office Facilities) is an ESPRIT II project which began in January 1989 and is due to end in mid-1992. The project looks at the issues of interworking between existing Ethernet LANs and the ISDN in the distributed office environment.

PROOF has reached a point where most of the system investigation and architectural definition has been completed and the scope of the work ahead is better defined. The aim of the project is to demonstrate OSI applications, supported by interworking Primary Rate Integrated Services Digital Network (PRISDN) and Local Area Networks (LANs), in a distributed office environment.

This report details some of the areas of investigation within the PROOF project.

1. Introduction

PROOF is a 3,5 year project, funded by the Commission of the European Communities (CEC) under the ESPRIT II work program; it began on the 1st January 1989. The project involves four European partners, three industrial and one academic. 3 Net from the UK, is the prime contractor, the others are Nixdorf Computer AG in West Germany, System Wizards in Italy, and University College London. The project will culminate in a demonstration of distributed office applications supported by interworking Primary Rate Integrated Services Digital Network (ISDN) [1] and Local Area Networks (LANs). Two important applications, message and document handling services, will be used to prove its effectiveness.

To accomplish its objectives, PROOF is divided into five distinct Work Packages: Gateway Development (WP 1), ISDN Usage and Interfacing (WP 2), LAN Integration (WP 3), Application Support (WP 4), Application Integration and Demonstration (WP 5). All work packages lead up to the final integration and demonstration of the applications contained in WP 5 and each is considered in turn below.

2. WP 1 - Gateway Development

This work package is concerned with the development of the PROOF gateway between the LAN and the ISDN. This involves both hardware and software specification and design. The work package takes into account work from other work packages; in particular Task 2.1 'ISDN Environment Study', Task 3.1 'LAN Gateway Facilities', Task 3.2 'Protocol Design' and Task 4.4 'Management Services'.

The work undertaken so far first concentrated on the overall architecture of the gateway and then progressed to the hardware and software design and implementation. A report [2] has been delivered detailing the architecture of the gateway, an overview of which is given in figure 1.

The factors considered in the design were:

- Modularity and flexibility
- Performance
- OSI standard conformance
- Class of service
- Layer of interworking
- Addressing and routing
- Management
- Security

The gateway will operate a Connection Oriented (CO) service based on X.25 and will perform as a Connection Orientated Network Service (CONS) relay. It will interface to ISDN via the Primary Rate interface at the S/T reference point. This is defined, in Europe, as a 2.048 Mbps interface with 30 "B" Channels intended to carry a wide variety of user information, a control ("D") Channel intended to carry signalling information and a synchronisation/error channel. The bit rate of each of the channels is 64 Kbps. This primary rate interface is commonly known as "30B + D" and is outlined in CCITT I. series recommendations [1].

The gateway will also interface to an Ethernet LAN. This interface is defined by ISO specification 8802-3 [3] and is known as Carrier Sense Multiple Access with Collision Detect (CSMA/CD).

In terms of hardware the gateway will be based on Motorola 68000 series microprocessor technology and the Siemens' Primary Rate interface chip set. The gateway will use the industry standard bus VMEbus [4].

The PROOF gateway will consist of two modules; a Router Module and an Ethernet Service Module.

The Router Module contains the interface between the Primary Rate ISDN (PRISDN) and the Service Module. It provides the following features:

- Primary Rate interface
- Trunk monitoring, maintenance and initialization functions
- Initial call control functions
- Routing of calls from the Primary Rate ISDN network to the Service Module
- Management functions

The Ethernet Service Module is the main work unit of the system and will be designed around the VMEbus standard at REV C.1 (IEEE P1014).

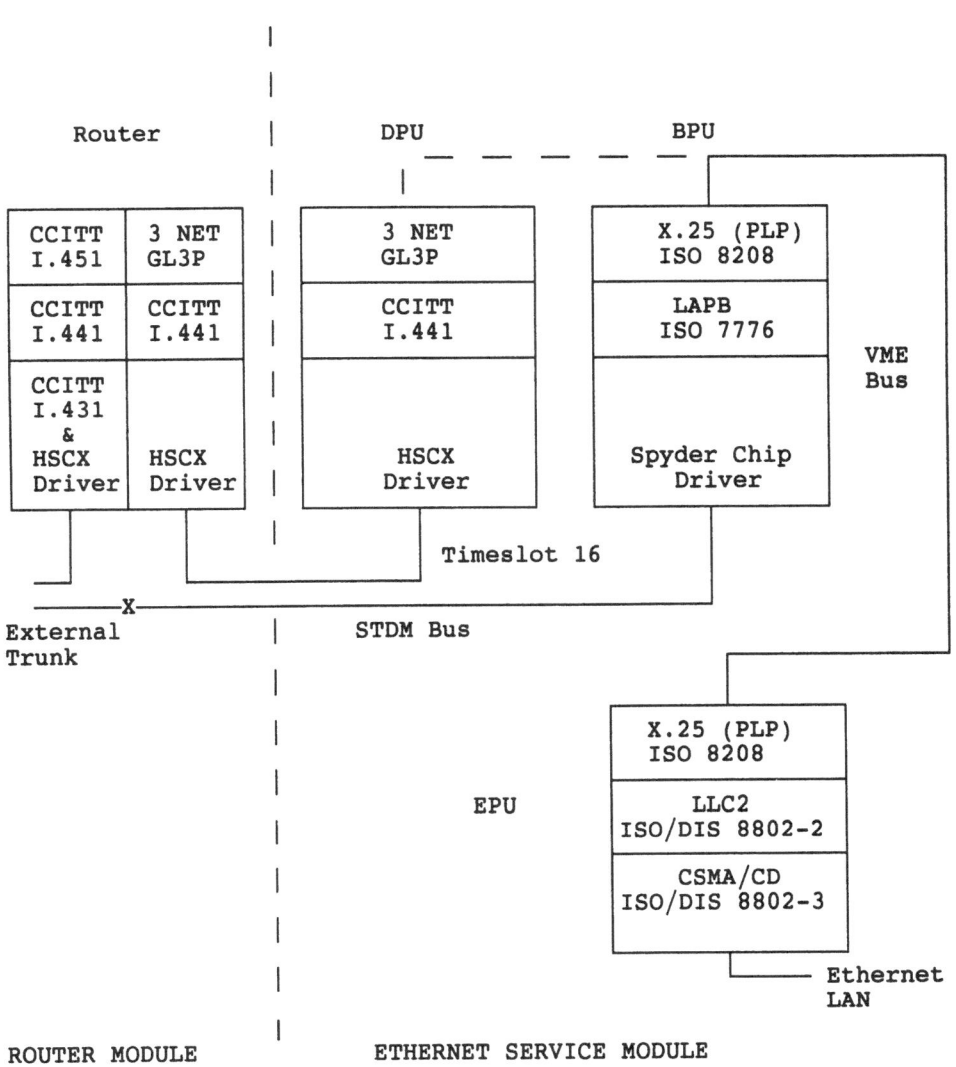

Figure 1 Gateway Architecture

The Ethernet Service Module consists of 3 separate processing units :

- D Channel Processing Unit (DPU)
- B Channel Processing Unit (BPU)
- Ethernet Processing Unit (EPU)

and two bus interfaces

- Serial TDM bus
- Parallel VME bus

The DPU's primary function is to access the D Channel signalling information present in timeslot 16 of the ISDN Primary Rate trunk, enabling the software to interpret and act on the ISDN protocols present. Once set up, the BPU/EPU combination will run independently until told to break the connection by the DPU or by inband signalling.

Similarly, the BPU is responsible for all B Channel data processing. It supports a 32 Channel HDLC controller for data to be transmitted in the 64 Kbps B Channels over the ISDN. The BPU interface to the EPU is via the VME bus. The BPU formats the data arriving from the EPU into protocol data units which are then transmitted to the Router over the STDM bus.

The EPU provides the interface to the Ethernet LAN and in the initial development the EPU board will be off-the-shelf.

Above this hardware foundation there will exist a suite of software ranging from the operating environment to the supporting communication protocols.

At the heart of the software system there is a real-time operating environment called 3 NOS [5]. This performs services such as task, memory, time and event management and allows for inter-task communication.

The communication software which operates in this environment is based on the OSI 7 layer model [6] and the interfaces comply with CCITT and ISO standards.

The ISDN processing software is layered and conforms to the I Series recommendations I.431, I.441 and I.451 [1], for Layers 1, 2 and 3 respectively.

The B Channel processing software will be connection oriented and based on the X.25 protocol. This consists of the physical layer, LAPB for Layer 2 as defined by ISO 7776 [7] and the network layer as specified by ISO 8208 [8].

The Ethernet processing software is layered with the physical, MAC and network layers conforming to ISO 8802-3, ISO 8802-2 [9] and ISO 8208 specifications respectively.

Design and implementation is continuing with special emphasis on routing and addressing, management and bandwidth issues for transmission rates greater than 64 Kbps.

3. WP 2 - ISDN Interfacing

The subject of this work package was the investigation and determination of the directions to be taken in the area of PRISDN within PROOF.

The factors which had to be considered when proposing feasible solutions were:

- compliance with international standards
- constraints implied by the PABX (The problem of digit sequence integrity has to be solved)

709

– flexibility in coping with various bandwidth demands

Even though the interface structures recommended by the international standard bodies allow only for a mix of B, H0 and H11 Channels with or without D Channel, the solutions finally adopted should provide enough flexibility to meet future requirements of multiple-slot connections in general, i.e. to meet bandwidth demands of an integer number of B Channels (n * B Channels).

Investigation into the actual situation of National, European and International standards related to the primary rate access interface has shown that the CCITT D Channel protocols allow the handling of n * B Channels. Furthermore, ideas for possible Channel structures and bandwidth allocation schemes for this interface have been developed.

One major problem associated with the switching of n * B Channels is how to make sure that the information within the n time slots forming an n * B Channel remains unchanged during switching operation. This problem arises due to the internal characteristics of the time/space switches used in the switching fabrics. Investigation of ideal switches, i.e switches without time delay, was made to identify solutions for multi-Channel switching. A companion paper [8] covers this work detail.

The final results were that

– digit sequence integrity for n * B Channels can be guaranteed, if the incoming and outgoing time slots concerned follow special time conditions
– the time conditions in an ideal switch can easily be taken into consideration
– the complexity increases when considering real switches and the component delay which has to be dealt with

The functionality of a real switch was defined in terms of a matrix which allows an easy decision as to how Channels must be switched in order to maintain data sequence integrity.

Concerning the PABX two solutions have been proposed for further consideration.

They are based on the assumption that an n * B Channel could occupy any time slots available (called arbitrary grouping) and that empty time slots are sequentially hunted. Whether or not allocation strategies which could compensate for different grades of service will be applied has yet to be decided, but could be adopted in both solutions.

In the first solution the information sequence is maintained through all internal switches involved by observing certain time conditions. In the second solution the peripheral entities of the PABX are responsible for restoring the correct sequence.

Another task to be done in Work Package 2 is to develop a primary rate interface for a server computer and to install the necessary protocol software.

The protocols to be supported by the server are:
D Channel: The normal I.420 set will be used.

B/H Channel: layer 4:   ISO TP Class 0 or 2 (ISO 8073)
                          layer 3:   X.25 PLP (ISO 8208)
                          layer 2:   LAPB (ISO 7776)
                          layer 1:   CCITT I.431

First the hardware will be developed based on the following hardware concept:

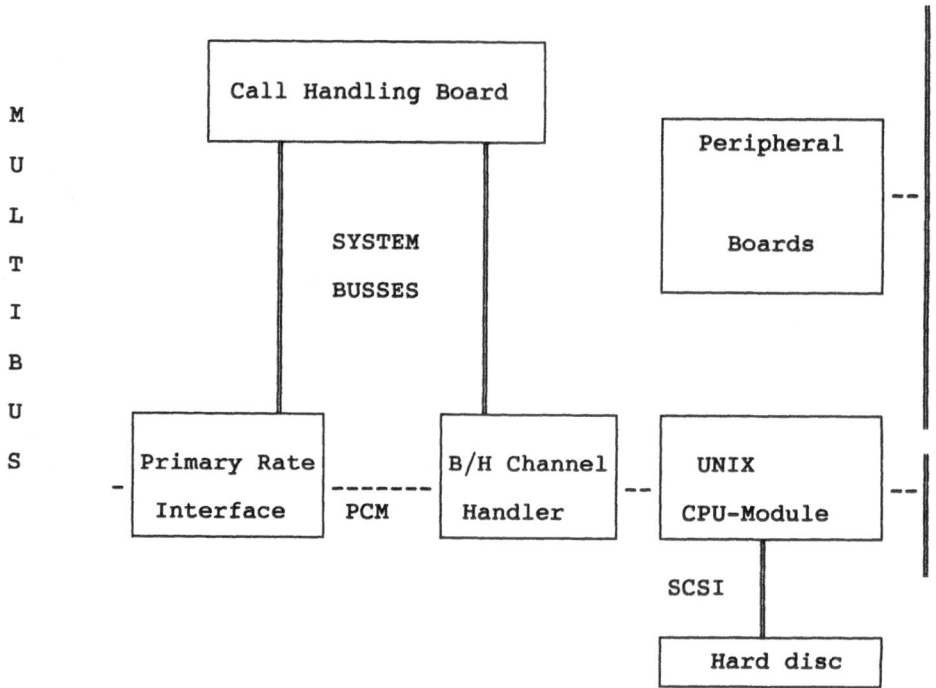

At present, the work is focussing on the B/H Channel handler which is a plug-in board on top of an application independent CPU module. This processor module has a hard disc interface and runs UNIX. The X.25 PLP and the upper protocols will be handled on this board.

4. WP 3 - LAN Integration

This work package is concerned with the LAN side of the gateway, its specification and design. As part of this work package, we have considered how the gateway should fit with the OSI and CCITT reference models, the protocols that should be used, and how such issues as naming, addressing and routing should be tackled. The main PROOF gateway will employ CO protocols, however, in order to gain some insight into the alternative Connectionless (CL) architecture, some experimental work has been done and a pilot (CL) version of the PROOF gateway has been built.

4.1 LAN Gateway Architecture

The main conclusions for the LAN gateway architecture are:

i) The gateway should handle packet-mode traffic only and the architecture should enable existing LAN applications to gain transparent access to the ISDN.
ii) The gateway should terminate all the ISDN signalling protocols.
iii)The gateway should make the decisions about when ISDN circuits should be opened and closed.
iv) It should be assumed that the only service available from the ISDN will be the 64 Kbps circuit-switched one - a Layer 1 service as far as OSI is concerned.
v) From the point of view of the OSI model, the gateway should operate as a CO Network Service (CONS) relay. Ideally, both Connection Oriented and Connectionless working should be supported, together with interworking by means of a transport service bridge.

Unfortunately, there are only sufficient resources available to implement the CONS version at present.

4.2 Protocols

The protocols selected for use in the CONS gateway are:-

i)   On the LAN:

       ISO 8878 [10]   Use of X.25 to provide the CONS
       ISO 8881 [11]   Use of X.25 PLP in local area networks
       DIS 8208   X.25 PLP

       DIS 8802-2   LLC-2
       DIS 8802-3   CSMA-CD MAC Layer

       ISO 8802-3   CSMA-CD Physical Layer
       ----

ii)   on the ISDN (B Channels):

       ISO 8878   Use of X.25 to provide the CONS
       DIS 8208   X.25 PLP

       ISO 7776   HDLC

| CCITT I.431 | Primary Rate Layer 1 |
| CCITT G.703 | Physical/Electrical characteristics |

----

iii)   On the ISDN (D Channels):

| CCITT Q.931 | ISDN User-Network Interface Layer 3 |
| | Basic Call Control (=I.451) |

| CCITT Q.921 | ISDN User-Network Interface Data Link |
| | Layer (=I.441) |

| CCITT I.431 | Primary Rate Layer 1 |
| CCITT G.703 | Physical/Electrical characteristics |

The project has now begun the detailed discussion of addressing and routing. The use of the CCITT format NSAP addresses will be encouraged with the IDP containing the public network address of the gateway and the DSP being allocated according to ECMA 117 [12]. This will enable routing to be done at least partly through an algorithmic usage of the NSAP address, thus reducing the size and complexity of routing tables.

4.3 Connectionless Gateway Experiments

Many LAN applications are based upon a CL Network Service. Typical of these is "transparent file access" in which a disc is mounted remotely and accessed across the LAN. In order to assess the feasibility of extending such applications so that they work across the circuit-switched ISDN, two experiments [13] have been undertaken.

In the first, measurements were taken of file-access traffic on a LAN. The experimental set-up consisted of a single-user workstation and a file server; the times and lengths of all packets transmitted over a period of one hour were recorded. Subsequently, the timing information was modified to take account of the additional delays that would be imposed by transmissions at 64Kbps. Where there was a prolonged burst of network activity, the delays built up to an unacceptable degree. This was especially noticeable in the case of traffic resulting from random access to a remote software library; here, the random access defeats the caching strategy. However, it appeared that less intensive activities, such as remote access to a mailbox, are perfectly feasible. In these cases, 91% of the packets were delayed by less than one second.

In a second analysis of the measured data, the effects of different strategies for timing out idle ISDN circuits was considered. If circuits were closed after an "idle timeout" of 10 seconds then, typically, about 16 disconnects would occur in an hour's session. The cost of a one hour session based upon British Telecom's current peak-rate telephone tariff would then be around $L = 2.20$. If the idle timeout was increased to 60 seconds then the number of disconnects dropped to 6 and the cost increased to about $L = 3.50$. With no disconnects, the cost would have been $L = 8.80$.

The second CL experiment is still continuing and involves the construction of a VME-based Primary Rate ISDN interface for a powerful Unix workstation. The workstation already has the capability of acting as a CL Network relay (a "router") and so the ISDN interface has only to manage ISDN circuits and perform mappings between Network addresses and ISDN numbers.

Most of the software for this gateway is now complete and hardware testing is underway. We expect that the complete system will become operational during the second quarter of 1990. The system will be used to verify some of the conclusions from the experimental study and will act as a platform for the testing of circuit-management strategies.

5. WP 4 - Application Support

This work package is aimed at supporting tools which are required to be able to run the applications planned for WP 5.

The main activity lines of this work package are:

- Provision and usage of the appropriate hardware and software for mounting the applications (the so called ISO infrastructure) which is a pre-requisite for most of the applications testing and for all the interworking. This requires, of course, provision of the appropriate Multi-user Servers, Single-user workstations, Local Area Networks, and gateways. In addition, interworking accordance with the OSI standards is required although the partners are adopting different implementations of the OSI stacks as part of their on-going equipment development. An example of such an ISO infrastructure is the ISODE system used and distributed by UCL, but, in general, the provision of the basic OSI stacks depends mainly on background software produced in other projects.
- Provision and usage of Directory Services, because several of the applications (e.g. Message Services, and Document Services) have a requirement for Directory Services. Several partners have developed, or have on-going development projects, in Directory Services. They provide these Directory Services as background to the work package, and they will upgrade the implementation in the light of the experience of the applications (including authentication and security).
- Study and investigation in the Authentication and Security Services fields in order to identify other applications for the appropriate facilities, namely:
  - end to end services (e.g. message origin authentication, content confidentiality, non-repudiation of origin, and replay detection)
  - connection-based services (e.g. access control and connection confidentiality).

- Assessment of the suitability of the OSI standards for Network Management which are gradually maturing for the management of a gateway such as the PROOF one. Integration of rudimentary OSI network management tools available as background in the PROOF gateway.
- Continuous work throughout the project of studying, participation and possibly monitoring of the Standards activities in the PRISDN area, in the Directory Services area and in the Security and Authentication and Network Management areas.

In fact, demonstrations of working applications on top of ISO Infrastructures have been given at both the UCL and System Wizards sites during two reviews of the PROOF project. In particular, during the October 1989 review X.400 [14], X.500 [15] and Network Management applications have been demonstrated at UCL, whilst during the March 1990 review X.400, X.500 and FTAM applications have been demonstrated at System Wizards interworking with X.400, X.500 and FTAM applications at UCL; System Wizards and UCL have different ISO Infrastructure and different implementations for X.400 and X.500.

Furthermore, since the beginning of 1990 X.400 has been heavily used as the E-mail service within the PROOF project.

In the area of the Directory Services, System Wizards has distributed to the partners the specification and the implementation of a general purpose DUA programming interface at the end of 1989. In the meanwhile, UCL has included some developments in its QUIPU Directory Service such as enhanced security, faster searches, extended attribute handling, algorithmic choice of which DSA to contact and alignment to the latest standards.

In the Security and Authentication areas, a study has been completed which considered the security requirements and mechanisms for PROOF. Emphasis was given to the special characteristics of the ISDN. Broadly, the conclusions were:

- Security requirements normally relate to applications; transport media such as the ISDN do not alter these. This means that security services and mechanisms which have already been deemed suitable for OSI should be suitable for PROOF. Essentially, this means the adoption of the X.509 [16] scheme.
- Notwithstanding the above, the ISDN does provide some assistance with authentication by providing identification of the communicating parties. This may be used for access control at the boundary with the public network. Operation normally requires subscription to the appropriate ISDN supplementary services. Where these are available, they should be used by PROOF systems.

The analysis of PROOF requirements, and the conclusions reached have been set out in the PROOF deliverable D4.3a, titled, 'Security and Authentication Facilities In The PROOF Project'.

In the Management Services area, the existing UCL OSI MIS software has been more fully integrated with ISODE. Studies have begun on the management of the

PROOF gateway and a discussion document has been prepared. Further discussion documents [17] have been prepared defining a Management Information Base (MIB) for some of the protocols to be used in the PROOF and UCL gateways. Studies have been made of draft ISO MIB proposals as well as the proposals related to the SNMP [18] and CMOT work in the US Internet community.

Finally, in respect of Standards Activities, each partner has participated in and tracked standards activities which are most relevant for its work in PROOF.

3 NET has been actively involved with standards bodies working on interfacing between ISPBXs and other installations, and on interworking between DPNSS [19] and the CCITT ISDN protocols.

Nixdorf is attending German working groups and making investigations regarding terminal connections to ISDN, provision of data services over ISDN and interfacing ISDN-PABX.

There has been project participation in the Directory Services area and documents on Access Control have been submitted to meetings by both SW and UCL.

UCL represents the project at standard meetings in the areas of Security and Management.

6. WP 5 - Application Integration and Demonstration

This work package brings together the work done in the project and results in a demonstration of a number of applications over an integrated ISDN Primary Rate network. One of the principles behind the PROOF gateway architecture is that it should provide transparent connectivity at the network level. This means that its introduction should have little or no impact on the applications. Consequently, work at this stage of the project is concentrating on enhancements to the applications themselves. As the PROOF demonstrator network becomes available, the applications will be ported to it.

6.1 The Proof Demonstrator

The uncertainties surrounding the plans of the carriers, the different standards applying in the different countries, and the costs of approvals and network connection mean that it is not possible to commit to use of public ISDN services in the PROOF demonstrator. Thus, the demonstrator will be based on one site and will be built around the Nixdorf PABX which will act as an ISDN emulator.

The demonstrator will include at least two PROOF gateways, a Primary Rate server, several LANs complete with hosts and workstations and possibly, some Basic Rate workstations. The applications demonstrated will include FTAM and X.400 mail integrated with X.500 directory services; some aspects of X.509 style security services will be integrated with the mail. Remote network management based on OSI Network Management standards will also be shown.

6.2 Basic Rate Capability

There has been little activity on Basic Rate ISDN from the carriers in the UK to date. However, this seems likely to change and UCL is now actively pursing ways of achieving a Basic Rate capability from a Unix workstation. It is hoped that such a system may soon be available as background and that it can then be integrated into the PROOF environment.

6.3 Interpersonal Message Services

The partners have access to several X.400 mail systems and a part of this work package consists of upgrading one or more of these systems and ensuring that interworking is possible. In the absence of suitable public ISDN services, most of the initial testing is being done over international X.25 networks.

As a first task, one or more of the existing X.400 1984 implementations is being upgraded to 1988 specifications, and both security and the directory usage is being added. The main security and directory features will be implemented in the User Agents.

There will be particular emphasis on how the message task should be distributed in the light of the idiosyncrasies of the ISDN. For example, where a basic rate ISDN workstation is employed, it could mount either a full Message Transfer Agent, a User Agent with or without a remote message store, or could simply access a mailbox via remote login or transparent file access.

The UCL and Systems Wizards X.400 mail systems have been shown to interwork, and this connection is now being used for service mail within the project. Version 4.0 of the UCL 'PP' X.400 system has been released to beta test sites. Although this uses the 1984 version of the protocol, the service offered conforms to the 1988 standard. This will protect systems using it from disruption when the proper 1988 protocols are introduced. It is expected that full X.400 (1988) working will be available within the project by mid 1990.

System Wizards will use PP for the purposes of the PROOF demonstrator, re-hosting it on their target system, and integrating it with their DirWiz X.500 product.

6.4 Document Services

One advantage offered by the ISDN over existing public switched services is an increase in bandwidth. This will facilitate the transfers of much larger quantities of data in real-time. This small work package will investigate the use of ISDN to support multi-media document services based around the ODA standards. Most of this work will take place towards the end of the project. One possibility is the mounting of a database of technical documents stored in ODA format on an ISDN server. These documents could be accessible remotely via X.400 or FTAM. Clearly the security implications of such a service will require careful consideration.

References

[1] CCITT Blue Book Volume III - Fascicle III.8, "Integrated Services Digital Network [ISDN] Overall Network Aspects and Functions, ISDN User-Network Interfaces", Melbourne 1988

[2] Patel B., J. Martin and J. Wratten, "Ethernet-ISDN Gateway Overall Architecture", 3 NET PROOF Note 2402/3 NET/89/15, 3 NET, July 1989.

[3] ISO 8802-3, "Information Processing Systems - Local Area Networks - Carrier Sense Multiple Access with Collision Detection - Access Method And Physical Layer Specification, Part 3, November 1987.

[4] VMEbus Specification Manual Rev C.1 - VMEbus International Trade Association (VITA), October 1988

[5] Teather R., "3 NOS Operating System Outline Design", 3 NET Ref: OD-01-007-001.011, June 1988.

[6] ISO 7498-2 "Information Processing Systems - Open Systems Interconnection - Basic Reference Model", Part 1, 1988.

[7] ISO 7776 "Information Processing Systems - Data Communications - High Level Data Link Control Procedures - Description of the LAPB compatible DTE Data Link Procedures", 1984.

[8] ISO 8208 "Information Processing Systems - Data Communications - X.25 Packet Level Protocol For DTE's", 1984.

[9] ISO 8802-2, "Information Processing Systems - Local Area Networks - High Level Data Link Control Procedures - Address Resolution/Negotiation In Switched Environments", 1984.

[10] ISO 8878, "Information Processing Systems - Data Communications - Use Of X.25 To Provide The OSI Connection Mode Network Service", 1985.

[11] ISO 8881, "Information Processing Systems - Data Communication - Use Of The X.25 Packet Level Protocol In Local Area Networks", October 1988.

[12] Standard ECMA 117, "Domain Specific Part Of The Network Layer Addresses", ECMA, June 1986.

[13] KNIGHT G. and WALTON S., "Connectionless ISDN-LAN Gateway Experiments", UCL PROOF Note : 2404/UCL/90/09, UCL, March 1990.

[14] KILLE S.E., PP - A Message Transfer Agent, Proc. IFIP WG 6.5 Conference on Message Handling Systems and Distributed Applications, October 1988.

[15] SIROVICH F. and ANTONELLINI M., The THORN X.500 Distributed Directory Environment. Published in IESNEWS, Issue No. 19, December 1988.

[16] CCITT Draft Recommendation X.509 (ISO DIS 9594-8) "The Directory Authentication Framework", Gloucester, November 1987.

[17] PAVLOU G., "Management Information For The Gateway Managed System Objects", UCL PROOF Note : 2404/UCL/90/04, UCL, January 1990.

[18] CASE, J.D., FEDOR, M., SCHOFFSTALL, M.L., DAVIN, C. "Simple Network Management Protocol (SNMP), (DARPA RFC 1098), April 1989.

[19] British Telecom Network Requirement No. 188, Digital Private Network Signalling System (DPNSS), Issue 05, BT Computer Code IS18SE, December 1989.

Project No. 2569

THE EUROWORKSTATION PROJECT AND ITS FUTURE EXTENSIONS

Ch. Müller-Schloer, Michael Geiger
Siemens AG
Otto-Hahn-Ring 6
D-8000 München 83
F. R. Germany

Summary.

This paper gives an architectural overview of the EuroWorkStation (EWS) project. First it shows a brief overview of the existing workstation arena in order to derive the goals of the EWS project. Then it discusses the coprocessor architecture and its backbone, the Coprocessor Communication System (CCS), which allows easy and flexible adaptation of the workstation to the specific requirements of different users. Migration of applications between coprocessors, as well as cooperation of applications is made possible by adapting a distributed UNIX Operating System to the special needs of coprocessor communication.

The special modules Basic Workstation/Floating Point Accelerator, Graphics Engine, Symbolic Engine and Electronics Simulation Accelerator are discussed in some detail.

Special emphasis is given to the future aspects of extending the present architecture in response to the rapid change of technology.

The last chapter deals with the current status of our implementations with regard to the project goals.

1. Motivation for a New High Performance Technical Workstation

1.1. Overview of the Workstation Arena

Scientific Workstations will play a major role in enhancing the productivity of engineers and scientists in the near future. Considering the West German market alone, it is estimated that the total number of UNIX machines in use escalates from about 8,000 in 1987 to 51,000 in 1993 [4]. In 1989, the workstation market is dominated by US companies. Japan shows growing interest in this area. The EuroWorkStation project aims to provide a competitive European response to this situation.

The Workstation market covers a wide spectrum of different products ranging from low cost monoprocessor workstations to expensive high end multiprocessor systems.

In Figure 1 several workstations are mapped with respect to their performance and the year of availability. It is obvious that, very roughly, the performance / time function follows Joy's law which predicts a two-fold performance increase per year. Also, it seems reasonable to use Joy's performance curve as a divider between a performance driven and a price driven segment of workstations. Price driven machines can typically be characterized by attributes like: monoprocessor, standard CMOS technology, motherboard / desktop, limited extensibility. Performance driven workstation features are: mono- or multiprocessor, bipolar or CMOS, backplane / tower system, extensibility.

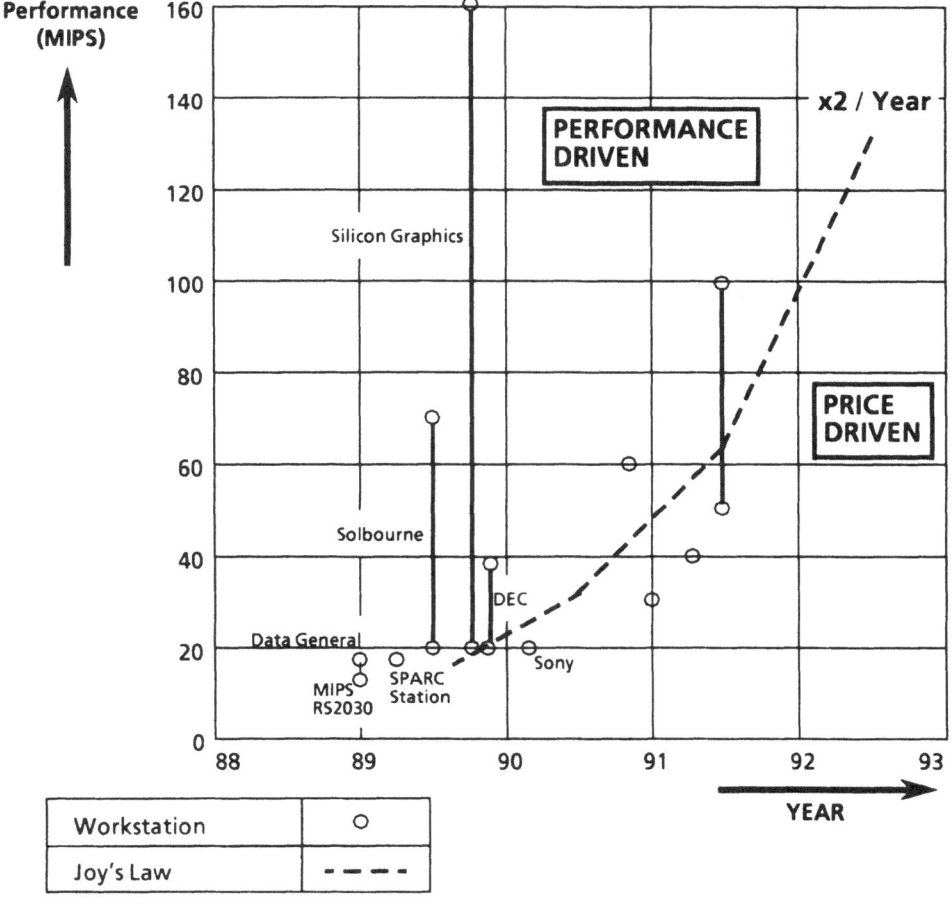

Figure 1 : The Workstation Arena

A modular workstation family allows a seamless migration from low cost entry systems upwards to configurable multiprocessor systems with coprocessors for special applications.

1.2. Basic Architectural Features of the EuroWorkStation

As a result of the above considerations, the EuroWorkStation is designed to be a modular, extendable workstation which can be easily transformed from a general purpose low cost machine into a high performance technical workstation for special purpose applications. This goal is achieved by the following architectural principles:

– *Modular open architecture.* This feature allows the tuning of the workstation to specific user requirements. Modularity also makes the workstation suitable for further developments in other ESPRIT projects. The flexibility of the hardware architecture is reflected in a modular distributed UNIX operating system.

– *High computing power provided by specialized modules.* The project provides competitive performance levels in the following special application areas:
   - numerical computation,
   - simulation,
   - graphics,
   - symbolic computation for AI,
   - high capacity archiving and retrieval.

– *Support of high level languages* such as C, Pascal, Fortran, Lisp

– A *high resolution imaging system* and *advanced user - system interface,* including
   - monochrome or colour display
   - 2D and 3D processing
   - an object oriented user interface management system.

– *Network integration* and *distributed operation*

The realisation of these basic architectural features leads to technical principles guiding the activities of the EWS project, which concentrate on 3 central requirements: *performance, modularity* and *standards.*

1.2.1. Performance.

Each EWS workstation contains a Basic Workstation platform which provides a standard high performance integer and floating point computing environment to the user and, in addition, provides certain central services, such as network access, to the specialized modules. The current version of the basic workstation is based on the SPARC processor. For future versions, the MIPS processor is considered. For applications with higher performance requirements, special purpose coprocessors are added. Within the ESPRIT project, coprocessors for 3D graphics, Lisp and mixed-mode simulation are developed.

1.2.2. Modularity.

Modularity is provided mainly by a high level coprocessor communication system (CCS, see Section 2.5) which allows the flexible combination of the special purpose coprocessors tailored to the needs of the user. A CCS interface, almost identical for each coprocessor, minimizes the effort of adapting special architectures for use in the EWS.

1.2.3. Standards.

The most important standard for technical workstations is UNIX. EWS uses a distributed UNIX System V (CHORUS), which is adapted to the special needs of a fast coprocessor coupling. The graphics engine will support X-windows and later also PHIGS. Within CCS, in the present version Multibus II and the Intel transport protocol (data link and transport level) are used. The Basic Workstation supports a SCSI interface, Ethernet (TCP/IP and ISO/OSI) and FDDI.

2. Coprocessor Architecture of the EWS

The EuroWorkStation (EWS) is a personal computer system for scientific and technical applications. It is basically conceived as an element of a network distributed system, and as a locally configurable platform, allowing additions of special purpose coprocessors. Figure 2 shows the overall coprocessor architecture of the EWS workstation.

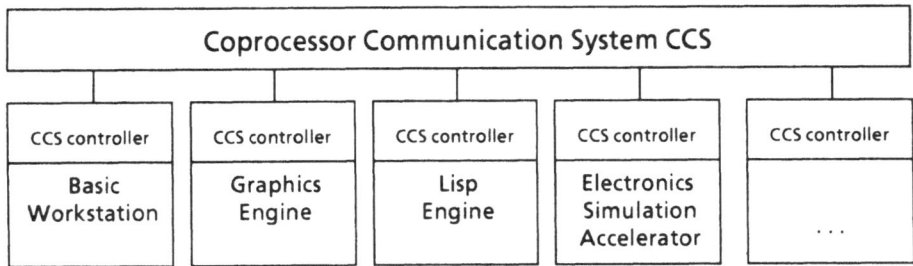

Figure 2: Overall Architecture of the EWS

The essential components are:
- The *Basic WorkStation (BWS)* provides a standard Unix-SPARC environment for high performance scientific computing and networking applications.

- The *Floating Point Accelerator (FPA)* extends the BWS to a symmetric tightly coupled parallel SPARC architecture. Threading and parallel execution of Fortran programs are mainly managed by the FPA Fortran Parallelizer, and the Parallel Fortran Compiler.

- The *special purpose coprocessors* extend the BWS to an heterogeneous loosely-coupled multiprocessor architecture:
  - GRACE: Graphics engine for acceleration of 3D graphic processing
  - COLIBRI: Lisp accelerator
  - ESimAC: Electronics simulation accelerator

- The physical link between these coprocessors and the BWS is given by application of the Multibus II message passing concept, thus providing a loosely coupled multiprocessor system.

- The *Coprocessor Communication Subsystem (CCS)* defines the protocol set used for communication between processors over Multibus II and a generic programming interface for the development of special applications.

2.1. Basic Workstation and Floating Point Accelerator

2.1.1. Basic Workstation.

The BWS architecture (see Figure 3) can be viewed as two interconnected domains (implemented on two boards) communicating through a shared memory space.

The first domain is dedicated to external communications and is implemented on the IO board. It is organised around four coupling units, allowing the external world to exchange data and messages with the central memory of the BWS via SCSI, Ethernet, FDDI and/or Multibus II.

Each of these units can access the BWS main memory by physical DMA channels, provided by means of an ASIC solution (CMOS 1 μm gate array technology). IO board and Central Memory are connected by the Communication Bus (CBUS).

The second domain, implemented on the CPU-board, is dedicated to central processing and includes a 33MHz Cypress SPARC RISC, a Weitek FPU, a MMU with virtual copy back cache controller and a virtual cache (64KB SRAM). The board is completed by a 16 MByte interleaved DRAM main memory and two internal busses, the MBUS (CPU to MMU) and the scaleable PBUS for IO space access.

All boards of the EWS fit the triple Eurocard format.

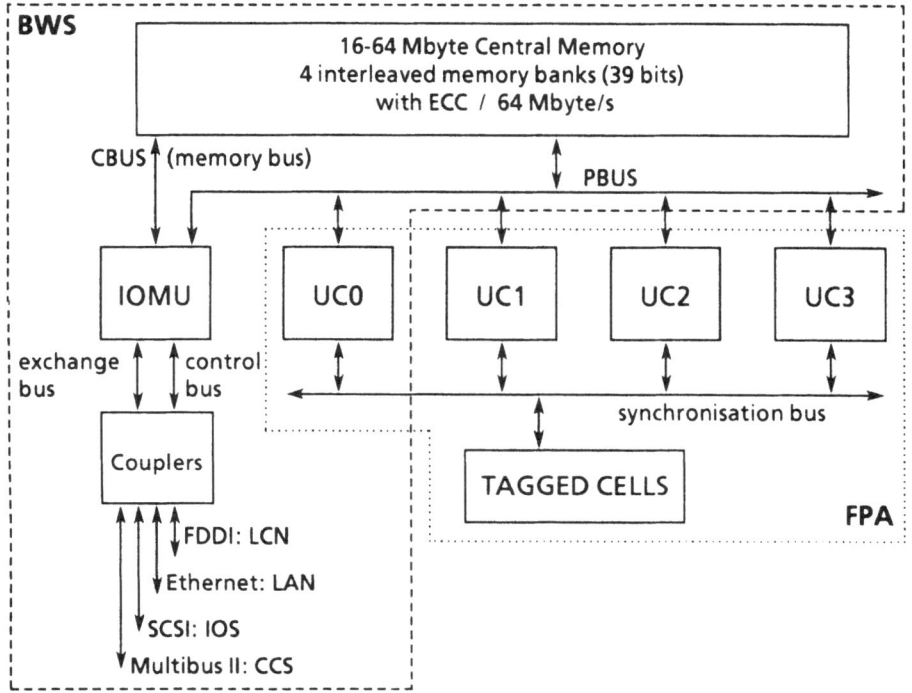

Figure 3 : Architecture of BWS and FPA

2.1.2. Floating Point Accelerator FPA.

The floating point accelerator's function is obtained by a symmetric tightly coupled parallel extension of the BWS (see Figure 3).

The main goal of the FPA is to provide a substantial speed-up of compute-bound application programs on an architecture which is basically the one of a personal computer. While designing the FPA, the main accent has been put on defining a simple set of basic hardware and certain software features which are easily manageable at the compiler level.

The FPA architecture uses both the Central Memory board and the IO board of the BWS. A specific CPU board integrates the SBUS and four complete SPARC units (IU, FPU, CMU, Cache). The memory space for IO and the Central Memory are shared among the four processing units. The MMU's and the cache controllers of the BWS are enhanced in this FPA architecture to support the multiprocessor cache coherency (Cypress 605 CMU).

2.1.3. High Capacity Network Storage.

In order to provide large storage capacities (several GBytes) for data intensive applications (e.g. VLSI design), the BWS is equipped, in addition to the traditional Winchester disk, with an optical storage server.

724

Following the modular architecture of the project, the HCNS will be accessible from both local and remote workstations by standard file access protocols. The HCNS will provide mechanisms for versioning, replication and atomic operations which will be integrated in the operating system.

2.2. Graphics Engine GRACE

The Graphics Coprocessor Engine (GRACE) within the EuroWorkStation constitutes the interface between the human users and their applications. It serves mainly as a means to display the computational results of other coprocessors and as an interface for the human machine interaction.

The hardware of GRACE is divided into a Gerneral Purpose Processor Board (GPPB) and the 3D Geometry and Rendering Pipeline (3DGRP) as shown in Figure 4.

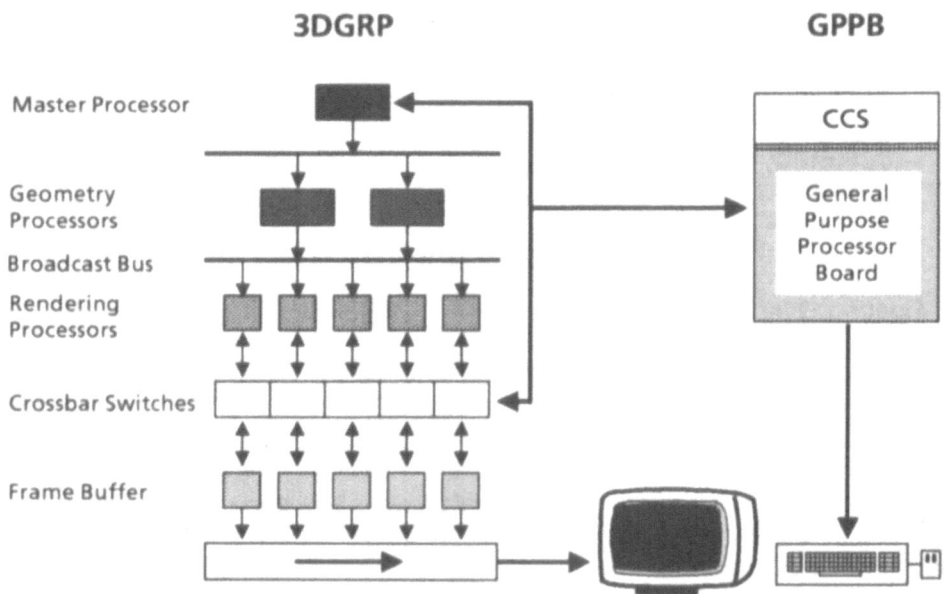

Figure 4: GRACE Coprocessor Architecture

The GPPB is used in serveral coprocessors and administrates the interface to the Basic Workstation via the Coprocessor Communication System (CCS). Within the GRACE coprocessor, the GPPB performs the overall control and provides the platform for porting existing software to the graphics system and for the User Interface (UI).

The 3DGRP performs the calculation of the 3D graphics objects and the access to the frame buffer. The frame buffer consists of an on-screen part (having 24 bit true colour, double buffering, 24 bit Z-buffer, transparency, window control, cursor) and an off-screen memory with 64 bit planes. Additional planes are used to support the window manager.

The access to the frame buffer is parallelized by a distributed architecture. There are five different parts each being exclusively used by one rendering processor. This increases the bandwidth to the frame buffer memory and therefore enhances the speed of image generation. Every part of the distributed frame buffer is completed by its own rendering processor which performs the access to the frame buffer and processes the rendering parts of the graphics algorithms. These tasks comprise the conversion of geometry objects to pixels and the shading of these objects (interpolation of surface colour). The processing step that has to be done prior to the rendering is the geometry calculation which involves transformation and clipping computations, splitting of complex objects (e.g. curved surfaces) into planar sets, and computation of lighting models in order to obtain a realistic view of a scene.

The throughput of the 3DGRP is calculated at 40 MPixels/s. This rate may be achieved by bitblock operations. The speed for line drawing is estimated at about 400,000 vectors/s (including shading/depth cueing, 100 pixels/line), and about 50,000 shaded triangles/s will be processed (100 pixels/triangle).

The Interaction Framework (IF) is a subtask of the UI design and provides the runtime environment where interaction software executes within the intelligent controller of the GPPB board. The IF supports an extensible set of interaction and feedback techniques, and an extensible set of modelling facilities which can be used by application programmers and user interface designers to optimize the use of the intelligence of the graphics engine for their purposes.

A detailed description of the GRACE workpackage is presented in [3].

2.3. Symbolic Engine COLIBRI

Within the COLIBRI (Coprocessor for LISP based on RISC) workpackage a symbolic engine is developed with a LISP performance comparable to or better than special purpose LISP machines, but at a considerably lower cost. The COLIBRI coprocessor board consists of two semicustom integrated circuits (CPU and MMU), two external caches, 16 MB of main memory and a CCS bus interface (see Figure 5).

2.4. Electronics Simulation Accelerator ESIMAC

This workpackage aims to supply the EWS with a hardware accelerator which allows substantial speed-up for a wide range of CAD algorithms used for the simulation of large circuits and systems covering

726

- the *high level simulation* comprising the system level and the register transfer level,

- the *logic simulation* (intended magnitude of speed: a few tens of Mgev/s (gev = equivalent gate evaluation))

- the *electrical circuit simulation* (intended magnitude of speed: a few tens of MFlops).

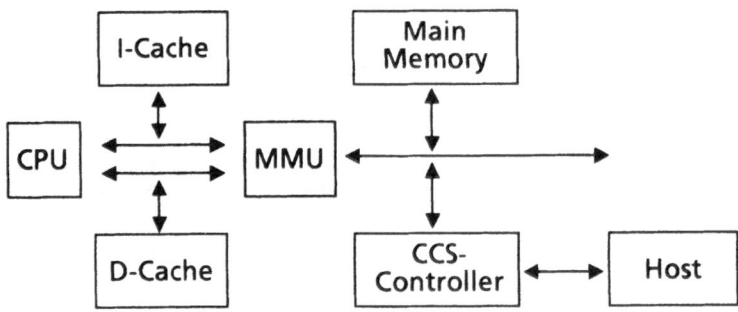

Figure 5: COLIBRI Coprocessor Architecture

Following the technical development towards complex digital-analogue ICs (especially ASICs), accelerating hardware support should also be supplied to a mixed-mode technique which integrates the abovementioned simulation levels. As preliminary studies show [2], the event-driven algorithm template yields a suitable and efficient approach for simulation acceleration on both the algorithmic and the hardware area, and is applicable to all the simulation levels mentioned above as well as to additional design problems like fault simulation, test pattern generation and routing. Hence the architecture of the accelerator ESimAc covers the following criteria:

- the internal communication concept is well fitted to the event-driven algorithm template,

- there is a sufficient amount of effective memory bandwidth provided,

- specialized units for event evaluation are included.

Moreover the ESimAc architecture is intended to be scalable and modifiable with respect to the mentioned simulation levels and to an appropriate system size depending on the user's application. Finally, this engine is embedded (on the board level) into the basic workstation by a suitable coprocessor concept, the CCS (see Section 2.5.).

A detailed description of the ESimAc workpackage is presented in [1].

2 5. Coprocessor Communication Subsystem CCS

The Coprocessor Communication System forms the backbone of EWS (see Figure 6).

Each coprocessor "sees" a full UNIX System V interface towards other applications and at the same time each application is capable of using the special purpose processor hardware of each individual coprocessor. If a standard bus, such as Multibus II, forms the physical link between the coprocessors, then it is the task of the CCS controller to cover the intermediate communication levels to offer a UNIX interface to the applications. The function "CCS controller" can be implemented either completely in software or as a separate controller hardware with CCS software running on it.

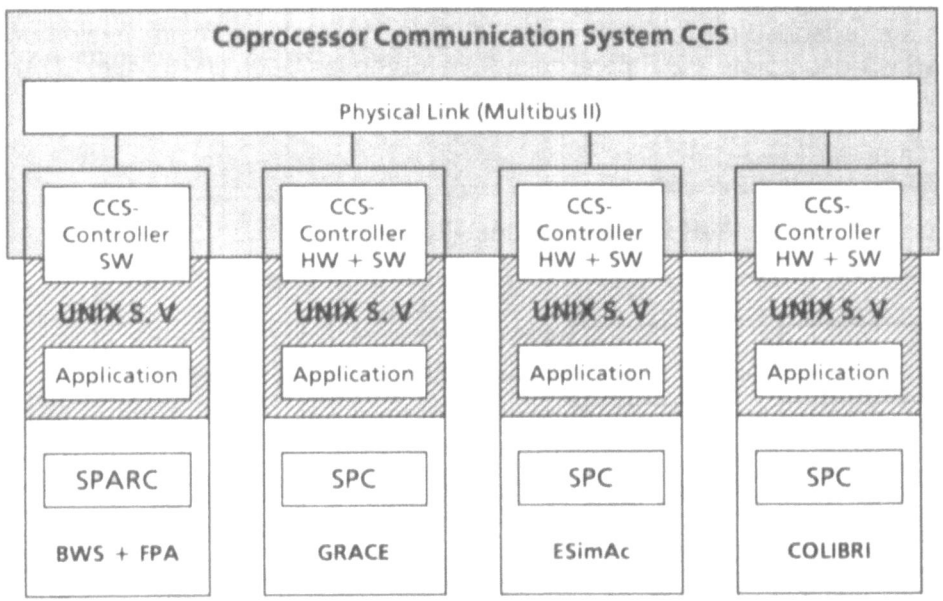

Figure 6: CCS System Architecture

3. Future Development Goals of EWS

While current development is mainly guided by the three goals **performance, modularity** and **compatibility,** future actvities will enhance these features but will at the same time preserve the architectural basis of EWS. In fact, the result will be a general workstation based on a shared-memory, multiprocessor architecture, completed by specialized coprocessors connected via a communication system utilizing the advanced features of the distributed UNIX operating system (see Figure 6). Performance intensive tasks like graphics and simulation are handled by special hardware, connected to the system by a common controller. The general workstation acts as a common high performance computational platform.

Present work has given already some hints for future improvements (see Figure 7):

- GPPB performance is now limited by the performance of the used microprocessor, a i80386. This processor is sufficient as long as it is responsible only for handling CCS. In the cases of GRACE and ESimAc, the controller runs also extensive application specific software. A slow machine can become the bottleneck. Therefore a RISC solution will be chosen for the general purpose controller (GPC).

- The increased hardware capabilities of the GPC allow for a future decentralization, including e.g. local I/O. This demands a more fully developed operating system, extended beyond the mere CCS functions. Hence a future version of EWS will incorporate a full operating system (European Operating System EOS) which contains, of course, the full CCS functionality.

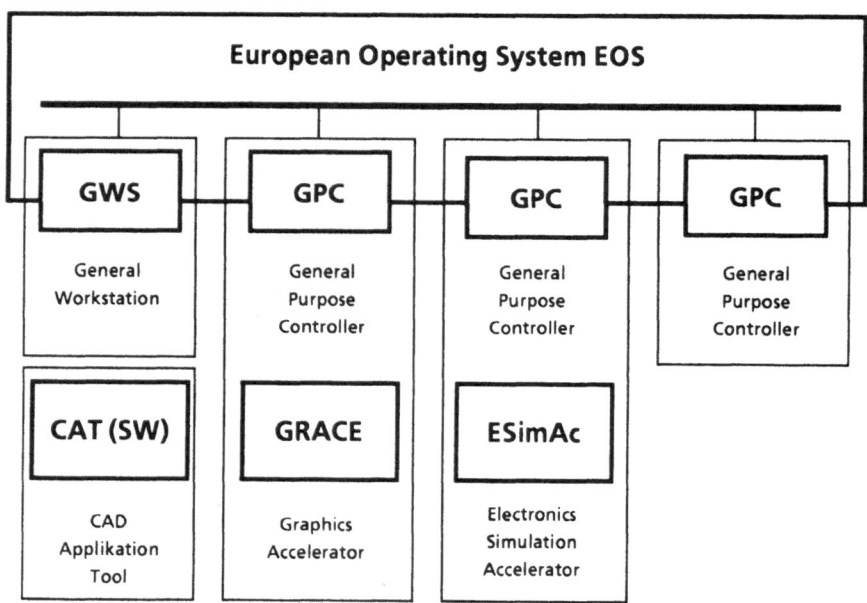

Figure 7: The Future Design of the EuroWorkStation

- Multibus II as workstation backbone might not be necessary if bandwidth requirements are kept low by allowing only short high level messages between the coprocessors. It will be determined if a fiber link (FDDI) is sufficient for this internal communication. In this case the difference between workstation internal and external communication is further decreased. A fiber link between coprocessors is much more flexible than a bus. It can be hoped that eventually it will also be faster and less expensive.

Hence, an extension of the present work on EWS should comprise the following workpackages:

– *European Operating System EOS.* In order to guarantee a fast and secure operating system for EWS, the extension of the CHORUS operating system (including the Coprocessor Communication System CCS) into a fully distributed UNIX OS supporting shared memory multiprocessor architectures as well as each of the coprocessors while still keeping full compatibility with the X/Open standard, will be an issue of central importance. Backbone of this "European Operating System" EOS will be either a Multibus II (already in use) or FDDI.
FDDI allows a transparent and fast extension of a workstation of cooperating coprocessors into a network of cooperating workstations.

– *General Workstation GWS.* The General Workstation platform is built around a high performance, high capacity memory unit and on high throughput network communication facilities. It supports a shared-memory multiprocessor architecture. The processors will be state-of-the-art RISC processors. Also, the GWS supports high resolution 2-D colour graphics and adequate input devices.

– *General Purpose Controller GPC.* Each coprocessor is controlled by a general purpose processor which has the following tasks:
  - platform for UNIX application processes (i.e. EOS environment)
  - communication with other coprocessors
  - control of special purpose coprocessor hardware
The GPC workpackage will achieve a much higher performance level than the current i80386 solution by using a state of the art RISC processor which can be extended into a closely coupled multiple processor configuration. Performance of the GPC will start from around 50 MIPS for a single processor system.
Besides acting as a controller for special purpose hardware, the GPC can work as a high performance computational platform similar to the GWS.

– *Graphics Accelerator GRACE.* The graphics subsystem is aiming at providing a high performance environment for graphics applications and a comfortable user interface. Not every user needs high resolution and true colour. Because of this, a scaleable concept will be developed within a family range of configurations, reusing and extending a good part of the hardware already developed. The display facilities will range from smaller black and white screens up to large high resolution displays providing true colour output. The scaleability lies in the number of processors used on the one hand and in the size of the frame buffer (colour and spatial resolution) on the other. The high end version will be able to support up to 1 million Gouraud shaded 3D triangles per second. Another important issue is the integration of a multimedia extension. The video interface will allow the integration of graphics and video, as well as picture processing. On the software side, the X11 standard and one rendering standard will be implemented as well as some real applications.

- *Electronics Simulation Accelerator ESimAc.* Towards the end of 1990, ESimAc will achieve an acceleration of logic gate level simulation and high level (behavioral) simulation separate from each other. Future development will add circuit level simulation and mixed-mode capability over all three levels. Also, industry standard interfaces for netlist format and model libraries (such as VHDL) will be supported.

4. Industrial Importance of the EWS Project

The modular architecture framework of EWS allows for an easy expansion and fast adaptation to a changing environment. This is examplified already now by incorporating new technologies (like high performance RISC or optical fiber links) while keeping the standard EOS / CCS interfaces.

Referring to Figure 1 the EuroWorkStation covers the whole spectrum of the workstation market. In the price driven area, the BWS as a standalone machine is a promising competitor. Due to the special needs of computational intensive application programs (graphics, simulation etc.), the extension of the BWS with appropriate coprocessors transforms the system to a high performance workstation. CHORUS has gained much visibility in the European computer industry and is used in a variety of industrial and research projects. The GRACE approach, being a high end solution, will be transformed to a scaleable architecture and find its entry into product designs. ESimAc is a very promising candidate for a simulation server; an adaptation to a VHDL simulation accelerator is considered.

5. Current Status of the Project

Currently all newly developed hardware is available, at least as prototypes. The Coprocessor Communication System is completely implemented and guarantees quick and transparent communication over the Multibus II. Software developed for the special purpose coprocessors is mostly available, but in some cases not yet fully integrated. Therefore, some demonstrations shown at the exhibition are still running on the development platforms. The final integration of all components is planned to take place until the end of 1990.

References

1. Coleman, N., Jacoby, K., Pfeuffer, E., Thinschmidt, H., (1989). The Electronics Simulation Accelerator ESimAc. This issue.

2. Jacoby, K., (July 1988). Ereignisgesteuerte Verfahren als Grundlage für die HWSimulationsbeschleunigung. Siemens AG, internal report no. ZT ZTI SYS 1-67/Jac, 57 pages, Munich, FRG.

3. Mehl, M., (1989). GRACE - the Graphics Coprocessor Engine within the EuroWorkStation. This issue.

4. VDI Nachrichten, (1989). Vol. 43, no. 18, pp. 1.

List of Consortium Members

Siemens AG
GIPSI S.A.
CM Bull
INRIA
INESC
CHORUS Systèmes
Fraunhofer Gesellschaft-AGD
Transtools
Rutherford Appelton Laboratories
Brunel University

INFORMATION EXCHANGE SYSTEM

HIGH SPEED NETWORKING IN EUROPE
AND THE EBIT PROJECT

Paul VAN BINST
Université Libre de Bruxelles
CP 230, bd. du Triomphe
1050 Brussels
Belgium

Summary.

The RARE association (Réseaux Associés pour la Recherche Européenne), in close collaboration with the Commission of the European Communities (CEC), has adopted a strategy for European high speed networking (2 Megabits/second and higher rates) for the academic and research community; technical studies are being conducted in collaboration with the CEC and major European PTT's in the scope of the EBIT project and RACE programme. A good part of the work done so far was made within RARE Working Group 6, of which the author of this paper is the deputy chairman.

The current European High Speed Networking Initiative (EHSNI) plans for a European high speed backbone linking regional or disciplinary high speed networks. Data rates of 2 Mbps are feasible across Europe right now and active planning has started to define such an infrastructure and its evolution to 140 Mbps within a few years and to the range of Gigabits/second by the end of the decade.

A number of regional networks, such as BERKOM, and scientific communities, such as High Energy Physics or Aerospace research, have the capabilities and needs to use such high speed networks right away.

1. Context and Needs: A Rapidly Growing Demand for European High Speed Networking.

In European countries, the data communication world is rapidly evolving towards very high speed capabilities, as a result of efforts of European PTTs and of large European projects such as RACE. At the same time, supercomputers, scientific and graphic workstations, and data sources (both scientific and industrial) are individually producing or consuming data at a rate which grows regularly by a factor of ten about every five years. The "big science" researchers, and the most advanced sectors of industrial research, are continually incorporating all these tools in their day-to-day work; international research initiatives such as CERN or ESPRIT have induced them to collaborate all over Europe. This is producing a rapidly growing demand for advanced Europe-wide communication services.

The European High Speed Networking User Meeting, which RARE and the CEC organized in 1989, brought together representatives of many branches of academic and industrial research, which are more and more getting organized in well-coordinated, international disciplinary user groups. They confirmed that, for carrying on their normal work, they urgently need powerful long-distance data communication facilities all over Europe: they referred to high speed today, very high speed tomorrow and ultra high speed in the longer term. These requirements are not utopian, since they correspond, for long distance communication, to what has been available within large research centres for some years, or is just appearing there today. Several user groups have already well worked out specifications for volumes of data and transfer speeds; others can see the same needs arising in only a few years time.

These speeds are explicitly required for source data transmission, for accessing large remote data bases, for accessing and graphically interacting with supercomputers, for remotely controlling complex experiments, for transmitting images, whether for studying earth resources, or for medical research, or for document distribution, etc. Distance learning is already an important consumer of high speed communications.

In fact, installing a very high speed networking facility with complete European coverage would lead to a completely new way of interacting between researchers over Europe, and would significantly boost their efficiency. Those who will have access to it will benefit from a quality and diversity of communication, over the whole of Europe, equivalent to what they use today within each campus or research centre.

It was also clearly shown that user groups are starting initiatives of their own to satisfy these needs; however, they are running into technical and financial difficulties, since for most of them it is not their normal job to deal with forefront networking technology, and also due to the fact that, today, the European PTTs marketing strategy tends to disfavour high speed international links. Users will welcome any initiative that could help them getting the services they need.

Another important element in this context is the situation and plans in the US: the development of data communication tools for the research community (academic and industrial) is coordinated by the Federal Government, in view of providing for an interconnected set of networks to support research and other scholarly pursuits. A major part of these efforts goes through the National Science Foundation: its network, which provides communications between the main research centres of the US, has been upgraded in 1988 to high speed capabilities; plans are to give it very high speed capabilities starting in 1990, ultra high speed before the end of the century, and to use it as the main communication stream between regional networks, which means hundreds and later thousands of institutions. This will provide the US research community with a very powerful tool for its day-to-day work; furthemore, the US data communication industry will see its know-how tremendously increased by its active participation in developing and running this network.

The US networking evolution shows clearly two points:

– the need for high speed communications appeared simultaneously, and grows at similar rates, within research communities on both sides of the Atlantic. Many of these communities are in fact worlwide collaborations;

- while Europe was hampered by its structure and could not react quickly, the USA have been able to draw conclusions switfly, and to start building within short delays a powerful data communication facility. A bill that went before the Congress in 1989 proposes to set up a 3 Gbps network in 1996.

2. The Strategy: A European High Speed Networking Initiative.

In view of the above context and needs, RARE has proposed to launch a European High Speed Networking Initiative (EHSNI), aiming at satisfying user requirements for long distance data communication services, in the short term and long term, by providing the European research community academic and industrial - with a powerful common communication facility based on forefront networking technology, with access speeds well above the conventional ones.

Furthermore, this action would benefit the European data communication industry, by giving it the possibility of gaining first hand experience in transferring results of its R & D projects to the real world and testing them within a large base of competent and demanding users.

The EHSNI will provide, for the most demanding user groups within European industrial and scientific research, very high speed networking services as soon as it becomes technically feasible, even accepting a flavour of "pilot action" with new services, products and technologies.

The EHSNI will be constituted of a central component, a European backbone, and of national, regional or disciplinary networks. It is organized in three main phases:

- Phase 1 (high speed): is based on current Megabits/second wide area technology. It can start right now;

- Phase 2 (very high speed): will be based on 140 Mbps technology. It could start in 2 to 4 years from now;

- Phase 3 (ultra high speed): will use, probably before the end of the decade, Gbps technology.

The EHSNI provides a general framework for the evolution of scientific very high speed networking in Europe for the next 10 years.

Each phase will be based on a set of rules for:

- setting up a European backbone using the relevant technology. This will have to be centrally managed and coordinated. It will be done gradually, through several successive steps;

- building upon national, regional or disciplinary networks and linking European backbone. How to realize this will be studied, in each case, by the coordinators of these networks, in relation with those of the EHSNI.

In reference to this model, the situation today can be summarized as follows:

– regional or disciplinary networks using the 2 Mbps technology of Phase 1 are being installed, due to user initiatives; rules concerning a European backbone are urgently needed, to ensure a minimum of coherency. The backbone itself must be started with very little delays;

– a few regional networks based on the 140 Mbps technology of Phase 2 are already being planned or installed.

3. The Expected Users of the EHSNI and How They Will Benefit.

Potential users of this EHSNI will mainly be found among the "big science" research institutes, and the largest industrial research centres.
Those mentioned at the User Meeting include the following disciplines:

space research,
high energy physics research,
medical research,
fusion research,
earth resources research,
aeronautics and space industry,
car engineering industry,
combustion analysis research,
supercomputer centres,
document distribution institutes (libraries, etc.),
distance learning initiatives (EuroPACE, DELTA, etc.)

This list is clearly not exhaustive.
A crude estimate of the number of sites, institutes, laboratories needing high speed - and hence potential EHSNI users - would start at around 30, and would grow steadily up to the range of 200 - 500; they represent the 5% most demanding sites within the RARE community.
What they intend to use the EHSNI for has been described in paragraph 1 above. In fact, they will most probably find that, as it becomes more powerful, the EHSNI can support many more applications: one will think of person-to-person or group communication techniques, such as videoconferencing on a wide scale, high definition facsimile transmission, multi-media mail (including graphics, images, voice and video as well as text and computer data), etc.

4. First Practical Steps : Relation with the EBIT Project.

In 1989, thirteen European PTT's signed a Memorandum of understanding about the launch of the EBIT (European Broadband Interconnect Trial) project, in close synergy with the RACE programme of the CEC.

EBIT aims at the promotion and rapid realization of switched high-speed services in Europe; the first phase is based on available 2 Mbps services, making use of various satellite and ground based facilities which are presently available in a number of European countries.

It was soon realized and formally agreed, between RARE, the CEC and the European PTT's, that the use of the EBIT facilities would potentially be an ideal first step in the realization of the EHSNI. Indeed, in the presently changing European telecommunications landscape, international high speed leased lines are not attractively priced while switched services offer much more versatility and are in line with the principles of future Integrated Broadband Communications services, as developed for instance under the RACE programme or in the BERKOM project.

Feasibility studies have been conducted and plans are being made to define in which way the EHSNI will make use of the EBIT facilities. First practical realizations and tests could take place this year within the High Energy Physics and Areospace research communities.

5. Conclusions.

European "big science" researchers and the most advanced sectors of industrial research urgently need powerful long distance communication facilities all over Europe. To satisfy these requirements, RARE has launched a European High Speed Networking Initiative, the EHSNI, constituted by a European backbone and national, regional or disciplinary networks. This will benefit both the users and the European data communication industry, by giving it first hand experience with new technologies in an operational environment.

The Commission of the European Communities and major European PTT's have taken the first steps towards achieving these goals.

6. Acknowledgments.

We wish to thank Mr. Jacques PREVOST, chairman of RARE Working Group 6 and all the members of the Working Group for their active collaboration in the preparation of this paper.

THE COSINE PROJECT

H. E. DAVIES
iCPMU
c/o RARE Secretariat
Postbus 41882
NL 1009 DB Amsterdam
The Netherlands

Abstract.

The main purpose of COSINE, Eureka Project No. 8, is to create a computer networking infrastructure based on the use of OSI protocols which will provide services to the whole research and development community (academic and commercial) throughout Europe. A set of sub-projects and services is being established to provide such things as information, directory and electronic mail services available throughout the 19 member states as well as gateways to equivalent services in the USA. The COSINE sub-projects are largely coordinating and pump-priming activities which aim to ensure that services already provided or being planned at a national level can be extended to give international coverage.

1. COSINE Background

COSINE (Cooperation for Open Systems Interconnection Networking in Europe) is Eureka Project No. 8 and is funded by a total of 19 European countries as well as the Commission of the European Communities. Its main purpose is to create a computer networking infrastructure, based on the use of OSI protocols, which will provide services to the whole research and development community (academic and commercial) throughout Europe. The COSINE services will be of particular interest to the Esprit community since they will extend the kinds of data communications and messaging facilities which are already available locally and, in some cases nationally, to the whole of Europe in a systematic and consistent way.

Following preparatory work started in 1986, a contract for the COSINE Specification Phase was awarded to RARE (Réseaux Associés pour la Recherche Européenne). The result of this Phase, completed in July 1988, was a set of technical reports [1,2,3,4,5,6,7] summarised in [8,9] which analysed the requirements for pan-European networking services in terms of the size and needs of the user community, the suitability of defined and emerging OSI standards, the likely availability of commercial systems and services, and the operational problems of providing a unified service to the whole community. In May 1989, the COSINE Policy Group published the COSINE Implementation Phase (CIP) Project Proposal [10] which specifies a number of pilot sub-projects and services which will be set up and operated during the three year lifetime of this phase of COSINE.

RARE has been asked to undertake the management of the CIP and has created the COSINE Project Management Unit to carry out this task. Detailed arrangements are defined in a contract between RARE and the CEC, acting on behalf of the COSINE Policy Group. Work on the CIP had already started in August 1989 through a Memorandum of Understanding between the CEC and RARE. The COSINE Funding Arrangement which defines the terms of the financial contributions from member states came into force on 1 January 1990, triggering the start of the three year CIP clock.

The ways in which COSINE sub-projects and services will continue after the end of the CIP are as yet undefined but one of the tasks of the CPMU in the later stages of the project will be to make the successful services self-funding, for example by getting them taken on by commercial operators or by implementing some other kind of charging scheme.

2. Aims and Objectives

The objectives of COSINE, as defined in the CIP Project Proposal, are as follows:

(a) to create a common operational OSI interworking infrastructure on the basis of federated research networks to support all European research.
(b) to establish and integrate on the required scale all the functions and support services necessary to allow the users to take full advantage of the infrastructure.
(c) to take steps to ensure that the infrastructure remains available to European researchers after completion of the project.
(d) to thereby contribute to the market pull for OSI.

The principal aim, therefore, is to create a set of OSI-based networking services for the European research community which will outlast the 'pump-priming' period which COSINE funding supports. An important subsidiary aim is to involve the industrial community, not only as network users alongside researchers from the academic community, but as partners in the development and operation of the necessary projects and services.

COSINE has no intention of trying to provide an exclusive set of networking services. On the contrary, COSINE forms part of a wide spectrum of networking activities. It will have links to existing networks such as EARN (European Academic Research Network), EUNET (European Unix Network) and HEPNET (High Energy Physics Network); there will be close cooperation with other new projects such as Y-net in order to ensure that as far as possible the sets of services are complementary; and the progress of high speed initiatives such as EBIT and EASYNET will be closely followed so that users can benefit from the results that they are expected to produce. Through RARE and the CCIRN (Coordinating Committee for International Research Networking, for which RARE provides one of the two co-chairman), COSINE will be able to follow and participate in the coordination of networking services between Europe and the United States.

3. Planned Sub-Projects and Services

The list of sub-projects and services planned for COSINE is shown in Table 1. A sub-set of these (as indicated in the Table) is covered by the first year budget for COSINE and work has already started on them; more details of active projects are given in section 5 below. The 'core set' includes all the first year sub-projects and services plus one further sub-project, P8. The CPMU will complete the specification of Activity Plans (ie detailed sub-project specifications) for the core set by the end of the first year though not all of the sub-projects will actually start by then.

Where the technology and expertise in a particular area is already sufficiently well developed, services are being set up immediately. To meet other user requirements, further work is necessary either to develop suitable products or provide additional support to service developments already being undertaken in the community. In these cases, pilot sub-projects, some of them with less than the three year lifetime of COSINE, are being set up to provide additional information before full scale services can be started. The detailed specification of some services (S3, S4, S5 and S6) cannot be made until further experience becomes available from the corresponding pilot sub-projects. Although the set of sub-projects and services defined in the CIP Project Proposal forms the contractual as well as the technical basis of the all current COSINE activity, it is expected that changes will be introduced (in a controlled fashion and with the agreement of the COSINE Policy Group) to take account of technological developments, changing user needs, and experience gained as the Project progresses.

Initially, the arrangements for Service S1.1, X.25 (1984) Service Provision, were different from that of the other COSINE sub-projects and services. The International X.25 Infrastructure (IXI) Pilot Project was initiated in 1989, the pilot service being provided by the Netherlands PTT Telecom under a contract with the CEC, with RARE providing technical management supervision via the IXI Project Team. When the CIP Execution Contract between the CEC and RARE was signed, IXI formally became COSINE Service S1.1 and the IXI Project Team came under the management control of the CPMU. The basic data transmission services provided by IXI, which provides X25 based connections at 64 kbit/sec accessible from all COSINE member states, are fundamental to many of the other planned services. IXI is the subject of a separate paper [11] and full details of the service will be found there.

The total budgeted cost of the projects in the core set, including the costs of the CPMU and other management activities, is 22.9 MECU. The estimated total cost of all the COSINE sub-projects and services is 35.7 MECU but it is expected that other parties, including for example equipment and service suppliers, will be willing to contribute their own funds in some cases.

4. Organization and Method of Working

COSINE Policy is determined by the COSINE Policy Group (CPG) on which all member states are represented. A sub-set of this group, the COSINE Policy Bureau, acts as an Executive Committee and deals with the more urgent matters that arise between meetings of the CPG.

The CEC, besides making its own financial contribution to COSINE, manages the Project's funds and monitors the progress of the sub-projects. In order to do this, it provides a COSINE Project Officer (CPO) and small Secretariat.

The sub-projects and services themselves are carried out by companies or other suitable organisations (including, in some cases, groups based in Universities) who are selected as a result of an open tender. Activity Plans are (or will be) defined for each sub-project and service and form the basis of an Invitation to Tender. The tendering process, including tender evaluation, is based on that used by the CEC for Esprit Projects. The principal changes that have had to be made result from the fact that COSINE sub-projects and services are not research projects but are organised as commercial sub-contracts with payments to sub-contractors made only on delivery of agreed product or service items.

Management of the sub-contracts, including preparation of Activity Plans, issue and evaluation of tenders, and supervision of sub-contractors, has been entrusted to RARE acting through the CPMU. The CIP Execution Contract between RARE and the CEC defines procedures for reporting, accounting and approval of major expenditures which enable the COSINE Project Officer to carry out his monitoring functions. The CPO acts as the formal channel of communication between the CPMU and the CPG; a RARE/CPB Liaison Group meets from time and provides another channel of discussion of problems and proposals for solutions.

The role of the CPMU is strictly one of project (sub-contract) management within the terms specified by the CIP Project Proposal. The CPMU is, however, able to call on the resources of the RARE Working Groups which are a valuable source of expertise in many areas of computer networking and whose members have a particular awareness of the additional difficulties of operating international services. Besides supplying people who, acting individually, can work as experts in tender evaluation teams, RARE Working Groups can be asked to investigate technical problems that arise. In their normal role, which extends beyond COSINE, they can investigate and analyse new networking techniques and strategies, make contributions to the development of OSI standards, and prepare further proposals for sub-projects and services which could be incorporated into COSINE. A diagram representing the inter-relationship between all the parties involved is shown as Fig. 1.

There is no intention in COSINE to provide services directly to end users; any attempt to do so would require enormously larger resources in order to provide the necessary geographical coverage and would in case not be very productive. Instead, COSINE services will be provided via existing national organizations. Where these do not exist or cannot offer suitable support (for example, because of regulatory constraints), alternative access methods will be provided, in the case of access to IXI via the national Public Switched Packet Data Network (PSPDN). Much of the COSINE activity, including that supplied by sub-contractors, will therefore be in the form of coordination, support and encouragement of national activities with

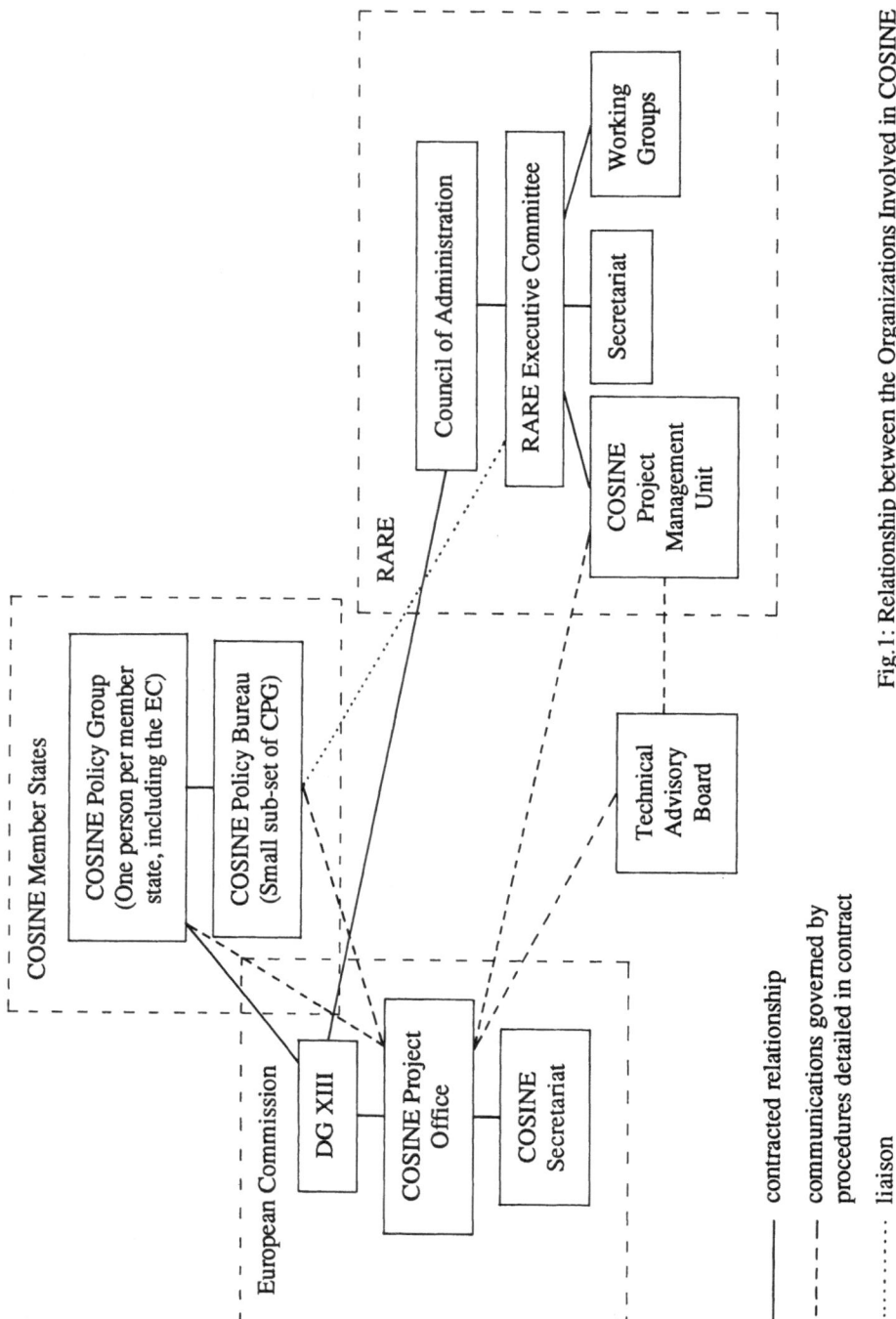

Fig.1: Relationship between the Organizations Involved in COSINE

—— contracted relationship

– – – communications governed by
procedures detailed in contract

·········· liaison

the aim of ensuring that incompatibilities are minimised and that all member states are fully informed about progress which is being made elsewhere.

5. Active Projects

Activity Plans for the first set of COSINE sub-projects and services were prepared before the MoU deadline of 10 January 1990; tenders were invited and evaluated during the first half of 1990. These sub-projects and services are as follows:

(a) P1.1 Pilot FTAM Gateway Service to the USA: selected File transfer, Access and Management (FTAM) products which can pull and send files to and from North American systems on DECnet and the Internet will be used to mount a service which will be available to all European users who have IXI access. The pilot service will be openly available for trial purposes to European FTAM end-user systems and will provide experience of operating such a gateway so that the CPMU can plan a self-sustaining gateway service which can be contracted beyond the pilot project.

(b) P2.1 Pilot International Directory Service: this sub-project will encourage the establishment of a Europe wide X.500 based directory service and will interconnect the national X.500 pilot directory projects. It will include the North American pilot X.500 directory service and ensure that there are no significant problems of interworking at the international level between different X.500 implementations. In so doing, it will remove one of the major barriers to the spread of other OSI applications, in particular electronic mail and the transfer of files, by providing directory information for both applications and for human users. At the end of the pilot project, the CPMU will have all the information needed to issue an invitation to tender for the full directory service (Service S5) and national networking organisations will also be able to place invitations to tender for a directory service that will be able to operate internationally with other national services. Cooperation with Public Telecommunications Operators (PTOs) is being sought; the pilot service should interwork with those PTO X.500 services that are available.

(c) P2.2 Support and Information Services: this sub-project will establish a COSINE information service which will be available to all target users of OSI networks in the COSINE countries. It will coordinate national information service provision initiatives so that a consistent access is provided to all support services for information providers. It will provide initially a focal point for users to get information about networking products, projects and services, and will be used as a vehicle for promoting OSI and COSINE. Special Interest groups will be able to run their own information services for group communication and information dissemination. The service will be reachable by a variety of methods, including message handling, file transfer and terminal connections through both private research networks and PSPDNs.

(d) S1 Provision of X.25 (1984) Infrastructure: described separately in [11], IXI in combination with national networks and PSPDNs will provide the carrier service for data transferred as part of the other COSINE sub-projects.

(e) S2.1 Interworking of Existing X.400 Administrative Domains: this service will enable national end-users to use an international message handling service (MHS) based on the X.400 (1984) set of protocols. It will provide national MHS managers with information on other national MHS services and with an error reporting service at the international level. It will link the COSINE community with similar communities in other parts of the world and ensure full connectivity with existing international RFC mail networks. Cooperation with public X.400 services will be encouraged and the requirements for a transition to X.400 (1988) will be studied. The COSINE MHS service is likely to provide the vehicle to support a number of other applications, for example the Pilot Information Service, sub-project P2.2.

(f) S2.2 Message Handling Gateway Services to the USA: this service complements service S2.1 by providing one of its essential components, the transmission via a gateway of messages exchanged between Europe and North America. This service will also be used to define tests that can be used to ensure that an operational gateway is performing to adequate levels of service and functionality and will provide the CPMU with the information needed to specify a self-sustaining gateway service for the longer term.

Activity Plans have also been prepared for sub-projects P6 and P9.2. Sub-project P6 aims to demonstrate interworking of equipment from different suppliers and, in particular, to investigate the effectiveness of protocol conformance testing services and the degree of assurance they provide to users who need to know whether products which have received a conformance certificate will in practice interwork.

Sub-project P9.2 will be in two phases; the specification of a machine independent implementation of the ISO Virtual Terminal Protocol and the installation of software conforming to this specification on a number of widely used computer systems.

6. The User Community

COSINE services will be made available to the whole R&D community in all the member states. In those countries which already have an extensive network linking, for example, university and government funded research laboratories, it will be relatively easy to make COSINE services accessible to a large community of end users. In other countries, national networking organisations may need assistance in mounting new services and in making them available to research workers working within their boundaries. Across Europe, the degree of integration of academic, government funded and commercial research organisations varies widely. Generally speaking, there is still a lack of easy intercommunication between workers in commercial R&D laboratories and those who are government funded, directly or indirectly. Although PSPDNs can in principle be used to link users of all kinds, no matter where they are located, regulatory and tariff issues continue to impose obstacles to rapid progress, particularly to the provision and use of services which require medium to high bandwidth. COSINE will attempt to assist user groups to remove such obstacles.

Sub-project P3 will give support to international users groups of research workers in particular disciplines. The requirements and existing patterns of use of selected groups will be studied and analysed. Other user groups will be encouraged to experiment with networking techniques to confirm that the full range of characters, data formats and other requirements that are particular to them can be made available through the proposed COSINE services. Subsequently, all these groups will be helped to make use of those services which are appropriate to their requirements and the lessons learnt in providing this support will be passed on to other groups.

7. Transition from Existing Network Services

A large proportion of potential users of COSINE services are already network users. They will not wish to give up their existing working methods until they are convinced not only that any alternative service is better but also that the cost in time and trouble of switching to it is outweighed by subsequent gains in the quality and nature of the new service. Sub-project P4 therefore aims to ease the process of migration to OSI based services of a number of selected groups.

In some cases, some aspects of a transition can be handled in a way which is transparent to users; for example, traffic for a message handling system which makes use of a proprietary data transmission system can be moved to an OSI network such as IXI without changing any of the facilities which are available to users at the application level. More frequently, changes will be visible to users, especially if they are to take advantage of improved functionality of services, including for example the ability to reach a larger number of other users and services.

Sub-project P4 will investigate how existing services can be mapped on to the set of OSI protocols and the use of gateways as a means of passing traffic between proprietary and OSI based systems. Close cooperation will be established with existing networks such as EARN and EUNET which already provide services to a large part of the COSINE community.

The large, and still growing, community of users of services based on the TCP/IP protocols deserves special attention. Outside the framework of COSINE, RARE is providing a forum for the discussion of TCP/IP services through its support of RIPE (Réseaux IP Européens). COSINE will accept proposals for new projects for pilot services based on based on ISO IP protocols once the standards and the equipment needed to implement them have reached the required level of maturity. The CPMU is working with the relevant RARE Working Groups to ensure that steps towards the provision of OSI connectionless services are taken as soon at is practical to do so.

8. Access to Services

Because of the way COSINE services are organised, users will normally access them through the appropriate national service. COSINE pilot information services, including detailed information on access methods, are expected to be available through these channels from early in 1991.

9. Conclusion

COSINE aims to set up and offer a number of OSI based data communications services to members of the R&D community throughout Europe. The first pilot projects and services are under way; other services will be initiated and it is planned that they should be transformed into longer lasting, self-funding services by the time the Project formally terminates at the end of 1992.

10. References

[1] Cornillie-Braun, Anne et al (1988) COSINE Specification Phase Report 1: The Scope of COSINE, RARE, Amsterdam

[2] Hutton, James (1988) COSINE Specification Phase Report 2: Protocol Profiles, RARE, Amsterdam

[3] Fluckiger, François (1988) COSINE Specification Phase Report 3: Future Services, RARE, Amsterdam

[4] Brinkhuijsen, Rob (1988) COSINE Specification Phase Report 4: Public Services, RARE, Amsterdam

[5] Bauerfeld, Wulfdieter (1988) COSINE Specification Phase Report 5: Operational Aspects 1, RARE, Amsterdam

[6] Bauerfeld, Wulfdieter (1988) COSINE Specification Phase Report 6: Operational Aspects 2, RARE, Amsterdam

[7] Fluckiger, François (1988) COSINE Specification Phase Report 7: Migration Strategies, RARE, Amsterdam

[8] Hutton, James et al (1988) COSINE Specification Phase Results, RARE, Amsterdam

[9] COSINE Policy Group (1988) COSINE Specification Phase Overview Report, RARE, Amsterdam

[10] COSINE Policy Group (1989) CIP Project Proposal

[11] Devoil, John (1990) 'COSINE Service S1, Provision of X.25 Infrastructure (IXI)', ESPRIT Conference 1990

749

Table 1: COSINE Sub-Projects and Services

Sub-Projects
P1	Pilot Gateway Services to USA	
	P1.1	FTAM	*
	P1.2	Remote Access to Computing Services
P2	Pilot Information Services	
	P2.1	International Directory Services	*
	P2.2	Support and Information Services	*
P3	International User Group Support	*
P4	Pilot Projects on migration of existing networks or user groups	*
P5	Tools and techniques of OSI adoption and migration	
P6	Pilots for implementations/demonstrations of multivendor interworking	
P7	Pilot procurement exercises for OSI-based products	
P8	Security mechanisms: Study and pilots	*
P9	Future facilities	
	P9.1	OSI over ISDN
	P9.2	Full screen terminal services
	P9.3	Job transfer and manipulation
	P9.4	High speed networking
P10	LAN/WAN interworking: investigation and testing	

Services
S1	Provision of X.25(1984) infrastructure	
	S1.1	X.25(1984) service provision	*
	S1.2	Preparation, monitoring and evaluation	*
S2	Message handling services	
	S2.1	Interworking of existing X.400 administrative domains	*
	S2.2	Gateway services to the USA	*
S3	Information services	
S4	Other gateway services to the USA	
	S4.1	FTAM
	S4.2	XXX
S5	International Directory Services	
S6	Security key management	

* sub-project or service included in the 'core set'

PROVISION OF AN INTERNATIONAL X.25 INFRASTRUCTURE (IXI)

J. DEVOIL
IXI Project Team
$^c/_o$ RARE Secretariat
Postbus 41882
NL-1009 DB Amsterdam

Abstract.

The IXI project has established an international packet switched network which provides a backbone service to interconnect public networks and private research networks within Europe. This report provides an overview of the connectivity, quality of service, and traffic observed during the first three months of the pilot IXI service.

1. Project Overview

The project to provide the International X.25 Infrastructure (IXI) for COSINE began in December 1988 with the release of a documented set of requirements for a pan-European backbone network service. This was a consequence of both the work done in the COSINE Specification Phase and ongoing discussions with CEPT concerning the provision of pan-European managed data network services.

The requirements, which were developed from the results produced by working parties comprised of representatives of national research networks, academic institutes, governmental agencies (including the CEC) and PTTs, can be summarised as the provision of X.25 network services for the interconnection of private research networks and public data networks (PSPDN) in the COSINE countries. The CCITT recommendations for X.25 packet switching of 1980 and 1984 had to be supported at the interfaces to the connecting networks which operate these different versions of the recommendations.

A contract was negotiated between the CEC (acting on behalf of COSINE) and PTT Telecom for the provision of a pilot IXI service for an initial period of 12 months. This contract was signed in September 1989 and the preparation and commissioning of the network started immediately. Réseaux Associés pour la Recherche Européenne (RARE) were requested by COSINE to set up a project team responsible for preparing for the introduction of the IXI service (including acceptance testing), and to perform ongoing monitoring and evaluation of the IXI service.

Following successful acceptance testing performed by an independent testing organisation, which included performance tests on an unloaded network and tests of conformance to CCITT recommendations (by using CTS/WAN and NET2 test profiles), the IXI Pilot operation started in late April 1990.

2. IXI Connectivity

The Pilot IXI service has been operational since late April 1990 and, by the end of July, provided connections to the following networks:

Private Networks:
> ACONET (Austria)
> RES-ULB (Belgium)
> CEC (Belgium)
> CIRCE (France)
> DFN (Germany)
> ARIADNE (Greece)
> EARN (Ireland)
> HEANET (Ireland)
> GARR (Italy)
> JRC/Ispra (Italy)
> EARN (Netherlands)
> Nikhef/HEPNET (Netherlands)
> SURFnet (Netherlands)
> RCCN (Portugal)
> ARTIX (Spain)
> SWITCH (Switzerland)
> JANET (U.K.)
> NORDUNET (Denmark, Finland, Iceland, Norway and Sweden)
> CERN (Switzerland)

Public Networks:
> DATAPAC (Denmark)
> DCS (Belgium)
> DN1 (Netherlands)
> Telepac (Portugal).

The original requirements for the IXI service included connectivity to other public data networks which, up to the end of July 1990, has proved impossible to provide due to various regulatory and tariff reasons. They are:

> DATEX-P (Austria)
> Hellaspac (Greece)
> Ierpac (Ireland)
> Itapac (Italy)
> Luxpac (Luxembourg)
> PSS (U.K.)
> Jupak (Yugoslavia)

Negotiations are ongoing with the Telecommunications Administrations of these countries in an attempt to provide the required PSPDN connectivity.

3. Performance

In addition to the monthly network management reports supplied by PTT Telecom there is an ongoing project to monitor and evaluate the IXI Service. The availability of the service is monitored by establishing X.25 connections from test equipment located in the RARE Secretariat (close to the Nikhef Access Point) to "Echo Points" within the IXI Network. This provides a measure of the availability of the IXI backbone service as seen from an Access Point, and by calculating the average call set-up times a measure of network performance is obtained.

However, during the second part of June (1990) a problem was detected in the network switching equipment. This appeared as an apparently random failure of individual Access Points to set-up new outgoing calls, while still permitting incoming calls to be established. This gradually eroded the validity of the monitoring programme which requires access to the backbone network with the ability to set-up (and then break down) X.25 connections. From week 25 onwards insufficient data could be gathered to provide meaningful measure of the availability of the backbone service, although it was still a large enough sample to derive valid average call set-up times. Monitoring was discontinued after week 28 as it was believed that the test traffic it generated could be a contributing factor in provoking the failure of the switching nodes. The data provided in TABLE 1 are, therefore, an indication of the availability and performance of the IXI service during the first two months of operation. The data in TABLE 2 gives an indication of the network performance represented by the average call set-up time measured by the monitoring system. TABLE 3 provides a measure of the traffic on the network measured in data segments of sixty-four bytes.

The problem had not been resolved by the end of July, but had been diagnosed as a software design or implementation error within the switching equipment.

TABLE 1. An indication of the IXI Backbone availability (as a percentage) measured from a test station in the RARE Secretariat (in Nikhef, Amsterdam) setting-up calls to "echo points" contained in the switching equipment at the network nodes.

| Network | Week No. | | | | | | | | | |
Node	19	20	21	22	23	24	25	26	27	28
Amsterdam	100.00	100.00	100.00	99.41	100.00	82.02	*	*	*	*
Berne	100.00	100.00	100.00	99.70	100.00	100.00	*	*	*	*
Madrid	98.85	100.00	100.00	96.30	99.88	100.00	*	*	*	*
Athens	97.69	84.33	100.00	98.81	91.65	99.90	*	*	*	*
Lisbon	100.00	100.00	100.00	99.41	100.00	100.00	*	*	*	*
Brussels	84.42	100.00	100.00	99.56	100.00	100.00	*	*	*	*
Dublin	100.00	100.00	100.00	99.56	100.00	100.00	*	*	*	*
Vienna	99.81	98.81	98.07	99.26	100.00	100.00	*	*	*	*
Bologna	100.00	98.98	100.00	99.56	99.52	100.00	*	*	*	*
Average	97.86	98.01	99.79	99.06	99.01	97.99	*	*	*	*

Note: The data for weeks 25 through 28 has been recorded as part of the monitoring programme, but the disturbances introduced by the software problem in the switching nodes invalidate it as a measure of network availability.

TABLE 2. An indication of IXI backbone performance given by the average call set-up times (in seconds) as recorded by the program used to measure network availability.

Network Node	Week No.									
	19	20	21	22	23	24	25	26	27	28
Amsterdam	0.26	0.26	0.32	0.26	0.25	0.24	0.25	0.35	1.15	0.30
Berne	0.70	0.70	0.73	0.71	0.73	0.80	0.83	0.66	0.89	0.70
Madrid	0.70	0.66	0.67	0.89	0.68	0.75	0.76	0.98	0.87	0.71
Athens	0.82	0.92	0.87	0.81	0.85	0.88	0.94	0.98	1.10	1.17
Lisbon	0.77	0.79	0.83	0.79	0.79	0.84	0.86	0.88	1.06	1.13
Brussels	0.75	0.64	0.67	0.63	0.66	0.71	0.71	0.99	0.89	0.93
Dublin	0.60	0.65	0.67	0.65	0.66	0.68	0.73	0.69	1.40	0.81
Vienna	0.78	0.63	0.75	0.64	0.65	0.70	0.79	0.84	0.92	0.90
Bologna	0.61	0.73	0.65	0.63	0.65	0.70	0.69	1.09	0.80	0.90
Average	0.67	0.66	0.68	0.67	0.66	0.70	0.73	0.83	1.01	0.84

TABLE 3. IXI service usage measured as the number of 64-byte segments transmitted in both directions during calls set-up from each of the Access Points. (Data supplied by PTT Telecom).

Connected Network	Segments Transmitted/Received		
	May	June	July
ACONET	11,444	2,619	1,108
RES-ULB	4	565	97
CEC	0	208,666	48,835
DCS	177,269	553,850	841,712
DATAPAK	420	11,563	930
REUNIR	2,067	12,159	29,407
DFN	41,410	80,287	154,988
ARIADNE	13,080	644,977	210,595
HEANET	431	1,037	1,182
EARN-IE	0	9,490	138,582
GARR	16,633	56,609	15,456
JRC/Ispra	0	540	112,850
SURFNET	11,465	190,200	130,073
HEPNET-Nikhef	793,690	582,073	153,185
EARN-NL	0	0	0
DN1	4	1,627	20,888
TELEPAC	2	826	17,818
RCCN	541	6,555	14,339
ARTIX	32,321	65,698	48,314
NORDUNET	2,720	29,873	30,829
CERN	2,428	14,211	42,443
SWITCH	193,904	101,218	82,934
JANET	247,813	182,773	1,337,485
Total	1,547,646	2,175,343	3,434,050

4. Accessibility and use.

Use of the Pilot IXI Service is restricted to the "COSINE community" of researchers, including industrial research departments, as well as governmental agencies for their programmes concerned with research and development. The traffic carried by the IXI network is non-commercial.

All organisations which can connect to a private research network with an IXI Access Point are authorised to interconnect via the IXI backbone service. To do so they must register their end-systems (DTEs) and receive a "COSINE Address" for them which will be recognised by IXI. This is done via the network operations and management procedures of the private research networks. For organisations which are connected to one of the public data network services with an IXI Access Point there is an authorisation and registration process by which they can apply for access to the IXI service.

5. Summary

The IXI backbone is providing a 64 Kilobit per second packet switched data interconnection service between research networks in the COSINE countries, but software problems within the switching equipment have reduced both the usability of the network service and the ability of the research network operators to commit production traffic to it. The unwillingness (or in few cases the inability due to legal restrictions) of several PTTs to provide connection to IXI from their public data networks reduces the potential connectivity of this pan-European research network. This is to the particular disadvantage of industrial researchers, and of academic institutes in countries without a widespread national research network.

BASIC RESEARCH

Perspectives in Supramolecular Chemistry
Towards Molecular Devices

Jean-Marie LEHN
Université Louis Pasteur,
4, rue Blaise Pascal, 67000 Strasbourg and
Collège de France,
11, Place Marcelin Berthelot, 75005 Paris, France

Molecular and supramolecular devices may be defined as structurally organized and functionally integrated chemical systems built into supramolecular architectures. The development of such devices requires the design of molecular components (effectors) performing a given function and suitable for incorporation into an organized array such as that provided by the different types of polymolecular assemblies. The components may be photo-, electro-, iono-, magneto-, thermo-, mechano-, or chemoactive, depending on whether they handle photons, electrons, or ions, respond to magnetic fields or to heat, undergo changes in mechanical properties, or perform a chemical reaction. A major requirement would be that these components, and the devices that they bring about, perform their function(s) at the *molecular* and *supramolecular* levels as distinct from the bulk material.

The nature of the mediator (substrate) on which molecular devices defines fields or molecular photonics, molecular electronics, and molecular ionics. Their development requires the design of effectors that handle these mediators and the examination of their potential use as components of molecular wires.

Much interest has been shown in the possibility of designing *electronic devices* that would operate at the molecular level. This involves several steps. It is first necessary to imagine a molecule that may possess the desired features, synthesize it and study its properties. The second step is to incorporate it into supramolecular architectures, such as membranes or other organized structures, and to investigate whether the resulting entity possesses the required properties. The third step requires to connect the basic unit to other components, in order to address it via relay molecules or with an external physical signal. Of course, the goal of designing molecular electronic circuitry may have many spin-offs along the way, in addition to asking novel questions about the handling of molecules. It is probably premature to try to define clearly the routes to the goal, let alone that the definition of the goal itself may change under the influence of progress made along a given path.

The possibility of designing devices such as molecular rectifiers, transistors, switches, photodiodes has been envisaged and some of the required features are present in compounds such as metal complexes or Donor-Photosensitizer-Acceptor systems that lead to photoinduced charge separation at the level of the isolated molecule.

Among the various devices and components performing molecular-scale electronic functions that may be imagined, a crucial one is a *molecular wire*, which might operate as a connector permitting electron flow to occur between the different elements of a molecular electronic system.

Our first approach towards the design of molecular wires was based on the *caroviologens* (CV^{2+}) that combine the structural features of carotenoids with the redox properties of methylviologen: long, conjugated polyolefinic chains bearing pyridinium groups at each end. Such compounds were incorporated into phospholipid vesicles and electron transfer experiments were performed between an external reducing phase and an internal oxidizing phase (see Figure).

Thus, the caroviologen approach does produce *functional molecular wires* that effect electron conduction in a supramolecular scale system.

Conjugated polyolefinic chains bearing an electron acceptor group on one end and a donor on the other end represent polarized molecular wires that should display preferential *one way electron transfer* and act as a *rectifying component*.

Such donor-acceptor carotenoids D-C-A were prepared and shown to possess very pronounced *nonlinear optical properties*. Oriented incorporation into vesicles and electron transfer experiments might reveal one way conduction by polarized molecules of this or similar type.

Modifications on the original CV^{2+} type of molecular wire may bear on the terminal groups or on the conjugated chain or on both. The carotenoid chain might be replaced for instance by extended dye systems, oligopyrroles or oligothiophenes, strings of redox centers allowing electron "hopping", etc.

Thus, symmetric and dissymmetric carotenoid chains bearing ferrocene, 2,2'bipyridine, pyridine as terminal groups have been prepared as well as metal complexes derivatives. These *metallo-carotenate* type of molecular wires combine the electrochemical and photochemical activities of metal complexes (ferrocene, or rhenium, ruthenium complexes etc.) with the long range conjugation properties of carotenoid chains, thus leading the way towards systems capable of performing electro- or photo-induced very long range electron transfer (VLReT).

These types of compounds may be considered to incorporate electro- and photosensitive switches at the end of or inside the conjugated chain and thus represent *switchable molecular wires* responding to external stimuli.

Endowing photo-, electro- and iono-active components with recognition elements opens perspectives towards the design of programmed molecular and supramolecular systems capable of molecular recognition directed selfassembling into organised and functional supramolecular devices. Such systems may be able to perform highly selective operations of recognition, reaction, transfer and structure generation for signal and information processing at the molecular and supramolecular levels.

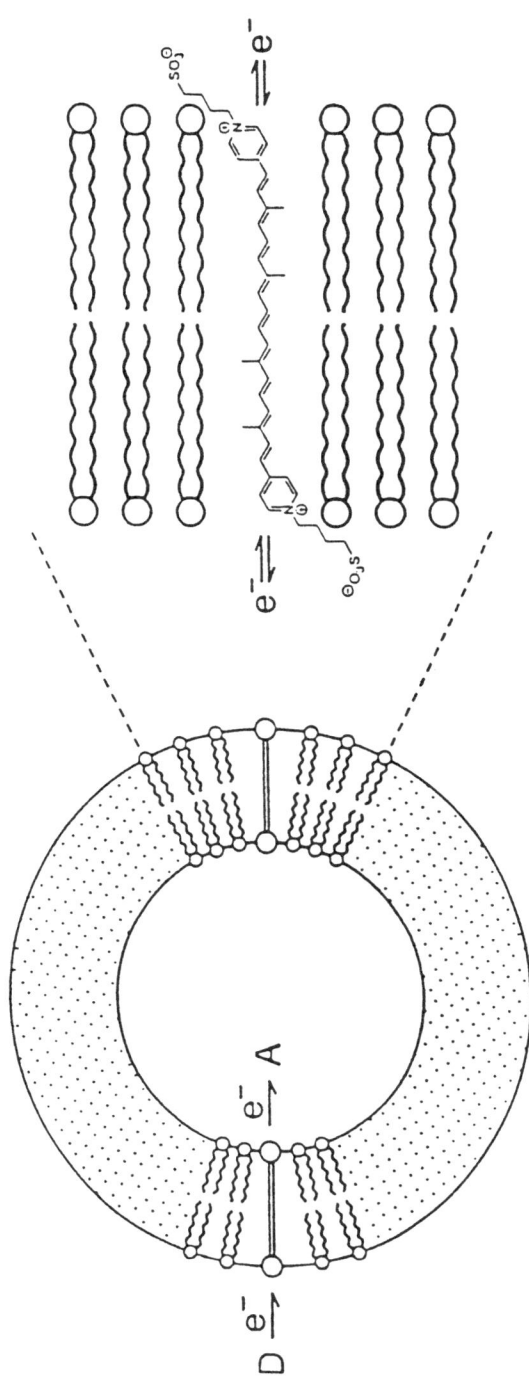

Perspectives of Local Probe Methods

Dr. Rohrer
IBM Research Division
Forschungslaboratorium Zürich
Säumerstrasse 4
CH-8803 Rüschlikon

Over the past years a variety of local probe methods developed from scanning tunneling microscopy, using various interactions between probe and object. These methods play a central role in advancing our understanding of individual functional units of nanometer dimensions and our capabilities of their manipulation and control. First attempts to work with biological macromolecules while in their functional state are presented.

EMERGING RELATIONSHIP BETWEEN OPTICS AND ELECTRONICS

ALAN HUANG
AT&T Bell Laboratories
Holmdel, New Jersey 07733

ABSTRACT

Optical logic gates and free space interconnects have been used to implement optical systems modules. These modules have been assembled to implement a circular optical pipeline. The design, performance, and motivation behind this optical pipeline are discussed.

BASIC STRATEGY

In the field of computers there is the concept of a platform. A platform is a set of standards used to minimize risk and maximize flexibility. An IBM personal computer is an example of a platform. By setting **standards** for the hardware backplane and software operating system many smaller vendors could with relatively little risk develop boards and programs for this computer. The users were also pleased by the variety and lower cost. We have actively been trying to establish such a platform for optics.

A DIGITAL INTERFACE

The most important **standard** which we have been trying to establish has been a digital interface. Optical systems have traditionally been analog. The problems with this approach is that it is increasingly difficult to build complexity and there was very little reuse of sub-systems because of their specialized nature. Our approach was digital. This initially made things more difficult however we have already noticed several benefits. It allowed us to decouple work on the logic gates from the optics and architecture. The architecture didn't depend on some novel aspect of physics while the devices weren't applicable to only one architecture. This minimized the risk for the device people and increased the flexibility for the system designers.

We did have some difficulty with the initial optical logic gates based on optical bistability. They were basically threshold gates and we were forced to provide a certain number of pico joules of energy at a certain location in our system. This was a problem in that it coupled the performance of our logic gates with our optical systems design. As a result, we adopted dual rail rather than an intensity dependent digital standard. This let us use the relative intensity of two signals rather than the absolute intensity of a signal. This let us decouple to a further degree the performance of the optical gates from the design of the optical system.

OPTICAL LOGIC

We are currently pursuing several types of optical logic gates [1,2,3,4]. The SEED (self-electrooptic effect device) [1,2] is an integrated electro-optical device which is functionally equivalent to a latching NOR gate.

An opto-electronic integrated circuit approach to implementing optical logic gates is also being pursued. CMOS optical input pads have been demonstrated [5]. We are currently trying to implement optical output pads with either SEED modulators [1] or microlasers [6]. See Figures 1 and 2.

We are also working on optical logic gates based on the QWEST effect [3] and polymer optical nonlinearities [4].

All of these optical logic gates adhere to a strict digital **standard**.

OPTICAL INTERCONNECTS

Our approach to random connectivity relies on multi-stage network theory. One approach is based on a perfect shuffle [7]. A more optically efficient approach is based on crossover networks [8]. We are currently using a split and shift network which is related to the previous networks. All of these approaches are capable of supplying thousands of very high bandwidth, low energy, constant latency interconnections. We have used these **standard** interconnects to construct various optical logic circuits.

OPTICAL SYSTEMS MODULES

We decided to concentrate on a **standard** optical interconnection module to reduce our engineering. The availability of optical logic gates to regenerate the signals both spatially and in terms of intensity greatly simplified our efforts since it bounded our light budget and aberration problems.

Our original attempt at building one stage of an optical pipeline took three, 4 by 12 foot optical benches. Our current approach takes about 1 square foot [9].

We are now investigating using three dimensional planar optics to reduce this setup to several square inches [10]. See Figures 3 and 4.

We are trying to provide optics with a mechanical platform. In our current system we removed as many degrees of mechanical freedom as possible in order to simplify the mechanical interface between modules. We hope to use three dimensional planar optics and its electron beam photolithographic based manufacturing techniques to reduce this tolerance from 10's of microns to microns.

In perspective, a great step in the development of electronics was the integrated circuit. It allow some people to concentrate on the development of the chips while others worked on using the chips to make more and more complicated systems. What optics has been missing has been **standard** "sockets" and components with a mechanical tolerance of microns. This would expand the market for both components and systems while reducing the cost.

ALGORITHMS

Software based on *symbolic substitution* has been developed to map an arbitrary combinatoric circuit onto a **standard** multi-stage, regular interconnect [11].

A technique called *computational origami* is used to fold a virtual array of these combinatoric circuits onto a **standard** optical pipeline architecture [12].

ARCHITECTURES

The intent is to use these virtual combinatoric circuits to implement a regular array of state machines. These state machines would be connected together in a large loop in a cytoarchitecture inspired by DNA biological information processing mechanisms. These state machines work on a symbol stream generated by a high level, general purpose functional language. A birth and death paradigm is introduced to facilitate a dynamic allocation of the state machines.

SYSTEM EXPERIMENTS

A simple optical pipelined processor has been demonstrated using four arrays of S-SEEDs, the photonic equivalent of an R-S flip flop. See Figure 5. We are currently using a 6 by 8 array of devices [9]. We are extending this to 64 by 32 and larger. Individual gates have been pushed to 1 Gigihertz but this requires a considerable amount of laser power so we cannot presently drive a large array at this rate. Lower power and faster devices are being pursued. Higher power and higher rep-rate lasers are being developed. The optical pipeline has been programmed to implement two, unary counters. Other circuits have been designed and are being tested.

THE FUTURE

In terms of raw speed, it is believed that the output of an electronic chip will be limited to around a Gigibit. The use of optical output pads should be able to extend this limit to around 10 Gigibits at which time the speed will be limited by the wires on the integrated circuit itself. By modifying the architecture and giving each logic gate an optical input and output capability this limit should be able to be pushed to around 100 Gigibits at which time the speed will be limited by some of the intrinsic properties of the semiconductor. Faster optical non-linearites exist. They are weak but they react in the order of femtoseconds. Their use is speculative at present but they are being studied.

In terms of parallelism, it is believed that optics can easily achieve over 50 times more connectivity. This should open up some of the architectural bottlenecks. Ideally, this should have a direct effect on throughput.

CONCLUSION

We are pursuing a **standard** signal interface, a **standard** mechanical interface, a **standard** optical interconnect, a **standard** optical logic module, a **standard** optical circuit design technique, and a **standard** architecture.

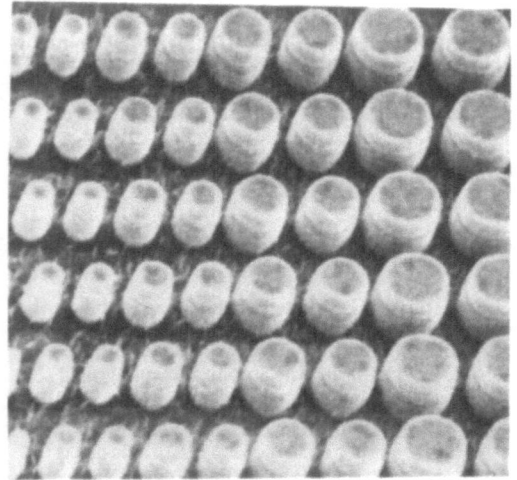

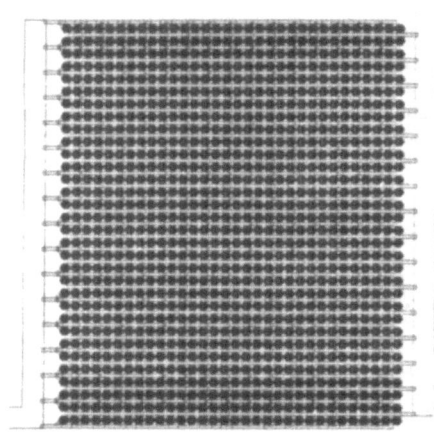

Figure 1 An array of microlasers.

Figure 2 An array of S-SEED optical logic gates.

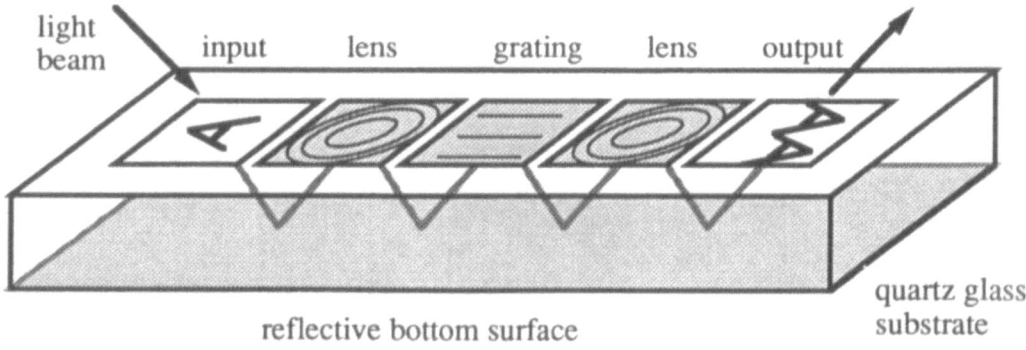

Figure 3 The conceptual view of a three dimensional planar optics experiment.

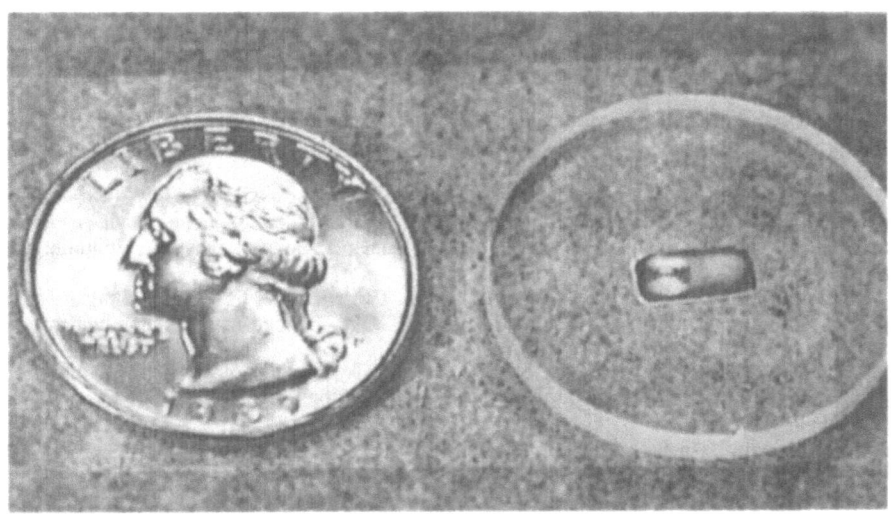

Figure 4 A view of a three dimensional planar optics experiment.

Figure 5 An optical pipeline processor.

REFERENCES

[1] D. A. B. Miller, J. E. Henry, A. C. Gossard, and J. H. English, "Integrated Quantum Well Self-Electroopptic Effect Devices: 2 x 2 Array of Optically Bistable Devices," *Applied Physics Letters*, vol 49, pp 821-823, 1986.

[2] A. L. Lentine, H. S. Hinton, D. A. B. Miller, J. E. Henry, J. E. Cunningham, and L. M. F. Chirovsky, "Symmetric Self-Electro-Optic Effect Device: Optical Set-Reset Latch," *Applied Physics Letters*, vol 52, pp 1419-1421, 1988.

[3] L. C. West, "Picosecond Integrated Optical Logic," *Computer*, pp 34-46, Dec. 1987.

[4] M. C. Gabriel, "Transparent nonlinear optical devices," Optoelectronic Materials, Devices, Packaging, and Interconnects, Ted E. Batchman, Richard F. Carson, Robert L. Gallawa, Henry J. Wojtunik, Editors, *Proc. SPIE* 836, pp.222-227 (1987).

[5] A. Dickinson, M. Prise, N. Craft, A. D'Asaro, and L. Chirovsky, "A Free Space Optical Data Link Using GaAs Multiple Quantum Well Modulators and Si CMOS Photodetectors," submitted to *Applied Optics*.

[6] J. L. Jewell, A. Scherer, S. L. McCall, Y.H. Lee, S. J. Walker, J. P. Harbison, and L. T. Florez, "Low Threshold Electrically-Pumped Vertical Cavity surface Emitting Micro-Lasers," to be published in *Electronics Letters*.

[7] K.H. Brenner and A. Huang, "Optical Implementation of Perfect Shuffle Interconnections," *Applied Optics*, vol. 27, pp 135-137, 1988.

[8] J. Jahns, "Crossover Networks and Their Optical Implementation," *Applied Optics*, vol 27, pp 3155-3160, 1988.

[9] "A Module for Optical Logic Circuits Using Symmetric Self Electrooptic Effect Devices," M. Prise, R. LaMarche, N. Craft, M. Downs, S. Walker, L.Chirovsky, and A. D'Asaro, submitted *Applied Optics*, July 1989.

[10] J. Jahns and A. Huang, "Planar integration of free-space optical components," *Applied Optics*, vol 28, pp 1602-1605, 1989.

[11] M. Murdocca and T. Cloonan, "Optical Design of a Digital Switch," *Applied Optics*, vol. 28, no. 13, pp 2505-2517, July 1989.

[12] A. Huang, "Computation Origami - The Folding of Circuits and Systems," *Conference Proceeding of the 1989 Optical Computing Topical Meeting of the Optical Society of America*, Salt Lake City, pp. 132-135, Feb. 1989.

Problems and Prospects for Computer Vision

Michael Brady
BP Professor of Information Engineering
Oxford University

Recent advances in techniques of image processing and image understanding, allied with rapid advances in sensor technology and parallel architectures, are making visually-guided processing and control an increasingly realistic and attractive option for industry. This lecture presents an overview of the current state of computer vision, and an assessment of how well current techniques work in practice.

We describe advances in a number of areas of computer vision ranging from early spatiotemporal processing, through model-based recognition, and the combination of vision systems with ideas from modern control theory. We will show examples of systems that compute three-dimensional shape and perform path planning. We describe parallel architectures that aim to make real-time vision applications a reality. The history of computer vision is a cycle of close attention to individual processes (eg edge detection) and the synthesis of entire systems that horizontally and vertically integrate such processes. The field is currently in a system building phase. Several examples will be sketched.

There is a continuous debate between advocates of "bottom-up" and "top-down" approaches to vision. Of course, neither are sufficient of themselves. There has been recent progress in top down approaches after they had been in the doldrums for a decade. We review the progress critically.

Most vision algorithms require representation of shape of objects and the surrounding environment. Different tasks make different demands on a shape representation. Determining the position and orientation of a polyhedral or simple object can be achieved by histogram representations such as the (generalised) Hough transform or extended Gaussian image. (EGI). Other representations are based on generalised cylinders, view spheres, symmetry sets, or, most recently algebraïc invariants.

The assessment of current performance is important for it is the practical unreliability of techniques prematurely pressed into regular use on complex, uncontrolled imagery after encouraging laboratory performance in simpler, controlled conditions that has, as much as cost (processor power or memory requirements), dissuaded production engineers from wholesale adoption of image-based solutions in the past. The general-purpose vision system capable of dealing with a complex changing environment with the same facility as humans is many years away. It is now clear that many practical applications can be successfully fielded now so long as the application and techniques are properly matched.

COMPUTATIONAL LOGIC

Robert Kowalski
Imperial College
London SW7, UK

The term "computational logic" has no generally agreed definition. Within the Compulog basic research action we use the term to mean the use of logic as a broad spectrum language for programs, program specifications, databases, and "knowledge bases" in artificial intelligence. We take logic programming as our starting point but seek to enhance it by incorporating related techniques for the use of logic in computing.

Historical background

Historically, the field of computational logic has developed from work on automated theorem proving. This work began in the 1950's with the efforts of logicians like Martin Davis, Paul Gilmore, and Hilary Putnam to devise efficient, automated proof procedures for symbolic logic. It had already been shown that mathematics could be represented in symbolic logic; and the goal of this early work on automated theorem proving was to mechanize the proofs of mathematical theorems by mechanizing the proof procedures of mathematical logic.

Significant advances in the automation of symbolic logic were made in the 1960's, especially with the development of the resolution principle by J. Alan Robinson and of model elimination by Donald Loveland. However, the ambitious goal of automated theorem proving turned out to be extraordinarily difficult to achieve.

It was in artificial intelligence and computer science that the first applications began to be made, most notably by Cordell Green in the late 1960's . In the early 1970's, Alain Colmerauer and Robert Kowalski showed how to use resolution logic to represent and execute computer programs. The computer language Prolog and the field of logic programming were born. In parallel, related techniques were used for deductive databases and for knowledge representation and problem solving in artificial intelligence.

The goal of computational logic

Although they have much in common, the three fields of programming, databases, and artificial intelligence have remained distinct. It is the goal of the Compulog basic research action to unify the techniques of these fields.

Historically the emphasis in automated theorem proving has been on the use of logic to represent a fixed state of knowledge and the use of mechanical proof procedures for problem solving. In artificial intelligence and computer science, however, it is also important to manage the way in which programs, databases, and knowledge bases change over time. As we see it in Compulog, therefore, the ultimate goal of computational logic is both to extend logic programming into a broad spectrum language for computing and artificial intelligence and to develop theorem-proving techniques for managing change.

To simplify terminology, we use the term "theory" to refer to a collection of statements written in logical form. Depending on the application, such a theory can be viewed as a program, database, or artificial intelligence knowledge base.

Logic programming

Logic programming is the starting point for our work. It also serves as the basis for research and development activities in several major centres, the most significant of which are probably ICOT in Tokyo, SICS in Stockholm and ECRC in Munich. All of these centres carry out a wide range of work including language and computer architecture development. ICOT has directed much of its effort towards concurrent logic programming computer architectures and towards artificial intelligence theory and applications; SICS has focussed on parallel extensions of PROLOG; and ECRC has concentrated on parallel logic programming architectures, constraint logic programming, and deductive databases. ECRC is a major partner in Compulog. Therefore the research activities of Compulog and ECRC are closely linked and complementary.

Compulog is concentrating on language development and on theory representation and management. It is not addressing matters of parallelism, because these are a major concern of ECRC and of other projects in ESPRIT. The Compulog work will be compatible with work on parallelism carried out elsewhere.

Compared with some other logic programming based projects, Compulog has somewhat broader aims of incorporating related developments in the use of logic in such fields as computer algebra, database systems, and artificial intelligence, as well as developments in mathematical logic itself.

The Compulog work areas

Within Compulog, the work has been organised into four main areas:

(1) language extensions, concentrating on constraint logic programming and structured types (Marseille, Kaiserslautern, Rome, Tübingen),
(2) knowledge assimilation, concerned with techniques for incorporating new knowledge into a knowledge base (ECRC, Bristol, Edinburgh, Lisbon, London, Passau),
(3) metalevel reasoning (Bristol, London, Pisa, Rome), and
(4) program development, analysis and transformation (Leuven, Edinburgh, London, Pisa, Uppsala).

Improved computer languages

The first work area and parts of the second and third contribute to the goal of improving the expressiveness of logic programming languages. Work on constraint logic programming, drawing on algorithms developed in computer algebra, significantly enhances the naturalness with which problems can be expressed, while increasing the effectiveness of available problem solving methods. A constraint logic programming language, Prolog III, is being developed within Compulog by Marseille.

Work on structured types draws on developments in hybrid knowledge representation formalisms, programming language types, and unification theory, and is related to work on constraint logic programming. It also combines logic programming with some of the features of object oriented knowledge representation.

Work within the area of knowledge assimilation, on default reasoning and abduction, contributes to enhancing naturalness of expression, by allowing explicit negative statements and the separation of rules and exceptions.

Work on metalevel reasoning extends logic programming languages by metalevel constructs. These can be used for specifying and implementing both knowledge assimilation and program development analysis and transformation strategies. They can also be used to represent theories about theories, to construct theories from other theories (programming in the large) and to reason about interactions between theories. A metalogic programming language, GÜdel, is being developed within Compulog by Bristol.

Improved theory management techniques

The second, third and fourth areas contribute to the goal of improving theory management techniques. Work on knowledge assimilation approaches this problem from the database and artificial intelligence points of view, whereas work on program development, analysis and transformation approaches it from the programming point of view. Work on metalevel reasoning provides tools for specifying and implementing theory development and maintenance techniques.

Knowledge assimilation generalises the problem of updating databases to the problem of assimilating new sentences into a theory, viewed as a deductive database or knowledge base. The two main aspects being considered are generalisations of integrity checking and belief revision. A third subarea concerns the connection between default reasoning and the revision of previous beliefs in the light of new ones. Belief revision and default reasoning are related to one another by abduction, which is a variety of hypothesis formation.

Metalevel reasoning enhances both expressiveness and problem solving power. It facilitates programming in the large as well as reasoning about several theories and their interaction. It is an essential tool for specifying and implementing knowledge assimilation, program development, and program transformation.

Work on program development, analysis and transformation is aimed at developing the techniques of abstract interpretation, compiling control, metaprogramming, proof plans, and program verification and synthesis. This work is related to techniques for proving the integrity of database updates and transactions. It both uses metaprogramming techniques and can be applied to metaprogramming to make it more efficient.

Achievements of Compulog

Although Compulog began only in April 1989, a number of achievements have already exceeded our original expectations. For example, in the area of language extensions, several numerical problems previously unsolved by computer algebra techniques alone have been solved using Prolog III; and several previously open complexity problems have been solved for structured typed languages. In the area of knowledge assimilation, previously disconnected methods for view updates, belief revision, default reasoning and abduction have been shown to be closely related; and certain bottom-up query evaluation inference engines developed within the datalog research community have been shown to be identical to top-down methods developed within the logic programming community. In the area of metalevel reasoning, first steps have been taken to clarify the semantics of different ways of combining object language and metalanguage in logic programming. In the area of program development, analysis, and transformation, connections have been found between program transformation methods using specialisation of metainterpreters, compiling control, and proofs as programs.

A number of new research opportunities have also been identified. For example, the possibility of developing object-oriented deductive databases using the work on structured types has begun to be investigated. Links between the theory construction techniques developed in Pisa, the transformation of rules and exceptions into ordinary logic programs, and theory construction defined by meta interpretation have been identified. Links between inductive proof methods used in proof plans and the methods used in program transformation and in program derivation by integrity checking have also been discovered.

Acknowledgements

Compulog is supported by the ESPRIT Basic Research programme. The principal investigators are John Lloyd (Bristol), Franìois Bry (ECRC, Munich), Alan Bundy (Edinburgh), Maurice Bruynooghe (Leuven), Luis Moniz Pereira (Lisbon), Chris Hogger and Robert Kowalski (London), Alain Colmerauer (Marseille), Mathias Jarke (Passau), Giorgio Levi and Franco Turini (Pisa), Luigia Aiello (Rome), Rudiger Loos (Tübingen), and Sten-Åke Tärnlund (Uppsala).

Publications

Other papers about computational logic and the Compulog action can be found in the proceedings of the Computational Logic Symposium held during the ESPRIT Conference:

J. Lloyd (editor) 1990, "Proceedings of the Computational Logic Symposium", Springer Verlag.

Compulog papers already published or accepted for publication include:

Baader, "A formal definition of expressive power for knowledge representation languages". Proceedings of ECAI 90.

Baader, "Terminological cycles in KL-ONE based knowledge representation languages". Proceedings of AAAI 90.

Brogi, Mancarella, Pedreschi, and Turini, "Hierarchies through basic metalevel operators". Proceedings of Meta 90, MIT Press.

Bry, "Query answering in recursive databases: bottom-up and top-down reconciled". Proceedings International Conference on Deductive and Object-oriented Databases, Kyoto, Japan, December 1989.

Bry, "Intensional updates: abduction via deduction". Proceedings of ICLP 90, MIT Press.

H.-J. Bückert, "A resolution principle for clauses with constraints". Proceedings of CADE 90.

Colmerauer, "An introduction to Prolog III". CACM, 1990

De Schreye, Martens, Soblon, Bruynooghe, "Compiling bottom-up and mixed derivations into top-down executable logic programs". To appear in "The Journal of Automated Reasoning", 1990

Donini, Lenzerini, Nardi, "An efficient method for hybrid deduction". Proceedings of ECAI 90.

Gallagher and Bruynooghe, "The derivation of an algorithm for program specialisation". Proceedings of ICLP 90, MIT Press.

Guessoum and Lloyd, "Updating knowledge bases". New Generation Computing, 1990.

Hill, Lloyd, and Shepherdson, "Properties of a Pruning Operator". To appear in "Journal of Logic and Computation".

Hollunder, Nutt, Schmidt-Schauss, "Subsumption algorithms for concept description languages". Proceedings of ECAI 90.

Kakas and Mancarella, "Generalised stable models: a semantics for abduction". Proceedings of ECAI 90

Kakas and Mancarella, "Database updates through abduction". Proceedings of VLDB 90.

Kim and Kowalski, "An application of amalgamated logic to multi-agent belief". Proceedings of Meta 90, MIT Press.

Kowalski and Sadri, "Logic programs with exceptions". Proceedings of ICLP 90, MIT Press.

Wiggins, "The improvement of Prolog program efficiency by compiling control: a proof-theoretic view". Proceedings of Meta 90, MIT Press.

COMPUTER SCIENCE
THE PAST AND THE FUTURE

John E. Hopcroft
Department of Computer Science
Cornell University
4130 Upson Hall
Ithaca NY 14853

ABSTRACT

This paper examines the milestones in Computer Science and proposes that future progress will not be in terms of "bigger and better," but rather in terms of qualitative differences in our abilities to process information. Achievements in artificial intelligence, programming languages and systems, computer architecture, numerical analysis, and complexity and algorithmic theory are discussed, and new avenues of research in these areas are suggested. We believe that there will be an exciting and symbiotic relationship between the way we think and the systems we build.

THE PAST

Although the history of modern computing began around 1945 with the construction of ENIAC at the University of Pennsylvania, the discipline of academic computer science did not begin until the sixties. In the early years, university research focused on the foundations of computer science and on various technologies such as compiling that were essential to using computers. The major areas defining the discipline were artificial intelligence, programming languages and systems, computer architecture, numerical analysis, and complexity and algorithmic theory.

Artificial Intelligence

Early work in artificial intelligence investigated such problems as robotics, and vision and speech recognition; tasks that were easy for humans but which were extremely difficult to automate. It was felt that progress in these areas would lead to an understanding of intellectual processes. Such was not to be the case. Progress in understanding these areas was made by advances specific to these domains and provided little or no insight into intelligent behavior. However, the complexity of the programming tasks undertaken by early AI researchers forced them to develop programming environments. Many of the early contributions to programming environments came from this community. Other contributions included LISP, expert systems, search techniques and logics. The effort in AI was undoubtedly too early to make

774

progress on the general problem of intellectual activity, but the importance of the subject matter argues strongly for continued effort. The area has attracted some of the best minds, many of whom have made solid contributions in other areas.

Programming Languages and Systems

Work in programming languages focused on language design. Important contributions included languages such as ALGOL, functional languages and object-oriented programming. Numerous constructs such as encapsulation and modularity also came from this work. Programming methodology concentrated primarily on programming in the small. A central theme was the development of a correctness proof along with a program. Little fundamental work has been done on understanding programming in the large. Recognizing that software is an evolving entity and that its architecture can extend or curtail its evolvement is critical. Practitioners have made major advances in the size and complexity of systems they produce but there is little codified knowledge with which to educate future software engineers. MacIntosh-like environments have extended computers to a much broader segment of the population, dictating a broader view of programming that encompasses user interfaces. Systems work has its biggest impact in the area of distributed computing, with important advancements in synchronization and fault tolerance. Relational data bases have provided a framework for payroll-like files but work remains to be done on spatial, scientific and design data bases.

Computer Architecture

Past research in computer architecture had little impact on computing technologies. Rapid advances in solid state physics, packaging, memories and disk drives drove the hardware field. Research in the VLSI area lagged behind the technology and had little or no impact on machines. This situation may be changing with parallelism. The increase in computing speeds evolved by reducing the size of components. Now we may be reaching a stage where further reduction of component size is becoming more difficult and costly. It appears likely that future increases in speed will come from executing programs in parallel. Until recently, the von Neumann model of computing provided a paradigm where a computer was universal and any increase in computing speed applied to all problems. However, with parallelism it is not clear how to map arbitrary problems onto a given architecture. In fact, each parallel architecture may bring significant speed-ups only to certain classes of problems. For small amounts of parallelism, automatic techniques may be found to extract the parallelism from existing sequential programs. For high-grained parallelism where tens of thousands of processors are applied to a problem, it appears necessary to write programs in a language for parallel execution. It seems that the fundamental algorithms for a given problem may well depend on the architecture used to solve the problem.

Numerical Analysis

     Numerical analysts were originally concerned with developing numerically stable algorithms for solving linear equations, integration of differential equations, optimization and other important engineering problems. Of particular importance was the notion of backward error analysis that related round-off error to the size of perturbation of the input data that would produce the same error. More recently, significant progress has been made in handling sparse systems. The most recent thrust has been to consider the impact of parallel architecture on scientific routines. Over the past 30 years, the area has changed from a narrow focus on numerical analysis to the broad discipline of scientific computation. In this process, there has been a combining of numerical and symbolic techniques that is likely to be of importance in the future for generation of efficient engineering codes. The interplay between numeric and symbolic computation also comes up in robustness of algorithms.

     Many classes of problems, e.g., graph theoretic problems, use only symbolic computation. Entities are represented by symbols rather than numbers. There is no notion of approximation or round off. If an algorithm is correctly implemented, it will work correctly on all input data. Similarly, other algorithms such as computing a square root, involve only numeric data. For these algorithms, a continuity argument can be used to show that the output is an approximation to the real output.

     A difficulty arises if an algorithm converts numeric data to symbolic. This can occur, for example, when the program branches if the result of a numerical computation is zero. The difficulty is that there is no way to determine for a quantity that is close to zero if it is indeed zero. If several such conversions from numeric data to symbolic are made, then they may be made in an inconsistent fashion. The solution is to make all conversions explicit and restructure the code so that all decisions are independent; then no inconsistency can occur.

     Consider the problem of intersecting two squares ABCD and EFGH. Suppose line EF is so close to vertex B that numerically one has trouble determining on which side of EF B lies. In that case, it probably does not matter whether one concludes that the squares intersect or not since a slight perturbation of the input data can cause either situation to occur. However, one must not structure their algorithm so that a numerical test is used to determine if EF intersects AB between A and B and another numerical test is used to determine if EF intersects BC between B and C. One might end up in the inconsistent situation where one decides that EF intersects AB between A and B but EF intersects BC above B. The solution of course, is to use a numerical test to determine on which side of EF B lies and then use symbolic computation to intersect the line segment AB with EF and the line segment BC with EF. This solution will insure that the two intersections are consistent.

The area of communications has received little attention except for packet switching and network protocols. Today it is impossible to separate computing and communications. What impact this will have on research remains unclear.

Complexity of Algorithmic Theory

The most significant research achievement came in theoretical foundations and in algorithm design, with perhaps the greatest progress occurring in our understanding of algorithms. In the mid sixties, a researcher would publish an algorithm, along with the execution time of a specific implementation, on a small set of sample problems. Another researcher, working on a new approach, would subsequently publish his or her results. Invariably the running times on the sample problems for the second algorithm would be better. However, it was never clear how much of the improvement was due to increases in computing speeds, better techniques for code optimization or simply more clever programming. In fact, it was possible that the apparent improvement came about because the algorithm was tuned to the sample data in the first paper. Conceivably, if both algorithms were tried on new problem instances, the original algorithm would do better. What was needed was a mathematical criterion with which to compare algorithms. A major advance came when asymptotic complexity was adopted as the measure of performance.

With a mathematical criterion for comparing algorithms, good algorithms were developed for a vast range of problems. It was quickly discovered that a small set of design paradigms gave rise to these algorithms. One of the techniques used was divide and conquer. In assembling a jigsaw puzzle, one partitions the pieces by color or some other distinguishing feature and then subdivides the piles again and again until one gets a small number of pieces which one can fit together. This technique is the divide and conquer approach. Using it, the time to assemble a jigsaw puzzle grows as $nlogn$ where n is the size of the puzzle measured by the number of pieces. Assembling the puzzle by brute force would require time that grows as n^2. The divide and conquer approach can be used to construct algorithms for a wide variety of problems. Consider multiplication of two n bit integers. The traditional algorithm taught in elementary school requires time on the order of n^2. Another approach is to break the integers into two $n/2$ bit integers and carry out the multiplication by four multiplications of $n/2$ bit integers. However, by some simple algebraic manipulation, one can actually carry out the multiplication of the two n bit integers with only three multiplications of $n/2$ bit integers. This gives rise to an algorithm whose complexity is only $n^{1.67}$.

Techniques such as divide and conquer provided a way of structuring the field that allowed non-experts to understand what algorithm to use under various circumstances.

Success in the algorithm area led to algorithm research for many problem domains. One particularly important area was computational geometry. An example of an advance in this domain is the plane sweep algorithm. Suppose one wished to intersect n line segments. The obvious solution is to intersect each segment with every other segment, an n^2 process. A faster way is to sweep a vertical line, called the sweep line, across the plane and to maintain a list of the line segments that intersect the sweep line at any given instance. The list is sorted by height of intersection with the sweep line. As the sweep line is moved across the plane, the set of line segments intersecting the sweep line and their vertical ordering changes only when: a new line segment is encountered, two line segments intersect, or the end of a line segment is encountered. The beginning of a line segment, the intersection of two line segments and the end of a line segment are called events. An event list is maintained and sorted by increasing horizontal position. Initially, the start and end of each line segment are placed on the event list. The sweep line is moved across the plane from event to event. Each time the sweep line encounters the start of a new line segment, the new segment is added to the list of lines intersecting the sweep line. The new line segment is tested to see if it intersects the line segments immediately above and immediately below it. If an intersection is discovered, it is placed on the event list. As the sweep line encounters the end of a line segment, the segment is removed from the list of lines intersecting the sweep list. When this happens, two new line segments intersecting the sweep list become adjacent and they are tested to see if they intersected. If so, the intersection is placed on the event list. When the sweep line encounters an intersection event, the order of the intersection of the two line segments with the sweep line are interchanged, thus creating two new adjacencies among line segments intersecting the sweep line which are then tested for intersection. The critical observation is that if two line segments intersect, their intersection with the sweep line will at some point be adjacent. At that time, the intersection will be discovered. By intersecting line segments only when their intersection with the sweep line are adjacent, many pairs of line segments need never be tested for intersection. The time of the algorithm is order $n+i$ where i is the number of intersecting line segments.

The major area of interest today, in the area of algorithms, is parallelism. We will return to this topic later.

Complexity theory concerned itself with the intrinsic resources in terms of time and space necessary to solve a given problem. Out of this work came the notion of a complexity class. Two classes of great importance are called P and NP. The class P is the set of all problems that can be solved on a computer in time bounded by a polynomial of the size of the problem description. This class is generally recognized as the class of problems that are computationally feasible. If a problem is not in this class, it is presumed that the needed computing resources grow too fast for its solution to be obtained in a practical amount of time. The class NP is the set of all problems that could be solved in polynomial time if computers had the power to guess solutions and only needed to verify that these solutions were indeed valid. For example, there is no polynomial time algorithm to determine if a graph has a Hamilton circuit, i.e., a circuit that traverses each vertex once and returns to its starting point. If a computer could guess the circuit, the problem of verifying that the circuit was indeed Hamiltonean would be easy to do in polynomial time. The importance of the class NP is that it includes many problems such as integer linear programming .

Researchers in disciplines such as Operations Research demonstrated that various problems were computationally reducible to others. One of the major accomplishments of complexity theory was to show that every problem in NP was computationally reducible (in a certain technical sense) to the problem of determining if a Boolean formula in conjunctive normal form (CNF) was satisfiable. A formula is satisfiable if there is an assignment of values TRUE and FALSE to its variables such that the formula evaluates to TRUE. Thus in some sense, the satisfiability problem for CNF was as hard as any problem in NP. For this reason it was said to be *complete* for NP. If there existed a polynomial time algorithm for the satisfiability problem, then the classes P and NP would be identical and there would be efficient algorithms for many important problems. The surprising fact is that many other problems were shown to also have this completeness property; in fact almost all important problems in searching, matching and covering were included in this category.

The proof that the satisfiability problem for CNF is complete for the class NP is actually quite simple. However, it requires the development of a notation such as the Turing machine for describing computations. Thus we shall take the result as proven and show how to establish some other problem complete. Consider the question of whether a graph can be colored with three colors so that no two adjacent vertices are the same color. The problem is in NP since once one has guessed a coloring of the vertices, and it is easy to check that no two adjacent vertices are the same color. To show that 3-colorability is complete, we show how to transform or reduce an instance of the satisfiability problem to an instance of the colorability problem. This is done by constructing a graph from a Boolean formula f in 3-CNF in such a way that the graph can be colored with three colors if and only if there is an assignment of TRUE and FALSE values to the variables of the formula for which the formula evaluates to TRUE.

Suppose the colors are RED, TRUE, and FALSE. Create a vertex which we can assume will be colored RED. For each variable xi in the formula f, construct two vertices connected by an edge and add an edge from the RED vertex to each of these new vertices. The two vertices corresponding to a variable xi must be colored, one TRUE and one FALSE. However, there is a choice of which is colored TRUE and which is colored FALSE. Thus it f has n variables, there are $2n$ assignments of TRUE and FALSE to the variable of f.

Next construct a graph that simulates an OR-gate. This graph consists of three edges. Call the vertices IN1, IN2 and OUT. Connect OUT by an edge to the RED vertex. Connect IN1 and IN2 to the vertices that must be colored TRUE and FALSE such as those vertices corresponding to the variables of f. Note that OUT can be colored TRUE only if a vertex connected to IN1 or IN2 is colored TRUE. Using these graphs that simulate OR gates, we can build a simulation of the function f in such a manner that the total graph can be colored with three colors with no two adjacent vertices the same color if and only if f is satisfiable. This simulation reduces the 3-CNF satisfiability problem to the coloring problem for graphs.

Returning to the issue of parallelism, there is another class of problems called log-space which is a subset of the class P. Log-space problems have the property that they can be solved very quickly (time log in problem size) on a parallel computer with a number of processors that grows as a polynomial of the problem size. Thus the log-space class consists of precisely those problems for which there exist efficient parallel algorithms. However, there are many problems in the class P not known to be in log-space for which we seek efficient algorithms. Just as the class NP had complete problems, so does the class P. These P-complete problems have the property that if any one of them has an efficient parallel algorithm (is in log-space), then every problem in P has an efficient parallel algorithm (P equals log-space). Thus complexity theory has provided us with an important foundation for practical work in parallelism.

Given that some problems were intrinsically too difficult to compute, researchers turned to other approaches such as approximation or nearly correct algorithms. Another direction was to look at randomized algorithms. As an example, consider determining if a number p is prime. If p is prime then

$$\left(\frac{a}{p}\right) = a^{\frac{p-1}{2}}$$

for all a where $\left(\frac{a}{p}\right)$ is the Legendre function. If p is composite, then the above equation is satisfied for at most half of the a's between 1 and p. This suggests a randomized algorithm for determining if p is prime. Guess an a and test if

$$\left(\frac{a}{p}\right) = a^{\frac{p-1}{2}}$$

If the answer is no,then p is composite. If the answer is yes, then either p is prime or we guessed a bad value for a. The probability that we guessed a bad value for a is less than 50%. Thus the probability that p is prime is at least a half. If we repeat the experiment 100 times and no a established that p is composite, we can safely conclude that p is prime. In fact, the probability of error is less than 2^{-100}. This observation made the area of randomized algorithms important.

The notion of a randomized algorithm raised important philosophical questions. For example, how does one generate a random number. So called random number generators generate numbers by a completely deterministic means and thus are not random at all. That is why we call them pseudo random. Although pseudo random numbers seem to make random algorithms work, this is only true if primality testing is in P, an unlikely event. Such observations have led to significant insights into the nature of randomness, how much randomness is needed and a theory of random algorithms.

One outcome of this effort is the notion of a zero knowledge proof. Suppose one has a valuable piece of knowledge and wishes to sell it. A buyer may be unwilling to pay money until he is sure that the seller indeed possess the knowledge. However, if the seller allows the buyer access to the knowledge before paying, the buyer may back away from the deal because he now possesses the knowledge. Thus for the transaction to take place, the seller needs a method of proving he has the knowledge without exposing the knowledge. Such a method is called a zero knowledge proof. Theoretical developments such as these, along with techniques for electronic signatures that can be attached to agreements, purchase orders or checks in such a way that any tampering will invalidate the message, are vital if electronic transactions are to replace paper documents.

THE FUTURE

One is always on shaky ground when trying to predict the future. Nevertheless, I believe that information science will develop in the next century, much as physical science developed in the past century. Computer and communication technology will allow man to expand his creative and intellectual capabilities by providing tools that process and assimilate information in unique ways. Today, computers process data which represents knowledge or information. Tomorrow, computer technology will enable us to assimilate data simultaneously across many domains, thus providing an enormous depth and scope to the information we are processing. And in the future, computers will capture more intelligent-like behavior and process the knowledge that data represents rather than the data itself.

The means to this future vision is still unclear. As mentioned earlier in my discussion of AI, the nature of intelligent activity continues to puzzle some of our finest minds. I can only suggest an approach that has been successful in the past. Because I have found that the first step towards solving a problem is asking the right questions, I choose a modest example of the problem that I am trying to solve and begin working. In a short time, the "right" questions start evolving. Thus, I will suggest two examples of problems whose solution should move us in the direction of our vision.

Consider a cellular communication system and a portable workstation technology that allows us to carry our workstation in our pocket like a billfold and keeps us in constant communication with an international computer network. I am on my way home on a December evening and I walk to the corner bus stop. I arrive at 5:58 p.m. The last bus is scheduled to reach the corner stop at 6:00 p.m. I wait until 6:10 p.m. and no bus. What do I do? Should I start to walk home or should I just wait patiently? Suppose the company is using a computer to track the location of delivery vehicles, buses, taxis, etc. for more efficient operation. I get out my workstation and dispatch a program to the bus company's database to get the current location of the bus and find it is 10 blocks away. I wait.

The technology, with the exception of the software, exists for this scenario. However, if I need to write a program to locate the bus company's computer and then search its database, there would be no way that an average citizen would use this technology. Furthermore, I have no way of knowing how the bus company's data base is organized, the name of the company or how access their files. The *program* must be sufficiently sophisticated to extract this information, starting from public directories, to make this application possible. During the next decade, advances in research will be made in user interfaces and software so that access to information such as that illustrated by this scenario could become a reality. The specific example is not what is of interest. Rather, this example captures a prototypic access to information which will change the way companies and people use information. Research in user interfaces, electronic media, information access and electronic protocols will change in a fundamental manner the way in which we interact with information.

A second example is a manufacturer who wants to design a new product; say a multi-fingered gripper. Instead of designing with paper and pencil, an engineer could build a computer model or electronic prototype of the gripper and explore the design by simulating gripping and manipulation of various objects. Since one is using a computer model rather than a physical prototype, one could easily change the location, dimensions or even the number of fingers, thereby rapidly exploring large design spaces. Providing engineers with software environments and computing power to bear on engineering problems, promises the potential of new products from medicine to jet aircraft. Parallelism has the potential to provide the 1000-fold increase in computing power needed for such applications. However, much foundational work needs to be done. With tens of thousands of processors, not every processor can communicate directly with every other processor; the intercommunication pattern will be critical. A given architecture is likely to effectively speed up only certain classes of problems. Thus research into parallelism is a critical component of making this example a reality.

The computing and communication technology exists or is on the horizon to create the information age. However, this technology is not currently accessible due to the lack of a science base to support the generation of the needed software. It remains to educate policy makers, researchers and the general public to the opportunities that are within our reach and to develop the research programs that will make the benefits of this technology a reality. And it further remains to keep our minds and imaginations open to the new possibilities and unique directions that "right" questions and new technologies lead us.

Recent Development in Database Systems

François Bancilhon
Altaïr

To many outsiders, the database field might look (or might have looked) concerned with the design and implementation of relational systems for business type applications.

It is actually concerned with the management of *large amounts of persistent, reliable and shared data*. "Large" means too big to fit in a conventional main memory. "Persistent" means that data persists from one session to another. "Reliable" means recoverable in case of hardware or software failures. "Sharable" means that several users should be able to access the data in an orderly manner. These four adjectives characterize *the* database problem and they define the specificity of the field.

It is therefore possible to find new solutions outside of the world of relational systems to this problem, and apply them to areas other than business applications.

Three new phenomena are emerging, which combine themselves with the two more traditional requirements for better programmer productivity and better system performance:

1. New non business-type users are feeling the need for large amounts of reliable, sharable and persistent data (CAD, CASE, office automation, CIM). These new customers of database technology bring in new requirements.
2. The main memory cost and memory-to-disk cost ratio are changing, thus modifying the assumptions behind current database system designs, and parallel machines offer new possibilities for database design performance.
3. Limitations of current relational systems have been realized, and new technologies are being imported in the database field.

These new areas have recently emerged in the database field, which try to answer to these requirements:

1. deductive database systems,
2. database programming languages,
3. object-oriented database systems.

In this talk, I will briefly explain the state of advancement of each one of these three approaches and try to assess its possible impact on future database systems and applications.

Some reflections on software research

C.A.R. Hoare
Oxford University Computing Laboratory

The history of computing has been dominated by continued explosive growth in the cost/effectiveness of hardware, whose benefits have been increasingly delayed by problems with programming and software. Software problems have been the subject of intensive study and research, but the results have been very slow in trickling through to practice. I have been wondering why this is, and what we might do about it.

HARDWARE- AND SOFTWARE-FAULT TOLERANCE

J.C. LAPRIE, J. ARLAT, C. BEOUNES, K. KANOUN
LAAS-CNRS
7, Avenue Colonel Roche
31077 Toulouse
France

Abstract

The paper is devoted to the methods and architectures aimed at tolerating hardware faults and software faults. The paper is composed of three parts. The first part presents in a unified way the methods for software-fault tolerance by design diversity; the faults to be tolerated are discussed from two viewpoints: their independent and their persistence. Hardware- and software-fault tolerant architectures are defined and analyzed in the second section; implementation issues involved in design diversity as well as structuring principles are discussed; several architectures are proposed, aimed at tolerating single faults and consecutive faults. These architectures are analyzed from both the reliability and the cost viewpoints in the third part

The text which follows is an extended summary of the paper "Definition and Analysis of Hardware- and Software-Fault-Tolerant Architectures", which has appeared in the July 1990 issue of IEEE Computer (Special Issue on Fault-Tolerant Systems, pp. 39-51).

Introduction

Computer systems able to tolerate physical faults have been an industrial and commercial reality for some years. Tolerance to design faults has also, although on a much more reduced scale, become a reality as witnessed by the real-life systems and the experiments reported in [Voges 88]. Design faultswhere the term "design" is to be considered in a broad sense, from the system requirements to realization, during initial system production as well as in possible future modificationsare of crucial importance for safety-critical systems, as they are indeed a source of *common-mode* failures. Such failures defeat fault tolerance strategies based on strict replication (thus intended to cope with physical faults), and have generally catastrophic consequences. From a broader perspective, the emergence of hardware-fault tolerant commercial systems will increase the user's perception of the influence of design faults, due to these systems' tolerance of physical faults. As a foreseeable consequence, software-faults tolerance aimed at providing service continuity to each system user, so necessitating design diversity (as opposed to the software-fault tolerance currently implemented in some current commercial fault-tolerant systems, aimed at preserving the system core integrity through the termination of erroneous tasks [Gray 86]), is likely to spread out beyond its currently privileged domain, i.e. safety-related systems.

In spite of the current and projected importance of computer systems able to tolerate both hardware and software faults, there does not exist in the published literature papers of a tutorial character dealing with hardware-and-software fault-tolerant systems. This paper is aimed at filling this gap.

Software-fault tolerance methods

This section is devoted to a unified presentation of the methods for software-fault tolerance by design diversity, i.e. the *separate* production of two or more variants of a system aimed at delivering the same service. The most well documented methods are the *Recovery Blocks* (RB) [Randell 87] and the *N-Version Programming* (NVP) [Avizienis 85]. A third type of method is identified from the careful examination of current, real-life systems (e.g. the computerized interlocking system of the Swedish Railways [Hagelin 88] and the flight control system of the Airbus A-320 [Traverse 88]): *N self-checking programming* (NSCP). The hardware-fault tolerance approaches equivalent to RB, NVP, and NSCP are respectively standby sparing, N-modular redundancy, and active dynamic redundancy.

The classes of faults to be considered are then identified, and discussed according to two viewpoints: independence (independent faults lead to separate failures whereas related faults lead to common-mode failures) and persistence. A distinction is made between *soft* software faults and *solid* software faults: a soft software fault has a negligible likelihood of recurrence and is recoverable, whereas a solid software fault is recurrent under normal operation or cannot be recovered.

Hardware- and software-fault-tolerant architectures

This section begins with a discussion of two keypoints when implementing design diversity: the number of variants to be generated, and the level (application software, executive software, hardware) where to apply fault tolerance.

Sets of hardware-and-software fault-tolerant architectures are then presented, aimed a implementing the three software-fault tolerance methods. The presentation of these architectures is based on the notion of error-confinement areas, and more precisely on hardware areas and software areas. The first set of architectures is aimed at tolerating one hardware fault and one software fault. Three basic architectures arc described, supporting respectively the recovery block approach with two variants, the N self-checking programming approach with four variants, and the N-version programming with three variants; a fourth architecture is also described, which implements N self-checking programming with three variants. The second set of architectures is aimed at tolerating two consecutive hardware faults and two consecutive software faults; the architectures which are presented are derived from the architectures for tolerating one fault, and account for the distinction between soft and solid software faults (tolerance to one single solid fault enables tolerating several consecutive soft faults). The presentation of all the architectures encompasses the analysis of such properties as the fault diagnosis and the ability to detect and tolerate sequences of faults.

Analysis and evaluation of the architectures

An analysis and an evaluation of the reliability of three of the defined architectures is performed. The probabilities of failure of the three software-fault tolerance methods are first characterized with respect to a) their ability to detect failures, and b) the distinction between separate failures and common-mode failures. Markov models encompassing the failure behavior with respect to hardware and software faults for these architectures are then derived.

Finally, the cost issues related to software fault tolerance are addressed. Estimates are provided, which enable the quantification of the usual qualitative statement according to which an N variant software is less costly than N times a non fault-tolerant software.

Conclusion

The results reported in this paper contribute to the field of fault tolerance in at least the following aspects :

1) The identification of N self-checking programming as an industrially significant fault tolerance scheme with some important and interesting differences, compared to recovery blocks and N-version programming.
2) The integration of the topics of hardware and software fault tolerance, and the detailed analysis of the effectiveness of various architectures involving the above three fault tolerance schemes.
3) The modelling and evaluation of the dependability and of the cost of these architectures, and the relating of these estimates to such experimental data as available.

Acknowledgement

The authors gratefully acknowledge the continuous support of Brian Randell, from The University of Newcastle upon Tyne.

References

Avizienis 85 A. Avizienis, "The N-version approach to fault-tolerant systems", *IEEE Trans. on Software Engineering*, vol. SE- 11, no. 12, Dec. 1985, pp. 1491-1501.

Gray 86 J.N. Gray, "Why do computers stop and what can be done about it?", in *Proc. 5th Symp. on Reliability in Distributed Software and Database Systems*, L,os Angeles, January 1986, pp. 3-12.

Randell 87 B. Randell, "Design fault toleran", in *The Evolution of Fault Tolerant Computing*, A. Avizienis, H. Kopetz, I.C. Laprie, eds, Springer-Verlag, Vienna, 1987, pp. 251-270.

Voges 88 U. Voges, ed., *Application of design diversity in computerized control systems*, Vienna: Springer Verlag, 1988.

Hagelin 88 G. Hagelin, "ERICSSON safety systems for railway control", in [Voges 88], pp. 11-21.

Traverse 88 P. Traverse, "Airbus and ATR system architecture and specification", in [Voges 88], pp. 95-104.

Structure and Behaviour of Concurrent Systems: Selected Results of the Esprit Basic Research Action No. 3148: DEMON (Design Methods Based on Nets)

Eike Best

Institut für Informatik, Universität Hildesheim, D-3200 Hildesheim
Tel. + +5121 883 741/0, FAX + +5121 869 281
and: GMD-F1.P, Schloß Birlinghoven, D-5205 St. Augustin 1

1. Motivation and Focus

The Esprit Basic Research Action DEMON (Design Methods Based on Nets) has been set up with the following aim [1]:

> In order to ensure the correct and efficient functioning of concurrent systems, effective formal reasoning is indispensable during their design. Suitable formalisms must properly describe concurrency and provide appropriate means (structuring techniques, algebra, proof rules) in order to facilitate such reasoning.
> Petri net theory is amongst the most mature formalisms capable of describing concurrency. This Action proposes to undertake foundational work needed for the eventual development of an effective design calculus for concurrent systems based on net theory. The envisaged calculus would comprise structuring techniques, algebras, proof rules, appropriate notions of equivalence and implementation, and analysis techniques.

The project is structured into two strands of activity.

The first strand may be seen as the 'core' part of the project, located properly within the theory of Petri nets. It is concerned with

(i) the development of classes of Petri nets together with structuring concepts, that is, an algebra of nets,
(ii) the definition, classification and evaluation of equivalence / simulation notions,
(iii) and the investigation of proof techniques based on indigenous analysis and synthesis techniques.

The second strand is intended to guide, bear upon and profit from these developments. The experience stemming from related abstract models and approaches that have in-built compositionality; the wish to describe existing programming languages; insights that have been gained with existing specification methods; and the necessities of practical design derived from a number of case studies; all of these are intended to be influential in the development of the net

792

classes / algebra of strand 1. Furthermore, sight is not to be lost of such practically important (but all too often neglected) concepts as priorities and dynamic structure.

This paper describes a small selection of the results obtained in the project so far. It shows how an indigenous proof technique of Petri nets can be brought to bear on an existing programming language, namely occam [18]. A Petri net semantics of occam is described. Both the benefits and the problems associated with such a semantics are sketched. The project has given a semantics for a large subset of occam including all operators, variables, and the priority constructs PRI ALT and PRI PAR [6,5].

Section 2 introduces and motivates our investigations by means of a simple example. Section 3 briefly surveys some of the investigations undertaken to relate the structure and the behaviour of (subclasses of) Petri nets to each other. Section 4 describes part of the problems that arise when the priority constructs of occam are provided with a Petri net semantics.

It should be emphasized that the work described in this paper strictly pertains only to a small part of the project. Work on modularity, abstract models and specifications, for instance, will have to be ignored for lack of space.

2. Introductory Example

Figure 1 shows an occam-like program. It can be verified that the two processes of this program cooperate in such a fashion that the following computation ensues:
Input stream: 0,1, 2, 3, 4, 5, . . .
Output stream: 0,1, 3, 6,10,15, . . .

That is, the program computes the triangular numbers $\frac{n.(n-1)}{2}$ from the sequence of natural numbers n.

Besides correctness in the usual sense, one might wish to prove or to postulate other properties of interest. One is the property of a state being a home state.

Home states are 'ground' states that can always be reached again. It is particularly important that such states exist if the system is stuck in some unwanted or unforeseen state. There is often a provision for reaching a home state forcibly: the reset button or power switch off. Such an inelegant method may, however, be undesirable; during the transmission of a large file, for instance, it may lead to unacceptable loss of performance to switch off the receiver if the transmission protocol is hung up in an unwanted state. In such a case it is important to know whether or not a home state can be reached during normal operation.

```
channel chl,ch2;
parallel integer x;
                while true do
                        read(x);
                        chl!x;
                        ch2?x;
                        print(x)
                endwhile
                integer y,z (initially 0);
                Boolean to2 (initially false);
                while true do
                        z := y + z;
                        while to2 do
                                ch2!z;
                                to2 := false
                        endwhile;
                        chl?y;
                        to2 := true
                endwhile
endparallel.
```

Figure 1: An occam-like program

The program shown in Figure 1 illustrates the notion of a state not being a home state. Once the two processes have started their execution, there is no way for them to return to their initial states simultaneously, even though each one of them can do so individually. Hence the initial state of the system is not a home state.

There are several formal ways of proving this fact. We shall illustrate one of them by using a Petri net translation of the program. The non-home-state property of the initial state will be related to certain substructures of the Petri net. Figure 2 shows the Petri net we associate with the program of Figure 1.

The set of places

$$Q = \{p_4, q_3, q_4, q_5, q_6\}$$

is a so-called t-set or trap. Formally, this means that every output transition of the set is also an input transition of the set, or, in shorthand form:

$$Q^\bullet \subseteq {}^\bullet Q.$$

The salient general property of a trap is the fact that once it carries at least one token, it will always continue to carry at least one token in every future marking. This is because by its structural property, there is no transition by which it might properly lose tokens.

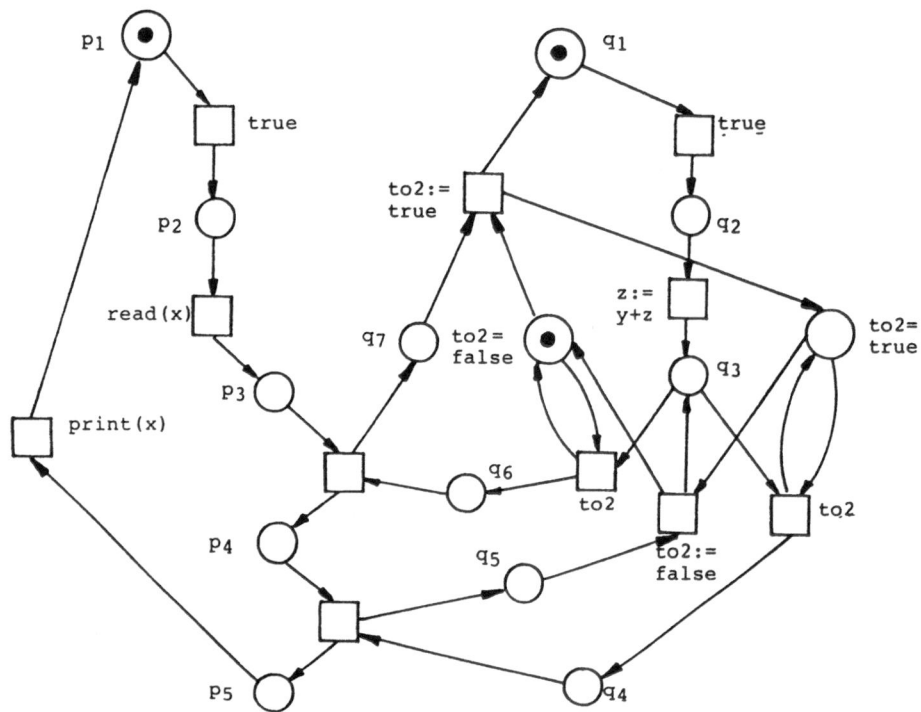

Figure 2: A Petri net associated with the example program

Lemma 2.1

Let M be a marking, Q a trap with $M(Q) = 0$ and $t \in {}^{\bullet}Q$ a transition such that t is enabled by some successor marking of M. Then M is not a home state. ■ 2.1

Hence, the property that $Q = \{p_4, q_3, q_4, q_5, q_6\}$ is unmarked initially and can be marked during the execution of the program proves that the initial state cannot be a home state.

Traps are very much the structural correspondents of what is known as 'inductive invariants': assertions that remain true once they are true, but are not necessarily true at the beginning of an execution. Indeed, the property that the initial state of the example program is not a home state can also be proved by means of an inductive invariant which corresponds exactly to the trap Q.

To see this, we introduce auxiliary integer variables p, q for the two processes of the program.

They are assigned values from 1 to 5 and from 1 to 7, respectively, in such a way that $p = i$ iff a token is on place i and $q = j$ iff a token is on place j. The inductive invariant in question is

$$I = (p = 4) \vee (3 \le q \le 6).$$

I is not valid initially, but once I is true it remains true.

Despite the fact that its initial state is not a home state, the system of Figure 1 is live, in the sense that every action remains executable in some future state. If the Boolean variable to2 is incorrectly initialised to true instead of to false, this property changes. Then, after the execution of the first read(x) and after execution of the empty inner loop of the second process, a deadlock occurs.

This deadlock is modelled in the associated Petri net by a marking which has tokens on $p3$, $q4$, the place to2 = true, and nowhere else. The set of places

$$D = \{p_4, q_1, q_2, q_3, q_5, q_6, q_7\}$$

forms a so-called d-set or siphon, that is, a set of places whose input transitions are also output transitions; in shorthand notation:

$$ {}^{\bullet}D \subseteq D^{\bullet}$$

The salient property of such sets is that once they are empty, they remain empty.

Lemma 2.2

If a marking M is in deadlock, then there is a d-set which is empty of tokens at M. ■ 2.2

The existence of the set D which can be made empty hence is indicative of the deadlock that may occur if to2 is incorrectly initialised to true.

Lemmata 2.1 and 2.2 only give weak necessary conditions for the existence of bad or undesired behaviour. What is much more interesting is to search for sufficient conditions, because one may then hope that bad behaviour can be characterised structurally. Not surprisingly, such characterisations cannot be achieved for general systems. Therefore, we have taken the approach to investigate certain simpler, but still nontrivial and interesting, subclasses of systems. One such class is the class of free choice systems. There is a conjecture about the relationship between free choice systems and certain subclasses of occam programs; we will describe this conjecture at the end of the next section.

3. Structure / Behaviour Results for Free Choice Systems

3.1 Free Choice Systems

The free choice property postulates that a choice (between two transitions t_1 and t_2) may not be influenced by the environment (i.e. we always have $^\bullet t_1 = {}^\bullet t_2$ in the case of choice). Figure 3 shows two allowed situations (3(i)) and an excluded situation (3(ii)).

(i) Allowed   (ii) Excluded

Figure 3: Illustration of the free choice property

In a marked Petri net, a marking is called bounded iff every place can only contain a bounded number of tokens. The translation of occam programs into Petri nets generates only bounded nets. A marking is called live iff no (partial) deadlock is possible.

3.2 Characterisation of Home States

The free choice Petri net shown in Figure 4 has a live and bounded marking which is not a home state; from any other reachable marking, it is impossible to reach the initial marking again. The net also has an unmarked trap, namely $\{s_0, s_2, s_3, s_5, s_6\}$. We have recently succeeded in proving that the non-existence of an unmarked trap actually characterises the home state property. Hence the non-existence of a

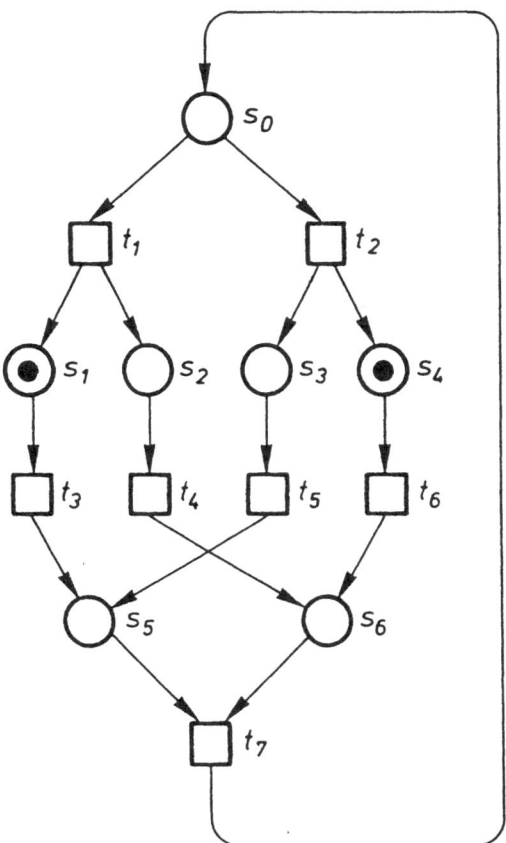

Figure 4: A free choice example

marked trap is a structural criterion for a state being a home state in (live) free choice nets.

More formally, let $\Sigma = (S, T, F, M_0)$ be a net with an initial marking M_0. We call a marking M a home state of Σ iff

$$\forall M' \in [M_0): \ M \in [M'),$$

where $[M\rangle$ denotes the set of markings reachable from M. Let $\Sigma = (S,T,F,M_0)$ be a live and bounded system. The main result proved in [2] states that a reachable marking M is a home state of Σ if and only if M marks all traps of (S, T, F):

Theorem 3.1

Let $\Sigma = (S, T, F, M_0)$ be a live and bounded free choice system. Let M be a marking reachable from M_0. Then M is a home state of Σ if and only if every trap $Q \subseteq S$ is marked under M. ∎ 3.1

The necessity of this characterisation is very easy to prove. It has been illustrated in the previous section (Lemma 2.1). The difficult part is the sufficiency.

We state three consequences of Theorem 3.1. The first one shows that any execution sequence which contains every transition at least once must necessarily produce a home state. This property has been conjectured in [13] where it has been phrased in terms of well-behaved bipolar schemata[1] which (under the translation given there) correspond to a class of live and bounded free choice nets a class which can be more precisely characterised by the results of [7].

Corollary 3.2

Let $\Sigma = (S, T, F, M_0)$ be a live and bounded free choice system. Let M be reachable from the initial marking with a transition sequence $\tau \in T^*$ such that every transition of T occurs at least once in τ.
Then M is a home state of Σ. ∎ 3.2

The second corollary shows that the home state property is monotonic.

Corollary 3.3

Let $N = (S, T, F)$ be a free choice net and let M be a live and bounded marking of N. If M is a home state then every marking $M' \geq M$ is also a home state.
∎ 3.3

This follows together with the well-known monotonicity property of liveness (see, e.g., [17]).

---

1 Actually, the conjecture in [13] is weaker, in that it concerns only the reproducibility of the marking in question.

The last corollary states that the home state property is polynomially decidable:

Corollary 3.4

Let $\Sigma = (S, T, F, M_0)$ be a live and bounded free choice system. Let $M \in [M_0\rangle$ be a reachable marking. Then it is decidable in polynomial time whether M is a home state of Σ.

Proof: (Sketch.) Starting with the set of all unmarked places, it is possible to reduce this set until either the empty set remains (meaning that the marking is not a home state) or a nonempty trap remains (meaning that the marking is indeed a home state). This algorithm is of the order $O(|S|^2 \cdot |T|^2)$ in the worst case. ■ 3.4

3.3 Characterisation of Well-Behavedness

A central and well known result of free choice net theory is the following:

Theorem 3.5 [14]

A marking M of a free choice net is live iff every nonempty d–set contains a trap which is marked under M.

In the project, further results have been obtained to characterise the class of structurally live and bounded free choice nets more precisely, that is, by means of the absence of certain bad structure [9,8,10].

Moreover, we have also obtained a linear algebraic characterisation of structural boundedness and liveness [12].

These results have led to the development of synthesis algorithms to generate all well-behaved (bounded and live) free choice systems from a very simple initial net. These algorithms take a variety of shapes.

The synthesis rules described in [9] consist of a set of two rules, one for appropriate additions to an already existing net, and another one for the expansion of parts of an already existing net. The rules are shown to preserve well-behavedness and to generate all well-behaved free choice systems. These rules are non-local, in the sense that a single element of a net may be replaced by an arbitrarily large net (which is of course restricted in structure).

The four synthesis rules described in [7], on the other hand, are local ones; that is, they allow only a bounded part of a net to be replaced or to be modified in a bounded way. They also preserve well-behavedness. However, not all well-behaved free choice systems are thereby generated, but only those that enjoy an additional property known as the absence of frozen tokens [4]. (The absence of frozen tokens indicates that all the resources of the system are necessary in every sufficiently large run of the system.) Along the way, these results imply another open conjecture (namely that live and bounded free choice nets without frozen tokens exactly characterise the class of well-behaved bipolar schemata of [13].)

Amongst other consequences, these results lead to a polynomial procedure to decide the liveness of bounded free choice systems. (The NP-completeness of deciding the liveness of free choice systems in general has long been known [16]; however, boundedness is no severe restriction in our case, since all nets derived as the semantics of occam programs are automatically bounded.)

The project has collected all of these results as part of an overview article on structure theory [12].

3.4 A Conjecture on Free Choice occam Programs

Let O be the class of occam programs satisfying the following two-properties:

(i) There are no nested loops.

(ii) Every clause alt chl?... ch2?... ... chj? endalt satisfies $chl = ch2 = \ldots = chj$, and communication is always with the same process. That is, alternatives may only distribute over the same channel and the same process .

We have the following conjectures:

Conjecture 3.6

The Petri net corresponding to an occam *program in O can be simulated – in a weak sense – by a free choice Petri net. Moreover, if the initial state is live then it is also a home state.*

Conjecture 3.7

Every live and bounded free choice Petri net such that the initial state is a home state can be simulated by a program in O.

Together, the two conjectures would imply a strong relationship between the class O and the class of bounded free choice systems and, moreover, the initial marking is a home state if it is live. (That this is not true in general has been shown in Figure 4.)

In the 'core' part of the project (see Section 1), the appropriate notions of simulation to prove these conjectures are being developed.

Our initial example (Figure 1) falls properly outside the class O, since it contains nested loops and its initial state is live but not a home state.

If the above essential equivalence turns out to be true, then all the nice structural properties of free choice nets carry over to the class O. Otherwise, one may still expect that methods and results can be transferred from one class to the other. In particular, in both cases there would be polynomial time algorithms to check liveness and home state properties of programs in this class.

4. Concluding Remarks

4.1 Potential exploitation

Note: The items described below are not part of the project DEMON, unless stated otherwise.

- Forging existing Petri net tools to fit an occam package.
  To date, there exist several tools to analyse Petri nets of almost all varieties. Tools that combine indigenous methods with state space reduction look particularly promising. The aim is to integrate such tools into a verification package for occam programs.

- Development of a new concurrent programming language.
  The aim here is to develop a concurrent programming language together with its formal (Petri net) semantics. The experience gained in giving the semantics of occam can be useful, for instance, in: (i) omitting non-compositional operators; (ii) devising flexible general constructs; (iii) setting up the language with an eye to as easy as possible verification of programs.
  The development of a prototype notation of this kind is actually one of the longer term aims of DEMON. Initial work has been done [15].

- Development of a design and verification package for occam-like concurrent programs.

- Standardisation of the formal semantics for full occam-2. (This could be a spin off project.)

4.2 Technical Innovation

We see the main technical innovation in our work in the attempt to make those formal methods that have been proved to be adequate for the description of concurrent systems (viz., partial orders, states and actions, etc., see [1]) available for the actual design and verification of concurrent systems, particularly (in the work described in this paper) concurrent programs.

4.3 Teamwork and synergy

The material condensed in this paper is contained in about 10 papers that will form part of the first annual progress report of the project. These papers are based on a collaboration between GMD, University of Milano, University of Zaragoza, Technical University of Munich and the University of Newcastle upon Tyne. This cooperation (a very amicable one, it may be added) has clearly been prompted by the project.
    Other parts of the project cooperate more intensively with other Basic Research

Actions, for instance with Cedisys.

Synergy is expected - and already partly experienced - between the European 'Schools' on concurrency, of which Petri net theory is but one. Later, we also envisage increased industrial interest in our results.

4.4 Comparison with competing efforts

The following discussion relates only to the part of the project outlined in this paper.

There are perhaps two main distinctions between the approach we follow and other related approaches.

(i) Compositionality for a more or less complete language.

Many papers deal with compositional semantics of abstract languages which may contain operators such as occam (and more). However few of them attempt to carry compositionality through to deal with other important aspects such as data, priorities, and timers.

While our approach consciously falls short of giving a semantics for full occam-2, we still have selected a subset which is large enough to allow extensive nontrivial systems to be expressed.

(ii) Concurrency semantics.

Many other approaches exist which deal with the formal interleaving semantics of the constructs we consider. However, some problems show up in the formalism (as well as at run-time) only when concurrency is properly taken into account. Such is the case, for instance, with the priority construct. We know of no other approach which attempts to describe priorities formally, composition-ally, and by means of partial orders (i.e., in a so-called 'true concurrency' approach).

4.5 Technology transfer

The results of the project will be made known mainly through published papers. However, we are also planning a Workshop in the Spring of 1991 (together with the Basic Research Action Cedisys) at which part of our results will be publicised. Also, there are some local activities in publishing our results, for instance a Summer School organised jointly with GMD, the University of Hildesheim and the Humboldt-University of Berlin in August 1990.

References

[I] Various Authors: Technical Annex of the Esprit Basic Research Action DEMON (Design Methods Based on Nets). GMD-Arbeitspapiere Nr. 435 (March 1990) .

[2] E. Best, L. Cherkasova, J. Desel and J. Esparza: Traps, Free Choice and Home States (extended abstract). Semantics for Concurrency, Leicester 1990 (eds. M.Kwiatkowska, M.W.Shields and R.Thomas), Springer-Verlag, Workshops in Computing, pp.16-20 (1990). Also: Hildesheimer Informatik-Berichte Nr.7 (August 1990).

[3] E. Best and J. Desel: Partial Order Behaviour and Structure of Petri Nets Formal Aspects of Computing Vol.2, pp.123-138 (1990).

[4] E. Best and C. Fernández: Nonsequential Processes: A Petri Net View. Springer EATCS Monographs (1988).

[S] E. Best and M. Koutny: Partial Order Semantics of Priority Systems. Hildesheimer Informatik-Berichte Nr.6/90 (June 1990).

[6] 0. Botti and J. Hall: A Petri Net Semantics of occam. DEMON Technical Report (June 1990).

[7] J. Desel: Reduction and Design of Well-Behaved Concurrent Systems. Report, Institut für Informatik, Technische Universitat München (1990). To appear in: Proc. of CONCUR'90, Springer Lecture Notes in Computer Science.

[8] J. Esparza: Structure Theory of Free Choice Nets. PhD Thesis, Universidad de Zaragoza (1990).

[9] J. Esparza: Synthesis Rules for Petri Nets, and How they Lead to New Results. Informatik-Berichte Nr.4/90, Institut für Informatik, Universität Hildesheim (1990). To appear in: Proc. of CONCUR'90, Springer Lecture Notes in Computer Science.

[10] J. Esparza and M. Silva: Circuits, Handles, Bridges and Nets. Proc. of the 10th Int. Conf. on Theory and Applications of Petri Nets, to appear in Advances in Petri Nets, Springer Lecture Notes in Computer Science (1990).

[11] J. Esparza and M. Silva: A Polynomial Time Algorithm to Decide Liveness of Bounded Free Choice Nets. Technical Report, Universidad de Zaragoza (1989).

[12] J. Esparza and M. Silva: On the Analysis and Synthesis of Free Choice Systems (survey paper). Submitted to the Advances in Petri Nets, Springer Verlag.

[13] H.J. Genrich and P.S. Thiagarajan: A Theory of Bipolar Synchronization Schemes. Theoretical Computer Science 30, 241-318 (1984).

[14] M. Hack: Analysis of Production Schemata by Petri Nets. MIT 1972/74.

[15] R. Hopkins and J. Hall: Towards a Petri Net Programming Notation. Draft DEMON Report (June 1990).

[16] N. D. Jones, L. H. Landweber and Y. E.Lien: Complexity of Some Problems in Petri Nets. TCS 4, 277-299 (1977).

[17] W. Reisig: Petri Nets - an Introduction. Springer EATCS Monographs in Theoretical Computer Science Vol.4 (1985).

[18] The occam-2 Reference Manual. INMOS Ltd. (1988).

Project no. 3104   BRA ProCoS project: Provably Correct Systems

AN ALGEBRAIC APPROACH TO VERIFIABLE COMPILING SPECIFICATION AND PROTOTYPING OF THE PROCOS LEVEL 0 PROGRAMMING LANGUAGE

C.A.R. Hoare   He Jifeng   Jonathan Bowen   Paritosh Pandya

Oxford University Computing Laboratory
Programming Research Group
11 Keble Road
Oxford OX1 3QD
England

Phone: +44-865-273838   Fax: +44-865-273839
E-mail: procos@prg.oxford.ac.uk

SUMMARY. A compiler is specified by a description of how each construct of the source language is translated into a sequence of object code instructions. The meaning of the object code can be defined by an interpreter written in the source language itself. A proof that the compiler is correct must show that interpretation of the object code is at least good (for any relevant purpose) as the corresponding source program. The proof is conducted using standard techniques of data refinement. All the calculations are based on algebraic laws governing the source language. The theorems are expressed in a form close to a logic program, which may used as a compiler prototype, or a check on the results of a particular compilation. A subset of the occam programming language and the transputer instruction set are used to illustrate the approach. An advantage of the method is that it is possible to add new programming constructs without affecting existing development work.

1.   Introduction

Compilation is specified as a relation between a source program p and the corresponding object code c. Further details of compilation are given by a symbol table Ψ, mapping the global identifiers of p to storage locations of the target machine. This compilation relation will be abbreviated as a predicate

$$\mathcal{C} \, p \, c \, \Psi \quad .$$

The internal structure of p, c and Ψ will be elaborated as the need arises.

Improvement is a relation between a product q and a product p that holds whenever for any purpose the observable behaviour of q is as good as or better than that of p; more precisely, if q satisfies every specification satisfied by p, and maybe more. For example, in a procedural language, a program is better if it terminates more often and/or gives a more determinate result. This relation is written

$$p \sqsubseteq q \quad,$$

where $\sqsubseteq$ is necessarily transitive and reflexive (a preorder).

If p and q are program operating on different data spaces, they cannot be directly compared. But if r is a translation from the data space of q to that of p, we can then compare them before and after the translation

$$r \,;\, p \sqsubseteq q \,;\, r \quad,$$

where the semicolon denotes sequential composition. The relation r is known as a simulation or refinement; such simulations are the basis of several modern development techniques (e.g., VDM and Z).

To define compiler correctness precisely, we need to ascribe meanings to p, c and Ψ. Let $\hat{c}$ be a formal description of the behaviour of the target machine executing the machine code c. Let $\hat{p}$ similarly be an abstract behavioural definition of the meaning of the program p. Finally, let $\hat{\Psi}$ be a transformation which assigns to each global identifier x of the source program the value of the corresponding location Ψx in the store of the target computer. In order to prove that the object code c is a correct translation of p, we need to show that

$$\hat{\Psi} \,;\, \hat{p} \sqsubseteq \hat{c} \,;\, \hat{\Psi} \quad;$$

or in words, that the source program is improved by execution of its object code. A compiler is correct if it ensures the above for all p, c and Ψ. We therefore define this to be the specification of the compiler

$$\mathcal{C} \, p \, c \, \Psi \stackrel{def}{=} \hat{\Psi} \,;\, \hat{p} \sqsubseteq \hat{c} \,;\, \hat{\Psi} \quad.$$

The task of proof-oriented compiler design is to develop a mathematical theory of the predicate $\mathcal{C}$. This should include enough theorems to enable the implementor to select correct object code for each construct of the programming language. It may be that the theory allows choice between several different codes for the same source code; this gives some scope for optimisation by selecting the most efficient alternative.

As example of the kind of theorem by which the design specifies a compiler, consider the following possible theorems (using $\langle$ and $\rangle$ as sequence brackets and $\frown$ for catenation):

$$\mathcal{C} \, (x := y) \, \langle \text{load}\Psi y, \text{ store}\Psi x \rangle \, \Psi$$
$$\mathcal{C} \, \text{SKIP} \, \langle \rangle \, \Psi$$
$$\mathcal{C} \, (p1; p2) \, (c1 \frown c2) \, \Psi \quad \text{whenever} \quad \mathcal{C} \, p1 \, c1 \, \Psi \quad \text{and} \quad \mathcal{C} \, p2 \, c2 \, \Psi$$

When these have been proved, the implementor of the compiler knows that a simple assignment can be translated into the pair of commands shown above; that a SKIP command translates to an empty code sequence; and that sequential composition can be translated by concatenating the translations of its two components.

The form of these theorems is extraordinarily similar to that of a recursively defined Boolean function, which could be used to check that a particular compilation of a safety critical program has been successful. It is also quite similar to a logic program, which could be useful if a prototype of the compiler is needed, perhaps for bootstrapping purposes. It is also similar to that of a conventional procedural compiler, structured according to the principle of recursive descent. Finally, it is quite similar to an attribute grammar for the language; and this permits the use of standard techniques for splitting a compiler into an appropriate number of passes. Thus the collection of theorems serve as an appropriate interface between the designers of a compiler and its implementors.

In this paper, we give a compiling specification for the **ProCoS** project language PL_0 [12], following the algebraic approach described above. In this case the simulation $\hat{\Psi}$ has a parameter Ω which denotes the free locations available to the machine program. We also present a prototype Prolog compiler based on this specification. PL_0 is a subset of the programming language occam [10], and the machine language ML_0 is a subset of the transputer instruction set [11, 13].

The paper is organised as follows. Section 2. briefly describes the programming language PL_0, and gives some algebraic laws for process refinement. Section 3. outlines the interpreter for ML_0 programs. Section 4. deals with the specification of a compiler and the proof of its correctness. Thus, Section 4.1. gives the formal meaning of correctness of compiling specification. The compiling specification itself is given in Section 4.3. and it is proved correct in Section 4.4.. Section 5. discusses a strategy for organising a compiler based on the specification given in this paper. A Prolog implementation of a prototype compiler which is very close to the compiling specification, is presented in Section 6.. Finally we discuss some advantages of our approach.

2. The Programming Language and its Process Algebra

The programming language PL_0 is a sequential subset of occam. It consists of the occam constructs, SKIP, STOP, assignment, SEQ, IF and WHILE. Constructs input?x and output!e permit a program to interact with the external environment. Rather than giving formal syntax [12], we present an example program (see Figure 1). Note that the concrete syntax is selected such that the program may easily be read directly by Prolog by defining each keyword and symbol as a Prolog operator with some suitable priority, associativity, etc. This obviates the need to write a parser in Prolog.

The semantics of PL_0 is well studied [5, 7]. The specification-oriented semantics of PL_0 given in [7] formally defines the relation between p and $\hat{p}$ mentioned in the introduction: $\hat{p}$ is just the strongest behavioural specification satisfied by p. However, in mathematics it is permissible (and indeed universal practice) to use notations directly to stand for what they mean. We can profit enormously by this convention if we just allow the program text itself to mean its strongest specification:

$$\hat{p} \stackrel{def}{=} p$$

The PL_0 language is enhanced with the following features to facilitate coding of a machine interpreter for ML_0. The assignment construct is generalised to multiple assignment and the use of array variables is permitted (for modelling machine memory). A special process Abort is included which models completely arbitrary behaviour. assert b ensures that either b holds or the behaviour is completely arbitrary. The scope of variables may be terminated dynamically using the end construct. We refer to this language as PL_{0+}.

```
int cs:  int ps:
int x:   int y:

seq[
  cs := 0,
  ps := 0,
  while(true,
    seq[
      input?x,
      ps := cs,
      cs := x,
      if[
        cs = 0  -> y := 0,
        ps = 0  -> y := 2-cs,
        ps <> 0 -> y := cs
      ],
      output!y
    ]
  )
].
```

Figure 1: Example PL_0 program

2.1. REFINEMENT ALGEBRA

A number of algebraic laws for proving refinement relation $p \sqsubseteq q$ between PL_{0+} processes have been developed. These are similar to the laws of occam [17].

In the following we present some simple laws as illustration. A more comprehensive set of laws may be found in [6]. A mathematical definition of the relation $\sqsubseteq$, and the consistency of the laws with respect a specification-oriented semantics of the language are explored in [7].

The program Abort represents the completely arbitrary behaviour of a broken machine, and is the least controllable; in short for all purposes, it is the worst:

Law 1 Abort $\sqsubseteq p$

The SEQ constructor runs a number of processes in sequence. If it has no argument it simply terminates.

Law 2 SEQ[] = SKIP

Otherwise it runs its first argument until that terminates and then runs the rest in sequence.

Law 3 SEQ[p] $= p$

SEQs may be unnested as follows. (The notation $\underline{p}$ denotes a list of processes.)

Law 4 SEQ[$\underline{p}$, SEQ[$\underline{q}$], $\underline{r}$] $=$ SEQ[$\underline{p}$, $\underline{q}$, $\underline{r}$]

The following law is used to describe the compositionality of WHILE-loops:

808

Law 5 SEQ[WHILE(b, p), WHILE($b \vee c$, p)] = WHILE($b \vee c$, p)

3. Interpreting Machine Programs

The machine code, a subset of the **transputer** instruction set [11], is interpreted by a PL_{0+} program $\mathcal{I}sfmT$, where s and f stand for the start and finish address of the ML_0 program in memory m. Thus, $m[s] \ldots m[f-1]$ is the ML_0 code to be executed. The set of memory locations T represents the data space available to the ML_0 program. For PL_0, T is defined as $ran\Psi \uplus \Omega$ where $ran\Psi$ is used to hold global variables and Ω includes the temporary stack used for evaluation of expressions. Any access to a location beyond that set is regarded illegal, and will allow completely arbitrary behaviour of the interpreter (which is modelled by the construct **Abort**).

The machine state consists of registers A, B and C, an instruction pointer P and a boolean *ErrorFlag*.

$$\mathcal{I}sfmT \stackrel{def}{=} \text{SEQ}[\ \langle P, ErrorFlag \rangle := \langle s, 0 \rangle, \text{WHILE}(P < f, mstep_T),$$
$$\textbf{assert}\,(P = f \wedge ErrorFlag = 0)\ ]$$

Here, $mstep_T$ is the interpreter for executing a single ML_0 instruction starting at location $m[P]$ using the given data space T [6]. The program **assert**($P = f \wedge ErrorFlag = 0$) assures that if the execution of the interpreter terminates, it will end at the finish address f with *ErrorFlag* cleared.

Properties of machine behaviour such as the following lemma can be proved using the laws of process refinement.

Lemma 1 (Composition Rule) If $s \leq l \leq f$ then $\mathcal{I}sfmT \sqsupseteq$ SEQ[$\mathcal{I}slmT, \mathcal{I}lfmT$]

Proof:

$$LHS$$
$$= \text{SEQ}[\langle P, ErrorFlag \rangle := \langle s, 0 \rangle, \text{WHILE}(P < j, mstep_T),$$
$$\text{WHILE}(P < f, mstep_T), \textbf{assert}(P = f \wedge ErrorFlag = 0)] \quad \{\text{by law 5}\}$$
$$\sqsupseteq \text{SEQ}[\mathcal{I}slmT, \langle P, ErrorFlag \rangle := \langle j, 0 \rangle,$$
$$\text{WHILE}(P < f, mstep_T), \textbf{assert}(P = f \wedge ErrorFlag = 0)] \quad \{\text{by law 2}\}$$
$$= RHS$$

4. Compiling Specification and its Verification

The compiling specification of PL_0 is defined as a predicate relating a PL_0 process and the corresponding ML_0 code. Section 4.1. gives a formal meaning to the compiling specification predicate. Section 4.2. presents the algebra of the weakest specification for the object code with respect the given symbol table, and Section 4.3. states many theorems of the compilation predicate $\mathcal{C}$. The aim is to is include enough theorems to enable the implementor to select correct ML_0 code for each construct of PL_0. Each theorem can be proved using the algebraic laws of process refinement. A sample proof is given in Section 4.4..

4.1. Compiling Specification Predicate

The compiler of the programming language PL_0 is specified by predicate $\mathcal{C}\,p\,s\,f\,m\,\Psi\,\Omega$ where

- p is a PL_0 process.
- s and f stand for the start and finish address of a section of ML_0 code to be executed.
- $m[s]\ldots m[f-1]$ is the ML_0 code for p.
- The symbol table Ψ maps each identifier (global variables and channels) of p to its address in the memory M, where we assume that m (used to store the code) and M (used to store data) are disjoint.
- Ω is a set of locations of the memory M, which can be used to store the values of local variables or the temporary results during the evaluation of expressions. Here we assume that

$$ran\Psi \cap \Omega = \emptyset$$

i.e., Ω only contains those addresses which have not been allocated yet. It is the responsibility of the compiler designer to ensure that this is so.

The machine code is interpreted by the program $\mathcal{I}\,s\,f\,m\,(ran\Psi \uplus \Omega)$.

The compiling specification predicate $\mathcal{C}$ is correct if the interpretation of ML_0 code has the same (or better) effect as PL_0 source code with appropriate translation from the data space of target code to that of source code, i.e.

$$\mathcal{C}\,p\,s\,f\,m\,\Psi\,\Omega \stackrel{def}{=} \text{SEQ}[\hat{\Psi}_\Omega, p] \sqsubseteq \text{SEQ}[\mathcal{I}sfm(ran\Psi \uplus \Omega), \hat{\Psi}_\Omega]$$

The relation $\hat{\Psi}_\Omega$ translates from the machine state to program state and then forgets the machine state consisting of the memory locations and machine registers:

$$\hat{\Psi}_\Omega \stackrel{def}{=} \text{SEQ}[\ \langle x, y, z, \ldots \rangle := \langle M[\Psi x], M[\Psi y], M[\Psi z], \ldots \rangle,$$
$$\textbf{end}_{ran\Psi \uplus \Omega \uplus \{A,B,C,P,ErrorFlag\}}\ ]$$

where $\langle x, y, z, \ldots \rangle$ contains all the program variables in the domain of Ψ.

Similarly, the compilation predicate $\mathcal{E}\,e\,s\,f\,m\,\Psi\,\Omega$ relates a PL_0 expression e to its ML_0 code, whose execution must leave the value of e in the register A.

4.2. Weakest Specification of the Object Code

This section shows how to derive the weakest specification of the target code when given the source program p and the symbol table Ψ.

Because the simulation $\hat{\Psi}_\Omega$ given in the previous section is a surjective mapping from the data space of the machine language ML_0 to that of the programming language PL_0, we can thus define its *inverse* as follows:

$$\hat{\Psi}_\Omega^{-1} \stackrel{def}{=} \text{SEQ}[\ \langle M[\Psi x], M[\Psi y], M[\Psi z], \ldots \rangle := \langle x, y, z, \ldots \rangle,\ \textbf{end}_{x,y,z,\ldots}\ ]$$

Algebraically, the inverse of the simulation $\hat{\Psi}_\Omega$ is uniquely defined by the following laws:

$$\text{SEQ}[\hat{\Psi}_\Omega^{-1}, \hat{\Psi}_\Omega] = \text{SKIP}_{PS} \qquad \text{SEQ}[\hat{\Psi}_\Omega, \hat{\Psi}_\Omega^{-1}] \sqsubseteq \text{SKIP}_{MS}$$

where PS and MS stand for the data spaces of PL_0 and ML_0 respectively. From these laws it is not difficult to show that

$$\text{SEQ}[\hat{\Psi}_\Omega, p] \sqsubseteq \text{SEQ}[q, \hat{\Psi}_\Omega] \quad \text{iff} \quad \text{SEQ}[\hat{\Psi}_\Omega, p, \hat{\Psi}_\Omega^{-1}] \sqsubseteq q$$

Consequently, the compilation predicate C can be redefined by

$$C\, p\, s\, f\, m\, \Psi\, \Omega \overset{def}{=} \text{SEQ}[\hat{\Psi}_\Omega, p, \hat{\Psi}_\Omega^{-1}] \sqsubseteq \mathcal{I}sfm(ran\Psi \uplus \Omega)$$

which claims that the weakest specification of a correct ML_0 implementation of the source program p is

$$WS_\Psi(p) \overset{def}{=} \text{SEQ}[\hat{\Psi}_\Omega, p, \hat{\Psi}_\Omega^{-1}]$$

The weakest specification operator WS_Ψ enjoys a number of algebraic properties:

Lemma 2 $WS_\Psi(\text{Abort}) = \text{Abort}$

Lemma 3 $WS_\Psi(\text{STOP}) = \text{STOP}$

Lemma 4 $WS_\Psi(\text{SKIP}) = MS \setminus M[ran\Psi] := anyvalue$
where $MS \setminus M[ran\Psi]$ denotes the remaining data space of ML_0 after removal of the subset of the memory M being allocated for global identifiers of the source program.

Lemma 5 $WS_\Psi(\text{SEQ}[p_1,\ldots,p_n]) = \text{SEQ}[WS_\Psi(p_1), WS_\Psi(\text{SEQ}[p_2,\ldots,p_n])])]$

Now we want to extend the definition of WS_Ψ to treat expressions:

$$WS_\Psi(e) \overset{def}{=} e[M[\Psi x]/x, M[\Psi y]/y, M[\Psi z]/z,\ldots]$$

where $e[t/x]$ is the expression obtained by replacing all free occurrences of x in e by t.

Lemma 6 $WS_\Psi(x := e) = \text{SEQ}[M[\Psi x] := WS_\Psi(e), WS_\Psi(\text{SKIP})]$

Lemma 7 $WS_\Psi(\text{IF}[b_1 \to p_1,\ldots,b_n \to p_n]) =$
$\text{IF}[WS_\Psi(b_1) \to WS_\Psi(p_1), WS_\Psi(\neg b_1) \to WS_\Psi(\text{IF}[b_2 \to p_2,\ldots,b_n \to p_n])])]$

Lemma 8 $WS_\psi(\text{WHILE}(b, p)) = \text{SEQ}[\text{WHILE}(WS_\Psi(b), WS_\Psi(p)), WS_\Psi(\text{SKIP})]$

4.3. THEOREMS OF COMPILING SPECIFICATION

We present some of the theorems of compiling specification for PL_0 to ML_0 translation from [6]. Note that ML_0 instructions are of variable length; each such instruction is implemented as a *sequence* of simpler single-byte **transputer** instructions. In an ML_0 program the argument of a jump instruction is the byte offset from the *end* of the jump instruction to the *start* of the target instruction. The function $mtrans(minstr)$ translates an ML_0 instruction into a sequence of **transputer** instructions and the function $Size(minstr)$ gives the length of $minstr$ in bytes [15]. The notation $m[s:t]$ denotes the sequence $m[s],\ldots,m[t]$.

(1) $C(\text{SKIP})sfm\Psi\Omega$ if
        $f = s$

(2) $C(\text{STOP})sfm\Psi\Omega$ if
        $m[s : f - 1] = mtrans(\text{stopp})$

(3) $C(x := e)sfm\Psi\Omega$ **if**

    $\exists l_1.\quad l_1 \leq f$

      $\mathcal{E}(e)sl_1m\Psi\Omega \quad \wedge$

      $m[l_1 : f-1] = mtrans(\mathtt{stl}(\Psi x))$

(4) $C(\mathtt{SEQ}[])sfm\Psi\Omega$ **if**

    $C(\mathtt{SKIP})sfm\Psi\Omega$

(5) $C(\mathtt{SEQ}[p_1,\ldots,p_n])sfm\Psi\Omega$ **if**

    $\exists l_1.\quad l_1 \leq f$

      $C(p_1)sl_1m\Psi\Omega \quad \wedge$

      $C(\mathtt{SEQ}[p_2,\ldots,p_n])l_1fm\Psi\Omega$

(6) $C(\mathtt{IF}[])sfm\Psi\Omega$ **if**

    $C(\mathtt{STOP})sfm\Psi\Omega$

(7) $C(\mathtt{IF}[b_1 \to p_1,\ldots,b_n \to p_n])sfm\Psi\Omega$ **if**

    $\exists l_1,l_2,l_3,l_4.\quad l_1 \leq l_2 \leq l_3 \leq l_4 \leq f$

      $\mathcal{E}(b_1)sl_1m\Psi\Omega \quad \wedge$

      $m[l_1 : l_2-1] = mtrans(\mathtt{cj}(l_4 - l_2)) \quad \wedge$

      $C(p_1)l_2l_3m\Psi\Omega \quad \wedge$

      $m[l_3 : l_4-1] = mtrans(\mathtt{j}(f - l_4)) \quad \wedge$

      $C(\mathtt{IF}[b_2 \to p_2,\ldots,b_n \to p_n]l_4fm\Psi\Omega$

(8) $C(\mathtt{WHILE}(b,p))sfm\Psi\Omega$ **if**

    $\exists l_1,l_2,l_3.\quad l_1 \leq l_2 \leq l_3 \leq f$

      $m[s : l_1-1] = mtrans(\mathtt{j}(l_2 - l_1)) \quad \wedge$

      $C(p)l_1l_2m\Psi\Omega \quad \wedge$

      $\mathcal{E}(\mathtt{NOT}\ b)l_2l_3m\Psi\Omega \quad \wedge$

      $m[l_3 : f-1] = mtrans(\mathtt{cj}(l_1 - f))$

Theorems of expression compilation follow a similar form.

4.4. CORRECTNESS OF THE COMPILING SPECIFICATION

We give a sample proof of correctness for the theorems of the SEQ construct.

Proof of Theorem 4   Direct from law 2.

Proof of Theorem 5

    $\mathcal{I}sfm(ran\Psi \uplus \Omega)$

  $\sqsupseteq$ $\mathtt{SEQ}[\mathcal{I}slm(ran\Psi \uplus \Omega), \mathcal{I}lfm(ran\Psi \uplus \Omega)]$ {by law 5 and lemma 1}

  $\sqsupseteq$ $\mathtt{SEQ}[WS_\Psi(p_1), WS_\Psi(\mathtt{SEQ}[p_2,\ldots,p_n])]$ {by the antecedent}

  $=$ $WS_\Psi(\mathtt{SEQ}[p_1,\ldots,p_n])$ {by lemma 5}

5. Compilation Strategy

Section 4.3. presented a number of theorems about the compiling specification predicate $\mathcal{C}$. In this section we discuss how these theorems may be used in actually generating code for the PL_0 programs. These theorems may directly function as clauses of a logic program implementing the compiler. To make such an approach practicable, we transform the compiling specification (using logic) to derive theorems which may be efficiently 'executed'.

5.1. RELOCATABILITY OF MACHINE CODE

A predicate $movem(s', f', m', s, f, m)$ is used in specifying relocation of machine code.

(9) $movem(s', f', m', s, f, m)$ if $m'[s' : f' - 1] = m[s : f - 1]$

(10) $(f' - s') = (f - s)$ if $movem(s', f', m', s, f, m)$

The following theorem, stating that code generated by the compiler is relocatable, is useful in implementing this strategy. A similar theorem applies to expressions. The theorems may be used to find the size of code even when its position in memory is unknown.

(11) $\mathcal{C}(P)sfm\Psi\Omega$ if
$$\exists s', f', m'.$$
$$\mathcal{C}(P)s'f'm'\Psi\Omega \quad \wedge$$
$$movem(s', f', m', s, f, m)$$

The theorem below can be derived from the above and Theorem 7.

(12) $\mathcal{C}(\mathrm{IF}[b_1 \to P_1, \ldots, b_n \to P_n])sfm\Psi\Omega$ if
$$\exists l_1, l_2, l_3, l_4, l_{23}, l_{34}, l_{4f}, m', m'', m'''. \quad l_1 \le l_2 \le l_3 \le l_4 \le f$$
$$\mathcal{E}(b_1)sl_1m\Psi\Omega \quad \wedge$$
$$\mathcal{C}(P_1)0l_{23}m'\Psi\Omega \quad \wedge$$
$$\mathcal{C}(\mathrm{IF}[b_2 \to P_2, \ldots, b_n \to P_n])0l_{4f}m''\Psi\Omega$$
$$m'''[0 : l_{34} - 1] = mtrans(\mathbf{j}(l_{4f})) \quad \wedge$$
$$m[l_1 : l_2 - 1] = mtrans(\mathbf{cj}(l_{23} + l_{34})) \quad \wedge$$
$$movem(0, l_{23}, m', l_2, l_3, m) \quad \wedge$$
$$movem(0, l_{34}, m''', l_3, l_4, m) \quad \wedge$$
$$movem(0, l_{4f}, m'', l_4, f, m)$$

5.2. DESIGN OF *BackJump* FUNCTION

The implementation of the WHILE construct requires backward jump. We first formulate a scheme for optimising the backward jumps. Let s be the target address of backward conditional jump, and let l be the start address for the cj instruction. We design a function *BackJump* which has the following properties.

$$s \le l \le f$$
$$Size(\mathbf{cj}(s - f)) = (f - l)$$
$$BackJump(s, l) = (s - f)$$

These can be solved to give the following specification of *BackJump*:

$$l - s = 0 \Rightarrow BackJump(s, l) = -(2 + (l - s))$$
$$0 \le i \land (16^i - i) \le (l - s) < 16^{i+1} - (i + 1) \Rightarrow$$
$$BackJump(s, l) = -(i + 2 + (l - s))$$

We may now formulate a compiling specification for the WHILE construct as follows.

(13) $\mathcal{C}(\texttt{WHILE}(b, P))sfm\Psi\Omega$ **if**
$$\exists l_1, l_2, l_3, l_{12}, m'. \quad l_1 \le l_2 \le l_3 \le f$$
$$\mathcal{C}(P)0l_{12}m'\Psi\Omega \quad \land$$
$$m[s : l_1 - 1] = mtrans(\texttt{j}(l_{12})) \quad \land$$
$$movem(0, l_{12}, m', l_1, l_2, m) \quad \land$$
$$\mathcal{E}(\texttt{NOT } b)l_2 l_3 m\Psi\Omega \quad \land$$
$$m[l_3 : f - 1] = mtrans(\texttt{cj}(BackJump(l_1, l_3)))$$

The compiling specification theorems for WHILE and the IF constructs outlined in this section have been used to implement a compiler in the logic programming language Prolog. Extracts of this compiler are included in the next section.

6. Prolog Implementation

The idea of using Prolog [4] for the construction of compilers has been accepted for some time [18]. Advantages include the fact that the code for the compiler can be very close to the compiling specification and thus the confidence in its correctness is increased. It can be used both for a prototype compiler and even for a 'real' compiler since the Prolog code itself may be compiled for increased efficiency [16].

The following sections include parts of a Prolog compiler from the PL_0 language to the ML_0 instruction set which follows the compiling specification outlined in Section 4.3. as closely as possible. The strategy presented in Section 5. is followed to produce a working compiler.

An interesting feature of the compiler is that compiled code is generated by using Prolog assertions to add clauses which specify the program memory contents dynamically. The compiler produces code by reading in a PL_0 source program and asserting ML_0 object code values in memory locations. A more standard approach is to include the actual object code generated as a parameter and to concatenate generated code sequences together. This approach is adopted in a companion paper [3] which may be compared with the code presented here.

Alternatively it would be possible to convert the program into a compiler 'checker' so that program source and object code produced by another compiler could be checked to be correct by ensuring that previously set memory locations are consistent with the results from the compiler. This could increase the confidence that two compilers produce the same results (e.g., whilst boot-strapping a compiler written in its own target language).

6.1. PROCESS COMPILATION

Each program construct is compiled using a separate Prolog clause. Individual instructions are assembled using m. This asserts that one or more consecutive byte locations in memory

814

contain the object code for the instruction. Alternatively, memory locations could be returned as lists and concatenated together at the end of each compilation clause as in [3].

A Prolog cut ('!') could be included at the end of each clause if only the first solution which the Prolog program finds is required. This would make the program more efficient by avoiding subsequent searching once a solution has been found. However the absence of such cuts allows the possibility of non-deterministic compilation (perhaps allowing multiple strategies and then chosing the 'best' for example). Most constructs are straight-forward and follow the original specification almost exactly:

```
c(skip,S,S,_,_,_).

c(stop,S,F,M,_,_) :-
        m(stopp,S,F,M).

c(X:=E,S,F,M,Psi,Omega) :-
        psi(Psi,X,PsiX),
        e(E,S,L1,M,Psi,Omega),
        m(stl(PsiX),L1,F,M).

c(seq[],S,F,M,Psi,Omega) :-
        c(skip,S,F,M,Psi,Omega).

c(seq[P|R],S,F,M,Psi,Omega) :-
        c(P,S,L1,M,Psi,Omega),
        c(seq R,L1,F,M,Psi,Omega).
```

The `if` and `while` constructs involve variable-length jump instructions which must be handled slightly differently from the specification in order to produce an executable program. See later for more details on the `movem` clause used below. `backjump` implements the *BackJump* function and `exists` instantiates a unique name.

```
c(if[],S,F,M,Psi,Omega) :-
        c(stop,S,F,M,Psi,Omega).

c(if[B->P|R],S,F,M,Psi,Omega) :-
        e(B,S,L1,M,Psi,Omega),
        exists(MP), c(P,0,L3_L2,MP,Psi,Omega),
        exists(MR), c(if R,0,F_L4,MR,Psi,Omega),
        m(j(F_L4),L3_L2,L4_L2,MP),
        m(cj(L4_L2),L1,L2,M),
        movem(0,L4_L2,MP,L2,L4,M),
        movem(0,F_L4,MR,L4,F,M).

c(while(B,P),S,F,M,Psi,Omega) :-
        exists(MP), c(P,0,L2_L1,MP,Psi,Omega),
        m(j(L2_L1),S,L1,M),
        movem(0,L2_L1,MP,L1,L2,M),
        e(~B,L2,L3,M,Psi,Omega),
        backjump(L1,L3,L1_F),
        m(cj(L1_F),L3,F,M).
```

Expressions are handled separately and straight-forwardly [6].

6.2. TRANSPUTER INSTRUCTIONS

Each instruction Instr is located at a particular byte address S in a memory M. The sequence of basic byte instructions, including all necessary prefixed nfix and pfix instructions must be calculated; the position of any following instructions F is then known. Each instruction is assembled into a list of byte values Code which is subsequently set in memory.

```
m(Instr,S,F,M) :- mtrans(Instr,CodeSeq), setm(CodeSeq,S,F,M).
```

setm takes a list of byte values, a start address and a memory, sets the values in the memory at that address, and then returns the finish address following the set locations. setmemory takes a single byte value instead of a list and actually adds a Prolog clause at the end of the database of clauses using the built-in Prolog assertz clause:

```
setm([],S,S,_).
setm([Code|R],S,F,M) :-
        setmemory(Code,S,SuccS,M), setm(R,SuccS,F,M).

setmemory(Code,S,F,M) :-
        Loc =.. [M,S], assertz(value(Loc,Code)), F is S+1.
```

Thus setm([33,245],0,F,bytemem) would assert

```
value(bytemem(0),33)   and   value(bytemem(1),245)
```

as two byte locations in memory and return F=2, for example. (Note that the code Loc =.. [bytemem,0] simply sets Loc to the Prolog 'functor' bytemem(0), etc.)

A compiler checker may be implemented by first asserting the bytemem values to be checked and then using a version of setmemory without the assertz clause":

```
setmemory(Code,S,F,M) :- Loc =.. [M,S], value(Loc,Code), F is S+1.
```

6.3. CORRESPONDENCE BETWEEN VARIABLES AND MEMORY LOCATIONS

The memory location PsiX for a particular variable X may be retrieved from the symbol table Ψ using the following Prolog code:

```
psi([X->PsiX|_],X,PsiX).
psi([_->_|Psi],X,PsiX) :- psi(Psi,X,PsiX).
```

6.4. RELOCATION OF MACHINE CODE

Compilation may proceed in a straight-forward sequential manner following the original specification directly except where forward jump instructions (j and cj) are involved. This occurs in the case of if and while constructs. In these cases the size of the relative jump and hence the size of the jump instruction itself are not known in advance.

The solution adopted in the Prolog program is to first compile the code which is to be jumped over into a separate piece of memory (uniquely named by the identifier returned by the exists clause and starting at location 0 for convenience). This temporary memory may subsequently be relocated into the actual position in memory once the jump instruction involved has been calculated and the real location (following the jump instruction) is known.

This is possible because all the instructions in the ML_0 language are relocatable; that is to say, they have the same effect wherever they are in memory.

The following Prolog code relocates a list of instructions from S1 up to (but not including) F1 in memory M1 into M2, starting at position S2:

```
movem(S1,S1,_,S2,S2,_).
movem(S1,F1,M1,S2,F2,M2) :-
        Loc =.. [M1,S1], value(Loc,Code), SuccS1 is S1+1,
        setmemory(Code,S2,SuccS2,M2),
        retract(value(Loc,Code)),
        movem(SuccS1,F1,M1,SuccS2,F2,M2).
```

Here the built-in Prolog clause retract also removes the temporarily asserted memory values. This is not strictly necessary, but keeps the number of asserted memory values to a minimum.

6.5. PERFORMANCE PROFILE

The example program given earlier compiles to object code in memory locations in about a second using compiled Quintus Prolog [16] on a Sun SPARCstation. This is acceptable for use in practice on small programs. However the performance of the prototype compiler is not linear with the size of input, but this could be alleviated by using *difference lists* [14] for the manipulation of lists.

An interpreter based on the mathematical interpreter definition, and written in PL_{0+} also also been implemented using Prolog. A compiled program can be interpreted at a rate of about 10 instructions per second. Alternatively the original program can be interpreted directly (and considerably more quickly). The results from the machine code program should of course be the same (or 'better' if the program is non-deterministic).

7. Discussion

In this paper, we have outlined a compiling specification for the a simple subset of the occam programming language and its correctness proof using the algebraic technique presented in [9]. The compiling specification is given as a set of theorems. The theorems are proved using the algebraic laws of process refinement for PL_0. The complete specification as well as the full correctness proof may be found in [6]. Further work is in progress to investigate more complicated language constructs such as recursion [8].

There are several advantages in following this approach:

- Each theorem and its proof is independent of the other theorems. This modularity is important if the verification method is to be practicable. Specification and its proof can be developed one theorem at a time. New theorems can be added to capture different ways of compiling the same construct. For example, the specification may be extended with the following theorem.

(14) $\quad C(\text{SKIP})sfm\Psi\Omega \qquad\qquad \textbf{if}$
$$\exists l_1. \quad s \le l_1 \le f \quad \wedge$$
$$m[s:l_1 - 1] = mtrans(\text{j}(f - l_1))$$

The compiler algorithm can then generate code using any of the alternative theorems; or possibly using several of them, choosing the 'best' (for example, the smallest) code. For example, the following theorem applies to IF clauses where one of the guards is TRUE:

(15) $\quad C(\text{IF}[\text{TRUE} \to p_1, \ldots, b_n \to p_n])sfm\Psi\Omega \qquad\qquad \textbf{if}$
$$C(p_1)sfm\Psi\Omega$$

- The compiling specification for PL_0 and its correctness proof are envisaged to be valid even for a larger language such as occam. The proofs given here will remain valid provided that the algebraic laws of PL_0 continue to hold for the full language. Also, the interpreter for the machine programs should only be extended with more instructions such that the behaviour of the existing ML_0 instructions remains unchanged (or is refined). Since illegal instructions are modelled as Abort, new instructions can only improve the machine.

The form of compiling specification is very similar to a logic program, with each theorem corresponding to a clause. However, such literal translation of the specification into a logic program may be inefficient. Hence, a strategy for executing the specification has been devised and a prototype Prolog compiler has been developed following this strategy. We have not given here a formal proof that the compiler satisfies the compiling specification. This should, however, be simple as the Prolog compiler is very close to the compiling specification. Of course, the resulting compiler will generate 'verified code' only if the compiler itself is executed on a trusted implementation of Prolog ... running on trusted hardware ...

Acknowledgements

The work was supported by the ESPRIT BRA **ProCoS** [1] and the UK IED safemos collaborative projects and we acknowledge the help of partners on both these projects. Copies of **ProCoS** project documents are available from: Annie Rasmussen, Department of Computer Science, Technical University of Denmark, Building 344Ø, DK-2800 Lyngby, Denmark.

References

[1] Bjørner, D. (1990) *ESPRIT BRA 3104: Provably Correct Systems ProCoS Interim Deliverable*, **ProCoS** doc. id. [ID/DTH DB 8].

[2] Bowen, J.P. and Pandya, P.K. (1990) *Specification of the ProCoS level 0 instruction set*, **ProCoS** doc. id. [OU JB 2].

[3] Bowen, J.P., He, Jifeng and Pandya, P.K. (1990) An Approach to Verifiable Compiling Specification and Prototyping, *Proc. PLILP90*, Workshop on Programming Language Implementation and Logic Programming, Linköping, Sweden, 20–22 August 1990.

[4] Clocksin, W.F. and Mellish, C.S. (1981) *Programming in Prolog*, Springer-Verlag, Berlin.

[5] He, Jifeng and Hoare, C.A.R. (1989) *Operational Semantics for ProCoS Programming Language Level 0*, **ProCoS** doc. id. [OU HJF 1].

[6] He, Jifeng, Pandya, P.K. and Bowen, J.P. (1990) *Compiling Specification for ProCoS level 0 language*, **ProCoS** doc. id. [OU HJF 4].

[7] He, Jifeng (1990) *Specification oriented semantics for the ProCoS level 0 language*, **ProCoS** doc. id. [OU HJF 5].

[8] He, Jifeng, Bowen, J.P. (1990) *Compiling Specification for ProCoS Language PL_0^R*, **ProCoS** doc. id. [OU HJF 6].

[9] Hoare, C.A.R. (1990) *Refinement algebra proves correctness of compiling specifications*, Technical Report PRG-TR-6-90, Programming Research Group, Oxford University, (also **ProCoS** doc. id. [OU CARH 1]).

[10] **INMOS** Limited (1988) *Occam 2 Reference Manual*, Prentice Hall International Series in Computer Science, UK.

[11] **INMOS** Limited (1988) *Transputer Instruction Set: A compiler writer's guide*, Prentice-Hall International, UK.

[12] Løvengreen, H.H. and Jensen, K.M. (1989) *Definition of the ProCoS Programming Language Level 0*, **ProCoS** doc. id. [ID/DTH HHL 2].

[13] Nicoud, J-D. and Tyrrell, A.M. (1989) The Transputer T414 Instruction Set, *IEEE Micro*, pp. 60–75.

[14] Nilsson, U. and Małuszyński, J. (1990) *Logic, Programming and Prolog*, John Wiley & Sons, Chichester, UK.

[15] Pandya, P.K. and He, Jifeng (1990) *A simulation approach to verification of assembling specification of ProCoS level 0 language*, **ProCoS** doc. id. [OU PKP 3].

[16] Quintus Computer Systems, Inc. (1990) *Quintus Prolog – Sun 3 & Sun 4 User Manual*, Release 2.5, Mountain View, California, USA.

[17] Roscoe, A.W. and Hoare, C.A.R. (1988) The Laws of Occam Programming, *Theoretical Computer Science*, **60**, pp. 177–229.

[18] Warren, D.H.D. (1980) Logic programming and compiler writing, *Software—Practice and Experience*, **10**, pp. 97–125.

Characterising structural and dynamic aspects of the interpretation of visual interface objects

Jon May, Martin Böcker
Standard Elektrik Lorenz AG,
Hirsauer Str. 210,
7530 Pforzheim,
West Germany,
Tel. +49 7231 71041,

Philip J. Barnard, Alison J. K. Green
MRC APU,
15 Chaucer Road,
Cambridge, CB2 2EF,
U.K.
Tel. +44 223 355294

Abstract.

This paper reports work carried out under the amodeus project (BRA 3066), which seeks to develop interdisciplinary approaches to studying interactions between users and systems. One of the main objectives is to extend the scope of user and system modelling techniques, and this paper describes the approach being taken towards the implementation of an expert system which would be able to evaluate interface designs, by producing predictions about user behaviour and consequent 'ease of use'. It is proposed that the problem of constructing an expert system which is able to generalise to novel interface designs can be solved by eliciting information about the structural and dynamic aspects of the interface. This paper describes the central ideas behind the approach, those of the *psychological subject* of an interface object and its *psychological predicate*, and of the *thematic transitions* that occur in interactions when the psychological subject changes.

1. Introduction

The AMODEUS (Assimilating Models of Designers, Users and Systems) project seeks to develop interdisciplinary approaches to studying interactions between users and systems. It has three general objectives: a) to extend the scope of user and system modelling techniques; b) to establish bridges between the behavioural and computing sciences; and c) to bridge from theory to the practicalities of design. This paper reports work under the first of these objectives and is specifically concerned with the problem of extending cognitive modelling techniques to deal with the nature, organisation and dynamic properties of "visual objects" at the human interface.

One strand of the cognitive modelling work makes use of a high level model of the overall organisation and functioning of the resources underlying human cognition. This model, called Interacting Cognitive Subsystems (Barnard, 1985,1987), provides the basic constructs around which an expert system is being developed. The aim is for the expert system to build descriptions of human cognitive activity that can predict user behaviour, and so provide advice about interface designs that is based upon firm psychological foundations (Barnard, Wilson & MacLean, 1988).

In order to build the models of cognitive activity, the expert system elicits descriptions of the users, their tasks, and the properties of the actual or envisaged user interface. Rules embodying psychological principles, derived from experiments, then reason about the nature and quality of the cognitive activity that is likely to occur and its consequences for smooth and effective user learning and performance.

So far, both within this theoretical approach and others (e.g. Card, Moran & Newell, 1983; Kieras & Bovair, 1984; Polson, 1987), by far the greatest analytic effort has been directed towards analysing the nature and properties of user tasks and their conceptual models of the system and its operation. More often than not, task analysis focuses upon the conceptual structure of the tasks and the decomposition of action sequences required to carry them out. It is not hard to understand why most current analyses focus largely on the structural description of tasks. Early human-computer dialogues mostly involved simple command entry or menu selection. The core problem for the user was to enter information at a prompt or to scan a menu and make a selection. Commands and their sequencing were the prime constituents of overt interaction.

With modern graphic interfaces, the screen environments and the visual objects present at the interface are not only more diverse, they are also structurally very much more complex and their detailed properties are a major influence on usability. A particular feature of current and future window based interfaces is that they allow the user to manipulate objects represented upon the screen directly, rather than through the medium of a textual command language. For "wimp" interfaces, and displays carrying multimedia information, it is becoming increasingly clear that information in the visual structures present in a screen environment plays a crucial role in determining the course of an interaction, what is known about it and how learning progresses (eg. see Mayes, Draper, McGregor & Oatley, 1988).

It is therefore apparent that in order to analyse interactions effectively, we must develop a method for describing and analysing the properties of visual objects at an interface, and their visuo-spatial behaviour within an interaction sequence. This paper explores the problem of producing structural descriptions of the visuo-spatial properties of screen environments and the objects contained within them. The immediate objective is to produce a psychologically motivated means of describing screen environments. These descriptions will form the basis for rules of inference to enable an expert system to reason about the usability of screen based interfaces.

It is beyond the scope of this paper to describe our underlying theoretical framework of Interacting Cognitive Subsystems in any great detail (see Barnard, 1987). We nevertheless assume, in line with certain other theoretical approaches (eg. Hinton, 1979), that human visual perception involves the active construction of a representation of the objects present in visual space. These representations are not simply recodings of the visual information collected by the retina, but are constructions derived from this information in the course of visual perception. Their content not only depends upon properties of perceptual processing activity, but is also based upon the perceiver's expectations, their knowledge and memories, the task they are performing, and so on. Our approach to describing screen environments will therefore go beyond simple visual properties to incorporate means of representing the psychological status of information.

A key aspect of the approach will involve the construction of hierarchical descriptions of the entities present in a screen environment. In addition to this we shall present distinctions intended to capture the psychological status of information within those hierarchical descriptions. Two of these latter distinctions, relating to the concept of a psychological subject and predicate and to thematic sequencing of the visual analysis of screen environments, are introduced in sections 2 and 3 by reference to multimedia displays and automatic bank teller machines. Section 4 then takes icon search as a simple, but prototypical case to relate the form of analysis offered to empirical evidence concerning user performance. Section 5 concludes the paper by outlining the form of principles based upon the analysis that can be further tested, refined and embodied within an expert system that models cognitive activity.

2. Structural description of a screen environment

Figure 1 illustrates some objects that might be found in a window based interface. There are three "passive" windows superimposed upon a neutral background, together with one active window, a pop-up menu, and the pointer that enables the user to manipulate objects.

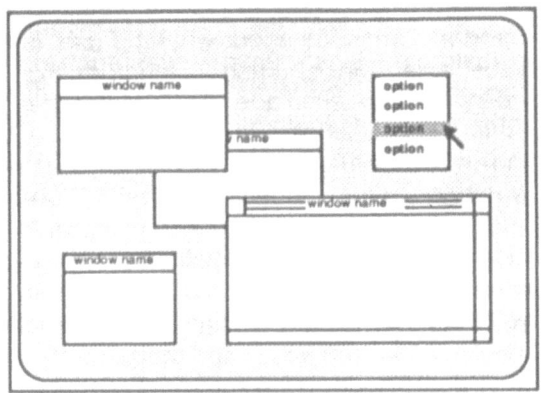

Figure 1: A typical window based interface

822

In a structural sense, the **Screen environment** can be said to consist of instances of objects, and each of these objects has a spatial position and attributes. One form of decomposition is shown in figure 2. In this decomposition some of the spatial relationships remain implicit in the locations of the descriptions themselves while others, for example occlusion, are itemised explicitly. Clearly, it would be possible to provide increasingly detailed descriptions of this interface that would ultimately capture the exact nature, content and positioning of the information on the screen. In a multimedia display one window might include a dynamically changing video image, whilst others contain static pictures, diagrams or text.

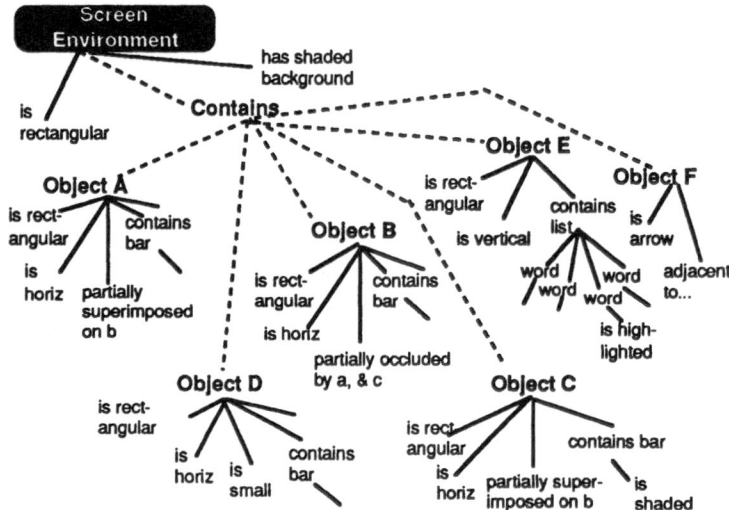

Figure 2: A structural decomposition of the hypothetical window based interface

However, our target is not so much to capture completely the 'screen image' but to capture the psychologically salient aspects of screen content in terms of its elements and organisation with respect to human cognitive processing. This will obviously depend upon assumptions about what the user is looking at and what information is being abstracted. Since human visual information processing is highly selective we need to orient our technique for structural description to take into account both what the user is attending to and what kind of visual information is being abstracted and used. From the point of view of what is represented in the head of the user, we shall distinguish between different *levels* of decomposition, and also between key information that forms the *psychological subject* of the current analysis and that which forms the *psychological predicate* .

Consider the two windows shown in figure 3, which are hypothetical examples of windows that might appear on the screen of a multimedia workstation. Both windows display moving video images, and incorporate a volume control in the bottom edge of the window to control the audio channel associated with the image. However, the video image in one window is being relayed in real time from a camera, while the image in the other is coming from a recorded source - a video recorder, for example. Since the user has an extra degree of control over the contents of the video window as opposed to the camera window (they can stop, rewind, or fast forward the image), an array of icons corresponding to standard video recorder controls is incorporated within the lower border of the window. For the user looking at the screen, this is the only salient feature of the window frames that can be used to distinguish them.

Figure 3: Two windows from a hypothetical multimedia workstation

Psychological theories universally acknowledge that the human visual attention is selective, and that the focus of our attention shifts over time from object to object or from one attribute of an object to another. There are, in the general cognitive science literature, many ways of talking about visual attention (eg. see Norman, 1976). Computer scientists seeking to model the interactive behaviour of computer systems have also proposed constructs such as a 'template' function (Harrison, Roast and Wright, 1989) to capture the fact that the user will only be attending to a part of the information displayed upon a VDU. Here, by direct analogy with systemic linguistics (Halliday, 1970), we propose that it is helpful to think about the selective nature of visual attention in terms of the part being played by a particular visual entity in the encoding and the subsequent *semantic interpretation* of relationships among entities within the visual field.

The concept of the *Psychological Subject* and *Predicate* relates the content of visual representations to the cognitive activities involved in the encoding and interpretation of objects, their inter-relationships and attributes. In systemic linguistics, the psychological subject of a clause is its first element (referred to as its 'theme') and this constituent fulfils a textual function. As Halliday puts it "it is, as it were, the peg on which the message is hung" (170, p.161). The body of the message (the 'rheme') is the remainder of the information expressed in a clause or sentence. In our visual application of this kind of idea, the *Psychological Subject* represents a particular point of departure for the mental processing of the visual information, in this case, present in a screen environment. The *Psychological Predicate* represents the structure and content of the visual information abstracted from the visual image and encoded *in relation* to the psychological subject. In this context, the psychological predicate is the equivalent of that part of an object specification that will come to form the "body" of a particular semantic interpretation of the visual scene. This predicate can contain *superordinate* information which refers to *other* visual objects within the visual image, or *subordinate* information which refers to *constituent elements* of the object itself. As with systemic linguistics, our emphasis is on the objects and their structural inter-relationships *per se*, but upon how information in those structures *is used* within the process of visual interpretation.

Typically, just as the psychological subject of a sentence is its first constituent, so those entities that can fulfil the role of the psychological subject of a particular visual object will be constrained by the properties of the early visual encoding of that object. When we look at or "attend to" a decorated dinner plate, its psychological subject would be a bounded shape specification (large or small; oval, circular), and the psychological predicate would specify properties of the decoration.

In the specific context of human-computer interaction, suppose a user has returned from a coffee break and needs to re-orient to the display shown in figure 1 to identify the video window. This is basically a task in which the array of objects on the screen will be scanned to identify the one with the appropriate attributes. At a given point in the scanning sequence the video window may become the psychological subject and relevant visual information encoded as a predicate structure. This is illustrated in figure 4.

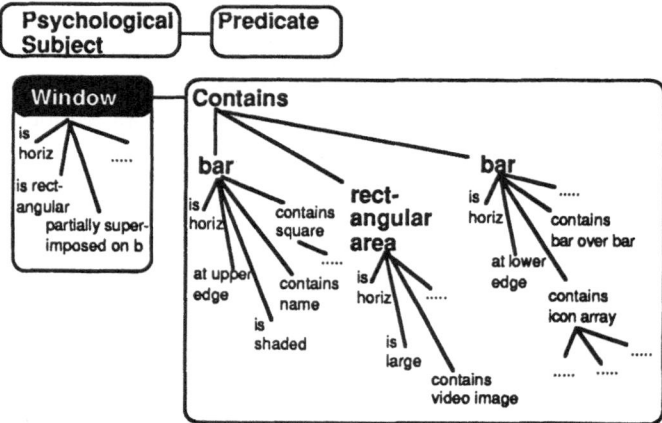

Figure 4: Psychologically motivated structural description of the video window, considered as part of the screen environment

The particular target window is actually a constituent of a spatial array of other similar rectangular objects. It is distinguished from these by a property of the horizontal bar at the foot of the window frame. In effect, the target is embedded in a *superordinate* visual structure, and a *subordinate* constituent of the target object itself, the linear array of icons, conveys crucial information concerning the window identity. The structural description shown in the figure should not be directly equated with a user's actual mental representation of visual objects. Rather, it should be understood as a means of capturing key properties of that mental representation. The underlying theoretical model assumes that these object-oriented descriptions capture key products of early visual processing activity, and form the basis for higher level semantic interpretation of what is "in" the visual scene, as well as providing the basis for the control of overt action.

In the process of object interpretation, the psychological subject would be the visual entity that is the current focus of what is being encoded. The subject-predicate relationships characterise the relevant cognitive *Information Structure* (as opposed to computer system information). So, for example, not all of the information shown in figure 4 would necessarily be incorporated into the semantic representation. Only part of the predicate structure may be represented in semantic form (eg. the description {Window {contains bar(horizontal, lower edge...) {contains icon array} } } might be sufficient to distinguish a window within the screen environment). Similarly, it should not be regarded as an entirely static concept, since visual processing involves continuous dynamic change in the representation of a visual scene.

3. Dynamic aspects of visual interactions

In visual processing activity, the psychological subject of any object based mental representation will change continuously, even when the full visual scene is static. If a user were processing information in a video window, many different types of changes may occur from moment to moment, both to the objects shown and to their visual frame of reference. A video image might well zoom in from showing two people in conversation to showing a close up of a particular face. In either case, the psychological subject may remain the same (a person's face) but the details of the relevant structural descriptions would need to be changed. Alternatively, visual attention may shift from one face to another here attention shifts from one object to another of the same class. In interactive systems, selection of an item in a pull down menu (like that shown in figure 1) may lead to a second pull down menu appearing offset to the right and slightly down. Again the psychological subject would be replaced by a visual object of the same class.

In purely visual terms, then, there is a *thematic structure* to the sequence of visual events in much the same way as there is a thematic structure in speech or text (cf Halliday, 1970). The concept of *Thematic structure* represents the psychological status of information in terms of the transitions that occur over time. So, for example, a constituent that is part of a predicate structure at one moment may become the psychological subject of the next step in much the same way as the object of one verbal sentence may become the subject of the next.

Some of the simpler aspects of such transitions can be illustrated by reference to the behaviour of an automatic teller machine (ATM). In contrast to the earlier multimedia illustration, the dynamic visual behaviour of the system is largely confined to text, and it can be used to provide a simple exemplification of thematic transitions. Such a machine typically consists of a dynamic VDU with selection buttons adjacent to it, a slot for card insertion, a keypad for number entry and a slot for dispensing money.

Figure 5 shows the sequence of screen displays for part of a cash withdrawal transaction. This part of the interface contains text, a numeric array, list structures for menu presentation, and iconically labelled buttons. Detailed structural descriptions of each of these components would, of course, be extensive, but in figure 6 the sequence is shown in a high level form, without attempting to break down the structure of the components into their respective constituent structures, or to present the relationships between them. During the sequence of transactions with the ATM, the focus of attention shifts from one component to another. At each stage different visual objects can, given certain constraints, assume the role of psychological subject, with the others forming the relevant superordinate and subordinate predicates. Progression from stage to stage is marked, psychologically, by changes in the subject, that is, thematic transitions.

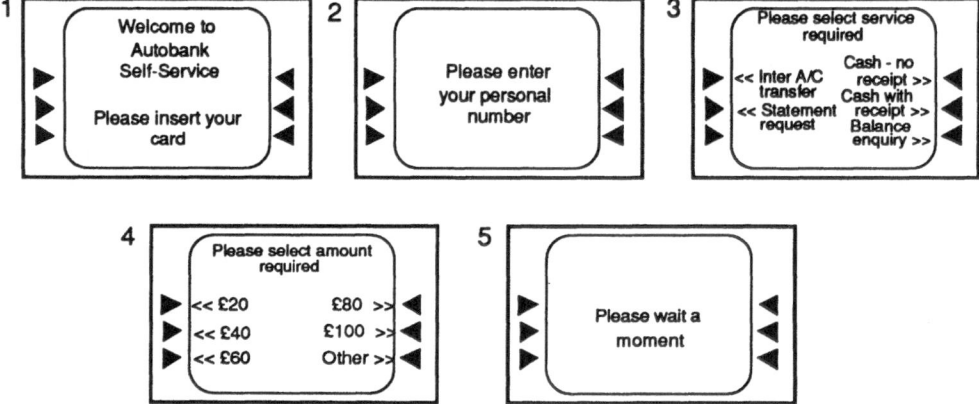

Figure 5: Transitions on the VDU display component of a typical Automatic Teller Machine (ATM)

When someone comes to use the ATM, the first thing that they encounter is a text screen, to the left of a card slot, which is above some number keys and a cash slot, which are to the right of a set of option keys. This textual description has been written to reflect the psychological salience of the various elements of the ATM interface, taking into account the user's intentions, the current state of the interaction and the design of the interface itself. It is clear that this description could be written in the form of a subject-predicate hierarchy without much difficulty, as shown in the lefthand column of figure 6.(See next page)

Some assumptions have been made, of course, about the user, the interface and their interaction. The text screen has been described as the first thing they encounter because of two factors - the role of the text screen in regulating encounters (and thus the user's knowledge about which part of the ATM interface to look at to check that it can be used), and its size and obviousness within the interface design, which make it stand out from the other elements (a brightly coloured card-slot, for example, might lead a user to focus on it as the subject rather than on the text screen). Similarly, if the person approaching the ATM was not a customer wanting to use it, but a service engineer coming to inspect the keypad, they would have different intentions about their task, and would direct their attention toward the keypad rather

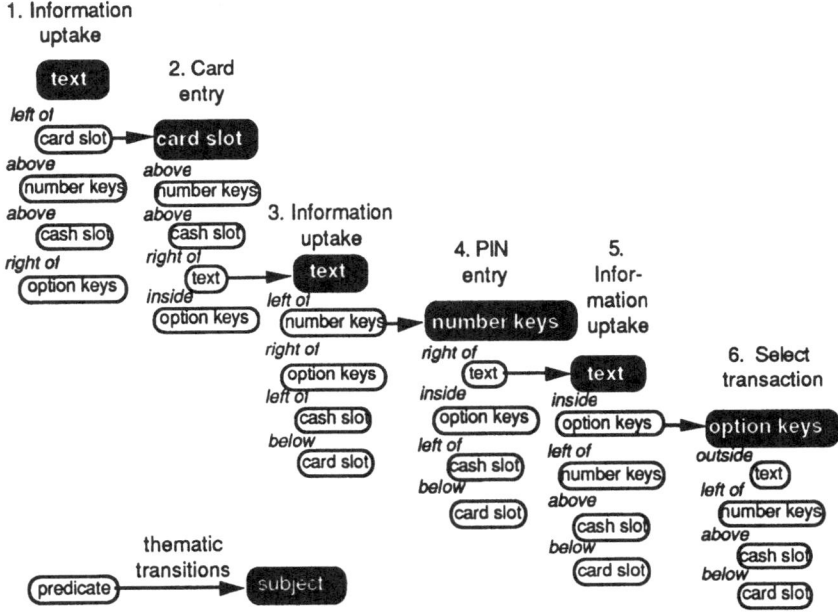

Figure 6: The thematic transitions between subjects in the course of an ATM interaction

than the text screen. These distinctions serve to illustrate the importance of knowing the psychological context of any interaction as well as details about the interface.

Once the interaction has begun, and the user has read the text screen to check that the ATM is in use, they have to insert their bankcard. This part of the interaction requires them to attend to the card slot, and so there is a change in the psychological subject of the interaction. The card slot must replace the text screen as the subject of the description, and the predicate of the superordinate structure must be re-ordered to reflect this thematic transition (as shown in the second column of figure 6). Likewise, the subordinate predicate would change from being a representation of the constituent structure of the textual material to a representation of the visual structure of the card slot (for simplicity, these subordinate predicates are not elaborated in figure 6). Having entered the card, attention switches back to the text screen to determine the next stage in the interaction (the third column). The text screen now displays a message requesting that the user's PIN number is entered and so the third thematic transition occurs, with the numeric keypad becoming the subject (the fourth column). The interaction continues in this way, with each stage of the interaction being marked by a thematic transition in the structural description of the interface.

4. Using structural descriptions to analyse and predict user performance with icon arrays

In the previous two sections we introduced three basic contrasts for capturing the structural and dynamic aspects of visual interface objects: the notion of *levels* involving subordinate and superordinate visual structures, the idea that the actual mental information structure is organised with respect to the current psychological subject and its predicates; and the concept of thematic structure which captures, within and between levels, the dynamic changes to the actual objects that form subject and predicate information. While these notions were introduced with reference to very different interactive settings, those of multimedia displays and an automatic bank teller machine, the basic concepts are quite general and can be very widely applied. In this section, we take a very simple interactive task, that of searching for a particular icon embedded within an array of other icons, to illustrate how general principles of human performance can be related to those ideas and explicitly motivated by established empirical phenomena.

Icon arrays are a common feature of many direct manipulation interfaces. Figure 7 illustrates two rather different kinds of icons, those that essentially have an abstract form and seek to convey (in this case) text editing operations, through the use of symbolic elements and semantic determiners (e.g. a cross). Other types of icon draw upon more depictive or representational constituents that visually map onto objects in the domain.

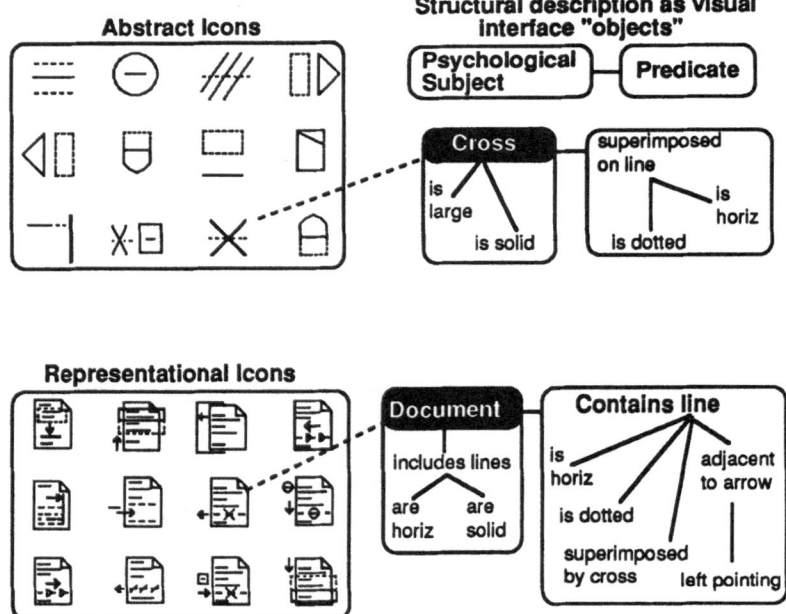

In this case, the 'representational' icons make use of document shapes to contextualise and convey aspects of the meaning of the text editing operations.

Although the representational forms are often thought to have a distinct advantage, and are perhaps the most common form in current use, two recent research studies have provided data showing that 'target' operations can be located and selected faster within arrays with the abstract forms rather than those with the representational form (Arend, Muthig and Wandmacher,1987; Green & Barnard, 1990).

In both studies there were not only overall differences between the two types of icon, but there were also substantial differences in the times to locate and select different icons of the same type. Furthermore, both studies showed that effects on search performance with the alternative types was not only a function of their visual form, but was also related to properties of the array such as the number of icons within the array to be searched (Arend, Muthig and Wandmacher,1987) or the extent to which individual icons remained at the same location within the array from one occasion to the next (Green & Barnard, 1990). Increasing the number of items in the array had a more detrimental effect on search performance with representational icons than with abstract ones, whilst maintaining icons in a constant spatial position within the array substantially reduces, over the course of learning, differences in search performance between the two types. The explanation of these kinds of results cannot be built around simple notions of confusability, since overlap in search times for the two types has been observed. The fastest 'representational' icons are within the same performance range as the slowest 'abstract' ones. What is required is not an explanation of 'Abstract/Representational differences' but a unified account of how objects with particular properties are readily located in an array of other objects and responded to, whilst others with different properties take longer.

The performance differences can be understood in a unified way by reference to the example structural descriptions also shown in figure 7. The overall visual form of the members of the representational set are indeed *similar* in the sense that a document shape is a prominent constituent of all of them. If, for the sake of argument, we now assume that the mental mechanisms of object encoding tend to operate on the basis of principles like "from the outside in"; "from large to small"; "from top to bottom" and "from left to right", then most of the representational icons will have a structural description in which the 'document shape' serves as the psychological subject of the representation. The discriminating constituents will be embedded within the subordinate predicate structure, in a manner that depends on the detailed properties of the individual icons. In marked contrast, for the abstract array, most of the items will be encoded with different visual forms serving as the psychological subject of the mental representation. For example, figure 7 illustrates the decomposition of the 'delete word' icon across both the icon sets. The psychological subject of the representational version, the 'document' shape, is shared with several other icons within the set. In the abstract set, the subject is the 'cross' and this is shared with only one other icon.

When the user can not take advantage of any prior knowledge about the spatial location of icons within the array, as is the case with newcomers to a system, or with arrays whose content and layout varies from occasion to occasion, the superordinate predicate of an icon's structural description cannot play any role in its detection. Instead, the arrays must be searched, and the subordinate make-up of individual icons evaluated in relation to mentally encoded properties of the target item, which will have been generated from the user's *semantic* representation of the icon.

In a search task, then, the user will initially orient to the array, locate an item, evaluate the appropriateness of its constituent elements with respect to a stored representation of the target and, if there is a mismatch, move on to the next item until the correct one is found. As with the ATM example discussed earlier, there are thematic transitions *between* levels (eg. from array to icon) and *within* levels (eg. from icon to icon). If we now assume that this evaluation starts with the psychological subject of the incoming visual information, and proceeds through the subordinate predicate structure until a decision can be made, the overall performance differences between the two types of icon and between individual icons of the same type can not only be understood, but can subsequently be predictively modelled.

So, for example, with a novel or randomly organised array, the empirical evidence suggests that the structural descriptions of the icons have a number of systematic influences on search performance. Representational icons not only take longer on average to locate than their abstract counterparts, but detailed analysis of the Green & Barnard (1990) data shows that individual icons within both sets vary substantially in the times taken to locate them.

Within the abstract set, those icons that are slower to locate either contain a constituent that would be prominent in the semantic representation of *other* icons within the set, or require the evaluation of two constituents of the predicate to discriminate the target from semantic representations of other icons. Likewise, *within* the representational set, search times for individual icons appear to increase quite systematically, according to the number of constituents that must be extracted from the predicate and evaluated.

Of course, the representational icons not only have to be evaluated to a greater depth within the predicate. Whereas only a few members of the abstract set contain constituents that would be prominent in the semantic representation of other members of the set, the 'document shape' is a prominent constituent of all of the representational icons. This form of evidence suggests that we can model search performance not on the basis of the 'type' of icon, such as their 'abstract' or 'representational' character, but on the more general basis of properties of their structural descriptions, and that such modelling can also be applied to other objects like windows, keypads and even bank card slots.

The presentation of a detailed psychological model of the search process is beyond the scope of this paper, but we can illustrate its broad form. We assume that the structural description of the superordinate array of icons presents the principal constituents of the individual icons as first level predicates, but does not include representations of detailed subordinate elements within those icons. In making a thematic transformation from the array to an individual icon or from one icon to another (cf figure 7) the movement can be restricted to items within the array that contain the pertinent target property as a first level predicate, for example, all those whose principal feature is a vertical rectangle. With the abstract set, which has a high proportion of discriminative information at this level, the number of icons that will need to be individually scanned and evaluated will be relatively small, and performance should be relatively unaffected by the number of icons actually in the local visual neighbourhood (cf Arend et al, 1987). By contrast, the set of representational icons contain little discriminative information within the first level predicates. Not only will more icons have to be scanned, but they will also have to be evaluated to a greater predicate depth to determine whether the incoming visual subject and predicate structures match those of the internally generated target.

In terms of human information processing activity, the search times will naturally increase as a function of the number of icons that are scanned and as a function of the number of predicate elements within each icon that need to be examined and evaluated. Both the number that are likely to get scanned and the depth to which they need to be evaluated can be estimated by analysing the properties of the structural descriptions of the complete set of icons. Furthermore, as noted in the previous section, the precise course of the evaluation is also likely to be influenced by the user's underlying knowledge of the set of icons. For both the abstract and the representational sets there are, for example, icons representing related actions. As the user learns about the set, they are likely to form related semantic representations of the target icons. So, for example, the *semantic* representations of the abstract icons for scroll forward and scroll backward (in figure 7, the second item on the middle row, and the last item on the bottom row in the 'Abstract' set) may both be organised around a common component eg:

scroll forward: {rectangle(horizontal, dotted) abuts above polygon(down pointing)}
scroll backward: {rectangle(horizontal, dotted) abuts below polygon(up pointing)}

One consequence of this form of hypothesised semantic conceptualisation of related icons is that the psychological subject of the incoming visual object for scroll forward readily matches that of the semantic representation that has been used to construct the target. With scroll back, there is a mismatch between the item that dominates the semantic representation (a rectangle) and the psychological subject of the incoming visual object (an upward pointing polygon). Indeed, the evidence from the two empirical studies cited earlier suggest that icons with otherwise equivalent visual complexity tend to be located faster in random arrays when their semantic and object based representations have matching initial constituents.

In summary, structural description of visual interface objects enables us to understand key determinants of human information processing activity during the process of scanning for icons in an array. Within this analysis the concepts of the psychological subject and predicate play a key role - not only in determining the form and content of thematic transitions between objects and their constituents, but also in relation to the user's semantic knowledge of the underlying set of operations and what is required to confirm or reject an icon in the context of the search task.

Obviously, other forms of theoretical expression could have been called upon to represent the key determinants of search performance. In that respect, our broader objective in this research is not simply to add "yet another model"of human search performance. Rather, our objective is to develop a set of concepts and techniques, based around structural description, that enable us to model a broad range of different tasks within a general theory of the organisation and functioning of cognitive resources (Barnard, 1985; 1987). The descriptive techniques so developed are intended to permit the formulation of explicit yet *general* heuristics for incorporation into an expert system that can be applied over different mental codes and tasks.

5. Implementing an expert system design tool

To be of any value, an expert system design tool must be able to aid designers by making predictions about and assessments of completely novel designs and task structures. A system constructed from accumulated "know-how" about existing design solutions would, like handbooks of interface guidelines and design principles, be able to provide only limited assistance outside its area of derivation, rapidly becoming dated and inapplicable as technology and ideas progress. The challenge is to provide a method of abstracting, from any situation, information about the factors which are psychologically relevant, and to have in place an inferencing system which can use this information to model the users' cognitive behaviour.

It is a relatively small step to move from the kind of analysis of empirical evidence, outlined in the previous section, to an expert system implementation which embodies rules that will generalise to settings other than icon search. It is, for example, not hard to develop questioning techniques that elicit the kind of general properties at an interface in terms of their shapes and other visual attributes. Rules that assign psychological status to the information in terms of their subject or predicate depths in a particular task context are also directly implied by the examples described in this paper. Obviously, this basic terminology applies equally readily to the multimedia and ATM examples with which they were originally introduced.

Most importantly, the analysis suggests general principles that will support predictive inferences about user behaviour and 'ease of use' of interface designs. For instance, we have argued that the psychological status of information within the structural description of a particular icon influences how many icons get examined. An icon is likely to be examined if its structural description contains a first level predicate that matches the dominant constituent of the user's conceptual representation of the target. As the proportion of such matches increases within a set of concurrently displayed objects, so the number of objects that will (on average) have to be evaluated also increases. This in turn implies an increase in the average number of thematic transitions between objects at the same level (icons in this case), and consequently an increase in average search time.

For the user, the ease with which they will be able to make a thematic transition depends upon several factors. First, of course, they must be aware that a transition is required and which element of the interface they must attend to next (or, for example, after approaching the ATM they would continue to stare at the text-screen, waiting for something to happen). This relies upon both users' knowledge of the overall structure of the interaction, and the cues provided by the interface. Mismatches between what the user expects to happen next and what the interface does next will cause problems, since the user will be attending to the wrong element, while explicit stimuli (flashing messages on the ATM screen, for instance, or clicks and whirrs from the card and cash slots) would help by orienting their attention directly.

Having realised that they must redirect their attention, the user must work out where the relevant target element is positioned on the interface, relative to their current focus of attention. The question is how deep within the predicate the next element to be attended to is embedded, for the further 'down' it is the more spatial relationships must be parsed in order to derive its location. The implication of this is that a design in which the thematic transitions occur between subjects and their most salient neighbours (ie those most likely to be at the 'start' of a predicate) are easier to use than those where attention must alternate between elements that are not spatially related.

It is important to note that the ordering of elements within the predicate is not solely governed by spatial structure or visual salience, although these play an important role, but also by the user's knowledge and expectations about the task. This can be seen within the ATM example illustrated in figure 6 by comparing the three descriptions of the task 'information uptake'. Although the same element of the interface is the subject in each case (ie the text screen), the predicates differ, with the element the user expects to use next starting the predicate, easing the anticipated thematic transition.

In an analysis of a dynamic task sequence, then, the subject-predicate descriptions of interface elements will vary according to the subsequent thematic transition, which in turn depends upon the user's knowledge of the task sequence.

The actual implementation upon which we are currently working does not solely rely upon the concept of structural description. It also calls upon a detailed model of human cognitive processes and memory structures (Barnard, 1987). Nonetheless, structural descriptions of visual interface objects and their dynamic properties play a crucial role in enabling us to extend the application of this particular form of cognitive modelling to current and future interface designs.

6. References

Arend, U., Muthig, K., and Wandmacher, J. (1987). Evidence for global feature superiority in menu selection by icons. *Behaviour and Information Technology, Vol. 6, No. 4*, 411-426.

Barnard, P. J. (1985). Interacting cognitive subsystems : A psycholinguistic approach to short-term memory. In A. Ellis, (Ed.), *Progress in the Psychology of Language, Vol. 2*, Chapter 6, pp 197-258. London, UK : Erlbaum.

Barnard, P. J. (1987). Cognitive resources and the learning of human-computer dialogues. In J. M. Carroll, (Ed.), *Interfacing Thought: Cognitive aspects of human-computer interaction*, pp 112-158. Cambridge, MA. : MIT Press.

Barnard, P. J., Wilson, M. and MacLean, A. (1988). Approximate modelling of cognitive activity with an expert system: A theory-based strategy for developing an interactive design tool. *The Computer Journal, Vol. 31, No. 5,* 445-456.

Card, S. K., Moran, T. P., and Newell, A. (1983). *The Psychology of Human-Computer Interaction.* Hillsdale, NJ: Erlbaum.

Green, A. J. K. Green, and Barnard, P. J. (1990). Iconic interfacing : The role of icon distinctiveness and fixed or variable screen locations. In B. Shakel, (Ed.), *Proceedings of Interact '90 : The 3rd IFIP Conference on Human-Computer Interaction.*

Halliday, M. A. K. (1970). Language structure and language function. In J. Lyons, (Ed.), *New Horizons in Linguistics.* Middlesex, England: Penguin.

Harrison, M. D., Roast, C. R. and Wright, P. C. (1989) Complementary methods for the iterative design of interactive systems. In Salvendy, G. and Smith, M. J. (eds.), *Designing and Using Human-Computer Interfaces and Knowledge Based Systems,* Elsevier Scientific, 651-658.

Hinton, G. E. (1979). Some demonstrations of the effects of structural descriptions in mental imagery. *Cognitive Science, Vol. 3, No. 3,* 231-250.

Kieras, D. E., and Bovair S. (1984). The role of a mental model in learning to operate a device. *Cognitive Science, 8,* 255-273.

Norman, D. A. (1976) *Memory & Attention: An introduction to human information processing (2nd edition).* New York, Wiley.

Mayes, J. T., Draper, S. W., McGregor, A. M., and Oatley, K. (1988). Information flow in a user interface : the effect of experience and context on the recall of MacWrite screens. In B. M. Jones and R. Winder, (Eds.), *People and Computers IV.* Cambridge, UK : CUP

Polson, P. G. (1987), A quantitative theory of human-computer interaction. In J. M. Carroll, (Ed.), *Interfacing Thought: Cognitive aspects of human-computer interaction,* pp 112-158. Cambridge, MA. : MIT Press.

Esprit Basic Research Action 3228: SPRINT

Speech Recognition using Connectionist Approaches

Khalid Choukri [*]
Project Coordinator
CAP GEMINI INNOVATION
118 rue de Tocqueville
75017 Paris. France
phone (33) 1 40 54 66 27
e-mail: choukri@csinn.uucp

Summary

SPRINT (ESPRIT Basic Research Action BRA-3228) aimed at tackling various problems that remain unsolved in speech recognition by exploring the particularities of neural networks (e.g. non-linearity, self-organization, parallelism) to upgrade Automatic Speech Recognition Systems (ASR) performance. It focussed on the use of connectionism paradigms to investigate some of the problems in relationship with speech variabilities. The work carried out covers adaptation to new speakers and/or new environments, noise immunity, classification of speech parameters using a set of "phonetic" symbols, and classification of a sequence of feature vectors by lexical access. Some results are given to illustrate neural network performance. The topics covered are :

First, theoretical studies have been conducted to establish various neural network capabilities, to generate any spectral transformation (Speaker-to-speaker Mapping) and their classification capabilities to discriminate between several phonetic classes. Speaker adaptation procedures based on learning spectral transformation with neural networks have been implemented, evaluated, and compared with well-established methods.

Considerable effort has been expended on the examination of network techniques for spectra classification and isolated word robust recognition. It mainly focused on the evaluation of various structures of multi-layer perceptrons and neural nets with different topologies (e.g. scaly or fully connected, locally connected, shared weights (TDNNs)). These experiments showed the need for specifically designed networks. Evaluations with LVQ alone, a TDNN-derived network alone and combined TDNN-LVQ architectures proved the combined architecture to be the most efficient.

An information theoretic distance metric together with a multilayer perceptron which has outputs with a probabilistic interpretation has been studied, unifying the probabilistically formal Hidden Markov Modeling (HMM) techniques and Multilayer Perceptron approaches (MLP), and leading to the development of an HMM--MLP hybrid.

---

[*] Authors and contributors list is given at the end of this paper.

1. Introduction

The aim of the work carried out within this project was to examine, both theoretically and experimentally, whether connectionist techniques can be used to improve the current performance of automatic speech recognition systems (ASR) by expanding their capability profile in different directions, in particular towards speaker independence and noise insensitivity.

The approach adopted allowed us to better understand the behaviour of neural networks when applied to speech, to appreciate their usefulness, and in the future to efficiently implement and use them in speech recognition devices in order to tackle speech variabilities (inter-speakers, noise).

Speech is a complex phenomenon but it is useful to divide it into levels of representation (Figure 1). Thus the performance of neural networks at each level has been evaluated with the aim to solve problems that occur at that specific level. The work described below concerns speaker adaptation, noise immunity, classification of speech parameters using a set of "phonetic" symbols, investigation of relevant graphemic symbols, and classification of a sequence of feature vectors by lexical access.

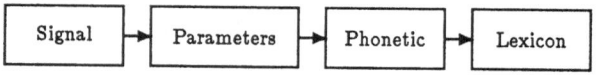

Figure 1: *Speech recognizer and SPRINT Tasks*

The tasks we consider are: Signal-to-Parameters and Parameters-to-Parameters, Parameters-to-Phonetic, Phonetic-to-Sub-lexical, and Parameters-to-Lexical.

2. Signal-to-Parameters and Parameters-to-Parameters

The main objective of this task was to provide the speech recognizer with a set of parameters leading to the best recognition performance, in particular for new speakers and noisy environments. The approaches we implemented, had to cope with the large inter-speaker variability, due to the speaker specific characteristics and the varying speech acquisition conditions, which creates difficulties in Automatic Speech Recognition. Two contributions are described below concerning speaker adaptation procedures and dimensionality reduction of feature space.

2.1 Parameter Transformations for Speaker Adaptation

2.1.1 Technical approaches

Spectral parameters corresponding to the same sound uttered by two speakers (S and N; S for Standard, N for New) are generally different. Speaker-independent recognizers usually take this variability into account, using stochastic models and/or multi-references. An alternative approach consists in learning spectral mappings to transform the original set of parameters into another one more adapted with respect to the characteristics of the current user and the speech acquisition conditions. The new parametric space (called herein interpretation space) may be of low dimension to ensure dimensionality reduction.

In the most general form, transformations could be applied to spectra of the reference speaker and the new one as depicted in the following figure:

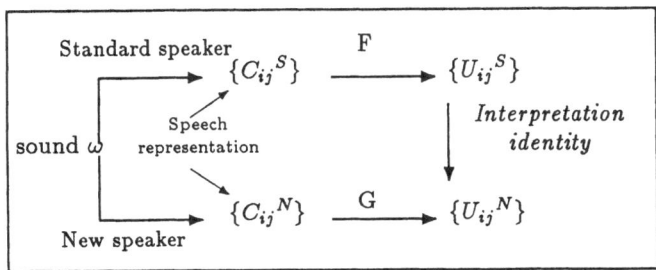

Figure 2: General form of spectral transformations

Two transformations F and G are found such that the averaged error (E_i) is minimum over a representative set of sounds: $G(C_{ij}^N) = F(C_{ij}^S) + (E_{ij})$. It should be noted that the transformation F is used to adapt the reference spectra at once (at the end of the adaptation phase) while G is applied as a normalization procedure during the recognition phase. The way to proceed is identical and can be summed up as follows:

– Speech acquisition for new speaker and Parameter extraction.

– Each new utterance is time-warped against each corresponding reference utterance. Thus temporal variability is softened and corresponding feature vectors are available.

– The spectral transformations are learned from these associated vectors.

The best approaches we experimented with were based on linear data analysis techniques, namely Canonical Correlation Analysis (CCA) and Linear Multivariate Regression (LMR) (G = identity), which allow the computation of linear transformations [5]. Our purpose was to built this transformation by non-linear functions using Multi-layer Perceptrons (MLP).

The mathematical formulation is based on a very important result, regarding input-output mappings, and demonstrated by Funahashi[7] and Hornik, Stinchcombe & White[9]. They proved that a network using a single hidden layer (a net with 3 layers) with an arbitrary squashing function can approximate any Borel measurable function to any desired degree of accuracy. Although any input-output transformation can be done with three layers, many might be done more easily with four layers or more with respect to the number of hidden units and the error surface.

Comparison with classical non-linear methods derived from polynomial filtering [12] was carried out. Experiments are described in the next section.

2.1.2 Results

A speech isolated word database consisting of 20 English words (including the digits) recorded 26 times by 16 different speakers (TI data base [6]) was selected. The first repetition of the 20 words formed reference templates, tests were conducted on the remaining 25 repetitions. The spectrum was parametrized using 8 MFCCs (Mel-based cepstral coefficients), computed every 20 ms. Recognition was carried out by comparing the input template to each reference template using an euclidean metric on the MFCC coefficients once both considered patterns were temporally aligned by a dynamic time warping algorithm.

The new user utters an adaptation vocabulary (words in isolation extracted from the data base): usually the first utterance of five words, "one", "two", "five", "six", "no". Preliminary experiments showed that using only 5 words to estimate all quadratic filtering parameters was insufficient. In that case ten words form the adaptation vocabulary.

Technique	Score	Confidence
Baseline system	68.25	± 0.29
Linear Multi-variate Regression or MLP 8-8	83.25	± 0.20
Canonical Correlations	76.82	± 0.25
Quadratic Filtering (10 adaptation words)	86.35	± 0.21
MLP 8-6-8	78.8	± 1.6
MLP 8-8-8	83.9	± 1.5
MLP 8-8-8-8	73.4	± 1.7
MLP 8-10-8	73.5	± 1.7

The results obtained show that MLP and LMR are slightly equivalent when used to learn a spectral transformation. Evaluation with larger adaptation vocabularies are under way to take into account more examples for the training phase of MLPs.

2.2 Transformations for Noise Robustness

The second contribution of this task was an attempt to train networks to carry out transformations of the speech parameters so as to provide recognition that is robust to contamination of the speech signal by background noise. It is established that in a Hidden Markov Model based recogniser it is possible to attain "acceptable" recognition performance at signal-to-noise (SNR) as low as -4dBs segmental SNR. MLPs can be viewed as extensions of digital filters, with the possibility of the use of non-linearities. Such networks can carry out a more general set of transformations than, for example, finite impulse response (FIR) filters. Preliminary experiments on the use of MLPs for isolated digit recognition in noise indicated that it was hard to train such a network to work well at varying signal-to-noise ratios. Work is still in progress and robustness to noise was tackled from other points of view as described in section 4.3.

2.3 Dimensionality reduction

The auto-associative structure can be trained using existing data as both input and target. Once trained these networks could provide dimensionality reduction simply by removing the output layer and using the outputs of the hidden layer directly.

A linear auto-associative network (e.g. a bottle-neck network with 27 input and output units and 8 hidden units) has been shown to be equivalent in performance to a singular value decomposition. It has further been shown that such a network with non-linear units can do no better (in terms of mean squared error) than the linear singular value decomposition[4].

The task is the transformation of filter bank analyzed speech data through the bottle-neck. The first aim here is to train linear networks with varying width bottle-necks. The input and target data are thus identical 27 element vectors of SRUbank analyzed data [*]. Four different width bottle-necks were trained, i.e. widths of 4, 8, 12 and 16 units. These were trained using 2000 vectors, i.e. 20 seconds of speech. The conjugate gradient learning technique was used[16]; various random starts were tried and found to make no difference. The following table gives an example of the results in terms of sum squared error over all the test data.

Number of Updates	100	200	300	400	500	600
Number of hidden units						
4	266	197	177	177	177	177
8	296	207	181	176	171	171
12	350	200	175	171	168	166
16	407	218	182	170	168	163

The second aim was to train networks with varying width bottle-necks using logistic rather than linear units, and the same training data. When using only conjugate gradients learning the results are too dependent on the random start. If however some steepest descent is used to prime the network, more consistent results are obtained (see the following table). The networks were trained using steepest descent (SD) followed by conjugate gradients (CG) for various random starts.

In the cases examined the learning characteristics of these transformational networks appear to be similar to those of discriminating networks. It can be seen from the previous tables that though broadly similar, the performance of linear and non-linear networks is not identical. It is still to be confirmed that training the

---

[*] The "SRUbank" filter-bank analysis system[8], is a 27 band-pass filter bank followed by; squaring, low-pass filters, and down-sampling to a frame rate of 100 frames per second; the filters are roughly critical-band spaced.

Number of Updates	10000(SD)	100(CG)	200(CG)	400(CG)	
Number of hidden units					Random start
8	463	181	179	179	1
8	450	181	181	181	2
8	437	183	183	183	3
12	424	171	171	171	1
12	456	176	175	175	2
12	381	174	174	173	3

[1]The "SRUbank" filter-bank analysis system [8], is a 27 band-pass filter bank followed by; squaring, low-pass filters, and down-sampling to a frame rate of 100 frames per second; the filters are roughly critical-band spaced.

auto-associative networks can achieve results as good as singular value decomposition and to ascertain the effect on recognition performance of the various width bottle-necks and the different final errors achieved.

2.4 Conclusions

Learning spectral transformations for speaker adaptation or noise robustness has been proved to be as efficient as the conventional methods. The advantage of Neural nets is that they can very easily account for the context of each speech frame through Context sensitive multi-layer perceptrons and/or Recurrent nets to build more complex spectral transformations. Application of NNs to speech feature extraction and dimensionality reduction is still in progress.

3. Parameters-to-phonetic

The objectives of this task were to assess various neural network topologies, and to examine the use of prior knowledge in improving results, in the process of acoustic-phonetic decoding of natural speech. These results were compared to classical pattern classification approaches such as nearest neighbour classifiers, dynamic programming, and k-means.

3.1 Preliminary Experiments

A preliminary experiment was carried out to evaluate the performance of a one-hidden-layer feed-forward automaton, for classifying spectral events obtained from a Temporal Decomposition [11] (a technique that gives a representation of speech, in terms of overlapping acoustic events). Three successive targets were used as the input data, in order to take into account some contextual information. The task of the perceptron was to identify the middle target, out of phonetic and diphonic labels. Different architectures with a single hidden layer and different numbers of hidden nodes were compared. The 3 highest output nodes were selected to give a 3 best candidates lattice, for phoneme identification. The results of these experiments are comparable to a classical 3 Nearest Neighbours approach[1] (recognition score of 70%).

The kind of MLPs that we initially investigated gave scores comparable to classical techniques, though it can be shown that the connectionist approach provides a slight reduction in computational requirements. This suggests that one must better account for acoustic and phonetic properties of speech, and design a specific network for speech pattern identification.

3.2 Databases

Two speech data bases were recorded and labeled to provide a larger data set for future experiments. The speech was uttered by one male speaker in French. A first database useful for training (referred to as DB_1) is made of isolated non-sense words (logatomes) and contains 6672 phonemes, in a quite uniform distribution. The second database (referred to as DB_2) was provided by the recording of 200 phonetically balanced sentences and gathered 5270 phonemes. DB_2 was split equally into training and test sets (2635 data each). 34 different labels were used: 1 per phoneme (not per allophone) and one for the silence. For each phoneme occurrence, 16 frames of signal (8 on each side of the label) were processed to provide a 16 Mel-scaled filter-bank vector; this representation can also be viewed as a 16x16 pixels spectrogram-like representation.

3.3 Classical Classifiers

Both to check the sufficient consistency of the data and to have some baseline reference scores, experiments using classical classifiers were conducted. The two strategies that were investigated are k-nearest neighbours and k-means.

A first protocol considered each pattern as a 256-dimension vector, and achieved k-nearest neighbours with the euclidean distance between references and tests. Time misalignments may cause major errors, thus a second protocol attempted to decrease these influences by carrying out some Dynamic Time Warping between references and tests and taking the sum of distances along the best path, as a distance measure between patterns. The same data were used in the framework of a k-means classifier, for various values of k (number of representatives per class). The best results are:

Classification method	Classification score
K-means alone ($K \geq 16$)	61.3 %
K-nn alone (K=5)	72.2 %
K-nn + DTW (K=5)	77.5 %

3.4 Neural Classifiers

3.4.1 LVQ classifiers

A first neural experiment was conducted using Learning Vector Quantization technique (LVQ). A study of the importance of the weights initialization procedure proved to be an important parameter for the classification performance. We have compared three initialization algorithms: k-means, LBG, Multiedit. With k-means and LBG, tests were conducted with different numbers of reference vectors, while for Multiedit, the algorithm discovers automatically representative vectors in the training set, the number of which is therefore not specified in advance.

Initialization by LBG gives better performance for self-consistency (evaluation on the training database: DB_1), whereas test performance on DB_2 (sentences) are similar for all procedures and very low. Further experiments were carried out on DB_2 both for training and testing. LBG initialization with 16 and 32 classes were tried (since they gave the best performances in the previous experiment). Even though the self-consistency for sentences is slightly lower than the one for logatomes, recognition scores are far better as illustrated here:

nb ref per class	16	32
K-means	60.3 %	61.3 %
LBG → LVQ	62.4 % → 66.1 %	63.2 % → 67.2 %

This experiment and some others (not presented here) [3] confirm that the failure of previous experiments is more due to a mismatch between the corpora for this recognition method, than an inadequacy of the classification technique itself.

3.4.2 The Time-Delay Neural Network (TDNN) Classifiers

A. Waibel [15] introduced TDNN as a specific architecture of MLP that can take into account the "dynamic nature of speech". A typical TDNN is composed of an input layer, two hidden layers, and an output layer. Successive TDNN's layers are not fully interconnected, and shift invariance imposes extra constraints on connection weights. Therefore, the number of connections in a TDNN is much lower than the one for a similar fully connected perceptron.

The input layer typically receives N_0 successive speech frames, represented by a M_0 dimension vector of parameters (M_0 x N_0 input pattern). The first hidden layer is composed of a M_1 x N_1 matrix of units. The second hidden layer is a K x N_2 matrix, where K represents the number of classes in the output. Only groups of P0 successive columns in the input layer are connected to 1 column in the first hidden layer, this structure being repeated with a shift of S_0 nodes. Time-invariance forces all sets of connections to be identical from one group to another. In the same way, only P_1 columns of the first hidden layer are fully interconnected to 1 column in the second hidden layer, with all groups of weights being identical. In the following a "TDNN-derived" network has a similar architecture, except that M_2 is not constrained to be equal to K, and the connectivity between the last 2 layers is full. Each output node takes for input all the nodes in a same line of the previous layer, the connection being either constrained to be equal or left independent from each other, according to the variant. A TDNN can then be fully described by its set of typological parameters, i.e.: M_0xN_0 / P_0,S_0 - M_1xN_1 / P_1,S_1 - M_2xN_2 - Kx1.

Various TDNN-derived architectures were tested on recognizing phonemes from sentences (DB_2) after learning on the logatomes (DB_1). Best results are given below:

TDNN-derived structure	self-consist.	reco score
16x16 / 2,1 - 8x15 / 7,2 - 5x5 - 34x1	63.9 %	48.1 %
16x16 / 2,1 - 16x15 / 7,4 - 11x3 - 34x1	75.1 %	54.8 %
16x16 / 4,1 - 16x13 / 5,2 - 16x5 - 34x1	81.0 %	60.5 %
16x16 / 2,1 - 16x15 / 7,4 - 16x3 - 34x1	79.8 %	60.8 %

The first net is clearly not powerful enough for the task, so the number of free parameters should be increased. This increase results immediately in better performance as can be seen for the other nets. The third and fourth nets have equivalent performance, they differ in the local windows width and delays. Other tested architectures did not increase this performance. The main difference between training and test sets is certainly the different speaking rate, and therefore the existence of important time distorsions. Though TDNN-derived architectures seem more able to handle this kind of distorsions than LVQ, as the generalisation performance is significantly higher for similar learning self-consistency, but both fail to remove all misalignment effects.

These experiments led to an "optimal" architecture and further experiments were then performed using it. Despite this poor performance, we can draw several conclusions about the relative performance of the different nets.

In order to upgrade classification performance we have changed the cost function which is minimized by the network by introducing weights in this function: the error term corresponding to the desired output is multiplied by a constant H superior to 1, the terms of the error corresponding to other outputs being left unchanged which compensates for the deficiency of the simple mean square procedure. We obtained our best results with the best TDDN-derived net we experimented for $H = 2$:

Database	Net :	self-consist.	reco score
DB_1	16x16 / 4,1 - 16x13 / 5,2 - 16x5 - 34x1	87.0 %	63.0 %
DB_2	16x16 / 4,1 - 16x13 / 5,2 - 16x5 - 34x1	87.0 %	78.0 %

The too small number of independent weights (too low-dimensional TDNN-derived architecture) makes the problem too constrained. A well chosen TDNN-derived architecture can perform as well as the best k-nearest neighbours strategy, though the recognition process is by a few orders of magnitude faster with the network, which is a very important result in itself. Performance gets lower for data that mainly differ by a significant speaking rate mismatch which could indicate that TDNN-derived architectures do not manage to handle all kinds of time distortions.

So it is encouraging to combine different networks and classical methods to deal with the temporal and sequential aspects of speech.

3.4.3 Combination of TDNN and LVQ

A set of experiments using a combined TDNN-derived network and LVQ architecture were conducted. For these experiments, we have used the best nets found in previous experiments. The main parameter of these experiments is the number of hidden cells in the last layer of the TDNN-derived network which is the input layer of LVQ [2].

Evaluation on DB_1 with various numbers of references per class gave the following recognition scores:

refs per class	4	8	16
TDNN +k-means	76.2 %	78.1 %	79.8 %
TDNN +LBG	77.7 %	79.9 %	81.3 %
TDNN +LVQ (LBG for initialization)	78.4 %	82.1 %	81.4 %

Best results have been obtained with 8 references per class and the LBG algorithm to initialize the LVQ module. The best performance on the test set (82.1%) represents a significant increase (4%) compared to the best TDNN-derived network alone.

Other experiments were performed on TDNN + LVQ by using a modified LVQ architecture, presented in [2], which is an extension of LVQ built to automatically weight the variables according to their importance for the classification. We obtain a recognition score of 83.6% on DB_2 (training and tests on sentences).

3.5 Conclusions

Experiments with LVQ alone, a TDNN-derived network alone and combined TDNN-LVQ architectures proved the combined architecture to be the most efficient with respect to our databases (improvement of 6% vs k-nn + DTW and 5% vs TDNN alone). The results are summarized below (training and tests on DB_2):

method	score
k-means	61.3 %
LVQ alone	67.2 %
k-nn alone	72.2 %
k-nn + DTW	77.5 %
TDNN alone	78.0 %
TDNN + LVQ	83.6 %

We also proposed to use low dimensioned TDNN's for discriminating between phonetic features[3], assuming that phonetics will provide a description of speech that will appropriately constrain a priori a neural network, the TDNN structure warranting the desirable property of shift invariance.

Phonetics describe speech events in term of features: oppositions are usually binary (voiced versus unvoiced, grave versus acute, vocalic versus non-vocalic) but they may be ternary or even based on 4 levels. Features can be defined according to acoustical properties (voiced vs unvoiced), articulatory properties (front vs back) or even linguistic properties (vowel vs consonant). The first experiments investigated for what kind of oppositions TDNN's are best suited. This has been done by evaluating individual performance for different feature extraction tasks.

While modes of articulation were very efficiently detected, those related to place of articulation provided less satisfying results. Actually, place of articulation is a more abstract notion, whereas acoustic correlates for modes of articulation are much more directly rendered on the spectrum[3].

The feature extraction approach can be considered as another way to use prior knowledge for solving a complex problem with neural networks. The results obtained in these experiments are an interesting starting point for designing a large modular network where each module is in charge of a simple task, directly related to a well-defined linguistic phenomenon.

4. Parameters-to-lexical

The main objective of this task is to use neural nets for the classification of a sequence of feature vectors by lexical access (transformation of speech parameters into lexical items: words). Many factors affect the performance of automatic speech recognition systems. They have been categorised into those relating to multi-speaker (or speaker independent) recognition, the time evolution of speech (time representation of the neural network input), and the effects of noise.

- multi-speaker (or speaker independent) recognition
  Work has been carried out on: speaker dependent, multiple speaker and speaker independent systems. The aim has been to examine the ability of various network paradigms to "learn" to generalise in order to deal with intra and inter--speaker variability.

- Time Evolution of Speech
  The problem of recognising time--varying speech patterns by neural networks has been approached from two directions: appropriate transformation of the speech signal to fit the fixed size network input layer; or suitable architectures and systems to deal with the temporal nature of speech. Initially, work used the fixed size input windows of various static networks by placing an utterance randomly within the window. Work was then carried out investigating various approaches to scaling the duration of observed data to fit the input window of static networks. Shift invariant networks were investigated and finally an HMM--MLP hybrid was developed.

- Noise and Distortion
  Work here investigated the recognition performance of "basic" MLP topologies
  on speech contaminated with background noise.

4.1 Comparison of Various Network Topologies

Experiments were carried out to examine the performance of several network
topologies such as those evaluated in section 3 for phonetic classification task. Time
delay neural networks (TDNN) offer a simple position invariant network that is
related to a single state Hidden Markov Model. A TDNN can be thought of as a
single Hidden Markov Model state spread out in time. The lower levels of the network
are forced to be shift-invariant, and instantiate the idea that the absolute time of an
event is not important. Scaly networks are similar to time delay neural networks in
that the hidden units of a scaly network are fed by partially overlapping input
windows. As reported in previous sections, LVQ proved to be efficient for the
phoneme classification task and an "optimal" architecture was found as a combination of a TDNN and LVQ. It will be reused herein.

It is interesting to compare the networks in terms of performance, the number of
weights and learning time. From experiments reported in detail in [14] there seems
little justification for fully-connected networks with their thousands of weights when
TDNNs and Scaly networks with hundreds of weights have very similar performance. This performance is about 83% (the nearest class mean classifier gave a
performance of 69%) on the E-set database (a portion of the larger CONNEX
alphabet database which British Telecom Research Laboratories have prepared for
experiments on neural networks). The first utterance by each speaker of the "E"
words: "B, C, D, E, G, P, T, V" were used. The speakers are divided into training
and test sets, each consisting of approximately 400 words. There were different
speakers in the training and test sets, approximately 50 speakers in each set.

Other experiments were conducted on an isolated digits recognition task,
speaker independent mode (25 speakers for training and 15 for test), using
networks already introduced. A summary of the best performance obtained is:

K-means		TDNN		LVQ		TDNN+LVQ	
train	test	train	test	train	test	train	test
97.38	90.57	98,90	94.0	98.26	92.57	99.90	97.50

Performance for training is roughly equivalent for all algorithms. For generalization, performance of the combined architecture is clearly superior to other techniques. This is not an isolated result and it has been observed on other tasks as
well. Although independent speaker recognition is usually a very difficult task, the
excellent results obtained with the multi-module architecture suggest that this is not
true for this small vocabulary task. To compare with other tasks on the same data,
we give below some results which were obtained in different experiments on
speaker dependent and multi-speakers tasks[13]. For speaker dependent isolated
digit recognition, Peeling and Moore [13] obtained 99.8% with HMM and 99.05%

with MLP. For multi-speaker recognition, performance was of 99.17% with HMM and 98.01% with MLP. Although it is difficult to compare these different tasks, performance of our TDNN, which is 94%, gives an idea of the difficulty of the speaker independent experiment compared to the multi-speaker experiment. In view of these results, the 97.5% we have obtained with the combined architecture can be considered as a very good performance and this two modules net appears as a very powerful classifier.

The initial comparison of network topology work found little difference in the performance of the various network topologies (i.e. fully connected nets, scaly nets and TDNNs) and a slight advantage to a combination of nets. To draw accurate conclusions more "sophisticated" experiments are need.

4.2 Time Evolution of Speech

The problems to be tackled here involve the search for appropriate input signal coding and the best choice of network structures. The digitized speech signal of successive time intervals is usually transformed into a vector sequence, obtained by the extraction of particular features. Such vectors, their number depending on word duration, are projected onto the fixed-size collection of N times M network input elements (number of vectors times number of coefficients per vector). Different projection methods have been selected. These are:

Linear Normalization: the boundaries of a word are determined by a conventional endpoint detection algorithm and the N' feature vectors linearly compressed or expanded to N by averaging or duplicating vectors.

Time Warp: word boundaries are located initially. Some parts of a word of length N' are compressed, while others are stretched and some remain constant with respect to speech characteristics.

Noise Boundaries: the sequence of N' vectors of a word are placed in the middle of or at random within the area of the desired N vectors and the margins padded with the noise in the speech pauses.

Trace Segmentation: it provides another means to make patterns equal in length. The procedure essentially involves the division of the trace that is followed by the temporal course in the M-dimensional feature vector space, into a constant number of new sections of identical length.

These time normalization procedures were used with the scaly neural network [10]. It turned out that three methods for time representation - time normalization, trace segmentation with endpoint detection or with noise boundaries - are well suited to solve the transformation problem for a fixed input network layer. With respect to their recognition rates, they are in the 98.5% range (with +/- 1% deviation) for 10 digits or 57 words vocabularies in speaker independent mode. There is no clear indication that one of these approaches is superior to the other ones. The time warp method and the inclusion of noise boundaries of the unwarped speech signal don't provide adequate recognition results.

4.3 Robustness to NOISE

The work examining the robustness of "basic" MLPs to contamination of speech with noise found that the networks examined offered little inherent robustness to the contamination of speech with background noise. However, the networks demonstrated graceful degradation when noise was added to the data at the input of the network.

The recognition performance for the DTW approach is in the range from 96.5% to 99.5% which outperform the tested neural networks by 3%. It should be emphasized that these comparison results have been achieved for the same input parameter set, manually corrected for errors of the endpoint detection algorithm. Using fully automatic endpoint detection, the results are quite different from those presented here. In these investigations, a considerable performance degradation of a DTW approach with increasing vocabulary size has been observed in comparison to neural networks results. Therefore, it seems that neural networks are less sensitive to errors with respect to endpoint detection, whereas DTW performs better with noisy environment data for a perfect endpoint determination.

4.4 Conclusions

The network techniques investigated have delivered comparable performance over other techniques. It is now well agreed that Hybrid systems (Hidden Markov Modeling and MLPs) yield enhanced performance. Initial steps have been made towards the integration of Hidden Markov Models and MLPs. Mathematical formulations are required to unify hybrid models. The temporal aspect of speech has to be carefully considered and taken into account by the formalism. Nonetheless, the network formalism offers enticing properties for enhancing the capabilities of automatic speech recognitions systems.

5. Conclusions

This first phase of SPRINT allowed us to gain know-how and tools for connectionist based approaches applied to speech processing. Baseline networks were investigated and compared for several tasks and then compared with conventional approaches.

It has been established that basic MLPs are efficient tools to learn speaker-to-speaker mappings for speaker adaptation procedures. We are expecting more sophisticated MLPs (recurrent and context sensitive) to perform better. For phonetic classifications, sophisticated networks, combinations of TDNNs and LVQ, revealed to be more efficient than classical approaches or simple network architectures; their use for isolated word recognition offered comparable performance. Various approaches to cope with temporal distortions were implemented and demonstrate that combination of sophisticated neural networks and their cooperation with HMM is a promising research axis.

With respect to each task the main results are summarized below:

We established that neural nets are powerful tools for isolated word recognition. On a very difficult database (E-set) we obtained 83% recognition score versus 69% with K-means approach. We also obtained 97.50% on an isolated digit database in speaker independent mode while HMM gave 99.8% but in multi-speaker mode (tests are still on-going). For noise robustness we obtained similar results using both neural nets and DTW. We obtained encouraging results for phoneme classification, 83.6%, when compared to the best conventional approach we tested (K-nn + DTW), 77.5%. Application to speaker-to-speaker transformation proved that neural nets can perform as well as classical methods.

6. Consortium

The partners involved in SPRINT are:

CSInn: Cap Sesa Innovation, (Project coordinator), France

ENST: Ecole Nationale Superieure des Telecommunications, France

IRIAC: Institut de Recherche en Intelligence Artificielle et Connexionisme, France

RSRE: Royal Signals and Radar Establishment, England

SEL: Standard Elektric Lorenz, Germany

UPM: Universidad Politecnica de Madrid, Spain

References:

[1] Y. Bennani. Decodage acoustico-phonetique par reseaux connexionnistes. Technical Report 88D016, E.N.S.T, 1988.

[2] Y. Bennani, N. Chaourar, P. Gallinari, and A. Mellouk. Comparison of Neural Net models on speech recognition tasks. Technical Report, LRI, 1990.

[3] F. Bimbot. Speech processing and recognition using integrated neurocomputing techniques: esprit project sprint (bra 3228), first deliverable of task 3. June 1990.

[4] H. Bourlard and Y. Kamp. Auto-association by multilayer perceptrons and singular value decomposition. Biological Cybernetics, 59:291--294, 1988.

[5] K. Choukri. Several approaches to Speaker Adaptation in Automatic Speech Recognition Systems. PhD thesis, ENST (Telecom Paris), Paris, 1987.

[6] K. Choukri. Speech processing and recognition using integrated neurocomputing techniques: ESPRIT Project SPRINT (Bra 3228), First deliverable. October 1989.

[7] Ken-Ichi Funahashi. On the approximate realization of continuous mappings by neural networks. in Neural Networks, 2(2):183--192, march 1989.

[8] J.N. Holmes. The jsru channel vocoder. IEE Proc. F, 127(1):53--60, Feb. 1980.

[9] K. Hornik, M. Stinchcombe, and H. White. Multilayer feedforward networks are universal approximators. Neural Networks, vol. 2(number 5):359--366, 1989.

[10] A. Krause and H. Hackbarth. Scaly artificial neural networks for speaker-independent recognition of isolated words. IEEE ICASSP (Glasgow), pp. 21--24, 1989.

[11] C. Montacie, K. Choukri, and G. Chollet. Speech recognition using temporal decomposition and multi-layer feed-forward automata. IEEE ICASSP, S1:409--412, 1989.

[12] W. J. Rugh. Non linear system theory The volterra and Wiener approach. The John Hopkins University Press, Baltimore, MD, 1981.

[13] R. Moore S.M. Peeling. Isolated digit recognition using mlps. NATO -ASI ON SPEECH RECOGNITION AND UNDERSTANDING, 1987.

[14] A. Varga. Speech processing and recognition using integrated neurocomputing techniques: ESPRIT Project SPRINT (Bra 3228), First deliverable of Task 5. June 1990.

[15] A. Waibel, T. Hanazawa, G. Hinton, K. Shikano, and K. Lang. Phoneme recognition using Time-Delay Neural Networks. Technical Report, CMU / ATR, Oct 30, 1987.

[16] A.R. Webb, D. Lowe, and M.D. Bedworth. A comparison of nonlinear optimisation strategies for feed-forward adaptive layered networks. Memorandum 4157, RSRE, July 1988.

Authors and contributors

Younes BENNANI	John BRIDLE	Nasser CHAOURAR
Khalid CHOUKRI	Lorraine DODD	Francoise FOGELMAN
Patrick GALLINARI	David HOWELL	Manfred IMMENDORFER
Anders KRAUSE	Ken McNAUGHT	Abdelhamid MELLOUK
Claude MONTACIE	Roger MOORE	Olivier SEGARD
Hélène VALBRET	Andrew VARGA	Alexandre WALLYN

Action No. 3042

PERFORMANCE AND PHYSICAL LIMITS
OF HETEROSTRUCTURE FIELD EFFECT TRANSISTORS

D.R. Allee, and A.N. Broers
Department of Engineering, Cambridge University
Trumpington Street, Cambridge CB2 1PZ, U.K.
Tel: 0223 332675, Fax: 0223 332662

M. Van Rossum, S. Borghs, and W. De Raedt
IMEC, Lueven, Belgium

H. Launois, B. Etienne, and Y. Jin
L2M, Bagneux, France

R.Adde, and R. Castagne
IEF, Orsay, France

A. Antonetti, and D. Hulin
LOA, Palaiseau, France

ABSTRACT: The technological progress made in the first 9 months of this ESPRIT basic research action is reviewed. The primary goal of this project over a 2.5 year period is the study of the physical limits of carrier transit time and hence device switching speed as heterojunction field effect transistors (HFETs) are scaled down from submicron to ultra-submicron dimensions (0.2-0.02μm gate length). Since the switching time of heterostructure field effect transistors only approach the carrier transit time when the device parasitics are minimized, an analysis of various techniques to reduce gate resistance for sub-100nm gate lengths is made including the consideration of superconducting gate electrodes. The most promising approach is the use of multi-fingered gates or multiple gate feeds with a normally conducting gate metal, combined with gate aspect ratios as large as is practical to fabricate. Nanometer scale structures have been fabricated on bulk substrates with PMMA liftoff, and contamination resist and ion milling down to 16nm. A 35nm by 25μm gate electrode has been fabricated in 0.4μm source drain gap of a device structure, and at L2M a 0.15μm gate length AlGaAs/GaAs HFET with transconductance of 694mS/mm and f_{max} of 132GHz has been achieved. Working pseudomorphic AlGaAs/InGaAs HFETs with InAs grown ohmic contacts and nanometer scale gate electrodes are expected in the next trimester.

1. Introduction

Since the invention of the planar integrated circuit in 1959, there has been a dramatic increase in the performance and decrease in the cost per function of integrated circuits. These advances have been achieved primarily by reducing the minimum feature size. The primary goal of this project over a 2.5 year period is the study of the physical limits of carrier transit time and hence device switching speed as heterojunction field effect transistors (HFETs) are scaled down from submicron to ultra-submicron dimensions (0.2-0.02μm gate length). Intensive investigation through Monte Carlo simulations, high frequency s-parameter characterization, and ultra-fast laser spectroscopy will be made of the conditions where the carriers might exceed the saturation velocity due to near ballistic transport. The ultimate goal is to reach carrier transit times in the 1ps regime and to minimize the corresponding device switching time by simultaneously reducing the device parasitics. AlGaAs/GaAs HFETs will be used for these studies for gate lengths above 0.1μm whereas pseudomorphic AlGaAs/InGaAs HFETs will be used for gate lengths down to 0.02μm. The results of this basic research action will provide invaluable guidelines for scaled down devices in the next generations of HFETs for information technology.

There are five members of this consortia. IMEC is developing fabrication processes suitable for ultra-submicron HFETs with the exception of the gate level. Cambridge University is responsible for the fabrication of the nanometer scale gate electrodes. CNRS-Laboratoire de Microstructures et de Microelectronique (L2M) will investigate and optimize the influence of the epitaxial layer structure of AlGaAs/GaAs HFETs for gate lengths down to 0.1μm. CNRS-Institute D'Electronique Fondamentale (IEF) is developing the design, modeling and electrical characterization tools including Monte Carlo simulations and high frequency s-parameter characterization. The Laboratoire d'Optique Appliquee (LOA) will use time resolved optical spectroscopy to study the carrier dynamics.

In this paper we will review the technological progress that has been made in the first 9 months of this research effort. We will begin with an analysis of various techniques to reduce the parasitic gate resistance for sub-100nm gate lengths. The fabrication of various nanometer-scale gate electrodes and structures down to 16nm with both PMMA liftoff and contamination resist will be presented as well as the optimization of MBE layer growth and advanced source drain metallizations for short gate length HFETs. DC and high frequency characteristics have just been obtained for 0.2μm and 0.15μm gate length AlGaAs/GaAs HFETs.

2. Gate Resistance

2.1. NORMALLY CONDUCTING GATES

Although the fundamental limit to device switching speed is the carrier transit time determined primarily by the gate length, the device parasitics, particularly the gate and source resistance, must also be minimized in order to approach this limit. Maintaining a low gate resistance as the gate length is reduced below 100nm is particularly difficult because of the unrealistically large gate aspect ratios required. Measured gate resistances for typical sub-100nm gate lengths fabricated with PMMA liftoff are several hundred ohms[1], two orders of magnitude too high for high performance devices. These extremely high gate resistances offset the principal advantage of reduced carrier transit time. As a result, the fastest FETs to date have had larger gate lengths between 0.1μm and 0.15μm[2].

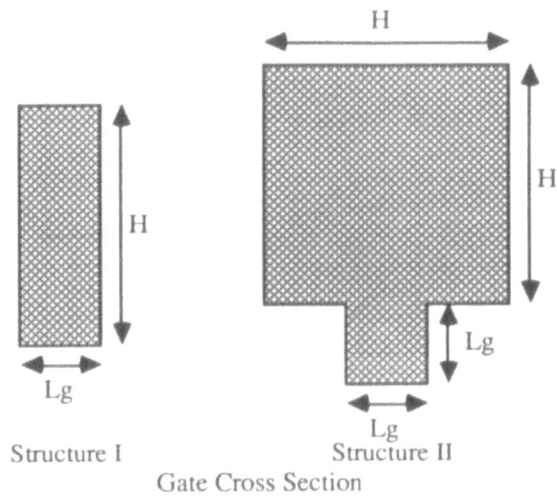

a)

Gate Cross Section

Structure I   Structure II

b)

Figure 1: a) Approximate cross sections of two common gate structures. Structure I is a simple rectangular gate with aspect ratio H/Lg. Structure II is a T-gate. Lg is the gate length. b) multi-fingered gate and multiple gate feeds can be used to reduce gate resistance by n^2 by breaking the gate into n equivalent segments.

First we will analyze various approaches to reduce the gate resistance for normally conducting gate metals. In the next, superconducting gate electrodes are considered. There are several possible techniques to reduce the gate resistance to a few ohms. The most obvious though often impractical approach is to increase the thickness of the gate metallization forming a rectangular high aspect ratio electrode (fig. 1a, structure I). Another approach is to fabricate a mushroom gate structure (fig. 1a, structure II) with a very small footprint[3,4]. Both of these techniques are limited primarily by the mechanical stability of the gate electrodes and secondarily by the parasitic sidewall gate to source capacitance. A third approach is to break the gate electrode into several segments (fig. 1b).

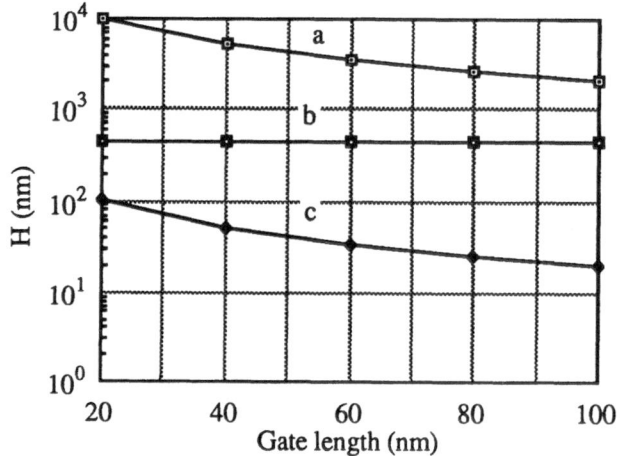

Figure 2: The size of H necessary to achieve a total gate resistance of 10Ω as a function of gate length, Lg, for various gate structures. The gate metal is assumed to be Au at 300K ($\rho_n = 2.04\mu\Omega$ cm). The total device width is 100μm. a) a simple rectangular gate (structure I), b) a T-gate (structure II), c) ten simple rectangular gates (structure I) in parallel; each gate is 10μm wide to maintain 100μm total device width.

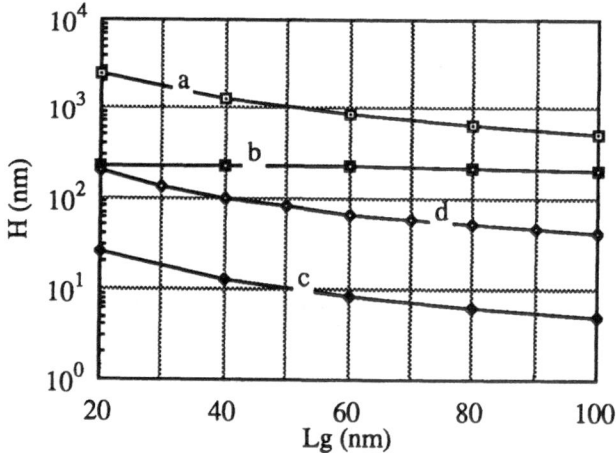

Figure 3: The size of H necessary to achieve a total gate resistance of 10Ω as a function of gate length, Lg, for various gate structures. The total device width is 100μm. For structures a-c, the gate metal is assumed to be Au at 77K ($\rho_n = 0.5\mu\Omega$ cm). a) a simple rectangular gate (structure I), b) a T-gate (structure II), c) ten simple rectangular gates (structure I) in parallel; each gate is 10μm wide to maintain the total device width of 100μm. d) a simple rectangular gate (structure I) made of superconducting YBaCuO at 77K. The superconducting material properties are assumed to be : $T_c = 90K$, $\rho_n(100K) = 150\mu\Omega$ cm, $\lambda_L(0K) = 140nm$.

Each segment is placed in the interstice between an interdigitated source and drain forming a multi-fingered gate. While maintaining the same total device width, the gate resistance is reduced by a factor of n^2 where n is the number of segments. One factor of n arises from each gate segment being shorter, the second factor of n from the segments being in parallel. Alternatively, n equivalent gate segments can be formed with multiple gate taps.

The dimensional parameters H and Lg are defined for the two gate structures in figure 1. Lg is the gate length. H is the thickness of the gate metallization for the rectangular gate (structure I) and the vertical and horizontal dimensions of the upper portion of the T-gate (structure II). The value of H necessary to achieve a total DC gate resistance of 10Ω is plotted as a function of gate length for a rectangular gate (fig. 2a), a T-gate (fig.2b), and 10 rectangular gates in parallel (fig. 2c). The calculations assume the gate metal is gold at room temperature with a bulk resistivity of $2.04\mu\Omega$cm. The actual resistivity and magnitude of H would be somewhat higher because of the extremely small dimensions of these structures. The total device width is assumed to be 100μm; a device width typical of commercial quarter micron microwave HFET's. A single rectangular gate is clearly implausible for gate lengths below 100nm because of the difficulty of fabricating gates with at least 20:1 aspect ratios. For the T-gate, H must be around 40nm almost independent of the gate length below 100nm because most of the conductivity is in the upper portion of the "T". These dimensions are just plausible for 100nm gate lengths but not for 20nm gate lengths. The only technique that appears promising for gate lengths all the way down to 20nm is placing ten 10μm gate segments in parallel. For a gate length of 20nm, the aspect ratio need only be 5:1.

These calculations are repeated assuming device operation at 77K where gold has a resistivity of $0.5\mu\Omega$cm (fig. 3a-c). A single rectangular gate still requires an unreasonably large aspect ratio for gate lengths less than 100nm. The T-gate might be possible to fabricate down to a gate length of 50nm with an H of about 200nm. Again the parallel gate segments is the most promising technique with a gate length of 20nm requiring an aspect ratio of only 1.5:1.

Although DC resistance has been calculated, the data in figure 2 and 3 are valid up to the frequency at which the current is localized at the conductor surface due to the skin effect. The most limiting case is the T-gate where uniform current is assumed in the large upper portion. This assumption will break down when the skin depth is approximately equal to H/2. Using the well known formula for skin depth[5] ,

$$\lambda_s = \frac{1}{\sqrt{\frac{1}{2}\omega\mu_o\sigma_n}}$$

where ω is the frequency in rad/sec, μ_o is the permeability of free space, and σ_n is the conductivity, the current in the T-gate will be uniform for frequencies up to 129GHz at 300K and 127GHz at 77K.

2.1. SUPERCONDUCTING GATES

With the advent of high T_c superconductors[6] , it may soon be possible to reduce the gate resistance with a superconducting gate electrode operating at liquid nitrogen temperature. The gate resistance, however, is strictly zero only at DC and increases with the square of frequency. We are particularly interested in frequencies deep into the millimeter wave band (30-300GHz) beyond the f_{max} of longer gate length HFETs. This phenomenon can best

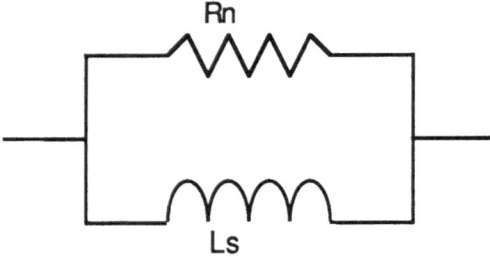

Figure 4: The London two-fluid model of a superconductor. A superconducting inductance, L_s, whose physical origin is the inertia of the Cooper pairs is in parallel with a normal channel resistance, R_n.

be understood with the London two fluid model[7]. The superconductor is modeled as a parallel combination of a superconducting channel and a normally conducting channel. The superconducting channel has no resistance but has inductance due to the inertia of the Cooper pairs. The normal channel has resistance and inductance due to the inertia of the electrons. This inertial or kinetic reactance of the normal channel is negligible compared the resistance and is usually ignored (fig. 4). At DC, the inductance of the superconducting channel shorts out the normal channel resistance resulting in a perfectly conducting wire. As the frequency increases, the reactance of the superconducting channel increases causing current to flow in the normal channel and resulting in a net series resistance for the superconductor.

In order to determine the magnitude of this resistance at microwave and millimeter wave frequencies, lets assume we have a gate electrode made of c-axis oriented $YBa_2Cu_3O_{7-\delta}$, the most thoroughly studied high Tc superconductor. As a thin film, it has a critical temperature, T_c, of 90K, a c-axis coherence length, ζ, of 1.2nm, an energy gap at 0K of 54meV, a normal state resistivity at 100K of $150\mu\Omega cm$, and a DC critical current density at 77K greater than $2\times10^6 A/cm^2$ [8]. In bulk form the London penetration depth at 0K, $\lambda_L(0)$, has been measured to be 140nm[9].

There are several assumptions in the London two-fluid model. The normal electron mean free path and the coherence length of the superconducting electrons must be less than the London penetration depth in order to use the local theory, $J = \sigma E$. Since the penetration depth is proportional to $\left(1-t_r^4\right)^{-1/2}$ where t_r is the temperature normalized to T_c, λ_L at 77K is 205nm. This is much larger than the coherence length and also larger than the mean free path of the normal electrons if we assume sufficiently "dirty" material. Furthermore, the cross section dimensions of the gate electrode must be equal to or less than twice the penetration depth in order to assume uniform current density. This is also true since the largest gate length we are considering is 100nm. Finally, the operating frequency must be less than the energy gap frequency at which the electric field will break the Cooper pairs. Since the energy gap is proportional to $(1-t_r)$, the energy gap at 77K is 20.7meV corresponding to a frequency of 5.02THz. The largest frequency we will be

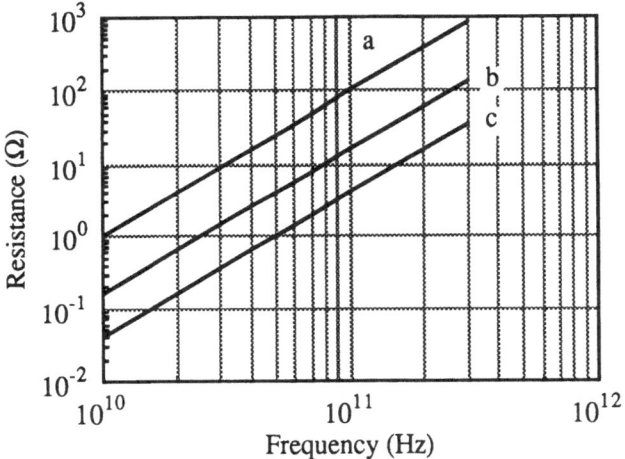

Figure 5: The gate resistance as a function of frequency for superconducting YBaCuO at 77K. The superconducting material properties are assumed to be : T_c = 90K, $\rho_{normal}(100K)$ = 150$\mu\Omega$ cm, $\lambda_L(0K)$ = 140nm. The total device width is 100μm. The three gate electrodes shown all have square cross sections (structure I, Lg = H). a) Lg = H = 20nm, b) Lg = H = 50nm, c) Lg = H = 100nm.

considering is 300GHz, the upper limit of the millimeter wave band. Under these conditions, the London two-fluid model is applicable.

The total current through the superconductor can be written as

$$J = (\sigma_n + \sigma_s)\, E$$

where σ_n and σ_s are the normal channel and superconducting channel conductivities respectfully. J is the current density and E is the electric field. These conductivities can be conveniently expressed in terms of measured parameters as follows.

$$\sigma_n = \sigma_{no}\, t_r^4 \qquad \sigma_s = \frac{1 - t_r^4}{j\omega\mu_o\lambda_L^2(0)}$$

σ_{no} is the normal channel conductivity just above Tc; j is the square root of negative 1. The remaining variables maintain their previous definitions. The normal channel resistance, R_n, and the superconducting inductance, L_s, are

$$R_n = \frac{Z}{L_g H \sigma_{no} t_r^4} \qquad L_s = \frac{Z \mu_o \lambda_L^2(0)}{L_g H \left(1 - t_r^4\right)}$$

where Z is the device width. A rectangular cross section of length, L_g, and height, H, is assumed (structure I). The equivalent series resistance, R_{series}, is

$$R_{series} = \frac{R_n \omega^2 L_s^2}{R_n^2 + \omega^2 L_s^2}.$$

The end to end gate resistance for a 100μm wide device as a function of frequency for H=Lg=20nm, 50nm, and 100nm are plotted(fig. 5). While the gate resistances are all less than 1Ω at 10GHz, they reach 1000Ω at 300GHz. At 100GHz, the value of H required for a total gate resistance of 10Ω as a function of gate length is also plotted for a superconducting gate electrode (fig. 3d). It is significant that H for a superconducting electrode at 100Ghz is still an order of magnitude larger than the H required for the ten normally conducting gate electrodes in parallel. While the superconducting gate electrode could be fabricated in a parallel structure as well, there isn't much motivation to do so because of the sufficiently low aspect ratio of the normally conducting gate electrodes in parallel. This is particularly true in view of the current processing difficulties associated with high T_c superconductors: the difficulty of depositing films on GaAs substrates and the high temperature anneal that is incompatible with GaAs.

For completeness, the current carrying capabilities of high T_c superconductors at millimeter wave frequencies must also be considered. To the author's knowledge, this measurement has not yet been made. Kwon has estimated the AC critical current to be 50mA/μm based on the low critical magnetic field of the high Tc superconductor[10]. This current density would be sufficient for small signal applications.

In conclusion, the most promising approach to reduce the gate resistance for sub-100nm gate lengths is the use of multi-fingered gates or multiple gate taps with a normally conducting gate metal, and gate aspect ratios should be as large as is practical to fabricate.

3. Ultra-high resolution electron beam lithography

At Cambridge, the central piece of equipment for nanometer-scale lithography is a JEOL-4000EX, 400kV transmission electron microscope with a LaB_6 source. Previous modifications enabled the use of this instrument as a lithography tool on thin membranes. The minimum beam diameter at the sample was measured to be 0.4nm. Such a small beam diameter is possible because the high beam voltage reduces diffraction effects and the small focal length (a few mm) of the objective lens reduces the spherical and axial chromatic aberrations. An Aharonov-Bohm ring was previously fabricated on a Si_3N_4 membrane using contamination resist and ion milling of AuPd; the ring diameter was 0.16μm and the wire width was 15nm[11]. Although this ring is one of the smallest and most accurately

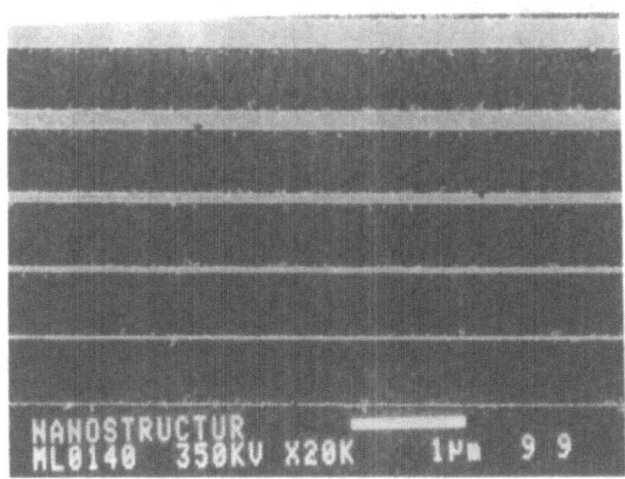

Figure 6: PMMA liftoff of Au on GaAs substrate. The lines from top to bottom are 270nm, 170nm, 94nm, 56nm, 31nm, and 16nm.

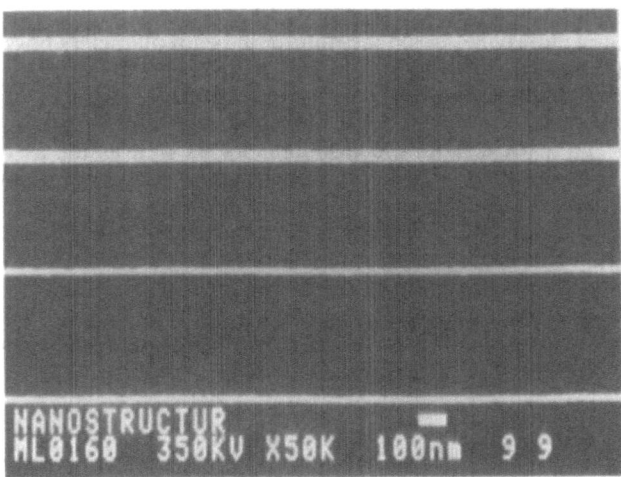

Figure 7: Ion milled Au:Pd on Si substrate using contamination resist. The two upper lines are 50nm; the two lower lines are 25nm. Note the excellent edge uniformity.

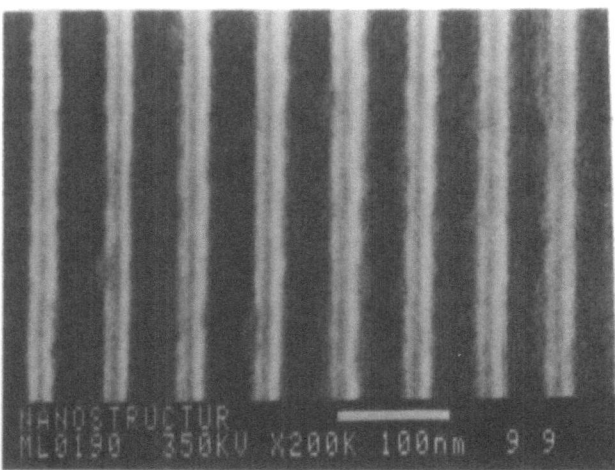

Figure 8: Ion milled AuPd on Si substrate using contamination resist. The lines are 25nm on 65nm period.

defined structures ever made, it is still an order of magnitude larger than the beam diameter because of the resolution limitations of the pattern transfer process. We are currently investigating alternative lithographic processes in an attempt to discover pattern transfer techniques that more closely approach the minimum beam diameters possible in this state-of-the-art electron optical column.

Many additional column modifications have been made as a part of this contract to enable the patterning of nanometer-scale structures on bulk substrates. A secondary electron detector and a backscatter electron detector as well as the associated scanning electronics have been installed. The backscatter electron detector provides mass contrast images and is particularly important for alignment and focusing on ohmic contact marks beneath a layer of e-beam resist, whereas the secondary electron detector provides high resolution topographic images. For completeness, a scanning transmission electron detector was also installed. Pattern rotation electronics and an electrical x and image shift now allow nanometer-scale gate electrodes to be precisely aligned in a sub-micron source drain gap. Beam blanking plates and a Faraday cup for beam current measurement have been designed and are ready for their final installation.

4. Experimental Results

The unusually high beam voltage of the JEOL-4000EX is particularly advantageous for the fabrication of low resistance nanometer-scale gate electrodes. The high beam voltage reduces forward scattering in the resist. As a result, thicker resist can be used while still maintaining linewidths close to the resist's resolution limit. Typically PMMA is used and has a resolution limit of about 10nm. The thicker resist allows the fabrication of a larger

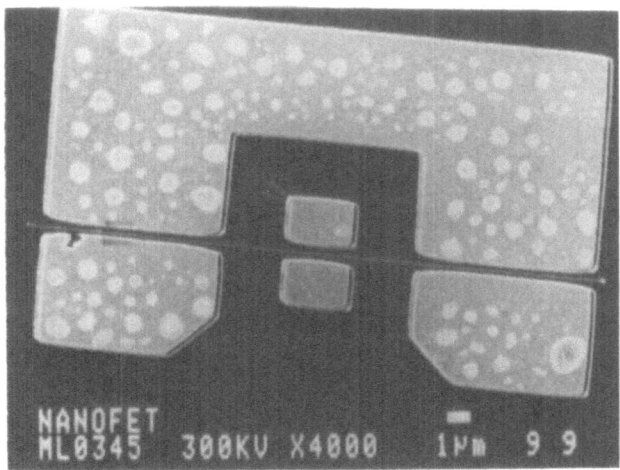

a)

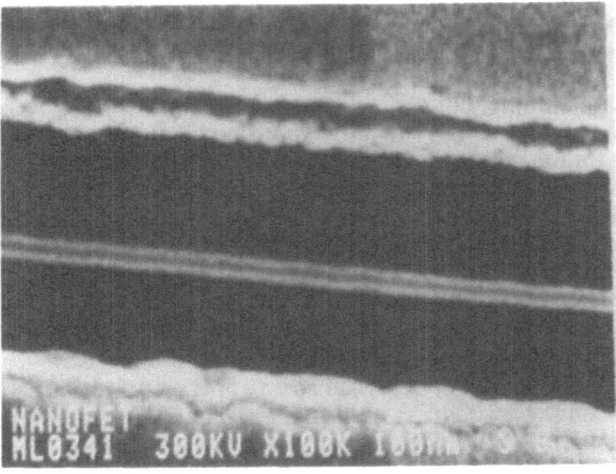

b)

Figure 9: a) 35nm gate electrode aligned in the source drain gap of a device structure. The electrode was fabricated using contamination resist and ion milling. b) a high magnification view of the gate in the source drain gap.

862

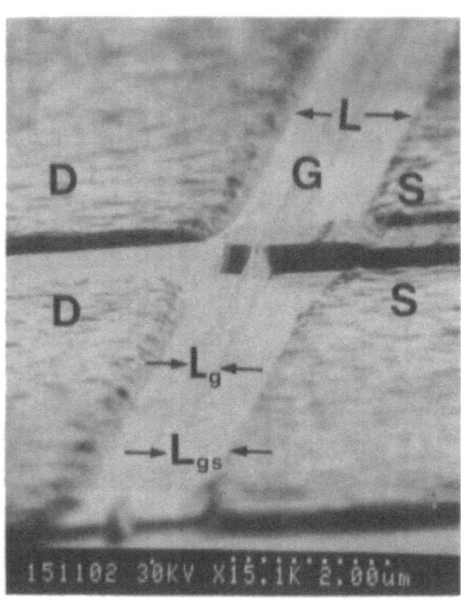

Figure 10: A planar doped 0.2μm AlGaAs/GaAs HFET. The transconductance is 625mS/mm; f_t and f_{max} are 60 and 100 GHz respectively.

gate aspect ratio either through liftoff of a evaporated gate metal or through electroplating on a predeposited plating base. The latter approach is being extensively investigated at other laboratories for the fabrication of high aspect ratio Au absorbers for X-ray masks[12]. To our knowledge, this technique has not been applied to the fabrication nanometer scale gate electrodes possibly because of concern over contaminants from the plating solution. We feel this approach may have significant advantages over liftoff. The tearing of metal at the gate edges during liftoff can be avoided possibly resulting in greater edge uniformity and reduced likelihood of a discontinuity. The thin plating base can be removed after the resist with a short chemical etch. A third approach to nanometer-scale fabrication uses contamination resist. In this process, discovered and developed by A.N. Broers[13,14], a finely focused, high current density electron beam polymerizes residual vacuum pump oil on the specimen forming a good ion milling mask. The LaB$_6$ source and the superior electron optics of the JEOL-4000EX greatly facilitate the formation of nanometer-scale contamination patterns. We are currently pursuing all three approaches.

Several nanometer-scale test structures have been fabricated on bulk substrates, both GaAs and Si. Using PMMA and liftoff, Au lines down to 16nm have been fabricated on GaAs (fig. 6). Using contamination resist , lines have been ion milled in a Au:Pd film on Si substrate (prepared at IMEC). The smallest line is 25nm and has outstanding edge uniformity (fig. 7). Also using contamination resist, a periodic AuPd grating was formed on a Si substrate (fig. 8). The metal lines are 25nm on a 65nm period. To our knowledge, this is the densest metal grating ever fabricated using contamination resist. Using a device structure prepared at IMEC, a 35nm Au gate electrode was fabricated in the source drain gap with contamination resist and ion milling (fig. 9). The source is broken into two segments; a central gate contact (not shown) will divide the gate into two equivalent segments and hence reduce the gate resistance by a factor of four.

Both bulk doped and planar doped AlGaAs/GaAs HFETs have been fabricated at L2M with gate lengths of $0.2\mu m$ (fig.10). The transconductances were 370 and 625 mS/mm with gm/gd (transconductance / output conductance) ratios of 13 and 16 respectively. Both of these devices had an f_t of 60GHz and an f_{max} of 100GHz. $0.15\mu m$ planar doped gate length devices were also recently fabricated both with a double and single gate recess. The single gate recess device had a transconductance of 612mS/mm, a gm/gd of 8.3, f_t of 81GHz and f_{max} of 102GHz. As expected the double gate recess device was superior having a transconductance of 694mS/mm, a gm/gd of 13, f_t of 112GHz and f_{max} of 132GHz.

At IMEC, pseudomorphic AlGaAs/InGaAs layer growth conditions are being optimized for bulk and delta doping. For bulk doping, room temperature mobilities of $28000cm^2/Vs$ and sheet carrier densities of $1.95 \times 10^{12} cm^{-2}$ have been achieved. For the fabrication of sub-100nm gate length HFETs and the corresponding reduction of source drain spacing to a few tenths of a micron, the ohmic contact edge definition becomes critical. The common alloyed AuGeNiAu contacts are inadequate. IMEC has developed InAs grown ohmic contacts with contact resistances less than 0.1 Ohm-mm. The contact metal was optimized to be PdGeTiPt which also provides high contrast in the scanning electron microscope greatly facilitating gate alignment. The first working nanometer-scale gate length pseudomorphic HFETs are expected in the next trimester.

5. Conclusion

In the first 9 months of this ESPRIT basic research action, several technological advances have been made. After an analysis of various techniques to reduce gate resistance for sub-100nm gate lengths, the most promising approach is the use of multi-fingered gates or multiple gate feeds with a normally conducting gate metal, and gate aspect ratios should be as large as is practical to fabricate. Nanometer scale structures have been fabricated on bulk substrates with PMMA liftoff, and contamination resist and ion milling down to 16nm. A 35nm by $25\mu m$ gate electrode has been fabricated in $0.4\mu m$ source drain gap of a device structure. A $0.15\mu m$ gate length AlGaAs/GaAs HFET with transconductance of 694mS/mm and f_{max} of 132GHz has been achieved. Working pseudomorphic AlGaAs/InGaAs HFETs with InAs grown ohmic contacts and nanometer scale gate electrodes are expected in the next trimester.

[1] Allee,D.R., P.R.de la Houssaye, D.G.Schlom, J.S.Harris Jr., and R.F.W.Pease, "Sub-100nm gate length GaAs MESFETs and MODFETs fabricated by molecular beam epitaxy and electron beam lithography," *J. Vac. Sci. Technol. B* **6**, No. 1, 328-332 (Jan./Feb. 1988).

[2] Lester,L.F., P.M.Smith, P.Ho, P.C.Chao, R.C.Tiberio, K.H.G.Duh, and E.D.Wolf, "$0.15\mu m$ gate length double recess pseudomorphic HEMT with f_{max} of 350GHz," *IEDM Technical Digest*, 172-175 (1988).

[3] Weitzel,C.E., and D.A.Doane, "A review of GaAs MESFET gate electrode fabrication technologies," *J. Electrochem. Soc.* **133**, 409C (Oct. 1986).

[4] Chisholm,A., S.Sainson, M.Feuillade, A.Clei, "0.15μm e-beam T-shaped gates for GaAs FETs," *Microcircuit Engineering Abstracts*, pp.-36 (Sept. 1989).

[5] Ramo,S., J.R.Whinnery, T.van Duzer, *Fields and Waves in Communication Electronics*, Wiley, (1984)

[6] Wu,M.K., J.R.Ashburn, C.T.Torng, P.H. Hor, R.L.Meng, L.Gao, Z.J.Huang, Y.Q.Wang, and C.W.Chu, "Superconductivity at 93K in a new mixed-phase Y-Ba-Cu-O compound system at ambient pressure," *Phys. Rev. Lett.* **58**, 908 (1987)

[7] Gittleman,J.I., B.Rosenblum, "Microwave properties of superconductors," *Proc. IEEE* **52**, 1138-1147 (1964)

[8] Kapitulnik,A., K.Char, "Measurements on thin film high-Tc superconductors," *IBM J. Res. Develop.* 33, No.3, 252-261 (1989)

[9] Cava,R.J., B.Batlogg, R.B.van Dover, D.W. Murphy, S.Sunshine, T. Siegrist, J.P.Remeika, E.A.Rietman, S.Zahurak, and G.P. Espinosa, "Bulk superconductivity at 91K in single-phase oxygen-deficient perovskite $Ba_2YCu_3O_{9-d}$," *Phys. Rev. Lett.* **58**, No. 16, 1676-1679 (1987)

[10] Kwon,O.K., *Chip-to-chip interconnections for very high speed system level integration*, Ph.D. Thesis, Stanford University (1986)

[11] Broers, A.N, A.E.Timbs, and R.Koch, "Nanolithography at 350kV in a TEM," *Microelectronic Engineering*, Vol. 9, 187-190 (1989).

[12] Windbracke,W., H. Betz, H.L.Huber, W. Pilz, S. Pongratz, "Critical dimension control in X-ray masks with electroplated gold absorbers," *Microelectronic Engineering*, Vol. 5, No. 1-4, 73-80 (1986).

[13] Broers,A.N., "Combined electrical and ion beam processes for microelectronics," *Microelectron. and Reliabil.* **4**, 103 (1965).

[14] Broers,A.N., W.W.Molzen, J.J.Cuomo, and N.D.Wittels, " Electron beam fabrication of 8nm metal structures," *Appl. Phys. Lett.* **29**, 596 (1976).

Growth and Characterization of Ultrathin Si_mGe_n Strained Layer Superlattices

Dr. Hartmut Presting, H. Kibbel, E. Kasper
Daimler Benz Research Institute
Wilhelm-Runge-Str. 1 1
D - 7900 Ulm, F.R.G.
T.: 49-731-505-2049
Fax:-505-4 102

M. Jaros
University of Newcastle upon Tyne
NEI 7RU, U.K.

G. Abstreiter
Technical University of Munich
D-8046 Garching, F.R.G

Summary:

Growth of Ultrrathin Si_mGe_n (m monolayers (ML) Si, n ML Ge) strained layer superlattices (SLS) by molecular beam epitaxy (MBE) is reported. Diode structures (doping sequence p^+-n-n^+ on n^+-substrate and n^+-n-p^+ on p^+-substrate) were grown for optical device applications with strain symmetrization of the SLS by a thin homogeneous buffer layer serving as a virtual substrate. The concept of zone folding is described and the transition matrix elements of the folded bandstructure as a function of period length are calculated. The concept of a virtual substrate consisting of the actual Si-substrate and a thin buffer layer is explained. The strain adjustment of the SLS by choice of the buffer layer composition and the strain symmmetrization is outlined. Different structural (TEM- and X-ray analysis, Rutherford Back Scattering) and optical (Raman, Photoluminescence, modulation spectroscopy) characterization methods of the SLS are discussed. A Raman spectrum of a Si_6Ge_4 SLS is shown together with an $Si_{0.6}Ge_{0.4}$ alloy sample, the occuring peaks are discussed.

The performance of future microelectronic circuits will be strongly enhanced by the monolithic integration of superlattice devices with conventional integrated circuits on top of a silicon substrate. With this material concept the mismatch between the superlattice materials and the silicon substrate has to be accomodated. The SinGem SLS is a model system for the study of mismatch effects because of similar chemistry and well pronounced strain effects.

1. Introduction

Ultrathin strained layer superlattices (SLS) can be considered as man made semiconductors with novel optical and electronic properties. Recent progress in molecular beam epitaxy (MBE) has made the growth of such superlattices consisting of layers with atomic dimensions possible. Silicon/Germanium (Si/Ge) superlattices grown on a silicon substrate are presently subject to considerable experimental and theoretical interest because spectacular modifications of the optical properties of silicon are predicted for this system. The hope is to create a direct band gap silicon type semiconductor on a silicon substrate which can be used as an optical device (e.g. LED or photodiode) for data transmission purposes. The possible application could range from intrachip communication within a silicon electronic circuit chip to interchip communication in a complex computer board.

The performance of future microelectronics will be strongly enhanced by the monolithic integration of superlattice devices with conventional integrated circuits on top of a silicon substrate. With this material concept the mismatch between the superlattice materials and the silicon substrate has to be accomodated. The Si_mGe_n SLS is a model system for the study of mismatch effects because of similar chemistry and well pronounced strain effects.

2. Monolithic Integration of Silicon Based Superlattices in Conventional Integrated Circuits

Completely new and tailorable properties are expected from classes of semiconductor devices which are based on superlattice structures. A superlattice structure is produced by arranging alternate layers of different elements with a period length which superposes the natural period of the lattice (Figure 1). New electronic and optical properties arise from this synthetic semiconductor if the period of the superlattice is small (three to hundred atomic layers). The properties of the synthetic semiconductor are controllable by the geometrical dimensions, the chemical composition and the strain inside of the superlattice structure. Alternating layers of lattice mismatched materials can be grown with high perfection if the individual layer thickness is small. With very thin layers the lattice mismatch is accomodated by tetragonal distortion of the lattice cell leading to strained superlattices. Since this discovery /1, 2, 3/ the SLS's gained increasing interest for several reasons. By preparation of lattice mismatched semiconductor components a wider range of superlattice systems is available. The strain can be used as an additional factor influencing electronic and optical properties not available in unstrained material. The technologically important silicon based superlattices are strained because the silicon lattice constant is smaller than that of the most other semiconductors.

Todays microelectronics has reached a very high level of performance and complexity mainly by applying techniques for shrinkage of the lateral dimensions of integrated circuit (IC) structures down to the submicron level. The vast majority of high complexity IC's is manufactored with high yield and astonishingly low costs on silicon substrates. The performance of these conventionals IC's is furtheron increasing (Figure 2) with no reliable prediction of saturation effects. Future circuits based on new more general material concepts should utilize the tailorable properties of heterostructure - or superlattice - devices for high performance cores, high speed interlinks and novel sensor application on the chip. But the tremendous complexity of conventional IC's and the technical, economical and environmental advantages

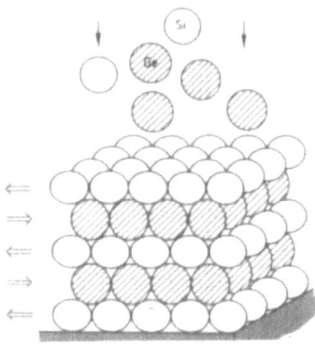

Figure 1 : Strained Layer Superlattice (SLS) grown from condensing Si and Ge molecular beams. The thin layers are strained because of lattice mismatch between Si and Ge.

of silicon substrates should not be missed. A material concept combining silicon based superlattices with conventional IC's (as shown in Fig. 2) can fulfill the requirements if progress in mismatch accomodation, low temperature stacking of different device levels and production technology continues.

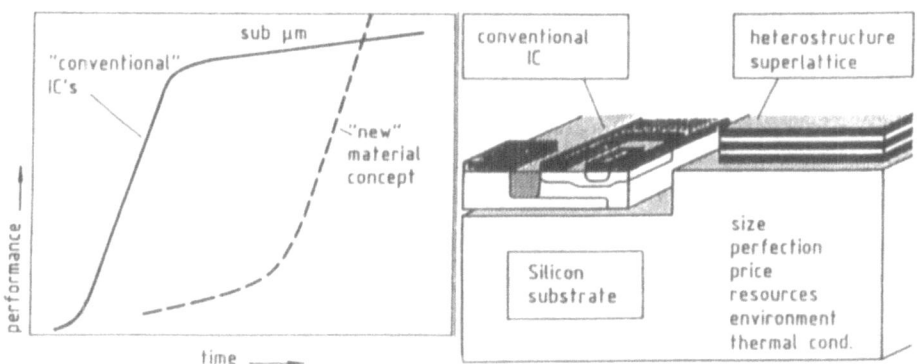

Figure 2 : New material concept for future microelectronic integrated circuit (IC's). Improved performance (left side) is expected for silicon based monolithic integration of conventional IC's with heterostructure or superlattice devices.

Silicon substrates are favoured for a broad industrial application because of their superior properties regarding wafer size, crystal perfection and thermal conductivity (Table 1), handling and price. The large resources (silicon is the second most frequent element of earth crust) and the environmental harmlessness are additional factors favouring broad usage.

A realization of this concept could start with the conventional IC on the silicon substrate. The high complexity conventional part covers the main part of the chip with the exception of core regions, interlink regions and sensor regions. The conventional IC-part is nearly completely processed up to the last metallization level. Then the superlattice structure is defined by a low temperature process within the predetermined regions and connected by the last metallization level.

3. Theoretical Concept

3.1 Zone Folding

Folding of the Brillouin zone (momentum space) of the superlattice is a direct consequence of the new introduced super-period in real space. It has been predicted that this can change the indirect character of the Si-type band gap /4/. One expects minizones in the momentum space not observed in the host crystal as depicted in Figure 3 for a superlattice with a period length of $L = 5a_0$ ($a_0 = 0.5431nm$, bulk lattice constant of silicon). More detailed calculations of the zone folding effects underline the essential effect of strain in this material system /5, 6/. The material will change its character from the indirect band gap of Si-type semiconductors to a direct band gap semiconductor if the lowest transition in the folded bandstructure is direct which means that the valence band maximum and the conduction band minimum are at the same k-value in momentum space. According to the calculations carried out by us and others /5, 7/ a direct band gap is only predicted for strained superlattices on top of a virtual substrate offering the same or a greater in-plane lattice constant $a_{//}$ than the in-plane lattice constant of the superlattice as a whole. Following these arguments a Si_mGe_n superlattice grown directly on a silicon substrate would be of indirect band gap character.

3.2 Transition Matrix Elements

The strength of the transition matrix elements between valence band maximum and conduction band minimum (oscillator strength) in a superlattice structure depends basically on the intermixing of conduction and valence band extrema states (X and G states) due to the potential introduced by the superlattice periodicity. This means that the oscillator strength becomes stronger the smaller the superlattice period is because the potential introduced by the superlattice becomes stronger with smaller separation of the atoms in adjacent supercells. Other factors are the energy separation of the folded valence and conduction band states, the strain and the position in the Brillouin zone where the conduction band minimum is folded to

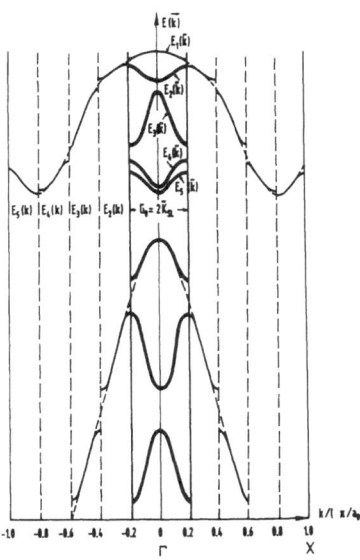

Figure 3 : Brillouin Zone Folding of a Si-like bandstructure introduced by a superlattice with a period of L=5a$_0$ (a$_0$=0.5431nm, lattice constant of Si). The bandstructure is folded into the first minizone (dark line) which has the extension of G = 2KSL = 2 (/5a$_0$), 1/5th of the original zone. By this process the original bandstructure with the indirect bandgap transforms into a bandstructure with a direct bandgap at k=0.

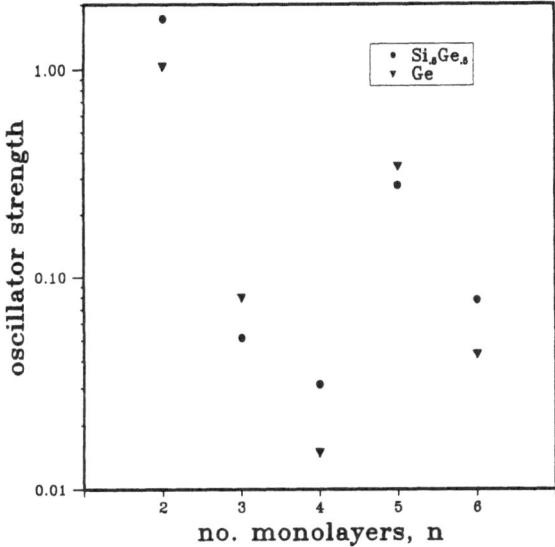

Figure 4 : Calculated transition matrix elements (oscillator strength) as a function of period length for a SinGen superlattice on a Si$_{0.5}$Ge$_{0.5}$ alloy substrate (note that the actual period length is 2n).

/7/. **Within the work carried out during the ESPRIT BASIC RESEARCH (EBR)
action** we have calculated the oscillator strength for a symmetrical Si_nGe_n superlattice on a $Si_{0.5}Ge_{0.5}$ and a Ge substrate (Figure 4). The oscillator strength can be
seen to vary strongly as a function of period length of the superlattice but little as a
function of substrate (note the logarithmic scale in Fig. 4). According to the
arguments above the smallest 4 monolayer period superlattice (Si_2Ge_2; note that
the SLS period is twice the number of monolayers shown in Figure 4) shows the
highest oscillator strength while there is a relative maximum occuring for a 10 ML
superlattice (Si_5Ge_5).

4. Virtual Substrate

The strain can be adjusted by a virtual substrate /8/ consisting of the the substrate
and the $Si_{1-y}Ge_y$ buffer layer between substrate and superlattice (Fig. 5). Often a
rather thick and graded buffer layer is used which is inconvenient for MBE-growth
methods and for device technology. We instead proposed a thin, homogeneous
buffer and gave design rules for that buffer layer /9, 10/.

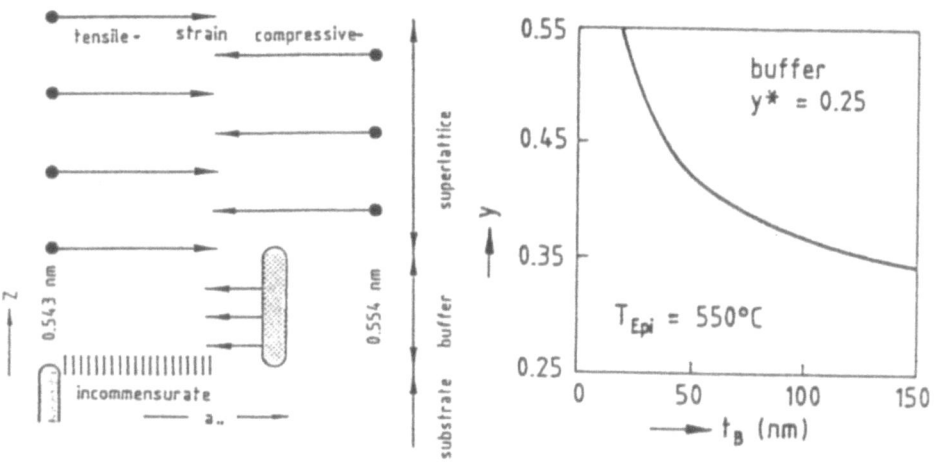

Figure 5 : Strain symmetrization in a Si_mGe_n SLS by a virtual substrate consisting of a
Si substrate with a covering incommensurate $Si_{1-y}Ge_y$ buffer layer. Left side : in plane
lattice constants of the layers as influenced by chemical composition (dots) and strain
(arrows). Right side : design chart of the corresponding buffer layer with effective Ge
content of 25% (y*=0.25). Shown is the actual Ge content y versus thickness of tB of
the buffer. Experimental strain values of $Si_{1-y}Ge_y$ layers grown at 550C (Tg) were used
for calculation of this design chart /10/.

4.1 Effective Ge-content

The thin homogeneous $Si_{1-y}Ge_y$ buffer layer with thickness t_B and the Ge-content y is compressively strained (negative value of ε_B) by the substrate (tetragonal distortion of the cubic lattice cell). The in-plane lattice constant $a_{//}$ of the buffer layer is somewhere between the natural lattice constants of Si and the $Si_{1-y}Ge_y$ buffer. For the following superlattice the virtual substrate offers the same $a_{//}$ as a $Si_{1-y}Ge_y$ substrate with an effective Ge-content y*. Applying Vegard's law for the natural lattice constants of $Si_{1-y}Ge_y$ alloys results in

$$a_{//} = a_0 (1 + 0.042y + \varepsilon_B) \qquad (1)$$

with the Si bulk lattice constant $a_0 = 0.543nm$. The effective Ge-content y* of the virtual substrate is given by

$$y^* = y + 24\,\varepsilon_B \qquad (2)$$

For a design chart (Figure 5) of virtual substrates the strain ε_B has to be measured as a function of growth temperature T_g, Ge-content y and thickness t_B of the buffer layer. We have constructed such a design chart /10/ using J.Bean's experimental data for growth at 550 °C /11/. For a virtual $Si_{1-y_B}Ge_{y_B}$ substrate with 25% effective Ge content (y* = 0.25) the design chart requires a $0.25\mu m$ thick buffer layer with 33.5% Ge content, or a $0.1\mu m$ thick buffer with 37.5% Ge, or a 50nm thick buffer with 42% Ge, or a 10nm thick buffer with 66% Ge. As this example shows, a wide range of thin, homogeneous buffer layers can be used for a virtual substrate with a definite in-plane lattice constant $a_{//}$.

4.2 Strain Distribution and Adjustment

Growing a SLS on different virtual substrates will result in different strain energies stored in the superlattice. Only the superlattice structure with the lowest energy content will be stable up to infinite thickness. All other structures will be unstable above a critical thickness of the whole superlattice, although the individual layers of the SLS are thinner than the critical thickness t_c for layer growth. Dislocations tend to move to the interface between virtual substrate and superlattice within these unstable SLS's.

Assuming equal elastic constants for the Si and Ge layers (for unequal elastic constants see /12/) the elastic energy E_h stored in a SLS with N periods and period length L is given by /10, 11/

$$E_h = 2\mu\,(1+\nu)\,/\,(1-\nu)\,N\,(\varepsilon_1^2 t_1 + \varepsilon_2^2 t_2) \qquad (3)$$

where ε_i and t_i are the strain and thicknesses of the individual layers of the superlattice period

$$L = t_1 + t_2 \qquad (4)$$

and μ and ν are the shear modulus and the Poisson's number respectively. The strain values are connected by the lattice mismatch η between the layers

$$\eta = 2\,(a_2-a_1)\,/\,(a_2+a_1) \tag{5}$$

where a_2 and a_1 are the unstrained bulk lattice constants of the superlattice component materials. The strain ε_1 in the silicon layers of the SLS can be chosen deliberately by an appropriate choice of the virtual substrate. The strain ε_2 in the germanium layers is then connected by the lattice mismatch η ($\eta = 0.042$) between silicon and germanium. Neglecting second order terms we obtain /14/

$$\eta = \varepsilon_1 - \varepsilon_2 \tag{6}$$

Minimizing the elastic energy E_h in (3) with respect to ε_1 and using (6) gives a condition for the stable strain symmertrized SLS

$$\varepsilon_1 = \eta\,(t_2/L)$$
$$\varepsilon_2 = -\eta\,(t_1/L) \tag{7}$$

For the most important case of equally thick layers ($t_1 = t_2 = L/2$) this condition reduces to

$$\varepsilon_1 = -\varepsilon_2 = \eta/2 \tag{8}$$

The symmetrical strain is of special importance because it marks the lowest energy state of the strained layer superlattice (SLS). Only in this strain situation the superlattice structure has the lowest energy content and and can be grown up to infinite thicknesses. In all other strain situations the SLS thickness is limited to a critical superlattice thickness t_c. Also the thickness of the individual layers can be larger than in the case of unsymmetrical strain.

If we discuss strain symmetrization in context with SLS stability we always refer to equ. (7) if unequally thick layers are used. Unsymmetrically strained layers are unstable above the critical superlattice thickness tc. This critical thickness is roughly the same as that for an alloy layer of the same integral composition. In case of the symmetrical strain situation of equ. (8) there is no limiting critical thickness of the superlattice as a whole. The situation changes if the strain is unsymmetrically distributed. Then the superlattice can gain energy by the introduction of misfit dislocations at its base if the thickness of the superlattice raises above the critical thickness t_c (N L > t_c). The critical superlattice thickness will decrease with increasing asymmetry of the strain distribution. By the introduction of misfit dislocations a completely unsymmetrical superlattice structure will change a// and shift the strain distribution to higher symmetry.

The strain in the SLS is adjusted by the choice of the virtual substrate with its in-plane lattice constant a//

$$\varepsilon_i = (a_{//}-a_i)\,/\,a_i \tag{9}$$

with a_i being the lattice constant of the ith layer. Symmetrizing the strain in a Si_nGe_m superlattice is obtained when the in-plane lattice constant of the buffer layer is chosen to be the lattice constant of the alloy with the same integral composition as the subsequent Si_mGe_n superlattice.

5. Growth by Molecular Beam Epitaxy

The basic idea of molecular beam epitaxy is atomic layer by layer growth from condensing molecular beams of that constituents which should create the thin crystal on top of a substrate. Principles, equipment, processes and industrial approaches of Si-MBE are described in /13/. For the growth of the Si_mGe_n superlattice we used an MBE-apparatus also commonly used for homoepitaxial device work. Details of this equipment are described elsewhere /16/ and the details of the growth process are outlined in /12/.

Within the work of the EBR action we have grown a p^+-n-n^+ doped diode structure (on n^+ substrate, series PIN II, sample No. B1856-B1873) with a n-doped Si_mGe_n SLS for device applications. For characterization purposes we have also provided samples from earlier grown serieses, such as series PIN I (No. B1585-B1590, n^+-n-p^+ on p^+-substrate) and a series with undoped superlattice layers grown on a semi-insulating substrate (No B1760-B1768, B1775-B1784). Figure 6 shows a sketch of the grown layer sequence of the two PIN serieses together with the used process temperatures during the MBE process. To achieve device relevant thicknesses of the SLS layer ($\sim > 0.25\mu m$) we used the concept of strain adjustment of the superlattice by a $Si_{1-y}Ge_y$ alloy buffer layer which was explained above (see chapter IV). In series PIN II we started with a 50nm thick Si buffer layer on top of which we grew a thin Ge buffer layer. Then a 60nm thick $Si_{1-yB}Ge_{yB}$ alloy layer follows where the Ge content y_B is adjusted in order to achieve strain relief of the subsequently following SLS by adjusting the strain near the strain symmetrized state (chapter IV). The substrate together with these three buffer layers can be considered as the virtual substrate with an effective Ge content y_{B*}. The Si_mGe_n SLS consists of N periods of alternating Si and Ge-layers (m ML Si + n ML Ge stacked N times above each other) and was chosen to be around 200nm thick. The buffer layers and the SLS-layers were n-doped by Sb-incorporation from a predeposited Sb-adlayer. On top of the SLS layer follows a 20nm thick $Si_{1-yc}Ge_{yc}$ alloy contact layer which is p^+-doped and where the Ge-content y_c is adjusted to provide the in-plane lattice constant $a_{//SLS}$ of the underlying SLS-layer. This important fact guarantees that there is no misfit dislocation network created at the p-n junction. A thin (1 nm) silicon cap protects the alloy layer and provides a Si-surface for post-epitaxial process steps.

874

$Si_n Ge_n$ SLS series PIN I (B1585-B1590)

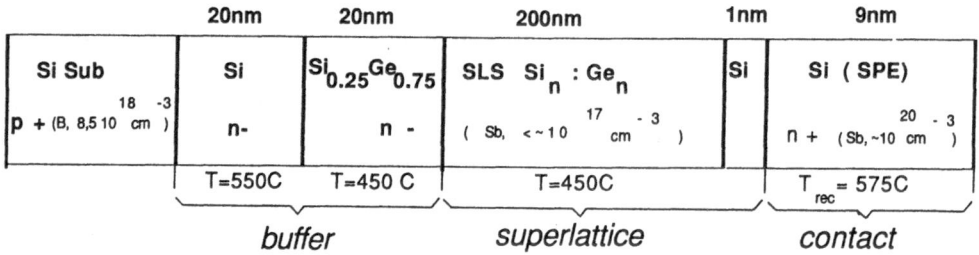

$Si_m Ge_n$ SLS series PIN II (B1856-B1873)

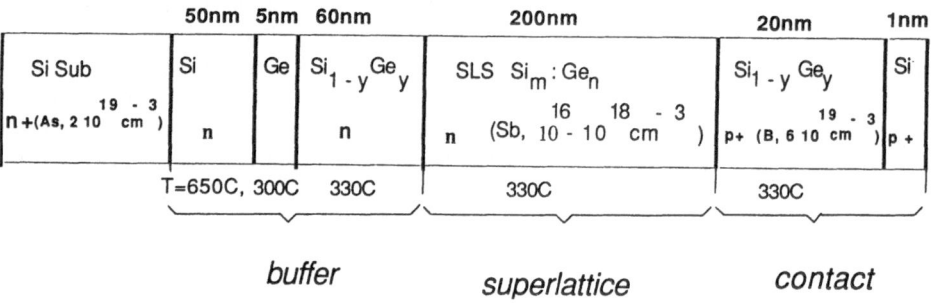

Figure 6 : Layer structure and MBE process temperatures Tg of the diode type samples of series PIN I (B1585-B1590, on p^+-doped substrate) and PIN II (B1856-B1873, on n^+- substrate). Series PIN II has been grown at lower temperatures and has been terminated by a 20nm thick $Si_{1-yc}Ge_{yc}$, p^+-doped alloy layer where yc is equal to the average Ge content of the SLS. Series PIN I has a very thin Si contact layer (thickness limited by the critical thickness) which has been grown by Solid Phase Epitaxy (SPE, amorphous deposition at room temperature, recrystallization at Trec=575C). Thus the p-n junction is situated near the misfit dislocation network (buffer-substrate interface) in series PIN I and away from it in series PIN II.

The different SLS compositions of series PIN II include a "6:4" subseries (B1857-B1859, B1873), a Ge-rich subseries (B1861, B1863) and two pseudomorphically grown SLS samples (B1864, B1872). Table II lists the sample numbers, buffer and SLS composition, the number of SLS periods, the total epitaxial thicknesses, growth temperature and growth rates of the samples in series PIN II. The "6:4" subseries ($m/n = 3/2$) consists of a $Si_{24}Ge_{16}$ SLS (B1857), a $Si_{12}Ge_8$ SLS (B1858), and two Si_6Ge_4 SLS's (B1859 and B1873) together with a reference $Si_{0.6}Ge_{0.4}$ alloy sample (B1856). For this 6:4 SLS samples the Ge-content of the buffer layer is $y_b = 0.38$ besides for sample B1873 which has $y_b = 0.55$. The different y_b alters the biaxial strain of the Si and Ge-layers within the SLS as well as the total strain of the SLS in sample B1873 compared to sample B1859 which has the same SLS period, composition and thickness. The different strain state leads to a different curvature of the 3 inch wafer which can be measured by X-ray curvature measurement (chapter IV.1). The two Ge-rich SLS samples B1861 (Si_3Ge_7) and B1863 (Si_2Ge_6) have a Ge-content of $y_b = 0.75$ in the buffer layer to achieve the strain symmetrized state of the SLS. In addition to the SLS layers grown on a $Si_{1-yb}Ge_{yb}$ buffer layer we have also grown two samples in a pseudomorphic (or commensurate) growth mode (B1864 and B1872) where the SLS is directly grown on a Si substrate (i.e. Si-buffer layer) and strained by the underlying Si lattice ($a_{//SLS} = a_{0,Si} = 5.431$Å). For not creating misfit dislocations at the buffer- SLS interface the SLS thickness has to be kept below the critical thickness. Sample B1864 has a $Si_{17}Ge_3$ SLS with $N = 50$ periods and a thickness of $d_{SLS} = 140$nm which according to the data in reference /16/ is below the critical thickness ($\eta = (a_{//SLS}-a_0)/a_0 = 0.0063$, $t_c \approx 350$nm at $T_g = 550°C$). The SLS of sample B1872 has $N = 20$ periods of a Si_4Ge_4 SLS where each SLS is interleaved by a 20nm thick Si layer which serves as a virtual buffer for each SLS. Thereby the restriction of a critical thickness is removed and the sample can be grown in a device relevant thickness with a good crystalline quality (without misfit dislocations). From this type of sample optoelectronic devices with impressive results have been fabricated by T.P.Pearsall et al. /23/. Also a Si homodiode (B1862) with a Si-homo layer instead of the Si_mGe_n SLS has been grown for reference purposes in this series where the same doping and the same growth conditions have been applied.

6. Characterization

The strain distribution in SimGen SLS can be determined by X-ray analysis /17, 18/, Raman scattering /19/ and Rutherford Backscattering (RBS) /20/. The growth quality, dislocation densities, superlattice composition and period can be studied by Transmission Electron Microscopy (TEM) /21,22/.

6.1 X-ray Diffraction

X-ray diffraction is a versatile tool for analyzing the superlattice material because

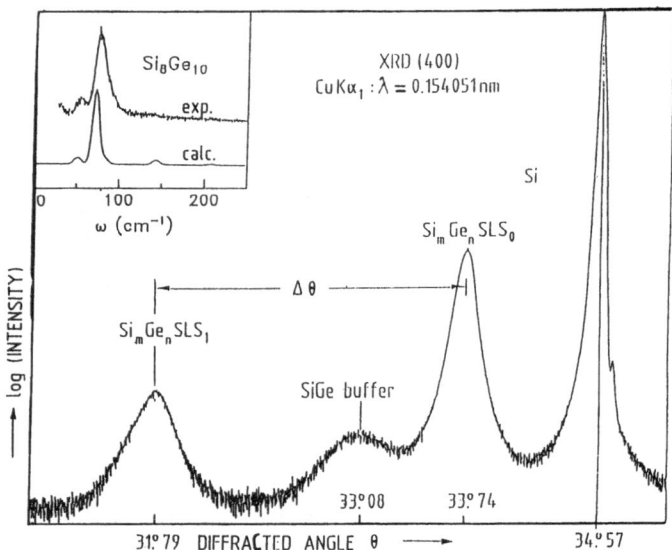

Figure 7 : (400) X-ray diffraction plot from a Si_8Ge_{10} SLS (log of diffracted intensity versus diffraction angle Q). Besides the substrate and the rather weak buffer peak the zeroth (marked SLS_0) and the first order (SLS_1) satellite peaks of the SLS can be seen. The inset shows the folded acoustic mode in the Raman spectrum measured and calculated (curve below).

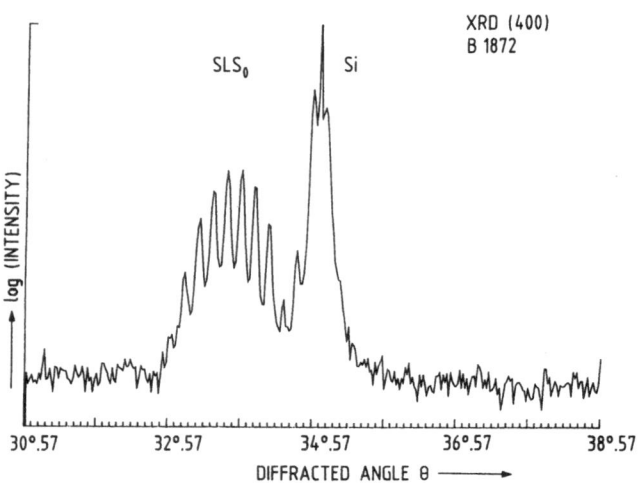

Figure 8 : (400) X-ray diffraction plot of the pseudomorphically grown sample B1872. The complicated pattern of the SLS peak arises from the two periods of the SLS in this sample. The smaller period of the Si_4Ge_4 SLS (L=8ML) itself is superposed on the large period of 20nm $Si+Si_4Ge_4$ (N=20) which leads to the fine structure on the SLS0 peak.

it is a nondestructive analysis method which reveals the strain distribution, the composition and the periodicity of the SLS at the same time. The tension (dilatation or compression) between the superlattice and the buffer and/or between the $Si_{1-yb}Ge_{yb}$ and the Si-buffer layer (and vice versa) leads to a bending of the 3 inch wafer which can be measured by the X-ray curvature measurement. Hereby the wafer is scanned stepwise through the X-ray beam and adjusted in angle to the Bragg position. The so measured curvature is related to the strain distribution of the individual layers in the SLS. Thus we can experimentally determine if strain symmetrization of the SLS by the buffer layer has been achieved and also by recording the sign of the curvature to which side the Ge-content y_b of the buffer has to be corrected for the strain symmetrized state of the SLS. Figure 7 shows a X-ray diffraction plot (logarithm of intensity versus diffraction angle Θ) of a Si_8Ge_{10} ultrathin superlattice taken by a double crystal diffractometer in the Si (400) reflection. The main peak at $\Theta = 34.57$ is caused by the (400) reflex of the Si substrate and serves as reference signal. A very low and rather broad intensity peak is caused by the thin buffer layer. The superlattice itself produces a zeroth order reflection (marked SLS_0) and satellites (the first marked SLS_1). The angular distance $\Delta\Theta$ between the satellites is given by

$$\Delta\Theta = \lambda \sin\Theta / (\sin(2\Theta_B) L) \qquad (10)$$

where Θ_B is the Bragg angle, Θ is the diffraction angle, and λ the used X-ray wavelength (CuKα radiation $\lambda = 0.154nm$). Thus the period length L of the SLS is related to the angular spread of the XRD satellite peaks. However, we like to stress that equation (10) is an approximation (linearized form) which for big angular distances DQ (small period length L) does not hold. Figure 8 shows the XRD plot of the pseudomorphically grown sample B1872 (series PIN II, see table II). The spectrum shows clearly the two periods of this sample. The large period consists of the Si_4Ge_4 SLS + 20nm Si which leads to the broad zeroth order satellite peak (SLS_0) while the small period of the Si_4Ge_4 SLS itself is superposed on the coarse period and gives rise to the fine structure on the broad peak. Generally the angular position and the amplitudes of the X-ray diffraction peaks reveal information on the layer thickness, composition and strain distribution while the linewidth and form is related to the structural perfection (interface sharpness, homogeneity, etc..) of the SLS. Due to interference and superposition effects from individual layers in the SLS it is necessary for a quantitative assessment of X-ray rocking curves from SLS's **to** compare the experimental data with a computer simulation.

6.2 Optical Characterization Tools

The most important optical characterization methods which have beeen performed for superlattices so far are Raman scattering /19/, photoluminescence /24/ and modulation spectroscopy measurements /23/. While Raman scattering reveals information on strain, periodicity , material quality and composition of the superlattice, photoluminescence and modulation spectroscopy (e.g. electroreflectance spectroscopy) probe the electronic bandstructure of the superlatttice directly. Recent photoluminescence measurements on 10 ML period superlattices (Si_6Ge_4 or Si_4Ge_6) /25/ give strong indications of direct bandgap transitions in the near infrared spectral range ($h\omega \sim 0.8eV$) for these superlattices which has been confirmed by Kronig-Penney type calculations. Here we will focus on Raman scattering furtheron.

Raman scattering by phonons is a local probe of the lattice dynamics and thus leads to information on the local crystalline structure, orientation, composition, built in strain etc. The probed sample is determined by the diameter of the laser focus which can be as small as $1\mu m^2$ and the penetration depth of the laser light into the semiconductor which can be easily varied from a few m down to a few tens of nm. It is therefore an ideal tool to investigate thin epitaxial layers grown on various substrates. Figure 9 shows a typical $Raman_n$ spectrum of a Si_6Ge_4 superlattice in comparison with a $Si_{0.6}Ge_{0.4}$ alloy sample both grown on a strain symmetrized buffer layer. The epitaxial layers have a total thickness of about 200nm. The Raman spectrum of sample B1856 is typical for an alloy with higher Si than Ge concentration, superimposed on the spectrum of the silicon substrate. The substrate gives rise to a small optical phonon mode of bulk Si at about $520cm^{-1}$, the other three main features at about $300cm^{-1}$, $400cm^{-1}$ and $500cm^{-1}$ of the alloy spectrum correspond to Ge-Ge, Si-Ge and Si-Si vibration in the alloy. The Ge-content y and the built-in strain of the alloy can be extracted from the exact positions and intensity ratios /26, 27/. The spectrum of the Si_6Ge_4 SLS sample B1778 is qualitatively different due to the ordered layer structure. In the wave number range at about $150cm^{-1}$ an additional strong phonon peak occurs due to the so called folded acoustic longitudinal modes. The exact position of this mode is strongly dependent on the superlattice period and can be seen as an experimental proof of the zone folding. The width of the folded mode is sensitive to interface roughness and thickness fluctuations. The optical phonon peaks of the SLS are also drastically different from those of the alloy spectrum. The Ge mode is increased in intensity compared to the Si-Ge mode. The energetic positions of all three modes are sensitive to the built-in strain. The Si and Ge modes also depend on the individual layer thicknesses due to confinement of the optical vibrations. In addition, higher order confined modes are observed on the low energy side especially of the Si mode. The intensity of the Si-Ge mode at about $400cm^{-1}$ can be seen as a probe of the interface sharpness of the SLS.

7. Conclusion

Ultrathin Si_mGe_n superlattices ($n + m < \sim 40$ ML) have been grown on silicon substrates by Molecular Beam Epitaxy. The concept of strain symmetrisation by a

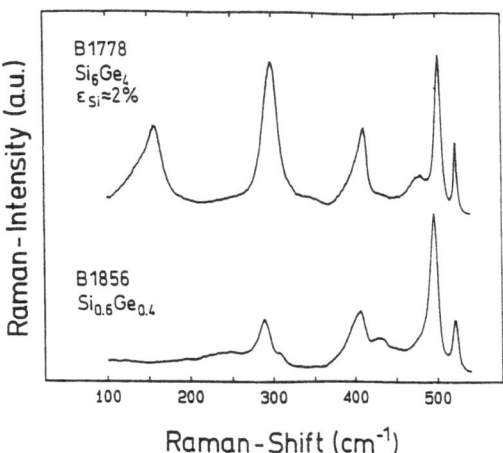

Figure 9 : Raman spectra of a Si₆Ge₄ SLS (B1778) and a Si₀.₆Ge₀.₄ alloy (B1856). The inset shows the in-plane strain eSi of the silicon layers in the superlattice.

thin homogeneous buffer layer has been applied and the strain distribution by the choice of the buffer layer composition is analyzed. The concept of Brillouin Zone folding is explained and the transition matrix elements between valence and conduction band of the superlattice bandstructure are calculated. Different characterization tools (X-ray analysis, TEM, Raman spectroscopy, photoluminescence and Raman spectroscopy) reveal information of the composition, built-in strain, periodicity and electronic band structure of the superlattice.

The concept of a heterostructure superlattice monolithically integrated with a conventional integrated circuit (IC) on top of a silicon substrate is outlined. This new and promising possibility has drawn a lot of scientific attention because of the predicted new optical properties of the Si_mGe_n SLS. In this concept the superlattice regions could for example form optical data links between conventional parts of the IC.

Acknowledgement : The assistance of H.-J. Herzog and U.Menczigar in the X-ray measurements and in Raman spectroscopy is acknowledged.

Table 1: Room temperature thermal conductivity of the semiconductors Si and GaAs compared with the best heat conductor diamond and the metal copper.

Material	Si	GaAs	diamond	copper
κ (W/mK)	145	46	2000	384

Table 2 : Si$_m$Ge$_n$ SLS series PIN II (B1856-B1873)

Sample number, buffer layer, Si$_m$Ge$_n$ SLS (m:n, N) and contact layer composition of the samples of series II. Also given are the total epitaxial thickness and growth temperature and growth rates of the SLS layers. The samples were grown on a n$^+$-substrate (n = 2 10^{19} As/cm^3, Φ = 76.2mm) on top of which a rather complicated buffer layer was grown ("6:4" series and "Ge-rich" series only, for a sketch of the layer structure see Fig. 4). The strain symmetrizing buffer consists out of a 50nm Si, 5 nm Ge and a 60nm thick Si$_{1-yc}$Ge$_{yc}$ alloy contact layer on top of which the 200nm thick Si$_m$Ge$_n$ SLS has been grown. The SLS of samples B1864 and B1872 is grown directly on the Si buffer layer in the pseudomorphic growth mode. There the SLS as a whole is strained because of the mismatch of the underlying silicon lattice, its thickness is below the critical thickness not to create any misfit dislocations at the substrate-buffer interface. On top of the SLS follows a 20nm thick Si$_{1-yc}$Ge$_{yc}$ alloy contact layer (y$_c$ = y$_{SLS}$ = n/(n + m)) which is terminated by a 1nm Si cap. Buffer and SLS-layer are n-doped (Sb) the contact layer has been p-doped. A Si homo-diode (Si homolayer instead of Si$_m$Ge$_n$ SLS) is grown under the same conditions for reference purposes.

No	buffer yb	SLS n:m, N	contact yc	epitax. thickness meas./ calc. [nm]	TG,SLS / R$_{Si}$, R [C°] / [Å/s]
"6:4 series"					
B1856	SiGe$_{0.38}$	Si$_{0.6}$Ge$_{0.4}$	0.4	348 / 336	330 / 0.5 (R$_{SiGe}$)
B1857	0.38	24:16, 36	"	354 / 335	300 / 0.41, 0.25
B1858	"	12:8, 73	"	372 / 338	" / " , "
B1859	"	6:4, 150	"	390 / 343	" / " , "
B1873	0.55	6:4, 145	0.55	347 / 336	" / 0.54, 0.27
"Ge-rich"					
B1861	0.75	3:7, 145	0.75	342 / 339	300 / 0.27, 0.28
B1863	0.75	2:6, 192	0.75	365 / 351	" / 0.27, 0.27
"pseudomorphic"					
B1864	-	17:3, 50	Si	198/ 208	" / 0.55, 0.26
B1872	-	4:4+20nm Si, 20	100nm Si	660/ 670	" / 0.27, 0.27
"Si-Homo"					
B1862	-	Si, 200nm	"	264/ 271	" / 0.55 (R$_{Si}$)

References

/1/ G.C.Osbourne, J.Appl. Phys. 53, 1985 (1982)

/2/ J.W.Matthews and A.E.Blakeslee, J. Cryst. Growth 27, 118 (1974)

/3/ E.Kasper, H.-J.Herzog and H.Kibbel Appl.Phys. 8, 199 (1975)

/4/ U.Gnutzmann and K.Clausecker, Appl. Phys. 3, 9 (1974)

/5/ S.Froyen, D.M.Wood and A.Zunger, Phys Rev. B 36,4547 (1987) and A.Zunger, 2nd Int. Symp. Si-MBE, Honolulu, Oct. 87

/6/ I.Morrison, M.Jaros and K.B.Wong, Phys Rev. B 35, 9693 (1987)

/7/ M.Jaros in Proceedings of Workshop I, Esprit Basic Research Action, Action No 3174, Ulm October 1989

/8/ E.Kasper, H.-J.Herzog, H.Dämbkes and er, Mat. Res. Soc. Proc. Vol. 56, ed. by J.M.Gibson, G.C.Osbourne and R.M.Tromp, p.347, Materials Research Society, Pittsburgh (1986)

/9/ E.Kasper in Physics and Applications of Quantum Wells and Superlattices, ed. by E.E.Mendez and K. von Klitzing, p. 101, Nato ASI Series B, Vol. 170, Plenum Press, New York (1987)

/10/ E.Kasper, H.-J.Herzog, H.Jorke and G.Abstreiter, Superlattices and Microstructures 3, 141 (1987)

/11/ J.C.Bean, "Silicon Based Heterostructures" in Silicon Molecular Beam Epitaxy, ed. by E.Kasper and J.C.Bean, CRC Press, Boca Raton USA, (1989)

/12/ E.Kasper, H.Kibbel and H.Presting, Thin Solid Films 183, p.87-93, 1989

/13/ Silicon Molecular Beam Epitaxy, ed. by E.Kasper and J.C.Bean, CRC Press, Boca Raton, USA, (1989)

/14/ W.A.Brantley, Journal Appl. Phys. 44, 534 (1973)

/15/ H.Presting, H.Kibbel, E.Kasper and H.Jorke to be published in Journal of Appl. Physics

/16/ E.Kasper and K.Wörner, J.Electrochemical Soc. 132, 2481 (1985) and E.Kasper Surf. Science 174, 630 (1986)

/17/ J.-M. Baribeau, Appl. Phys. Lett. 52, 105 (1988)

/18/ B.K. Tanner, Journal Electrochem. Soc. 136, 3438 (1989)

/19/ G.Abstreiter, K.Eberl, E.Friess, W.Wegscheider and R.Zachai, Journal of Crystal Growth 95, p.432-38 (1989)

/20/ S.Mantl, E.Kasper and H.Jorke, Mat. Research Soc. Symp.Proc. Vol. 91, 305 (1987) and H.-J.Herzog, H.Jorke, E.Kasper and S.Mantl, Proc. 2nd Int. Symp. Si-MBE, ed. by J.C.Bean and L.Showalter, Electrochemical Soc., Pennington N.Y., p.58 (1988)

/21/ D.C.Houghton, D.C.Lockwood, M.W.C.Dharma-Wardana, E.W.Fenton, J.-M.Baribeau and M.W.Denhoff, Journal of Crystal Growth 81, 434-39 (1987)

/22/ R.Hull, J.C.Bean, D.J.Eaglesham, J.M.Bonar and C.Buescher, Thin Solid Films 183, p.117-132 (1989) and K.Eberl et al., ibido, p.95 (1989)

/23/ T.P.Pearsall, J.Bevk, J.C.Bean, J.Bonar, J.P.Mannaerts and A.Ourzmad, Phys. Rev. B 39, 3741 (1989) and Fred H. Pollak and H.Shen, Superlattices and Microstructures, Vol. 6, p.203 (1989)

/24/ R.Zachai, K.Eberl, G.Abstreiter, E.Kasper and H.Kibbel, Phys. Rev. Letter 64, 1055 (1990)

/25/ M.A.Renucci, J.B.Renucci and M.Cardona, Proc. of the 2nd Int. Conf. on Light Scattering in Solids, ed. by M.Balkanski, p. 326, Flammarion, Paris 1971

/26/ F.Cerdeira, A.Pinczuk, J.C.Bean, B.Ba

INDEXES

INDEX OF AUTHORS

INDEX OF PROJECT NUMBERS

INDEX OF ACRONYMS

INDEX OF KEYWORDS

890